19th International Conference on Applied Mechanics and Mechanical Engineering (AMME-19)

IOP Conference Series: Materials Science and Engineering
Volume 973

Cairo, Egypt
7-9 April 2020

ISBN: 978-1-7138-4792-2
ISSN: 1757-8981

Printed from e-media with permission by:

Curran Associates, Inc.
57 Morehouse Lane
Red Hook, NY 12571

Some format issues inherent in the e-media version may also appear in this print version.

Institute of Physics
Dirac House, Temple Back
Bristol BS1 6BE UK

Phone: 44 1 17 929 7481
Fax: 44 1 17 920 0979

techtracking@iop.org

Additional copies of this publication are available from:

Curran Associates, Inc.
57 Morehouse Lane
Red Hook, NY 12571 USA
Phone: 845-758-0400
Fax: 845-758-2633
Email: curran@proceedings.com
Web: www.proceedings.com

TABLE OF CONTENTS

Preface

Peer review declaration

Experimental investigation of star grains in dual thrust solid propellant motors 1
M El-Naggar, H Belal, H M Abdalla

Mitigation of irregular burning in a small solid propellant rocket motor 19
M Elkshen, H Belal, M A Al-Sanabawy

Validation of a simplified model for liquid propellant rocket engine combustion chamber design 36
M Hegazy, H Belal, A Makled, M A Al-Sanabawy

Validation of relations for the determination of pyrotechnic igniter mass 50
A Salama, H Belal, H M Abdalla

Simplified algorithm for determining injection duration to achieve fast response in SI engines 63
Omar Shawky

The effect of rectangulargurney flap on wells turbine performance 76
A T Kotb, M A Nawar, R M Abd El Maksoud

Numerical studies of turbulent premixed flame interaction with repeated solid obstacles 91
M Elshimy, S Ibrahim, W Malalasekera

Development of rotary heat exchanger model to investigate performance with integrating controller 100
A M Ghoneim, M H Aly, O E Mahmoud

Handling performance of an 8x8 combat vehicle 116
M Ahmed, M El-Gindy, H Lang

Evaluating the effect of shape on energy absorbing response of structures used in armored vehicles floors 129
Hisham Kamel

Test rig design for upper-limb socket prosthetics at transradial amputation level 137
M M Bondok, M A El-Sheikh, M A El-Hadek

Design and manufacturing of a test rig for measuring the torque required in soil drilling operations 146
M A Abdeldayem, M E Abo-Elnor, M H Mabrouk

Assessment of steel wire's fatigue life using finite elements modelling and experimental testing 159
H M Lotfy, A B El-Shabasy, T A Attia, HA Hassan

Interactions between microorganisms and composite material with marble & granite filler 172
H Ahmed, O Darwesh

Optical non linear characteristics of particles from natural resources 184
H Ahmed, A Salah, S Hasabelnabi, A Khalil

Improving water desalination efficiency by using carbon nanotubes as pre-treatment 194
A AbdAllah, W El-Sallamy, B Azzam, T Osman

Optimization of friction stir welding parameters using response surface methodology 206
I Sabry, N Gadallah, M Abu-Okail

Study of peak temperatures in friction stir spot welding of AA2024-Al/polycarbonate and AA2024-Al/polypropylene systems 217
R K Mohammed, S S Mohammed, T S Mahmoud

The effect of tire rubber particles on the mechanical and physical properties of polyester............................. 226
K F Abo Elenien, N A Azab, G Bassioni, M H Abdellatif

Effect of modified bouc-wen model parameters on the dynamic hysteresis of magnetorheological dampers 237
M Abdelhamed, W G Ata, A M Salem, A Z Khafagy

Experimental and computational dynamic structural analysis of free-flight rockets....................................... 250
M A Habaka, SS Farag, M Khalil, M Y Ahmed

Numerical investigations of bistable piezoelectric cantilever for enhanced energy harvesting........................ 257
M A Abdelnaby, M S Shiba, S Ali

Simulating ultrasonic vibration enhanced FSW process of AA6082 using finite element method 267
OM Mabrouk, M S El-Wazery, M A El-Desouky

Passive vibration attenuation: a comparison study 276
A A Abdelghany, M M Hegazy, A Badawy

Phase stability of mechanically alloyed Ti-Fe-Al alloys 288
S A Ibrahim, M. M Seleman, H M Ahmed, A E Hannora

Fracture toughness assessment of axial partially through cracks in X65 and X70 steel pipes 298
A E Said, H A Hassan, T G Abu-El-Yazied, A M Makady

Assessment of St. St. 316L joints welded by laser beam and metal active gas MAG processes...................... 305
A M Magdy, A B El-Shabasy, H A Hassan

Pre-Cracking techniques of polymeric materials: an overview 317
A Kamal, AH Elsheikh, E Showaib

Abrasive water jet machining of CFRPs: single response optimization using taguchi method optimization.. 326
I I Edriys, M Fattouh, R Masoud

An experimental study on mechanical and ballistic characteristics of different HTPB composite propellant formulations 338
A F NourEldin, WM Adel, YA Attai, M A Ismail

Penetration of long rods into semi-infinite metallic targets over wide range of impact velocities 348
M A Abdel-Kader, A M Riad, A Z Ibrahim

Transient heat transfer for GPMG's barrel 7.62x51mm... 361
M Ghanem, O Abdelsalam

Multi-fidelity drag prediction for base bleed projectile.. 369
M M Aziz, AZ Ibrahim, M Y M Ahmed, A M Riad

Experimental modal testing of a honeycomb sandwich plate.. 378
A Aborehab, M Kamel, A F Nemnem, M Kassem

A new artificial neural network model integrated with a cat swarm optimization algorithm for predicting the emitted noise during axial piston pump operation 387
A H Elsheikh, M Abd Elaziz, HA Babikir, D Wu, Y Liu

Passive vibration damping of a cantilever plate 394
A Mohamed, A Hassan, A A Omer

Modelling and experimental investigation of high – pressure common rail diesel injection system........ 404
W Niklawy, M Shahin, MI Amin, A Elmaihy

Performance of homogeneous charge compression ignition (HCCI) engine with common rail fuel injection........ 417
W Niklawy, M Shahin, M I Amin, A Elmaihy

Experimental investigation and numerical simulation of the thermosyphon heat pipe charged with R134a 429
S Y Zakaria, A G Ibrahim, A M Rashad, S E El-shamarka

Aerodynamic performance improvement using a micro-cylinder as a passive flow control around the S809 airfoil........ 442
W Mostafa, A Abdelsamie, M Mohamed, D Thévenin, M. Sedrak

Assistive exoskeleton hand glove using soft pneumatic actuators: design optimization........ 455
M N El-Agroudy, M I Awad, A Rehan, S A Maged

Design of gain scheduled fuzzy PID controller for multi-link robot manipulators 465
Maya Emarah

Design of bio-inspired muscle sarcomere structure using a hybrid hydrogel-IPMC actuator........ 471
H M Hamdi, SA Abdelwahhab, MI Awad, FA Tolbah

Performance investigation and control parameters choice for sliding mode control of coupled tanks system........ 483
M A Louis, M R Roman, O E Mahmoud, M F Sedrak

Effect of applying lean maintenance in oil and gas fields 499
M Nasr Hassan, A F Barakat, A S Sobh

A new integrated product and process development model 512
A Ayman, A B Alhamaki, M H Abdellatif

Ultrasonic-assisted drilling of nickel-based super alloy in conel 601: An experimental study 522
A M Abdelaziz, H Youssef, M Al-Makky, H El-Hofy

Computational fluid dynamic model for machining using minimum quantity coolant........ 533
W Abdelfattah, H Hegab, A Mohany, H A Kishawy

Optimizing ladle-refining performance during treating special steel melts for aviation technology........ 540
S A Elghazaly, K K Gyula, W S Elghazaly

Author Index

IOP Conf. Series: Materials Science and Engineering **973** (2020) 011001 doi:10.1088/1757-899X/973/1/011001

Arab Rebublic of Egypt
Ministry of Defence
Military Technical College

10th International Scientific Conference of The Military Technical College
7 - 9 July, 2020

المؤتمر الدولى العلمى العاشر للكلية الفنية العسكرية

A.M.M.E.

Conference Program of
The 19th International Conference on Applied Mechanics and Mechanical Engineering,
(AMME-19)

برنامج
المؤتمر الدولى التاسع عشر فى الميكانيكا التطبيقية والهندسة الميكانيكية

Cairo 2020

IOP Conf. Series: Materials Science and Engineering **973** (2020) 011001 doi:10.1088/1757-899X/973/1/011001

Military Technical College
Kobry El-Kobbah,
Cairo, Egypt

19^h Int. Conference on Applied Mechanics and Mechanical Engineering,
July 7-9, 2020.

CONTENTS

Subject	**page**
Preface	1
Controbutors	2
Conference Committees	4
Reviewers	5
Conference Program	6
Scientific Sessions	10

IOP Conf. Series: Materials Science and Engineering **973** (2020) 011001 doi:10.1088/1757-899X/973/1/011001

Military Technical College
Kobry El-Kobbah,
Cairo, Egypt

A.M.M.E.

19^h Int. Conference on Applied Mechanics and Mechanical Engineering,
July 7-9, 2020.

PREFACE

The Military Technical College has the pleasure to organize the 19th International Conference on Applied Mechanics and Mechanical Engineering, (AMME-19) which is sponsored by the Ministry of Defense in the period 7-9 July, 2020.

The main objective of this conference is to bring together scientific researchers and engineers of Egyptian armed forces and their colleages in the international academic and industrial institutions. This occasion provide a chance to exchange information in the following designated fields of interest

1- Mechanical Power Engineering	**MP**
2- Robotic, Automation and Control	**RC**
3- Dynamics and Vibration	**DV**
4- Material Science and Processing	**MS**
5- Production Technology	**PT**
6- Solid Mechanics	**SM**
7- Mechanical Design	**MD**
8- Automotive Engineering	**AE**
9- Mechatronics	**MC**
10- Ballistics	**BL**
11- Acoustics and Noise Control	**AN**

From the 70 manuscripts submitted to the conference secretary, 49 papers have been accepted. These selected papers will be presented during the conference interval 7-9 July 2020 in 11 sessions.

Finally, the conference committee hopes that the conference will achieve its planed mission and would like to acknowledge all contributors, membes of scientific committees, chairman and board members of the conference sessions.

Note:

Please remember that you have only 15 min. to present your paper and 5 min. for discussion

Prof. Dr. Shawky A. Hegazy
AMME-19 Conference Rapporteur

IOP Conf. Series: Materials Science and Engineering **973** (2020) 011001 doi:10.1088/1757-899X/973/1/011001

Military Technical College
Kobry El-Kobbah,
Cairo, Egypt

A.M.M.E.

19h Int. Conference on Applied Mechanics and Mechanical Engineering,
April 7-9, 2020.

CONTRIBUTORS

I **Armed Forces**

- Military Technical College
- Armed Forces units
- Technical Research and Develop. For Armed Forces
- Technical Research for Naval Forces

II **Egyptian Universities and Organizations**

- Faculty of Engineering, Cairo University
- Faculty of Engineering, Ain Shams University
- Faculty of Engineering, Helwan University,
- Faculty of Engineering, Al Minia University
- Faculty of Engineering , Menoufia University
- Faculty of Engineering at Shoubra, Benha University,
- Faculty of Petroleum and Mining Engineering, Suez University
- Faculty of Engineering, Port-Said University.
- Faculty of Engineering, Egyptian Russian University
- Production Engineering and Mechanical Design Department, Tanta University,
- Faculty of Science, Zagazig University, Zagazig
- Faculty of Engineering, Alexandria University
- School of Engineering, Badr University in Cairo
- Arab Academy for Science, Technology & Maritime Transport
- October University of Modern Sciences and Arts
- Modern Academy for Engineering and Technology
- Akhbar Elyom Academy
- Ahram Canadian University (ACU)
- Canadian International College, Fifth Settlement, New Cairo
- Egypt-Japan University for Science and Technology, Alexandria
- Production Engineering, Education Development Fund, Cabinet of Ministers
- Nile High Institute of Engineering and Technology – Mansoura
- Steel Technology, Metals Technology, CMRDI
- Materials sciences & materials applications,
- Higher Technological Institute, 10th of Ramadan City
- Higher Technology Institute, 10th of Ramadan City, 6th of October Branch
- High Institute of Engineering, City of Science and Culture, 6 of October City

IOP Conf. Series: Materials Science and Engineering **973** (2020) 011001 doi:10.1088/1757-899X/973/1/011001

Military Technical College
Kobry El-Kobbah,
Cairo, Egypt

A.M.M.E.

19^h Int. Conference on Applied Mechanics and Mechanical Engineering,
April 7-9, 2020.

- High Institute of Engineering and Technology – Elmahala Elkobra
- Agricultural Microbiology Department, National Research Centre, Giza,
- National Institute of laser enhanced Science, NILES

III <u>Arab and Foreign Universities and Organizations</u>

- Taibah University, Medina, KSA
- Umm Al-Qura University, KSA
- UAE University, Al-Ain, UAE
- College of Technological Studies, PAAET, Kuwait
- University of Ontario, Oshawa, Ontario, Canada
- Metallurgy, Miskolc University, Hungary
- Huazhong University of Science and Technology, Wuhan 430074, China.
- Loughborough University, Loughborough, UK
- Hamburg University of Technology, Germany
- University of Magdeburg, , Germany
- University of California- USA

IOP Conf. Series: Materials Science and Engineering **973** (2020) 011001 doi:10.1088/1757-899X/973/1/011001

Military Technical College
Kobry El-Kobbah,
Cairo, Egypt

19^h Int. Conference on Applied Mechanics and Mechanical Engineering,
April 7-9, 2020.

Conference Committees

Higher Committee

- Maj. Gen. Prof. Dr. Ismail Mohamed Kamal — *Chairman*
- Maj. Gen. Prof. Dr. Yehia Zakarya Mehaseb — *Vice - Chairman*
- Maj. Gen. (R) Prof. Dr.Mohamed T. Sallam — *General Rapporteur*
- Maj. Gen. (R) Prof. Dr. Shawky A. Hegazy — *Rapporteur*

Steering Committee:

- Brig. Gen (R) Assoc. Prof. Dr. Aly Elmaihy
- Brig. Gen. Assoc. Prof. Dr. Mohamed M. Elkhateeb

Organizing Committee:

- Brig. Gen. Assoc. Prof. Dr.Hesham M. Eltaher
- Lt. Col. Dr. Mohamed Atef

Scintific Committee:

- Maj. Gen. (R) Prof. Dr. Salman E. El-Shamarka - Former Command of M. T. C.
- Maj. Gen. (R) Prof. Dr. Ibrahim Saleh - Military Technical College
- Maj. Gen. (R) Prof. Dr. Ahmed Reyad - Military Technical College
- Maj. Gen. (R) Prof. Dr. Moustafa Adnan- Military Technical College
- Maj. Gen. (R) Prof. Dr. Mohamed Said Abo Elkhair - Military Technical College
- Maj. Gen. (R) Prof. Dr. Yehia Hendawy Hossam eldin – Future University
- Maj. Gen. (R) Prof. Dr. Nabil Gadallah – Modern Acadamy
- Maj. Gen. (R) Prof. Dr. Abd Elnaser Zayed – Nahda University
- Maj. Gen. (R) Prof. Dr. Adel Abdel Razek - Military Technical College
- Maj. Gen. (R) Prof. Dr. Al Hussain Sharaf - Military Technical College
- Maj. Gen. (R) Assoc. Prof. Dr. Ahmed Rashad - Military Technical College
- Maj. Gen. (R) Assoc. Prof. Dr. Amgad Kamel - Military Technical College
- Maj. Gen. (R) Assoc. Prof. Dr. Moatasem Shahin – Badr University
- Maj. Gen. (R) Dr. Mouatafa M Abdelwahab - Military Technical College
- Maj. Gen. (R) Dr. Mohamed Ibrahim Amin - Military Technical College
- Prof. Dr. M. A. El Nashar – Former Minister of Higher Education
- Prof. Dr. Moustafa El Gindy- Faculty of Engineering, Ontario Tech Universit
- Prof. Dr. Hossam Kishawy - Associate Dean, Ontario Tech University
- Prof. Dr. Corina Sandu - Virginia Tech

IOP Conf. Series: Materials Science and Engineering **973** (2020) 011001 doi:10.1088/1757-899X/973/1/011001

Military Technical College
Kobry El-Kobbah,
Cairo, Egypt

A.M.M.E.

19^h Int. Conference on Applied Mechanics and Mechanical Engineering,
April 7-9, 2020.

REVIEWERS

A. A. Zayed
A. A. Omer
A. El Shabasy
A. S. Elmosalamy
A. A Saad
A. M. Bahgat
A. M. Rashad
A. M. Salah
A. R. Hanora
A. M.a Sharaf
A. A. Elmaihy
A. M. Kamel
A. A. Roshdy
A. I. Fayed
A. M. Fekry
A. G. Mohamed
B. Iskander
G. A. Hassaan
H. Hebatalrahman
H. M. Abd Allah
H. El laban
H. M. Belal
H. shawky
H. M. Kamel
H. E. Ragheb
I. S. Moustafa
K. A. Shaban Gamry
K. M. Abd Elfatah
M. A. Shahin
M. E. Shamekh
M. Z. Shaheen
M. A. Kamel
M. A. Elgamil
M. A. Al-Sanabawy
M. H. Abdelshafy
M. H. Abdellatif
M. I. Amin
M. I. Bennaya
M. I. Tolba Salam
M. M. El-Sayed Seleman
M. S. eldin hassan
M.. S. Abo Elkhair
M. Wahba
M. Y. Mohamed
M. A. Ismail
M. S. Asfoor
M. A. Ahmed
M. El-Gindy
M. F. Mohamed
M. M. AbdElwahab
M. Zaki
N. A. Gadallah
O. M. Kamal
O. Mohamed Ramy
S. A. Elghazaly
S. A. Hegazy
T.. Abu EL-Yazied
T. S. AbdElzaher
T. Samir
T. M. Abd Elbadea
W. G. Kotob
W. S. Eldin Hamed
W. A. Eltalabawy
W. M. Hanna
W. M. Naseem
Y. Hendawy hossam
Y. Attia

IOP Conf. Series: Materials Science and Engineering **973** (2020) 011001 doi:10.1088/1757-899X/973/1/011001

Military Technical College
Kobry El-Kobbah,
Cairo, Egypt

19ʰ Int. Conference on Applied Mechanics and Mechanical Engineering,
April 7-9, 2020.

CONFERENCE PROGRAM

IOP Conf. Series: Materials Science and Engineering **973** (2020) 011001 doi:10.1088/1757-899X/973/1/011001

وزارة الدفـــــاع
الكلية الفنية العسكرية

Program of the 19th International Conference on Applied Mechanics and Mechanical Engineering (AMME-19)

	9:00 - 10:45		11:00 - 12:45		13:00 - 14:45		15:00 - 16:45
Tuesday 7 July 2020	Ballistics I (4)	**Break**	Mechanical Power Engineering I (4)	**Break**	Automotive (2) / Mechanical Design (3)	**Break**	*Material Science and Processing I* (6)
	001_AMME-19		005_AMME-19		009_AMME-19		014_AMME-19
	002_AMME-19		006_AMME-19		010_AMME-19		015_AMME-19
	003_AMME-19		007_AMME-19		011_AMME-19		016_AMME-19
	004_AMME-19		008_AMME-19		012_AMME-19		017_AMME-19
					013_AMME-19		018_AMME-19
							019_AMME-19

IOP Conf. Series: Materials Science and Engineering **973** (2020) 011001 doi:10.1088/1757-899X/973/1/011001

وزارة الدفـــــــاع
الكلية الفنية العسكرية

Program of the 19th International Conference on Applied Mechanics and Mechanical Engineering (AMME-19)

	9:00 - 10:45		11:00 - 12:45		13:00 - 14:45		15:00 - 16:45
Wedensday 8 July 2020	Dynamics and Vibration I (5)	Break	Material Science and Processing II (5)	Break	Ballistics II (4)	Break	Dynamics and Vibration II (4)
	020_AMME-19		025_AMME-19		030_AMME-19		034_AMME-19
	021_AMME-19		026_AMME-19		031_AMME-19		035_AMME-19
	022_AMME-19		027_AMME-19		032_AMME-19		036_AMME-19
	023_AMME-19		028_AMME-19		033_AMME-19		
	024_AMME-19		029_AMME-19				

IOP Conf. Series: Materials Science and Engineering **973** (2020) 011001 doi:10.1088/1757-899X/973/1/011001

وزارة الدفـــــــاع
الكلية الفنية العسكرية

Program of the 19th International Conference on Applied Mechanics and Mechanical Engineering (AMME-19)

	9:00 - 10:45		11:00 - 12:45		13:00 - 14:45
Thursday 9 July 2020	Mechanical Power Engineering II (4)	Break	Mechatronics (1) / Robotic Automation and Control (3)	Break	Production Technology (5)
	037_AMME-19		041_AMME-19		045_AMME-19
	038_AMME-19		042_AMME-19		046_AMME-19
	039_AMME-19		043_AMME-19		047_AMME-19
	040_AMME-19		044_AMME-19		048_AMME-19
					049_AMME-19

IOP Conf. Series: Materials Science and Engineering **973** (2020) 011001 doi:10.1088/1757-899X/973/1/011001

Military Technical College
Kobry El-Kobbah,
Cairo, Egypt.

19th Int. Conference on
Applied Mechanics and
Mechanical Engineering,
July 7-9, 2020.

SCIENTIFIC SESSIONS

IOP Conf. Series: Materials Science and Engineering **973** (2020) 011001 doi:10.1088/1757-899X/973/1/011001

Military Technical College
Kobry El-Kobbah,
Cairo, Egypt.

A.M.M.E.

19th Int. Conference on
Applied Mechanics and
Mechanical Engineering,
July 7-9, 2020.

Session No. 1
Ballistics I

Room: A
Date: Tuesday 7/7/2020 **Time:** 900 : 1045

Board:

Prof. Dr.	Ahmed Mohamed Reyad	M. T. C.
Assoc. Dr.	Mahmoud Yehia Mohamed	M. T. C.
Col. Dr.	Moustafa Samir Mohamed	M. T. C.

001_AMME-19	***Experimental investigation of star grains in dual thrust Solid propellent motors*** M El-Naggar, H Belal and H Abdalla
002_AMME-19	***Mitigation of irregular burning in a small solid propellent rocket motor*** M Elkshen, H Belal and M Allam Al-Sanabawy
003_AMME-19	***Validation of a Simplified Model for Liquid Propellant Rocket Engine Combustion Chamber Design*** M Hegazy , H Belal, A Maklad and M Allam Al-Sanabawy
004_AMME-19	***Validation of relations for the determination of pyrotechnic igniter mass*** A Salama, H Belal and H Abdalla

IOP Conf. Series: Materials Science and Engineering **973** (2020) 011001 doi:10.1088/1757-899X/973/1/011001

Military Technical College
Kobry El-Kobbah,
Cairo, Egypt.

19^h Int. Conference on
Applied Mechanics and
Mechanical Engineering,
July 7-9, 2020.

001_AMME-19

Experimental investigation of star grains in dual thrust solid propellant motors

M El-Naggar[1] H Belal[1] and H Abdalla[1]

Abstract

One of the great challenges in designing tactical solid missiles is to achieve high acceleration in the boost phase then maintaining constant speed during the sustain phase. This could be achieved by using a dual thrust solid-propellant rocket motor. Many of these tactical motors use a combination of star, tubular or finocyl grains to achieve this profile. The present study uses two tandem star grains with different design parameters and different transition geometry. Previous researches had consistently shown that the main advantage of star grain is the potential higher volumetric loading in addition to high tailorability. The pressure-time curve for the designed grains is calculated using a zero-dimensional internal ballistic module and a small-scale test motor is used to verify the calculated pressure-time curve. Different transition geometries are compared. Tapered transition is shown to give a comparable performance with the sharp transition with the advantage of higher volumetric loading.

[1] Egyptian Armed Forces

IOP Conf. Series: Materials Science and Engineering **973** (2020) 011001 doi:10.1088/1757-899X/973/1/011001

Military Technical College
Kobry El-Kobbah,
Cairo, Egypt.

A.M.M.E.

19h Int. Conference on
Applied Mechanics and
Mechanical Engineering,
July 7-9, 2020.

002_AMME-19

Mitigation of irregular burning in a small solid propellent rocket motor

M Elkshen[1], H Belal[1] and M Al-Sanabawy[1]

Abstract

An irregular burning phenomenon arises in some solid propellant motors (SRM), where the time-averaged pressure differs largely from that dedicated by instantaneous area burning. This phenomenon is generally considered the result of the interaction of pressure disturbances within the motor cavity with solid propellant deflagration. Hence, it is a combustion-dependent acoustic phenomenon. Historically, different suppression devices included the classical resonance rods and paddles, and arrays of acoustic resonators and various baffle arrangements were applied. In this paper, irregular burning which arises in a modification of a shoulder-launched rocket is investigated. Several techniques dealing with the irregular burning problem were attempted. In each case, the signal of pressure was analysed in the time domain through time average and AC pressure signal, and in the frequency domain through Fourier transform and power spectral density. The most successful technique was the drilling of holes through the grain, connecting the perforation to the external burning region of the cylinder. Next was the increasing of blocking factor, both methods lead to stabilizing the modified motor.

---

[1] Egyptian Armed Forces

IOP Conf. Series: Materials Science and Engineering **973** (2020) 011001 doi:10.1088/1757-899X/973/1/011001

Military Technical College
Kobry El-Kobbah,
Cairo, Egypt.

19^h Int. Conference on
Applied Mechanics and
Mechanical Engineering,
July 7-9, 2020.

003_AMME-19

Validation of a Simplified Model for Liquid Propellant Rocket Engine Combustion Chamber Design

M Hegazy[1], H Belal[1] , A Maklad[1] and M Al-Sanabawy[1]

Abstract
The combustion phenomena inside the thrust chamber of the liquid propellant rocket engine are very complicated because of different paths for elementary processes. In this paper, the characteristic length (L*) approach for the combustion chamber design will be discussed compared to the effective length (L_{eff}) approach. First, both methods are introduced then applied for real LPRE. The effective length methodology is introduced starting from the basic model until developing the empirical equations that may be used in the design process. The classical procedure of L* was found to over-estimate the required cylindrical length in addition to the inherent shortcoming of not giving insight where to move to enhance the design. The effective length procedure was found to be accurate within ± 10%.

[1] Egyptian Armed Forces

IOP Conf. Series: Materials Science and Engineering **973** (2020) 011001 doi:10.1088/1757-899X/973/1/011001

Military Technical College	A.M.M.E.	19h Int. Conference on
Kobry El-Kobbah,		Applied Mechanics and
Cairo, Egypt.		Mechanical Engineering,
		July 7-9, 2020.

004_AMME-19

Validation of relations for the determination of pyrotechnic igniter mass

A Salama[1], H Belal[1] and H Abdalla[1]

Abstract

Propellant ignition is an essential process in solid rocket motor operation as it is responsible of starting the combustion and to reliably self-sustain burning of the solid propellant. Despite the importance of igniter design and ignition process, the relations dealing with igniter mass determination are not well studied and not validated against actual motors with a systematic variation of parameters. The main objective of the current research is to validate the available equations in open literature for determination of the mass of igniter pyrotechnic charge. Data from seven real solid propellant rocket motors were used to validate the relations. These motors have different sizes, grain geometries and pyrotechnic compositions. The results show that the equation of Brayan–Lawerene has the smallest root mean square error (RMSE), and so, it is considered more convenient.

---

[1] Egyptian Armed Forces

IOP Conf. Series: Materials Science and Engineering **973** (2020) 011001 doi:10.1088/1757-899X/973/1/011001

Military Technical College
Kobry El-Kobbah,
Cairo, Egypt.

19^h Int. Conference on
Applied Mechanics and
Mechanical Engineering,
July 7-9, 2020.

Session No. 2

Mechanical Power Engineering I

Room: A

Date: Tuesday 7/7/2020 **Time:** 1100 : 1245

Board:

Maj.Gen.(r) Prof.	Salman E. El-Shamarka	Former Command. of M. T. C.
Maj. Gen.(r) Assoc. Prof.	Ahmed Mohamed Rashad	M. T. C.
Maj. Gen. (r) Dr.	Mohamed Ibrahim Amin	M. T. C.

005_AMME-19	***Simplified algorithm for determining injection duration to achieve fast response in SI engines*** Omer Shawky
006_AMME-19	***The effect of rectangular Gurney flap on Wells turbine performance*** A T M Kotba[1], M A A Nawarb and R M Abd El Maksoudc[3]
007_AMME-19	***Numerical Studies of Turbulent Premixed Flame Interaction with Repeated Solid Obstacles*** M Elshimy, S Ibrahim and W Malalasekera
008_AMME-19	***Development of rotary heat exchanger model to investigate performance with integrating controller*** A M Ghoneim, M H Aly and O E Mahmoud

IOP Conf. Series: Materials Science and Engineering **973** (2020) 011001 doi:10.1088/1757-899X/973/1/011001

Military Technical College
Kobry El-Kobbah,
Cairo, Egypt.

19^h Int. Conference on
Applied Mechanics and
Mechanical Engineering,
July 7-9, 2020.

005_AMME-19

Simplified algorithm for determining injection duration to achieve fast response in SI engines

Omer Shawky[1]

Abstract
For modern SI engines, the injection duration and its timing are one of the most important parameters which influence the mixture homogeneity, emissions, combustion efficiency, and engine response. In the present work, a simplified port fuel injection system with its driver algorithms was designed to control the injection duration and its timing. The Calibrated Dose method is used in this system instead of the stream mixing method. Therefore, the injected fuel mass is accurately computed according to the engine load, engine speed, engine operating condition, and the amount of air drawn into the engine. The system consists of a conventional port fuel injection system equipped with a micro-controller, a Throttle Position Sensor (TPS),a couple of Hall Effect Sensors (HES), load cell with its driver module, injector controller circuit, and a custom-made inlet port. The micro-controller is used to calculate the injection duration each engine cycle according to engine speed, amount of air drawn per cycle and engine load. The two HES are used to detect the injection timing and measure the engine speed. The TPS is used to monitor the throttle position angle so it can calculate the amount of air drawn into the engine at that time with respect to engine speed. The load cell and its driver module are used to measure the engine torque and consequently the brake power. The injector controller circuit is used to initiate the injection process at the perfect timing for the accurate duration.

---

[1] Mechanical Dept. Menoufia University

IOP Conf. Series: Materials Science and Engineering **973** (2020) 011001 doi:10.1088/1757-899X/973/1/011001

Military Technical College
Kobry El-Kobbah,
Cairo, Egypt.

19th Int. Conference on
Applied Mechanics and
Mechanical Engineering,
July 7-9, 2020.

006_AMME-19

The effect of rectangular Gurney flap on Wells turbine performance

A T M Kotba[1], M A A Nawarb[2] and R M Abd El Maksoudc[3]

Abstract

Wells turbine is very adept at converting pneumatic power from ocean waves into mechanical energy. However, such turbines incur from low performance, limited operating range and low efficiency. The present study introduces a method to enhance the power produced from Wells turbine by implementing rectangular Gurney flap (GF). The Gurney flap is placed on both pressure and suction sides of the trailing edge (TE) and perpendicular to the chord line without change chord length. GF increases the lift coefficient by altering the Kutta condition at the TE. The turbine performance is herein evaluated by solving numerically 3D incompressible Reynolds Averaged Navier–Stokes equations (RANS) by using ANSYS Fluent 19.0. The performance is demonstrated according to the flow coefficient, torque coefficient, total pressure loss coefficient, and the turbine efficiency, as well as the velocity and pressure fields around the turbine blades. The validation of the present work was achieved using previous experimental work and CFD work by using SST $k-\omega$ turbulence model. The present work showed that the Wells turbine augmented with GF increases the torque coefficient by 41.98%. Besides, the stall has been delayed compared with the conventional Wells turbine.

---

[1] Demonstrator, Mechanical Power Engineering Department, Faculty of Engineering, Mattaria, Helwan University, Egypt.
[2] Assoc. Prof., Mechanical Power Engineering Department, Faculty of Engineering, Mattaria, Helwan University, Egypt.
[3] Assoc. Prof., Mechanical Power Engineering Department, Faculty of Engineering, Mattaria, Helwan University, Egypt.

IOP Conf. Series: Materials Science and Engineering **973** (2020) 011001 doi:10.1088/1757-899X/973/1/011001

Military Technical College
Kobry El-Kobbah,
Cairo, Egypt.

19[h] Int. Conference on
Applied Mechanics and
Mechanical Engineering,
July 7-9, 2020.

007_AMME-19

Numerical Studies of Turbulent Premixed Flame Interaction with Repeated Solid Obstacles

M Elshimy[1], S Ibrahim[1] and W Malalasekera[2]

Abstract

This paper presents numerical simulations of hydrogen and propane turbulent premixed flames interaction with repeated solid obstructions. The laboratory-scale combustion chamber used in this study is equipped with three solid baffles which promote the generation of turbulence and a square obstacle located downstream from the ignition source. The test cases considered have two different area blockage ratios (ABR) of 24% and 50%, respectively. The large eddy simulation (LES) turbulence modelling technique is used. The numerical simulations are carried out using an in-house computational fluid dynamics (CFD) model. Two different flow configurations are examined, both using three consecutive baffles to identify the subsequent effects and the sensitivity of each fuel to increasing the ABR. These effects are studied using the nature of the flame-obstacles interaction, generated combustion overpressure and resultant flame speed. The modelling capability is confirmed by validating the numerical results against published experimental data. Conclusions are drawn that increasing the ABR increases the combustion overpressure, rate of pressure rise and flame speed. It is also concluded that the larger obstacle has a significant effect on the propagating flame structure and that hydrogen flames are more sensitive to an increased ABR and produce a significantly higher peak overpressure.

---

[1] Department of Aeronautical and Automotive Engineering, Loughborough University, Loughborough, LE11 3TU, UK
[2] Wolfson School of Mechanical, Electrical and Manufacturing Engineering, Loughborough University, Loughborough, LE11 3TU, UK

IOP Conf. Series: Materials Science and Engineering **973** (2020) 011001 doi:10.1088/1757-899X/973/1/011001

Military Technical College
Kobry El-Kobbah,
Cairo, Egypt.

19th Int. Conference on
Applied Mechanics and
Mechanical Engineering,
July 7-9, 2020.

008_AMME-19

Development of rotary heat exchanger model to investigate performance with integrating controller

A M Ghoneim[1], M H Aly[1] and O E Mahmoud[1]

Abstract

The heat wheel Considers one of the best heat recovery devices with very high effectiveness, which leads to great savings in energy and operating expenses. This research study a novel dynamic model of a rotary heat exchanger automatically control the temperature of outlet air despite the operating conditions are changed. The model is implemented under MATLAB/SIMULINK environment to investigate the performance of the heat wheel considering the air properties changing during the operation. The transient effect of air-flow rate, inlet air temperature and wheel rotational speed on the output temperature and effectiveness are discussed. The optimum rotational speed of a heat wheel was investigated for different operating conditions. The temperature distribution of air and matrix for one Full cycle of heat wheel were investigated. The model was validated based on data found in previously published experimental work. Results reveal that the effectiveness of the device is inversely proportional to the amount of flow, while it is directly proportional to the rotational wheel speed. The effectiveness is slightly affected by the inlet air temperature.

[1] Mechanical Power Engineering Department, Faculty of Engineering at Mataria, Helwan University, Cairo, Egypt

IOP Conf. Series: Materials Science and Engineering **973** (2020) 011001 doi:10.1088/1757-899X/973/1/011001

Military Technical College
Kobry El-Kobbah,
Cairo, Egypt.

A.M.M.E.

19th Int. Conference on
Applied Mechanics and
Mechanical Engineering,
July 7-9, 2020.

Session No. 3

Automotive Engineering / Mechanical Design

Room: A

Date: Tuesday 7/7/2020 **Time:** 1300 : 14.45

Board:

Prof. Dr.	Nabil A. Gadallah	Modern Academy
Maj. Gen. (r) Dr.	Moustafa M. Abdel-Wahab	M. T. C.
Prof. Dr.	Shawky A. Hegazy	M. T. C.

009_AMME-19	***Handling Performance of an 8x8 Combat Vehicle*** M Ahmed, M El-Gindy and H Lang
010_AMME-19	***Evaluating the effect of shape on energy absorbing response of structures used in armored vehicles floors*** Hesham kamel
011_AMME-19	***Test rig design for upper-limb socket prosthetics at transradial amputation level*** M M Bondok, M A El-Sheikh and M A El-Hadek
012_AMME-19	***Design and manufacturing of a test rig for measuring the torque required in soil drilling operations*** M A Abdeldayem, M E Abo-Elnor and M H Mabrouk
013_AMME-19	***Assessment of steel wire's fatigue life using finite elements modelling and experimental testing*** H M Lotfy, A B El Shabasy, T Attia and H A Hassan

IOP Conf. Series: Materials Science and Engineering **973** (2020) 011001 doi:10.1088/1757-899X/973/1/011001

Military Technical College
Kobry El-Kobbah,
Cairo, Egypt.

19h Int. Conference on
Applied Mechanics and
Mechanical Engineering,
July 7-9, 2020.

009_AMME-19

Handling Performance of an 8x8 Combat Vehicle

M Ahmed[1], M El-Gindy[1] and H Lang[1]

Abstract

In this research paper, the handling stability of an 8x8 combat vehicle will be assessed using two different control systems. The first technique utilizes a Torque Vectoring Controller (TVC) to control the vehicle yaw rate to meet the desired value. The second technique utilizes an Active Rear-axles Steering (ARS) to minimize the vehicle sideslip. TVC will be designed as a Single Input Single Output (SISO) control problem using a Sliding Mode Control (SMC) technique, while an Optimal Linear Quadratic Regulator (LQR) will be utilized to develop the ARS controller. The two controllers will be evaluated against a conventional vehicle with fixed rear axles. TruckSim software is used in corporation with Matlab/Simulink to implement and assess the controllers using Double Lane Change (DLC) over high and low coefficient of friction road surfaces at various speeds. The results give an insight into the driving conditions at which each controller is utilized and introduce a novel method to coordinate the integration between both controllers for integrated chassis applications.

---

[1] Automotive, Mechanical and Manufacturing Engineering, Faculty of Engineering and Applied Science, University of Ontario Oshawa, Ontario, Canada

IOP Conf. Series: Materials Science and Engineering **973** (2020) 011001 doi:10.1088/1757-899X/973/1/011001

Military Technical College
Kobry El-Kobbah,
Cairo, Egypt.

19th Int. Conference on
Applied Mechanics and
Mechanical Engineering,
July 7-9, 2020.

010_AMME-19

Evaluating the effect of shape on energy absorbing response of structures used in armored vehicles floors

Hesham kamel [1]

Abstract

Improvised explosive devices (IEDs) are a major threat to the lives and wellbeing of soldiers transported in armored vehicles. The blast of an IED under vehicle body can cause severe injuries especially to the lower extremities of vehicle occupants. Blast-mitigating structures are developed to absorb as much energy as possible to protect occupants from the effect of blast. These structures are placed on vehicle floor to soften the impact on occupants' lower extremities thus minimizing injuries. Shape plays an important role in determining the amount of energy absorbed in these structures. This paper presents a finite element study where nonlinear finite element (FE) models were used to simulate the motion of vehicle floor due to the blast of an IED. A benchmark structure was developed to mimic Skydex that is a trademark structure. Different shapes were then used and compared with Skydex. The study shows the potential of optimizing shape to maximize the performance of blast-mitigating structures.

[1] Egyptian Armed Forces.

IOP Conf. Series: Materials Science and Engineering **973** (2020) 011001 doi:10.1088/1757-899X/973/1/011001

Military Technical College
Kobry El-Kobbah,
Cairo, Egypt.

19[h] Int. Conference on
Applied Mechanics and
Mechanical Engineering,
July 7-9, 2020.

011_AMME-19

Test rig design for upper-limb socket prosthetics at transradial amputation level

M M Bondok[1] , M A El-Sheikh[1] and M A El-Hadek[1]

Abstract

Statistics show that the transradial level is the most common level of upper-limb amputation seen by the prosthetist. Amputee needs an intermediate part to wear and fit the prosthetic device with the residual limb firmly. The socket is the only channel between residual limb and prosthetic components and is the crucial part of the prosthesis influencing the amputee's acceptance considerably. The purpose of this design is making a test rig that can assess the socket while simulating arm movements that are common during activities of daily living such as:(Abduction, Adduction, Flexion, Extension, Supination and Pronation) in Euler angles: Yaw, Pitch and Roll. This test rig allows to predict when dislocation between the residual limb and the socket will be happened with any weight and at any plane.

[1]Production and Mechanical Design Department, Faculty of Engineering, Port-Said University.

IOP Conf. Series: Materials Science and Engineering **973** (2020) 011001 doi:10.1088/1757-899X/973/1/011001

Military Technical College
Kobry El-Kobbah,
Cairo, Egypt.

19^h Int. Conference on
Applied Mechanics and
Mechanical Engineering,
July 7-9, 2020.

012_AMME-19

Design and manufacturing of a test rig for measuring the torque required in soil drilling operations

M A Abdeldayem[1], M E Abo-Elnor [1] and M H Mabrouk[1]

Abstract
Soil drilling operation has become one of the most important interests to researchers due to its many applications in engineering systems. Such as construction industry, soil samples for geological sciences and space sampling. The dominant factor in determining drilling parameters based on drilling operations experience or in some times based on proposed modelling techniques. As a result, soil drilling process using auger drilling is studied to obtain drilling parameters and then optimize these parameters to improve drilling performance which enables proper selection of machine for a required job. One of the main challenges that faces researchers during using modelling techniques to define the soil drilling problem is the complex nonlinear behaviour of the drilled medium itself due to its discontinuity and heterogeneous formation. This paper presents a developed apparatus that has been designed and manufactured to be used in measuring and recording the total torque required during soil drilling operation. A simplified auger drilling machine is built in soil-tool interaction laboratory, Military Technical College, to obtain experimental results that can be used to verify the presented models. Data acquisition measuring system is established to analyse experimental results using a. The Labview® program enables recording and displaying the output data collected mainly from sensors planted in the test rig. Results of both analytical and numerical models are then compared to experimental results to aid in developing the presented parametric study that can be used to define the working parameters during drilling operations in different types of soils.

[1] Egyptian Armed Forces

IOP Conf. Series: Materials Science and Engineering **973** (2020) 011001 doi:10.1088/1757-899X/973/1/011001

Military Technical College
Kobry El-Kobbah,
Cairo, Egypt.

19^h Int. Conference on
Applied Mechanics and
Mechanical Engineering,
July 7-9, 2020.

013_AMME-19

Assessment of steel wire's fatigue life using finite elements modelling and experimental testing

H M Lotfy[1], A B El Shabasy[1], T Attia[1] and H A Hassan[1]

Abstract

Various applications consist of wire rope as one of their most important construction parts. Wire ropes formed of three parts as basis; wires, strands and core. to get a deep understanding of wire ropes and their mechanical properties, it is essential to study their main part; the wire. For this reason, steel wire has been studied in this work to get its fatigue life as the number of cycles to failure using finite element modeling. FEM results have been verified using experiments; Flex tester machine and fatigue apparatus which is created and structured for this study. Outcomes have proved that using of bigger bending diameters resulted in increasing the fatigue life of the steel wire and vice versa. It is noticed that the FEM results have small and reasonable value of deviation comparing to the experimental results. Other mechanical properties of the steel wire have been obtained from FEM such as; equivalent Von-Mises stress, equivalent elastic strain, equivalent plastic strain and equivalent total strain.

[1] Mechanical Design and Production, Faculty of engineering, Ain Shams University, Cairo, Egypt

Military Technical College
Kobry El-Kobbah,
Cairo, Egypt.

19h Int. Conference on
Applied Mechanics and
Mechanical Engineering,
July 7-9, 2020.

Session No. 4

Material Science and Processing I

Room: A

Date: Tuesday 7/7/2020 **Time:** 1500 : 1645

Board:

Prof. Dr.	Mazen Ibrahim Negm	Al-Azhar Univ.
Prof. Dr.	Mohamed Tolba Sallam	M. T. C.
Brig. Gen. Dr.	Walid Abdel-Hai El-Thalabawy	M. T. C.

014_AMME-19	***Interactions Between Microorganisms and Composite Material with Marble & Granite Filler*** H Ahmed and O Darwesh
015_AMME-19	***Optical Non-linear Characteristics of particles from Natural Resources*** H Ahmed, A Salah, S Hasabelnbi and A Khalil
016_AMME-19	***Improving water desalination efficiency by using carbon nanotubes as pre-treatment*** A AbdAllah, W El-Sallamy, B Azzam and T Osman
017_AMME-19	***Optimization of Friction Stir Welding Parameters Using Response Surface Methodology*** I Sabry , N Gadallah and M Abu-Okail
018_AMME-19	***Study of peak temperatures in friction stir spot welding of AA2024-Al/Polycarbonate and AA2024-Al/Polypropylene systems*** R K Mohammed, S S Mohammed, and T S Mahmoud
019_AMME-19	***The effect of tire rubber particles on the mechanical and physical properties of polyester*** K F Abo Elenien, N A Azab, G Bassioni and M H Abdellatif.

IOP Conf. Series: Materials Science and Engineering **973** (2020) 011001 doi:10.1088/1757-899X/973/1/011001

Military Technical College
Kobry El-Kobbah,
Cairo, Egypt.

19h Int. Conference on
Applied Mechanics and
Mechanical Engineering,
July 7-9, 2020.

014_AMME-19

Interactions Between Microorganisms and Composite Material with Marble & Granite Filler

H Ahmed [1] and O M Darwesh[2]

Abstract

It is promising to develop new composite materials which have antimicrobial activity for many applications; this is our goal in this study. All experiments were done to achieve this an important objective by studying the interaction between the new composites and some pathogenic reference microorganisms. The produced composite was manufactured from industrial wastes. Recently, many research groups in all fields have focusing work on removing the wastes and recycling it as useful materials. In this study, the preparing new composite from marble and granite wastes with thermoplastic matrix PMMA was investigated. This composite is considered as new type of micro- nano composite because it has fillers in the nano and micro sizes. Pathogenic microorganisms, used in this study, obtained from the American type culture collection (ATCC; Rockville, MD, USA), yeast and one pathogenic fungus were applied. The evaluation of analysis was done by spectrophotometer SEM and EDX techniques. The suitable applications of new composite material were mentioned based upon the interaction recorded in the study. The study ends with recommendation for future modifications in the new composite materials to improve its resistances to microorganism.

[1] Dr. Eng. Consultant in materials sciences & materials applications, Egypt
[2] Agricultural Microbiology Department, National Research Centre, Giza, Egypt

IOP Conf. Series: Materials Science and Engineering **973** (2020) 011001 doi:10.1088/1757-899X/973/1/011001

Military Technical College
Kobry El-Kobbah,
Cairo, Egypt.

19h Int. Conference on
Applied Mechanics and
Mechanical Engineering,
July 7-9, 2020.

015_AMME-19

Optical Non-linear Characteristics of particles from Natural Resources

H Ahmed[1], A Salah[2], S Hasabelnbi[2] and A Khalil [2]

Abstract

In these work some nano and micro particles such as granite, zeolite, diopside ferrous and sand glass which considered as waste or byproduct from mining industries. The micro and nano particles from natural resources were prepared by sieving and mixing with acidic solution at certain concentration, the particles are soaked and immerged in the acidic solution for long time to achieve reasonable hemogenity, sterring is done directly before measurements. The non linear characteristics of the prepared solution were measured by z-scan open aperture technique. The results explained in terms of the composition and structure of the micro and nano particles. Granite particles does not introduce any emission due to compact structure, while zeolite, diopside ferrous and sand glass have significant non linear spectrum related to their structure. The research introduces new applications of the ceramics materials micro and nano particles in non linear optical devices. The research ends with conclusions and recommendation for future work in this new field of applications. The new techniques and the new materials selected in the current study have good non linear optical characteristics; these are promising for practical application. It can be used for various optical applications such as optical modulators, sensors and other optical protection applications.

---

[1] Consultant in materials sciences & materials applications, EGYPT
[2] National Institute of laser enhanced Science, NILES, EGYPT.

IOP Conf. Series: Materials Science and Engineering **973** (2020) 011001 doi:10.1088/1757-899X/973/1/011001

Military Technical College
Kobry El-Kobbah,
Cairo, Egypt.

A.M.M.E.

19h Int. Conference on
Applied Mechanics and
Mechanical Engineering,
July 7-9, 2020.

016_AMME-19

Improving water desalination efficiency by using carbon nanotubes as pre-treatment

A AbdAllah[1], W El-Sallamy[2], B Azzam[3] and T Osman[4]

Abstract

Filters based on carbon nanotubes (CNTs) have been highlighted as an emerging technology for water desalination. In this paper, the weight percentage (wt.%) of CNTs required for desalination is discussed. The idea is based upon addition of CNTs to simple filters that is employed as adsorbent for dissolved solids in water (sodium chloride removal). Different CNTs filters (using CNTs Multi walled tube type) specimens were fabricated. Experimental work was done, through building a test rig, to show the effect of percentage of weight of CNTs on desalination of (Red Sea and Brackish) water. The effect of multiple filtration stages has been also shown. It was shown that the filters containing CNTs would be effective if used at the intake of modern RO (Reverse Osmosis) water desalination plants. Using filters with CNTs as pre-treatment of intake water for the RO units will reduce feeding pressure, specific energy and power consumption.

---

1 Production Engineering, Education Development Fund, Cabinet of Ministers, Cairo, Egypt
2 Production Engineering and Printing Technology Department, Akhbar Elyom Academy, 4th Industrial Zone, 6th of October City, 12563 Giza, Egypt
3 Mechanical Design and Production Department, Taibah University, Janadah Bin Umayyah Road, Tayba, 42353 Medina, KSA
4 Mechanical Designs and Production Department, Cairo University, 1 Gamaa Street, 12613 Giza, Egypt

IOP Conf. Series: Materials Science and Engineering **973** (2020) 011001 doi:10.1088/1757-899X/973/1/011001

Military Technical College
Kobry El-Kobbah,
Cairo, Egypt.

19ʰ Int. Conference on
Applied Mechanics and
Mechanical Engineering,
July 7-9, 2020.

017_AMME-19

Optimization of Friction Stir Welding Parameters Using Response Surface Methodology

N Gadallah[1] ,I Sabry[1] and M Abu-Okail[1]

Abstract

The aim of this article is to create a new technique for predicting discontinuity formation, its place and magnitude during aluminium alloy (AA6061) friction stir welding (FSW). The effectiveness of the technique was demonstrated using visual inspection, hardness and tensile test of the friction stir welded joints. The measured current was analysed through power calculations. In each of the FSW stages, the energy consumption is significantly varied, clearly distinguishing the penetration of the tool, its revolution, its traverse movement and its metal removal rate. The findings of tracking the energy consumption indicate that using power consumption means the significance of weld quality. FSW has been carried out based on two factors - two levels. Response surface methodology (RSM) is employed to develop a mathematical model. Analysis of variance (ANOVA) technique checks the adequacy of the developed mathematical model, which is used effectively at 95% confidence level. In contrast, tensile and hardness tests also showed that welds at high power usage failed continuously within the welding area, due to reduced welding temperature and absence of penetration in the welding zone.

---

[1] Manufacturing Engineering Dept., Modern Academy for Engineering and Technology, Cairo, Egypt.

IOP Conf. Series: Materials Science and Engineering **973** (2020) 011001 doi:10.1088/1757-899X/973/1/011001

Military Technical College
Kobry El-Kobbah,
Cairo, Egypt.

19[h] Int. Conference on
Applied Mechanics and
Mechanical Engineering,
July 7-9, 2020.

018_AMME-19

Study of peak temperatures in friction stir spot welding of AA2024-Al/Polycarbonate and AA2024-Al/Polypropylene systems

R K Mohammed[1], S S Mohammed[1], and T S Mahmoud[1]

Abstract
Compared to ordinary fusion welding processes, friction stir spot welding (FSSW) have significant advantages for example: joining of conventionally non-weldable alloys, improved mechanical properties and reduced distortion of weldable alloys joints due to the pure solid-state joining of metals. This welding technique gives the online feedback control and opportunity of automation, allowing automatic adaptation to geometrical and environmental variations of the component. However, several fundamentals of the process such as the temperature inside the stirred zone of the weld and its power on mechanical properties, are not yet fully understood. In the present study, FSSW was adopted to weld AA2024 aluminium alloy sheets with both polycarbonate (PC) and polypropylene sheets using two different tools having two different pin configurations. The two tools have conical pin with 2.5 mm height, however, tool (T1) has draft angle of ≈ 16.69° and the tool (T2) has a draft angle of ≈ 6°.The effects of FSSW parameters of the process, typically, the tool rotational speeds and dwell time, on temperature variation in the tool/work piece interface was also recorded, via an infrared camera. The results revealed that, increasing the tool rotational speed and/or the dwell time increase(s) the peak temperature in FSSW of both AA2024Al/polycarbonate and AA2024Al/polypropylene welded joints. For AA2024Al/polycarbonate joints, the peak temperatures recorded for tool T1 are higher than those recorded using tool T2. The peak temperature ranges were found to vary from 54 to 150 °C and from 45 to 136 °C, for joints welded using T1 and T2 tools, respectively. While, for AA2024Al/polypropylene joints, the peak temperatures recorded for tool T2 are higher than those recorded using tool T1. The peak temperature ranges were found to vary from 40 to 115 °C and from 61 to 143 °C, for joints welded using T1 and T2 tools, respectively.

---

[1] Mechanical Engineering Department, Faculty of Engineering at Shoubra, Benha University, Cairo, Egypt

IOP Conf. Series: Materials Science and Engineering **973** (2020) 011001 doi:10.1088/1757-899X/973/1/011001

Military Technical College
Kobry El-Kobbah,
Cairo, Egypt.

19h Int. Conference on
Applied Mechanics and
Mechanical Engineering,
July 7-9, 2020.

019_AMME-19

The effect of tire rubber particles on the mechanical and physical properties of polyester

K F Abo Elenien [1], N A Azab[2], G Bassioni[3] and M H Abdellatif[4]

Abstract
In this work, a novel composite of polyester with rubber particles -obtained from recycled wasted car tires -was developed. Worn tires generate a large amount of wastes that do not derogate resulting in environmental problems. This has the advantage of producing useful product that is economical and eco-friendly. The effect of the volume fraction of the rubber particles in the composites on the physical and mechanical properties of the composites was studied. Varying the volume fraction from 0-60% resulted in an increase in the impact properties. The addition of rubber particles had also the effect of reducing the density of the composites which is a desirable property

---

[1]PhD student, Mechanical Design and Production Engineering Department, Faculty of Engineering, Ain Shams University.
[2]Associate Professor, Mechanical Design and Production Engineering Department, Faculty of Engineering, Ain Shams University.
[3]Professor, Chemistry Department, Faculty of Engineering, Ain Shams University
[4]Professor, Mechanical Design and Production Engineering Department, Faculty of Engineering, Ain Shams University.

IOP Conf. Series: Materials Science and Engineering **973** (2020) 011001 doi:10.1088/1757-899X/973/1/011001

Military Technical College
Kobry El-Kobbah,
Cairo, Egypt.

19^h Int. Conference on
Applied Mechanics and
Mechanical Engineering,
July 7-9, 2020.

Session No. 5

Dynamics and Vibration I

Room: A

Date: Wensday 8/7/2020 **Time:** 900 : 1045

Board:

Prof. Dr.	Adel Abdel-Razek Omer	M. T. C.
prof. Dr.	Al Hussain M. Sharaf	M. T. C.
Assoc. Prof. Dr	Hesham Mohamed Eltaher	M. T. C.

020_AMME-19	***Effect of Modified Bouc-Wen Model Parameters on Dynamic Hysteresis of Magnetorheological Dampers*** M Abdelhamed, W G Ata, A M Salem, A Z Khafagy
021_AMME-19	***Experimental and Computational Dynamic Structural Analysis of Free-flight Rockets*** M Habaka, S S Farag , M Khalil and M Y Ahmed
022_AMME-19	***Numerical Investigations of Bistable Piezoelectric Cantilever for Enhanced Energy Harvesting*** M A Abdelnaby, M S Shiba, and S Ali
023_AMME-19	***Simulating Ultrasonic Vibration Enhanced FSW Process of AA6082 Using Finite Element Method*** O M Mabrouk, M S El-Wazery, M A EL-Desouky
024_AMME-19	***Passive Vibration Attenuation: A COMPARISON STUDY*** A A Abdelghany, M M Hegazy and A Badawy

IOP Conf. Series: Materials Science and Engineering **973** (2020) 011001 doi:10.1088/1757-899X/973/1/011001

Military Technical College
Kobry El-Kobbah,
Cairo, Egypt.

19h Int. Conference on
Applied Mechanics and
Mechanical Engineering,
July 7-9, 2020.

020_AMME-19

Effect of Modified Bouc-Wen Model Parameters on Dynamic Hysteresis of Magnetorheological Dampers

M Abdelhamed[1], W G Ata[1], A M Salem[1] and A Z Khafagy[1]

Abstract

Magnetorheological (MR) dampers are a semi active control device that becoming a promising actuators in vibration mitigation. The unique properties of MR fluids are used to control the damping force without consuming large power source. Despite MR dampers are fail-safe, efficient, and robustness devices, they poses significant nonlinear characteristics. A well-known Bouc-Wen models and their extension modified Bouc-Wen are the most accurate models that predict the hysteresis of the MR dampers. However, these models suffer from some complexity and limitations and they miss a good fitting to the measured data. In this paper, the existing modified Bouc-Wen (MBW) models are introduced and their dynamic characteristic are simulated using MATLAB Simulink environment. The robustness of the model is judged by a comparison between the experimental tests of the MR damper (on and off state) on the MTS damper test machine in MTC and the simulated characteristics. To better understand the damper nonlinearity, an investigation into the effect of the model parameters on the dynamic characteristics of the damper is studied. A parametric study is proposed to identify the optimal model parameters that best fits the experimental data. The response of the model with the obtained parameters is validated across the measured MR damper characteristics under different sinusoidal excitations and command current. The results show a good agreement between the simulated and measured responses of the MR damper.

[1] Egyptian Armed Forces.

IOP Conf. Series: Materials Science and Engineering **973** (2020) 011001 doi:10.1088/1757-899X/973/1/011001

Military Technical College
Kobry El-Kobbah,
Cairo, Egypt.

A.M.M.E.

19[h] Int. Conference on
Applied Mechanics and
Mechanical Engineering,
July 7-9, 2020.

021_AMME-19

Experimental and Computational Dynamic Structural Analysis of Free-flight Rockets

M Habaka[1], S S Farag[1], M Khalil[1] and M Y Ahmed[1]

Abstract
Aerospace vehicles' designers often strive to approach the lightweight structures in order to improve flight performance without compromising vehicles durability. However, this goal may be achieved on the expense of structural rigidity of the vehicle. A missile with high slenderness ratio is a typical vehicle that should be considered as flexible body if precise dynamic behaviour analysis is sought. This analysis is crucial to identify the flight performance of such missile with high accuracy. The present paper discusses the results of a side-by-side experimental and numerical analysis of the vibration characteristics of a full-scale free-flight missile having 70mm caliber. The experimental modal analysis is conducted on the missile with empty motor to represent its unpowered flight regime. Experimental modal analysis setup involved accelerometer sensors while vibration excitation is achieved using an impact hammer. Modal analysis is also conducted numerically using a well-used high-fidelity commercial tool. Results of experimental modal analysis have shown that the first four modes as bending modes with frequencies of 134.4, 400.6, 819.4 and 1173.6 Hz, respectively. The results also demonstrate the close agreement between numerical and experimental approaches.

---

[1] Egyptian Armed Forces.

IOP Conf. Series: Materials Science and Engineering **973** (2020) 011001 doi:10.1088/1757-899X/973/1/011001

Military Technical College
Kobry El-Kobbah,
Cairo, Egypt.

19^h Int. Conference on
Applied Mechanics and
Mechanical Engineering,
July 7-9, 2020.

022_AMME-19

Numerical Investigations of Bistable Piezoelectric Cantilever for Enhanced Energy Harvesting

M A Abdelnaby[1], M S Shiba[2] and S Ali[3]

Abstract.
Piezoelectric energy harvesting applications have been highly impacted research area recently. Numerical investigations forbistable piezoelectric cantilevers have great importance to understand its dynamic performance. This paper aims to develop comprehensive numerical models to simulate the dynamic response of a unimorph piezoelectric bender under bistability effect for enhanced energy harvesting. To this end, the equivalent stiffness, damping, electromagnetic coupling and capacitance coefficients of a bistable piezoelectric cantilever have been numerically developed. Besides, finite element models are developed to determine the bistability repulsion force between magnets. The resulting simulations are then implemented in an electromechanical model to predict the output voltage and power over a range of load resistance.

---

[1] Production Engineering and Printing Technology Department, Akhbar El Yom Academy, Egypt
[2] Higher Technology Institute, 10th of Ramadan City, 6th of October Branch, Egypt
[3] Department of Mechatronics, Canadian International College, Fifth Settlement, New Cairo, Egypt

IOP Conf. Series: Materials Science and Engineering **973** (2020) 011001 doi:10.1088/1757-899X/973/1/011001

Military Technical College
Kobry El-Kobbah,
Cairo, Egypt.

19th Int. Conference on
Applied Mechanics and
Mechanical Engineering,
July 7-9, 2020.

023_AMME-19

Simulating Ultrasonic Vibration Enhanced FSW Process of AA6082 Using Finite Element Method

O M Mabrouk[1], M S El-Wazery[2,3] and M A EL-Desouky[4]

Abstract
Several modifications have been made on friction stir welding process to overcome some certain limitations which have been reported. One of these modifications is to use ultrasonic energy as an assistance tool in FSW process. In the present paper, a mathematical model is developed to express the heat generation during ultrasonic vibration enhanced friction stir welding process (UVeFSW). A finite element model is built to perform a transient thermal heat transfer analysis using ANSYS mechanical APDL software package. The temperature contours, temperature distribution and the thermal cycles were predicted using the moving heat source technique. To validate the model, k-type thermocouples were used to measure the temperature and thermal cycles at five locations. The results showed a good agreement between the simulated and experimentally measured results.

[1,2]Production Engineering & Mechanical Design Department, Faculty of Engineering, Minoufia University, Shebin El-Kom, Egypt
[3]Current, Mechatronics Engineering Department, High Institute of Engineering and Technology – Elmahala Elkobra
[4]Civil Engineering Department, Nile High Institute of Engineering and Technology - Mansoura, Egypt

IOP Conf. Series: Materials Science and Engineering **973** (2020) 011001 doi:10.1088/1757-899X/973/1/011001

Military Technical College
Kobry El-Kobbah,
Cairo, Egypt.

A.M.M.E.

19[h] Int. Conference on
Applied Mechanics and
Mechanical Engineering,
July 7-9, 2020.

024_AMME-19

Passive Vibration Attenuation: A COMPARISON STUDY

A A Abdelghany[1], M M Hegazy[2] and A Badawy[3]

Abstract

A comparison study between two different approaches for passive vibration attenuation of solar array had been presented. A powerful finite element software "Ansys" was used to perform this study. Finite element models of a cantilever rectangular aluminum plate are created in order to model the satellite solar array. Stiffeners and circular patches (passive masses) are used separately in order to attenuate the vibration of the plate. The output results of the mode shapes and the FRF of the plate are investigated in order to figure out the best and adequate vibration attenuation technique for satellite solar array, the results showed that using stiffeners have a better effect on vibration attenuation more than using circular patches. A further investigation for vibration attenuation using stiffener is conducted to study the effect of changing the aspect ratio of stiffener.

---

[1] Egyptian Armed Forces
[2] Arab Academy for Science, Technology & Maritime Transport
[3] October University of Modern Sciences and Arts

IOP Conf. Series: Materials Science and Engineering **973** (2020) 011001 doi:10.1088/1757-899X/973/1/011001

Military Technical College
Kobry El-Kobbah,
Cairo, Egypt.

19th Int. Conference on
Applied Mechanics and
Mechanical Engineering,
July 7-9, 2020.

Session No. 6

Material Science and Processing II

Room: A

Date: Wedensday 8/7/2020 **Time:** 1100 : 1245

Board:

Prof. Dr.	Mohamed Tolba Salam	M. T. C.
Assoc. Prof. Dr.	Abdel-Hamid M. El-Mahallawy	M. T. C.
Brig. Gen. Dr.	Mohamed Elsayed Shamekh	M. T. C.

025_AMME-19	***Phase stability of mechanically alloyed Ti-Fe-Al alloys*** S A Ibrahim, M M Seleman, H M Ahmed and A E Hannora
026_AMME-19	***Fracture Toughness Assessment of Axial Partially Through Cracks in X65 and X70 Steel Pipes*** A E Said , H A Hassan , T G Abu-El-Yazied and A M Makady
027_AMME-19	***Assessment of St. St. 316L joints welded by laser beam and metal active gas MAG processes*** A M Magdy, A B El-Shabasy and H A Hassan
028_AMME-19	***Pre-Cracking Techniques of Polymeric Materials: an overview*** A Kamal, A H Elsheikh and E Showaib
029_AMME-19	***Abrasive Water Jet Machining of CFRPs: Single Response Optimization Using Taguchi Method Optimization*** I I Edriys, M Fattouh and R Masoud

IOP Conf. Series: Materials Science and Engineering **973** (2020) 011001 doi:10.1088/1757-899X/973/1/011001

Military Technical College
Kobry El-Kobbah,
Cairo, Egypt.

A.M.M.E.

19^h Int. Conference on
Applied Mechanics and
Mechanical Engineering,
July 7-9, 2020.

025_AMME-19

Phase stability of mechanically alloyed Ti-Fe-Al alloys

S A Ibrahim[1], M M Seleman[1], A E Hannora[2] and H M Ahmed[1]

Abstract
Iron alloying element addition to pure titanium can be used to enhance strengthening as well as β-phase stabilizer. However, the outcome depends on the processing route to some extent. Mechanical alloy technique is adapted in this work to produce titanium alloys with 7wt% iron content and 1wt% Aluminum. The obtained nano-size particles have two phases α and β structures and the formation of the second β-phase is mainly enhanced by the milling force. Heating the produced grain compacted powder to a temperature above the beta phase transus temperature followed by quenching to ambient temperature resulted in α-martensite partial transformation. This result can indicate that the obtained β-phase for the present alloys is not fully stabilized by the iron additions.

---

[1] Metallurgical and Materials Engineering Department, Faculty of Petroleum and Mining Engineering, Suez University, 43521 Suez, Egypt
[2] Department of Science and Mathematical Engineering, Faculty of Petroleum and Mining Engineering, Suez University, Suez 43521, Egypt

IOP Conf. Series: Materials Science and Engineering **973** (2020) 011001 doi:10.1088/1757-899X/973/1/011001

Military Technical College
Kobry El-Kobbah,
Cairo, Egypt.

19th Int. Conference on
Applied Mechanics and
Mechanical Engineering,
July 7-9, 2020.

026_AMME-19

Fracture Toughness Assessment of Axial Partially Through Cracks in X65 and X70 Steel Pipes

A E Said[1], H A Hassan[1], T G Abu-El-Yazied[1] and A M Makady[2]

Abstract

This paper presents analysis of fracture assessment methods of axial partially-through crack in X65 and X70 steel pipes with internal pressure. Two analytical methods Folin–Ciocalteu method (FC method) and Gauss–Seidel method (GS method)) are used to make assessment for two steel pipes (steel X65 and X70). Finite Element model of X65 and X70 steel full-scale pipes with axial part-through crack was established. In this work, a comparison is made between results obtained from FC and GS methods and finite element model with previous experimental results. The GS is more conservative assessment method as it provides smaller crack depth (a) corresponding to (Jcr). Finite Element model in case of steel X70 is more conservative than the analytical methods and its results close to the experimental values.

[1]Design and Production Eng. Dept., Faculty of Engineering, Ain Shams University, EGYPT.
[2]Department of Manufacturing Technology, College of Technological Studies, PAAET, Kuwait.

IOP Conf. Series: Materials Science and Engineering **973** (2020) 011001 doi:10.1088/1757-899X/973/1/011001

Military Technical College
Kobry El-Kobbah,
Cairo, Egypt.

19^h Int. Conference on
Applied Mechanics and
Mechanical Engineering,
July 7-9, 2020.

027_AMME-19

Assessment of St. St. 316L joints welded by laser beam and metal active gas MAG processes

A M Magdy[1,3], A B El-Shabasy[2,3] and H A Hassan[2,3]

Abstract
Study the welding of Stainless steel 316L sheet was carried out using two welding techniques, Laser beam and Metal Active Gas MIG. Strips of 1.5 mm thick were cut from the Stainless-steel sheet and then welded using nitrogen gas for shielding. Metallographic examination using optical microscopy and SEM was conducted to study the microstructure of the welded joints. The tensile stress for joints welded by Laser beam and MAG were ranged from 386 to 632 MPa and from 126.5 to 599 MPa respectively according to the parameters of each welding techniques. The average of microhardness values measured at successive distances for joints welded by Laser beam and MAG were 265 HV and 261 HV, respectively. The obtained results were analysed and discussed based on the recent research.

[1] Design and Production Engineering Dept Faculty of Engineering Ain Shams University

IOP Conf. Series: Materials Science and Engineering **973** (2020) 011001 doi:10.1088/1757-899X/973/1/011001

Military Technical College
Kobry El-Kobbah,
Cairo, Egypt.

A.M.M.E.

19th Int. Conference on
Applied Mechanics and
Mechanical Engineering,
July 7-9, 2020.

028_AMME-19

Pre-Cracking Techniques of Polymeric Materials: an overview

A Kamal[1], A H Elsheikh[1] and E Showaib[1]

Abstract

In fracture toughness tests, a number of notched specimens with identical artificial pre-cracks are essential to obtain accurate fracture parameters. The test results are critically depending on the initiation stage quality. Any slight variance in pre-cracks front shape, length and orientation could significantly affects the test results; therefore, producing identical pre-cracks is a critical issue to obtain accurate results. The pre-cracking technique should be selected carefully to fulfil controllably and repeatability requirements of the standard pre-cracks for a certain material while preserving the induced residual stresses at the crack tip at a minimal value. The notching and pre-cracking standards for metallic material have been well specified in ASTM E399. However, the case is more cumbersome for polymeric materials due their viscoelastic nature. ASTM D5045, ISO 13586:2000 and ASTM D6068 specified different procedures to prepare a sharp pre-crack for polymeric materials. Many pre-cracking techniques have been proposed in literature. The present work introduces an overview of the pre-cracking techniques for polymeric materials.

---

[1] Production Engineering and Mechanical Design Department, Tanta University, Tanta, 31527, Egypt

IOP Conf. Series: Materials Science and Engineering **973** (2020) 011001 doi:10.1088/1757-899X/973/1/011001

Military Technical College
Kobry El-Kobbah,
Cairo, Egypt.

19^h Int. Conference on
Applied Mechanics and
Mechanical Engineering,
July 7-9, 2020.

029_AMME-19

Abrasive water jet machining of CFRPs: single response optimization using taguchi method optimization

I I Edriys[1], M Fattouh[2] and R Masoud[21]

Abstract

Taguchi method was applied to assess and optimize machining parameters and their effect on kerf characteristics during abrasive water jet machining (AWJM) of carbon fiber reinforced polymeric composites (CFRPCs). The main responses selected for these analyses are kerf width, kerf taper, metal removal rate, and surface roughness, the consistent machining parameters focussed for this study are abrasive flow rate, pressure, traverse rate, thickness of the workpiece and standoff distance, each parameter has three levels. Twenty-seven experiments were conducted on a typical CFRP composite workpiece materials based on Taguchi L27 design. The response curves and response tables were used to assess the data obtained to control the major significant process factors statistically affecting the kerf characteristics. The optimal settings of process parameters for each response are set up. From the analysis, it was detected that the percentage contribution of the control factors affecting the kerf width is standoff distance, workpiece thickness, abrasive flow rate, traverse rate, and pressure correspondingly. The results exposed that the thickness, feed rate, and standoff distance were the most significant factors affecting the kerf taper, metal removal rate, and surface roughness respectively.

---

[1] Faculty of Engineering, Tanta University, Tanta, Egypt
[2] Faculty of Engineering, Minufiya University,ShebinElkom, Egypt
[21] Kaha for chemical industries (270 army factor), Egypt

IOP Conf. Series: Materials Science and Engineering **973** (2020) 011001 doi:10.1088/1757-899X/973/1/011001

Military Technical College
Kobry El-Kobbah,
Cairo, Egypt.

A.M.M.E.

19^h Int. Conference on
Applied Mechanics and
Mechanical Engineering,
July 7-9, 2020.

Session No. 7

Ballistics II

Room: A

Date: Wedensday 8/7/2020 **Time:** 1300 : 1445

Board:

Prof. Dr.	Mohamed Saied Abo Elkhair	M. T. C.
Assoc. Prof. Dr.	Amr Ibrahim Hassan Fayed	M. T. C.
Maj.Gen (r) Dr.	Osama Mohamed Kamal	M. T. C.

030_AMME-19	***An Experimental Study on Mechanical and Ballistic Characteristics of Different HTPB Composite Propellant Formulations*** A F Nour Eldin, W M Adel and Y A Attai
031_AMME-19	***Penetration of long rods into semi-infinite metallic targets over wide range of impact velocities*** M A Abdel-Kader, A M Riad and A Z Ibrahim
032_AMME-19	***Transient Heat Transfer for GPMG's Barrel 7.62x51*** M Ghanem and O Abdelsalam
033_AMME-19	***Multi-Fidelity Drag Prediction for Base Bleed Projectile*** M M Aziz, A Z Ibrahim, M Y M Ahmed and A M Riad

IOP Conf. Series: Materials Science and Engineering **973** (2020) 011001 doi:10.1088/1757-899X/973/1/011001

Military Technical College
Kobry El-Kobbah,
Cairo, Egypt.

A.M.M.E.

19^{h} Int. Conference on
Applied Mechanics and
Mechanical Engineering,
July 7-9, 2020.

030_AMME-19

An Experimental Study on Mechanical and Ballistic Characteristics of Different HTPB Composite Propellant Formulations

A F Nour Eldin[1], W M Adel[2] and Y A Attai[1]

Abstract

The main goal of this paper is to investigate the changing in the tensile behavior and the burning rate characteristics of hydroxyl-terminated polybutadiene (HTPB) propellantunder the variations of the crosslinking density, which was predominantly determined by the equivalent ratio of diisocyanate to total hydroxyl (NCO/OH ratio), which known as the curing ratio. Four variousbatches with different curing ratios (NCO/OH) percentage were produced in which the production processes were fixed. Uniaxial tensile tests were conducted at different temperatures (-40, 20 and 55°C), and different strain rates (0.000656 1/s, 0.0328 1/s) using a Zwick universal test machine. In order to measure the burning rate, cured solid propellant strands were tested using the acoustic wave emission method under different pressure ranging from 4 to 10 MPa. The experimental results indicate that the tensile behavior of HTPB propellant is remarkably influenced by curing ratio, strain rate, and temperature. It was observed that a great change on stress-strain curves affected various curing ratios and temperatures on the mechanical behavior of propellant composition. The results showed that high curing ratio leads to increase theultimatestress and decrease the strainat maximum stress, but higher temperatures lead to decrease theultimate stress and the strain at maximum stress.The curing ratios (NCO/OH) have an intense impact on mechanical characteristics, but slightly impact on ballistic characteristics for propellant. Furthermore, careful measurements of these parameters are important to control the production quality and to provide a reliable comparison between different propellant batches.

[1] Mechanical power engineering department,Faculty of Engineering, Helwan University
[2] Egyptian Armed Forces

IOP Conf. Series: Materials Science and Engineering **973** (2020) 011001 doi:10.1088/1757-899X/973/1/011001

Military Technical College
Kobry El-Kobbah,
Cairo, Egypt.

19^{h} Int. Conference on
Applied Mechanics and
Mechanical Engineering,
July 7-9, 2020.

031_AMME-19

Penetration of long rods into semi-infinite metallic targets over wide range of impact velocities

M A Abdel-Kader[1], A M Riad[1] and A Z Ibrahim[1]

Abstract

In this paper, the analytical model that describes the penetration process of a long rod into a semi-infinite metallic target over wide range of impact velocities has been presented. In this model, the target strength factor R_t is not considered as a constant but it is a function of penetration velocity whereas, the rod strength factor Y_P is represented by Hugoniot Elastic Limit (HEL) of its material. The penetration process of the present model consists mainly of primary phase during which the rod front is subjected to erosion and secondary (after-flow) phase. The main equations of the used model are presented, arranged and compiled into a computer program. The program is capable of predicting the time histories of the parameters associated with the penetration process. Autodyn hydrocode package is used herein to simulate the penetration processes of long rods into targets examined by the present analytical model. The hydrocode is fed with the same data used in the model for rods and targets materials. Both the model predictions and the obtained numerical results of Auto dynhydro code, respectively, are compared with the corresponding experimental measurements of other investigators. Good agreement is generally obtained. In addition, representative samples of the model predictions and their corresponding simulation results of Autodyn hydro code are presented with relevant analyses and discussions. The obtained results prove the predictive capabilities of both the present model and Autodyn hydro code, respectively, where each of them could be used as a quick tool for determining the main parameters associated with the studied penetration process.

---

[1] Egyptian Armed Forces.

IOP Conf. Series: Materials Science and Engineering **973** (2020) 011001 doi:10.1088/1757-899X/973/1/011001

Military Technical College
Kobry El-Kobbah,
Cairo, Egypt.

19th Int. Conference on Applied Mechanics and Mechanical Engineering, July 7-9, 2020.

032_AMME-19

Transient Heat Transfer for GPMG's Barrel 7.62x51

M Ghanem[1] and O Abdelsalam[1]

Abstract
In this paper, a modified analytical solution for time-dependent one dimension heat diffusion equation in cylindrical coordinate had been mathematically solved to calculate the temperature distribution through thick-walled weapon barrel subjected to successive shots. The analytical solution has been compared with experimental measurements for the temperature at the outer surface of the barrel at two different positions. The General Purpose Machine Gun (GPMG) 7.62X51 mm was considered as the case of study. The results show a good agreement between the analytical model and the experimental work with a difference less than 3 %. The main difference between them mainly as a result of the very short time of the heating phenomena as it takes about 6 (ms) only, although the used Data Acquisition System was able to take a high sampling rate and a high sensitive thermocouple type k with only 0.5 mm thickness had been used.

[1] Egyptian Armed Forces

IOP Conf. Series: Materials Science and Engineering **973** (2020) 011001 doi:10.1088/1757-899X/973/1/011001

Military Technical College
Kobry El-Kobbah,
Cairo, Egypt.

A.M.M.E.

19th Int. Conference on
Applied Mechanics and
Mechanical Engineering,
July 7-9, 2020.

033_AMME-19

Multi-Fidelity Drag Prediction for Base Bleed Projectile

M M Aziz[1], A Z Ibrahim[1], M Y M Ahmed[1] and A M Riad[1]

Abstract
A computational study is performed to predict the drag reduction for a base bleed projectile. 2-D and 3-D simulations are conducted to describe the flow past a projectile with base bleed. The calculated computational results are compared with measurements of total drag coefficients based on published results for 155mm K307 projectiles with live/dummy base bleed grain. In addition, the mean drag coefficient measurements based on live firing tests for a five K307 projectiles with live BB grain at different Mach numbersare used to validate results obtained from the computational results. Differences between the drag coefficient calculated from computational and published live firing measurements of the drag coefficient for K307 are less than 5.2% and 6.8% for projectile with dummy and live BB, respectively. On the other hand, the maximum deviation between computational results and real measurements from the live firing tests of live BB projectiles is 2.75%.

---

[1] Egyptian Armed Forces

IOP Conf. Series: Materials Science and Engineering **973** (2020) 011001 doi:10.1088/1757-899X/973/1/011001

Military Technical College
Kobry El-Kobbah,
Cairo, Egypt.

19h Int. Conference on
Applied Mechanics and
Mechanical Engineering,
July 7-9, 2020.

Session No. 8

Dynamics and Vibration II

Room: A

Date: Wedensday 8/7/2020 **Time:** 1500 : 1645

Board:

Prof. Dr.	Moustafa Adnan Ahmed	M. T. C.
Prof. Dr.	Moataz Mohamed Hegazy	Arab Academy for S.&T.
Assoc. prof. Dr.	Amgad Mohamed Kamel	M. T. C.

034_AMME-19	***Experimental Modal Testing of a Honeycomb Sandwich Plate*** A Aborehab, M Kamel M Kamel, A F Nemnem and M Kamel
035_AMME-19	***A New Artificial Neural Network Model Integrated With a Cat Swarm Optimization Algorithm for Predicting the emitted noise during Axial Piston Pump operation*** A H Elsheikh, M Abd Elaziz, H A Babikir, D Wu, and Y Liu
036_AMME-19	***Passive Vibration Damping of a Cantilever Plate*** A Mohamed, A Hassan, A A Omer

IOP Conf. Series: Materials Science and Engineering **973** (2020) 011001 doi:10.1088/1757-899X/973/1/011001

Military Technical College
Kobry El-Kobbah,
Cairo, Egypt.

19^h Int. Conference on
Applied Mechanics and
Mechanical Engineering,
July 7-9, 2020.

034_AMME-19

Experimental Modal Testing of a Honeycomb Sandwich Plate

A Aborehab[1], M Kassem[1], A F Nemnem[1] and M Kamel[1]

Abstract

Honeycomb structures have been widely used in several applications and industries in the past decades. Such structures have a significant role in aerospace engineering especially in building up the satellite structures. Honeycomb structures provide key benefits represented in high specific strength, high specific stiffness, and superior dimensional stability. Detailed finite element modelling of such structures is a great challenge as it involves high computational expense. Thus, equivalent modelling is mandatory. The detailed and equivalent finite element models of a honeycomb plate are introduced intensively throughout analyzing its dynamic behaviour using the modal analysis module in ANSYS workbench software. The equivalent modelling is carried out via the three-layered sandwich theory and leads to the calculation of the first four natural frequencies and the related mode shapes. An experimental modal testing of the honeycomb sandwich plate is implemented for the sake of validating the computational results. A good agreement is obtained when comparing both experimental and computational results with mean error not exceeding 5%. Finally, a parametric study is performed to relate the modal analysis results with different boundary conditions cases. The results show that the model can be a reliable basis for satellite honeycomb sandwich structures design.

---

[1] Egyptian Armed Forces

IOP Conf. Series: Materials Science and Engineering **973** (2020) 011001 doi:10.1088/1757-899X/973/1/011001

Military Technical College
Kobry El-Kobbah,
Cairo, Egypt.

19^h Int. Conference on
Applied Mechanics and
Mechanical Engineering,
July 7-9, 2020.

035_AMME-19

A New Artificial Neural Network Model Integrated With a Cat Swarm Optimization Algorithm for Predicting the emitted noise during Axial Piston Pump operation

A Elsheikh[1], M Abd Elaziz[2], H Babikir[3], D Wu[3] and Y Liu[3]

Abstract

This study presents a new artificial intelligence based methodology to predict the emitted noise of an Axial Piston Pump (APP). The suggested method depends on augmentation of conventional Artificial Neural Network (ANN) via integration with Cat Swarm Optimization (CSO). CSO is used to obtain the optimal structure of ANN. The training and testing of the approach were accomplished using experimental data sets considering six system operating pressures (0.5, 1.0, 1.5, 2.0, 2.5, and 3.0 MPa) and five speed levels (600, 900, 1200, 1500, and 1800 rpm). Two valve seat materials were investigated: polyetheretherket one (PEEK) and 316L stainless steel. A reasonable agreement was observed between the predicted results obtained by the developed method and the experimental data.

[1] Production Engineering and Mechanical Design Department, Tanta University, Tanta, 31527, Egypt.
[2] Department of Mathematics, Faculty of Science, Zagazig University, Zagazig, Egypt.
[3]The State Key Laboratory of Digital Manufacturing and Equipment Technology, School of Mechanical Science and Engineering, Huazhong University of Science and Technology, Wuhan 430074, China.

IOP Conf. Series: Materials Science and Engineering **973** (2020) 011001 doi:10.1088/1757-899X/973/1/011001

Military Technical College
Kobry El-Kobbah,
Cairo, Egypt.

A.M.M.E.

19h Int. Conference on
Applied Mechanics and
Mechanical Engineering,
July 7-9, 2020.

036_AMME-19

Passive Vibration Damping of a Cantilever Plate

A Mohamed [1], A Hassan [1], A A Omer[1]

Abstract

Vibration control of mechanical systems divided into two types active and passive, this case study illustrates other approach of passive control of the cantilever plate by using double layers of the viscoelastic material (Dyad 606) one in the upper face of the Al substrate (5051) and the second one in the lower face. This layers are affected by axial uniform distributed force by applying tension load, the damping level depend on the value of this axial force where the vertical component of this force consistently the other way of the vibrating element that reduce vibration amplitude in this direction by about (95%) of the original amplitude without damping layers. Computational analysis using ANSYS program are used firstly to verify that the results from the software are compatible with that obtained from analytical and experimental evaluation by comparing the values of natural frequencies. Then using ANSYS model to calculate the reduction occurs in the lateral vibration with changing of the axial loads on the viscoelastic plates. The benefits of this model are enhancing the reliability and reducing the cost.

[1] Egyptian Armed Forces

IOP Conf. Series: Materials Science and Engineering **973** (2020) 011001 doi:10.1088/1757-899X/973/1/011001

Military Technical College
Kobry El-Kobbah,
Cairo, Egypt.

19th Int. Conference on Applied Mechanics and Mechanical Engineering,
July 7-9, 2020.

Session No. 9

Mechanical Power Engineering II

Room: A

Date: Thursday 9/7/2020 **Time:** 900 : 1045

Board:

Prof. Dr.	Ibrahim Saleh Moustafa	M. T. C.
Brig. Gen. Prof. Dr.	Ahmed Salah Abo elazam	M. T. C.
Brig. Gen. Dr.	Mohamed Safaa Eldin	M. T. C.

037_AMME-19	***Modeling and Experimental Investigation of High Pressure Common Rail Diesel Injection System*** W Niklawy, M Shahin, M I. Amin, A Elmaihy
038_AMME-19	***Performance of Homogeneous Charge Compression Ignition (HCCI) Engine with Common Rail Fuel Injection*** W Niklawy, M Shahin, M I. Amin, A Elmaihy
039_AMME-19	***Experimental Investigation and Numerical Simulation of the Thermosyphon Heat Pipe Charged with R134a*** SY Zakaria, A G Ibrahim, A M Rashad and S E El-Shamarka
040_AMME-19	***Aerodynamic performance improvement using a micro-cylinder as a passive flow control around the S809 airfoil.*** W Mostafa, A Abdelsamie, M Mohamed, D Thévenin and M Sedrak.

IOP Conf. Series: Materials Science and Engineering **973** (2020) 011001 doi:10.1088/1757-899X/973/1/011001

Military Technical College
Kobry El-Kobbah,
Cairo, Egypt.

19h Int. Conference on
Applied Mechanics and
Mechanical Engineering,
July 7-9, 2020.

037_AMME-19

Modeling and Experimental Investigation of High Pressure Common Rail Diesel Injection System

W Niklawy[1], M Shahin[2], M I Amin[3] and A Elmaihy[3]

Abstract

The high-pressure Common Rail (HPCR) injection system was originally introduced for diesel engines to both reduce pollutant emissions and enhancement of performance. HPCR separates fuel pressurization and injection processes from each other. The high injection pressure generated by the common rail system provides better atomisation and evaporation of fuel spray, resulting in improved air inlet and fuel jet mixing, which is advantageous for lowering soot emission. In this paper, a mathematical common rail injection system model has been presented. A Simulink/Matlab code was developed to execute this simulation. This work does not only seek to validate the presented numerical model but to have more insight for understanding the overall common rail injection system diesel engine performance under different operating conditions. Some simulation results are illustrated to highlight modelling capability.The engine used is an HCCI turbocharged diesel engine, 2776 cc, 4-stroke, and 4-cylinder, water-cooled with overhead valve mechanism. The common rail pressure, fuel consumption, start and duration of each injection through one engine cycle are measured at various engine speed and loads. The measured common rail fuel pressure and consumption are used to validate the simulation results. The findings of the simulation show good consistency with the experimental results. At last, some simulation results, which highlight the modelling capability, are illustrated at certain values of engine speed and load.

---

[1] PhD candidate, Mechanical Power Engineering Dept., Military Technical College
[2] Mechanical Engineering Department, School of Engineering, Badr University
[3] Mechanical Power Engineering Dept, Military Technical College

IOP Conf. Series: Materials Science and Engineering **973** (2020) 011001 doi:10.1088/1757-899X/973/1/011001

Military Technical College
Kobry El-Kobbah,
Cairo, Egypt.

19^h Int. Conference on
Applied Mechanics and
Mechanical Engineering,
July 7-9, 2020.

038_AMME-19

Performance of Homogeneous Charge Compression Ignition (HCCI) Engine with Common Rail Fuel Injection

W Niklawy[1], M Shahin[2], M I Amin[3] and A Elmaihy[3]

Abstract

Homogeneous Charge Compression Ignition (HCCI) engine is a newly introduced technology. However, controlling the HCCI timing of ignition and theheat release rate, at all operating regimes, present the biggest challenge due to the complicated and highly coupled combustion problems. The successful implementation of the common rail injection system in diesel engines has made this possible. The main key for this success is the capability to optimize the most important fuel injection parameters, namely timing, rate, and duration.This work experimentally investigates the performance of an HCCI Diesel engine with a Common Rail Direct Injection (DI) fuel system. The engine used is a turbocharged 2776 cc, 4-stroke, 4-cylinder, water-cooled with overhead valve mechanism. The engine performance is evaluated and presented at different loads and speeds. The Apparent Heat Release Rate during combustion is evaluated from the measured cylinder pressure-crank angle data. The pressure in the fuel line is measured at the entry to different fuel injectors to investigate the effect of pressure wave propagation in the fuel line on the fuel injection system.

[1] PhD candidate, Mechanical Power Engineering Dept., Military Technical College
[2] Mechanical Engineering Department, School of Engineering, Badr University
[3] Mechanical Power Engineering Dept, Military Technical College

IOP Conf. Series: Materials Science and Engineering **973** (2020) 011001 doi:10.1088/1757-899X/973/1/011001

Military Technical College
Kobry El-Kobbah,
Cairo, Egypt.

A.M.M.E.

19^h Int. Conference on
Applied Mechanics and
Mechanical Engineering,
July 7-9, 2020.

039_AMME-19

Experimental Investigation and Numerical Simulation of the Thermosyphon Heat Pipe Charged with R134a

S Y Zakaria[1], A G Ibrahim[1], A M Rashad[1]
and S E El-shamarka[1]

Abstract

This work presents an experimental and numerical study of Two-Phase Closed Thermosyphon (TPCT) heat pipe filledwith R134a as a base fluid. For this purpose, a test rig was designed and developed toperform a series of tests on the TPCT heat pipe by applying different heat inputs at the evaporator section. The surface temperature along the TCPT heat pipe and temperature changes of the cooling water across the condenser section were measured. The influence of the TCPT heat pipe operating conditions on its thermal performance was reviewed. The considered parameters were; heat pipe transport capacity, conductance, both heat transfer coefficients of the evaporator and condenser sections and the thermal efficiency. Then, a CFD model was developed to investigate the two-phase flow and heat transfer mechanism during the transient and steady-state operation of the TPCT heat pipe. The results of the experimental work were used to validate the CFD model and acceptable agreements were noticed. The validated CFD model is utilized to predict features of the mass and heat transfer processes, as well as the nucleate pool boiling and the liquid film condensation phenomena during TPCT heat pipe operation.

---

[1] Egyptian Armed Forces.

IOP Conf. Series: Materials Science and Engineering **973** (2020) 011001 doi:10.1088/1757-899X/973/1/011001

Military Technical College
Kobry El-Kobbah,
Cairo, Egypt.

19h Int. Conference on
Applied Mechanics and
Mechanical Engineering,
July 7-9, 2020.

040_AMME-19

Aerodynamic performance improvement using a micro-cylinder as a passive flow control around the S809 airfoil.

W Mostafa[1,*], A Abdelsamie[2,3], MMohamed[2,4], D Thévenin[3] and M. Sedrak[2]

Abstract
Aerodynamic performance improvement of airfoils is the first step towards enhancement of the wind turbine performance in electricity generation and energy conversion in renewable energy applications. The flow behavior around wind turbine blades profile can be improved by introducing active and/or passive flow controls. This work numerically describes the impact of adding micro-cylinder, as a passive flow control around S809 airfoil, on aerodynamic performance under various operating conditions. A suitable combination of flow analysis and optimization technique has been used in the current work. The numerical simulation has been performed using ANSYS Fluent 18.2 software. The airfoil was numerically analysed in flow at Reynolds number of 106; aerodynamic coefficients (lift and drag coefficients) at different angle of attacks were validated with the experimental data reported by Somers in NREL. The Response Surface Method (RSM) is applied to obtain the optimum position of micro-cylinder to achieve maximum lift to drag ratio. It has been found that the total aerodynamic forces are sensitive to the location of the micro-cylinder. A significant enhancement of lift to drag ratio can be achieved by adding micro-cylinder in front of S809 airfoil especially at high Reynolds number.

[1] Egyptian Russian University
[2] Helwan University
[3] Umm Al-Qura University
[4] University of Magdeburg
[5] Faculty of Engineering at El-Mattaria, Helwan University

IOP Conf. Series: Materials Science and Engineering **973** (2020) 011001 doi:10.1088/1757-899X/973/1/011001

Military Technical College
Kobry El-Kobbah,
Cairo, Egypt.

A.M.M.E.

19^h Int. Conference on
Applied Mechanics and
Mechanical Engineering,
July 7-9, 2020.

Session No. 10

Mechatronics / Robotic Automation and Control

Room: A

Date: Thursday 9/7/2020 **Time:** 1100 : 1245

Board:

Prof. Dr.	Yehia Hendawy Hossam Eldin	Future University
Col. Dr.	Amr Ahmed Roshdy	M. T. C.
Col. Dr.	Moustafa Shokry Asfoor	M. T. C.

041_AMME-19	***Assistive exoskeleton hand glove using soft pneumatic actuators: design optimization*** M N El-Agroudy, M I Awad, A Rehan and S A Maged
042_AMME-19	***Design of Gain Scheduled Fuzzy PID Controller for Multi-Link Robot Manipulators*** Maya Emarah
043_AMME-19	***Design of bio-inspired muscle sarcomere structure using a hybrid Hydrogel-IPMC actuator.*** H M Hamdi, S A Abdelwahhab, M I Awad and F A Tolbah
044_AMME-19	***Performance Investigation and control parameters choice for sliding mode control of coupled tanks system*** M A Louis, M R Roman, O E Mahmoud and M F Sedrak

IOP Conf. Series: Materials Science and Engineering **973** (2020) 011001 doi:10.1088/1757-899X/973/1/011001

Military Technical College
Kobry El-Kobbah,
Cairo, Egypt.

19h Int. Conference on
Applied Mechanics and
Mechanical Engineering,
July 7-9, 2020.

041_AMME-19

Assistive exoskeleton hand glove using soft pneumatic actuators: design optimization

M N El-Agroudy[1], M I Awad[1], A Rehan and S A Maged [1]

Abstract
An assistive exoskeleton hand glove can be the solution to people who suffer from impairment in hand functionality. Soft pneumatic actuators (SPA) are used to actuate such glove. This paper focus on optimizing the design of the soft actuator. Finite element (FE) model is linked with the design optimization tool to find the optimal dimensions and materials of the SPA that give maximum displacement of the actuator with acceptable stresses on the actuator's body. Performance analysis shows the relation between different parameters and how each affect the performance of the actuator. Results show the dimensions and materials needed to have maximum actuator displacement without being subjected to high stresses.

[1] Mechatronics Department, Faculty of Engineering, Ain Shams University, Egypt

IOP Conf. Series: Materials Science and Engineering **973** (2020) 011001 doi:10.1088/1757-899X/973/1/011001

Military Technical College
Kobry El-Kobbah,
Cairo, Egypt.

19[h] Int. Conference on
Applied Mechanics and
Mechanical Engineering,
July 7-9, 2020.

042_AMME-19

Design of Gain Scheduled Fuzzy PID Controller for Multi-Link Robot Manipulators

Maya Emarah[1]

Abstract
Gain Scheduling (GS) controller is one of the most popular approaches to nonlinear control design. Commonly, Gain Scheduling (GS) controllers have better performance than robust ones. In this paper, the structure of a proposed gain scheduled PID fuzzy controller (GS-PIDF) designed and analyzed. The Fuzzy PID controller gains weighted through gain scheduling designed system to grantee both the stability and the best performance of a multi links robot manipulator system. The overall designed controller system is applied to two links robot manipulator system, which is subject to modelling using Lagrangian dynamics method and simulation vie MatLab and Simulink package to minimize the vibration and acquire best performance.

---

[1] Assistant Professor, the High Institute of Engineering, City of Science and Culture, 6th of October City, Giza, Egypt.

IOP Conf. Series: Materials Science and Engineering **973** (2020) 011001 doi:10.1088/1757-899X/973/1/011001

Military Technical College
Kobry El-Kobbah,
Cairo, Egypt.

19^{h} Int. Conference on
Applied Mechanics and
Mechanical Engineering,
July 7-9, 2020.

043_AMME-19

Design of bio-inspired muscle sarcomere structure using a hybrid Hydrogel-IPMC actuator

H M Hamdi[1], S A Abdelwahhab[2], M I Awad[3] and F A Tolbah[3]

Abstract

Biological muscle is considered a powerful actuator due to its flexibility, lightweight, and efficiency. The building unit of a muscle, sarcomere, and the study of its energy balance cycle is considered in researches due to its importance to mimic the micro-level muscle structure to improve the artificial muscle performance. In this work, a new design of a linear actuator based on the sarcomere behavior is developed. The design is inspired by studying the four steps adenosine triphosphate (ATP) hydrolysis cycle, which is the main source of the required energy for sarcomere contraction. A new developed hybrid hydrogel-polymeric material actuator is designed in this paper using a combination between Ionic Polymeric Metallic Composites (IPMC) and hydrogel to behave like the sarcomere. This new actuator proposes an autonomous cycle using the effect oscillatory Belousov–Zhabotinsky (BZ) reaction. The physical model is proposed, and the mathematical model of the actuator is derived and formulated and identified using MATLAB/ Simulink.

---

[1] Design and Production Engineering Department, Faculty of Engineering, Ain Shams University, Cairo, Egypt.
[2] Production Technology Department, Faculty of Industrial Education, Helwan University, Cairo, Egypt.
[3] Mechatronics Engineering Department, Faculty of Engineering, Ain Shams University, Cairo, Egypt.

IOP Conf. Series: Materials Science and Engineering **973** (2020) 011001 doi:10.1088/1757-899X/973/1/011001

Military Technical College
Kobry El-Kobbah,
Cairo, Egypt.

19th Int. Conference on
Applied Mechanics and
Mechanical Engineering,
July 7-9, 2020.

044_AMME-19

Performance Investigation and control parameters choice for sliding mode control of coupled tanks system

M A Louis [1], M R Roman[2], O E Mahmoud[2] and M F Sedrak[2]

Abstract

In several food processing and chemical industries, liquid is pumped and kept in interrelating coupled tanks. However, automatic regulation of the liquid level and flow control between these tanks is a challenging problem because of the complexity and high non linearity of such system. This paper deals with the liquid level control of two horizontal coupled tanks system. A comprehensive comparative study is made for most popular sliding mode control (SMC) algorithms found in literature, namely Proportional-Derivative Sliding Mode Control (PD-SMC), Proportional-Integral-Derivative SMC (PID-SMC), Fractional Order SMC and finally dynamic SMC. Special emphasis is put on the effect of the sensor noise on the controller performance. Simulated experiments including robustness to variation in plant parameters and step input disturbances are made. Control algorithms parameters are selected to optimize designed performance indices by using MATLAB optimization toolbox. Simulation results reveal that dynamic SMC is superior to other control algorithms in the presence of sensor noise and has a significant reduction in the actuator chattering phenomenon.

---

[1] Mechatronics department, Faculty of Engineering, Egyptian Russian University, Cairo, Egypt

[2] Mechanical power engineering department, Faculty of Engineering at Mataria, Helwan University, Cairo, Egypt

IOP Conf. Series: Materials Science and Engineering **973** (2020) 011001 doi:10.1088/1757-899X/973/1/011001

Military Technical College
Kobry El-Kobbah,
Cairo, Egypt.

19^h Int. Conference on
Applied Mechanics and
Mechanical Engineering,
July 7-9, 2020.

Session No. 11

Production Technology

Room: A

Date: Thursday 9/7/2020 **Time:** 1300 - 1445

Board:

Maj. Gen. (r) Dr.	Wael Salah Eldin Hamed	M. T. C.
Assoc. Prof. Dr.	Amr Mohamed Fekry	M. T. C.
Col. Dr.	Ahmed Selem El Mosalamy	M. T. C.

045_AMME-19	***Effect of applying Lean Maintenance in Oil and Gas fields*** M Nasr Hassan, A F Barakat and A S Sobh
046_AMME-19	**A New Integrated Process and Product Development model** A Ayman, A B Alhamaki and M H Abdellatif
047_AMME-19	**Ultrasonic-assisted drilling of Nickel-based superalloy In conel 601: An experimental study** A M Abdelaziz , H Youssef , M Al-Makky and H El-Hofy
048_AMME-19	**Computational fluid Dynamic model for machining using minimum quantity coolant** W Abdelfattah, H Hegabb, A Mohany and H A. Kishawy
049_AMME-19	**Optimization Ladle-Refining Performance During Treating Special Steel Melts for Aviation Technology** S A Elghazaly, K K Gyula and W Saied Elghazaly

IOP Conf. Series: Materials Science and Engineering **973** (2020) 011001 doi:10.1088/1757-899X/973/1/011001

Military Technical College
Kobry El-Kobbah,
Cairo, Egypt.

19[h] Int. Conference on
Applied Mechanics and
Mechanical Engineering,
July 7-9, 2020.

045_AMME-19

Effect of applying Lean Maintenance in Oil and Gas fields

M Nasr Hassan [1], A a Fathalla Barakat[2] and A Sobh[3]

Abstract

Oil and Gas Sector are facing a huge increase and delay in Operations & Maintenance tasks for all facilities beside continuous demand from management to improve the productivity of wells and the efficiency of facilities. The top management decides to use Lean Maintenance as a new tool, technique and methodology to improve and modify the current operations, production and maintenance systems as done before by many researchers in the oil and gas sector. This improvement will be done through three real case studies by integrating Building Information Modelling with lean concepts, monitoring Key Performance Indicators within a modern Computerized Maintenance Management Software and finally using some LM tools such as Total Productive Maintenance, Value Stream Mapping and 5S Methodology to increase productivity, efficiency, and quality of the output production and services. The results of implementation since July 2017 till June 2018 show average increase in crude oil production by 6.72% per day in addition to overcome natural decline of oil wells, reduced Technician's total Preventive Maintenance tasks by 52.63% and transfer the less difficult inspections to Operator to be responsible for making checks to all parameters of all equipment and to enable Technicians to perform other maintenance tasks, Saving about $724,770 in Water Injection Pumps Operation's Expense. Last but not least, accelerate the workflow of production and maintenance processes by 28% compared with the previous financial year.

---

[1]Mechanical Dept., Faculty of Engineering, Helwan University, Cairo, Egypt
[2]Professor Faculty of Engineering, Ahram Canadian University (ACU) & Faculty of Engineering, Helwan University
[3]Mechanical Dept., Faculty of Engineering, Helwan University, Cairo, Egypt

IOP Conf. Series: Materials Science and Engineering **973** (2020) 011001 doi:10.1088/1757-899X/973/1/011001

Military Technical College
Kobry El-Kobbah,
Cairo, Egypt.

19h Int. Conference on Applied Mechanics and Mechanical Engineering, July 7-9, 2020.

046_AMME-19

A New Integrated Process and Product Development model

A Ayman[1], A B Alhamaki[1] and M H Abdellatif[1]

Abstract

Currently, the demand for more robust, economic, and quality products has increased the complexity in the integrated process development and product design model. In the knowledge economy century, all nations are competing to innovate new technologies to develop a new system that produces goods in less time and cost with compatible quality. Hence, a new IPPD model is needed to offer support and information/ technology sharing during the entire product life cycle. The new IPPD model proposed that integrate all domains shall affect the product development. The model has a friendly computerized interface supported with internet enablers that offer support to the user to collect and share data easily. This model contains 6 domains. The Business Environment is the first domain, which analyzing the pre-feasibility of the product creation. The second domain is the market analysis and segmentation taking into consideration customer needs and competition. Managerial readiness of the product developing company is the third domain, including all managerial functions and the necessity of implementing the organizational change techniques if necessary. The fourth domain is Technological creation and deals with product concepts and robust quality design. Production/Launch is the fifth domain which consists of production processes and quality techniques to assure producing with the required quality and price levels. Finally, the sixth domain deals with the vital information of product distribution, maintenance and after-market service. The model is easily adapted to the industry need

[1] Mechanical department, Faculty of engineering, Ain shams university, Cairo, Egypt.

IOP Conf. Series: Materials Science and Engineering **973** (2020) 011001 doi:10.1088/1757-899X/973/1/011001

Military Technical College
Kobry El-Kobbah,
Cairo, Egypt.

19^h Int. Conference on
Applied Mechanics and
Mechanical Engineering,
July 7-9, 2020.

047_AMME-19

Ultrasonic-assisted drilling of Nickel-based superalloy Inconel 601: An experimental study

A M Abdelaziz [1], H Youssef [1], M Al-Makky [1] and H El-Hofy [1,2]

Abstract
In this study, a comparison between ultrasonic-assisted drilling (UAD) and conventional drilling (CD) is presented under different feed rates using thrust force, torque, and hole geometrical errors as output responses. The experiments were done on plates of Inconel 601 (nickel-based superalloy), which is classified as a difficult-to-cut alloy due to its high Nickel content (60%), high hardness (43 HRC), and low thermal conductivity. The experiments were performed using DMG Mori Ultrasonic 20 linear, which is equipped with ultrasonic tool holder oscillating at 20 kHz with 7 μm amplitude. A coated carbide single margin twist drill had been used in the experiments. Full factorial design of experiments approach was employed, and the results had been statistically analyzed to find the most significant factor affecting the process responses. The results showed that the ultrasonic assistance had reduced the thrust force, and torque compared to conventional drilling (CD). Also, a reduction in hole cylindricity error was detected during UAD, which improves the hole quality. In case of UAD, twist drills did not suffer from a physical wear, however notch wear was observed in CD drills. Chip morphology was also studied. Short segmented chips were obtained when using UAD which improved chip evacuation and reduced the chance of chip jamming in the drill flutes.

---

[1]Production Engineering Department, Faculty of Engineering, Alexandria University, Alexandria 21544, Egypt
[2]Industrial and Manufacturing Engineering Department (IME), School of Innovative Design Engineering (IDE), Egypt-Japan University for Science and Technology (E-JUST), Alexandria 21934, Egypt

IOP Conf. Series: Materials Science and Engineering **973** (2020) 011001 doi:10.1088/1757-899X/973/1/011001

Military Technical College
Kobry El-Kobbah,
Cairo, Egypt.

19h Int. Conference on
Applied Mechanics and
Mechanical Engineering,
July 7-9, 2020.

048_AMME-19

Computational fluid Dynamic model for machining using minimum quantity coolant

W Abdelfattah[1,2], H Hegabb[2], A Mohany[1] and H A Kishawy[1]

Abstract
The cooling applications during machining has significant effects on the production costs, surface quality, and the mechanical properties of the final product. In conventional flood cooling, a large amount of continuous cooling fluid is usually used, and that increases the cost of the product as well as the harmful effects on the environment and the machining operator. This study focuses on simulating alternative cooling system, called minimum quantity coolant (MQC), which used an optimal flow rate compared to classical flood cooling. The cooling fluid is directly provided to the cutting edge through the insert holder. In the current work, a computational fluid dynamic (CFD) model has been developed to study the effects of the cooling fluid velocity on the accessibility of coolant to the chip-tool interface area under using various types of cooling fluids. Three types of coolant are used (i.e. water, mineral oil, and nano-fluid). The results of the proposed CFD model have been classified into two phases. The first phase obtains the coolant accessibility percentage into the chip-tool interface (MA%) with different coolant velocities (i.e., 0.5, 1, 1.5, and 2 m/s) for the three studied coolants. The second phase discusses the heat transfer effectiveness for the employed coolants with different inlet velocities since it is an important aspect, especially when machining hard-to-cut materials. It was found that increasing the coolant velocity would increase the coolant accessibility percentage into the chip-tool interface area. However, no significant effect has been found after 1.5 m/s for all employed coolants.

[1]Faculty of Engineering and Applied Science, University of Ontario Institute of Technology, Oshawa, Canada.
[2]Faculty of Engineering, Cairo University, Giza, Egypt.

IOP Conf. Series: Materials Science and Engineering **973** (2020) 011001 doi:10.1088/1757-899X/973/1/011001

Military Technical College
Kobry El-Kobbah,
Cairo, Egypt.

19[h] Int. Conference on
Applied Mechanics and
Mechanical Engineering,
July 7-9, 2020.

049_AMME-19

Optimization Ladle-Refining Performance During Treating Special Steel Melts for Aviation Technology

S A Elghazaly[1,] K K Gyula[2] and W S Elghazaly[1]

Abstract

The ever increasing demands of consumers towards superior quality steel products urged steelmakers to consider the implementation of various relatively new metallurgical technologies into the classical operations. The necessity to produce high quality steels led to the use of two stage steel production processes, one is the primary steelmaking and the other stage is the outside furnace treatment through which many metallurgical functions can be achieved like degassing, stirring and inclusion removal, Inclusion modification, desulphurization, deoxidation, decarburization, heating and alloying. In this research a trial was done to optimize the performance and usage of 30tons-ladle refining system during production of X65-pipeline steel as final product-Aluminium killed steel melts through controlling the mass flow contour using optimized modelling, optimum usage of Al2O3 and Sulphide modifiers and enhancing removal of non-metallic inclusions by altering their morphologies and hence floatation speed. Assessment of fine clusters of inclusions in the final steel product has been industrially correlated to the cleanness of melt before refining, to the slag composition and to the parameters of materials flow rates as well as their effects on the mechanical properties of final X65-steel product. Scanning-EM+DX analyzing unit and metallurgical microscopes were used to emphasis qualitatively and quantitatively the characters of non-metallic inclusions.

---

[1] Steel Technology, Metals Technology, CMRDI, Egypt
[2] Metallurgy, Miskolc University

IOP Conf. Series: Materials Science and Engineering **973** (2020) 011002 doi:10.1088/1757-899X/973/1/011002

Peer review declaration

All papers published in this volume of *IOP Conference Series: Materials Science and Engineering* have been peer reviewed through processes administered by the Editors. Reviews were conducted by expert referees to the professional and scientific standards expected of a proceedings journal published by IOP Publishing.

- **Type of peer review: Double-blind, reviewers can not see the authors information**
- **Conference submission management system: iKNiTO cs https://amme.conferences.ekb.eg/**
- **Number of submissions received: 70**
- **Number of submissions sent for review: 70**
- **Number of submissions accepted:53**
- **Number of withdrawn papers due to the change of conference date because of covid-19: 7**
- **Acceptance Rate (Number of Submissions Accepted / Number of Submissions Received X 100):74 %**
- **Average number of reviews per paper: 2**
- **Total number of reviewers involved: 3**
- **Contact person for queries: Prof. Dr. Shawky Hegazy**

shawkyhegazy@mtc.edu.eg

IOP Conf. Series: Materials Science and Engineering **973** (2020) 012001 doi:10.1088/1757-899X/973/1/012001

Experimental investigation of star grains in dual thrust solid propellant motors

M El-Naggar[1], H Belal[2] and H M Abdalla[3]

[1] M.Sc. Student, Rocket Department, Military Technical College, Egypt
[2] Assistant Professor, Rocket Department, Military Technical College, Egypt
[3] Associate Professor. Rocket Department, Military Technical College, Egypt
hbelal@mtc.edu.eg

Abstract. One of the great challenges in designing tactical solid missiles is to achieve high acceleration in the boost phase then maintaining constant speed during the sustain phase. This could be achieved by using a dual thrust solid-propellant rocket motor. Many of these tactical motors use a combination of star, tubular or finocyl grains to achieve this profile. The present study uses two tandem star grains with different design parameters and different transition geometry. Previous researches had consistently shown that the main advantage of star grain is the potential higher volumetric loading in addition to high tailorability. The pressure-time curve for the designed grains is calculated using a zero-dimensional internal ballistic module and a small-scale test motor is used to verify the calculated pressure-time curve. Different transition geometries are compared. Tapered transition is shown to give a comparable performance with the sharp transition with the advantage of higher volumetric loading.

1. Introduction

The objective of Dual-Thrust Rocket Motors (DTRMs) is to generate two levels of thrust: booster-phase thrust and sustainer-phase thrust. The booster stage is used to accelerate the vehicle from zero velocity to a certain velocity, and then the sustainer stage is used to maintain a constant velocity. A typical thrust-time curve with all terminology for a DTRM is shown in figure 1. There are different methods to achieve this dual thrust profile such as two independent rocket motors, a single chamber with an intermediate nozzle, two separate grains with different geometries, different propellants.[1-3] In the case of two independent rocket motors, the booster motor is mechanically separated from the missile. For example, Russian air defence missiles SA-2 or SA-3 which give the benefit of reducing the final empty mass. However, this separation may have a negative impact on the environment and require some complex mechanism. In order to avoid separation problems, a dual-thrust motor composed of a single chamber is preferred. In this case, a grain with a certain design is required: the first phase with a large burning surface followed by a second phase with the smaller burning area. Some designs used the same grain configuration but with different propellant. In such a case, different thrust levels are achieved through a difference in burning rates as for example, M112 Motor in HAWK missiles [4] shown in figure 2. The most practical dual thrust motors use a single chamber with different grain geometries. For example, a radial burning grain for boost and end burning grain as sustain or radial star grain for boost and a tubular grain for sustain. As examples, tactical motors of Mk58 in AIM-7 Sparrow missile [5] shown in figure 3 or in Super-530 D air-to-air missile [6] shown in the figure 4, even though a few researchers are still using dual chamber rocket motor with the cartridge loaded double base grains like the rocket motor for the booster for the Penguin MK2 MOD7 [7] missile shown in figure 5

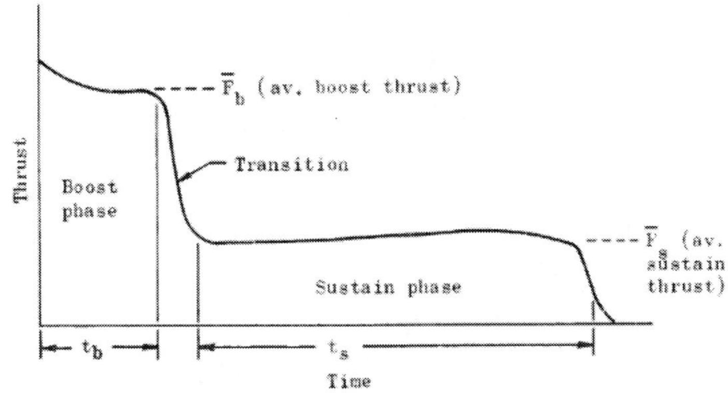

Figure 1. Boost- Sustain grain thrust-time curve [3].

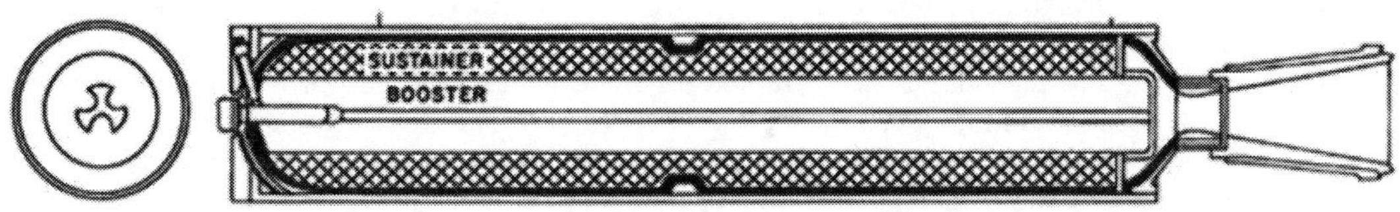

Figure 2. M-112 Hawk missile [4].

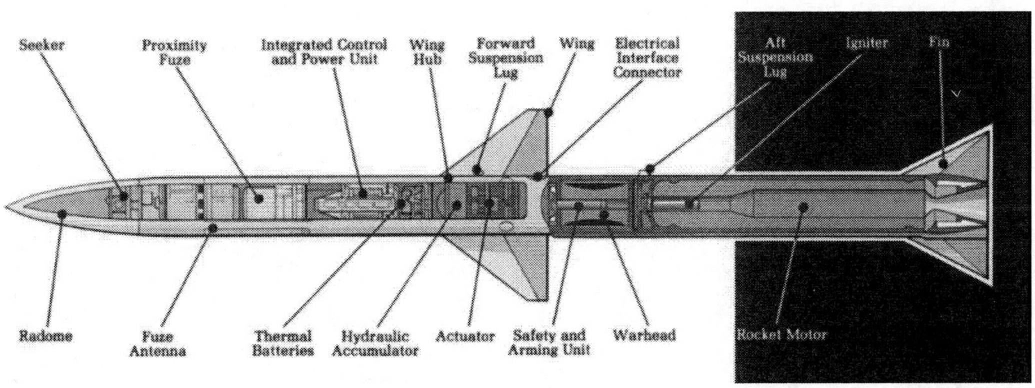

Figure 3. AIM-7 Sparrow missile [5].

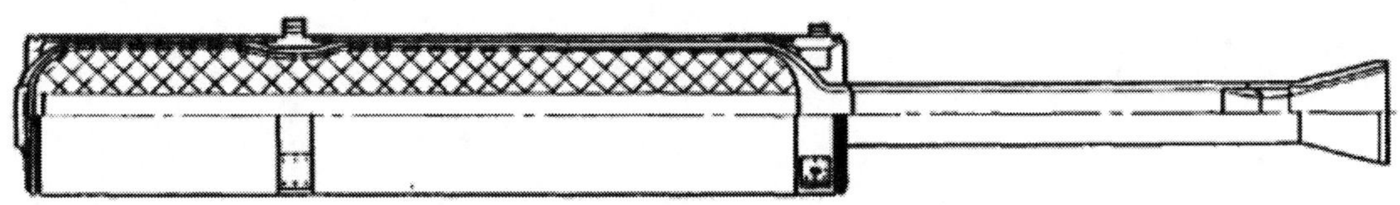

Figure 4. cross-section for super 530-D motor longitudinal [6].

IOP Conf. Series: Materials Science and Engineering **973** (2020) 012001 doi:10.1088/1757-899X/973/1/012001

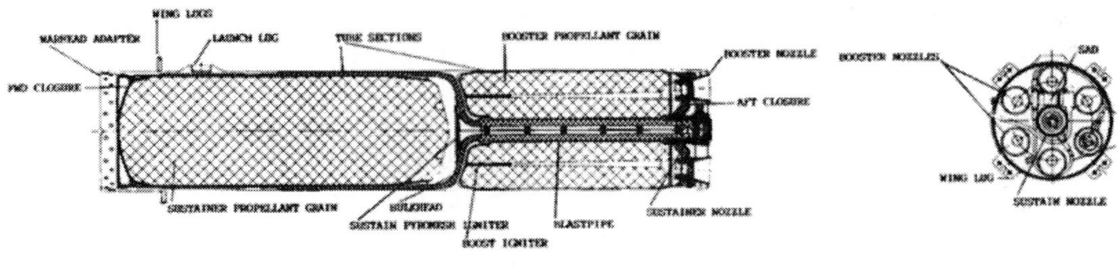

Figure 5. Penguin MK2-MOD7 rocket motor assembly [7].

Gawad [8] made two different methods to achieve DTRM'S, first by changing the burning area by single grain boost phase and second by changing the burning rate using dual end burning grain boost phase. He developed a mathematical model to estimate the pressure-time curve for DTRMs under geometric ballistic regression uncertainties, then made two experimental works: one for the 2-inch motor to get burning rate, and another for the 6-inch motor to get high thrust-to-weight ratio between booster and sustainer. This was made for two tubular grains with two different diameters along the grain. Ritchey and Anderson [9, 10] found that boost-sustain motors deliver specific impulse more effective than that delivered by boost ones, which makes the boost-sustain motors more suitable for certain applications.

Star grain, shown in figure 6, has been studied for more than 60 years. A previous survey showed that 40% of 129 operational motors used star grain [11] with number of star points ranging from 3-star points to the mighty 260 in SRM [12] the largest solid propellant ever built but did not fly, shown in figure 7. Grains with 16-star points [13] or 40-star points [14] shown in figures (8,9) respectively are used in the pyrogen for igniting the space shuttle solid booster separation motor. Star grain is characterized by several design parameters that provide good tailorability, in addition to higher volumetric loading than the traditional tubular grains.

The design variables of the star grain geometry are number of star points N, fillet radius f, angular fraction ε, star angle θ and maximum internal radius (s + f). Typical values are N =7, f = 3, ε = 0.5, θ = 75°, (s + f) = 60mm. The star grain has two main phases of burning; the first phase can be designed to be progressive, regressive or neutral depending on the star angle (θ), and second phase which is always progressive.

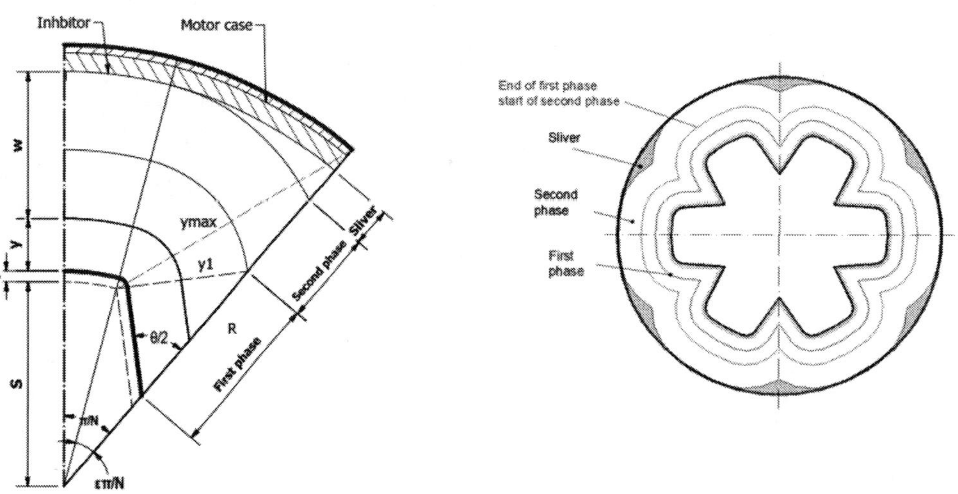

Figure 6. Star grain.

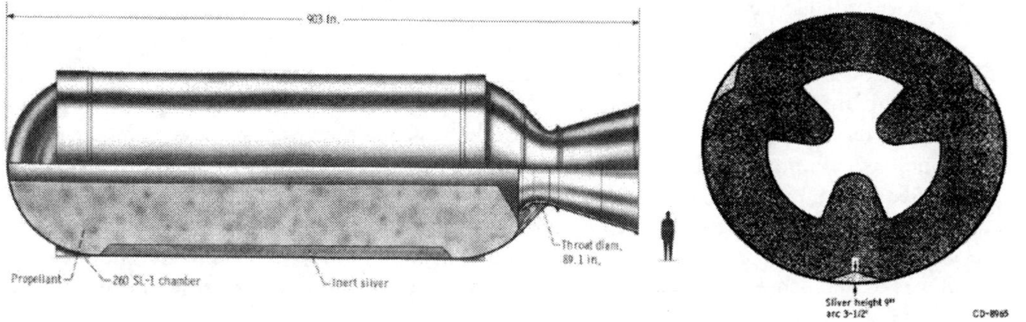

Figure 7. SRM with N=260 star points [12].

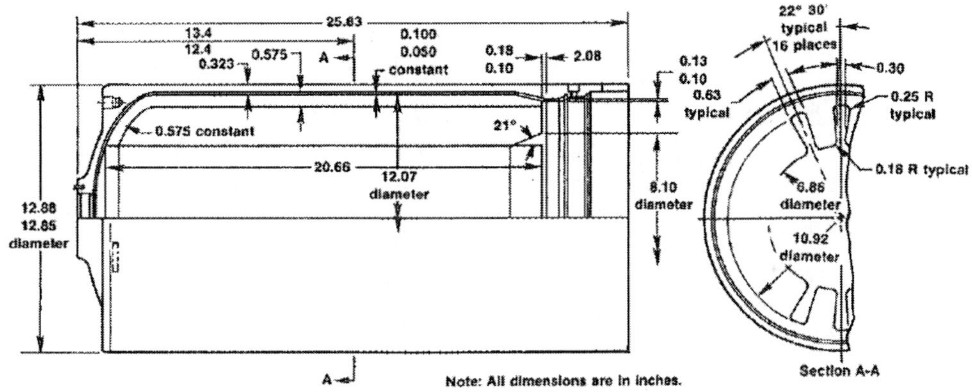

Figure 8. SRM with N=16 star points [13].

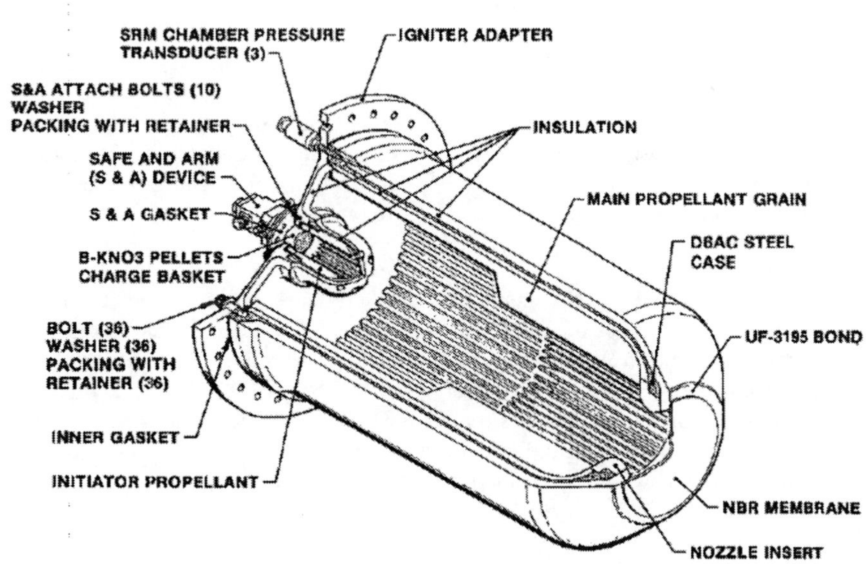

Figure 9. SRM with N=40 star points [14].

IOP Conf. Series: Materials Science and Engineering **973** (2020) 012001 doi:10.1088/1757-899X/973/1/012001

In 1992 Ricciardi [15] executed 16 conceivable configurations, 8 with convex and other 8 for concave points which have the boundaries $0 < r_2 < T_C$ and $(Y_t - \frac{T_C}{\sin\eta}) < r_2 < \sqrt{X_C^2 + Y_C^2}$ respectively as shown in figure 10. He described criteria and formulas adopted in developing a computer program capable of calculating the burning perimeter and evaluating the geometric evolution of a cylindrical star grain during its burning. If there is a variation of the seven geometrical parameters along the grain axis, the solutions have to be considered as approximated. If the cone semi angle is less than 10-15°, the error in area burning is less than 1-2%[15][15].

Stein [16] showed that star grain design increases volumetric loading efficiency (VLE) and specific impulse while keeping the pressure at a constant level in the combustion chamber. The idea of star grain design is to add more initial burn area that allows high starting up pressure and more neutral/ regressive burn to achieve maximum efficiency.

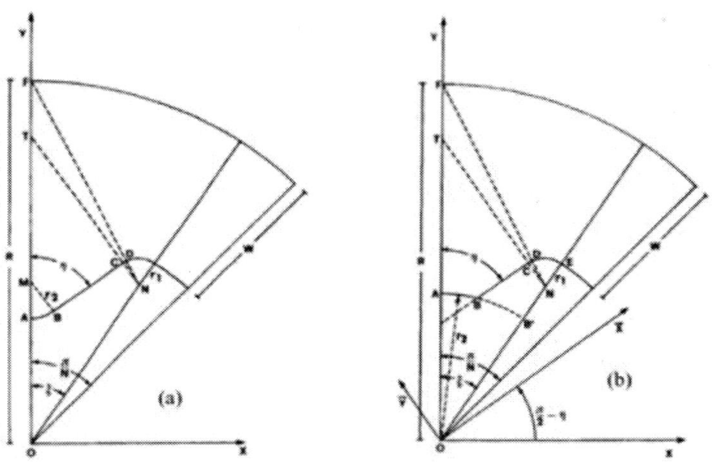

Figure 10. Star grain configuration for (a) convex and (b) concave [15].

Brooks[17] analysed and identified - for a given volumetric loading fraction - the most neutral-burning star with given web fraction, number of star points, and two small radii. He established a computer program that generates data for evaluating various other optimization criteria for star designs and to establish some universal limits of the capability of the star in terms of neutrality and sliver.

In this paper, a combination of two-star grains with different design parameters and different webs is used in order to achieve a dual thrust profile. The combination of design variables is calculated analytically followed by validation of the proposed design using a small-scale test motor. In addition, a genetic algorithm optimization module is used in order to tune the burning rate to fit the analytical data with experimental findings, and to predict the nozzle erosion rates as well.

2. Burn-back analysis of star grain

In all configurations studied, the grains have un-inhibited faces, thus the length of grain changes during burning. Two different types of transition geometries have been studied as shown in figure 11 where case (a) represents the sharp transition design and case (b) represents taper or conical transition design. The burn-back of the grains with sharp transition is shown in the figure 12 where the transition section is approximated using sharp edge to simplify the analytical equation rather than considering an arc of a circle with its center at the original transition plane. The case of taper transition is studied only experimentally and its analytical solution is beyond scope of this research.

IOP Conf. Series: Materials Science and Engineering **973** (2020) 012001 doi:10.1088/1757-899X/973/1/012001

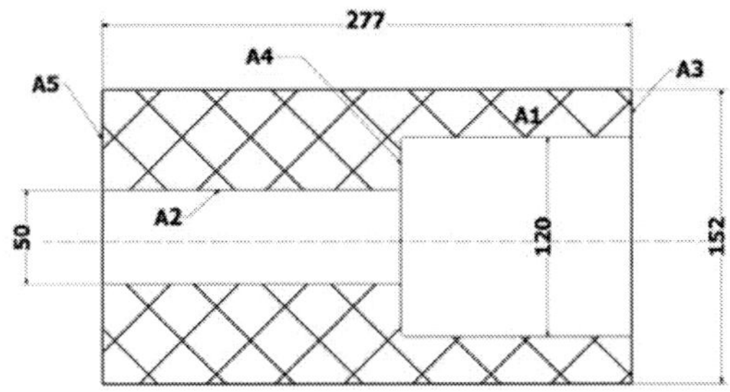

a) Sharp transition

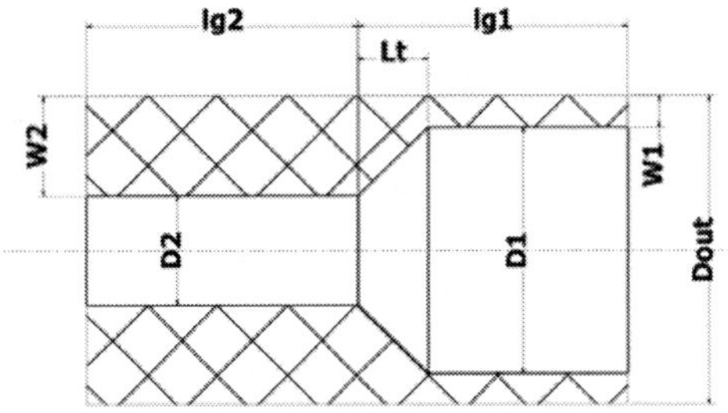

b) Taper transition

Figure 11. Longitudinal section for the different transition geometries.

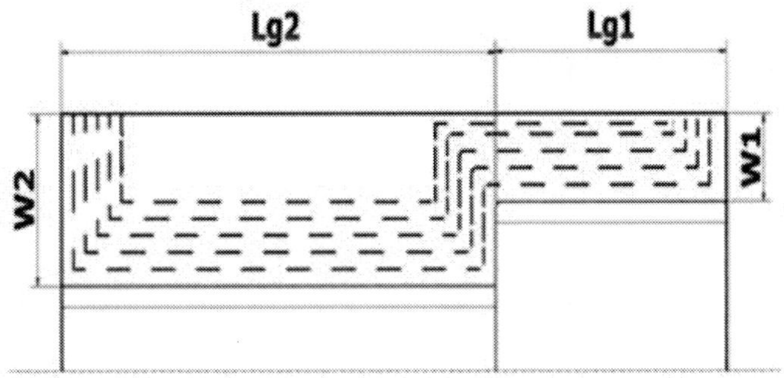

Figure 12. Burn-back of the grain with sharp transition.

In order to get neutral burning in the second phase, the grains will not have inhibited faces. Referring to Ricciardi [4], there are three types of configurations depending on web thickness (w), y_1 and y_{max} where y_1 is the condition of ending burning of the first phase as expressed in equation (1), and y_{max} is the maximum allowable burned distance of the propellant as in equation (2).

$$y_1 = s * \frac{\sin\frac{\varepsilon\pi}{N}}{\cos\frac{\theta}{2}} - f \qquad (1)$$

$$y_{max} = (R^2 + S^2 - 2RS\cos\frac{\varepsilon\pi}{N})^{0.5} - f \qquad (2)$$

Configurations 1, 2, and 3 correspond to ($y_1 < w$), ($w < y_1 < y_{max}$) and ($y_1 > y_{max}$) respectively. The details of these configurations are given below.

2.1. Configuration 1: **($y1 < w$)**, *figure 13*

This configuration is divided into three zones: zone 1: ($0 \leq y \leq y_1$), zone 2 ($y_1 < y \leq w$) and zone 3 ($w < y \leq y_{max}$).

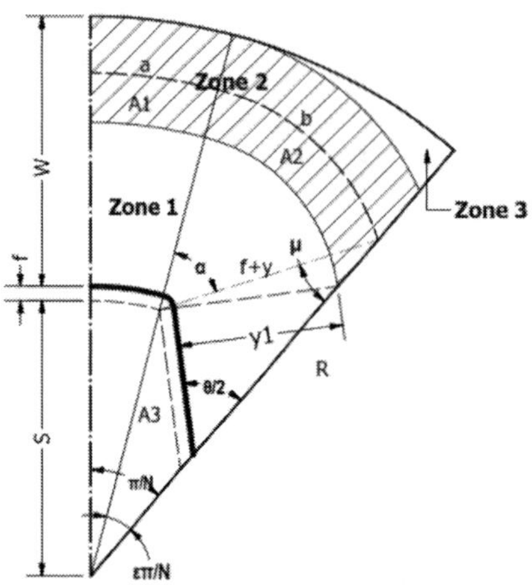

$y1 = 13.01mm$
$W1 = 16\ mm$
$S = 57\ mm$

Figure 13. The geometry of configuration 1.

Zone 1: ($0 \leq y \leq y_1$)

The burning area (A_b) of the surface as a function of burnt distance y is calculated by equation (3):

$$A_b = 2Nl_g(a+b+c)$$

$$=2Nl_g[(s+f+y)\left(\frac{\pi}{N}-\varepsilon\frac{\pi}{N}\right)+(f+y)\left(\frac{\pi}{2}+\varepsilon\frac{\pi}{N}-\frac{\theta}{2}\right)+s\frac{\sin\varepsilon\frac{\pi}{N}}{\sin\frac{\theta}{2}}-(f+y)\cot\frac{\theta}{2}] \quad (3)$$

The corresponding port area (A_p) of gases is found using equation (4):

$A_p = 2N(A_1 + A_2 + A_3 + A_4 + A_5)$

$$= N\begin{bmatrix}(s+f+y)^2\left(\frac{\pi}{N}-\varepsilon\frac{\pi}{N}\right)+(f+y)^2\left(\frac{\pi}{N}+\varepsilon\frac{\pi}{N}-\frac{\theta}{2}\right)+(f+y)^2\cot\frac{\theta}{2}+ \\ s^2\frac{\sin\varepsilon\frac{\pi}{N}}{\sin\frac{\theta}{2}}\sin\left(\frac{\theta}{2}-\varepsilon\frac{\pi}{N}\right)+(f+y)\left(s\frac{\sin\varepsilon\frac{\pi}{N}}{\sin\frac{\theta}{2}}-(f+y)\cot\frac{\theta}{2}\right)\end{bmatrix} \quad (4)$$

Zone 2: ($y_1 < y \leq w$)

The burning area of the surface in zone 2 as the surface progresses a distance y is calculated by equation (5):

$$A_b = 2NL_g\left[(s+f+y)\left(\frac{\pi}{N}-\varepsilon\frac{\pi}{N}\right)+(y+f)\left[\varepsilon\frac{\pi}{N}+\sin^{-1}\left(\frac{s}{y+f}\sin\varepsilon\frac{\pi}{N}\right)\right]\right] \quad (5)$$

The port area is found using equation (6):

$$A_p = N\begin{bmatrix}(s+f+y)^2\left(\frac{\pi}{N}-\varepsilon\frac{\pi}{N}\right)+(f+y)^2\left[\varepsilon\frac{\pi}{N}+\sin^{-1}\left(\frac{s}{y+f}\sin\varepsilon\frac{\pi}{N}\right)\right] \\ +s(f+y)\sin\left[\pi-\varepsilon\frac{\pi}{N}-\sin^{-1}\left(\frac{s}{y+f}\sin\varepsilon\frac{\pi}{N}\right)\right]\end{bmatrix} \quad (6)$$

Zone 3: (Sliver Area) ($w < y \leq y_{max}$)

In this zone, both segment b and segment c vanish, thus the burning area of the surface in zone 3 as the surface progresses a distance y is calculated using equation (7):

$$A_b = 2NL_g(f+y)(\phi-\delta) \quad (7)$$

where $\delta = 180 - \cos^{-1}\left[\frac{s^2+(f+y)^2-Rout^2}{2s*(f+y)}\right]$

The port area can be expressed as in equation (8)

$$Ap = N\left[R_{Out}{}^2\left(\frac{\pi}{N}-\frac{\varepsilon\pi}{N}\right)+ s(f+y)sin\left(\mu+\frac{\varepsilon\pi}{N}\right)+ 0.5\beta R^2 - 0.5Rs*sin\,\beta + (f+y)^2\Gamma\right] \quad (8)$$

where $\beta = \cos^{-1}\frac{R^2+s^2-(f+y)^2}{2Rs}$

2.2. Configuration 2: ($w < y_1 < y_{max}$), figure 14

Configuration 2 is divided into three zones, 1, 3 and 4. Zone 1 ($0 \le y \le w$) is the same as zone 1 in configuration 1, and zone 3 ($y_1 < y \le y_{max}$) is the same as zone 3 in configuration 1. So, zone 4 is the new zone in this configuration.

Zone 4: ($w < y \le y_1$)

The burning area of the surface in zone 4 as the surface progresses a distance y is expressed by equation (9):

$$A_b = 2NL_g(a+c) = 2NL_g\left[(f+y)*(\alpha-\delta)+\frac{y_1-y}{\tan\frac{\theta}{2}}\right] \quad (9)$$

Port area can be calculated using equation (10):

$$A_P = 2N*(A_1+A_2+A_3+A_4)$$

$$A_P = N\begin{bmatrix} R^2\left(\frac{\pi}{N}-\frac{\varepsilon\pi}{N}\right)+R^2\phi - Rs\sin\phi + (f+y)^2\Gamma \\ +s(f+y)\cos\left(\frac{\theta}{2}-\frac{\varepsilon\pi}{N}\right) - C(y_1-y) \end{bmatrix} \quad (10)$$

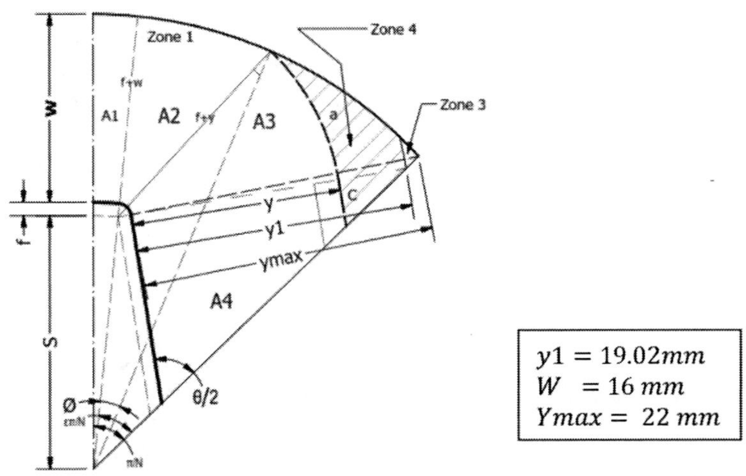

Figure 14. The geometry of configuration 2.

2.3. *Configuration 3:* $(y_1 > y_{max})$, *figure 15.*

This configuration can be divided into three zones: Zone 1 $(0 < y \leq w)$ is the same as zone 1 in configuration 1, and zone 4 $(w < y \leq CQ_O)$ is the same as zone 4 in configuration 2.

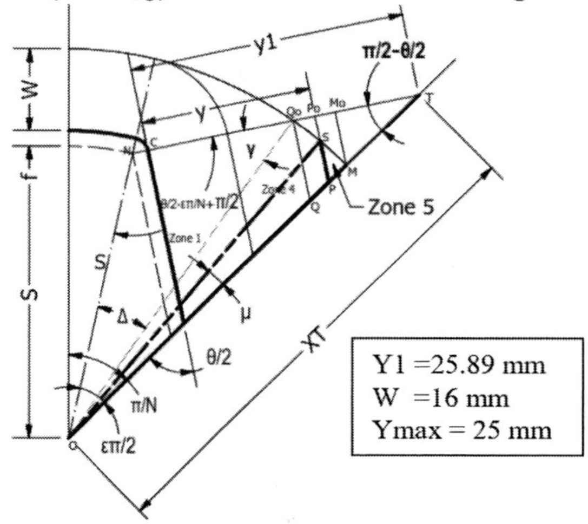

Figure 15. The geometry of configuration 3.

Zone 5: $(CQ_O < y < CM_O)$

The burning area of the surface in zone 5 as the surface progresses a distance y is calculated using equation (11):

$$A_b = 2Nl_g\left[X_P\cos\frac{\theta}{2} + \sqrt{R^2 - X_P{}^2\left(\sin\frac{\theta}{2}\right)^2}\right] \quad (11)$$

Port area is calculated by using equation (12)

$$A_P = N[(R^2\left(\frac{\pi}{N} - \mu\right) + R * \overline{O_P} * \sin\mu]] \quad (12)$$

where

$X_P = X_T - \frac{y_1 - y_C}{\sin\frac{\theta}{2}}$

$X_T = S\left(\cos\varepsilon\frac{\pi}{N} + \sin\varepsilon\frac{\pi}{N}\tan\frac{\theta}{2}\right)$

$\overline{O_P} = \frac{R*\sin(\frac{\theta}{2}-\mu)}{\sin\frac{\theta}{2}}$

The governing expressions for burning area in each phase are summarized in the following table 1. These areas are: AB1= the surface burning area of star 1., AB2 = the surface burning area of star 2, AB3 = the first surface burning area related to star 1, AB4 = the middle surface burning area between star 1, 2, and finally AB5 = the last surface burning area related to star 2. The resulting total burning area is shown in figure16 with individual burning areas for different surfaces. However, a long transition between booster and sustainer is observed. In addition, the booster time is too long, and the sustainer time is short. The desired objective is not achieved and hence SRM is to be modified to improve the results. After several trials, an acceptable solution for the transition between booster and sustainer, and the relative times of booster and sustainer have been reached. Design parameters for the accepted dual thrust profile are shown in figure 17.

IOP Conf. Series: Materials Science and Engineering **973** (2020) 012001 doi:10.1088/1757-899X/973/1/012001

Table 1. Summary for burning surfaces and how to calculate (refer to figure 1(a) for surface numbering).

	Y = 0	Y2 = 3.200	Y1 = 13	W1 = 16	Ymax1 = 21	W2 = 51	Ymax2 = 51.80
AB1	Zone1 $(Y > 0 \& Y < Y1)$	Zone2 $(Y > Y1 \& Y \le W1)$	Zone3 $(Y > W1 \& Y \le Ymax1)$	Zero	AB1	Zone1 $(Y > 0 \& Y < Y1)$	Zone2 $(Y > Y1 \& Y \le W1)$
AB2	First phase $Y \le Y2$	Second phase $Y \le W2 \& Y > Y2$	Sliver phase $Y > W2 \& Y \le Ymax2$		AB2	First phase $Y \le Y2$	Second phase $Y \le W2 \& Y > Y2$
AB3	$A_{C.CH} - A_{P1}$	$A_{C.CH} - A_{P1}$	$A_{C.CH} - A_{P1}$	$A_{C.CH} - A_{P1}$		Zero	
AB4	$A_{P1} - A_{P2}$	$A_{P1} - A_{P2}$	$A_{P1} - A_{P2}$	$A_{P1} - A_{P2}$	$A_{P1} - A_{P2}$	$A_{P1} - A_{P2}$	$A_{P1} - A_{P2}$
AB5	$A_{C.CH} - A_{P2}$	$A_{C.CH} - A_{P2}$	$A_{C.CH} - A_{P2}$	$A_{C.CH} - A_{P2}$	$A_{C.CH} - A_{P2}$	$A_{C.CH} - A_{P2}$	$A_{C.CH} - A_{P2}$

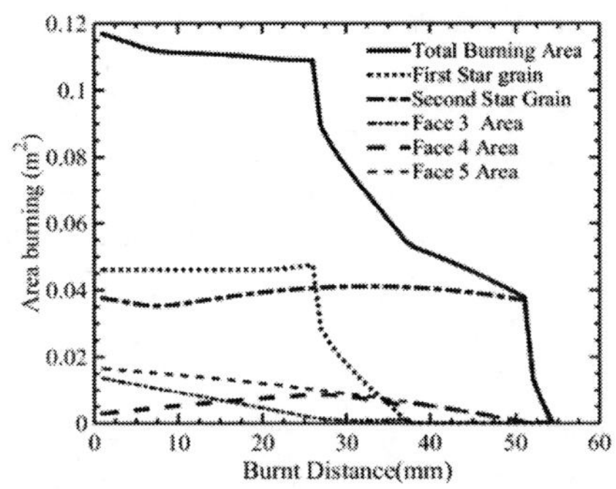

Figure 16. Calculated burning area for all surfaces, N=5, ε =0.8, f=5mm, w1=26 mm, w2=51 mm, lg1=97 mm, lg2=180 mm, s1=45mm, s2=20mm, D1=100 mm, D2=50 mm and θ (Neutral) =62.24°.

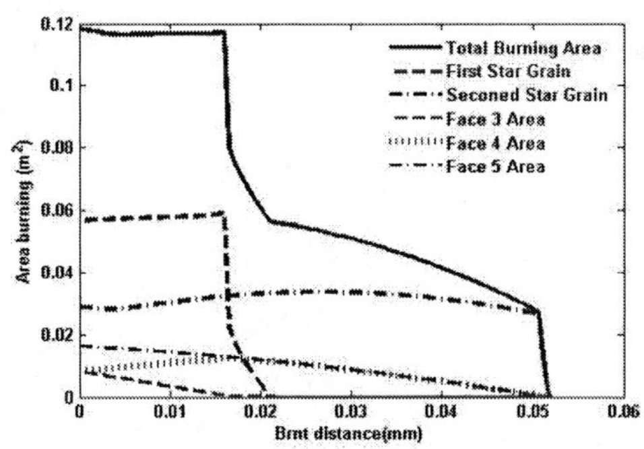

Figure 17. Calculated burning area for all surfaces, N=7, ε =0.5, f=3mm, w1=16 mm, w2=51 mm, lg1=120 mm, lg2=157 mm, s1=57mm, s2=22mm, D1=120 mm, D2=50 mm and θ =75°.

The flowchart of burn-back analysis for each configuration to get the area burning module $A_B(Y)$ is shown in figure 18 consequently.

IOP Conf. Series: Materials Science and Engineering **973** (2020) 012001 doi:10.1088/1757-899X/973/1/012001

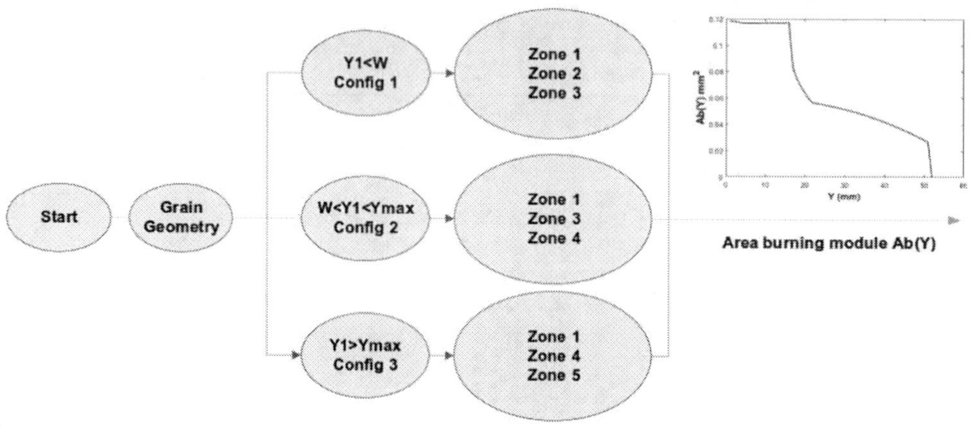

Figure 18. Flowchart of grain burn-back analysis.

3. Internal ballistics

In order to predict the pressure-time curve, the values of burning rate from composite propellant were obtained from the experiments using 2" motor of [8] which gave the preliminary data of the value for pressure exponent (n=0.021) and burning rate coefficient ($a=2.68\times10^{-4}$) for the proposed design, a 0-D internal ballistic prediction module (IBPM) was applied. The basic equation for this module according to [2] is:

$$V_c \frac{dP_c}{dt} = \rho_{sp} RT_c A_b a P_c{}^n - \Gamma P_c A_{cr} \sqrt{RT_c} \quad (13)$$

where V_c= chamber free volume, Pc = combustion pressure (stagnation pressure), ρ_{sp} = solid propellant density, R = gas constant, T_c = combustion temperature, A_b = burning area, and A_{cr}= nozzle critical area, according to the flowchart of input and output data for IBPM as shown in figure 19. In order to predict the pressure-time curve, the solution of equation 13 was performed according to the flow chart shown in figure 20. The resultant pressure time curve is shown in figure 21.

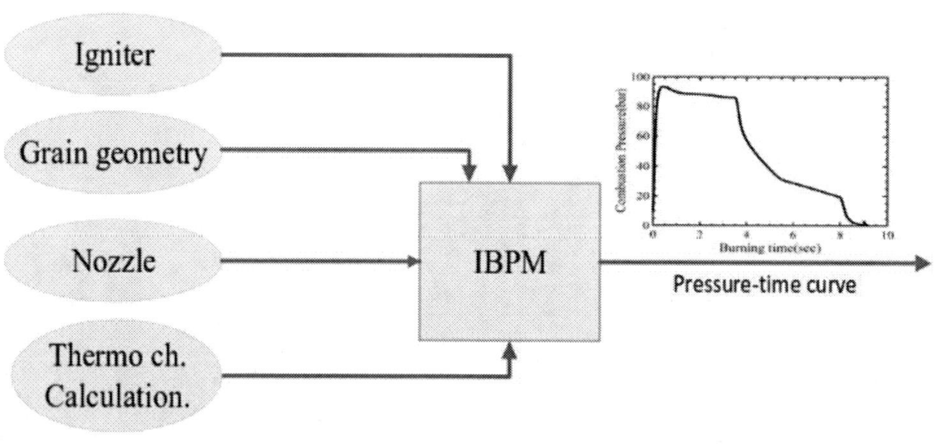

Figure 19 Flowchart of input and output data of IBPM.

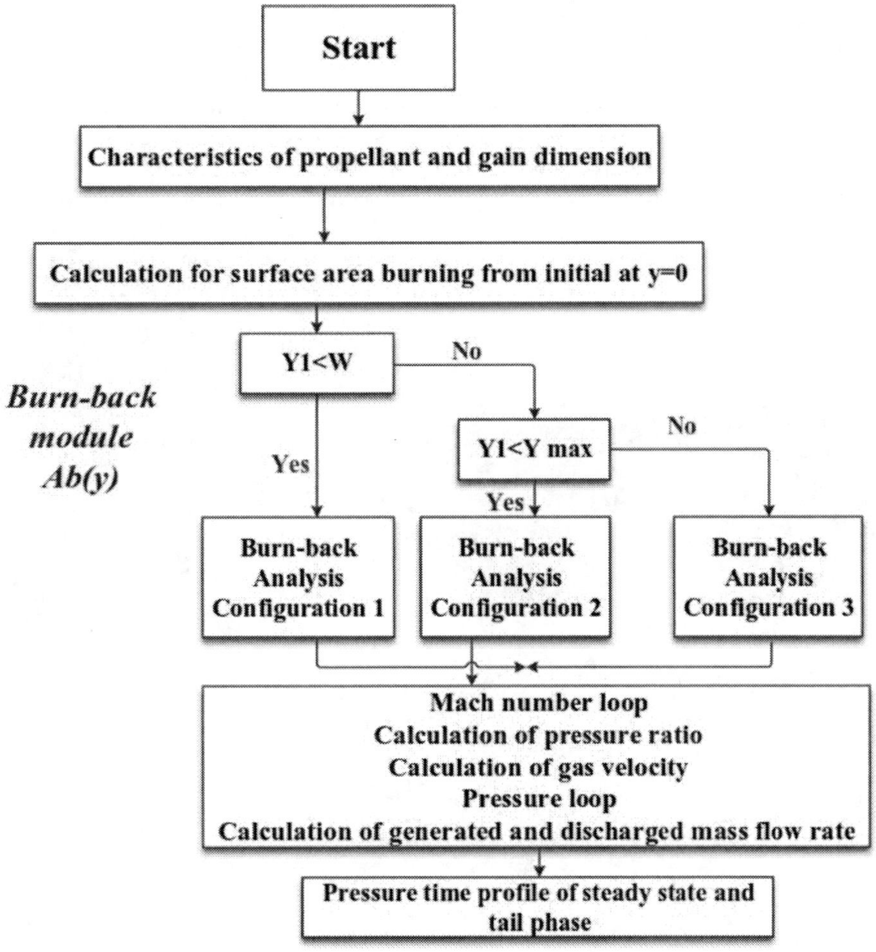

Figure 20. Flow chart of IBPM.

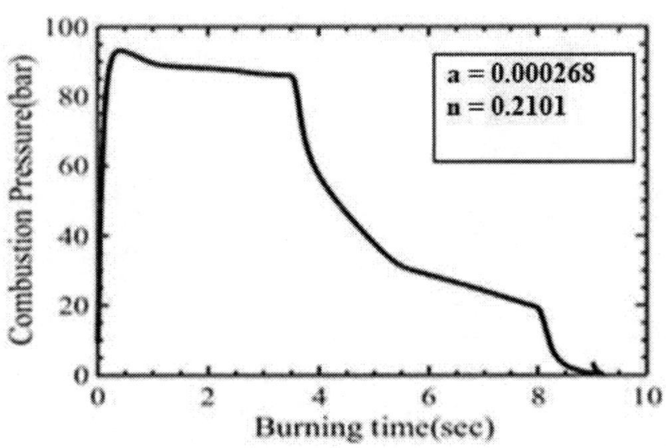

Figure 21. Predicted pressure-time curve for the proposal grain.

IOP Conf. Series: Materials Science and Engineering **973** (2020) 012001 doi:10.1088/1757-899X/973/1/012001

4. Experimental results and discussions

In order to validate the burn-back analysis of the proposed grain and the 0-D internal ballistic module, a small-scale test motor (6" motor) was used, figure 22. In order to study the effect of transition geometry on thrust time-curve, two mandrels were used in propellant casting in a small-scale test motor shown in the figure 23 with the geometry of resultant grains demonstrated in figures 24 and 25. The experimental results of sharp and taper geometry from factory 18 are shown in table 2. In order to obtain different operating pressures, different nozzles with different throat diameters were used. Table 3 shows test matrix.

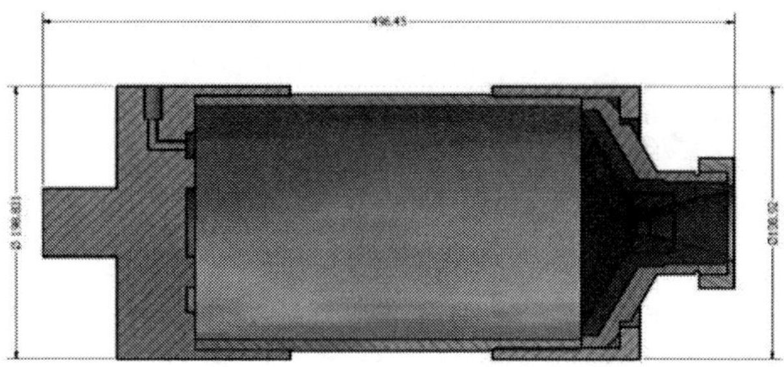

Figure 22. Small scale test motor used in experiments.

Figure 23. Taper and sharp mandrels used in SPRM.

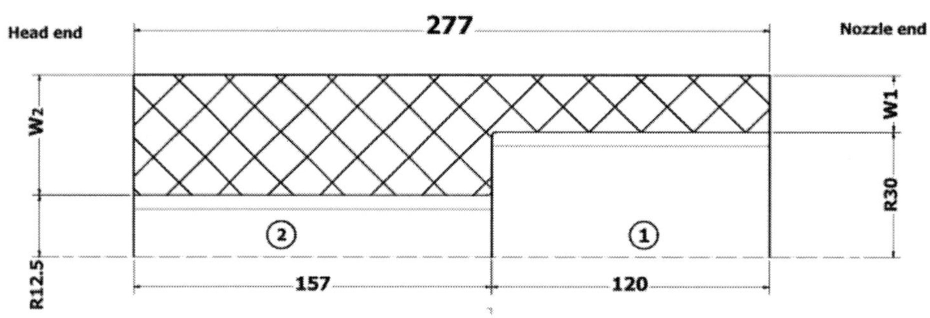

Figure 24. Main dimensions of sharp star grain.

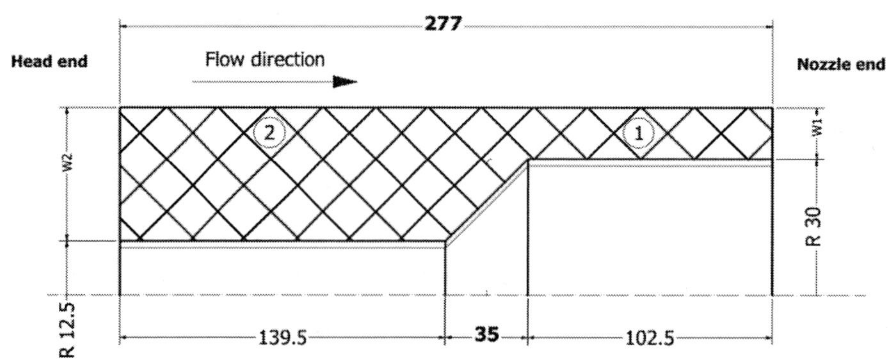

Figure 25. Main dimensions of taper star grain.

IOP Conf. Series: Materials Science and Engineering **973** (2020) 012001 doi:10.1088/1757-899X/973/1/012001

Table 2. Results for sharp and taper geometries.

Geometry	Temp (C)	Throat D_{cr} (mm)	m_p (kg)	T50% (Sec)	Pmax (BAR)	P50% (BAR)	R50% (mm/s)	C* (m/s)	Isp sec
Sharp	20	15	6.4	4.982	156.3	97.44	8.58	1357	229.5
	20	18	6.4	6.365	96.2	56.26	6.72	1460	220.0
Taper	21	15	6.9	5.940	140.8	80.33	8.59	1395	229.5
	21	18	6.9	7.000	91.9	50.57	7.29	1476	221.3

Table 3. Experimental test matrix.

	Sharp	Tapered
ϕ15	A1	B1
ϕ18	A2	B2

The following figures. (26-29) illustrates the effect of using different transition geometries on pressure-time and thrust-time traces in case of throat diameters 15 and 18 mm respectively.

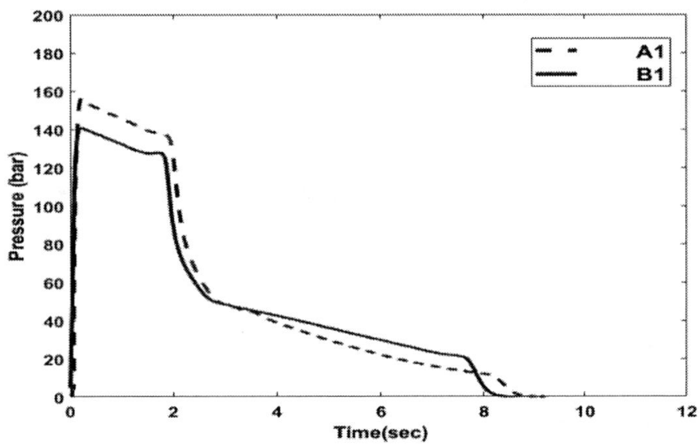

Figure 26. Pressure time curves for different transition geometries at the same throat.

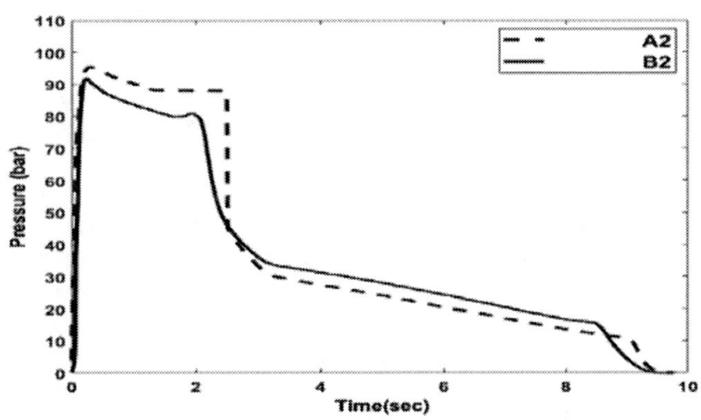

Figure 27. Pressure time curves for different transition geometry at the same throat.

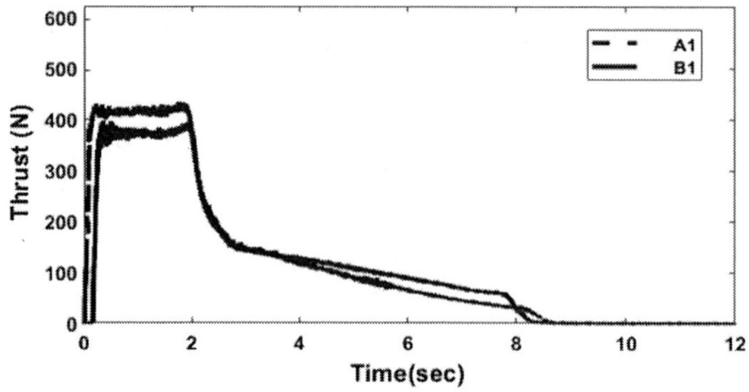

Figure 28. Thrust time curves for different transition geometry at 18 mm.

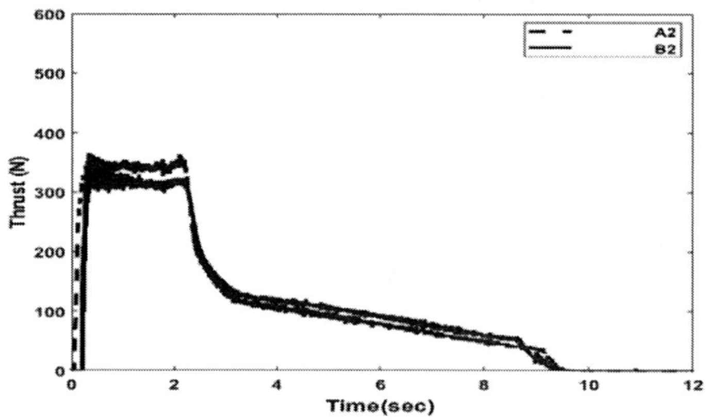

Figure 29. Thrust time curves for different transition geometry at 15mm.

From the previous figures, Basically, there is no difference in performance between taper transition and sharp transition. The tapered transition gives lower pressure in the booster phase due to lower initial burning surface. In addition, the taper configuration gives a clear advantage of higher volumetric loading, as the motor with the same size carries 6.9 kg in comparison to 6.4 kg for a case of sharp configuration.

5. Optimization module

In order to decrease the difference between the measured and predicted pressure-time curves and the predicted optimization module, genetic algorithm module in MATLAB, was used as illustrated by the flowchart in figure 30.

After many trials, the minimum error is reached with assuming different burning rates; one at lower pressure and one at higher pressure. As evident from [18], the composite propellant with bimodal ammonium perchlorate has a region of plateau burning with different burning rates before and after plateau region. The fitted pressure-time curve is shown in figure 31. The results of optimization for case A1 are as follows: n_h=0.2247, a_h=2.1727×10^{-4}, n_l=0.213, a_l=2.342×$^{10-4}$, P=62.19 bar. A summary for propellant data before and after tuning using optimization module is shown in table 4.
Comparing pressure-time curves which shows a progressive profile in the boost phase while the corresponding thrust-time curves show neutral profile, giving evidence of nozzle erosion. This erosion is validated by inspecting images of the nozzle throat after motors burnout as shown in table 5.

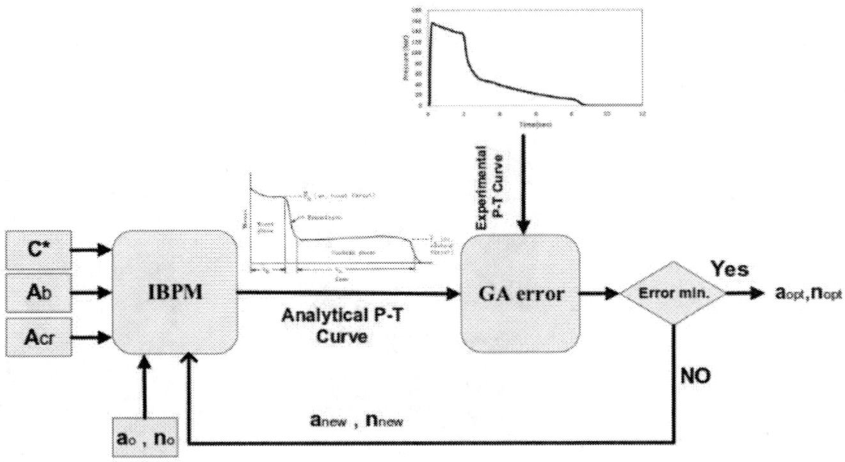

Figure 30. Flowchart of optimization module.

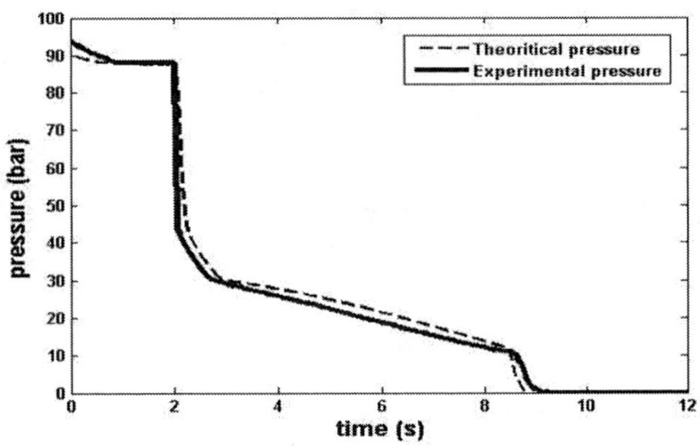

Figure 31. Pressure time curves for experimental and theoretical work at throat 18 mm.

Table 4. Experimental test matrix.

	Before tuning	After tuning	
		Low pressure region	High pressure region
a	2.68×10^{-4}	2.342×10^{-4}	2.1727×10^{-4}
n	0.213	0.213	0.2247

In addition to tuning the burning rate law parameters, an additional parameter- nozzle erosion rate- is included in the optimization module. A summary for the predicted and experimental erosion rates is shown in table 6. The comparison between nozzle erosion rates for different initial throat diameter shows that erosion rate for smaller throat diameter which gives a higher combustion pressure yield a larger erosion rate than that of initially large throat diameter. Such result is in agreement with previous researches [19]. From table 6, there is a good agreement between measured and predicted erosion rates. However, the result for case A2 shows a larger discrepancy between the measured average erosion rate and the erosion rate calculated from optimization module.

IOP Conf. Series: Materials Science and Engineering **973** (2020) 012001 doi:10.1088/1757-899X/973/1/012001

Table 5. Nozzles before and after static firing.

Cases	A1(ϕ 15)	A2(ϕ 18)
Before firing	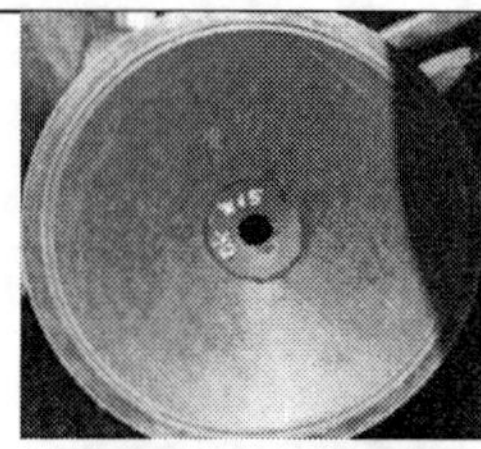	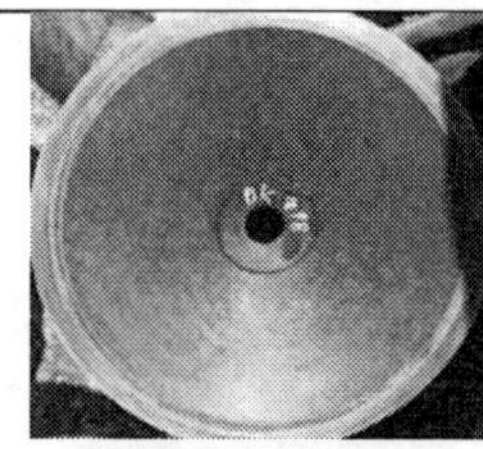
After firing	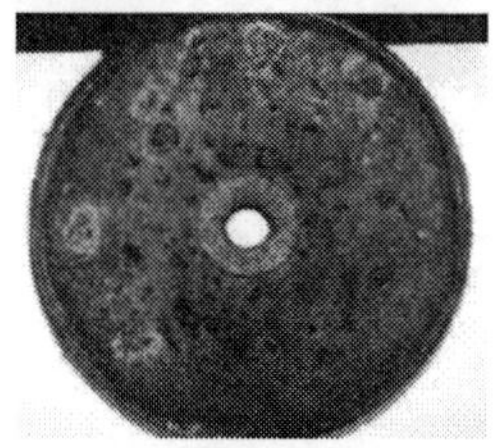	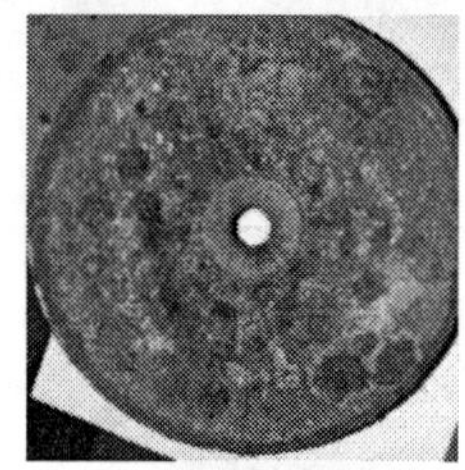
Graphite insert	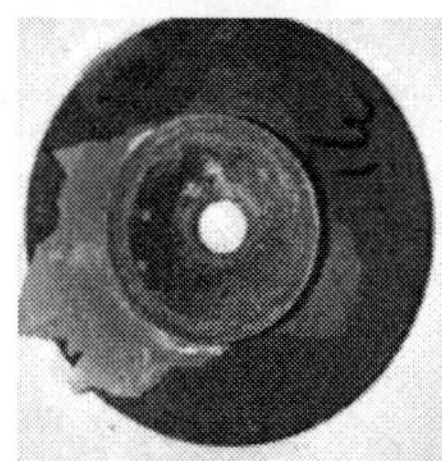	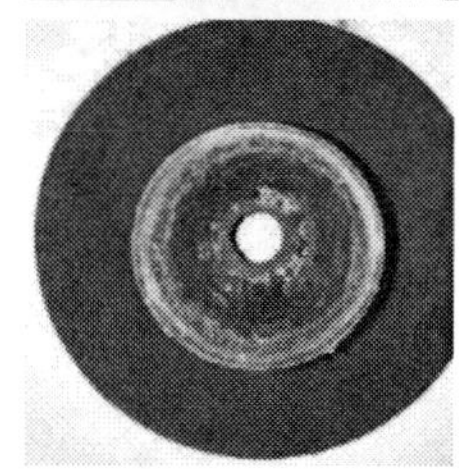

Table 6. Erosion rate from optimization module.

Case	A1	A2
D_{Cro} After burning	ϕ 15.6	ϕ 18.3
Average erosion rate (mm/s)	0.0375	0.01875
Predicted erosion rate (mm/s)	0.03358	0.0125

6. Conclusion

Star grain showed a good tailorability. As with only changes of star grain design parameters, there is good controllability on both relative durations of booster and sustainer phase, and, the transition between the booster phase and the sustainer phase. This tailorability comes at the cost of more complicated analytical burn-back where different configurations with each configuration have multi-zone with different burning regimes. Different transition (i.e. sharp transition vs. taper transition) were compared, with tapered transition is shown to give a comparable performance with the sharp transition with the advantage of higher volumetric loading.

The analytical procedure for predicting the pressure-time trace for a dual thrust rocket motor (0-D internal ballistic module) is derived and validated for sharp transition geometry at different throat diameter 0-D internal ballistic module showed to give very good fit to experimental data as long as the burning rate law is accurate, but in case of uncertainty in burning rate law, A GA-Module can be applied to find the more accurate data. In addition, combining the 0-D internal ballistic module with GA-module can be used to predict the erosion rates in a nozzle with a good fit with the experimental data, with higher combustion pressure leads to higher nozzle erosion rates as known from previous analytical and experimental researches.

IOP Conf. Series: Materials Science and Engineering **973** (2020) 012001 doi:10.1088/1757-899X/973/1/012001

References

[1] Barrere M, Jaumotte A, Fraeijs de Veubeke B and Vandenkerckhove J V 1960 *Rocket propulsion* (Liège: Belgium LTAS)

[2] Sutton G P and Ross D M 1976 *Rocket propulsion elements : An Introduction to the engineering of rockets* (John Wiley & Sons 4th ed)

[3] Bornstein L J 1979 Method of making dual-thrust rocket motor *Google Patent.*

[4] *Operator, organizational, direct and general support maintenance manual : intercept-aerial guided missile MIM-23B, MTM-23B, amd M18E2 :improved HAWK air-defense guided missile system 1984 (*Washington, D.C. Headquarters Dept. of the Army).

[5] Kopp C BAe Active Skyflash and AIM-120 AMRAAM 1994 *Australian Aviation* ***June* 1994**.

[6] TRICOT J 1988 Wrap-on composite case technology Super 530 D-Tactical motor application *24th Joint Propulsion Conference* p 3326.

[7] Haugen S 1990 Development and pre-qualification of the booster rocket motor for the Penguin MK2 MOD7 missile *26th Joint Propulsion Conference* p.1863.

[8] Abdel Gawad A R, Mohamed MY , Abdalla H M , and Elsenbawi M A 2015 Analytical Prediction of Dual-Thrust Rocket motors under Uncertainties *International Conference on Aerospace Sciences and Aviation Technology* pp 1-12.

[9] Ritchey H 1955 Tailor-Made Thrust *Jet Propulsion* **25** p55.

[10] Anderson S 1964 Design and Evaluation of a Propulsion Unit for a Single-Stage Hercules (U) *Rep. S-48, Redstone Arsenal Research Div., Rohm and Haas Co.(Huntsville, AL) July* **8**.

[11] Brooks W 1972 *Solid propellant grain design and internal ballistics* vol. 8076(WashingtonD.C National Aeronautics and Space Administratio).

[12] Crimmins P, Cousineau M, Rogers C, and Shell V 1999 The 260-The largest solid rocket motor ever tested *36th AIAA/ASME/SAE/ASEE Joint Propulsion Conference and Exhibit* p. 3690.

[13] SMITH G and CHASE C 1976 Space shuttle booster separation motor design *12th Propulsion Conference* p 772.

[14] Adams I and Call F 1979 *Space shuttle SRM development* (NASA Technical Reports).

[15] Ricciardi A 1992 Generalized geometric analysis of right circular cylindrical star perforated and tapered grains *Journal of Propulsion and Power* **8** pp. 51-58.

[16] Stein S. D. 2008 Benefits of the Star grain Configuration for a Sounding Rocket(United States: Air Force Academy Department of Astronautics, USAFA).

[17] Brooks W T 1982 Ballistic optimization of the star grain configuration," *Journal of spacecraft and rockets* **19** pp 54-59.

[18] Ghazimirsaied M S A , Audet A and Koch C R 2008 HCCI Engine Cyclic variation characterization using both chaotic and statisticalapproach *Proceedings of Combustion Institute-Canadian Section*, University of Toronto, Ontario.

Acknowledgments

The authors would like to thank the experts in the Abo-Zaabal factory and test facility for their generous support and valuable contribution.

IOP Conf. Series: Materials Science and Engineering **973** (2020) 012002 doi:10.1088/1757-899X/973/1/012002

Mitigation of irregular burning in a small solid propellant rocket motor

M E lkshen[1], H Belal[2] and M A Al-Sanabawy[3]

[1]M.Sc. Student, Rocket Department, Military Technical College, Egypt
[2]Assistant Professor, Rocket Department, Military Technical College, Egypt
[3]Associate Professor. Rocket Department, Military Technical College, Egypt
hbelal@mtc.edu.eg

Abstract. An irregular burning phenomenon arises in some solid propellant motors(SRM), where the time-averaged pressure differs largely from that dedicated by instantaneous area burning. This phenomenon is generally considered the result of the interaction of pressure disturbances within the motor cavity with solid propellant deflagration. Hence, it is a combustion-dependent acoustic phenomenon. Historically, different suppression devices included the classical resonance rods and paddles, and arrays of acoustic resonators and various baffle arrangementswere applied.In this paper, irregular burning which arises in a modification of a shoulder-launched rocket is investigated. Several techniques dealing with the irregular burning problem were attempted. In each case, the signal of pressure was analysed in the time domain through time average and AC pressure signal, and in the frequency domain through Fourier transform and power spectral density. The most successful technique was the drilling of holes through the grain, connecting the perforation to the external burning region of the cylinder. Next was the increasing of blocking factor, both methods lead to stabilizing the modified motor.

1. Introduction

Combustion instability has plagued SRM development for many years. Its occurrence has produced events ranging from unsatisfactory ballistic performance to tolerated pressure oscillations[1]. Irregular burning may be considered one of the exhibitions of combustion instability, in which the pressure varies from the predicted pressure trace dedicated by a balance of mass-generated from burning and mass discharged from the nozzle and stillnot oscillatory[2]. Usually, irregular burning is accompanied by a DC shift in pressure above or below the natural values.

In general, unstable burningin a solid propellant rocket motor is characterized by undesirable irregularities in the combustion pressure and the linear burning rate.Figure 1,presents the terminology pertinent to the different types of burning that have been observed together with brief descriptions of the principal phenomena associated with the principal types of unstable burning. The frequencies of such oscillations correspond to one or more of the acoustical modes relevant to the geometry of the cavity wherein the combustion arises, hence the term resonant burning.Figure**1** illustrates the pressure-time curves for the different types of burning.

IOP Conf. Series: Materials Science and Engineering **973** (2020) 012002 doi:10.1088/1757-899X/973/1/012002

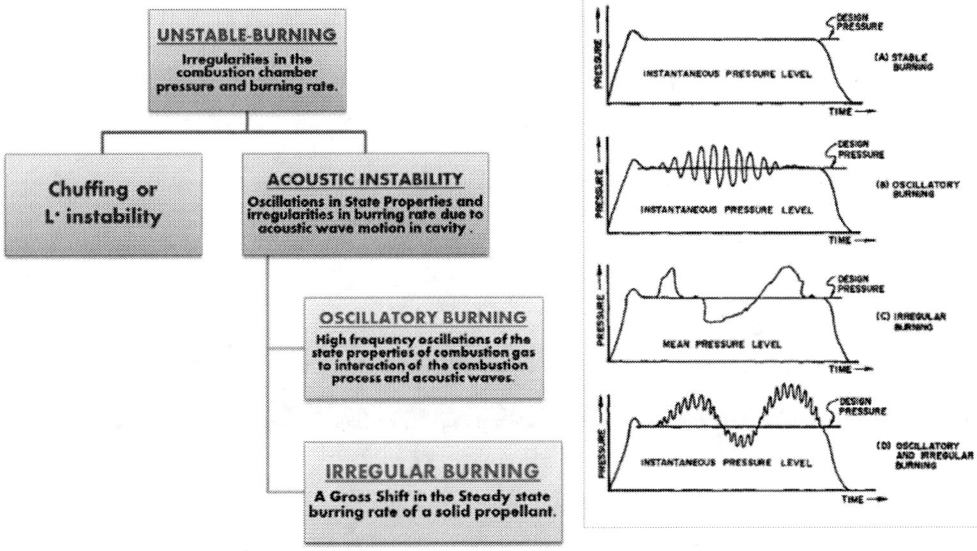

Figure 1. Unstable burning in a solid rocket motor[3].

Generally, the disturbances initially occur at a very small amplitude. If this state continued, combustion instability would be far less troublesome. However, the dynamic response of the combustion response to fluctuations in the internal flow field can result in the amplification of the initial disturbance and is often accompanied by a significant rise in base chamber pressure, called a DC pressure shift and limiting amplitudes of the oscillations[4],[5]. This occurs when the oscillatory gas motion is accompanied by a shift in the mean chamber pressure.In addition to fluctuating motor performance, a significant DCshift in chamber pressure may menace the structural integrity andpossibly lead to explosive failure of the motor case[6]–[8].

Historically, several techniques for dealing with the irregular burning problem were attempted. The best success of these techniques was radial holes, in which holes are drilled radially across the web of tubular grains, connecting the perforation to the external burning region of the cylinder. However, drilling toward the head end has been observed to be much more effective in eliminating secondary peaks than drilling the grain near the nozzle end. The stabilization which is obtained by drilling powder grains can also be accomplished by insertion of rods into the inner channel. Here also the effect on pitting of the inner surface of the grain is observed to be local, with head-end stabilization more important, than nozzle end stabilization in eliminating pressure peaks[9],[10].

The term "nozzle damping" represents the amount of the acoustic energy that isconvicted out of the combustion chamber through the nozzle. Factors such as the nozzlethroat size relative to the time-varying chamber port area, the throat erosion rate and howmuch the throat is submerged in the chamber have a major effect on this term.Additionally, nozzle damping is the largest factor for axial modes and is small for tangential andradial modes[11]–[13].

These techniques were supposed to affect the geometry of oscillating gas volume to change (usually increase) the frequencies and structure of the ordinary modes, thereby placing them outside the range in which the propellant will drive most strongly. Changes in the mode shapes relative to the distribution of burning surface may be such as to reduce the driving of combustion processes[14].

In this paper, a modification of a shoulder-launched rocket was attempted. The modification necessitated a change from forward multi-nozzle into a single central-nozzle keeping all other parameters unaltered. However, an irregular burning phenomenon emerged. This research attempts to mitigate these irregularities by applying different techniques in order to gain a better understanding of the phenomenon and its suppression methods.

IOP Conf. Series: Materials Science and Engineering **973** (2020) 012002 doi:10.1088/1757-899X/973/1/012002

2. The Motor

The original motor with 6 forward annular nozzles is shown in figure 2. Isometric views of the nozzle are shown in figure 3. In figure 4, the modified motor with central-nozzle is shown. The propellant used is a double-base propellant with the detailed composition given in table 1. The propellant grain was tubular without inhibited end.

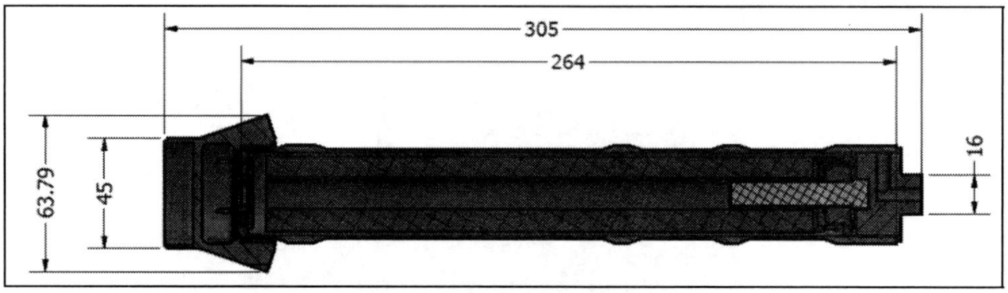

Figure 2. Test motor with multi-nozzle.

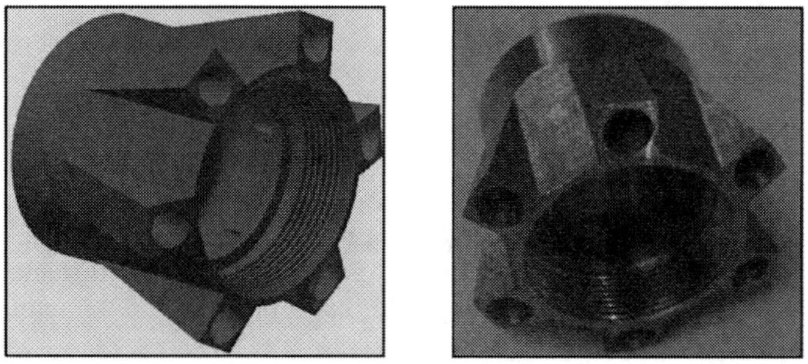

Figure 3. Multi-nozzle

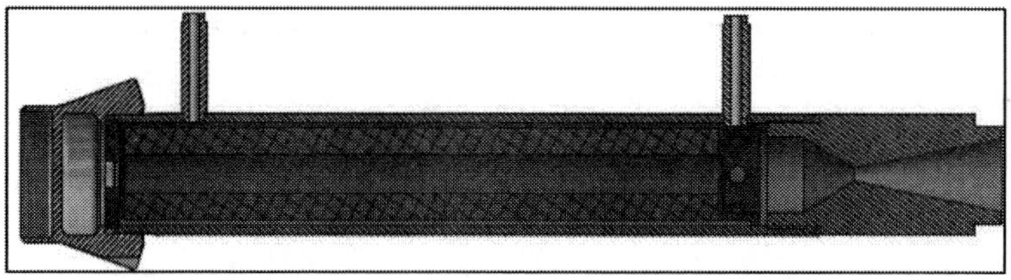

Figure 4. Test motor with central-nozzle.

Table 1. Composition of the double-base propellant.

Propellant	Density (g/cm^3)	Weight (g)
NC	1.653	59.0
NG	1.6	31.0
EC	1.14	5.0
$C_{16}H_{22}O_4$	1.05	2.0
$C_7H_6N_2O_4$	1.52	3.0
PBN_2O_6	4.53	2.0
C	2.26	.30

3. The problem

The static test for the original motor is shown in figure 5. The results show a typical pressure-time curve, without any instability or fluctuation in pressure. It is clear that the pressure-time trace has no oscillations except at ignitionand burnout.

The problem arises when replacing the multi-nozzle with a single central nozzle in the same motor, keeping the same throat area and using the same propellant grain. The modified motor shows a different pressure-time trace in static tests. The pressure-time trace (P-t trace) shows irregular burning phenomena, and where the pressure varies from the value predicted using the instantaneous burning area.

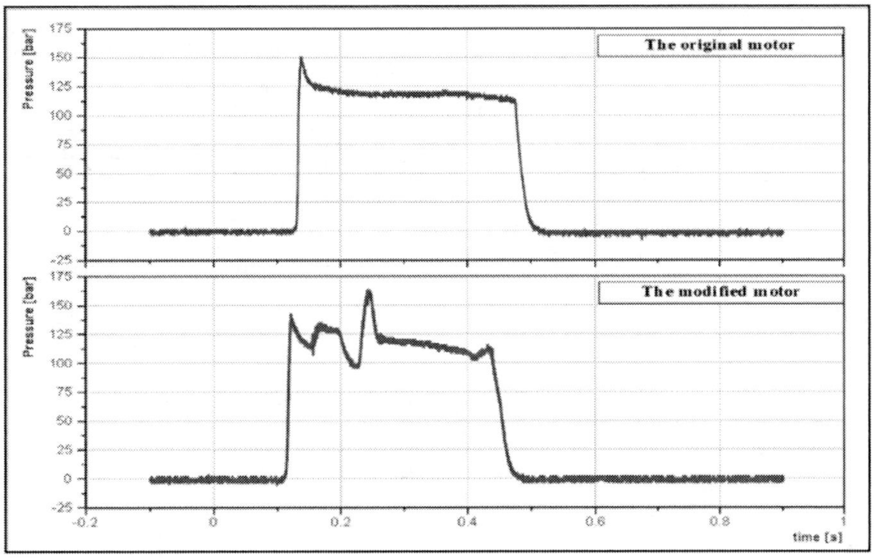

Figure 5. The (P-t) trace for the original motor and the modified motor.

4. Experimental technique and test matrix arrangement

Different methods are attempted to suppress the irregular burning: (a) using resonance rod, figure 6, (b) drilled holes[15], figure 7, and (c) nozzle admittance (damping) changes through using nozzles with differentdiameters[16]–[21].

The stabilizing effect of a hole is slightly decreased if the diameter of the hole is made very small in comparison with that of the axial perforation. However, increasing the diameter of the radial hole further than 0.4 times that of the axial perforationdoes not add to the stabilizing effect. So, the chosen diameter was 4 mm, (0.4 times the axial perforation). The burning area slightly increases but has no effect on the ballistic performance of the motor [15], figure 8.

Figure 6. Test motor with resonance rod.

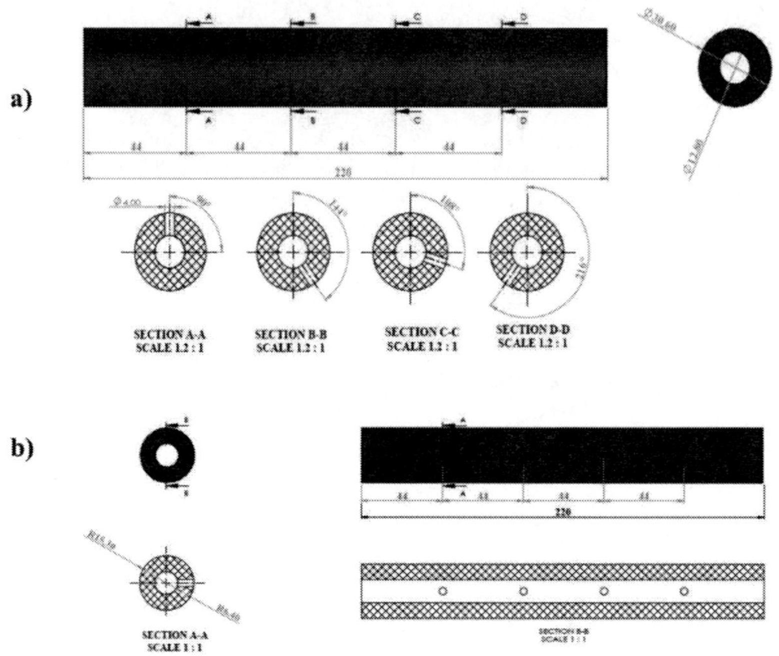

Figure 7. The double-base grain with (a) 4 radial holes (b) 4 inline holes.

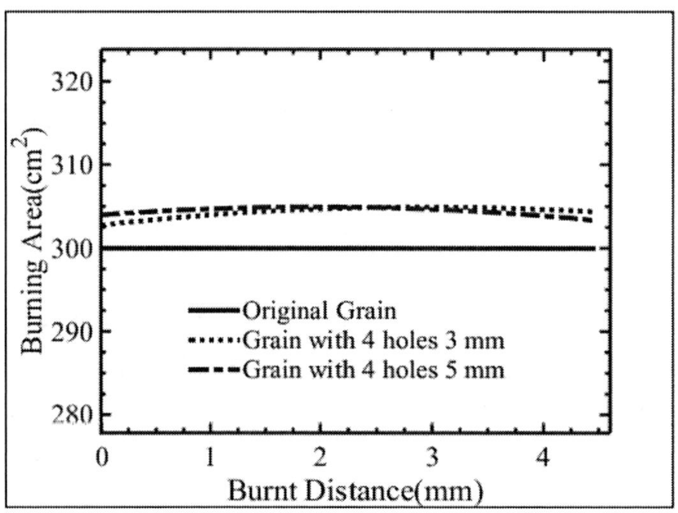

Figure 8. The changes of burning area without and with drilled holes.

Both time-averaged and high-frequency pressure measurements were taken at the motor head-end. The time-averaged measurements were made through one pressure port with one of HBM P3MB pressure strain-gauge transducer and one Kistler 6229A piezoelectric transducer covered with a thermal protective layer to record the high-frequency pressure oscillations.The test cases are described in table 2. In all of these tests, the pressure was measured at two positions, one was near to head and the other was near to nozzle, figure 9.

IOP Conf. Series: Materials Science and Engineering **973** (2020) 012002 doi:10.1088/1757-899X/973/1/012002

Figure 9. Test Stand.

5. Results

5.1. The P-t traces from the original motor and the modified motor

The static test results from the original motor are shown in figure 10 with the time-averaged pressure (in black) and the dynamic pressure (in green). It is clear that the pressure-time trace has no oscillations except at the ignition and burnout.

The modified motor shows a different pressure-time trace when statically fired, figure 11.First, a spectral analysis was done to investigate the frequency content of the original and the modified motor (it will be designated test #2). This analysis shows that there are exciting modes near 800 Hz and 1200 Hz, figure 12. The frequency for a given mode can be determined based upon chamber quantities of length, **L**, and the speed of sound inside a combustion chamber, **a**, as shown in equation (1)[2], where **n** is the mode number.The first longitudinal mode = 2165 Hz.

(1)

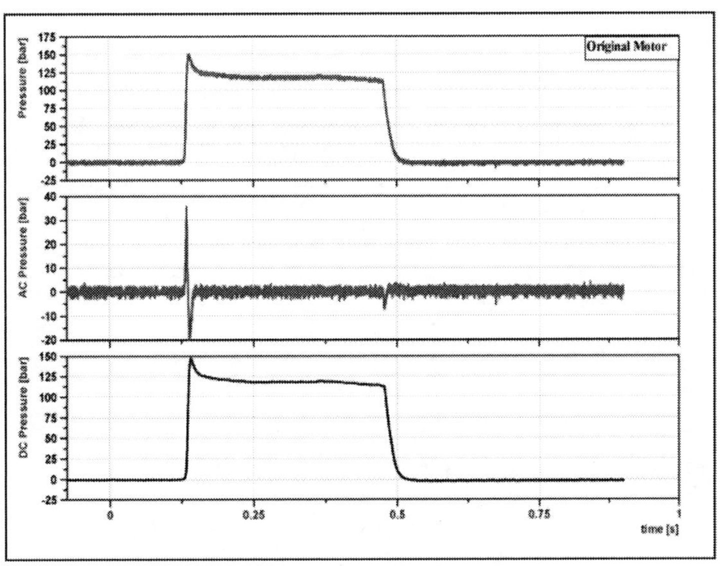

Figure 10. The (P-t) trace for original motor with multi-nozzle.

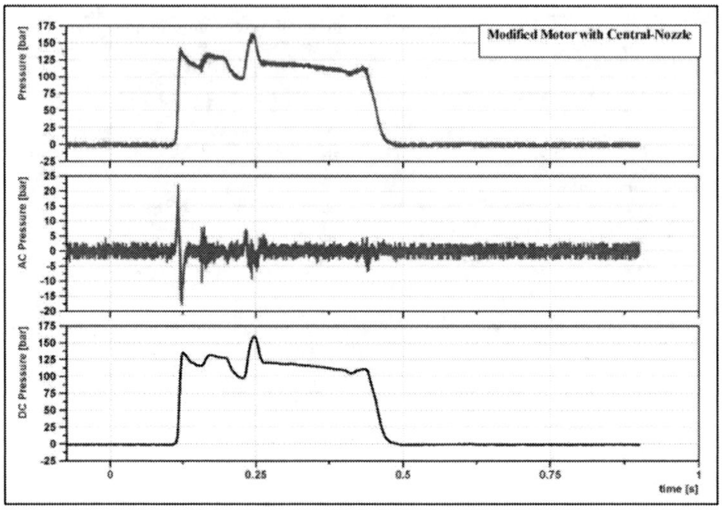

Figure 11. The (P-t) trace formodified motor with central-nozzle.

Figure 12,shows the power spectral density. it shows three broad peaks corresponding to the low-frequency instability. The bulk mode, and the first longitudinal modeare similar to those observed in a hybrid motor, figure 13[4]. It has been generally observed that the dominant mode involves low-frequency oscillations accompanied by lower amplitude higher frequency acoustic modes.

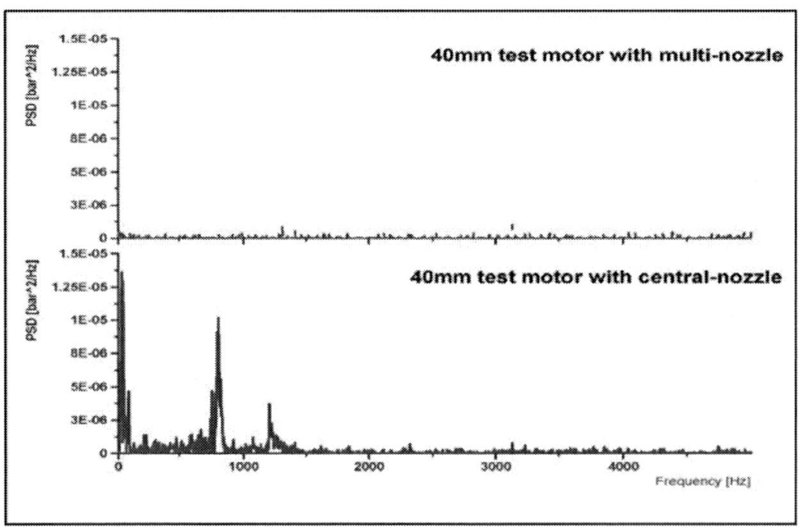

Figure 12. A spectral analysis performed for the two tests.

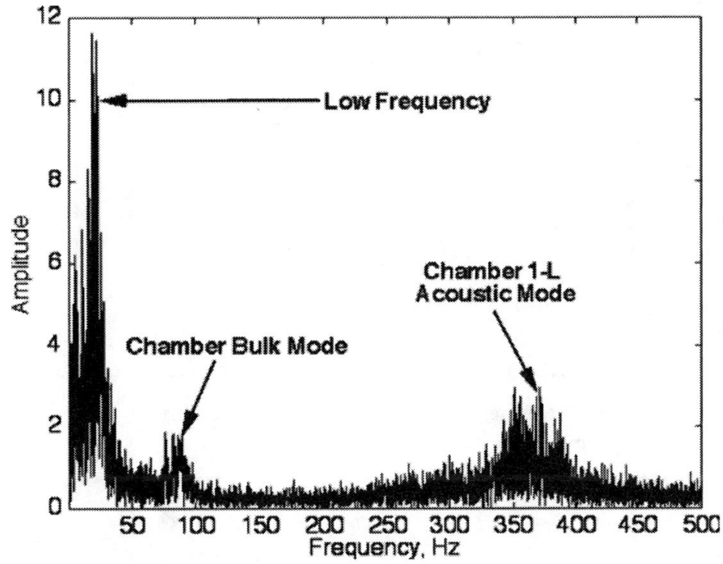

Figure 13. The Fourier transform corresponding to the low-frequency instability.

5.2. *The P-t traces for the other cases.*

Test #3 :(Resonance rod, central-nozzle dcr = 9.6 mm)
Resonance rods have been used to a much greater extent than any other device, the resonance rods were used to mitigate transverse modes of instability, with success. Resonance rods are installed in the grain perforation(s), either singly or in multiples, and are characterized by large length-to-width (transverse dimension) ratios but relatively small widths compared to the dimensions of the perforation. Various rod cross-sections have been used including rectangular, square,and circularcruciforms and shapes. Rods have been installed to extend either partially or over the full length of the grain perforation, and either cantilevered from one end or supported from both ends of the motor[22].

Resonance rod was applied successfully in control of instabilities in rocket motors MK40 and MK66 used in the American rocket Hydra (70 mm caliber), figure 14. For Talos cruise missile booster, figure 15, it is a good start to try to solve the problem of instability at expense of adding inert mass. Inthis case, the same resonance rod applied in MK66, figure 16, was tried. The P-t trace is shown in figure 17.The pressure spike at starting may be attributed to the decreased free volume inside the combustion chamber as the resonance rod is fitted inside the motor. The function of the resonance rod is to essentially disable the oscillations, meaning that the gas cavity is made acoustically random. However, this is often a difficult task since the mass of the rocket motor is required to be distributed symmetrically about the longitudinal axis. A common result of this requirement is that the suppressor, rather than creating randomness in the cavity, only "retunes" it. Another fundamental problem in the design of the mechanical suppressor is that the gas cavity is constantly changing, and the suppressor is often completely ineffective after the burn progressed partially, and this is what happened in this case as is evident from the figure.

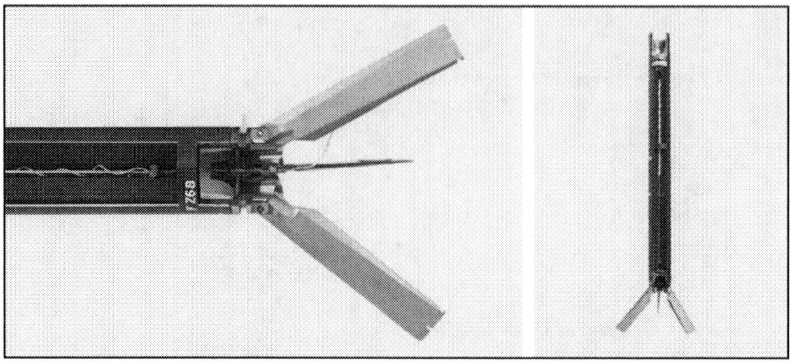

Figure 14. The American rocket Hydra (70 mm caliber)[23].

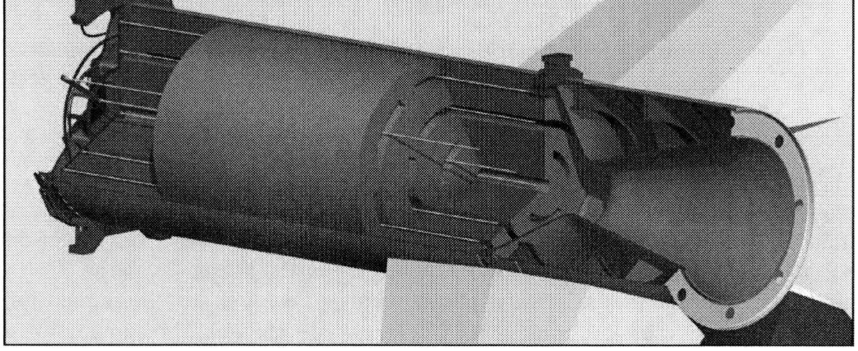

Figure 15. Talos cruise missile booster[24].

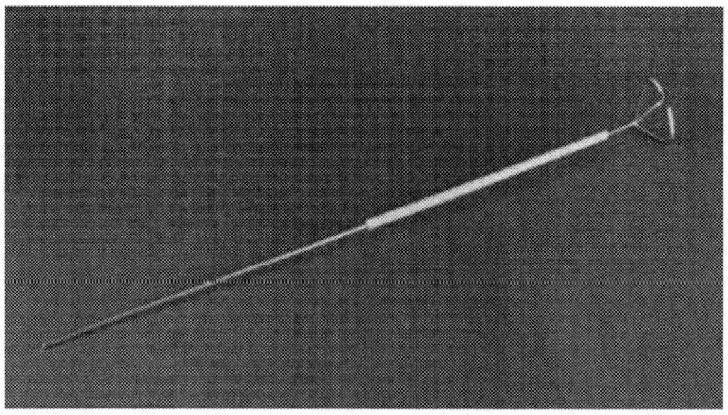

Figure 16. The resonance rod applied in MK66.

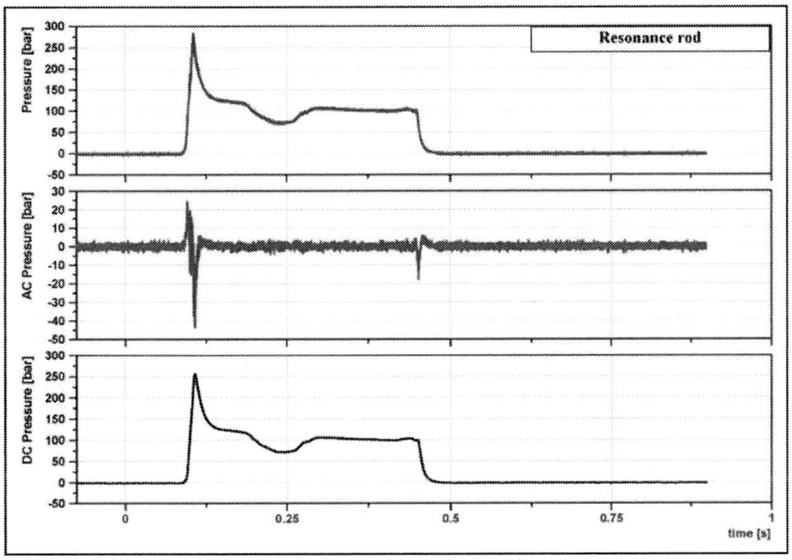

Figure 17. The modified motor tested with resonance rod (Test #3).

Test #4:(Grain with 3 holes radial, central-nozzle dcr = 9.6 mm)

Wimpress[15]has reported that by entirely empirical methods it has been found that several devices can be used to stabilize the burning in the central perforation, and the most common is the drilling of radial holes through the grain web at intervals along the axis. He reported also that two or more holes in a given plane at right angles to the axis of the grain have no more stabilizing effect than a single hole at the same point. The radial location of the holes is of no significance with respect to stabilizing effect, but it is general practice to avoid placing them in a single straight line because a radially unsymmetrical arrangement might affect the external ballistics of the rocket harmfully. In three-hole tubular grains, the radial holes are arranged in the form of a helix, with each hole in a plane passing through the axis of the grain but rotated 120° from the plane of the previous hole. Although it is apparent from figure 18, that irregularities gradually become less as the spacing between holes is decreased, it is not possible to establish definitely the maximum spacing compatible with stable burning.

The pressure-time curves obtained were more unstable than grain with three holes. Here, in this case, we confirmed that reducing the number of holes increases the problem of instability.

IOP Conf. Series: Materials Science and Engineering **973** (2020) 012002 doi:10.1088/1757-899X/973/1/012002

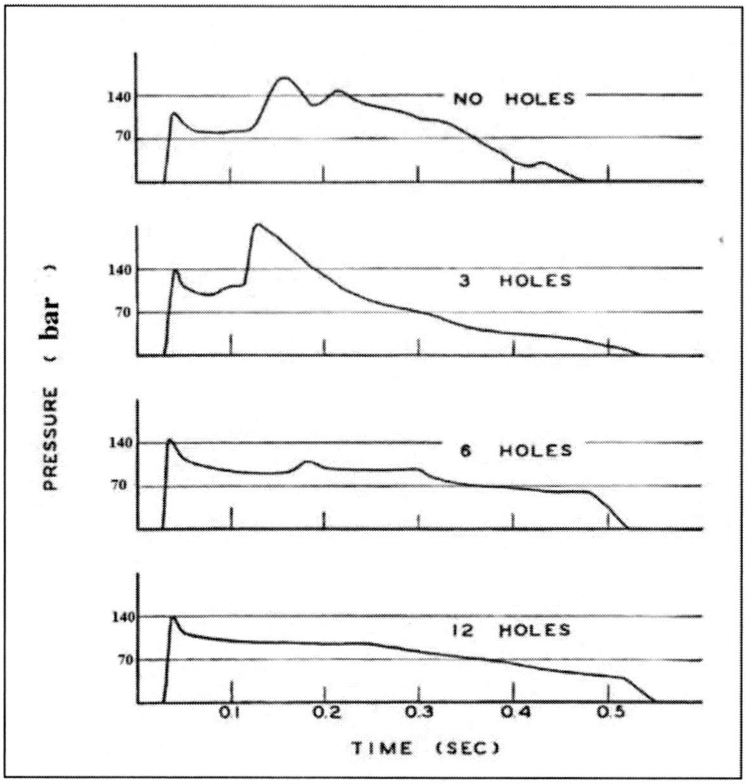

Figure 18. Stabilization of burning with radial holes taken from [14].

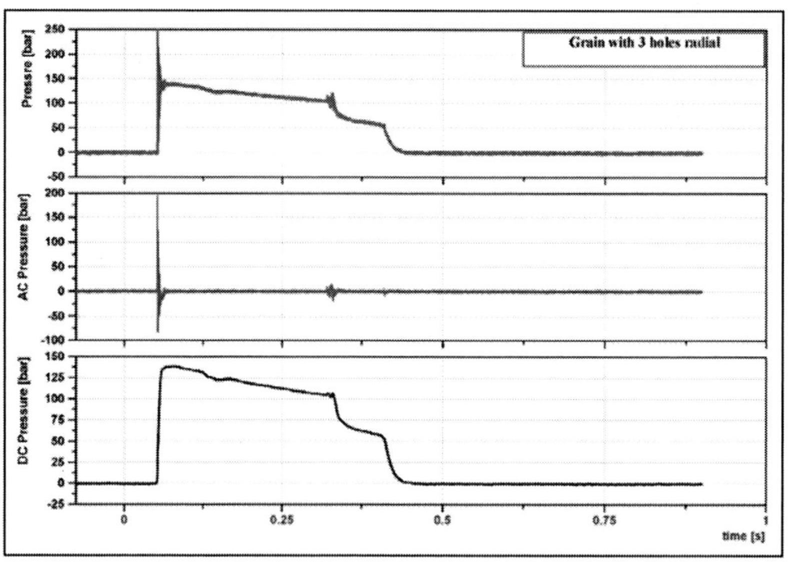

Figure 19. The modified motor tested with modified grain with 3 holes(Test #4).

Test #5 : (Grain with 2 holes radial, central-nozzle dcr = 9.6 mm)
The target of the holes was not achieved in this case, figure 20, as it appears that the number of holes was not sufficient to get rid of instability, but it was close to the target.

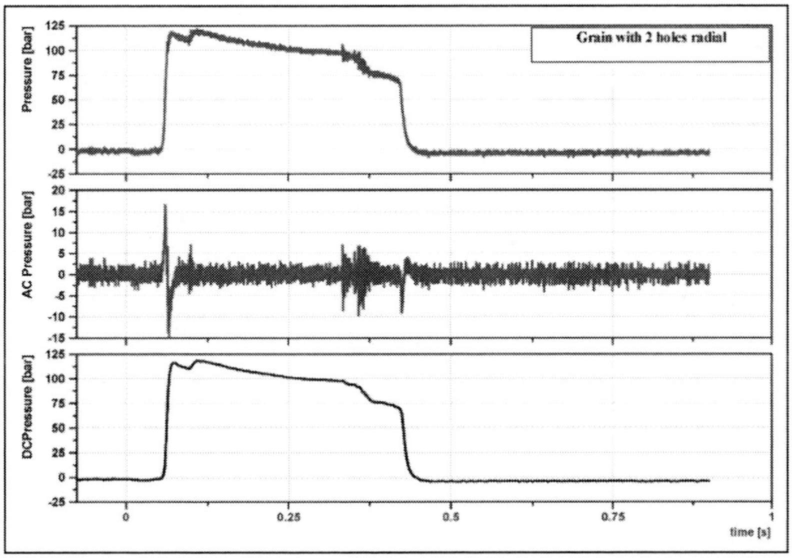

Figure 20. The modified motor tested with modified grain with 2 holes (Test #5).

Test #6:(Grain with 4 radial holes, central-nozzle dcr = 9.6 mm)
It seems that this is the most stable case. It is clear that the acoustic disturbances have decreased significantly, and this time the holes were effective in getting rid of the phenomenon of combustion instability, figure 21.

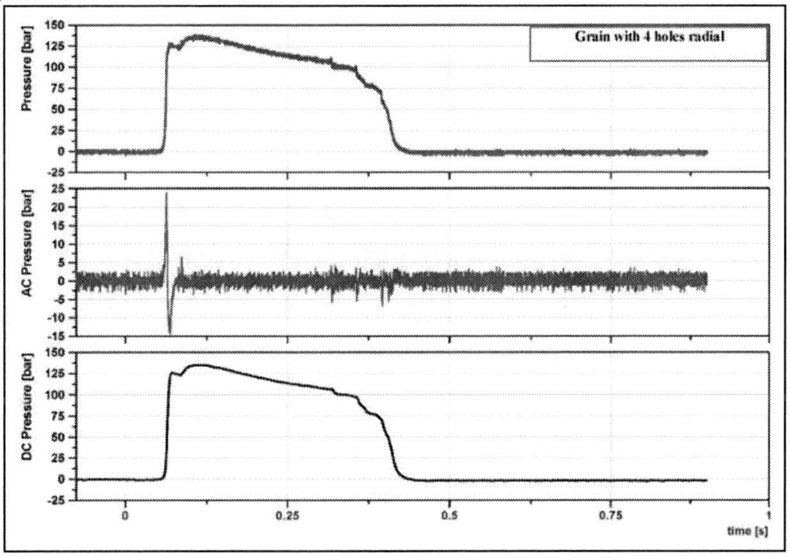

Figure 21. The modified motor tested with modifiedgrain with 4 holes(test #6)

IOP Conf. Series: Materials Science and Engineering **973** (2020) 012002 doi:10.1088/1757-899X/973/1/012002

Test #7 :(Grain with 4 inline radial holes - Central-nozzle dcr = 9.6 mm)
The holes made on the same line were not as efficient as the other, which were not made on the same line, although the same distance between the holes was preserved, figure 22.

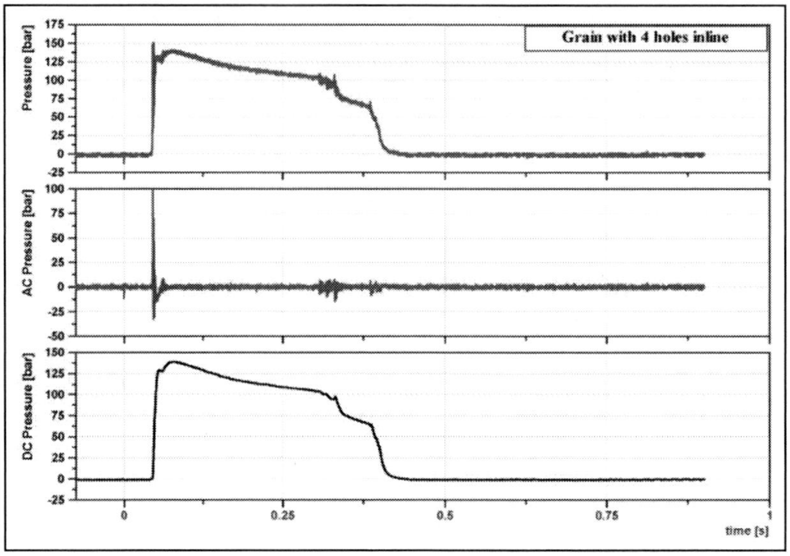

Figure 22. The modified motor tested with modified grain with 4 inline holes (Test #7).

Test #8:(Change blocking factor Kn, central-nozzle dcr = 10.6 mm)
The blocking factor $K_n$defined as ratio between burring area to throat area was reduced from 431 to 354, to test, The results showed that the problem has been worsened, figure 23

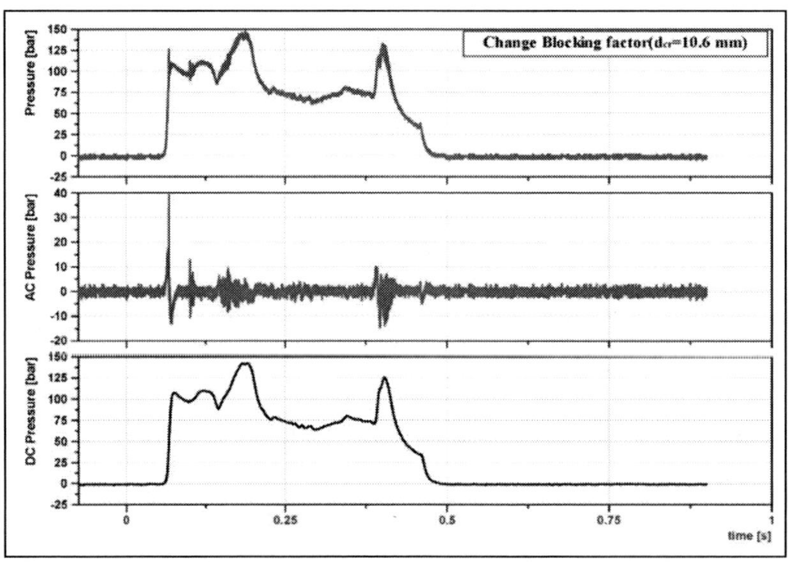

Figure 23. The modified motor tested witha throat diameter of 10.5 mm (Test #8).

Test #9:(Change blocking factor - central-nozzle d_{cr} = 8.5 mm)
The blockingfactorwas increased from 431 to 550, to mitigate combustion instability and it was positive. The effect of combustion instability was reduced, figure 24.

IOP Conf. Series: Materials Science and Engineering **973** (2020) 012002 doi:10.1088/1757-899X/973/1/012002

The results of the last two tests (increasing /decreasing blocking factor) show a decisive evidence for the phenomenon at hand to be an irregular burning rather than acoustic combustion instability. This is justified by referring to two researches.

According to Caprner et al. [25], they found that plotting the blocking factor K_n for both stable and unstable operations against combustion pressure, the intersection of the two plots being designated as the instability threshold, figure 25. From that figure, it is clear that the increasing pressure above the threshold pressure leads to instability, while from of Angelusetet al. [26] when he reported that the increasing pressure through increasing blocking factorleads to decrease irregular burning till all irregularities disappear,figure 26.

To summarize the results from these tests, all tests show irregular burning except for cases number 9 where the blocking factor was increased and case 6 of 4 drilled holes where the holes were not drilled on one line.

To have more investigation, a power spectral analysis was performed for the most stable case (test 6), figure 27. The curves show that even for such a stable case, there is still some acoustic or excited modes where there are relative peaks [26]. However, the powers contained in these peaks are too low to result in noticeable DC shifts as clear in other unstable cases.

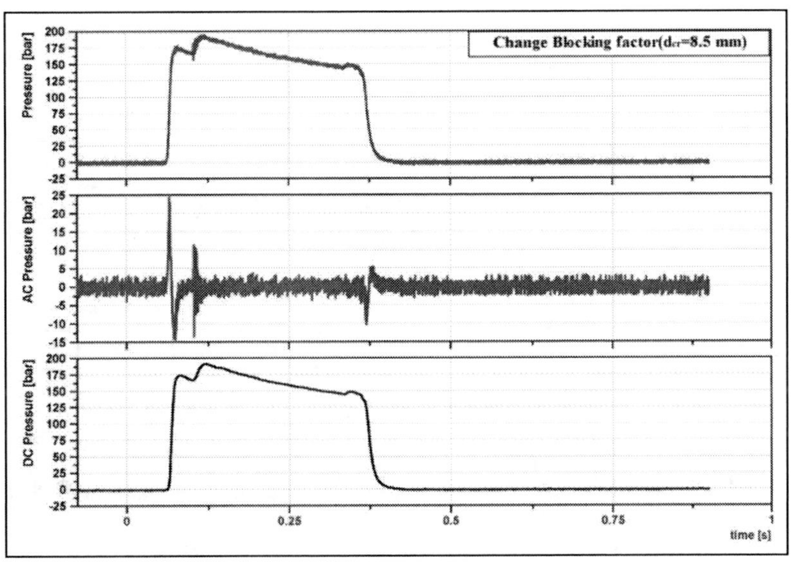

Figure 24. The modified motor tested with the throat diameter 8.6 mm(Test #9).

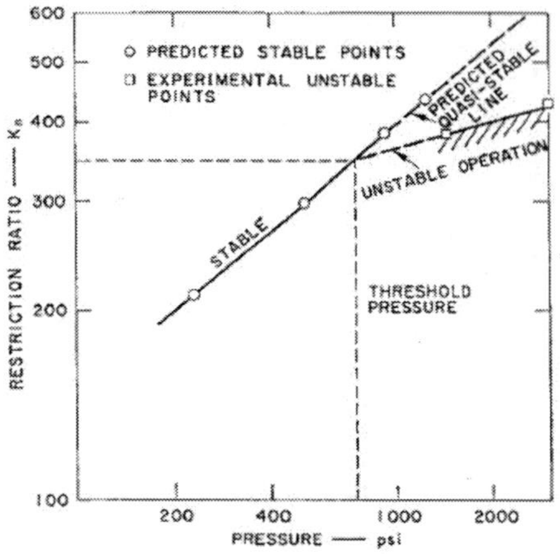

Figure 25. Selection of threshold point by extrapolating unstable data points back to $K_n$vs. P[25].

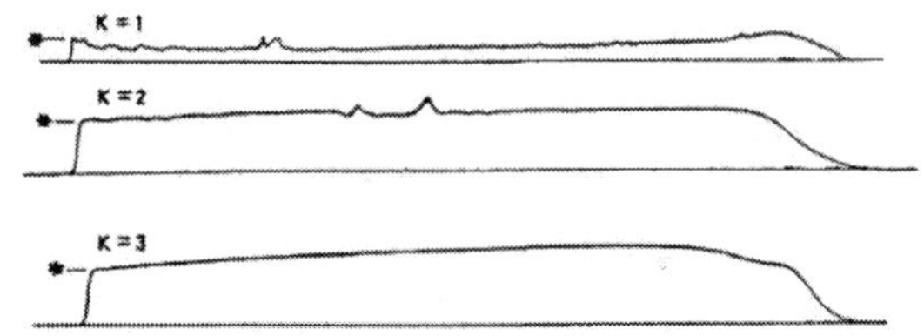

Figure 26.P-t traces for the same motor at different blocking factors, With increasing blocking factor, irregularities disappear.

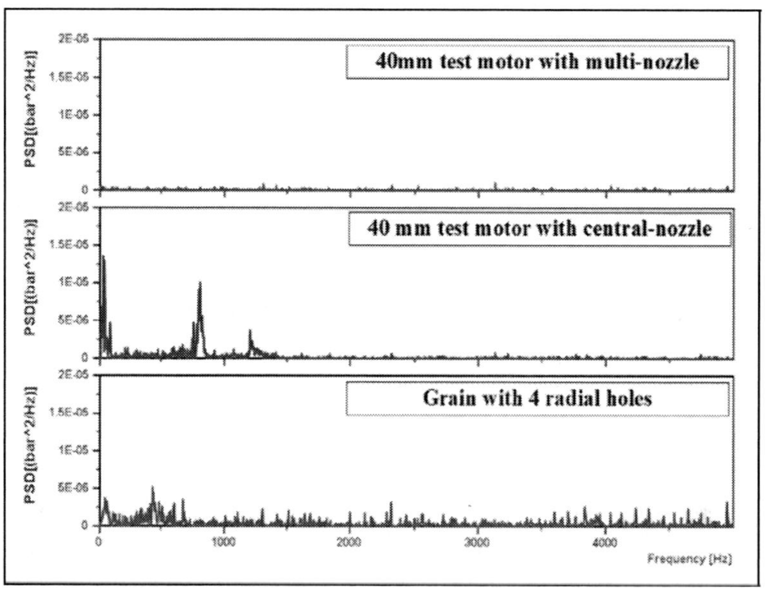

Figure 27. Power spectrum density.

The previous results are an experimental verification of the equation derived by Flandro[2]. It was shown that the change in time-average pressure is governed by the following equation (2)

$$\frac{d\bar{p}}{dt} = \left\{ \begin{array}{l} \left(\frac{\bar{a}^2}{\gamma V}\right)\left(\bar{\rho} S_b \left(b\bar{P}^n\right) - \right. \\ + \frac{\bar{a}(\pi R L)}{2\gamma^2 V} \bar{P} \sum_{m=1}^{\infty} R_m^2 \left\{ \bar{M}_b \left(A_b^{(r)} \right. \right. \end{array} \right. \quad (2)$$

where the classical first-order internal ballistics results are seen in the first term. S_b is the instantaneous burning surface area and A_{cr} is the nozzle throat area, C* characteristic velocity. The standard model for the propellant burning rate is displayed with n as the burning rate exponent and **b** as the coefficient. The second line shows the nonlinear correction indicating the effects of the pressure coupling where $\bar{a}$mean speed of sound, γratio of specific heats,$\bar{P}$mean chamber pressure,Lchamber length, $\bar{M}_b$Mach number in a combustion chamber, $\mathbf{A_b}$ and $\mathbf{B_b}$ are parameters of propellant admittance or pressure coupling terms, with $\mathbf{J_m(k_{mn})}$ is the Bessel functions of order m, with summing over acoustic modes. It is obvious that large mean pressure shifts are expected when higher-order harmonics are excited in the nonlinear growth. It is important to understand that in this type of modeling effort, it is necessary to account for all internal ballistics effects including the changes of chamber volume and burning surface area with time as well as the modification of the surface Mach number as the average pressure rises. Form spectral analysis of unstable case (Test #2) and stable case (Test #6), the reason for the DC shift is clearly due to the excited modes at 800 and 1200 Hz, while in stable cases the amplitude of such waves ($\mathbf{R_m}$ in equation 2) is very small , which results in too small time-averaged pressure[2].

Table 2. Summary of the test matrix

No.	Test name	Nozzle type	Irregular burning	
			Yes	No
1	Original motor 40 mm	Forward multi-nozzle		√
2	Modified motor 40 mm	Central-nozzle d_{cr} = 9.6 mm	√	
		Mechanical Mitigation		
3	Resonance rod	Central-nozzle d_{cr} = 9.6 mm	√	
4	Grain with 2 holes radial	Central-nozzle d_{cr} = 9.6 mm	√	
5	Grain with 3 holes radial	Central-nozzle d_{cr} = 9.6 mm	√	
6	Grain with 4 holes radial	Central-nozzle d_{cr} = 9.6 mm		√
7	Grain with 4 holes inline	Central-nozzle d_{cr} = 9.6 mm	√	
		Nozzle Damping		
8	Change blocking factor	Central-nozzle d_{cr} = 10.6 mm	√	
9	Change blocking factor	Central-nozzle d_{cr} = 8.5 mm		√

6. Conclusion

Combustion instability continues to hinder the development of solid propellant rocket motors from small motor such as the case discussed in this paper or for large motors used in the intercontinental ballistic missile. The problem with solving the combustion instability is that it cannot be extrapolated as the resonance rod which was applied successfully in a similar motor did not succeed in mitigating stability in this case.

Analysis of each case signal in time (time-averaged pressure and AC pressure) and frequency (Fourier transform amplitude and power spectral density) can explain how the instability happens. It was clear that exciting certain modes of chamber acoustic leads to instability. However, even for the stable cases, there are still some exciting modes but with a too-small amplitude that it cannot trigger instability or noticeable DC shift.

Despite that the general problem of acoustic instability had been studied for long times, the problem of irregular burning still needs more research in order to predict a priori the instability and predict the effectiveness of each fix to save the costs of experiments and labor works.

References

[1] Blomshield F 2007 Lessons Learned In Solid Rocket Combustion Instability *43rd AIAA/ASME/SAE/ASEE Joint Propulsion Conference & Exhibit*

[2] FlandroG 2007 Irregular Burning: On the Origin of the DC Shift*43rd AIAA/ASME/SAE/ASEE Joint Propulsion Conference & Exhibit*pp 1–17

[3] *Engineering Design Handbook,Elements of Aircraft and Missile Propulsion.* AMCP.
[4] Carmicino C and Pastrone D 2019 Acoustic Excitation as Triggering Mechanism of the 'DC Shift' *Hybrid RocketsAIAA J.***57** no. 11, pp 4845–4853
[5] Toong T, Richard F, John M, and Griffin Y 1965 Mechanisms of Combustion Instability*Proc. Combust. Inst.***10** pp 1301-1313
[6] Flandro G, Fischbach S, and Majdalani J 2007 Nonlinear rocket motor stability prediction: Limit amplitude, triggering, and mean pressure shift*Phys. Fluids***19**no. 9
[7] Flandro G, Perry E, and French J 2006 Nonlinear Rocket Motor Stability Computations: Understanding the Brownlee-Marble Observations*44th AIAA Aerospace Sciences Meeting and Exhibit*
[8] Lengellé J,DuterqueG,.TrubertJ 2002 Combustion of Solid Propellants (RTO-EN-023) *Intern. Aerodyn. Solid Rocket Propuls.* heldRhode-Saint-Genèse, Belgium, 27-31 May pp 27–31
[9] Gusachenko Land. ZarkoV 2008 Analysis of unsteady solid-propellant combustion models (review) *Combust. Explos. Shock Waves***44**no. 1 pp 31–42
[10] Green L 1956 Observations on the Irregular Reaction of Solid Propellant Charges *J. Jet Propuls.***26** no. 8, pp 655–659
[11] Gloyer P, Rice T, and Flandro G 2009 Ares i thrust oscillation mitigation via motor modifications*56th JANNAF Propulsion Meeting*
[12] French J, Flandro G, Gloyer P, and Jacob E 2010 UCDS Nozzle Acoustic Dynamics and Stability Modeling *46th AIAA/ASME/SAE/ASEE Joint Propulsion Conference & Exhibit*
[13] Larsen C, Gloyer P, and Flandro G 2008 Ares I Solid Rocket Booster Thrust Oscillation *NASA Eng. Saf. Cent. Tech. Report, RP-08-60*
[14] Grad H 1949 Resonance burning in rocket motors*Commun. Pure Appl. Math.***2**no. 1, pp 79–102
[15] Wimpress R andSage B1950 Characteristics of burning in perforations *INTERNAL BALLISTICS OF SOLID-FUEL ROCKETS*, McGraw-Hill New York
[16] Janardan B, Daniel B, and Zinn B 1974 Damping of Axial Instabilities by Small-Scale Nozzles
Under Cold-Flow Conditions *J. Spacecr. Rockets***11** no. 12 pp 812–820
[17] Javed A and Chakraborty D2014 Damping Coefficient Prediction of Solid Rocket Motor NozzleUsing Computational Fluid Dynamics*J. Propuls. Power***30** no. 1 pp 19–23
[18] ZINN B1973 Nozzle Damping in Solid Rocket Instabilities*AIAA J.***11** no. 11 pp 1492–1497
[19] Sun B,Li S, Su W, Li J, andWang N2016 Effects of gas temperature on nozzle damping experiments on cold-flow rocket motors*Acta Astronaut.***126**pp18–26
[20] Xing W,Wang N, Wei J, Dong Y, and Yan M 2013 Numerical Research on The Nozzle Damping Effect by A Wave Attenuation Method *Def. Technol.***9** no. 3 pp 162–166
[21] Janardan B and Zinn B1977 Rocket nozzle damping characteristics measured using different experimental techniques *AIAA J.* **15** no. 3 pp 442–444
[22] Oberg C,Haymes W, and Wong T 1972 Solid propellant combustion instability suppression devices *8th Joint Propulsion Specialist Conference*
[23] FFAR Rocket Systemhttps://fz.be/Mk40
[24] P. R. H. P. L. USNR-Rhttps://www.okieboat.com/Booster%20History.html
[25] Capener E,Dickinson L and Kier R 1964 Driving processes of finite-amplitude axial mode instability in solid-propellant rockets *AIAA J.***5**no. 5 pp 938–945
[26] Angelus T 1960 Unstable Burning Phenomenon Double-Base Propellants *Solid Propellant Rocket Research*, New York: American Institute of Aeronautics and Astronautics pp 527–559.

Validation of a simplified model for liquid propellant rocket engine combustion chamber design

M Hegazy[1], H Belal[2], A Makled[3] and M A Al-Sanabawy[4]

[1] M.Sc. Student, Rocket Department, Military Technical College, Egypt
[2] Assistant Professor, Rocket Department, Military Technical College, Egypt
[3] Associate Professor. Zagazig University, Egypt
[4] Associate Professor. Rocket Department, Military Technical College, Egypt
hbelal@mtc.edu.eg

Abstract. The combustion phenomena inside the thrust chamber of the liquid propellant rocket engine are very complicated because of different paths for elementary processes. In this paper, the characteristic length (L*) approach for the combustion chamber design will be discussed compared to the effective length (L_{eff}) approach. First, both methods are introduced then applied for real LPRE. The effective length methodology is introduced starting from the basic model until developing the empirical equations that may be used in the design process. The classical procedure of L* was found to over-estimate the required cylindrical length in addition to the inherent shortcoming of not giving insight where to move to enhance the design. The effective length procedure was found to be accurate within ± 10%.

1. Introduction

The combustion phenomena inside the thrust chamber of the liquid propellant rocket engine (LPRE) are very complicated because of different paths for elementary processes, figure 1 [1]. Previous research from the same group studied the vaporization controlled model for liquid propellant thrust chamber design. It was concluded that for some propellant combination i.e. LOX/Kerosene, vaporization is actually the rate-limiting process and can be used as to find the minimum length required for the combustion chamber, they also found that for other propellant combinations, the chemical reactions are slow that the vaporization is no longer the rate-limiting process [2]. In this paper, the L* approach will be discussed and applied to real case studies, next, the effective length approach will be introduced. This effective length design methodology is developed using the previously developed vaporization-limited model. Similarly, the effective length approach is applied for the same known liquid propellant motors in order to validate the applicability of the effective length and L* approaches.

The characteristic length approach is a classical approach to design a LPRE combustion chamber. This approach depends on tabulated values for L*, table 1 [3] which is defined as the ratio between the volume of the combustion chamber and the nozzle critical area, equation 1 [4]. The chamber volume is that volume enclosed between the injection head and the nozzle critical section as indicated in figure 2. Characteristic length is not a physical length but it is a parameter depending on the perfection of combustion processes.

IOP Conf. Series: Materials Science and Engineering **973** (2020) 012003 doi:10.1088/1757-899X/973/1/012003

$$L^* = V_c / A_{cr} \tag{1}$$

where V_c = Volume of the chamber (m^3), A_{cr} = Critical cross-section area (m^2)

Experimental data for the characteristic length is obtained through several testes using motors with different cylindrical lengths. The measurments involves the characteristic velocity efficiency defined as the ratio between experimental and theoretical characteristic velocities, equation 2, [3]. The characteristic velocity is known as the ratio of the product (chamber pressure × nozzle throat area) to the mass flow rate of gases leaving the the nozzle. Such parameters can be measured to determine the experimental characteristic velocity while the theoretical one can be obtained from any thermochemical calculation software for each motor.

$$\eta_{c^*} = c^*_{exp} / c^*_{th} \tag{2}$$

where c^*_{exp} = Experimental characteristic velocity (m/s), c^*_{th} = Theoretical characteristic velocity (m/s).

Similar curves can be obtained with this method by changing the propellant type, injection system, combustion chamber geometry (contraction area ratio), or the mixture ratio. From these curves, the optimum cylindrical length can be determined for the combustion chamber as showing in table 1 [3].

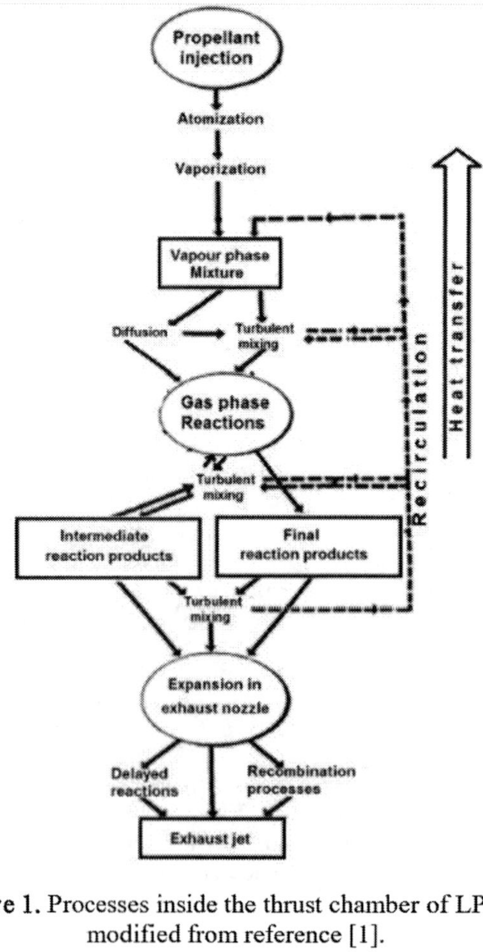

Figure 1. Processes inside the thrust chamber of LPRE, modified from reference [1].

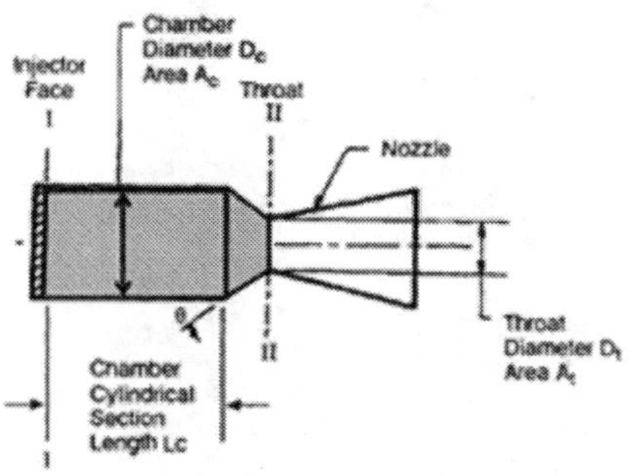

Figure 2. Definition of combustion chamber volume V_c [1].

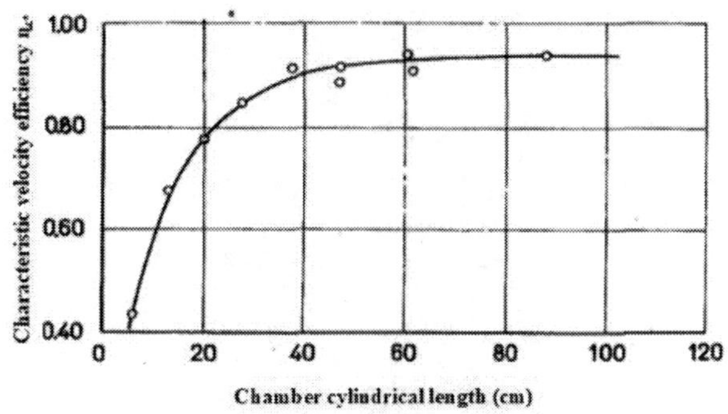

Figure 3. Variation in characteristic velocity efficiency with the chamber cylindrical length [3].

Table 1. Values of characteristic length for different propellant combinations [3]

Propellant Combination	L* (m)
Nitric-acid-Hydrocarbon	2-3
Nitric Acid-UDMH	1.5-2
LOX-Ethyl alcohol	2.5-3
LOX-Kerosene	1.5-2.5
Fluorine-Ammonia	1-1.5

The procedure starts by selecting a value of L* then find the volume of the chamber. This volume is used to evaluate the value of chamber length depending on volume shape (spherical, cylindrical, or ellipsoidal). Table 2 shows the data for 6 of actual LPREs, shown in figures 4-9, the table summarizes the main data needed for the analysis in this section, namely contraction area ratio a_c (ratio of the combustion chamber area to the critical area), convergent angle θ, critical area A_{cr} and cylindrical length L_c. Such values were obtained from available documentation.

Figure 4. Engine LR89-NA-7 [5].

Figure 5. Engine LR79-NA-11 [5].

IOP Conf. Series: Materials Science and Engineering **973** (2020) 012003 doi:10.1088/1757-899X/973/1/012003

Figure 6. Engine RD-0110 [5].

Figure 7. Engine H-1 [5].

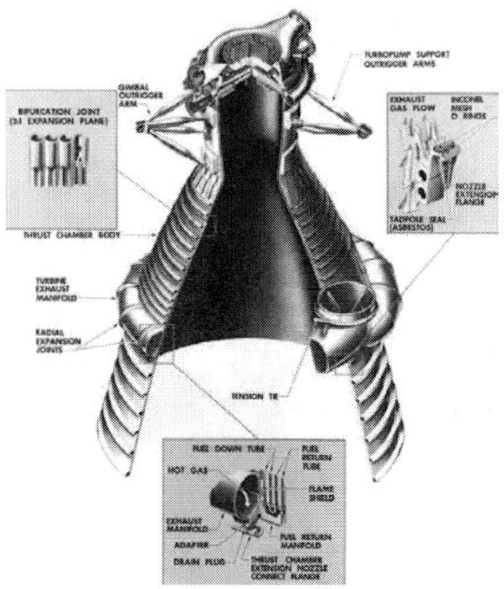

Figure 8. Engine F-1 [6].

Figure 9. Engine LR87-AJ-3 [5].

Table 2. Main parameters of used engines.

Engine	L, mm	a_c	A_{cr}, In^2	Convergent angle θ°
RD-0110[7]	204	4.538	8.69	30
F1[8]	944	1.3	961.4	13
LR79-NA11[8]	280	1.67	206	10
LR89-NA-7[8]	335	1.67	206	10
H1[8]	270	1.62	205	15
LR87-AJ-3[8]	412	2.17	182.56	25

The results of applying L* procedure are shown in table 3, choosing values for L* = 1.5 m and 2.5 m. These are the lower and upper values for the combination of LOX/ Kerosene, given in table 1. It is clear from the last three columns that the actual engines do not comply with the recommended ranges of L*. For the engine RD-0110, the actual length is far below the length calculated using the lowest value of L*. This also is true for the engine LR79-NA-11. This discrepancy adds to the disadvantages of L* methods where there is no measurable account for the operating parameters on calculated values. Increase in pressure, decrease of injection temperature, or alteration in the efficiency of the injection system are examples of operating parameters in question.

Table 3. Data for 6 LPREs with calculated values of cylinder length using L* method.

Engine	L_{act} mm	$V_{c\,(1.5)}$ m^3	$V_{c\,(2.5)}$ m^3	$L_{(1.5)}$ mm	$L_{(2.5)}$ mm
RD-0110[7]	204	0.0084	0.014	284	504
F-1[8,9,10]	944	0.931	1.552	905	1670
LR79-NA-11[8,9,10]	280	0.199	0.332	628	1227
LR89-NA-7[8,9,10]	335	0.199	0.332	628	1227
H-1[8,9,10]	270	0.198	0.331	759	1376
LR87-AJ-3[8,9,10]	412	0.177	0.295	605	1096

The previous data obligate the need for a different procedure to design a liquid propellant combustion chamber which can account for the process in order to find more accurate values and account for both operating conditions and design parameters.

2. Model of relevant chamber parameters

The model used to develop the equations needed for the new methodology is discussed in detail in [2]. The main ideas are reviewed in the following. The model is assuming a one-dimensional steady flow in which propellant vaporization is the rate-controlling process, considering reaction and mixing are infinite fast processes. In addition, there is no reaction in the liquid phase, neglecting secondary breakup. Applying these assumptions would result in the following equations:

Mass transfer
$$\frac{dm_d}{dt} = w = A_S\, K\, \alpha\, P_{a,s} \tag{3}$$

Heat transfer
$$q_v = h\, A_S \left(T_c - T_l\right) Z \tag{4}$$

Droplet-Heating rate
$$\frac{dT_d}{dt} = \frac{1}{m\, c_{p,l}}\left(q_v - w\lambda\right) \tag{5}$$

Droplet Acceleration
$$\frac{dV_d}{dt} = -\frac{3}{8} C_D \frac{\rho_m}{\rho_l} \frac{U^2}{r_S} \tag{6}$$

Gas velocity
$$\frac{u(x)}{u_{fin}} = 1 - \frac{\dot{m}_f(x)}{\dot{m}_{f,in}} \tag{7}$$

For the nomenclature and more details, interested readers can rely on reference [2]. The equations can be used to find the length required to vaporize the less-volatile component, considering the reaction takes place in gas phase between the more volatile component and the amount vaporized from the less-volatile one. The objective in this research is to develop an empirical equation relating the required length to the operating parameters (combustion pressure P_c, and injection temperature T_{in}), geometric parameter (contraction ratio a_c), spray parameter (mass median diameter MMD), and standard deviation σ_{LN}. The target is to illustrate the effect on the length required to vaporize 95% of propellant when using conditions other than those applied in a test case. The test matrix used for n-heptane as typical hydrocarbon propellant is considering the variation of injection temperature T_{in}, injection velocity V_{in}, MMD, σ_{LN}, Pc and a_c as shown in table 4.

Table 4. Operating and design parameters used in the parametric study.

T_{in}	V_{in} (m/sec)	MMD (μm)	σ_{LN}	P_c (bar)	a_c
220	15	100	1.25	10	1.65
280	30	150	2.3	20	3.15
380	60	450	3.6	40	12.6

2.1. Mass median droplet size effect

Table 5. Effect of MMD on $L_{95\%}$.

MMD (μm)	$L_{95\%}$ (mm)
100	110.73
150	196.10
450	946.88

The length required to vaporize a given fraction of the total mass of the spray is proportional to the 1.46 of the mass median droplet size. a result which is intuitionally accepted as the larger the size of a droplet, the longer the length it needs to vaporize.

2.2. Geometric standard deviation effect
For higher standard deviation, the percent of the spray vaporization and vaporization rate near the injector increases because of the large number of small droplets, and the required length to vaporize propellant is longer because of the larger droplets. The length required to vaporize a given fraction of the total mass of the spray is proportional to the 0.77 of the geometric standard deviation.

Table 6. Effect of geometric standard deviation on $L_{95\%}$.

Standard deviation (σ_{LN})	$L_{95\%}$ (mm)
1.25	113.23
2.3	196.10
3.6	335.47

2.3. Injection droplet velocity effect
The maximum vaporization rate is obtained with a low initial velocity. The higher the initial velocity, the more mass vaporized before the minimum point is reached. However, the larger the velocity, the larger the spray travel before complete vaporization. This yields the results that the length required to vaporize a given percentage of the total mass is proportional to 0.86 power of initial velocity.

Table 7. Effect of injection velocity on $L_{95\%}$

Droplet Injection velocity (V_{in}) (m/sec)	$L_{95\%}$ (mm)
15	126.28
30	196.01
60	353.80

2.4. Chamber pressure effect
The higher the chamber pressure, the higher the vaporization rate and amount vaporized in a given length. The length required to attain a given high percentage of mass vaporized is inversely proportional to 0.54 power of chamber pressure. This would not be attained if it was made use of the d^2-law of Spalding [11]. The pressure dependence is introduced via the droplet equation and vaporization equation. Experimentally, it was found that burning rate dependency on pressure is related to fuel (0.25 for furfuryl alcohol, 0.4 for benzene, 0.37 for hydrazine in air and 1 for hydrazine decomposition) [12].

IOP Conf. Series: Materials Science and Engineering **973** (2020) 012003 doi:10.1088/1757-899X/973/1/012003

Table 8. Effect of chamber pressure on $L_{95\%}$.

Gas pressure (bar)	$L_{95\%}$ (mm)
10	298.89
20	196.01
40	128.41

2.5. Contraction area ratio effect

It is the ratio of cylindrical chamber cross-section area to the nozzle throat area. Small contraction ratio yields high gas velocities at the end of the combustion chamber, which encourages vaporization. The higher the contraction ratio, the more mass is vaporized before the minimum point is reached.

Table 9. Effect of contraction ratio on $L_{95\%}$.

Contraction ratio	Final gas velocity (m/sec)	$L_{95\%}$ (mm)
1.65	382.06	160.50
3.15	188.20	192.40
12.6	46.51	373.59

Reduction in contraction ratio decreased the length of the combustor needed for complete vaporization. This effect, however, was not constant throughout the vaporization period. During the initial vaporization phase, an inverse effect of contraction ratio was noticed. Inversion occurs in the region where combustion gas velocity begins to exceed drop velocity. This analytical result is similar to the experimental effect, where efficiency improves with an increase in contraction ratio for the low-efficiency region and the inverse effect becomes evident as efficiency increases [13]. The calculation shows that the length required to vaporize 95% of the mass is proportional to 0.48 power of contraction ratio.

2.6. Droplet temperature effect

A higher initial liquid temperature is moderately beneficial, as droplet spends less time reaching the wet-bulb temperature which simply is the droplet temperature at which all heat flux goes to vaporize droplet and no more heating or droplet temperature increase. The calculation shows that the percentage of mass vaporized is inversely proportional to 0.51 power of initial droplet temperature.

Table 10. Effect of droplet injection temperature on $L_{95\%}$.

Droplet Temperature (K)	$L_{95\%}$ (mm)
220	212.57
280	196.10
380	165.35

For a summary, it is concluded that the chamber length required for a given percent of propellant vaporized increases with larger drop sizes and higher injection velocity, decrease with higher final gas velocity, higher chamber pressure, and higher initial temperature.

2.7. Empirical correlation

The above discussion can be used to derive a correlation relating the chamber length required to vaporize 95% of propellant. The resultant correlation for heptane can be expressed as:

$$L = 0.196 \frac{\left(\frac{MMD}{150}\right)^{1.46} \left(\frac{V}{30}\right)^{0.86} \left(\frac{\sigma_{LN}}{2.3}\right)^{0.77} \left(\frac{a_c}{3.15}\right)^{0.48}}{\left(\frac{P_C}{20}\right)^{0.54} \left(\frac{T_I}{280}\right)^{0.51}} \tag{8}$$

The above correlation has an error band of ±10% in comparison with the data from the solution of differential equations as in [2]. This correlation may be compared with the correlation given in [14] that expressed the empirical equation for the geometric standard deviation of 2.3 only. From the above equations, it may be evident that the data could be expressed by an "*effective length*" rather than the actual length. This length can be expressed by:

$$L_{eff} = L\frac{P_C^{\ 0.54}\ T_l^{\ 0.51}}{MMD^{1.46}V^{0.86}\sigma_{LN}^{\ 0.77}a_c^{\ 0.48}(1.1861\times 10^{8})} \tag{9}$$

The advantage of the effective length over actual length is that it combines the actual length, design and operating parameters. Effective length can be considered as a generalized or universal length. Effective length can be expressed as percentage propellant vaporized as in the following section.

2.8. The relation between mass vaporized and effective length

Variation percentage mass vaporized vs. effective length for different parameters is shown in figure 10. From that figure, it is clear that there is a stacking of data in a central stream except for a very la1rge change in standard deviation (the outlier dashed line in the figure). Nevertheless this outlier arises only in the very low-efficiency motors, those with low percentage mass vaporized. Taking advantage of such a stack to fit a generalized equation for variation of effective length with percentage mass vaporized, the following relation can be written.

$$L_{eff} = 5.7x\,10^{-8}\,M\,\%_{vap}^{\ 4} - 8x\,10^{-6}M\,\%_{vap}^{\ 3} + 3.17x\,10^{-4}\,M\,\%_{vap}^{\ 2} + 0.0017M\,\%_{vap} + 0.0347 \tag{10}$$

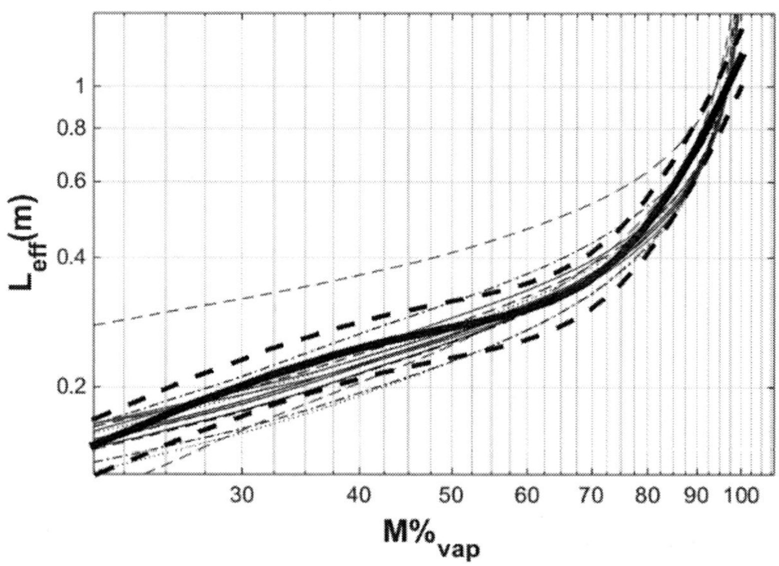

Figure 10. The spread of data and fitted mass % -effective length relation-dashed line indicates ±15 % variation from the fitted equation (10).

3. The Methodology

The methodology is described in the flow chart shown in figure 11, and can be summarized in the following steps:

- Start with targeting characteristic velocity efficiency (defined in equation 2).

IOP Conf. Series: Materials Science and Engineering **973** (2020) 012003 doi:10.1088/1757-899X/973/1/012003

- Using thermochemical calculations to find the theoretical characteristic velocity at both the theoretical mixture ratio and at vaporized mixture ratio.
- Find the required percentage mass vaporized (consider the propellants are one gas or volatile and the other has low volatility) using equation 11, which relates characteristic velocity efficiency to the percentage of vaporized propellant through two terms $\left(C^*_{th}\right)_{O_{vap}/F_{vap}}$ and $\%M_{vap}$. That dictates using look-up table technique to find $\%M_{vap}$ rather than direct substitution in figure 12 [2]

$$\eta_{C^*} = \frac{\left(C^*_{th}\right)_{O_{vap}/F_{vap}}}{\left(C^*_{th}\right)_{O/F}} \frac{\dot{m}_{vaporized}}{\dot{m}_{injected}} = \frac{\left(C^*_{th}\right)_{O_{vap}/F_{vap}}}{\left(C^*_{th}\right)_{O/F}} \%M_{vap} \qquad (11)$$

- Find the effective length using equation 10.
- The effective length equation represents a relation between chamber length, operating parameters, and injection system design
- From injection system cold flow test or from empirical correlations found in literature, find the MMD, injection velocity for the injector elements
- From the regenerative cooling or other cooling system constraints, find the injection temperature
- Choose the operating pressure according to structural constraints or feeding cycle constraints
- Use equation 9 to find the desired cylindrical length; this length satisfies the target characteristic velocity efficiency and takes into account the operating and design constraints

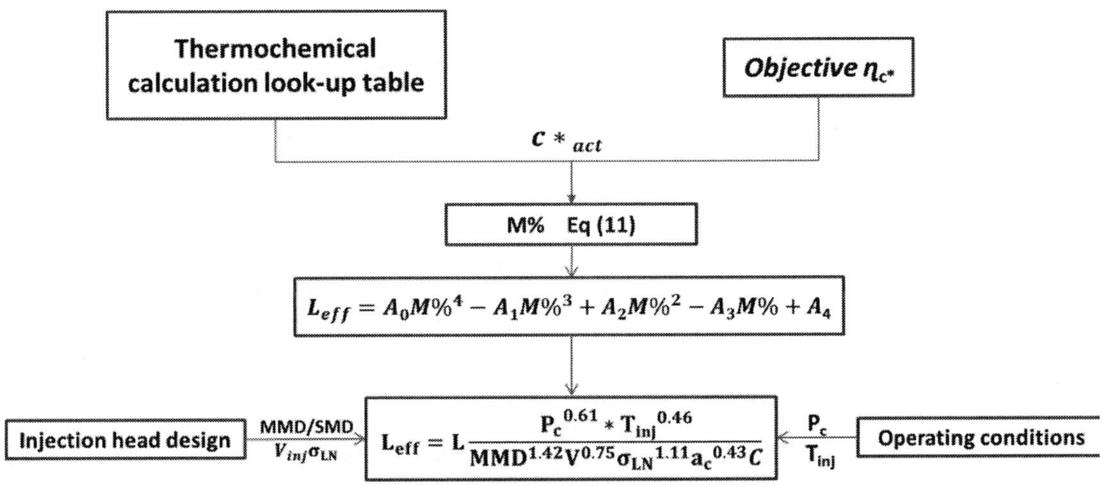

Figure 11. Flow chart for the proposed methodology.

4. Application

The effective length methodology is applied for the same test case mentioned in table 1. The data needed to find MMD for the injection head is shown in table 11. The temperature of injection liquid depends on the layout of cooling achieved by fuel or oxidizer. In some cases, temperature is known or otherwise, it will be assumed to be around 350 - 400 K. MMD is calculated for swirl injector [2,15] and for impingement injector [2].

IOP Conf. Series: Materials Science and Engineering **973** (2020) 012003 doi:10.1088/1757-899X/973/1/012003

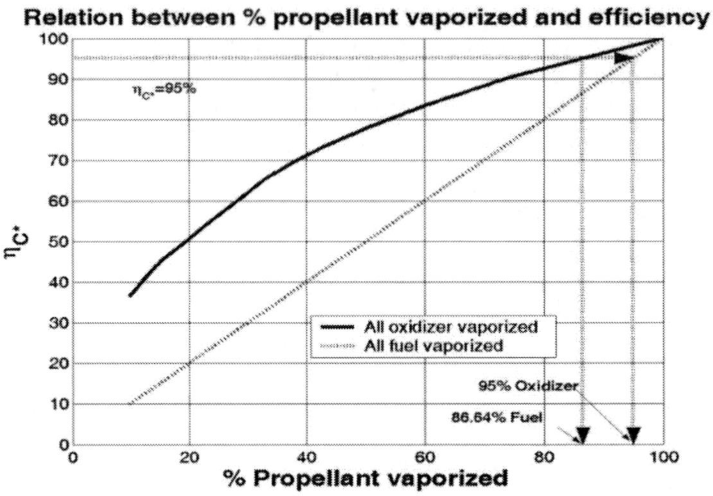

Figure. 12 Relation between C* efficiency and % propellant vaporized [2].

$$SMD = 2.25\sigma^{0.25}\mu_L^{0.25}\Delta P_L^{-0.5}\rho_g^{-0.25} \quad (12)$$

$$MMD = 1.2\ SMD \quad (13)$$

$$MMD = 8.41\times10^5 \frac{D_j^{0.57}}{V^{0.85}} \quad (14)$$

The details of the injector and injection head for the cases studied are shown in figures 13-18, the calculations of the needed data are summarized in table 12.

Table 11. Data for injection parameters for case studies.

Engine	Element type		Injector diameter *mm*		Number primary element		ΔP_{inj} bar		$\dot{m}$ *kg/s*	
	F	*O*	*F*	*O*	*F*	*O*	*F*	*O*	*F*	*O*
RD-0110[4]	bi-swirl injector	bi-swirl injector	13	5.41	-	-	6.96	4.27	6.6	15.7
F1[7]	Like doublet	Like doublet	6.15	7.14	702	714	58	40	1715	4070
LR79-NA11[7]	Like doublet &triplet	Like doublet &triplet	2.87	1.6	582	335	22.4	19.2	202	456
LR89-NA-7[7]	Like doublet &triplet	Like doublet &triplet	2.87	1.6	582	335	18	13.8	211	458
H1[7]	Like doublet &triplet	Like doublet &triplet	3.05	2.08	612	365	11.5	8.5	240	537
LR87-AJ-3[7]	Like doublet	Like doublet	3	2.08	610	560	27	11	183.3	412.7

IOP Conf. Series: Materials Science and Engineering **973** (2020) 012003 doi:10.1088/1757-899X/973/1/012003

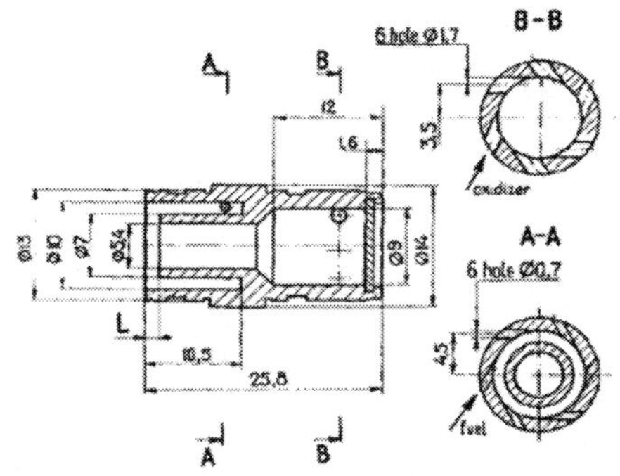

Figure 13. RD-0110 main injector [7].

Figure 14. Engine RD-0110 injection head [7].

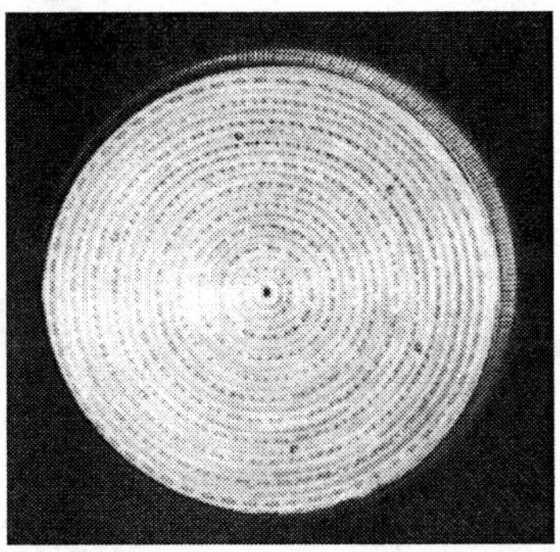

Figure 15. Engine H-1injection head [5].

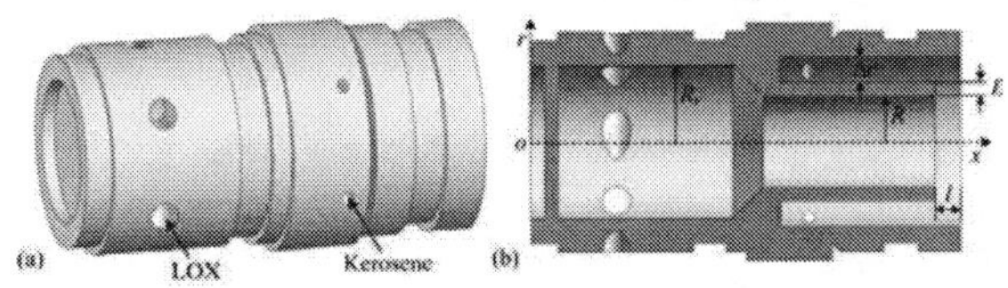

Figure 16. RD-0110 main injector [7].

Figure 17. F-1 injection head [6].

Figure 18. *LR87-AJ-3* injection head [5].

Table 12. Data needed for the L_{eff} methodology.

		Eq (11)	Eq. (10)	Operating Condition		Injection Parameters		Eq. (9)
Engine	η_{c^*}	M%	L_{eff} mm	P_c bar	T_{inj} k	MMD μm	V_{inj} m/s	L_{cal} mm
RD0110	95	95	1018.9	68.16	350	298	11.38	164
F1	95	95	1018.9	77.6	350	988	30.35	1011
LR79-NA11	95	95	1018.9	40.54	350	177.1	95.6	316
LR89-NA-7	94.7	95	1018.9	38.78	350	170.7	99.84	313
H1	94.8	95	1018.9	48.61	350	295.9	64.59	410
LR87-AJ-3	94.7	94	964	43.92	350	359.4	49.35	5۲9

The classical L* methodology is applied; in addition to the effective length methodology applied with different target characteristic velocity efficiency. Table 13 and figures 19-20 show the results. The comparison between the classical L* approach vs. the real data is shown in figure 19. As mentioned before, all the engines do not comply with that recommended range of L*. However, applying the effective length approach with different target characteristic velocity efficiencies (92.5%, 95% and 97.5 %) show a far better agreement with that of real values, figure 20.

Table 13. Comparison between the L* approach and L_{eff} approach.

Engine	L_{act} *mm*	$L_{(L^*=1.5)}$ *Mm*	$L_{(L^*=2.5)}$ *mm*	$L_{cal\ (92.5\%)}$ *mm*	$L_{cal\ (95\%)}$ *mm*	$L_{cal\ (97.5\%)}$ *mm*
RD0110	204	284	504	143	164	188
F1	944	905	1670	881	1011	1162
LR79-NA11	280	628	1227	275	316	363
LR89-NA-7	335	628	1227	273	313	360
H1	270	759	1376	357	410	471
LR87-AJ-3	412	605	1096	487	529	642

It is clear from the previous figure that the effective methodology shows better agreement with the real data, with the powerful advantage that it can propose insight on how to change design parameters to fit the design in a certain geometrical envelope, which is an inherent shortcoming in the classical L* methodology. However, the proposed methodology has an inherent assumption of vaporization limited engines which is an accurate assumption in a well-mixed propellant with a large difference in volatility. However, these assumptions are not valid for engines with mixing limited situations [16] such as the case in the gas generator, or low-cost engine with high flow rate injectors [17] or in cases where the propellants have similar volatility such combination MMH/NTO, but this is left for future work.

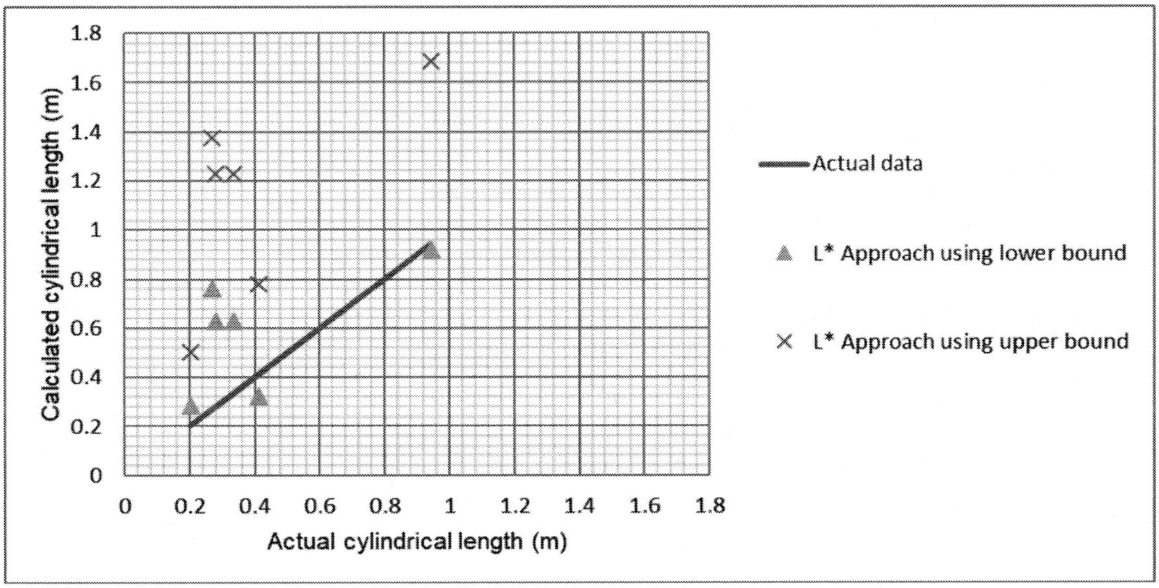

Figure 19. Comparison between L* procedure and the actual data.

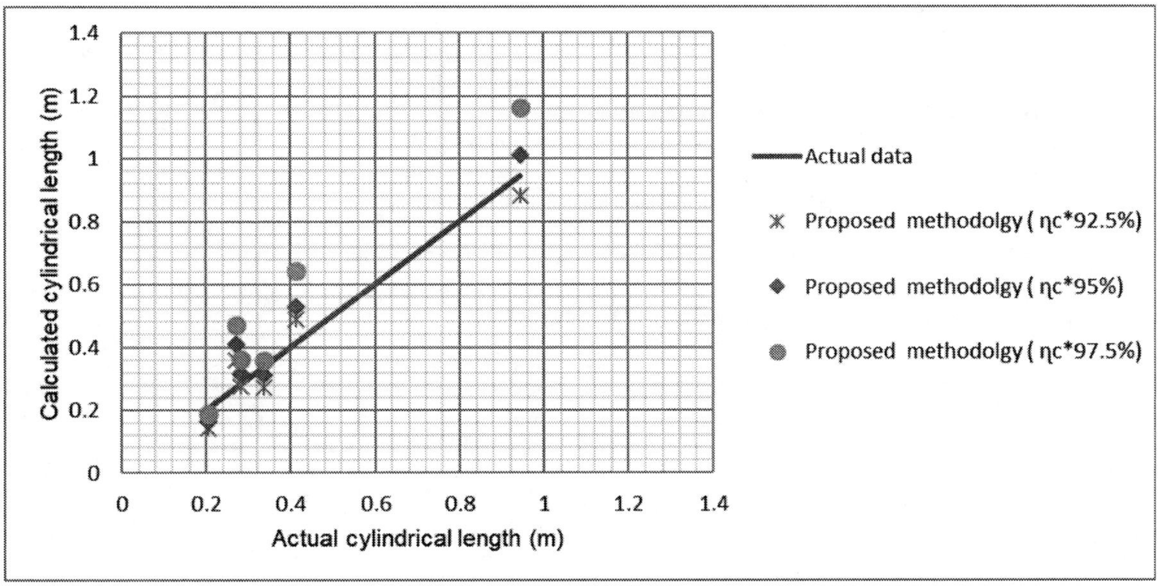

Figure 20. Comparison between L_{eff} procedure and the actual data.

5. Conclusion

The complete modelling of processes inside the liquid propellant thrust chamber is a formidable task. However, the insightful engineer can rely on the concept of the "rate-limiting process" in order to find

IOP Conf. Series: Materials Science and Engineering **973** (2020) 012003 doi:10.1088/1757-899X/973/1/012003

an engineering equation to design a motor. In this paper, the classical procedure of L* was found to over-estimate the required length. In addition, it does not give insight to where to move to enhance the design. An "effective length" procedure was proposed to find the cylindrical length of the thrust chamber. Validation of this methodology was performed on actual liquid motors, and the results showed a better agreement than the classical L* method.

References

[1] Harrje D and Reardon F 1972 Liquid Propellant Rocket Combustion Instability *NASA Special Publication-194*
[2] Belal H et al. 2019 *IOP Conf. Ser.: Master. Sci. Eng. 610 012088*
[3] Barrere M et al. 1960 *Rocket Propulsion* Elsevier Publishing Company, New York
[4] Joseph Himpan, 1950 The Calculation of the Volume of Rocket Combustion Chambers *Aircraft Engineering and Aerospace Technology, Vol. 22 Iss 7* pp 191–193
[5] http://www.astronautix.com/
[6] Anthony Young 2008 The Saturn V F-1 Engine Powering Apollo into History *Springer-Praxis Books In Space Exploration*
[7] Vigor Yang and William E Anderson 1995 Liquid Rocket Engine Combustion Instability Volume 169 *Progress In Astronautics And Aeronautics AIAA*
[8] Liquid rocket engine injectors 1976 I Liquid rocket engine nozzles *NASA Space Vehicle Design Criteria (Chemical Propulsion) SP-8120*
[9] Turbopump Systems For Liquid Rocket Engine 1974 *NASA Space Vehicle Design Criteria (Chemical Propulsion) SP-8107*
[10] Turns, S 2012 *Introduction to Combustion*, 3rd edition.
[11] Williams A 1973 Combustion of droplets of liquid fuels: A review *Combustion and flame* Vol. 21 pp 1-31
[12] Heidmann M 1959 Propellant vaporization as a criterion for rocket engine design; Experimental effect of chamber diameter on liquid oxygen-heptane performance *NASA TN D-65*
[13] Priem R 1958 Propellant vaporization as a criterion for rocket engine design; Calculation of chamber length to vaporize various propellants *NASA TN-3883*
[14] Simmons H 1977 The Correlation of Drop-Size Distributions in Fuel Nozzle Sprays-Part I *Journal of Engineering for Power*
[15] Yang V 2004 Liquid Rocket Thrust Chambers Aspects of Modelling, Analysis, and Design Chapter 20, *AIAA Progress in Astronautics and Aeronautics,* Vol **200**
[16] Berque J, Sion M, and Thomas JL 1999 Tricoaxial injector technology development *Progress In Astronautics And Aeronautics AIAA* 1999-2492

IOP Conf. Series: Materials Science and Engineering **973** (2020) 012004 doi:10.1088/1757-899X/973/1/012004

Validation of relations for the determination of pyrotechnic igniter mass

A Salama[1], H Belal[2] and H M Abdalla[3]

1 M.Sc. Student, Rocket Department, Military Technical College, Egypt.
[2] Assistant Professor, Rocket Department, Military Technical College, Egypt
[3] Associate Professor, Rocket Department, Military Technical College, Egypt.
hbelal@mtc.edu.eg

Abstract: Propellant ignition is an essential process in solid rocket motor operation as it is responsible of starting the combustion and to reliably self-sustain burning of the solid propellant. Despite the importance of igniter design and ignition process, the relations dealing with igniter mass determination are not well studied and not validated against actual motors with a systematic variation of parameters. The main objective of the current research is to validate the available equations in open literature for determination of the mass of igniter pyrotechnic charge. Data from seven real solid propellant rocket motors were used to validate the relations. These motors have different sizes, grain geometries and pyrotechnic compositions. The results show that the equation of Brayan–Lawerene has the smallest root mean square error (RMSE), and so, it is considered more convenient.

1. Introduction

Propellant ignition is an essential process for starting a solid propellant rocket motor. It can be simply defined as a fast and high heat release accompanied by large gas amount per unit mass of the igniter charge to raise the temperature of the propellant surface above the self-ignition temperature and partially raise the pressure in the motor free volume above the low deflagration limit for the propellant [1]. The ignition process is a complicated process, figure 1, which consists of different paths to transfer energy between different phases. The main problem is the failure to start the rocket motor or to reach the self sustaining state due to insufficient heat energy produced by the igniter. Many accidents have been encountered, caused by ignition failures as in the 3rd stage of Ariane-2 at 1986 [2] and in the 2nd stage Zefiro 23 motor of the Vega rocket at 53 km altitude.

There are two main types of igniters: pyrogen and pyrotechnics igniters [3]. Pyrogen igniter, figure 2, can be considered simply as a small rocket motor used to ignite a larger motor. It is not designed to generate thrust and its major objective is to produce sufficient energy output to ignite a large scale rocket motor. The igniter has to be kept as small and light as possible, and could have one or more nozzles; in most cases its nozzle is subsonic. The pelleted pyrotechnic [5] igniter, figure 3, uses a solid energetic material to ignite the solid rocket motor. Pyrotechnics give a large amount of energy at a very small time. With a proper design, no shockwave will be produced which results in avoiding pressure spikes during the ignition that may affect the propellant mechanical properties [6]. Different pyrotechnic compositions are used for the igniter, and their relevant properties are listed in table 1.

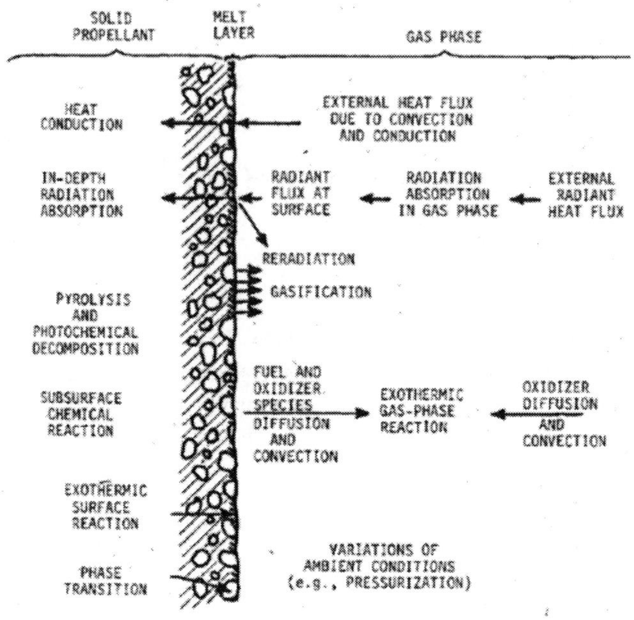

Figure 1. Physical processes involved in solid propellant ignition [4].

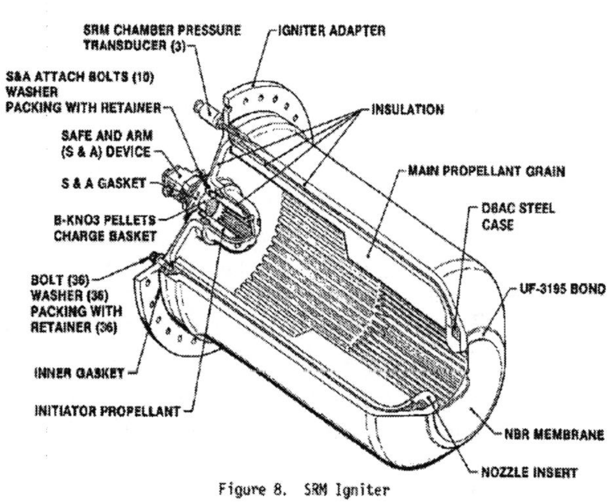

Figure 2. Pyrogen igniter [3].

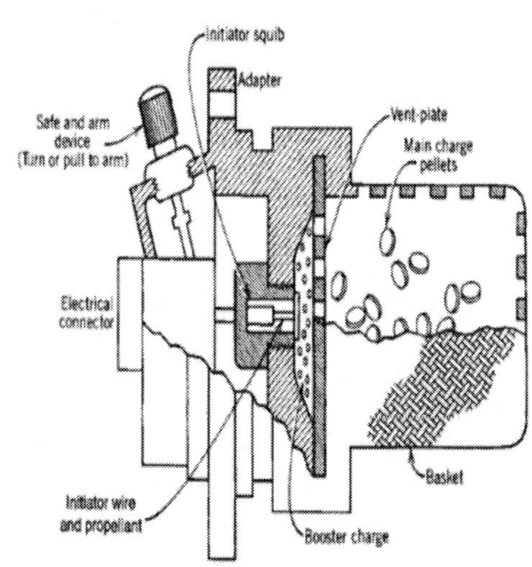

Figure 3. Pyrotechnic igniter [3].

IOP Conf. Series: Materials Science and Engineering **973** (2020) 012004 doi:10.1088/1757-899X/973/1/012004

Table 1. Main Pyrotechnic Composition Characteristics **[7].**

Composition	Heat of Combustion (cal/gram)	Burning rate at 1 atm (mm/s)	Characteristics
Black powder: 75% KNO_3, 15% charcoal, 10% sulfur	755	28.95	-Granular or consolidated configurations, figure 4. -Ease to ignite but produce low energy. -Quench at high altitudes, -Used in: C-5 57mm, SNEP 68mm, FROG-7 rockets
BKN 80% KNO_3, 19% B, 1% wax	1550	9.9	-Pellet form -Ease to ignite at low pressure and high altitudes. -High gas content -Used in Minuteman missile igniter.
MTV: Magnesium, Teflon, Viton (MTV)	2200	1.778	-Produce high energy. -Low hygroscopicity. -Low temperature and pressure dependence -Hard to ignite -Used in Pershing missile igniter, figure 5.

Figure 4. Black powder [8].

Figure 5. Magnesium/Teflon/Viton Pellet [9].

Five possibilities for the ignition pressure-time curve are shown in figure 6. The curves A and E represent a misfire for two contradictory reasons. The curve A shows a high energy release in a very short time. The misfire curve E is attributed to the heat being released through a very long ignition delay.

The factors affecting this ignition delay time are numerous. For example: imperfection in squib current, igniter charge temperature, pyrotechnic characteristics, charge packing or moisture may affect ignition time delay [8]. The curve B shows the ignition spike, giving an indication of over-weighted pyrotechnic charge. The curve (c) represents the proper pressure time ignition curve, while the curve D indicates a hang-fire as it failed to reach a self-sustained state. Figure 7 shows "***the ignition corridor***" [9] state in a logarithmic scale between ignition time and ignition energy represented in radiant flux. It is evident from the figure that start of combustion of solid propellant does not mean reaching to the sustained state. This ignition corridor defines the limits of energy imparted to propellant surface in order to get propellant ignited. As indicated in figure 4 lower values lead to case E while higher values lead to case A or "***over-driven combustion***".

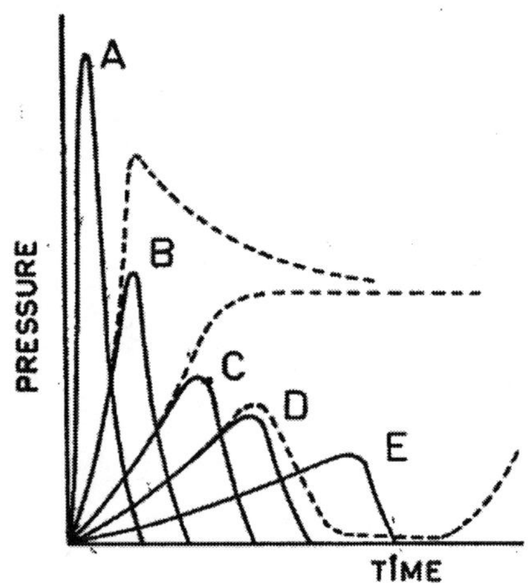

Figure 6. Possible pressure-time curves during ignition [6].

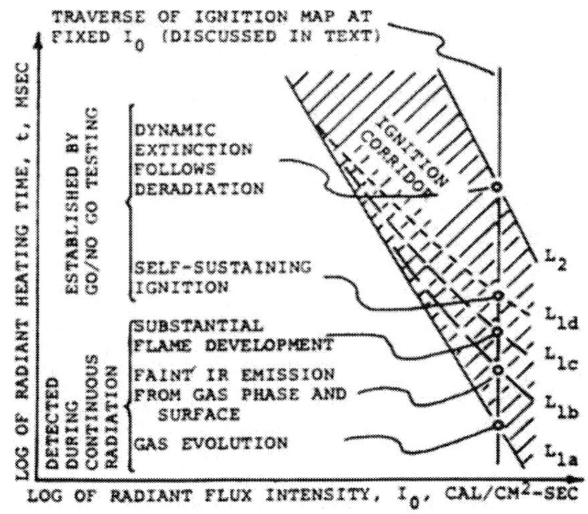

Figure 7. Ignition corridor [9].

2. Ignition charge mass equations

As it is discussed previously, ignition process is a complicated process, it is too hard to get analytical solution for the pyrotechnic mass required and hence all the equations available are empirical. The main contribution in this research is to validate these equations against real igniters in order to grip the main factors affecting the ignition process which may help acquire a better igniter design. Three empirical equations will be handled in the following sections.

2.1. Free Volume Equation [5]

In this relation the required mass depends only on free volume of the combustion chamber including the convergent section. This equation was developed based on more than 15 different rocket motors with varying propellant formulation, free volume, igniter composition and grain configuration, figure 8.

$$m_{ig} = 0.016944\,(V_c)^{0.7} \tag{1}$$

m_{ig}………pyrotechnic charge mass, gram.
V_C…………combustion chamber free volume including the convergent section, cm^3, figure 9.

2.2. Brayan–Lawerene equation [10]

This relation depends on calculating the total energy required to ignite propellant surface taking in consideration the thermodynamic properties of the propellant, the grain length and the total area exposed to ignition. This equation was developed based on 53 different rocket motors with varying propellant formulation, igniter composition and grain configuration, This equation states that:

$$Q = 38\left[A_s\, q_c \left(\frac{L_g}{A_s}\sqrt{4\pi A_P}\right)^{0.59}\right]^{1.06} \tag{2}$$

where
Q total energy required for ignition, cal.

IOP Conf. Series: Materials Science and Engineering **973** (2020) 012004 doi:10.1088/1757-899X/973/1/012004

A_s total area exposed to igniter products, figure 10, including the burning surface of propellant grain, the convergent area of combustion chamber, inhibited area of grain end and any other solid parts like grids, cm^2.
q_c experimental ignition energy per unit area of propellant, cal/cm^2.
L_g propellant grain length, cm.
A_p cross-sectional port area, cm^2.

The igniter mass can be calculated by defining the pyrotechnic charge composition and determination of its combustion heat using equation 3.

$$Q = m_{ig}\, \Delta H_{comb} \tag{3}$$

where
m_{ig} mass of igniter, gm.
ΔH_{comb} heat of combustion of igniter material, cal/gram. Calculated by NASA (CEA) [13] thermochemical calculations.

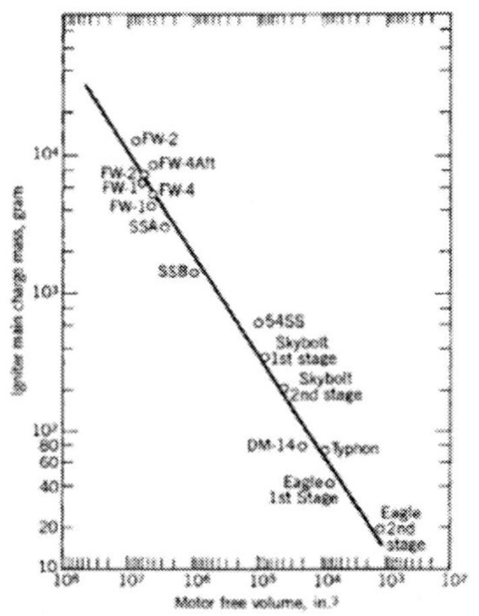

Figure 8. Igniter pyrotechnic charge weight to motor free volume [5].

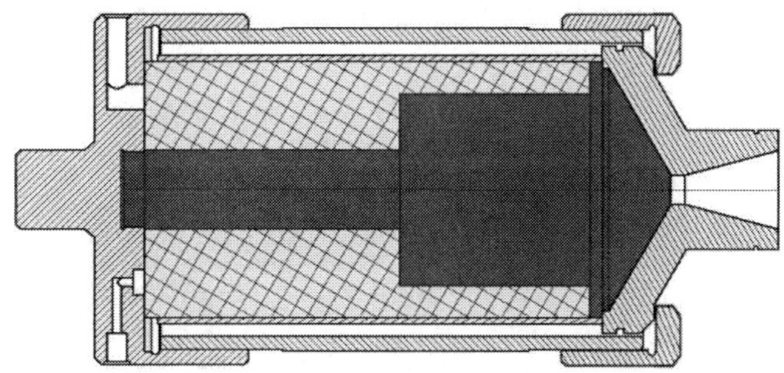

Figure 9. Free volume of combustion chamber (red volume).

IOP Conf. Series: Materials Science and Engineering **973** (2020) 012004 doi:10.1088/1757-899X/973/1/012004

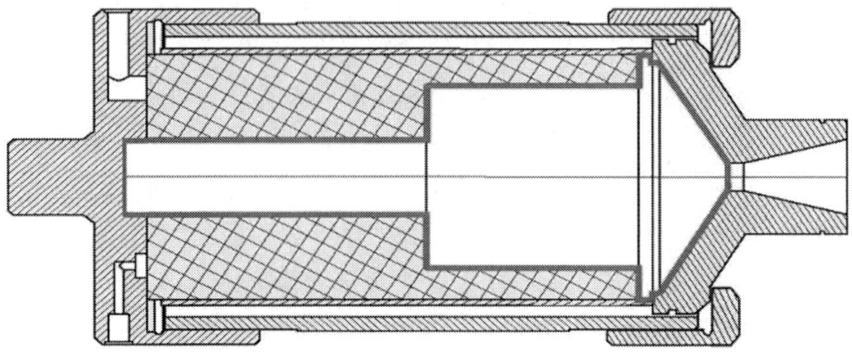

Figure 10. Area exposed including the grain surface inhibited surface and the convergent section.

2.3. Ideal gas equation [6]
This relation is based on the ideal gas law, it depends on different parameters: the initial free volume of the combustion chamber, the ignition pressure, isochoric temperature and the solid fraction in the combustion gases. The propellant and pyrotechnic properties are neglected

$$m_{ig} = \frac{1}{1-\varepsilon} \frac{P_{ig} V_{ig}}{\frac{R_u}{M} T_{ig}} \tag{4}$$

where

m_{ig}	calculated mass of ignition charge, kg.
V_{ig}	free ignition volume, m^3.
ε	fraction of condensed phase in combustion products of ignition charge,
R_u	universal gas constant, J/(mol-K);
m_{ig}	molar mass of combustion gas of ignition charge, kg/mol.
T_{ig}	combustion gas temperature of ignition charge, K.
P_{ig}	ignition pressure or the lower deflagration pressure, N/m^2.

Lower deflagration pressure is the minimum value for combustion chamber pressure to sustain the propellant combustion [11]. For double base propellant, it often has a value above 30 bar and for ammonium perchlorate based composite propellant it has a lower value above 20 bar [12], in spite of the fact that composite propellant is slightly more difficult to ignite than double base propellant [6].

3. Case studies
In order to validate the previous equations, the three equations are used to calculate the igniter mass for different motors. Most of these motors use double base propellant. The motors have a huge difference in size, table 2. Calculated igniter charge masses values are compared with those from real motors. In this study, different types of pyrotechnic charge with different compositions were used:

- Black-powder with composition Sulfur S 10%, Charcoal C 15%, potassium nitrate KNO_3 75%.
- SR 371C with composition Magnesium Mg 42%, potassium nitrate KNO_3,with acaroid resin binder [8]

The thermochemical characteristics for these types were calculated by NASA (CEA) [13] to get the required parameters.

Table 2. Case study motors.

	Free volume (cm^3)	Area exposed A_s (cm^2)	A_S/A_P	Grain length (cm)
Motor 1	207	165.82	322	45
Motor 2	1100000	133012.23	158	474
Motor 3	748	273.53	213	42.6
Motor 4	129089	90980.02	716	115
Motor 5	220683	152956.85	763	173.1
Motor 6	4058	3066.12	2310	89
Motor 7	37165	17806.49	138	287.

4. Results and discussion

The results of applying the three relations on motors in the case studies are shown in table 3. Comparison between the actual and calculated igniter masses using equations 1 through 3 are shown in figures 13-15 respectively. The following analysis can be drawn:

Table 3. Igniter pyrotechnic charge mass.

	Actual mass (gram)	Equation 1 (gram)	Equation 2 (gram)	Equation 3 (gram)
Motor 1	5	0.71	3.30	3.25
Motor 2	5066.4	287.08	2661.83	6476.84
Motor 3	5.9	1.74	5.28	1.17
Motor 4	3024	64.07	511.57	2027.70
Motor 5	5040	93.25	949.42	3466.43
Motor 6	63	5.69	23.61	63.74
Motor 7	100	26.80	77.61	136.75

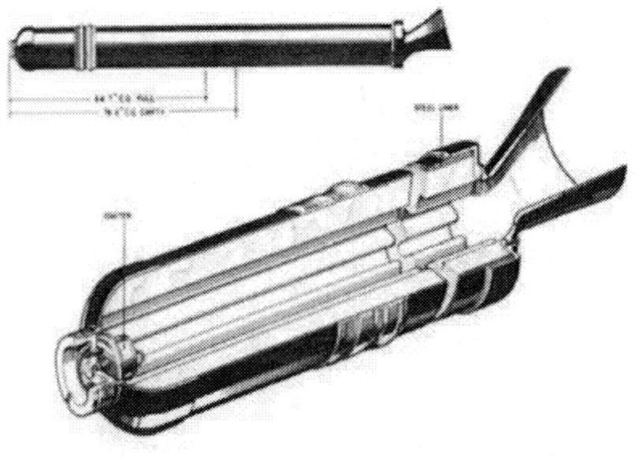

Figure 11. Motor No.7 [14].

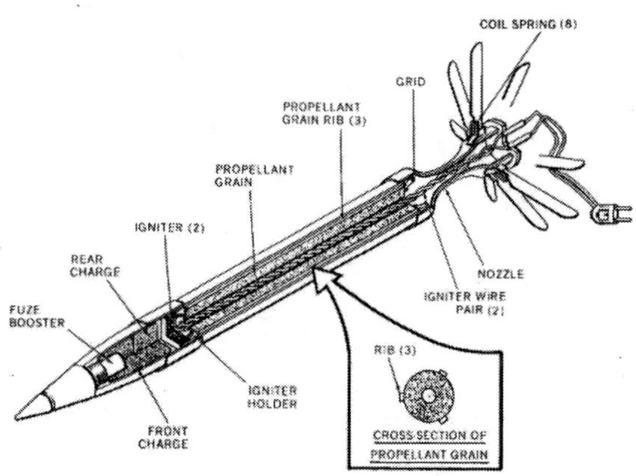

Figure 12. Motor No. 1 [15].

- Results from the first equation underestimated the igniter mass charge in all the cases. The best match was observed for the composite motor based on AP. This might be explained by the fact that this empirical equation was originally proposed for composite propellant [5].

- The second equation gives a slightly better result, especially with small scale motors. A larger scale motor has a large burning area, which justifies the notable discrepancy between actual and calculated pyrotechnic masses.
- For the third equation; a large deviation is observed for small motors.
- To have a quantitative judgement on the prediction of each equation, a root mean square error (RMSE) is calculated according to the following equation

$$RMSE = \sqrt{\frac{\Sigma\left(1-\frac{m_{calculated}}{m_{real}}\right)^2}{n}} \quad (5)$$

where

m_{cal}	the calculated igniter charge mass
m_{real}	the real igniter mass value
n	the total number of studied motors

On reviewing the pressure time curve for each motor, a presence of an ignition spike was observed, which gives a direct indication for an over-weighed igniter charge. This phenomenon was highly pronounced in motors 4, 5 and 6 as shown in figures 16-18. In motor #4 a pressure spike of 140 bar is noticed while the average pressure was targeted to be 70 bar; this implies that the igniter charge mass was nearly twice the proper value. Similarly, the figures show for the motor 5 an ignition spike of 110 bar and a steady state pressure of 60 bar, while for the motor 6, an ignition spike of 130 bar and steady state pressure 85 bar.

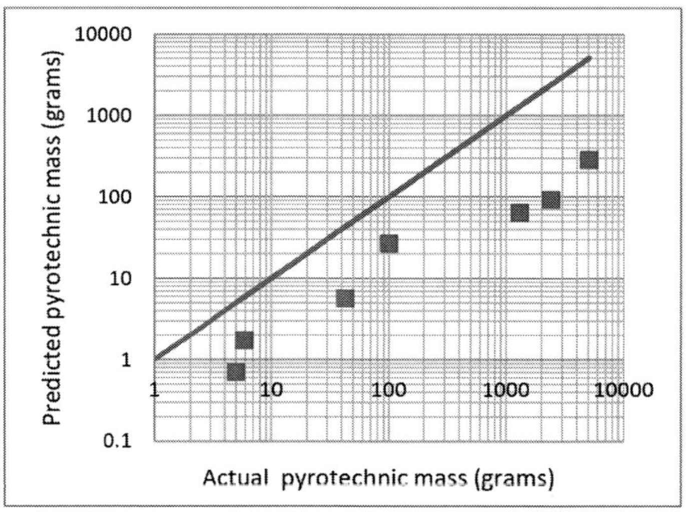

Figure 13. Comparison with results from equation 1.

IOP Conf. Series: Materials Science and Engineering **973** (2020) 012004 doi:10.1088/1757-899X/973/1/012004

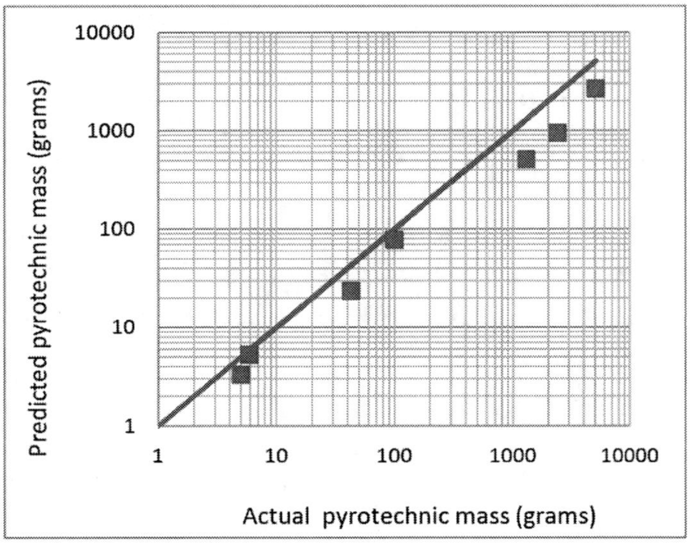

Figure 14. Comparison with results from equation 2.

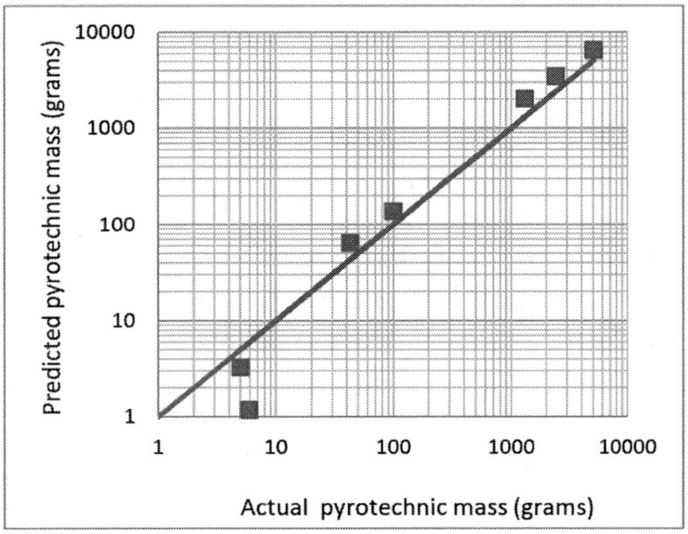

Figure 15. Comparison with results from equation 3.

Results are shown in table 4.

Table 4. Root mean square error.

	Root mean square error (%)
Equation 1	88
Equation 2	55
Equation 3	41

To remove spikes in the pressure-time curves it was necessary to modify the ignition masses for the motors 4, 5 and 6. The motors were fired with "*modified real values*" and no such spikes, were recorded. A comparison between the original values and "modified real values". versus re-calculated RMSE values are shown in table 5.

Table 5. Corrected igniter pyrotechnic masses.

	Original mass (gram)	Corrected mass (gram)
Motor 4	3024	1300
Motor 5	5040	2400
Motor 6	63	43

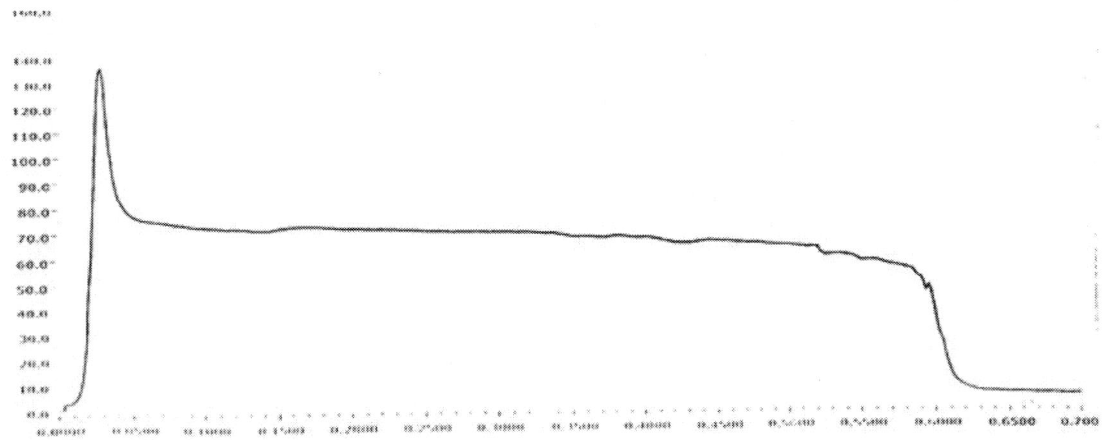

Figure 16. Pressure time curve for motor no.4

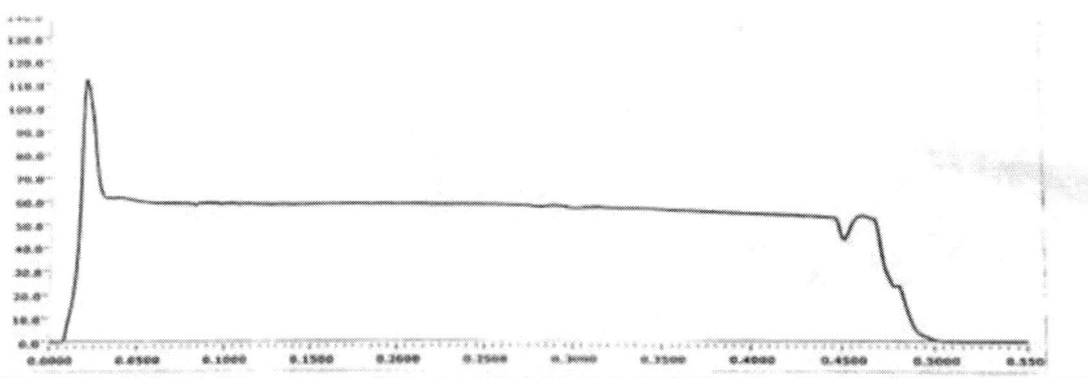

Figure 17. Pressure time curve for motor no.5

IOP Conf. Series: Materials Science and Engineering **973** (2020) 012004 doi:10.1088/1757-899X/973/1/012004

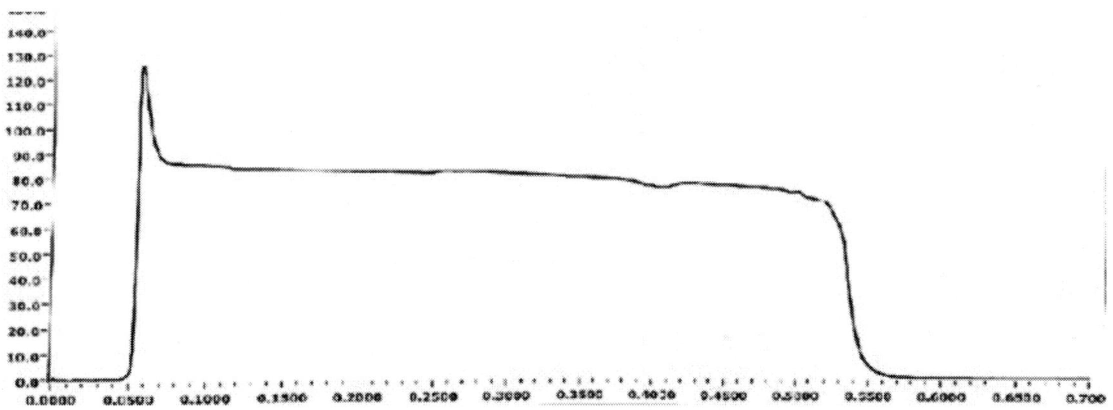

Figure 18. Pressure time curve for motor no.6.

After correcting the igniter pyrotechnic mass according to the pressure time curve of each motor the igniter pyrotechnic mass were changed and the results are shown in Figs 19, 20, 21.

- The first equation still has a large deviation between the calculated result and the real pyrotechnic mass,
- For the second equation the result is under estimated after the correction but the results make a better sense with the correcting masses.
- the third equation also give a better fitting with the correcting masses except motor number 1

Now the second equation give the minimum root square and the best fitting data and the third equation go a little higher by about 10%.

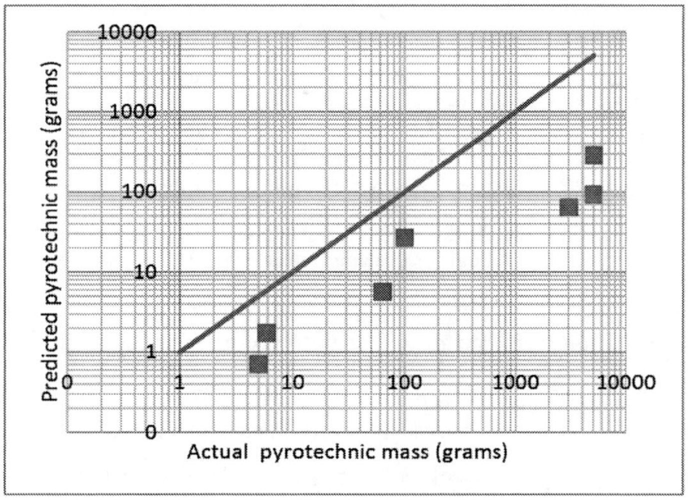

Figure 19. Equation 1 result vs. the corrected pyrotechnic igniter mass.

IOP Conf. Series: Materials Science and Engineering **973** (2020) 012004 doi:10.1088/1757-899X/973/1/012004

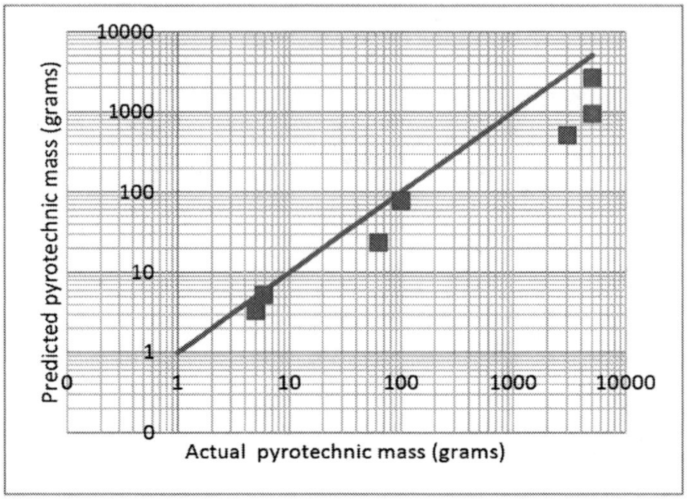

Figure 20. Equation 2 result vs. the corrected pyrotechnic igniter mass.

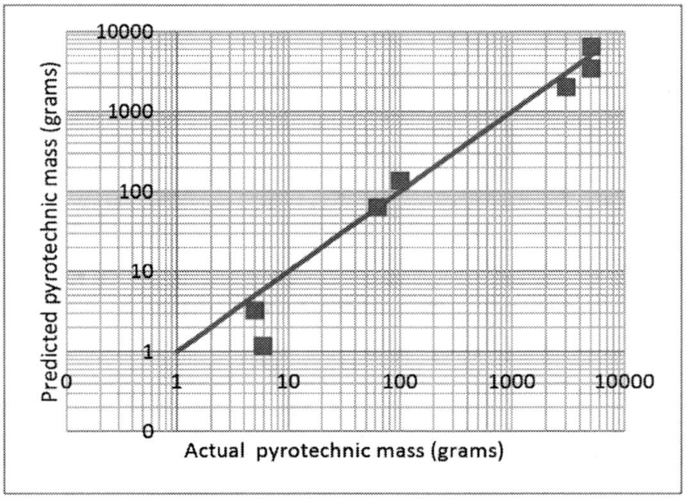

Figure 21. Equation 3 result vs. the corrected pyrotechnic igniter mass.

Table 6. Corrected- Root mean square error

	Root mean square Error (%)	Corrected Root mean square errors (%)
Equation 1	88	89
Equation 2	55	43
Equation 3	41	49

IOP Conf. Series: Materials Science and Engineering **973** (2020) 012004 doi:10.1088/1757-899X/973/1/012004

5. Conclusion

The equation of Brayan–Lawerene gives less root mean square error than other relations. Therefore, it could be considered more realistic as compared to other equations. The better resulting estimation could be argued to considering wider range of parameters as the exposed area, which is a dominant factor, the energy required for ignition of unit area of propellant, and the produced combustion energy from the igniter pyrotechnic. However, the effects of solid particles within the igniter gas products, the temperature and pressure evolving from these products have been neglected. The free volume equation is too simple as it considers only the effect of combustion chamber free volume, while neglecting the effects of composition and thermodynamic properties of propellant or even the pyrotechnics. The ideal gas equation takes into consideration the igniter composition and the free volume but it neglects composition and properties of the propellant except the deflagration pressure limit. However, all these equations did not consider neither the ignition moment nor the proper ignition delay that may affect the calculated ignition mass. As a future work, it is planned to finely tune Brayan–Lawerene equation through considering the effects of pyrotechnic charge particle size, the presence of solid phase in products, in addition to the relation between estimated mass and motor ignition delay.

References

[1] Von Elbe G 1966 Solid propellant ignition and response of combustion to pressure transients *2nd Propulsion Joint Specialist Conference* p 668
[2] Chang, I-Shih, Susumu Toda, and Seishiro Kibe 2000 European space launch failures *36th aiaa/asme/sae/asee joint propulsion conference and exhibit*
[3] Keller jr, r B 1971 *Solid rocket motor igniters NASA SP-8051*
[4] Kulkarni, Anil k Mridul Kumar, and Kenneth K Kuo 1982 Review of solid-propellant ignition Studies pp 243-244
[5] Sutton George P; Biblarz O 2016 *Rocket Propulsion Elements* John Wiley & Sons
[6] Barrere, Marcel, et al. 1960 *Rocket propulsion.* LTAS
[7] Robertson W 1972 Igniter material considerations and applications *8th Joint Propulsion Specialist Conference*
[8] Wimpress R 1950 *Internal Ballistics of Solid-fuels Rockets.* New York: mcgraw-Hill
[9] Deluca l, Caveny L, Ohlemiller T and Summerfield M 1976 Radiative ignition of double-base propellants. I-Some formulation effects *AIAA Journal 14(7)* pp 940-946
[10] Bryan, George J, and Evan K 1959 A method for predicting ignition energy requirements of Practical propellant systems. part 3 rockets *Naval ordnance lab white oak md*
[11] Zanotti C; Giuliani P 1994 Pressure deflagration limit of solid rocket propellants: experimental results *Combustion and flame 98 1-2* pp35-45.
[12] Atwood A et al. 2009 Radiant ignition studies of ammonium perchlorate based propellants *Progress in Propulsion Physics 1* pp121-140.
[13] Mcbride B, and Gordon S 1996 Computer program for calculation of complex chemical equilibrium compositions and applications ii *User's manual and program description, NASA reference pub. 1311* 1996.
[14] White C 1960 The development of the gosling ii solid propellant rocket motor *Rocket propulsion establishment wescott (united kingdom)*
[15] https://en.m.wikipedia.org/wiki/S-5_rocket#/media/File%3AS-5_57mm_rocket_cross-section. JPG retrieved 1st Dec 2019

IOP Conf. Series: Materials Science and Engineering **973** (2020) 012005 doi:10.1088/1757-899X/973/1/012005

Simplified algorithm for determining injection duration to achieve fast response in SI engines

Omar Shawky
DLECS - Cairo university
engomar582@gmail.com

Abstract. For modern SI engines, the injection duration and its timing are one of the most important parameters which influence the mixture homogeneity, emissions, combustion efficiency, and engine response. In the present work, a simplified port fuel injection system with its driver algorithms was designed to control the injection duration and its timing. The Calibrated Dose method is used in this system instead of the stream mixing method. Therefore, the injected fuel mass is accurately computed according to the engine load, engine speed, engine operating condition, and the amount of air drawn into the engine. The system consists of a conventional port fuel injection system equipped with a micro-controller, a Throttle Position Sensor (TPS),a couple of Hall Effect Sensors (HES), load cell with its driver module, injector controller circuit, and a custom-made inlet port. The micro-controller is used to calculate the injection duration each engine cycle according to engine speed, amount of air drawn per cycle and engine load. The two HES are used to detect the injection timing and measure the engine speed. The TPS is used to monitor the throttle position angle so it can calculate the amount of air drawn into the engine at that time with respect to engine speed. The load cell and its driver module are used to measure the engine torque and consequently the brake power. The injector controller circuit is used to initiate the injection process at the perfect timing for the accurate duration.

1. Introduction

For the sake of improving energy conversion efficiency in the transportation sector, many fuel systems were developed to increase combustion efficiency and engine response [1]. The transportation sector is strongly dependant on fossil fuels especially gasoline and diesel [2]. The passenger transportation sector tends to spark-ignition (SI) engines [3] which runs on gasoline. Therefore, engine developers always target to minimize fuel consumption and lower engine emissions [4]. For the SI engines specifically, one of the most effective parameters to decrease the fuel consumption is the accurate fuel injection duration [5] and its timing [6]. Also, the air to fuel ratio suitable to the engine operating conditions is so important to decrease fuel consumption [7]. SI engines use mainly two fuel injection strategies; the port fuel injection (PFI) strategy, and the gasoline direct injection (GDI) strategy [8-11]. The PFI systems are easier to manufacture and design than the GDI systems [12]. PFI technology is widely spread in the SI engines to reduce fuel consumption and increase the output power [13].

In this system, the engine is equipped with an electronic control unit consisting of a micro-controller, monitoring sensors and actuators [14]. The micro-controller, called as the engine control unit (ECU) [15], continuously detects what the driver is demanding out of the engine [16]. Accordingly, it processes the data received from the engine inputs such as airflow rate, fuel flow rate, throttle position and engine temperature [17]. The micro-controller uses this processed data to trigger the injection process at the accurate timing and for the suitable duration whereas the precise injection timing greatly influences the engine performance [18]. The time available for injection and mixing is very short also the size of droplets affects the combustion efficiency [19], therefore the faster the injection response, the better the combustion efficiency [20] which results in more power output [21]. The engine cold start condition also affects fuel consumption as the fuel atomization is reduced due to the engine's low engine temperature [22].

There were many attempts made to enhance the PFI for the small motorcycles to increase the compatibility, efficiency and decrease fuel consumption [23]. In the present work, a simplified port fuel injection system was designed and coded to achieve a fast-responsive injection process with variable load and speed.

2. Description of the proposed system

The proposed system aims to inject just the suitable amount of fuel at the accurate timing. The suitable fuel mass depends on many factors: air mass drawn, engine speed, required power, and air to fuel ratio. The control methodology is the Calibrated Doze method. This method handles each of these factors carefully to calculate the suitable fuel mass.

The present work was conducted on a carbureted internal combustion engine. The carburetor was replaced by the proposed port injection system. The engine used was a four-stroke SIE. The engine specifications are shown in table 1.

Table 1. Engine Specifications.

Model	GX390
Type	4-stroke, overhead valve, single-cylinder, horizontal shaft
Cooling system	Air cooling
Bore& Stroke	82·64 mm
Rated power	8.2/3600 (KW/rpm)
Maximum torque	27.1/2500 (N.m/rpm)
Displacement	389 cc
Compression ratio	8.0:1
Ignition system	Spark
Oil capacity	1.1L
Lubrication system	Splash
Starting	Kicking starting
Dimensions	405·450·443
Dry weight	31 kg

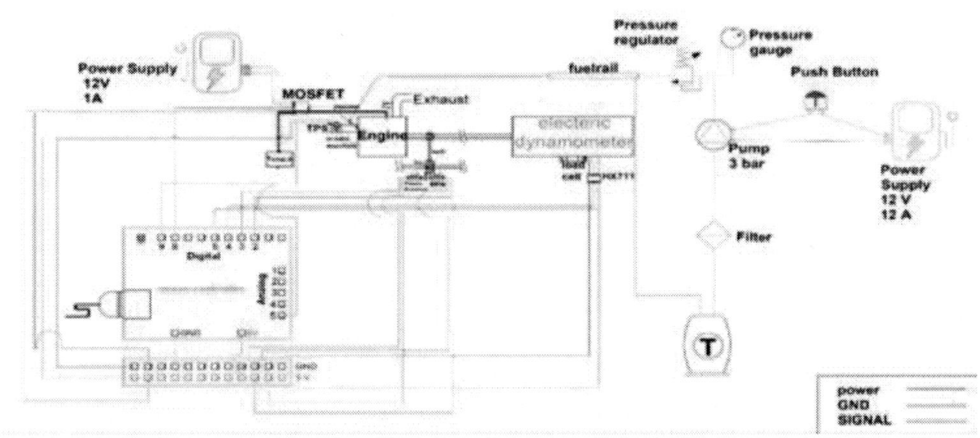

Figure 1. Detailed view for the proposed PFI system.

The proposed injection system consists of a conventional port injection system equipped with simplified control system. The control system is implemented using integrated electrical components for measuring engine inputs and outputs as shown in figure 1.

3. The proposed system methodology

The control system methodology is the Calibrated Doze method. This method consists of three steps to determine the suitable fuel mass. It begins with calculating the air mass, then it determines the air to fuel ratio suitable for engine operating conditions, at the end it determines the adequate fuel mass. The below points show how the method is implemented in the code. Each one of these points is explained in the coming subsections.

- Calculating the amount of air drawn per cycle per cylinder.
- Determining the air to fuel ratio according to the load and engine speed.
- Determining the accurate fuel mass suitable to the drawn air mass and the determined air to fuel ratio and then calculating the accurate injection duration to inject the determined fuel mass.
- Feedback and monitoring the outgoing parameters during engine operation. (mass of air - air to fuel ratio- calculated fuel mass - injection duration - engine speed in rpm - measured engine torque - brake power - engine temperature – engine throttle position).

3.1. Calculating air mass drawn into the engine cylinder

To calculate the amount of air drawn into the engine cylinder, the volumetric efficiency must be determined. There are two main factors greatly affecting the volumetric efficiency: the engine throttle position angle and the engine speed. The engine throttle position is determined using the TPS and the engine speed is computed using a HES. A function between the volumetric efficiency, engine throttle position, and engine speed was formulated by measuring the actual amount of air drawn into the cylinder by using an external anemometer then comparing these results to the output of the following equation:

$$\dot{m}_{calc} = \rho_{air} \cdot V_{st} \cdot \eta_v \tag{1}$$

Where $\dot{m}_{calc}$ the calculated amount of air, ρ_{air} the ambient air density, V_{st} the engine stroke volume and η_v the volumetric efficiency. By this comparison, the volumetric efficiency for all the engine speed range and different throttle position angles was formulated as shown in figure 2.

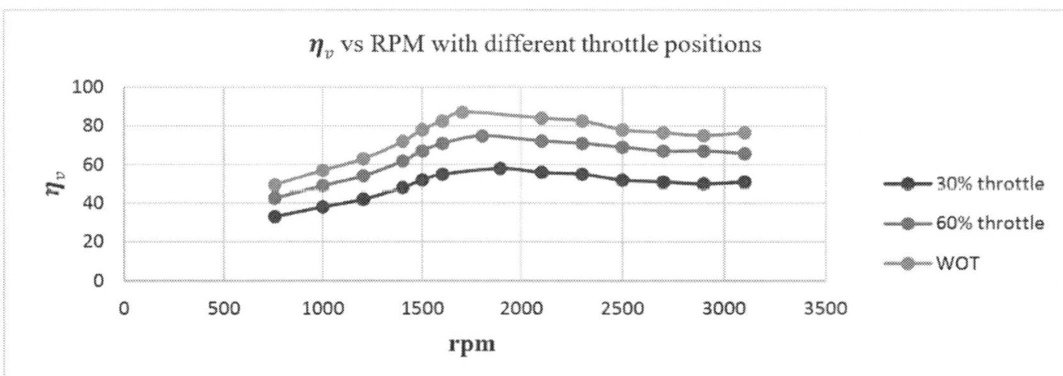

Figure 2. Curve fitting for volumetric efficiency vs RPM at different throttle position angles.

After this experimental calibration, three functions were developed at 30%, 60%, and WOT (Wide Open Throttle) as follow.

- At 30% throttle position angle:

IOP Conf. Series: Materials Science and Engineering **973** (2020) 012005 doi:10.1088/1757-899X/973/1/012005

$$\eta v = (-18 \cdot 10^{-3} \cdot RPM^6) + (-14 \cdot 10^3 \cdot RPM^5) - (\mathrm{RPM}^4 \cdot 10^{-2}) + (-7 \cdot 10^3 \cdot RPM^3) + (3 \cdot 10^{-3} \cdot RPM^2) + (0.1385 \cdot RPM) - 8.5545 \quad (2)$$

- At 60% throttle position angle:

$$\eta v = (-18 \cdot 10^{-3} \cdot RPM^6) + (-14 \cdot 10^3 \cdot RPM^5) - (RPM^4 \cdot 10^{-2}) + (-7 \cdot 10^3 \cdot RPM^3) + (3 \cdot 10^{-3} \cdot RPM^2) + (0.1935 \cdot RPM) - 13.709 \quad (3)$$

- At WOT:

$$\eta v = (-18 \cdot 10^{-3} \cdot RPM^6) + (-14 \cdot 10^3 \cdot RPM^5) - (\mathrm{RPM}^4 \cdot 10^{-2}) + (-7 \cdot 10^3 \cdot RPM^3) + (3 \cdot 10^{-3} \cdot RPM^2) + (0.204 \cdot RPM) - 12.346 \quad (4)$$

In equations (2), (3) and equation (4), there are four constant parameters and the last two parameters are changing with respect to the engine throttle position angle.So, a generic equation was formulated between the volumetric efficiency, engine speed, and engine throttle position angle. The generic function will be like equation (5).

$$\eta v = (-18 \cdot 10^{-3} \cdot RPM^6) + (-14 \cdot 10^3 \cdot RPM^5) - (RPM^4 \cdot 10^{-2}) + (-7 \cdot 10^3 \cdot RPM^3) + (3 \cdot 10^{-3} \cdot RPM^2) + ((\mathrm{A}) \cdot RPM) - (\mathrm{B}) (5)$$

where, A and B are functions of throttle position angle. Equation (6) and (7) were formulated to relate variable A and variable B with the throttle position angle.

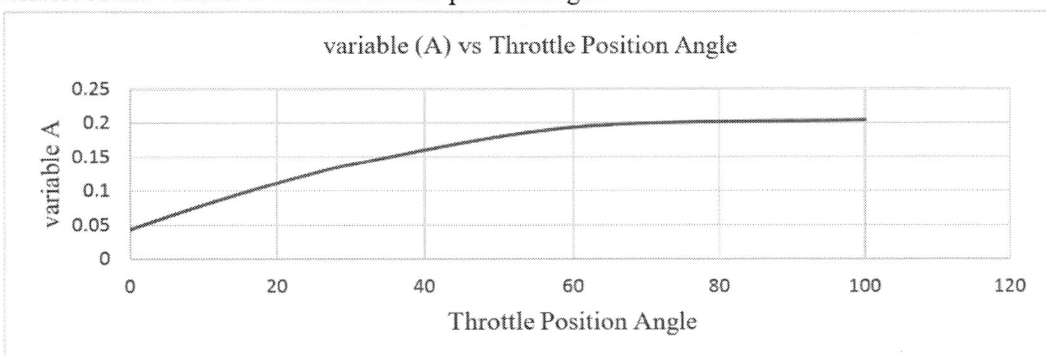

Figure 3. Curve fitting for the relation between variable A and TPS.

After the curve fitting for the experimental data, equation (6) was generated:

$$\mathrm{A} = \left(-0.0029 \cdot (\mathrm{TPS}^2)\right) + (0.4365 \cdot \mathrm{TPS}) - 1.894 \quad \textbf{(6)}$$

IOP Conf. Series: Materials Science and Engineering **973** (2020) 012005 doi:10.1088/1757-899X/973/1/012005

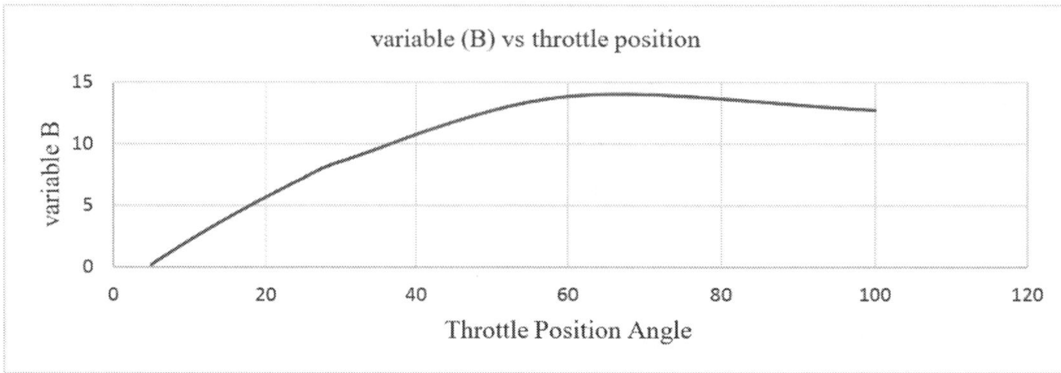

Figure 4.curve fitting for the relation between variable B and TPS.

After the curve fitting for the experimental data, equation (7) was generated:

$$B = \left((-2 \cdot 10^{-5}) \cdot (TPS^2)\right) + (0.0038 \cdot TPS) + (0.0433) \quad \textbf{(7)}$$

Using equation (5), (6), and (7), the air mass drawn is calculated accurately.

3.2. Determination of the air to fuel ratio

In order to relate the mass of air drawn to the fuel mass required, the accurate air to fuel ratio is required. This air to fuel ratio was obtained by measuring the amount of air drawn into the engine by an external anemometer and the amount of fuel consumed at a specific time and a specific throttle position with constant engine speed at 3000 RPM, Then these actual data were used to generate a function between the air to fuel ratio and the throttle position as shown in figure 5.

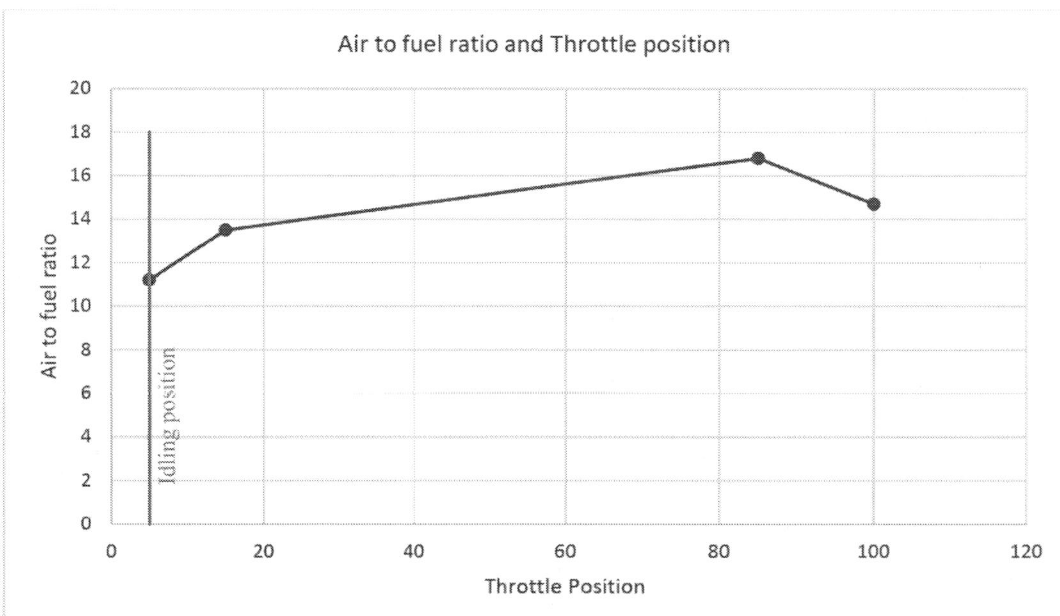

Figure 5. The air to fuel ratio at each throttle position with constant engine speed.

As shown in figure 5 the suitable air to fuel ratio was determined at each engine throttle position, and the ideal air to fuel ratio is 14.7.

3.3. Determination of the accurate fuel mass and injection duration

The accurate fuel mass is calculated using equation (8).

$$\textbf{fuel mass} = \frac{\textbf{calculated air mass}}{\textbf{assigned air to fuel ratio}} \qquad \textbf{(8)}$$

After the fuel mass calculation, the injection duration is to be determined. This is achieved by using a dedicated circuit developed for this process. This circuit is used to control the injection timing and duration as shown in figure ٦.An injector calibration must be done to determine the injection duration suitable for each fuel mass calculated. The used injector has 4 holes each one has a diameter of 0.1mm. As shown in figure 7 and table 2, the function between the injection duration and the fuel mass is nearly linear.

Table 2. Experimental calibration data showing the injection duration for each fuel mass.

fuel mass in grams per injection process (1/2000)	fuel mass in grams for the hole 2000 injection process	time (milliseconds) injection duration
0.002	4	2
0.0035	7	3
0.005	10	4
0.007	14	5
0.0085	17	6
0.0105	21	7
0.0125	25	8
0.014	29	9
0.016	32	10
0.0175	35	11
0.019	38	12

Calibration process:

- A code was developed to inject 2000 times.
- The injection duration was changed from 2 milliseconds to 10 milliseconds each injection cycle with constant injection time off as 10 milliseconds.
- The injected amount of fuel was weighed then divided by 2000 to get the weight of each injection process as shown in table 2.
- As shown in figure 7, formulating a function between the injection duration and the injected fuel mass according to the experimental results.

After this calibration, a function between the injection duration and fuel mass was generated as shown in equation (9).

$$\Delta t_{Inj} = 572.28 \cdot m_{fueL} + 0.9911 \qquad (9)$$

Where,$\Delta t_{Inj}$is the injection duration in milliseconds andm_{fueL}mass of fuel in grams.Equation (9) will determine the suitable injection duration for each fuel mass required.

IOP Conf. Series: Materials Science and Engineering **973** (2020) 012005 doi:10.1088/1757-899X/973/1/012005

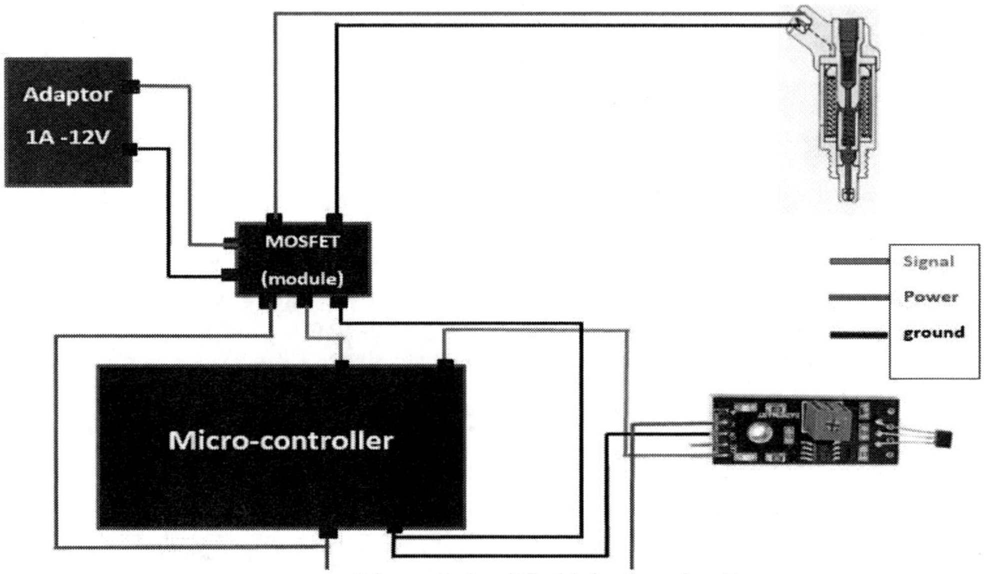

Figure 6.simplified injection circuit.

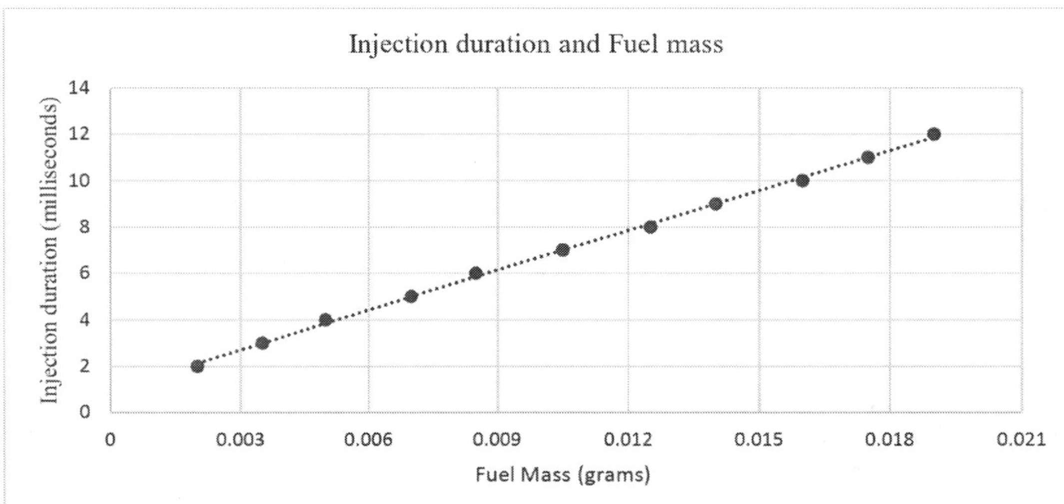

Figure 7. Curve fitting to formulate a function relating the amount of fuel and the injection duration.

After the injection duration is calculated, the injection timing is required. A HES is used to detect the beginning of the intake stroke from the red sign, magnet, on the external gear. The beginning of intake stroke is the beginning of the injection process. To represent all the cycle evens within a single revolution, an external gear is used as shown in figure 8. During the engine operation, the HES detects the suction stroke beginning each cycle. Then it sends a signal to the micro-controller and consequently to the injector driver circuit which will initiate the injection process for the determined injection duration. The injection process is initiated by opening the injector solenoid using the external voltage supply as shown in figure 6.

IOP Conf. Series: Materials Science and Engineering **973** (2020) 012005 doi:10.1088/1757-899X/973/1/012005

Figure 8. The reduction ratio between the external gear and the crankshaft gear. and the place of the magnet red sign.

3.4. Measuring the basic engine parameters

3.4.1. *Load measurement.* A load cell was used to measure the engine torque. A torque arm was used to transfer the engine torque to the load cell. One end of it was mounted on an electrical dynamo-meter body and the other was mounted on the load cell. Using this arrangement and the equation (10),

$$\mathbf{T} = \mathbf{F} \cdot \mathbf{L} \tag{10}$$

Where, *T*is the engine torque and*F* is the force acting on the iron arm by the electrical dynamo-meter body and *L* The iron arm length.

3.4.2. *Engine speed measurement.* Another HES was used to detect the end of each cycle and record its time in microseconds by using another sign 'magnet'. By comparing the end time of each cycle and the previous cycle, the periodic time is calculated. Substituting in the frequency formula as shown in equation (11).

$$\mathcal{F} = \frac{\mathbf{1}}{\mathcal{T}} \tag{11}$$

where*F* The frequency and *T* The periodic time.The engine frequency is obtained, so consequently the engine speed in revolution per minute is calculated by using equation (12).

$$\mathbf{RPM} = \mathcal{F} \cdot \mathbf{60} \cdot \mathbf{2} \tag{12}$$

The result was multiplied by 2 due to the reduction ratio done by external gear.

3.4.3. Monitoring engine temperature. The engine temperature was monitored by using a spark-plug temperature sensor, installed between the spark-plug and the upper surface of the engine cylinder head as shown in figure 9.a.The sensor itself is shown in figure 9.b

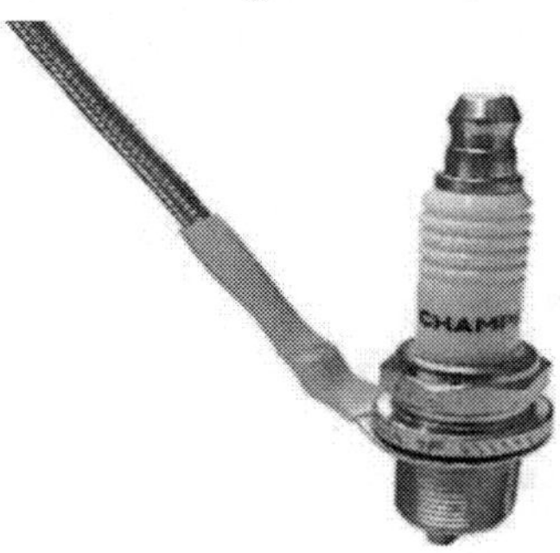

Figure 9.a. Spark plug temperature sensor installation.

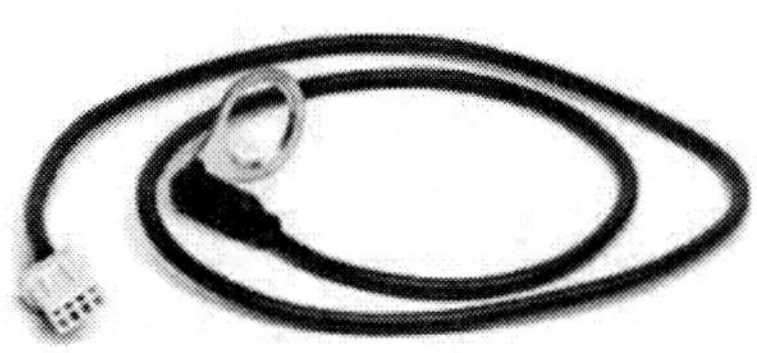

Figure 9.b. The spark plug sensor.

The temperature sensor is updated each millisecond. The temperature also controls the engine status. When the engine temperature is low, the air to fuel ratio is changed to be slightly rich until the engine reaches the ideal operating temperature. After reaching this condition, the engine status will be changed to warm status so the air to fuel ratio will be leaner than the cold status to achieve the least possible fuel consumption.

All the previous parameters are outputs of the used sensors except for the brake power so it will be calculated using equation (13).

$$\mathbf{P = T \cdot \omega} \tag{13}$$

where P the brake power, Tthe engine torque, obtained by the load cell, and ωthe angular velocity. The angular velocity is obtained using equation (14).

$$\omega = \left(\frac{\mathbf{2 \cdot \pi \cdot N}}{\mathbf{60}}\right) \tag{14}$$

where N is the engine speed in rpm, obtained by the HES.

3.5. Monitoring the outgoing parameters during engine operation

Now, it is easy to monitor the engine parameters every specified period. The mass of air drawn per cycle is calculated in grams. The air to fuel ratio is determined to be suitable for the current engine load and engine throttle position using the equation (8). The accurate fuel mass in grams is calculated with respect to the determined air to fuel ratio. The injection duration is determined using equation (9). Engine speed in rpm was measured using a HES. The engine output torque is measured through the load cell. Brake power was determined by engine torque and speed. Engine temperature was measured by a spark plug temperature sensor. These parameters were used to measure system efficiency compared with other systems.

4. Code flow chart

As shown in figure 10, a simple flow chart describes the developed code. First, the code reads the current engine speed, engine temperature, and throttle position angle. Then, it compares the engine temperature to the ideal engine temperature. After this comparison, the code decides the engine operating condition if it is a warm status or a cold start one. for the warm condition, the code calculates the accurate amount of air drawn into the engine then determines the air to fuel ratio according to the engine load, then it calculates the suitable fuel mass required and the suitable injection duration for it. At last, the code waits for the signal from the HES to start the injection process. After each injection process, the engine output torque is measured. For the cold start condition, the code opens the fuel injector for longer durations to make a rich mixture. This rich mixture is used to warm up the engine to the desired engine temperature. Every 30 seconds, all the engine parameters 'calculated or measured' are monitored.

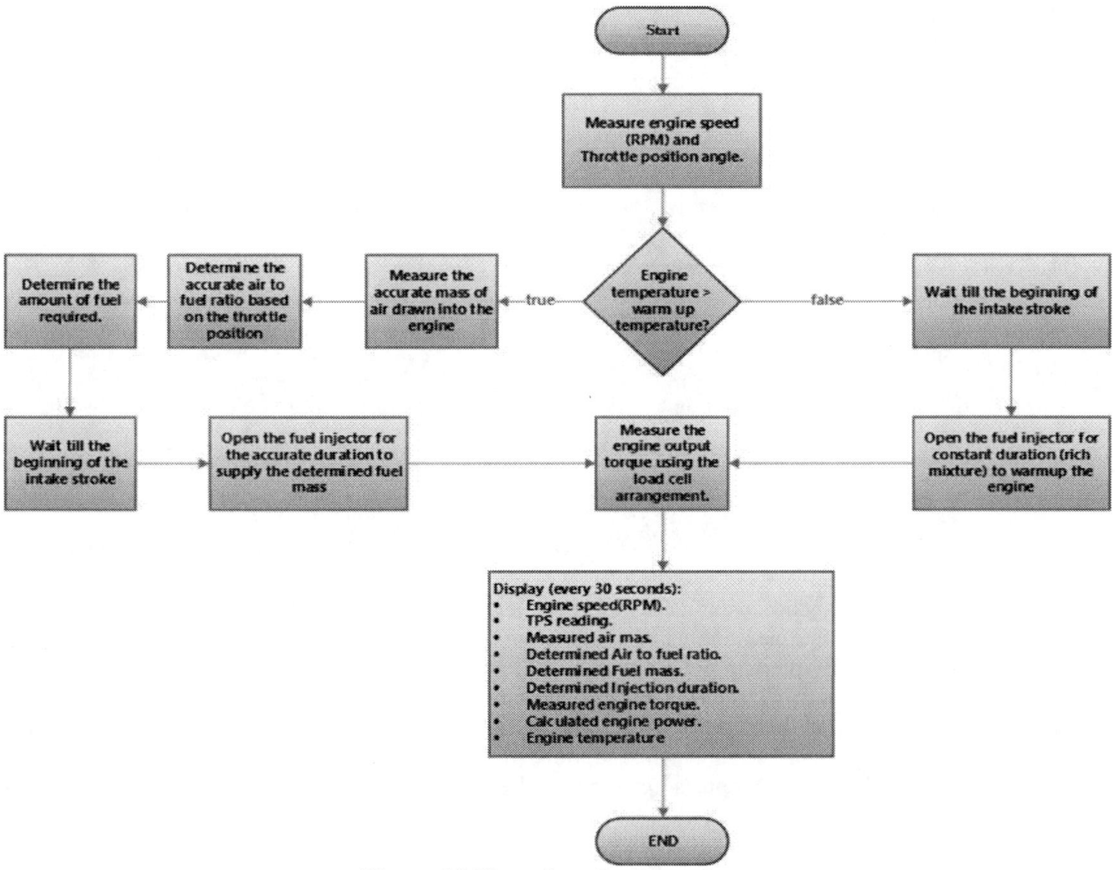

Figure 10. Flow chart for the code.

5. Results

As shown in figure 11, while using the PFI technique, the maximum brake torque is increased by 8%. It is also noticed that the torque in the conventional carburetor system reaches its maximum value at 3000 rpm while the proposed PFI system reaches its maximum torque at 2800 rpm which is more practical in different applications. The maximum brake torque is reached at 2800 rpm while using the PFI system because this system is based on the Calibrated Dose method which will inject just the accurate amount of fuel suitable for the drawn air. Also, the high-pressure injection just above the inlet valve results in a fast response to load compared to the system based on the stream mixing methods.

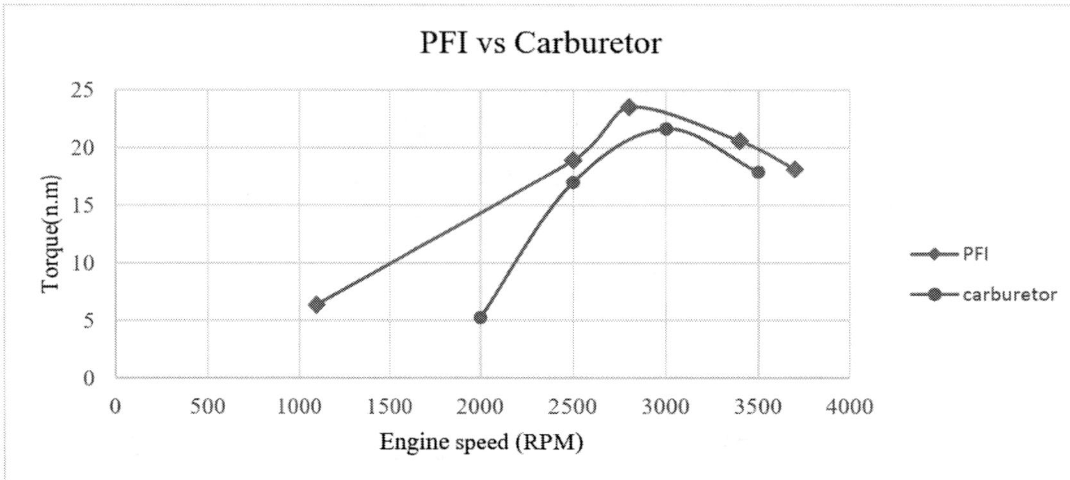

Figure11. Comparison between the output brake torque by using the proposed PFI system and the carburetor with respect to engine speed at (WOT).

As shown in figure 12, while using the PFI technique, the maximum brake power is increased by 7%. It is also noticed that the maximum brake power in the conventional carburetor system reaches its maximum value at 3000 rpm while the proposed PFI system reaches its maximum torque at 3500 rpm which gives the proposed PFI system a wider engine speed range to control. The PFI system achieves more brake power as it delivers the accurate fuel mass 'suitable for the drawn air just' above the inlet valve. This results in a more homogeneous mixture of fuel and air without losses due to friction of the inlet port walls. Also, the high-pressure injection helps to achieve the fuel atomization to increase mixture quality.

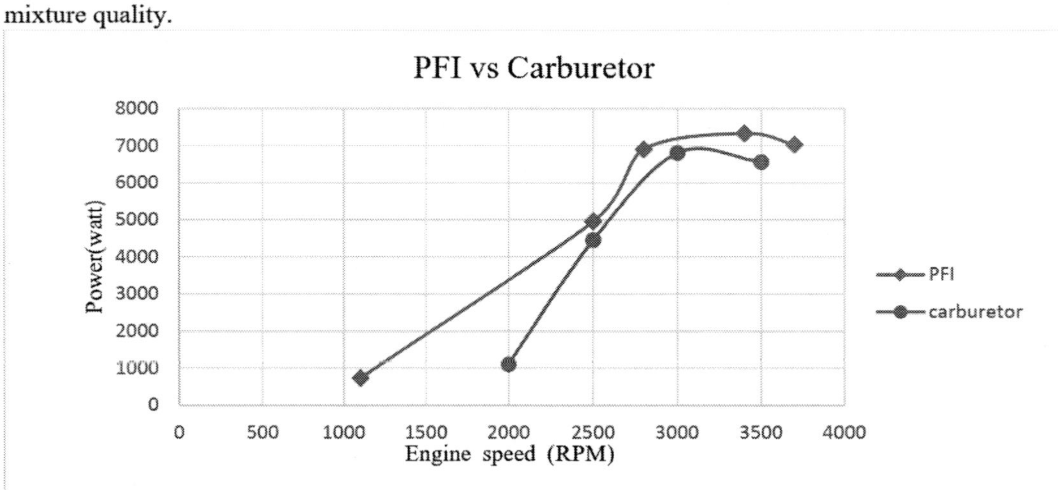

Figure 12. Comparison between the output brake power by using the proposed PFI system and the carburetor with respect to engine speed at (WOT).

6. Conclusion

The proposed algorithm for operating the PFI system has proven to be effective in increasing the specific power over wide range of speed. The proposed PFI system also results in higher brake torque for a wider engine speed range. Using the PFI technique, faster engine response is achieved compared to the carburetor system. Due to using the Calibrated Dose method instead of the stream mixing

IOP Conf. Series: Materials Science and Engineering **973** (2020) 012005 doi:10.1088/1757-899X/973/1/012005

method, the used algorithm for operating the PFI system results in lower fuel consumption. The designed algorithm for operating the PFI system could be used for increasing the output torque and power for the same engine block.

7. References

[1] Zhu R Hu J Bao X He L Lai Y Zu L and Su S 2016 Tailpipe emissions from gasoline direct injection (GDI) and port fuel injection (PFI) vehicles at both low and high ambient temperatures *Environmental Pollution* **216** pp 223–234

[2] Gupta T Kothari A Srivastava D K and Agarwal A K 2010 Measurement of number and size distribution of particles emitted from a mid-sized transportation multipoint port fuel injection gasoline engine *Fuel* **89** (9) pp 2230–2233

[3] Nakao Y, Sakurai Y, Hisano A, Saitou M, Kazari M, Murase T and Suzuki K 2016 Effects of Port Injection Specifications on Emission Behavior of THC *SAE International Journal of Engines* **9**(4) p 2427–2433

[4] Guzzella L and Onder C H 200 4 *Introduction to Modeling and Control of Internal Combustion Engine Systems* (Springer-Verlag Berlin Heidelberg) pp 173–221

[5] Movahednejad E Ommi F Hosseinalipour M and Samimi O 2006 Experimental and Theoretical Study of Injection Timing on Performance and Exhaust Emissions in a Port-Injected Gasoline Engine *ASME 2006 Internal Combustion Engine Division Spring Technical Conference (ICES2006)*

[6] Cowart J S and Hamilton L J 2006 Post Fuel Injection Fuel Vaporization Characteristics in the Intake Port of a Port-Fuel-Injected *Gasoline CFR Engine ASME Internal Combustion Engine Division Fall Technical Conference (ICEF2006)*

[7] Feng D, Wei H, Pan M, Zhou L and Hua J 2018 Combustion performance of dual-injection using n-butanol direct-injection and gasoline port fuel-injection in an SI engine *Energy* **160** pp 573-581

[8] *US6959693B2 Fuel injection system and method.* https://patents.google.com/patent/US6959693B2/en

[9] Mathivanan K Mallikarjuna J M & Ramesh A 2016 Influence of multiple fuel injection strategies on performance and combustion characteristics of a diesel fuelled HCCI engine *Experimental Thermal and Fluid Science* **77** pp 337-346

[10] Fanick R, Kroll S, Swarts A, and Quarderer S 2019 Effects of Dual Port Injection and Direct-Injection Technology on Combustion Emissions from Light-Duty Gasoline Vehicles *SAE Technical Paper* **2019-01-0999**

[11] Zabeu C, Camargos L, Marinsek L, Berti R et al. 2017 Influence of Injection Timing on the Engine Efficiency and Emissions in a SI PFI Engine Using Ethanol-Gasoline Blends with Several Water Content Levels *SAE Technical Paper* **2017-36-0147**

[12] Kim N ,Cho S ,Choi H, Song H H and Min K 2014 The Efficiency and Emission Characteristics of Dual Fuel Combustion Using Gasoline Direct Injection and Ethanol Port Injection in an SI Engine *SAE Technical Paper* **2014-01-1208**

[13] US6691021B2 *Failure determination system, failure determination method and engine control unit for variable-cylinder internal combustion engine.* https://patents.google.com/patent/US6691021B2/en

[14] US6763707B2 *Failure determination system and method for internal combustion engine and engine control unit.* https://patents.google.com/patent/US6763707B2/en

[15] Anand T and Ravikrishna R 2012 Modelling of mixture preparation in a small engine with port fuel injection Progress in Computational Fluid Dynamics, An International Journal (PCFD) **12** No. 6 pp375-388

[16] Balluchi A, Benvenuti L, Benedetto M D, Pinello C and Sangiovanni-Vincentelli A 2000 Automotive engine control and hybrid systems: challenges and opportunities *Proceedings of the IEEE* **88** Issue 7 pp 888-912 doi: 10.1109/5.871300.

[17] Movahednejad E, Ommi F, Hossinalipour M and Karbasforoushha M 2005 Experimental and Theoretical Study of Injection Parameters on Performance and Fuel Consumption in a Port-Injected Gasoline Engine *SAE Technical Paper* **2005-01-3805**

[18] Daniel R Wang C Xu H & Tian G 2012 Effects of Combustion Phasing, Injection Timing, Relative Air-Fuel Ratio and Variable Valve Timing on SI Engine Performance and Emissions using 2,5-Dimethylfuran *SAE International Journal of Fuels and Lubricants 5* Issue 2 pp 855-866.

[19] Huang Y Hong G and Huang R 2015 Numerical investigation to the dual-fuel spray combustion process in an ethanol direct injection plus gasoline port injection (EDI+GPI) engine *Energy Conversion and Management* **92** pp 275-286

[20] Patel R J and Brahmbhatt P 2019 Effect of Varying Fuel Injection Timing on Engine Performance of CNG Direct Injection in Retrofitted SI Engine *European Journal of Sustainable Development Research* **3** Issue 3 Article No: em0089

[21] Dernotte J, Mounaim-Rousselle C, Halter F and Seers P 2009 Evaluation of Butanol–Gasoline Blends in a Port Fuel-injection, Spark-Ignition Engine *Oil & Gas Science and Technology* **65**(65) pp 345-351

[22] Gupta T Kothari A Srivastava D K & Agarwal A K 2010 Measurement of number and size distribution of particles emitted from a mid-sized transportation multipoint port fuel injection gasoline engine *Fuel* **89**, Issue 9 pp2230-2233

[23] Shankar R Udayakumar G and Sasikumar K 2008 Advanced Port Injection Solution for Motorcycle Application *SAE Technical Paper* **2008-28-0031**

Acknowledgment
The author acknowledges Prof. Safwat Wilsonand Prof. GaberAsar for being good advisers and for supporting technically and morally.

IOP Conf. Series: Materials Science and Engineering **973** (2020) 012006 doi:10.1088/1757-899X/973/1/012006

The effect of rectangulargurney flap on wells turbine performance

A T Kotb[1], M A Nawar[2] and R M Abd El Maksoud[3]
[1]Demonstrator, Mechanical Power Engineering Department, Faculty of Engineering, Mattaria, Helwan University, Egypt.
[2]Assist.Prof., Mechanical Power Engineering Department, Faculty of Engineering, Mattaria, Helwan University, Egypt.
[3] Assoc. Prof., Mechanical Power Engineering Department, Faculty of Engineering, Mattaria, Helwan University, Egypt.
ahmedtarekmostafa15@gmail.com

Abstract. Wells turbine is very adept at converting pneumatic power from ocean waves into mechanical energy. However, such turbines incur from low performance, limited operating range and low efficiency. The present study introduces a method to enhance the power produced from Wells turbine by implementing rectangular Gurney flap (GF). The Gurney flap is placed on both pressure and suction sides of the trailing edge (TE) and perpendicular to the chord line without change chord length. GF increases the lift coefficient by altering the Kutta condition at the TE. The turbine performance is herein evaluated by solving numerically 3D incompressible Reynolds Averaged Navier–Stokes equations (RANS) by using ANSYS Fluent 19.0. The performance is demonstrated according to the flow coefficient, torque coefficient, total pressure loss coefficient, and the turbine efficiency, as well as the velocity and pressure fields around the turbine blades. The validation of the present work was achieved using previous experimental work and CFD work by using SST $k - \omega$ turbulence model. The present work showed that the Wells turbine augmented with GF increases the torque coefficient by 41.98%. Besides, the stall has been delayed compared with the conventional Wells turbine.

1. Introduction

Wells turbine is utilizedto convert ocean wave energy into a mechanical power.it is environmentally device friendly so researchers around the world are investigating on how to boost its performance. Different designs were experimentally studied by Gato et al. [1] and Curran and Gato [2] to investigate the effect of theirs designs on performance of turbine. Kim et al. [3] made a numerical simulation to study the blade sweep and its effect on the turbine performance. Setoguchi et al. [4,5] introduced influence of self-rectifying turbines and utilizing of inlet guide vanes on design parameters of turbine. Brito-Melo et al. [6] studied numerically different turbine design parameters. Dhanasekaran and Govardhan [7] introduced a numerical study for turbine performance. The effect of inlet conditions on the Wells turbine performance wasintroduced by Raghunathan et al. [8]. The previous studiesconfirmed that geometrical parameters such as blade geometry [9], blade setting angle [10], the use of end plates [11] and tip clearance [12] affect Wells turbine performance.

The previous studies showed that the optimum blade profile was found to be NACA0020 for small-scale Wells turbine [13], while the optimum blade profile was NACA0015for large-scale turbines [14]. The effect of guide vanes on the constant chord Wells turbine was showed by Thakker et al. [15]. Besides, Setoguchi and Takao [16] introduced its effect on the variable chord.A study by [14,17] had

IOP Conf. Series: Materials Science and Engineering **973** (2020) 012006 doi:10.1088/1757-899X/973/1/012006

been demonstrated that Wells turbine with guide vanes stalled before those without guide vanes. The effect of tip leakage flow on the Wells turbine performance was showed by Taha et al. [18,19] with uniform tip clearance [18] and also with non-uniform tip clearance [19]. Both studies showed that tip leakage flow effect on the performance of wells turbine and turbine stalling. The influence of geometry of turbine duct on its performancewas numerically studied by Shaaban and Hafiz [20]. Halder and Samad [21] showedthat the casing treatment influence the vortex pattern which result in stall limit change.

The effect of a number of rotor blades on the wells turbine performance had been studied through many researchers. Some of the previous studies recommended the use of eight rotor blades [1,2,12,22] and others recommended the use of six rotor blades[22,13].On the other side, using air turbines to be immersed in water gave a promising performance. Hamedet al. [23] proved that operating wells turbine as a hydraulic turbine achieves better performance. The circumferential casing groove (rectangular shape) effects on the turbine performance at different groove depth and width were demonstrated by Halder et al [24]. Moreover, two different cases, with and without a tip groove, were considered to guess the optimal turbine speed for the different flow velocities [25].

The design with grooved-casing performs better than that of the without grooved casing. Using variable blade sweep angles at the tip and mid sectionsenhancedWells turbine performanceand provided higher range of operating [26, 27]. Torres FR, etal [28] demonstrated a methodology to determine the optimal size of a Wells turbine taking into account hydro-aerodynamic model. Moreover, Nazeryan M, et al [29] investigated numerically theinfluence of blade thickness on both aerodynamic and entropy generation analysis of a Wells turbine. Besides, Gratton T, etal [30] performed for optimizing the blade profile in the Wells turbine for better performance, by the torque coefficient magnification. Kumar and Samad [31] showed the effect of rectangular flap and flap height on wells turbine performance.

In accordance toprevious researches, still much effort is needed to boost Wells turbine performance that incurs from low performance, limited operating range and low efficiency. Accordingly, the goal of the present paper is to overcome these disadvantages. The present study introduces a method to enhance the power produced from Wells turbine by implementing different geometries of Gurney flap (GF). The Gurney flap is placed on both pressure and suction sides of the trailing edge (TE) and perpendicular to the chord line without change chord length.

Wells turbine is investigated while having symmetrical NACA 0015 blades. The turbine performance is here in evaluated by solving numerically 3D incompressible Reynolds Averaged Navier–Stokes equations (RANS) by using ANSYS Fluent 19.0. The performance is demonstrated according to the flow coefficient, torque coefficient, total pressure loss coefficient, and the turbine efficiency, as well as the velocity and pressure fields around the turbine blades. The choice of the number of blades used is rationalized from previous studies.

2. Computational methodology

In this section, Wells turbine theory of operation, its gurney flapandits computational methodology areshowed.

2.1. Wells turbine

Wells turbine is one of the most important Ocean-wave energyconverters that convert Ocean-wave energy into mechanical power. Actually, in oscillation water column (OWC) wave energy converters, pressure fluctuations are generated due to motion of wave, inside a plenum, and therefore producing an oscillating air-flow, so the self-rectifying Wells turbine can efficiently convert whose energy into a unidirectional rotational motion; see figure.1a. In its basic arrangement, this is an axial-flow air turbine, consisting of a rotor with a number of symmetrical and untwisted airfoil blades, usually coding to the NACA00XX series, usually chord line is perpendicular to rotational axis(90 °stagger angle); see figure1b,[12].

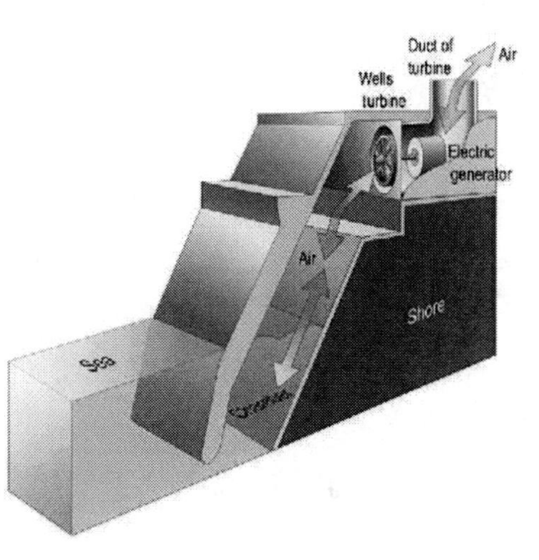

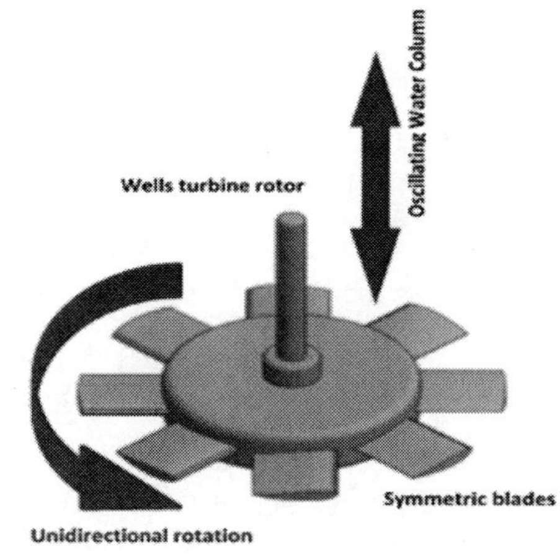

(a) **(b)**

Figure1. (a) Schematic cutaway of a Wells turbine, **(b)** The Wells air turbine.

2.2. Gurney flap scheme

The present study introduces a method to enhance the power produced from Wells turbine by implementing different geometries of Gurney flap (GF). The Gurney flap is placed on both pressure and suction sides of the trailing edge (TE) and perpendicular to the chord line without change chord length. In the present study,a rectangular Gurney flap (GF)isstudied, seefigure 2. In this paper a rectangular Gurney flap (GF) heightwas varied from 0.5-2%C and the width from 0.4-0.6%C.

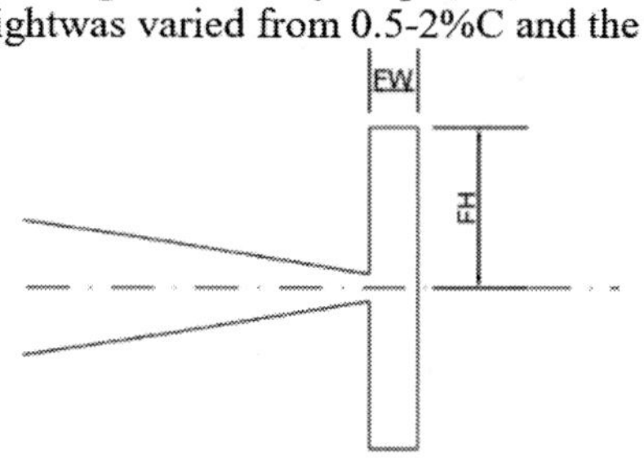

Figure2. Schematic of rectangular Gurney flap.

2.3. Computational equations

In figure3, the flow having an angle of incidence (α) will produce a lift force (L) and a drag force (D). The components of these forces in both tangential and axial directions are denoted by $F_t$and F_a, respectively. Accordingly, F_t generates torque on the turbine rotor, while F_a producesaxial stress on the rotor as a result of thrust forceon the turbine rotor axis [7]. Accordingly, for wells turbine in an oscillatingairflow, the angle of incidence (α)changes from positive to negative values in accordance with the upward and downward directions of flow, respectively; see figure 1. The tangential force F_t is acting in the same direction regardless of the positive or negative value of an angle of incidence (α) because of symmetry of the rotor blade.

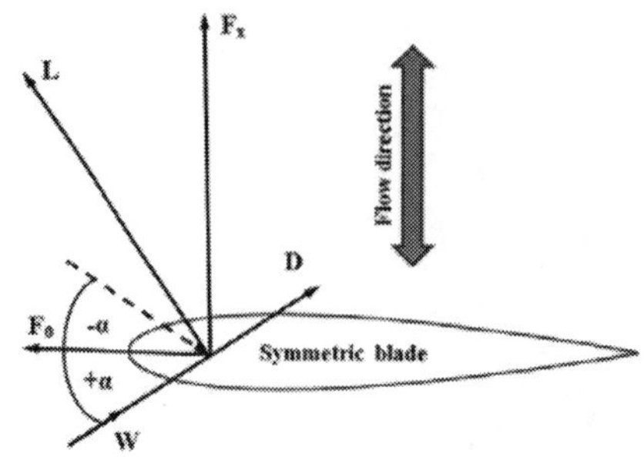

Figure 3. Forces acting on a Wells turbine blade.

In accordance with, torque has a unidirectional feature that is created from an oscillating flow [7]. The flow coefficient, φ, can be introduced as:

$$\phi = \frac{C_x}{U} \quad (1)$$

The flow coefficient is changedin accordance with the variation of the axial flow velocity C_x while keeping the rotor tip speed U constant. Thus, the torque coefficient C_T can be introduced as:

$$C_T = \frac{T}{\rho\omega^2 {R_T}^5} \quad (2)$$

whereTthe torque on rotor is, ρ is the working fluid density, ω is the angular velocity of turbine and$R_T$is the radius of rotor tip. Total pressure drop coefficient C_P is expressed by the following formula:

$$C_P = \frac{\Delta P_o}{\rho\omega^2 {R_T}^2} \quad (3)$$

WhereΔP_o is the total pressure drop.
The Turbine efficiency (η) is introduced as:

$$\eta = \frac{T\omega}{\Delta P_o Q} \quad (4)$$

whereQ is the volume flow rate through the turbine.

Rotor solidity is an important parameter that affects the performance of Wells turbines. The rotor solidityσ is defined as:

$$\sigma = \frac{Z * c * H}{\pi * ({R_T}^2 - {R_h}^2)} = \frac{Zc}{2\pi R_m} \quad (5)$$

whereZ is the number of blades, ***c*** is the chord length;***H*** is the blade height, R_T is the radius of rotor tip, R_h is the radius of rotor hub and $R_m$is the turbine mean radius. Eq. (5), shows that the rotor solidity σ can be varied by change the number of turbine blades.

In the present study, the influence of using different geometries of Gurney flap (GF) on performance of Wells turbine is numerically investigated by solving the 3D incompressible Reynolds-averaged Navier-Stokes equations RANS for wells turbine with Gurney flap (GF). The basic purpose of using any turbulence model is to bring results closer to the experimental results. In this study, the SST k–ωTurbulence model is used. In addition, Non-dimensional wall distance$y^+ < 1$ to guess flow pattern on buffer and sub-layer zones. The good prediction of the aerodynamic performance was achieved by the combination of the SST k–ωTurbulence model with non-dimensional wall distance$y^+ < 1$. Besides, the Moving Reference Frame method (MRF) is applied to simulate the turbine rotor, but the main governing equations are solved in the absolute frame. The finite volume method is appliedto appreciate the absolute frame using the computational fluid dynamics (CFD) code ''Fluent'' V19.0. The present simulationuses different discretization methods. These methods are shown in the following table; see table 1:

Table 1. Solution methods.

Pressure-velocity coupling	SIMPLEC
Gradient	Least squares cell based
Pressure	Standard
Momentum	The second order upwind
Turbulent kinetic energy(k)	The second order upwind
Specific dissipation rate(ω)	The second order upwind

The SIMPLEC algorithm applied for pressure-velocity coupling eliminates the influence of mesh skewness on the final solution. The maximum residual values for momentum, continuity, velocity x, velocity y, velocity z and turbulence equations of the solutions are10^{-4}. Moreover, the valve of turbulence intensity at inlet and outlet is 5% (The turbulence intensityis defined as the ratio of the root-mean-square of the velocity fluctuations to the mean flow velocity).

3. Results and discussions

In this part four parts are presented and they are a mesh size independence study, model validation, effect of different geometries of Gurney flap on turbine performance, and the field of flow around the blades.

3.1. Mesh independent study

Curran and Gato [2]wells turbine dimensionsare used in this investigation. The specifications of the Wells turbine that is under study are shown in table 1. The full cylindrical domain of eight-blade rotor is reduced to slice of domain angle $\theta \frac{360}{8} = 45°$ of the full domain, see figure 4. This minimized was made on account of the rotor of turbine symmetry around the axis of rotational and as a result, the computational time is reduced [20]. Besides, table 2 presentsdomain meshing and domain boundary conditions. Figure.4 and figure.5 describe the domain of computational with boundary conditionsand the mesh of computational, respectively.

The large cylindrical domain extended 4C and 8C upstream and downstream the rotor, respectively [20] to confirm that the boundaries doesn't influence on the flow at the turbine, Here, refinement grids with inflation are used near the blade surface to accurately capture the flow field inside the boundary layer. The grid is made up of unstructured tetrahedral mesh. The total number of cells in the computational domain is 2,614,335. The inlet and outlet conditions of computational domain are uniform velocity and outlet pressure, respectively. For the blades, hub and casing surfaces no slip wall conditions are used. Flow coefficientφ is changed from 0.1 to 0.325 and the fixedrotational speed of turbine (2000 rpm). Henceforth, the blade tip speed was kept constant.

Figure 6 shows the influence of number of cells in computational domain on the CFD results of turbine efficiency η at flow coefficient $\varphi = 0.15$. The number of cells was gradually increased in five steps from 1,015,551 cells to 3,953,000 cells, with constant non-dimensional wall distance $y+ < 1$ at the same φ. This study showed that the maximum percentage of the diffractions of torque coefficient C_T, total pressure loss coefficient C_p, and efficiency η are ± 0.99%, ±0.66% and ± 0.33%, respectively. Cells number of 2,614,335 is the best solution quality, where CFD results are independent of the number of cells in computational domain. Thus, this meshes number was utilized during the present study.

IOP Conf. Series: Materials Science and Engineering **973** (2020) 012006 doi:10.1088/1757-899X/973/1/012006

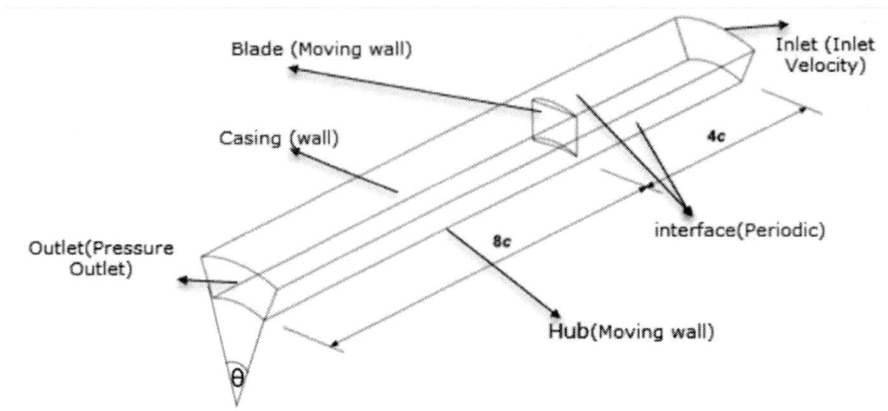

Figure 4. The computational domain with boundary conditions.

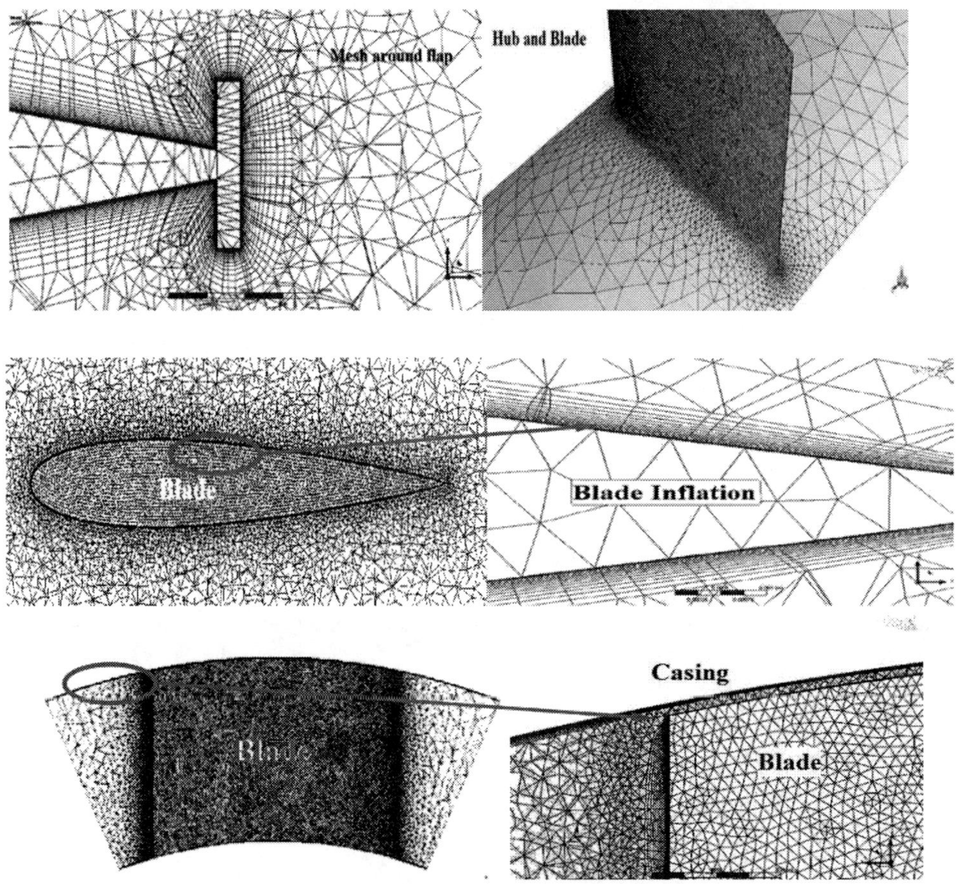

Figure 5. Unstructured/tetrahedral computational mesh.

Table 2. Specifications of turbine model.

Parameter	Dimension
Blade Profile	NACA0015
Number of blades, Z	8
Domain angle, θ	45°
Blade chord length, *c*	0.125 m
Blade thickness, t	0.01875 m (15% of *c*)
Hub diameter, $\boldsymbol{D_t}$	0.4 m
Casing diameter, $\boldsymbol{D_c}$	0.59 m
Tip clearance	0.001 m
Blade height, H	0.094 m
Solidity, $\boldsymbol{\sigma}$	0.644
Stagger angle	90°
Turbine rotational speed, N	2000 rpm

Table 3. Meshing and boundary conditions.

Parameter	Description
Flow domain	Single blade
Interface	Periodic
Mesh/Nature	Unstructured/tetrahedral
Elements	2,614,335
Fluid	Air
Turbulence model	SST $k - w$
Inlet	Velocity
Outlet	Pressure
Hub	Wall
Casing	Wall
Blade	Wall
Maximum Residual criteria	$1\text{x}10^{-4}$

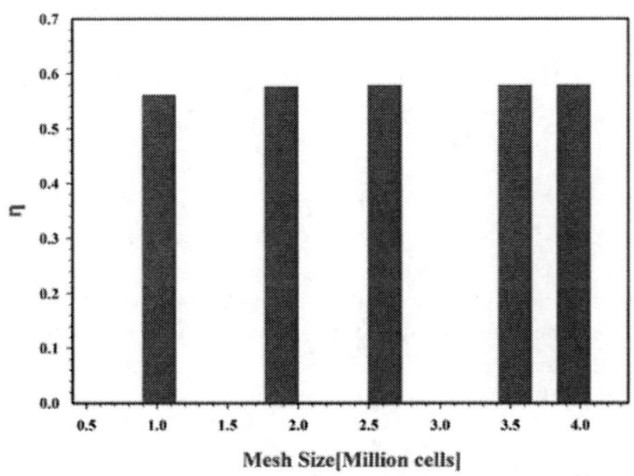

Figure 6. Influence of cells number on turbine efficiency.

IOP Conf. Series: Materials Science and Engineering **973** (2020) 012006 doi:10.1088/1757-899X/973/1/012006

3.2. Model validation

A validation curve of torque coefficient C_T, total pressure loss coefficient C_p, and efficiency η has been determined at different values of flow coefficientφ as shown in figures 7a, 7b,and 7c, respectively. The validation results were compared withboth the experimentalresults [2] and CFD results [12, 21, 23and 31]. The results of the present work showgood agreementwith the results of Curran and Gato [2].Accordingly, the present CFD model has been great admitted as acomputational tool for the performance of Wells turbine.

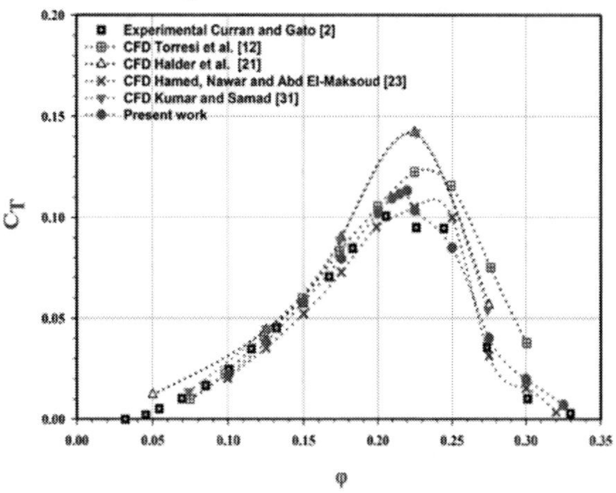

(a)

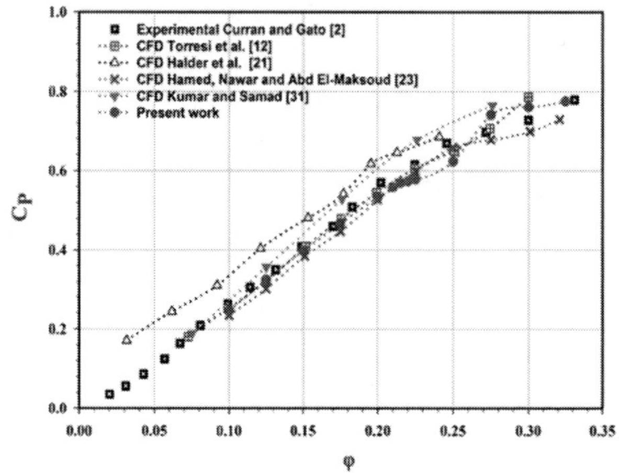

(b)

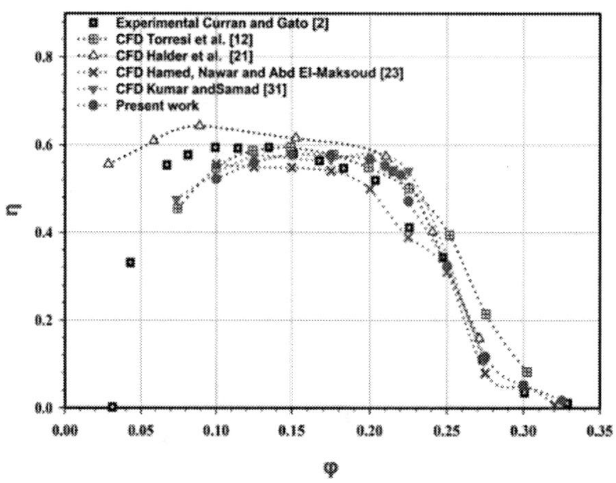

(c)

Figure 7. Validation with experimental and CFD results for:
(a) C_T, (b) C_p, and (c) η.

Figure 8 presents the error percentage of the present work compared with the experimentalresults of Curran and Gato [2]. As it is depicted from the figure that the maximum error percentage for torque coefficient, the total pressure drop coefficient and efficiency are 9.9%,7.3%, and 11.7%, respectively.It is notably to mention that the error percentage conserned the efficiency is dependent upon the errors of the torque and pressure drop coefficients.

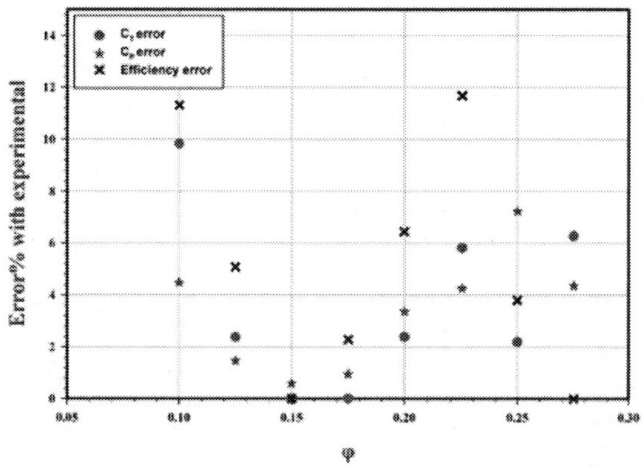

Figure 8. Error percentage with experimentalresults.

Validation of the rectangular flap at FW = 0.5 %C & FH = 1.5 %C is also achieved with Kumar and Samad [31] for more satisfaction. Figure 9 presents that the present CFD results have a good agreement with Kumar and Samad [31]. However, Kumar and Samad [31] studied only the effect of varying the rectangular flap height (FH) on the Wells turbine performance. So, in this paper, the effect of varying flap height (FH) and flap width (FW) on the Wells turbine performance has been considered.

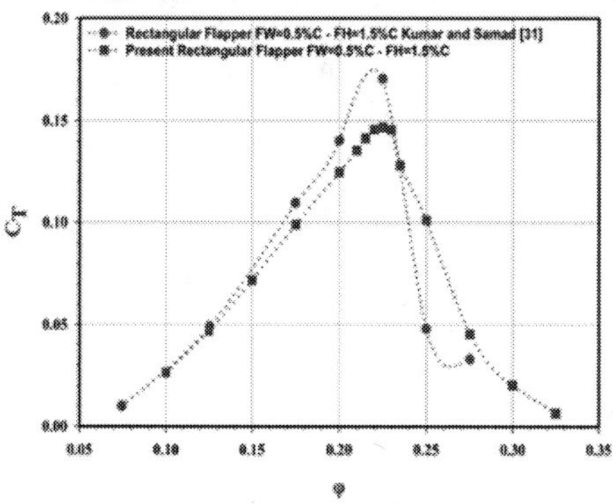

(a)

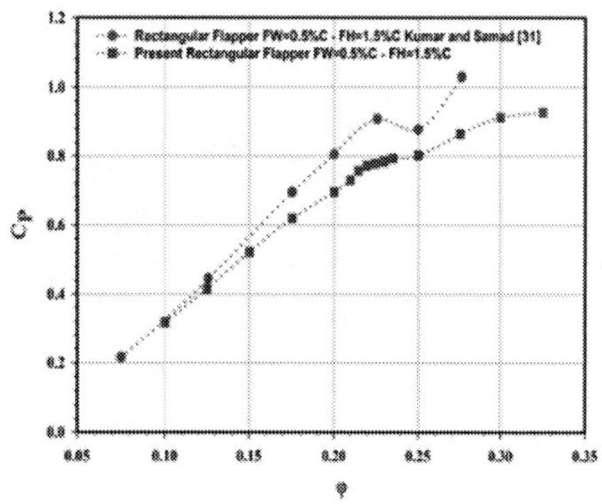

(b)

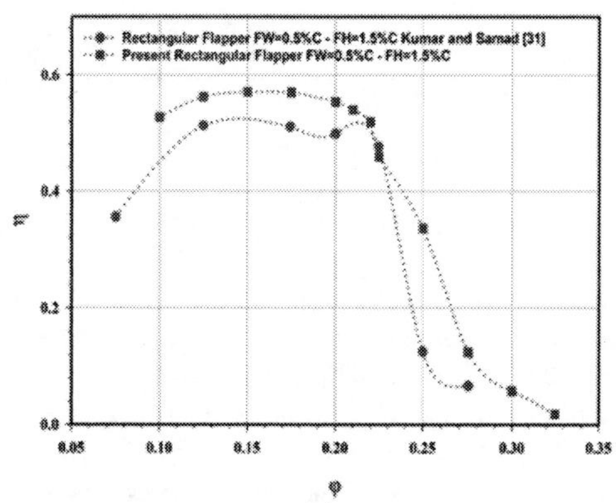

(c)

Figure 9. Verification of the present rectangular flap results. For:
(a) CT, (b) Cp, and (c) η.

IOP Conf. Series: Materials Science and Engineering **973** (2020) 012006 doi:10.1088/1757-899X/973/1/012006

3.3. Effect of rectangular flap height

Figure10 demonstrates the effect of rectangular flap height on the variations of C_T, C_P, and η at different values of φ. The results show thatthe rectangular flap of FW = 0.5 %C & FH = 1.5 %C gives a maximum value of C_T= 0.148at φ = 0.225.While, the maximum C_T = 0.113 at Φ = 0.22 for blade without flap (FH = 0 %C). Hence, the percentage increase in $C_T$is41.98% Φ = 0.225 at, see figure 10(a). On the other hand, C_P is increased as the flap height increases especially at the values of φ> 0.2 as shown in figure 10(b). Despite the decrease in the efficiency compared with the blade without flap (reference case), the amount of power generated had been enhanced, figure 10(c).

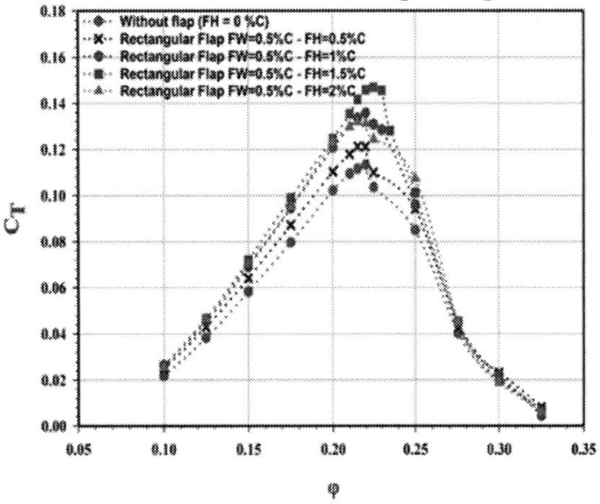

(a)

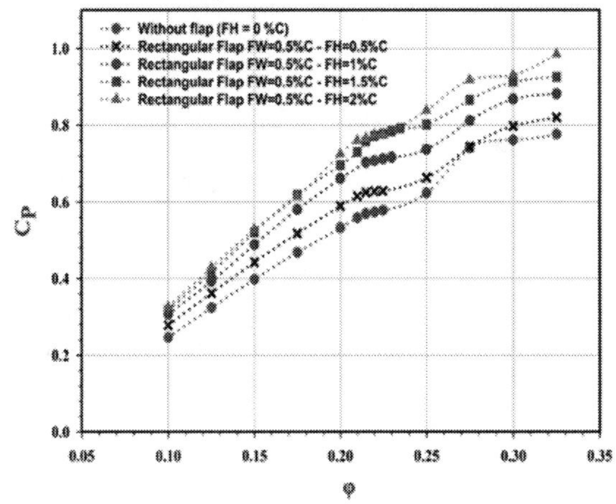

(b)

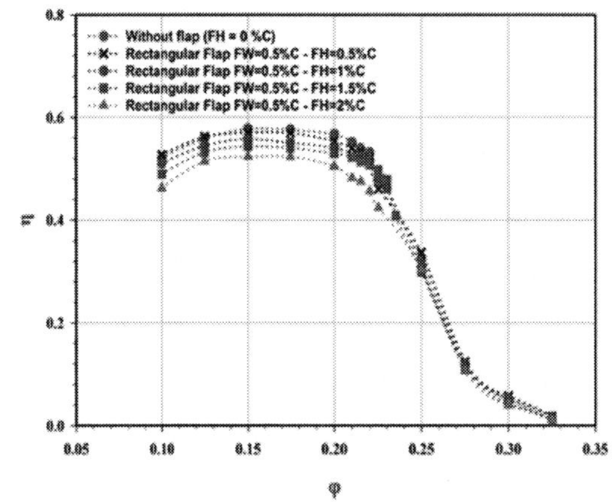

(c)

Figure10. Effect of rectangular flap height (FH) on: (a) C_T, (b) C_p, and (c) η.

The velocity streamlines are presented for Wells turbine blade without flap (FH = 0 %C) and with flaps of different heightsas shown in figure11. The comparison of velocity streamlines between blades with flap is established at FW = 0.5%C, H = 50%,φ = 0.2 andFH = (0%, 0.5%, 1%, 1.5%, and 2%). It is noticed that, at the blade without flap (FH = 0 %C); a vortex is formed at the end of the trailing edge. This vortex separates the flow from the surface of the blade, which reduces the power generated, see figure 11 (a). The flap position at the end of the blades generates to vortices regions; a forward and backward. The forward vortices size is very small compared with the backward vortices, which diminish their effect. Hence, the backward vortices, that is located downstream the blade, made a low-pressure zone. This reduces the chance of the boundary layer separation and give the opportunity to entrain more streamlines thatin turn enhance the power generated. Accordingly, thesize and location of

IOP Conf. Series: Materials Science and Engineering **973** (2020) 012006 doi:10.1088/1757-899X/973/1/012006

the vortices have a very vital effect on the performance of the blades. The effect of flap of FH = 0.5 and 1 is very small compared with the flapof FH = 1.5 and 2 as shown in figures 11 (b, c, d, and e) respectively.

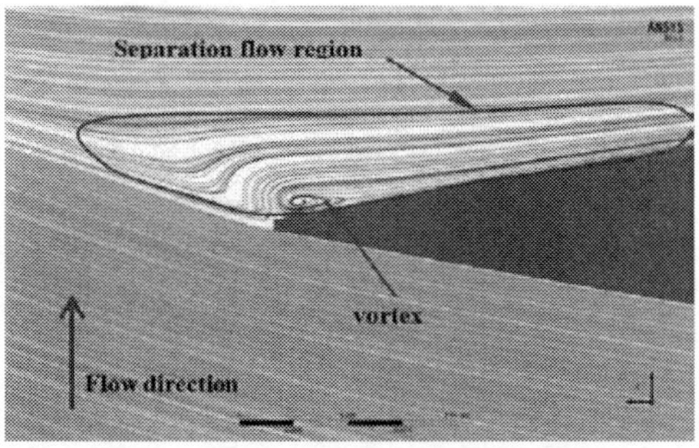

(a) FH = 0 %C

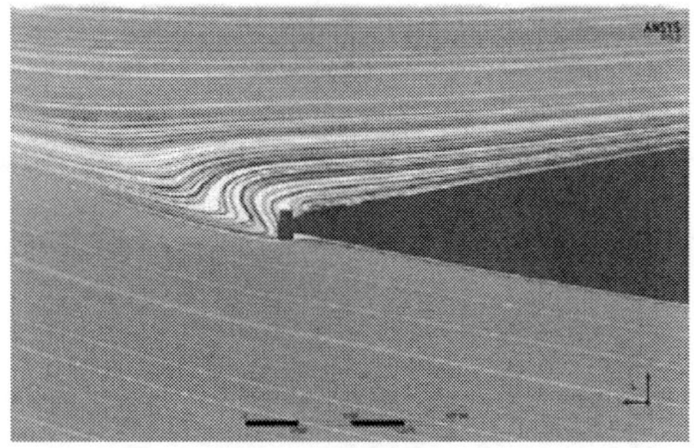

(b) FH=0.5%C

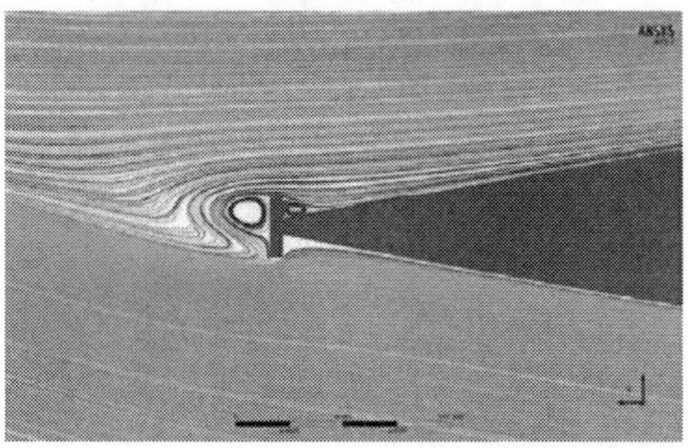

(c) FH=1%C

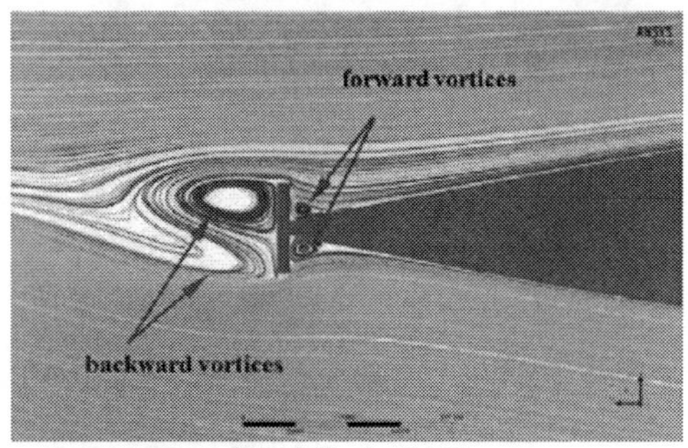

(d) FH=1.5%C

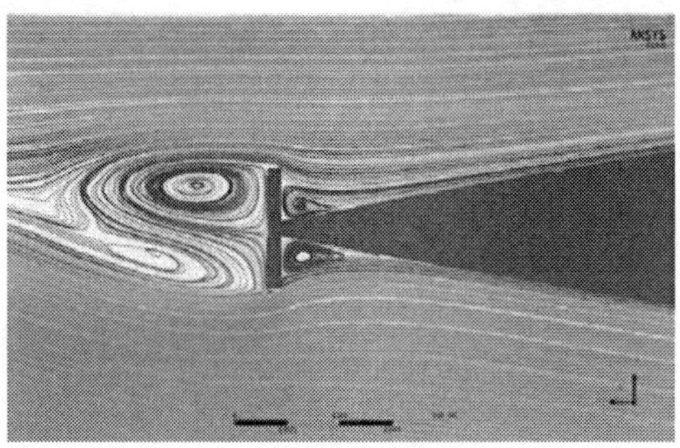

(e) FH=2%C

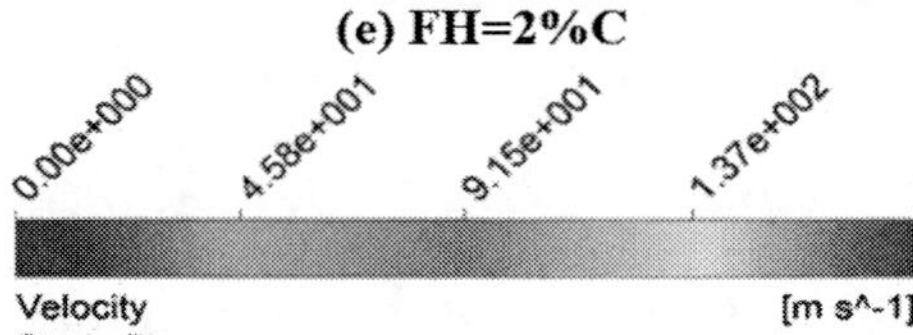

Figure 11. Velocity streamlines of rectangular flaps of different heights at FW=0.5%C, H=50% and Φ=0.2.

3.4. Effect of rectangular flap width

Figure12 illustrate the effect of rectangular flap width on the variations of C_T, C_P, and η at different values of φ. the rectangular flap of FW = 0.5 %C gives a maximum value of C_T at φ = 0.225 compared with the other geometries as shown in figure 12(a). On the other hand, the rectangular flap of width

(FW = 0.4, 0.5, and 0.6 %C) approximately the same values of $C_p$through the range of φ as shown in figure 12(b). Despite the decrease in the efficiency compared with the blade without flap (FH = 0%), the amount of power generated had been enhanced, figure 12(c). therefore, the optimum rectangular flap dimensions are FH = 1.5 %C and FW = 0.5 %C

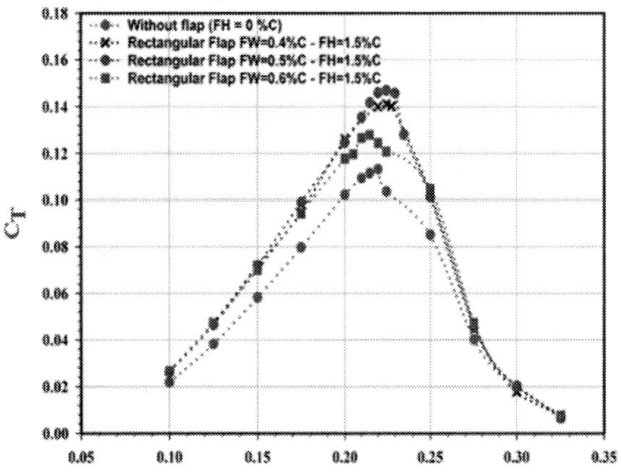

(a)

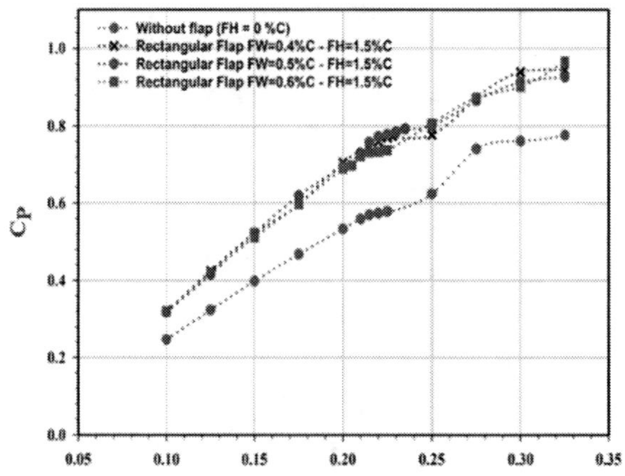

(b)

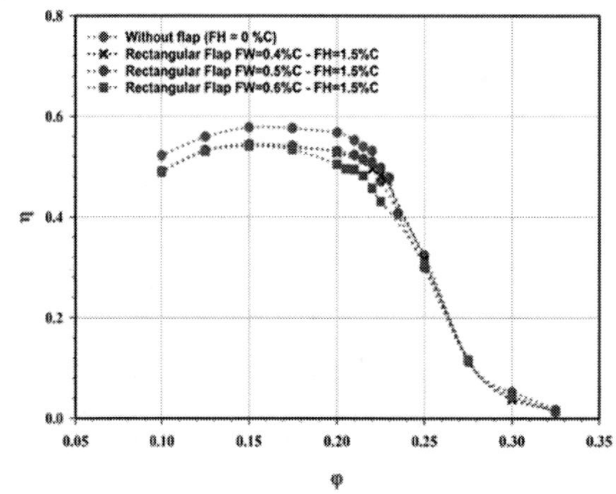

(c)

Figure12. Effect of rectangular flap width (FW) on: (a) C_T, (b) C_p, and (c) η.

The velocity streamlines are presented for Wells turbine blade without flap (FH = 0 %C) and with flaps of different widths as shown in Figure 13. The comparison of velocity streamlines between blades with flap is accomplished at FH = 1.5 %C, H = 50 %, φ = 0.2 and FW = (0%, 0.4%, 0.5%, and 0.6%). The backward vortices that is formed at FW = 0.5%C is very dense compared to others, see figure 13 (C).

IOP Conf. Series: Materials Science and Engineering **973** (2020) 012006 doi:10.1088/1757-899X/973/1/012006

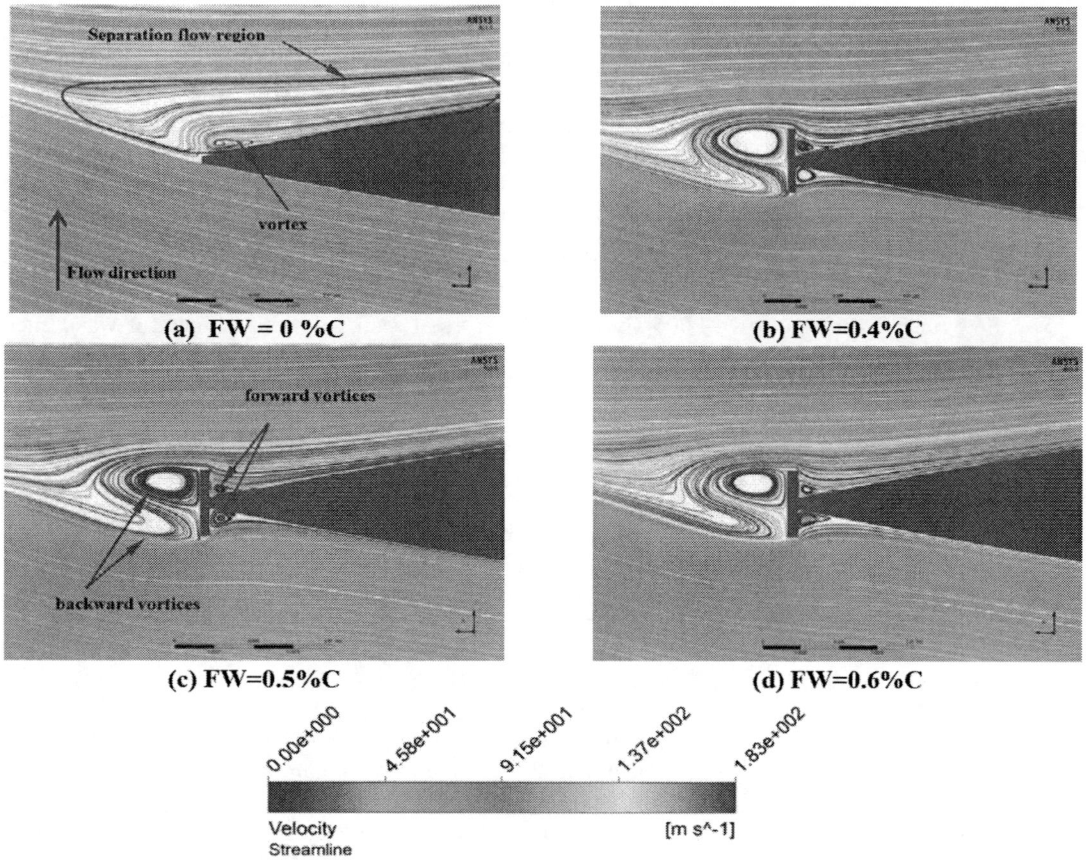

Figure 13. Velocity streamlines of rectangular flaps of different width at FH=1.5%C, H=50% and Φ=0.2.

4. Conclusions

The aim of the present work is enhancing the Wells turbine performance. Rectangular Gurney flap has been elaborated herein.The turbine performance is investigated under steady unidirectional flow conditions using SST $k - \omega$ model where its numerical results were validated. The model is based on 3D incompressible RANS equation in the single rotating reference frame.Accordingly, the following conclusions are taken:

1. The Gurney flap generates two types of vortices; forward and backward vortices. The forward vortices size is very small compared with the backward vortices that diminish their effect.
2. The blade without flap achieves maximum torque coefficient equals to 0.113 at $\Phi = 0.22$. Furthermore, the stall inception appears at flow coefficient equals to 0.22.
3. Compared with the blade without flap, theblade with rectangular flap (FH = 1.5 %C and FW = 0.5 %C) achieves maximum torque coefficient equals to 0.148.Moreover, the stall point is delayed to flow coefficient equals to 0.225. Furthermore, the maximum percentage of increasing in the torque coefficient is 41.98% at such flow coefficient
4. The Gurney flap width (FW) does not affect ontotal pressure drop coefficient through the whole range of operation for rectangle flap.

IOP Conf. Series: Materials Science and Engineering **973** (2020) 012006 doi:10.1088/1757-899X/973/1/012006

References

[1] Gato L, Warfield V and Thakker A 1996 Performance of a High-Solidity Wells Turbine for an OWC Wave Power Plant *Journal of Energy Resources Technology* **118** pp 263-268.
[2] Curran R and Gato L 1997 The energy conversion performance of several types of Wells turbine designs *Proceedings of the Institution of Mechanical Engineers Part A: Journal of Power and Energy* **211** pp 133-145.
[3] Kim T, Setoguchi T, Kaneko K and Raghunathan S 2002 Numerical investigation on the effect of blade sweep on the performance of Wells turbine *Renewable Energy* **25** pp 235-248
[4] Setoguchi T, Takao M and Kaneko K 2000 A Comparison of Performances of Turbines for Wave Power Conversion *International Journal of Rotating Machinery* **6** pp129-134.
[5] Setoguchi T and Takao M 2006 Current status of self rectifying air turbines for wave energy conversion *Energy Conversion and Management* **47** pp 2382-2396
[6] Brito-Melo A, Gato L and Sarmento A 2002 Analysis of Wells turbine design parameters by numerical simulation of the OWC performance *Ocean Engineering* **29** pp 1463-1477.
[7] Dhanasekaran T and Govardhan M 2005 Computational analysis of performance and flow investigation on wells turbine for wave energy conversion *Renewable Energy* **30(14)** pp 2129-2147.
[8] Raghunathan S, Setoguchi T and Kaneko K 1989 The Effect of Inlet Conditions on the Performance of Wells Turbine *Journal of Energy Resources Technology* **111(1)** pp 37-42.
[9] Kim T, Setoguchi T, Kinoue Y and Kaneko K 2001 Effects of blade geometry on performance of wells turbine for wave power conversion *Journal of Thermal Science* **10(4)** pp 293-300.
[10] Setoguchi T, Santhakumar S, Takao M, Kim T and Kaneko K 2003 A modified Wells turbine for wave energy conversion *Renewable Energy* **28(1)** pp 79-91.
[11] Takao M, Setoguchi T, Kinoue Y and Kaneko K 2006 Effect of end plates on the performence of a wells turbine for wave energy conversion *Journal of Thermal Science* **15(4)** pp 319-323.
[12] Torresi M, Camporeale S, Strippoli P and Pascazio G 2008 Accurate numerical simulation of a high solidity Wells turbine *Renewable Energy* **33(4)** pp 735-747.
[13] Setoguchi T, Kim T, Takao M, Thakker A and Raghunathan S 2004 The effect of rotor geometry on the performance of a Wells turbine for wave energy conversion *International Journal of Ambient Energy* **25(3)** pp137-150.
[14] Takao M, Thakker A, Abdulhadi R and Setoguchi T 2006 Effect of blade profile on the performance of a large-scale Wells turbine for wave-energy conversion *International Journal of Sustainable Energy* **25(1)** pp 53-61.
[15] Thakker A, Dhanasekaran T and Ryan J 2005 Experimental studies on effect of guide vane shape on performance of impulse turbine for wave energy conversion *Renewable Energy* **30(15)** pp 2203-2219.
[16] Setoguchi T and Takao M 2001 State of art on self-rectifying air turbines for wave energy conversion *Proceedings of the fourth international conference on mechanical engineering* Dhaka Bangladesh pp 117–126.
[17] Boake C, Whittaker T, Folley M and Ellen H 2002 Overview and initial operational experience of the LIMPT wave energy plant *Proceedings of the 12th international offshore and polar engineering conference* ISOPE Kitakyushu Japan vol 1 pp 586–594.
[18] Taha Z and Sugiyono T 2010 A comparison of computational and experimental results of Wells turbine performance for wave energy conversion *Appl Ocean Res* **32** pp 83–90.
[19] Taha Z, Sugiyono T, Ya T and Swada T 2011 Numerical investigation on the performance of Wells turbine with non-uniform tip clearance for wave energy conversion. *Appl Ocean Res* **33** pp 321–331.
[20] Shaaban S and Abdel HA 2012 Effect of duct geometry on Wells turbine performance *J. Energy Convers Manag.* **61** pp 51–58.
[21] Halder P and Samad A 2015 Casing Treatment of a Wave Energy Extracting Turbine *Aquatic Procedia 4*(Icwrcoe) pp 516–521.

IOP Conf. Series: Materials Science and Engineering **973** (2020) 012006 doi:10.1088/1757-899X/973/1/012006

[22] Thakker A and Abdulhadi R 2008 The performance of Wells turbine under bi-directional airflow *Renewable Energy* **33** pp 2467–2474.
[23] Hamed H, Nawar M and Abd El-Maksoud R 2017 A New Operating Concept to Enhance Wells Turbine Performance *17th International Conference on Aerospace Sciences & Aviation Technology* 11-13 April Proceedings of the 17th Int. ASAT Conference Military Technical College Kobry Elkobbah Cairo Egypt.
[24] Halder P, Samad A, Jinhyuk K, and Young C 2015 High performance ocean energy harvesting turbine design–A new casing treatment scheme *Energy* ***86*** pp 219-231.
[25] Halder P and Samad A 2016 Optimal Wells turbine speeds at different wave conditions *International journal of marine energy* **16** pp133-149.
[26] Halder P, Samad A and Thévenin D 2017 Improved design of a Wells turbine for higher operating range *Renewable energy* **106** pp122-134.
[27] Halder P, Rhee SH and Samad A 2017 Numerical optimization of Wells turbine for wave energy extraction *International journal of naval architecture and ocean engineering* **9(1)** pp 11-24.
[28] Torres FR, Teixeira PR and Didier E 2018 A methodology to determine the optimal size of a wells turbine in an oscillating water column device by using coupled hydro-aerodynamic models *Renewable Energy* **121** pp 9-18.
[29] Nazeryan M and Lakzian E 2018 Detailed entropy generation analysis of a Wells turbine using the variation of the blade thickness *Energy* **143** pp385-405.
[30] Gratton T, Ghisu T, Parks G, Cambuli F and Puddu P 2018 Optimization of blade profiles for the Wells turbine *Ocean Engineering* **169** pp 202-214.
[31] Kumar P and Samad A 2019 Introducing Gurney Flap to Wells Turbine Blade and Performance Analysis with OpenFOAM *Ocean Engineering* **187** 106212

NOMENCLATURE

Abbreviations

LE	Leading edge	TE	Trailing edge
GF	Gurney flap	FH	Flap height
FW	Flap width	PS	Pressure side
SS	Suction side	RANS	Reynolds-averaged Navier-Stokes
OWC	Oscillating water column	SST	Shear stress transport
MRF	Moving Reference Frame	SIMPLEC	Semi-Implicit Method for Pressure-Linked Equations-Consistent

Symbols

C	Chord length (m)	K	Turbulence kinetic energy (m^2/s^2)
H	Blade height (m)	y^+	Non-dimensional wall distance(–)
R_T	Rotor tip radius (m)	F_t	Tangential force (N)
R_h	Rotor hub radius (m)	F_a	Axial force (N)
R_c	Casing radius (m)	T	Torque (Nm)
R_m	Rotor mean radius (m)	Q	Volume flow rate (m^3/s)
T	Blade thickness (m)	U	Rotor tip speed (m/s)
Z	Number of blades	C_x	Axial velocity (m/s)
σ	Turbine solidity	W	Relative velocity (m/s)
C_T	Torque coefficient (–)	Δp_o	Total pressure drop (Pa)
C_P	Total pressure drop coefficient (–)	ω	Angular velocity (rad/s)
$\emptyset$	Flow coefficient (–)	ρ	Air density (kg/m^3)
η	Efficiency (–)	α	Incidence angle (°)
L	Lift force (N)	*	Non-dimensional parameter

Numerical studies of turbulent premixed flame interaction with repeated solid obstacles

M Elshimy[1], S Ibrahim[1] and W Malalasekera[2]
[1]Department of Aeronautical and Automotive Engineering, Loughborough University, Loughborough, LE11 3TU, UK
[2]WolfsonSchool of Mechanical, Electrical and Manufacturing Engineering, Loughborough University, Loughborough, LE11 3TU, UK

m.a.a.a.elshimy@lboro.ac.uk

Abstract. This paper presents numerical simulations of hydrogen and propane turbulent premixed flames interaction with repeated solid obstructions. The laboratory-scale combustion chamber used in this study is equipped with three solid baffles which promote the generation of turbulence and a square obstacle located downstream from the ignition source. The test cases considered have two different area blockage ratios (ABR) of 24% and 50%, respectively. The large eddy simulation (LES) turbulence modelling technique is used. The numerical simulations are carried out using an in-house computational fluid dynamics (CFD) model. Two different flow configurations are examined, both using three consecutive baffles to identify the subsequent effects and the sensitivity of each fuel to increasing the ABR. These effects are studied using the nature of the flame-obstacles interaction, generated combustion overpressure and resultant flame speed. The modelling capability is confirmed by validating the numerical results against published experimental data. Conclusions are drawn that increasing the ABR increases the combustion overpressure, rate of pressure rise and flame speed. It is also concluded that the larger obstacle has a significant effect on the propagating flame structure and that hydrogen flames are more sensitive to an increased ABR and produce a significantly higher peak overpressure.

1. Introduction

The increasing scarcity of fossil fuels combined with a negative environmental impact has forced the search for an alternative fuel as a source of energy. Hydrogen (H_2) has been identified as a potential replacement for hydrocarbons due to its high energy density. Further, the use of hydrocarbons such as LPG (C_3H_8) or CNG (CH_4) as a fuel produces harmful greenhouse gases, unlike hydrogen. This presents a significant opportunity; however, research needs to be carried out to investigate the safety related factors in the production, storage and transport of the fuel [1]. Due to hydrogen's low ignition energy and wide flammability range it can pose a significant safety risk in an event where accidental ignition occurs [2]. The peak combustion overpressure is one of the main concerns in the event of an explosion and previous experiments have found that when compared to alternative hydrocarbons, hydrogen burns at much higher velocities and produces significantly higher peak combustion overpressures [3], [4].

The numerical results are validated against experimental data published by The University of Sydney, Australia [5]. Experiments concluded that by increasing the ABR and reducing the separation distance between successive solid obstructions the peak combustion overpressure and flame speed are increased [5]. This is due to the obstructions increasing the level of turbulence in the chamber which promotes combustion [3], [5].

Computational Fluid Dynamics (CFD) has become increasingly accepted as a substitute to experiments, enabling the numerical modelling of tests which may potentially be dangerous in a safe manner [6]. Large eddy simulations (LES) is used by researchers to study turbulent premixed flames in various applications [7-10]. Direct numerical simulations (DNS) can produce results to a high degree of accuracy, however it typically requires a large computational expense to do so. Reynolds-Averaged Navier-Stokes (RANS) is usually more fitting for large-scale applications. For appropriate results to be produced using LES, a suitable sub-grid-scale (SGS) model needs to be implemented in order to produce accurate results over a range of flow length scales. The present work uses the LES technique with a dynamic flame density model (DFSD)[11] to study the flame structure as it interacts with baffles and a solid obstacle before it propagates through the vented chamber.

Previous research has used the DFSD model to make accurate predictions for turbulent premixed flames propagating through baffles and past a solid obstacle [8], [12]. This research is further developed by investigating turbulent premixed flames for a stoichiometric propane-air mixture and alean hydrogen-air mixture (Φ=0.7). A comparison is made between the mixture interaction with three consecutive baffles and a solid obstruction with an area blockage ratio (ABR) of 24% and 50%.

2. Experimental Setup

The numerical results are validated using experimental data obtained from a lab-scale chamber constructed at The University of Sydney [3-5]. The combustion chamber measures 50 x 50 x 250 mm for a total volume of 0.625 liters. The chamber can be equipped with up to three removable baffle and

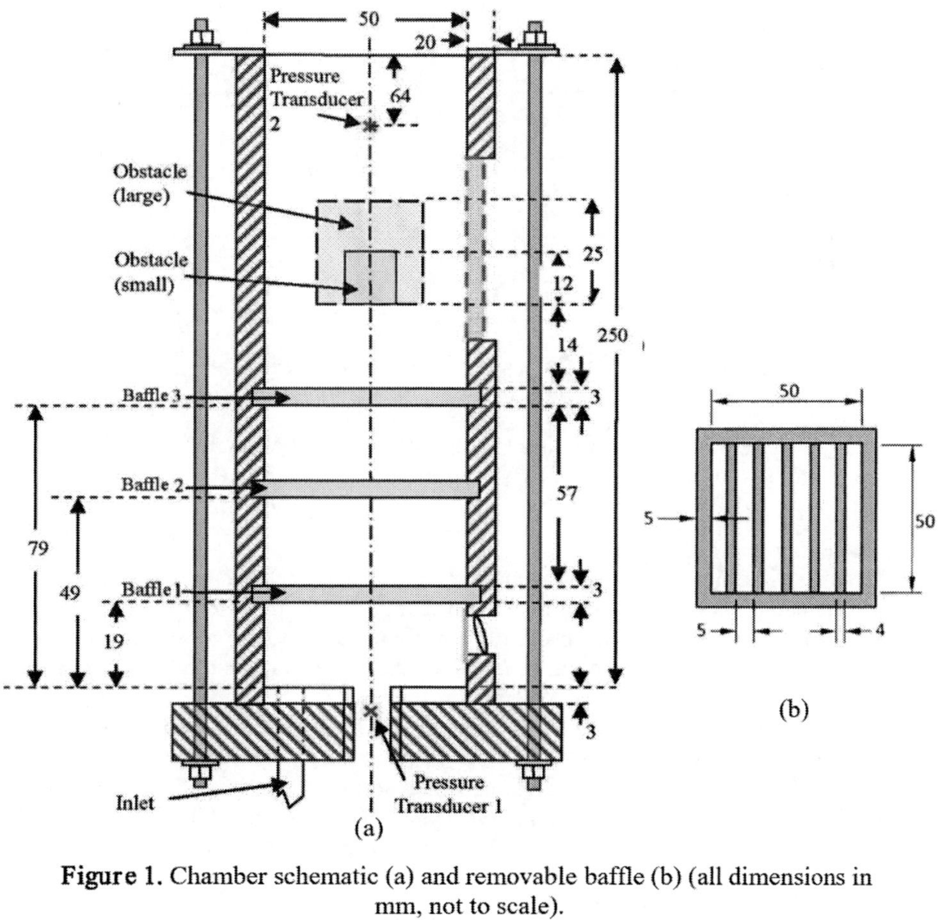

Figure 1. Chamber schematic (a) and removable baffle (b) (all dimensions in mm, not to scale).

a single solid obstacle as shown in figure 1. The solid obstruction is placed 96 mm from the base of the chamber with the option to use a small cubic obstacle with a side length of 12 mm and an ABR of 24% or a large cubic obstacle with a side length of 25 mm and an ABR of 50%. The baffles are located at 19 mm, 49 mm and 79 mm from the chamber base. Each baffle is made of five 4 mm wide and 3 mm thick strips evenly separated by 5 mm gaps. Each of the baffles provides an ABR of 40%.

The fuel-air mixture is introduced into the chamber using a non-return valve at atmospheric pressure and is allowed to settle before ignition occurs. Ignition is provided using an Nd:YA laser which produces an infrared output 2 mm above the base of the chamber, this identifies the start of the experiment. One second before ignition the vent at the top of the chamber is opened and remains open for the length of combustion. Experimental images are obtained using high-speed laser-induced fluoresce from OH at a rate of 5 kHz. Two Keller type PR21-SR piezo-electric pressure transducers attain pressure readings at a rate of 25kHz, with one transducer located in the base of the chamber and another in the wall of the chamber located 64 mm from the top vent.

Two different configurations are going to be studied utilizing three baffles and a small or large obstacle as shown in figure 2. Configuration BBBS uses baffles 1, 2 and 3 with the small square obstacle whereas configuration BBBL uses the three baffles combined with a large square obstacle.

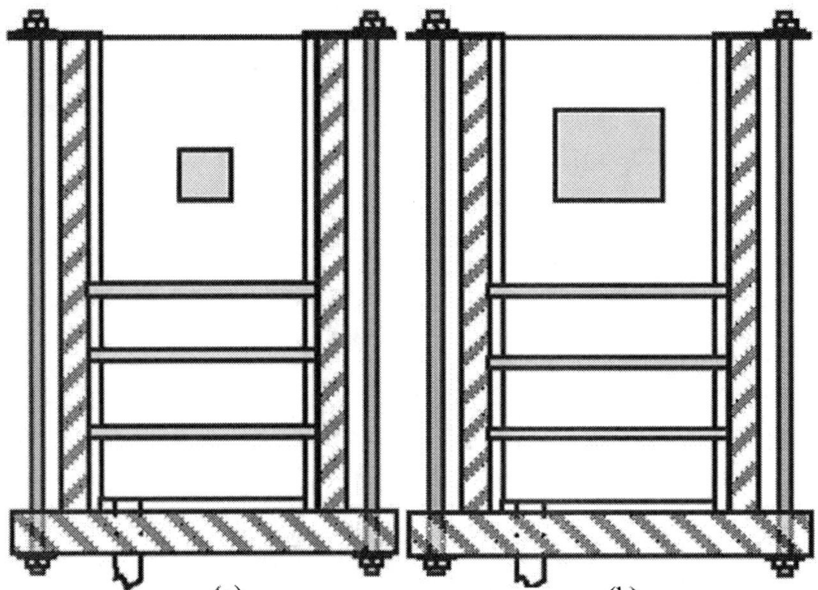

Figure 2. Configuration BBBS with the small obstacle (a) and configuration BBBL with the large obstacle (b).

3. Numerical Setup

Numerical results are obtained using an in-house code known as PUFFIN developed by Kirkpatrick et al. The complete computational domain measures 325 x 325 x500 mm whereas the chamber measures 50 x 50 x 250 mm. The domain uses non-reflecting boundaries to avoid any influence of reflected pressure waves from the extended domain. The computational grid has a total of 2.7 million cells with 90 cells in the x and y directions and 336 cells in the z direction. This grid refinement has been used in previous research and is kept consistent for simulations using the small and large obstacle for both fuels [13]. Grid refinement varies in the z direction with areas of increased refinement where the flame interacts with and propagates past the baffles and the solid obstacle providing a more representative modeling of the flame propagation. The lean hydrogen-air mixture has a laminar burning velocity of 1.25 m/s and a laminar flame thickness of 0.12 mm. The propane-air mixture is composed of 95%

C_3H_8, 4% C_4H_{10} and 1% of other hydrocarbons by volume and has a laminar burning velocity of 0.39 m/s and a laminar flame thickness of 0.37 mm.

The laminar flamelet approach is used to model the mean reaction rate ($\bar{\dot{\omega}}$) using the equation:

$$\bar{\dot{\omega}} = \rho_u u_L \Sigma \quad (1)$$

Where ρ_u is the unburned mixture density, u_L is the laminar burning velocity and Σ is the Flame Surface Density (FSD).The FSD is dynamically modelled using a DFSD model. Further details on the DFSD model can be found elsewhere [11].

4. Results and Discussion

In this section numerical results are presented and discussed for a stoichiometric propane-air mixture (Φ=1.0) and a lean hydrogen air mixture (Φ=0.7). Numerical flame images are compared to LIF-OH images from experiments. Comparisons will be made between numerical results and experimental data for the overpressure-time histories, peak overpressure magnitude timing and magnitude as well as the maximum rate of pressure rise. The data presented in this paper is an average value obtained from 50 experiments [5]. Flame speed-time traces will also be presented and discussed.

4.1. Flame Images

The images shown below convey the flame obstacle interaction as the flame propagates through the baffles and past the larger solid obstacle. Experimental images shown in figure 3 for the propane-air mixture convey the flame as it jets through the final baffle and begins to interact with the large solid obstacle. The large flow generated wake behind the obstacle burns gradually as also shown in the numerical images in figure 4 as the flame continues to propagate through the chamber. Experimental images are not available for the hydrogen-air mixture due to experimental limitations, however numerical images are shown in figure 5. Research investigating these experiments concluded that further advanced equipment would be required to capture hydrogen combustion when using a large obstacle due to the extreme flame speeds [5]. Note the difference in times after ignition between figure 3 and figure 5 due to the hydrogen-air mixture reactivity.

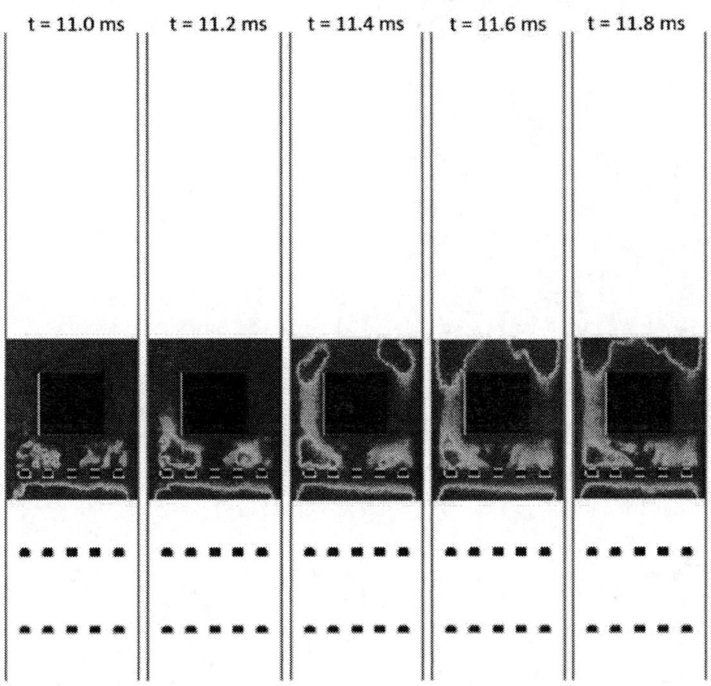

Figure 3. High Speed LIF-OH images of propane-air flame structure at different times after ignition.

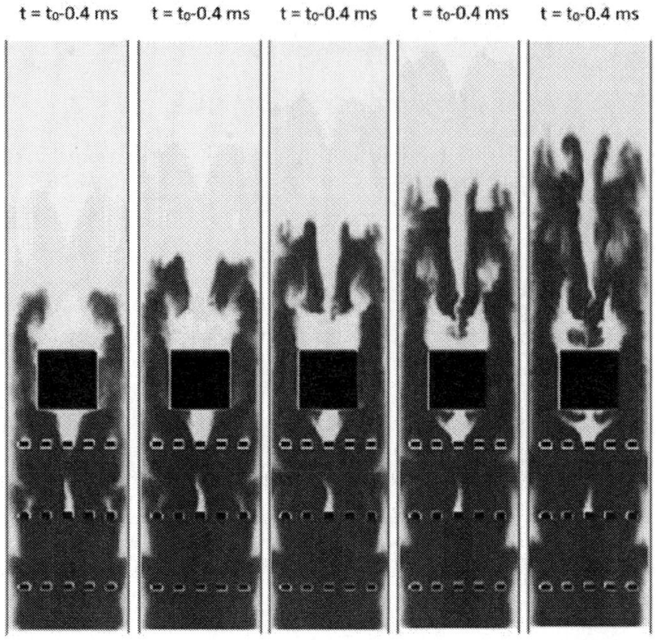

Figure 4. Numerical results showing Iso-lines of the reaction progress variable on the chamber midplane for propane-air flames.

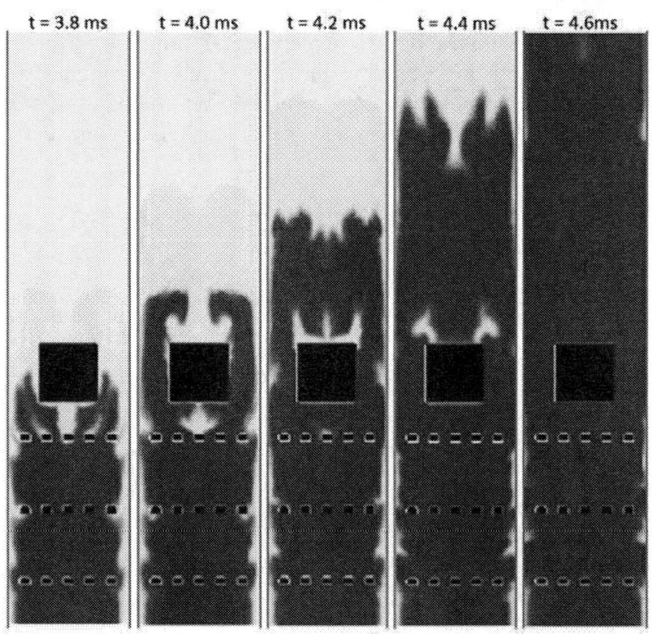

Figure 5. Numerical results showing Iso-lines of the reaction progress variable on the chamber midplane for hydrogen-air flames.

4.2. Overpressure Results

4.2.1. Peak Overpressure Timing. The time at which the maximum overpressure occurs is typically while the flame is propagating past the solid obstacle. This is due to the narrow passages on the sides of the obstacle followed by the flow generated wake behind the obstacle which provides a contribution to the overpressure magnitude when it is consumed. Tables 1 and 2 list the timing for the mixtures to reach the peak overpressure magnitude for the small and large obstacle. The lean hydrogen-air flames produce peak overpressure magnitude in less than half the time due to the mixture reactivity when compared to the stoichiometric propane-air flames.

IOP Conf. Series: Materials Science and Engineering **973** (2020) 012007 doi:10.1088/1757-899X/973/1/012007

Table 1. Comparison of the time peak overpressure magnitude occurs for hydrogen (Φ=0.7) and propane (Φ=1.0) using configuration BBBS.

	Hydrogen (ms)	Propane (ms)
Experimental Result	4.32	10.72
LES Result	4.15	10.52

Table 2. Comparison of the time peak overpressure magnitude occurs for hydrogen (Φ=0.7) and propane (Φ=1.0) using configuration BBBL.

	Hydrogen (ms)	Propane (ms)
Experimental Result	4.28	10.00
LES Result	4.44	10.37

4.2.2. Maximum Rate of Pressure Rise. The maximum rate of pressure rise will vary depending on the separation distance between obstructions, the fuel-air mixture strength and the ABR. Tables 3 and 4 below detail the maximum rate of pressure rise for the mixtures using the small and large solid obstacle. The maximum rate of pressure rise increases for both fuels with an increased ABR from 24% to 50%. Note that the lean hydrogen-air mixture has pressure that increases at a significantly higher rate when compared to the stoichiometric propane-air mixture.

Table 3. Comparison of the maximum rate of pressure rise for hydrogen (Φ=0.7) and propane (Φ=1.0) using configuration BBBS.

	Hydrogen (MPa/s)	Propane (MPa/s)
Experimental Result	320.51	8.53
LES Result	300.46	8.06

Table 4. Comparison of the maximum rate of pressure rise for hydrogen (Φ=0.7) and propane (Φ=1.0) using configuration BBBL.

	Hydrogen (MPa/s)	Propane (MPa/s)
Experimental Result	337.15	10.30
LES Result	349.56	13.85

4.2.3. Overpressure-time Traces. The pressure time histories presented convey various key events in the combustion process including the initial rise of pressure, the peak overpressure and the time at which the flame departs the chamber. When comparing figures 6 (a) and 7 (a) to figures 6 (b) and 7 (b) the hydrogen-air mixture produces a significantly higher magnitude of peak overpressure in a much shorter duration of time. Comparing figure 6 to figure 7 also conveys that increasing the ABR increases the peak overpressure magnitude for both fuels. The stoichiometric propane-air mixture produces a peak overpressure of 99.46 mbar using the smaller ABR which increases to 106.67 mbar. The lean hydrogen-air mixture is more sensitive to the increased ABR as it produces a peak magnitude of overpressure of 789.75 mbar which increases to 931.15 mbar using the larger ABR. The peak overpressure for the hydrogen-air mixture is almost ten-fold the peak propane-air mixture overpressure using an obstacle with an ABR of 50%.

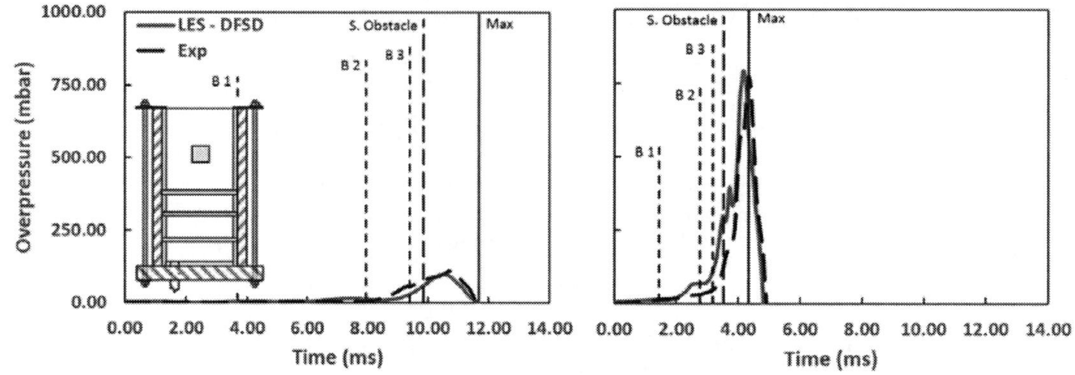

(a) (b)

Figure 6. Overpressure-time traces for propane-air at Φ=1.0 (a) hydrogen-air at Φ=0.7 (b) using configuration BBBS.

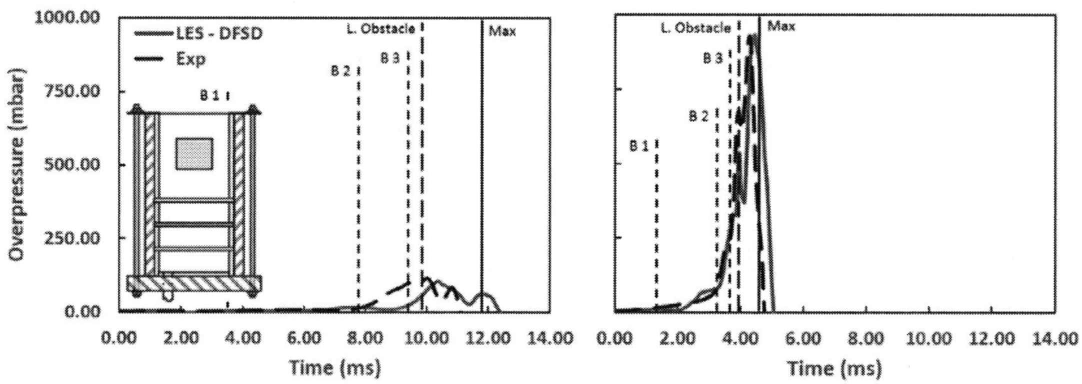

(a) (b)

Figure 7. Overpressure-time traces for propane-air at Φ=1.0 (a) hydrogen-air at Φ=0.7 (b) using configuration BBBL.

4.3. Flame Speed-Time Traces

Given the nature of the overpressure and rate of pressure rise results detailed it is expected that the combustion of the hydrogen-air mixture achieves significantly higher flame speeds in a shorter time period for both area blockage ratios when compared to the propane-air mixture.

figure 8 (a) clearly shows the increased speed with an increased ABR for the propane-air mixture. In figure 8 (b) this impact is less apparent for earlier phases of combustion, however the results presented convey that the increased ABR eventually increases the maximum flame speed as shown. Note that in figure 8 (b) the experimental speed-time data is not available for the configuration with the larger ABR due to the experimental limitations mentioned earlier.

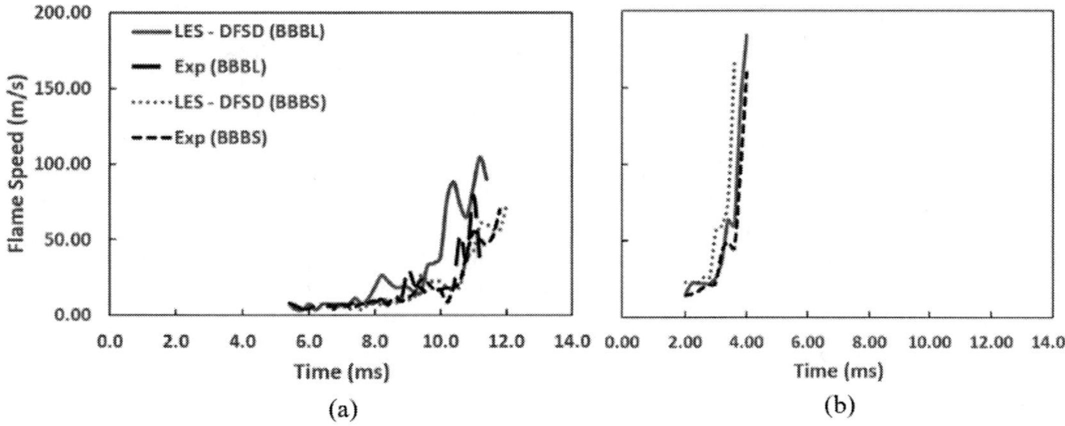

Figure 8. Speed-time traces for propane-air at Φ=1.0 (a) hydrogen-air at Φ=0.7 (b).

5. Conclusions

This paper compares between numerical results obtained for the deflagration of hydrogen and propane flames using an ABR of 24% and 50%. The numerical results are validated against published experimental data and the following conclusions are drawn:

- Increasing the ABR increases the generated peak combustion overpressure and the maximum rate of pressure rise for both fuel-air mixtures.
- The combustion of hydrogen flames produces a significantly higher combustion overpressure at much higher flame speeds.
- Hydrogen flame deflagration takes place in a reduced amount of time due to the high mixture reactivity even at leaner mixture strength.
- Numerical calculations for the overpressure and flame speed produce good agreement for both fuel-air mixtures when compared with experimental data for both area blockage ratios.

References

[1] Marban, G and Valdes-Solis T 2007 Towards the hydrogen economy? *Int. J. Hydrogen Energy* **32** pp 1625-1637
[2] Bjerketvedt D, Roar Bakke J and Wingerden K V 1997 Gas explosion handbook *J. of Hazardous Materials* **52** pp 1-150
[3] Masri A R, AlHarbi A, Meares S and Ibrahim S S 2012 A comparative study of turbulent premixed flames propagating past repeated obstacles *I&EC research* **51**pp 7690-703
[4] AlHarbi A, Masri A R and Ibrahim S S 2014 Turbulent premixed flames of CNG, LPG and H_2 propagating past repeated obstacles *ETF science* **56**pp 2-8
[5] AlHarbi A 2013 Turbulent premixed flames propagating past repeated obstacles Ph.D. Thesis *The University of Sydney*
[6] Versteeg H K and Malalasekea W 2007 *An Introduction to Computational Fluid Dynamics – the Finite Volume Method* Pearson 2nd edition pp 40-113
[7] Pitsch H 2006 Large-eddy simulation of turbulent combustion *Annu. Rev. Fluid Mech.* **38** pp 453-482
[8] Masri A R, Ibrahim S S and Cadwallader B J 2006 Measurements and large eddy simulation of propagating premixed flames *Exp. Threm. Fluid Sci.* **30** pp 687-702
[9] Knikker R, Veyante D and Meneveau C 2004 A dynamic flame surface density model for large

eddy simulation of turbulent premixed combustion*Phys. Fluids* **16** pp L91-L94

[10] Charlette F, Meneveau C and Veyante D 2002 A power-law flame wrinkling model for LES of premixed turbulent combustion part II: dynamic formulation *Combust. Flame***131** pp 181-197

[11] Gubba S R, Ibrahim S S, Malalasekera W and Masri A R 2007 Les modelling of propagating turbulent premixed flames using a dynamic flame surface density model *2nd ECCOMAS Thematic Conf. on Comp. Comb.* (Amsterdam: Delft)

[12] Ruipengyu Li, Malalasekera W and Ibrahim S 2018 Numerical study of vented hydrogen explosions in a small scale obstructed chamber *Int. J. Hydrogen Energy* **43** pp 16667-16683

[13] Gubba S R, Ibrahim S S, Malalasekera W and Mari A R 2011 Measurements and LES calculations of turbulent premixed flame propagation past repeated obstacles *Combust. Flame* **158** 12 pp 2465-2481

IOP Conf. Series: Materials Science and Engineering **973** (2020) 012008 doi:10.1088/1757-899X/973/1/012008

Development of rotary heat exchanger model to investigate performance with integrating controller

A M Ghoneim[1,4], M H Aly[2,4] and O E Mahmoud[3,4]
[1] Demonstrator. [2] Lecturer. [3] Assoc. Prof.
[4]Mech. Power Eng. Dept., Fac. Of Eng. Mataria, Helwan University, Egypt.
ahmedghoneim014@gmail.com

Abstract. The heat wheel Considers one of the best heat recovery devices with very high effectiveness, which leads to great savings in energy and operating expenses. This research study a novel dynamic model of a rotary heat exchanger automatically control the temperature of outlet air despite the operating conditions are changed. The model is implemented under MATLAB/SIMULINK environment to investigate the performance of the heat wheel considering the air properties changing during the operation. The transient effect of air-flow rate, inlet air temperature and wheel rotational speed on the output temperature and effectiveness are discussed. The optimum rotational speed of a heat wheel was investigated for different operating conditions. The temperature distribution of air and matrix for one Full cycle of heat wheel were investigated. The model was validated based on data found in previously published experimental work. Results reveal that the effectiveness of the device is inversely proportional to the amount of flow, while it is directly proportional to the rotational wheel speed. The effectiveness is slightly affected by the inlet air temperature.

1. Introduction

A large part of building energy is consumed in the process of air conditioning to achieve comfortable and healthy conditions. Energy loss in the ventilation process leads to an increase in energy costs needed to maintain the required indoor conditions. To reduce the losses, a device is needed that can transfer energy from the exhaust air to the fresh air [1]. Utilizing heat recovery facilities can improve the system efficiency by recovery the escaped energy. The estimations show that more than 70% of the escaped energy can be recovered [2]. Heat wheels have many applications such as air conditioning, gas turbine system, flue gas desulfurization, and thermal power plant systems [3]. One of the main advantages of the heat wheel is high effectiveness, exceeds 80% [4]. Figure 1 shows a schematic diagram of a heat wheel. While the wheel rotates, energy transfers from the hot airstream to the wheel, then energy transferred to the cold air stream from the stored energy in the wheel as the wheel rotates 180° [5]. Diverse experimental and theoretical investigations have examined the heat wheel to assess its effectiveness. Some investigations studied the heat transfer process within the wheel flow channels by applying the partial differential equation model under simplified assumptions [6,7]. However, due to solutions inaccuracy and complexity, the results were not used by rotary heat exchanger manufacturers and designers. Instead, most of the dependable results are in the form of correlations studied by Kays and London [8]. They proposed a correlation based on the results of a complete study on heat wheels done by Lamberston [9], Bahnke and Howard [10]. Kays and London's correlation gives dependable results for rotary heat exchanger performance to high rotational speeds only. Yilmaz and Buyukalaca provide an analytical solution for the performance of rotary heat exchanger [11,12] which can be useful for any wheel rotational speed. No one of the previous studies has presented the transient air temperature variations of the heat wheels. Also, the effect of integrating of control system target the outlet air temperature has not been studied too.

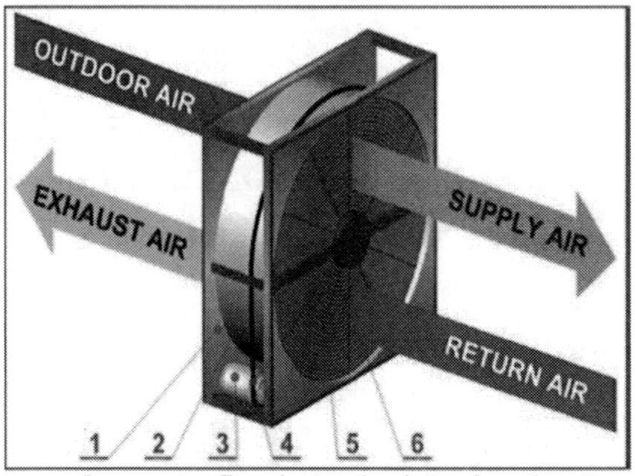

Figure 1. Schematic diagram of a counter flow heat wheel: 1—housing, 2—rotor, 3—electric motor, 4—belt, 5—return airside, 6—supply air side [13].

In this search, the modelling of the heat wheel was performed considering the changing of air properties during the operation. The model validation was done by comparison with previously published experimental work. The transient effect of air-flow rate, inlet air temperature and wheel rotational speed on the output temperature and effectiveness are discussed. Besides, an automatic control system is designed using the rotary heat exchanger model to regulate the outlet air temperature from heat exchanger despite operating conditions are changing.

2. Mathematical model

2.1 Describe the Heat Wheel and Simplifying Model Assumptions

This section briefly studied the mathematical model for the heat wheel dynamics. A rotary heat exchanger can be presented as a frequent flow type heat exchanger. Any channel in the wheel is periodically passed from the hot air to the cold air while the wheel is continuously rotating. The wheel is located between air channels air with counter flow directions. To simplify the problem, the following simplifying assumptions are made:

1. The flow is deemed as one-dimensional flow and is incompressible.
2. The channels that make up the heat wheel are equal with constant heat transfer surface areas.
3. The channels are treated adiabatic and impermeable.
4. The air heat conduction is negligible.
5. The effect of drop pressure along the channel is negligible.
6. The rotary heat exchanger includes no work exchanges so that potential and kinetic energy differences for gas are negligible.
7. The matrix thermodynamic and physical properties are considered as constant.

Based on the above assumptions, the model used in this paper is transient and one-dimensional. One of the channels is split into a finite number of equal discrete elements or channels perpendicular to the flow directions (as shown in figure 2) and each discrete element contains control volume from the air and matrix as shown in figure 2. A rotary heat exchanger can be presented as a finite number of channels along the circle that the wheel traverses during a whole cycle to investigate the performance of the heat wheel through a full cycle as shown in figure 3.

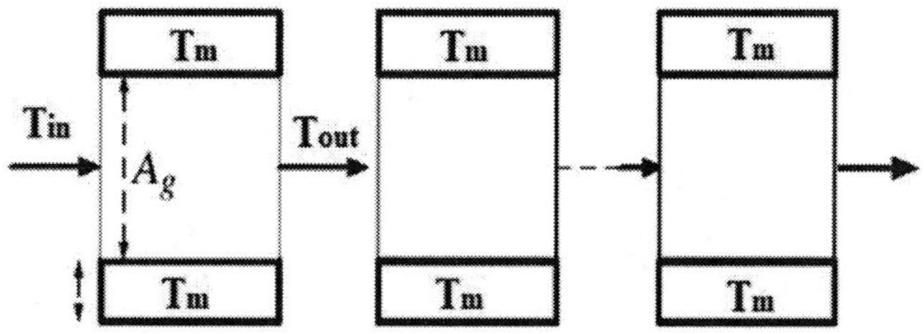

Figure 2. Schematic diagram of a discrete channel.

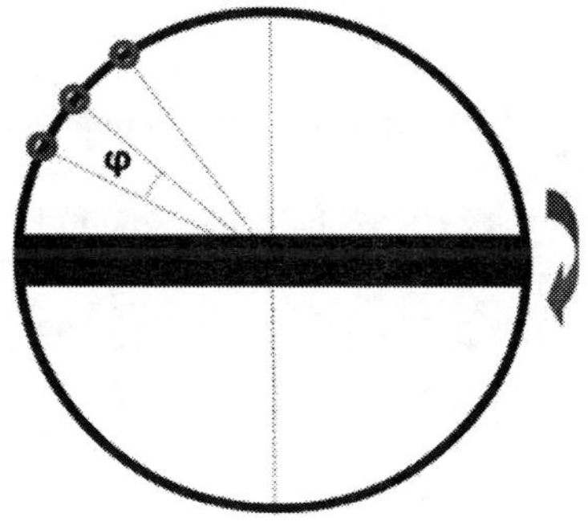

Figure 3. Schematic diagram of a finite number of channels along the cycle around the wheel.

2.2 Governing Equations

The following equations are applied to each discrete element of the channel: Conservation of energy in the air stream, conservation of energy in the solid wall.
Heat transfer equation for the airstream:

$$\rho_g A_g dz Cp_g \frac{dT_g}{dt} = U_g A_g \rho_g Cp_g T_{g_{in}} - U_g A_g \rho_g Cp_g T_{g_{out}} - h_{HT} p dz \left(T_g - T_m\right) \tag{1}$$

$$U_g = \frac{v_f}{\sigma}, \sigma = \frac{A_g}{A_g + A_m}$$

$$\frac{dT_g}{dt} = \frac{U_g}{dz}\left(T_{g_{in}} - T_{g_{out}}\right) - \frac{h_{HT} p}{\rho_g A_g Cp_g}\left(T_g - T_m\right) \tag{2}$$

Heat transfer for wall:

$$\rho_m A_m dz C_m \frac{dT_m}{dt} = \frac{kA_m}{dz}\left(T_{m_p} - T_m\right) - \frac{kA_m}{dz}\left(T_m - T_{m_f}\right) + h_{HT} p dz \left(T_g - T_m\right) \tag{3}$$

$$A_m = p.\frac{s}{n}$$

$$\frac{dT_m}{dt} = \frac{k}{\rho_m C_m dz^2}\left(T_{m_p} + T_{m_f} - 2T_m\right) + \frac{h_{HT} p}{\rho_m A_m C_m}\left(T_g - T_m\right) \tag{4}$$

The Nusselt number is used to calculate the heat transfer coefficient as follows:

$$h_{HT} = \frac{Nu.k_g}{D_h} \tag{5}$$

To consider the changing temperature and air velocity profiles along the channel, the local Nusselt number is calculated through the following correlation [14].

$$Nu = Nu_{FD} + \frac{0.0841}{0.002907 + Gz^{-0.6504}} \tag{6}$$

$$Gz = \frac{L / D_h}{\text{Re.Pr}}, \quad \text{Re} = \frac{\rho_g U_g D_h}{\mu_g}, \quad \text{Pr} = \frac{\mu.Cp_g}{k_g}$$

where Nu_{FD} is the Nusselt number for the fully developed flow and its value depends on the channel geometry. This is presented in Table 1 for typical axial symmetric flow channel geometries [15] and in figure 4 for sinusoidal channels [16].

In order to take the effect of temperature changing during the operation on the air properties, the thermodynamic properties of air are evaluated through the following correlations [17].

$$\rho_g = \frac{351.99}{T_g} + \frac{344.84}{T_g^{\,2}}, \quad k_g = \frac{2.3340*10^{-3} T_g^{\,3/2}}{164.54 + T_g}, \quad \mu_g = \frac{1.4592*10^{-6} T_g^{\,3/2}}{109.10 + T_g}$$

$$Cp_g = 1030.5 - 0.19975 T_g + 3.9734*10^{-4} T_g^{\,2}$$

Table 1. Fully developed Nusselt Numbers for Many Axial Symmetric Channels.

Channel Shape			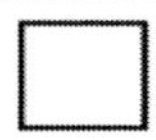	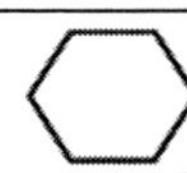	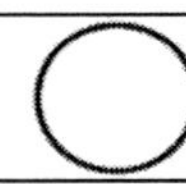
Nu_{FD}	7.54	2.35	2.89	3.44	3.66

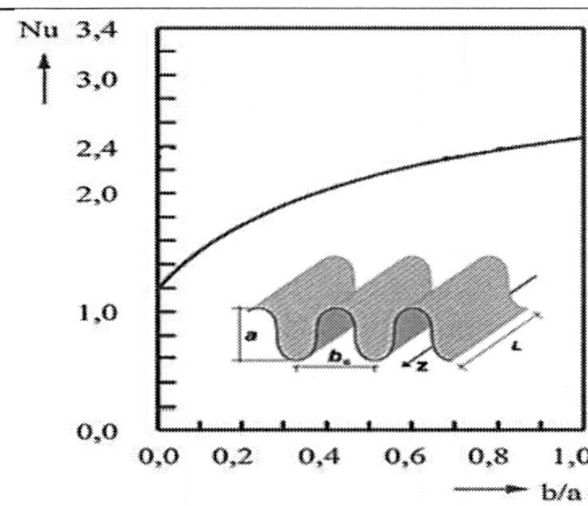

Figure 4. Nusselt Number at Fully Developed Air-flow in Sinusoidal Channels.

Finally, the heat wheel effectiveness should be calculated as the main parameter that was commonly used to evaluate the performance of the heat exchanger. The heat wheel effectiveness is:

$$\varepsilon = \frac{(T_{h,in} - T_{h,out}) + (T_{c,in} - T_{c,out})}{2(T_{h,in} - T_{c,in})} \tag{7}$$

IOP Conf. Series: Materials Science and Engineering **973** (2020) 012008 doi:10.1088/1757-899X/973/1/012008

2.3 *Implementation in Simulink*

Simulink is an environment integrated with MATLAB and is used for simulation of dynamic systems, which provides access to a wide range of tools for numerical computations [18]. Simulink has block libraries with numerous standard operations that allow you to build your own blocks. Blocks are linked through the input and the output ports to form dynamic system models. Simulink provides many solvers that support the developers in getting up the equations for the dynamic systems. Simulink is intuitive and flexible for the programmer by a very flexible graphical interface with respect to the systems of equations that are able to handle them [19].

The model of heat wheel has been built up based on the main framework for each segment shown in figure 2 that can be regarded as two main subsystems (air subsystem and matrix subsystem), that affected by each other. In figure 5, a block diagram shows the simulation of one segment from the air channel of the rotary heat exchanger. The segments are linked through input and output ports to represent an air channel consisting of several segments. Figure 6 displays a sample of a Simulink model for an air channel. For simulating the rotation of the wheel, the input air temperature and flow speed are periodically changed from the warm air to the cold air, so a block diagram has been built up and jointed to the main framework for each segment. Figure 7 shows a Simulink block for a periodically changed from the warm air to the cold air. The change between the hot period and cold period of all air channels is carried out sequentially by a delay equal to the difference in angle between the channels taking into account that the flow direction in the hot period is opposite the flow direction in the cold period. The time each channel passes through the air stream can be counted to estimate the output conditions in the various angels for the wheel.

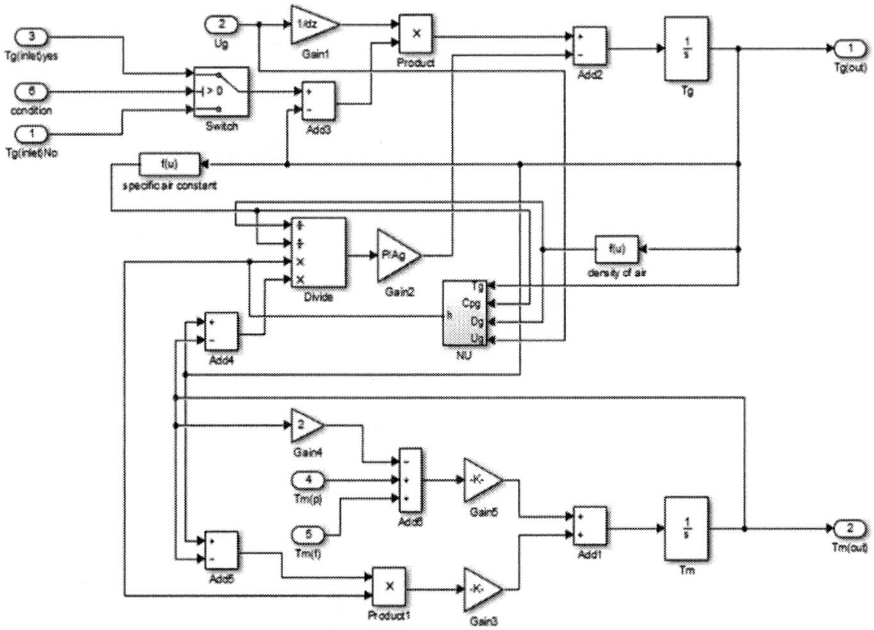

Figure 5. Block diagram for one segment.

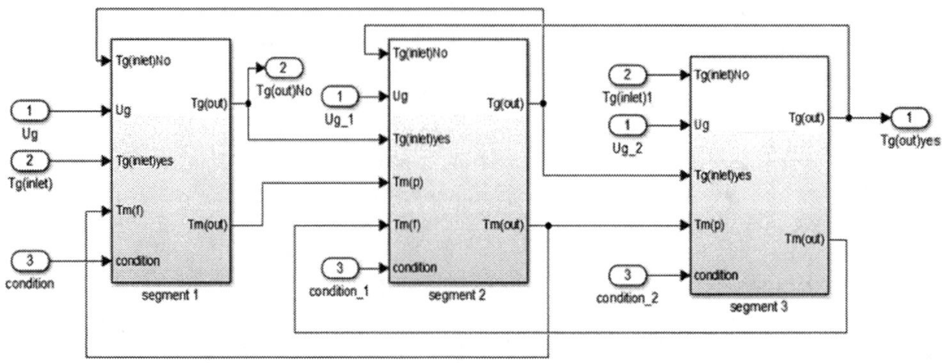

Figure 6. Block diagram of channel model

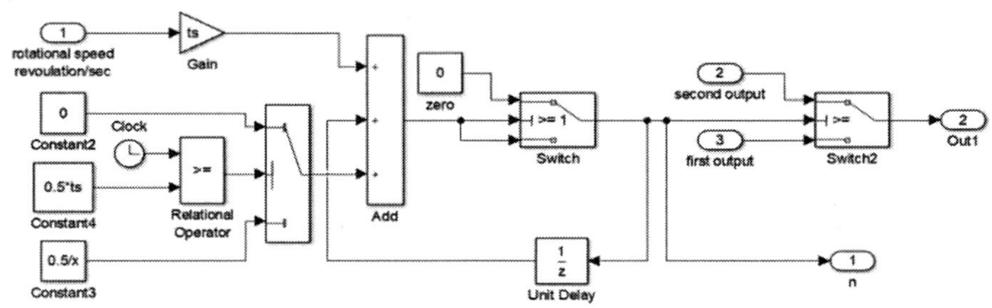

Figure 7. Block diagram for rotation motion.

2.4 Model validation

The numerical simulation has been compared with the published actual values of an experimental work [20-22]. The model has been verified in numerous working conditions with different air-flow rates and inlet air temperature of both air streams and revolution speed. The validation of the model results with the experimental data and a good agreement among model result and experimental data is achieved. In figure 8 the validation of the model results with the experimental actual results. The extreme relative error among the model results and the experimental data is continuously lesser than 5%. The response of the dynamic model based on changes in control inputs has been compared with the published actual values, in figure 9. The simulation results and the experimental actual dates are in good agreement in terms of steady-state values and temperature trends. Consequently, it is probable to say that the mathematical model can fine predict the performance of the heat exchanger.

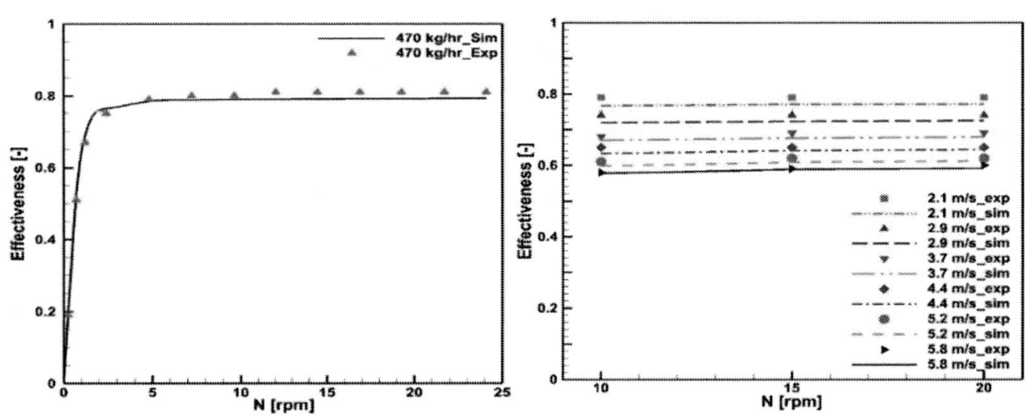

(a) (b)

Figure 8. Validation with experimental data.

IOP Conf. Series: Materials Science and Engineering **973** (2020) 012008 doi:10.1088/1757-899X/973/1/012008

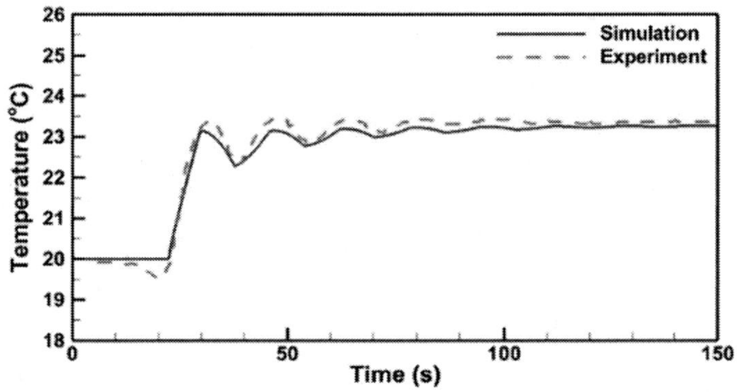

Figure 9. Outlet cold air temperature versus time with a step increase to wheel speed.

3. Results and discussion

The research has been displayed a heat wheel with corrugated sinusoidal ducts as it the most well-known in heat wheels due to having large heat transfer surface area and the simplicity of construction (figure 10).Geometric properties, the material for the matrix with thermophysical characteristics and operating conditions of the modelled are itemised in Table 2.

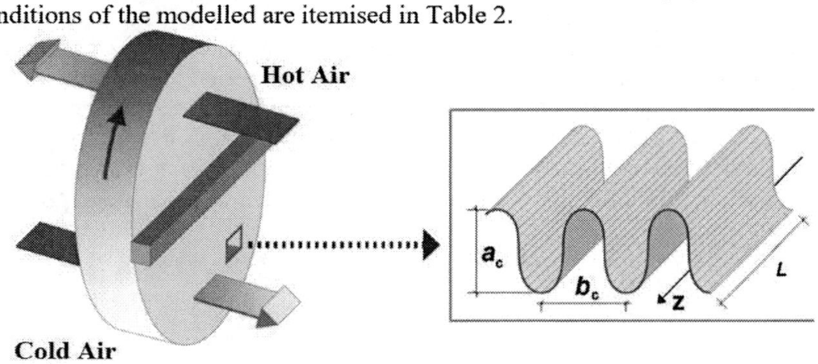

Figure 10. Scheme of heat wheel.

Table 2. The characteristics of the wheel and the heat wheel operating conditions.

Type	Value
Channel height (mm)	2
Channel width (mm)	3.8
Channel hydraulic diameter (mm)	1.6
Porosity	0.91
Rotor diameter (mm)	600
Shaft diameter (mm)	60
Rotor height (mm)	200
Purge Sector	None
Heatwheel material	Aluminium
Density (kg/m^3)	2700
Specific heat capacity (J/kg·K)	900
Thermal conductivity (W/m·K)	220
Matrix angular velocity (rpm)	0-25
Air-flow rate (cfm)	100-1800
Inlet air temperature (°C)	20-100

IOP Conf. Series: Materials Science and Engineering **973** (2020) 012008 doi:10.1088/1757-899X/973/1/012008

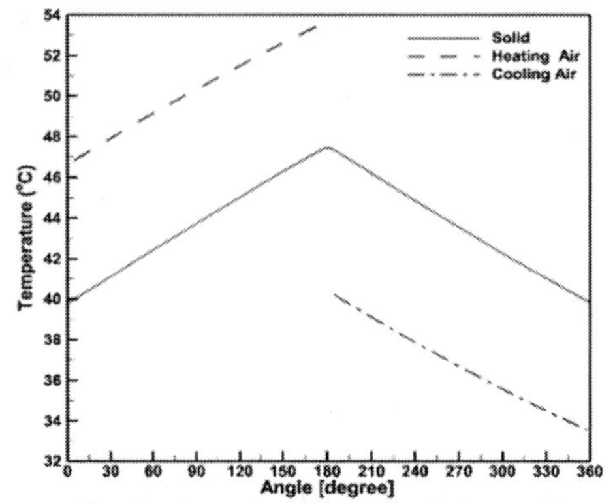

Figure 11. Temperature distributions for the air and the wall at middle of the wheel

Figure 11 displays the variation of the air and matrix temperature in the middle of the wheel for both the cold and hot periods at the steady-state. The variation of the air temperatures is almost linear for both air streams with the rotation angle. As the wheel rotates during the hot period, the air temperature rises. And when the wheel rotates at the cold period the air temperature reduces. The variation of the matrix temperature starts from the lowest wall temperature and is increased to a point where the wheel is shifted to the cold period then the matrix temperature is decreased. During the steady-state operation, the matrix temperature at the finish of the warm period must be equal to the matrix temperature at the start of the cold period. Figure 12 shows the mean temperature for hot air, cold air and matrix versus wheel width at steady state. In the rotary heat exchanger, the warm air-flow is opposite the direction of the cold air-flow therefore that the warm air exit face is the face of the cold fluid entry and vice versa. With this type of flow, the temperature variance among the warm and cold streams is almost constant along with the heat exchanger. The matrix represents the medium of energy transfer between the warm and cold air and therefore it possesses medium temperature among the warm and cold air streams along with the heat exchanger.

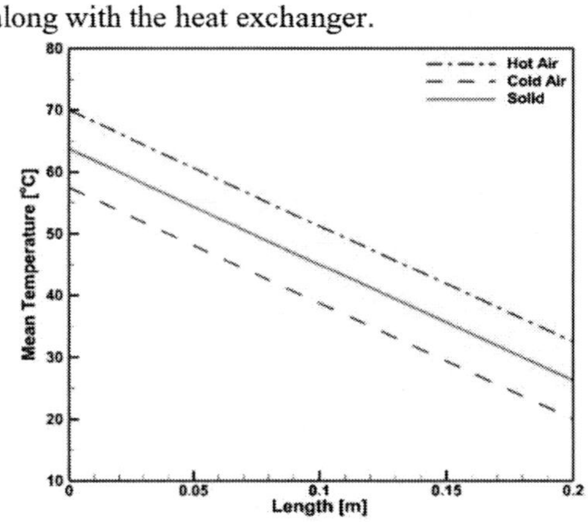

Figure 12. The mean temperature for the air and the wall along the wheel

3.1 The influence of rotational velocity on the Rotary heat exchanger

The wheel rotational velocity is a significant parameter on the heat wheel effectiveness. Figure 13 displays the effect of the rotational speed of the wheel on outlet air temperatures from the wheel and the effectiveness. As seen, the effectiveness of the heat wheel is greatly increased by increasing the wheel rotational speed until a specific value then, no important change was realized in the effectiveness. By rising the rotational velocity, the exit cold air temperature is raised while the exit hot air temperature is reduced. The trend of the exit air temperature is like that of the heat exchanger effectiveness. As displayed, after 6 rpm the change among temperatures is nearly fixed which is like the effectiveness curve. Increasing the rotation speed of the wheel from zero to 6 rpm led to increasing heat exchanger effectiveness from zero to 0.7. Rising the rotational speed due to increase heat transfer area available for energy transfer among hot and cold air which due to the rise in the heat exchanger effectiveness. Instead, it reduces the highest temperature of the wheel when it is in the warm period and rises the lowest temperature of the wheel when it is in the cold period because of the decrease in the time required for heat transfer between the wheel and air which lead to a decrease in the heat exchanger effectiveness. The equilibrium among these two reverse effects limits the movement of heat exchanger effectiveness curve. Three different values for wheel rotational speed are given as 3, 5 and 15 rpm, respectively (All-wheel channels have the same temperature at the beginning of rotation). Hot air temperature variations and cold air temperature differences from startup until the steady-state for various wheel rotational speed are shown in figure 14. Periodical fluctuation variations are shown in the hot and cold air temperatures as the wheel is periodic between the cold air and hot air. Additionally, extreme temperature variations among the wheel and the air have the lowest value in the high-speed case whereas they are higher in the small-speed case. The higher speed indicates that the wheel will have a lesser exposure time to either one of the warm and cold periods. Therefore, the average temperature variance among the matrix and the air remains bigger than those for the cases with lower speeds. Sustaining a high difference in temperature (between the air and the matrix) through alternating between hot air and cold air results in better heat transfer performance. Therefore, the steady-state is gotten quicker. Transient outlet air temperature variations from the beginning of the raising in rotational speed until the steady-state conditions for the rotational speeds are presented in figure 15. At the beginning the wheel isn't moving, the hot air passes in one half of the wheel and the cold air passes in the other half. Thus, one half of the wheel has the hot air temperature and the other half has the cold air temperature. When the wheel rotates the hot half of the wheel moves to the cold air area, which raises the air temperature fast, then the cold half of the wheel returns to the cold air area, which decreases the air temperature. Thus, causing frequency temperature between high value and low value. When the wheel rotates at low speeds, the time that the hot half of the wheel stays in the cold area increases, which reduces its temperature and at the same time the cold half temperature of the wheel increases, which reduces the frequency process between high and low-temperature values for the air.

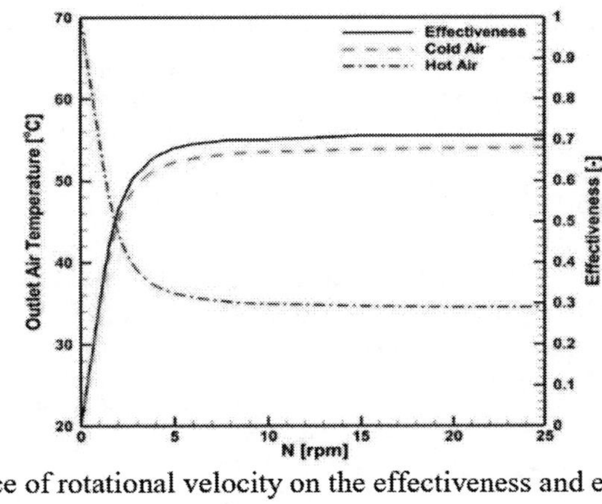

Figure 13.influence of rotational velocity on the effectiveness and exit air temperatures .

IOP Conf. Series: Materials Science and Engineering **973** (2020) 012008 doi:10.1088/1757-899X/973/1/012008

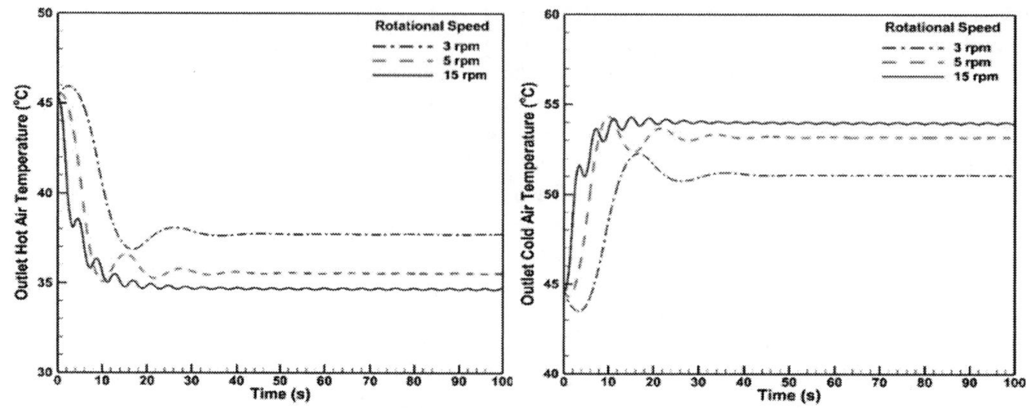

(a) (b)

Figure 14. Influence of rotational velocity on exit air temperatures.

3.2 *The influence of air-flow rate on the heat wheel*

The performance of the heat wheel is influenced by the amount of flow air. The increase in air-flow rate (from 300 to 1800 cfm) led to a decrease in the effectiveness of the heat wheel (from 0.82 to 0.59). The effectiveness of the rotary heat exchanger was decreased due to the inability of the wheel to transfer all the energy that increased as a result of increased flow. Figure 16 (a) shows the influence of the air-flow rate on the effectiveness and exit air temperatures and figure 16 (b) shows the influence of the rotational velocity of the wheel on the effectiveness of heat wheel to three different values for air-flow rate 100, 850 and 1600 cfm, respectively. Therefore, at the lowest flow rate of the air stream (100 cfm), the greatest value of heat exchanger effectiveness happened and with rising the flow rate of the air stream, the heat exchanger effectiveness reduced. This was caused by the longer residence time of the air in the heat wheel for low flow rates but short residence time of the air for high flow rates. the effectiveness of the heat wheel is increased by increasing the wheel rotational speed until specific value and after that, no important change is realized in the effectiveness. But, the increase in the rotational speed led to an increase in the energy consumption, in addition to increasing the leakage rate between the two fluids, the effect of carryover, which leads to the contamination of the supply fluid. It is clear from the above that the less rotation speed leads to reduce operating expenses and reduce the mixing rate between the two fluids but increase the rotation speed leads to increase the efficiency of the heat exchanger until specific value and after that, no important variation is realized in the efficiency and prefer to work at that speed to achieve the highest performance rate with the lowest operating expenses and the lowest leakage rate between the two fluids. The optimal rotational speed is depending on air-flow rates and heat exchanger geometry.

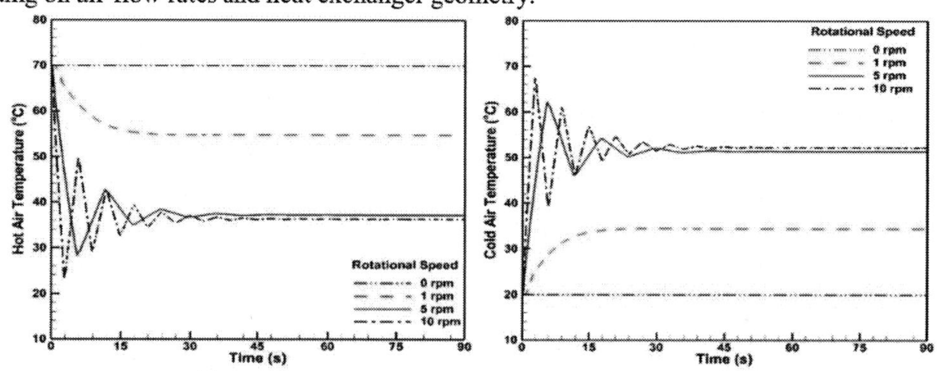

(a) (b)

Figure 15. Outlet Air Temperature Versus Time with a Step Increase to rotational speed.

IOP Conf. Series: Materials Science and Engineering **973** (2020) 012008 doi:10.1088/1757-899X/973/1/012008

As seen in figure 16 (b), the optimal rotational speed is increased (from 1.8 to 7 rpm) by increasing the air-flow rate (from 100 to 1600 cfm).

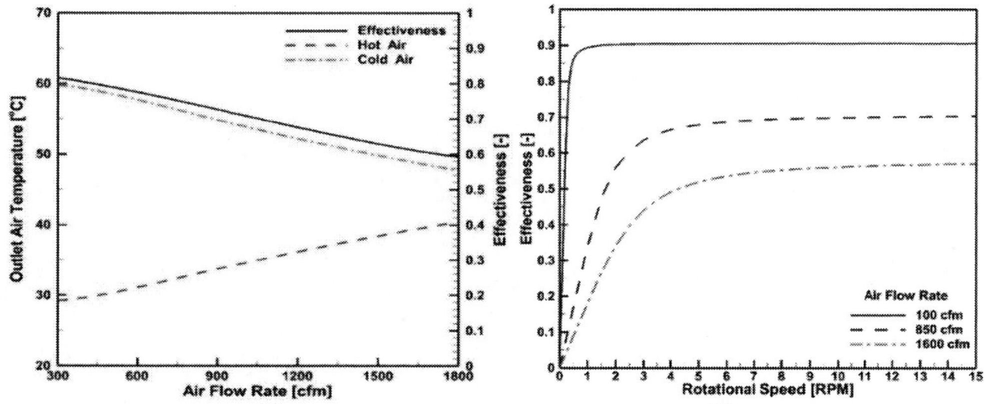

(a) (b)

Figure 16. Influence of air-flow rate on the effectiveness and exit air temperatures.

Transient exit air temperature variations from the start-up of the increase in air-flow rate until the steady-state conditions for various air-flow rates are displayed in figure 17 (a, b) outlet air temperature versus time with a step increase to flow for cold fluid and figure 18 (a, b) outlet air temperature versus time with a step increase to flow for hot fluid). The hot and cold air outlet temperatures observed small periodical variations that match with the rotary heat exchange periodic variations as the heat wheel is periodic among the cold period and hot period. Increasing the flow rate of hot fluid leads to arise in the temperature of the hot air more than the rise of cold air temperature, because when the fluid flow rate increases, the effectiveness of the heat exchanger decreases. Therefore, the heat wheel is less able to transfer heat from hot air to cold air.

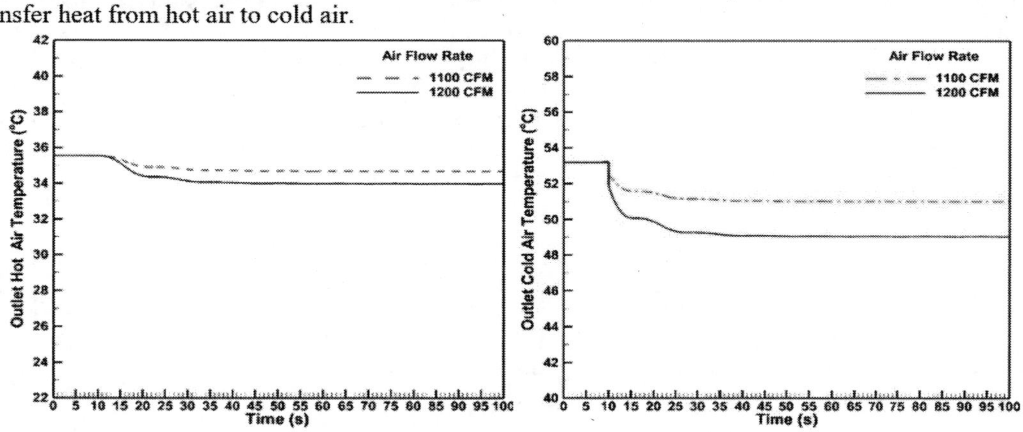

(a) (b)

Figure 17. Outlet air temperature versus time with a step increase to flow for cold fluid.

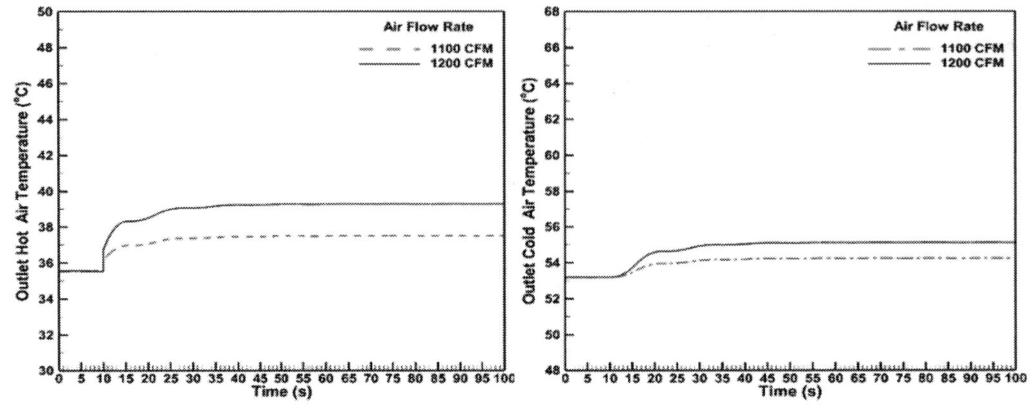

(a) (b)

Figure 18. Outlet air temperature versus time with a step increase to flow for hot fluid.

3.3 The influence of inlet air temperature on the performance of the Rotary heat exchanger

Figure 19 (a) shows the change of exit air temperature and the effectiveness with inlet hot air temperature, the effectiveness of the heat exchanger has a slight increase due to increase inlet air temperature. The small change in the rotary heat exchanger effectiveness causes by the change of the air properties with temperature. The specific heat capacity of the air at 1 bar changes from 1.006 kJ/kg K at 0°C to 1.012 kJ/kg K at 100°C, and the density of the air at 1 bar changes from 1.2942 kg/m3 at 0oC to 0.9464 kg/m3 at 100°C [17]. Figure 19 (b) shows the influence of the rotational velocity of the wheel on the effectiveness of the heat wheel matrix to three different values for inlet hot air temperature 25, 55 and 85°C, respectively. As seen, the optimal rotational speed is slightly decreased by increasing the inlet hot air temperature (the optimal rotational speed is unchanged approximation with air temperature about 6 rpm). The effect of outlet air temperature variation in inlet hot air temperature is displayed in figure 20. There are periodic changes seen in hot and cold air exit temperatures because of the fluctuating among the hot air and cold air. The effectiveness of the heat wheel is around 0.7 and it is approximately uniform for an extremely wide variety of temperatures. Therefore, when the temperature of the inlet hot air rises, most of the energy is transferred to the cold air and the increase of the cold air temperature is higher than the increase of the hot air temperature. The results presented that by rising the inlet hot gas temperature, the temperature of the exit cold air grew 2.2 times more than the temperature rise of the exit hot air. Steady-state conditions are achieved approximately with the similarly time range for all states.

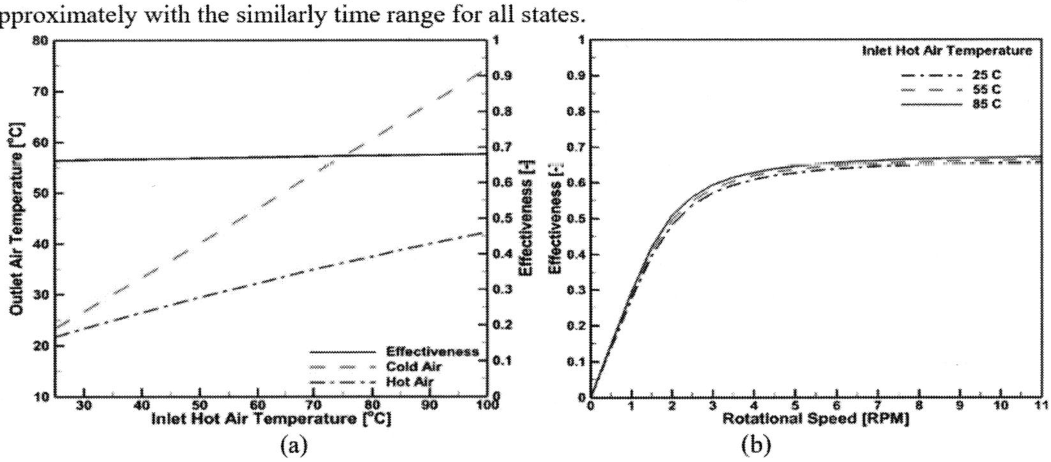

(a) (b)

Figure. 19. Influence of inlet gas temperature on the effectiveness of the rotary heat exchanger.

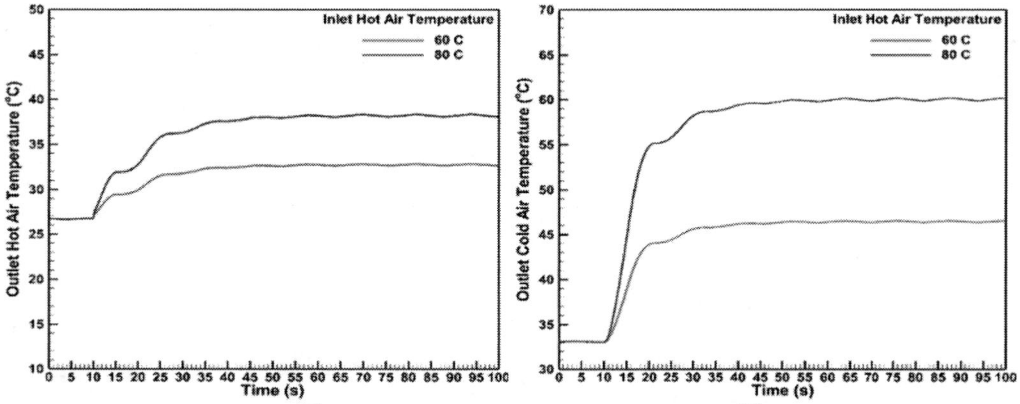

(a) (b)

Figure. 20. Outlet air temperature versus time with a step increase to inlet hot air temperature.

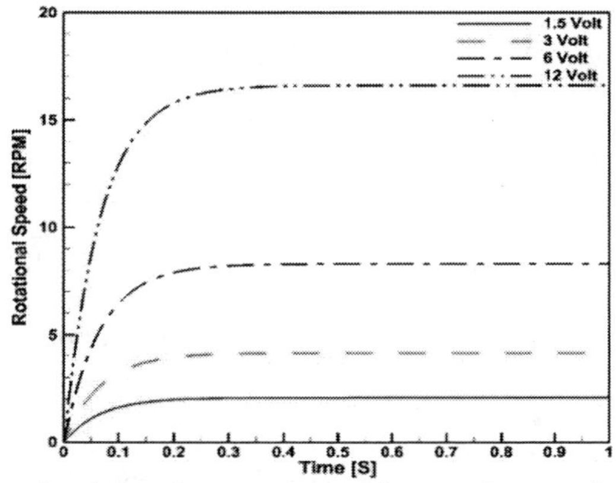

Figure. 21. Wheel rotational speed versus time with a step increase in the inlet voltage supply

3.4 Control outlet air temperature of the heat exchanger system

Controlling the performance of heat exchangers can be used to prevent overheating or overcooling of the supply air during part-load operating conditions of HVAC systems. Numerical investigations are implemented to control the energy transfer rates in the rotary heat exchanger by using wheel speed control. Variable speed DC electric motor was used to drive the heat wheel. The wheel rotational speed variations for different supply voltage (1.5, 3, 6 and 12 V) are shown in figure 21. In the time domain, requirements for a control system are stated in terms of the standard quantities on the settling time, rise time, peak time, overshoot, steady-state error, and overshoot of a step response. The parameters of the PI controller adjusted repeatedly through computer simulations until the closed-loop system behaviour as desired. Setpoint tracking and disturbance rejection of the feedback controller is shown in figure 22. The feedback PI controller shows 24.24% of overshoot and 11.5 seconds of settling time.

IOP Conf. Series: Materials Science and Engineering **973** (2020) 012008 doi:10.1088/1757-899X/973/1/012008

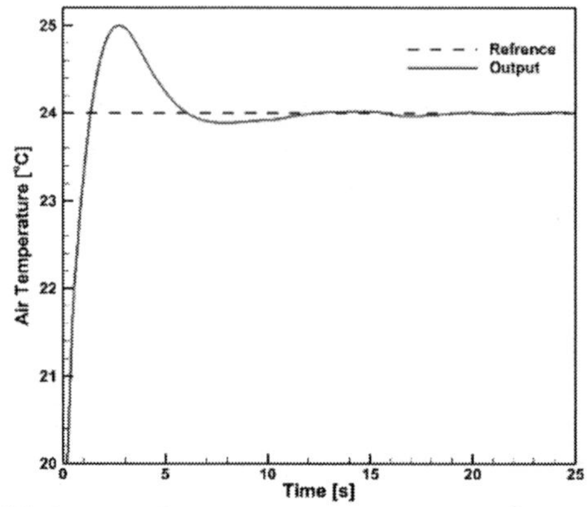

Figure 22. Output air temperature response using PI controller.

4. Conclusion

- The developed heat transfer model could predict the effectiveness of the heat wheel, the exit air temperature from the heat wheel and the optimum rotating velocity of the wheel for certain operating conditions.
- Accurate agreement among simulations and the published experimental data is accomplished within an acceptable margin of error (the greatest error is always smaller than 5%).
- The effectiveness of the rotary heat exchanger is greatly increased by growing the wheel rotational speed until a specific value and after that, no big change was achieved in the effectiveness. Growing the rotation speed of the wheel from zero to 6 rpm led to increasing heat exchanger effectiveness from zero to 0.7.The response of the outlet air temperature is like that of the heat exchanger effectiveness.
- Periodical fluctuation variations are shown in the hot and cold air temperatures while the wheel is a rotation between the cold airstream and hot airstream. Also, extreme temperature variations between the wheel matrix and the air have the lowest amount in the high-speed state as they are greater in the low-speed state.
- The increase in air-flow rate (from 300 to 1800 cfm) led to a decrease in the effectiveness of the heat wheel (from 0.82 to 0.59).
- The optimal rotational speed is depending on air-flow rates and heat exchanger geometry. The optimal rotational speed is increased (from 1.8 to 7 rpm) by increasing the air-flow rate (from 100 to 1600 cfm).
- The effectiveness of the heat wheel is very nearly constant for a very largevariety of temperatures.
- The optimal rotational speed is unchanged approximation with air temperature.
- The search proved that by rising the inlet hot air temperature, the temperature of the outlet cold air rise 2.2 times more than the exit hot air temperature for balanced air-flow.
- Controlling the performance of heat exchangers can be used to prevent overheating or overcooling of the supply air during part-load operating conditions of HVAC systems.
- Integrating the RHE with PI controller to achieve the required supply air temperature by modifying the rotating speed. The feedback PI controller shows 24.24% of overshoot and only 11.5 seconds to settling.

Nomenclature

A	Cross-section area for any channel, m^2
B	Friction coefficient, Nm s/rad
C_m	Specific heat constant of the matrix, J/kg K
C_p	Specific heat constant, J/kg K

IOP Conf. Series: Materials Science and Engineering **973** (2020) 012008 doi:10.1088/1757-899X/973/1/012008

D_h	Hydraulic channel diameter, m
Dz	Length of one part, m
G_z	Greatz number
h_{HT}	Convection coefficient, W/m^2 K
i_a	Armature current, A
J	Moment of inertia, kg m^2
K	Conduction coefficient, W/mK
K_e	back emf constant, V rad/s
K_i	Motor torque constant, Nm/A
L	The thickness of the wheel, m
L_m	armature inductance, mH
Nu	Nusselt number
P	channel perimeter, m
Pr	Prandtl number
R	armature resistance, Ω
Re	Reynolds number
T	Temperature of air, K
t	Time, s
U_g	The speed of air within the channel, m/s
V	Voltage, V
v_f	face air velocity, m/s
Greek symbols	
μ	Viscosity of air, $N.s/m^2$
ε	Effectiveness
ρ	Density of air, kg/m^3
Σ	wheel porosity
T	Torque, Nm
Φ	Angle between Discretization channel, rad
Ω	Angular velocity, rad/s
Subscripts	
A	armature
B	back
C	Cold
F	forward segment
FD	Fully developed
G	gas
H	hot
In	Inlet
L	Load
M	Motor
M	Channel Material
Out	Outlet
P	previous segment
Abbreviation	
RHE	Rotary heat exchanger
PI	A proportional–integral controller

References

[1] Seo J w, Lee D Y and Dong-Seon KimD S 2018 A Simple Effectiveness Model for Heat Wheels *International Journal of Heat and Mass Transfer* **120** pp 1358–64.

[2] Liddament M W 1996 *A Guide to Energy Efficient Ventilation- Air Infiltration and Ventilation Centre* (AIVC, Guide GV)

[3] Özdemir K and Serincan M F 2018 A Computational Fluid Dynamics Model of a Rotary Regenerative Heat Exchanger in a Flue Gas Desulfurization System *Applied Thermal Engineering* **143** pp 988–1002.
[4] Mardiana-Idayu A and Riffat S B 2012 Review on Heat Recovery Technologies for Building Applications *Renewable and Sustainable Energy Reviews* **16** (2) pp 1241–55.
[5] Fathieh F et al 2015 Determination of Air-to-Air Heat Wheel Sensible Effectiveness Using Temperature Step Change Data International *Journal of Heat and Mass Transfer* **87** pp 312–26.
[6] Romie F E 1988 Transient Response of Rotary Regenerators *Journal of Heat Transfer* **110** (4a)
[7] Baclic B S 1985 The Application of the Galerkin Method to the Solution of the Symmetric and Balanced Counterflow Regenerator Problem *Journal of Heat Transfer* **107** (1).
[8] Kays A L and London W M 1984 *Compact Heat Exchanger* (New York McGraw-Hill Inc).
[9] Lambertson T J 1957 *Performance Factors of a Periodic-Flow Heat Exchanger* (Monterey California: Naval Postgraduate School)
[10] Bahnke G D and C P Howard. 1964 The Effect of Longitudinal Heat Conduction on Periodic-Flow Heat Exchanger Performance *Journal of Engineering for Power* **86** (2).
[11] Büyükalaca O and Yılmaz T 2002 Influence of Rotational Speed on Effectiveness of Rotary-Type Heat Exchanger *Heat and Mass Transfer* **38** (4–5): 441–47.
[12] Yilmaz T and Büyükalaca O 2003 Design of Regenerative Heat Exchangers *Heat Transfer Engineering* **24** (4) pp 32–38.
[13] Paulina K, Jedlikowski A and Anisimov S 2019 The Influence of Geometrical Parameters on Heat and Mass Transfer Processes in Rotary Heat Exchangers *SN Applied Sciences* **1** (6): 526.
[14] Niu, J L and Zhang LZ 2002 Heat Transfer and Friction Coefficients in Corrugated Ducts Confined by Sinusoidal and Arc Curves *International Journal of Heat and Mass Transfer* **45** (3) pp 571–78.
[15] Bejan A 1984 *Convective Heat Transfer* (New York: John Wiley & Sons).
[16] Shah R K and London A L 1978 *Laminar Flow Forced Convection in Ducts Advances in Heat Transfer* (New York: Academic Press).
[17] McQuillan F G, Culham J R and Yovanovich M M1984 *Properties of Dry Air at One Atmosphere* (Waterloo Ontario: University of Waterloo Microelectronics Heat Transfer Lab)
[18] *Simulink 6* (United States. The MathWorks, Inc. n.d. 3 Apple Hill Drive, Natick, MA 01760-2098)
[19] Nielsen T R, Rose J and Kragh J 2009 Dynamic Model of Counter Flow Air to Air Heat Exchanger for Comfort Ventilation with Condensation and Frost Formation *Applied Thermal Engineering* **29** (2–3) pp 462–68.
[20] De Antonellis S, Intini M, Joppolo C and Leone C 2014 Design Optimization of Heat Wheels for Energy Recovery in HVAC Systems *Energies* **7** (11) pp 7348–67.
[21] Rabah A A, Fekete A and Kabelac S 2009 Experimental Investigation on a Rotary Regenerator Operating at Low Temperatures *Journal of Thermal Science and Engineering Applications* **1** (4): 041004.
[22] Zhuang Wu, Roderick VN Melnik and Finn Borup 2006 Model-Based Analysis and Simulation of Regenerative Heat Wheel *Energy and Buildings* **38** (5) pp 502–14.

IOP Conf. Series: Materials Science and Engineering **973** (2020) 012009 doi:10.1088/1757-899X/973/1/012009

Handling performance of an 8x8 combat vehicle

M Ahmed[1], M El-Gindy[1] and H Lang[1]
[1]Automotive, Mechanical and Manufacturing Engineering, Faculty of Engineering and Applied Science, 2000 Simcoe St. N, Oshawa, Ontario, Canada, L1H 7K4

Moataz.ahmed@uoit.net.

Abstract. In this research paper, the handling stability of an 8x8 combat vehicle will be assessed using two different control systems. The first technique utilizes a Torque Vectoring Controller (TVC) to control the vehicle yaw rate to meet the desired value. The second technique utilizes an Active Rear-axles Steering (ARS) to minimize the vehicle sideslip. TVC will be designed as a Single Input Single Output (SISO) control problem using a Sliding Mode Control (SMC) technique, while an Optimal Linear Quadratic Regulator (LQR) will be utilized to develop the ARS controller. The two controllers will be evaluated against a conventional vehicle with fixed rear axles. TruckSim software is used in corporation with Matlab/Simulink to implement and assess the controllers using Double Lane Change (DLC) over high and low coefficient of friction road surfaces at various speeds. The results give an insight into the driving conditions at which each controller is utilized and introduce a novel method to coordinate the integration between both controllers for integrated chassis applications.

1. Introduction

Ground vehicle handling stability had great attention during the last decades to stabilize the vehicle during a sudden maneuvers and cornering which can lead to rollover or loss of control. In general, improving the vehicle lateral stability is applied through two methods. The first is Direct Yaw Control (DYC) which can be designed by applying a corrective yaw moment to the vehicle to meet a desired signal. In addition, DYC can be implemented either by redistribute traction forces in case of Torque Vectoring Control (TVC) or braking forces in case of Differential Braking (DB)[1]. The second method to control the vehicle handling is Active Steering Control (ASC). This method aims to control the yaw motion by generating additional lateral forces at the front tires in case of Active Front Steering (AFS) [2]or at the rear tires in Active Rear Steering (ARS) as in [3]or by integrating both[4].

Despite of the effectiveness of the DYC and ASC systems, DYC systems are limited to the traction forces between the tires and the road surface, while the ASC systems are limited by the saturation of the generated lateral forces of the steered tires. Therefore, many researchers attempted to combine both controllers [5, 6]. This combination technique allows the vehicle to reach the maximum permissible stability.The combination of the controllers is known as Integrated Chassis Controller (ICC) which is developed using two approaches: top-down or bottom-up. In top-down approach all actuators are considered in one controller while in bottom-up method each controller is designed separately, and a low lever controller is used as a coordinator.

Several researches integrate ASC and DYC using top-down method such as in [7-9] where Linear Parametric Variables (LPV)-H_∞ was utilized. Also Predictive Controller (MPC) was used widely in the integration as in [10, 11]. Furthermore, optimization based control methods were introduced as in[12, 13]. Despite of the advantages of this method, it is not preferred in automotive research as adding or modify individual controller is not supported and required redesigning the whole controller. In contrast, bottom-up approach support this without affecting other controllers. Therefore, this method is recently used in research as in [14-17] , which require a coordinator between the controllers and often gave a preference to a controller over another. However, the criteria for giving priority to one controller with respect to another was not considered in previous studies and need more justification. Besides, still there is no scheme to follow to add prospective controllers.

Multi-wheel vehicles offer several advantages as it has higher stability and fault tolerant capabilities than two-axle vehicles. Despite its enormous advantages, most of previous work only considered studying the stability of two axles vehicle and fewer showed concerns to study multi axles vehicle as in [18-20], which was the motivation of applying this study on an eight Wheel Drive (8WD)-four axles combat vehicle.

In this paper, the influence of ARS and TVC on lateral stability nd handling performance of a Multi-wheeled combat vehicle is investigated. Virtual testing of the controllers over low and high coefficient of friction surfaces at low and high speed is performed. In addition, the paper aims to give an insight into the advantages and disadvantages of each controller and define a performance map that can illustrate a region of priority of each controller. In addition, this performance map can be used as a novel and general scheme to integrate both controllers and future controller.

The paper is arranged as follows: At first, an ARS system is designed introducing a new configuration by utilizing the active and independent steering of the 3rd and 4th axles to minimize the vehicle sideslip. The control signal consists of two terms: feed forward and feedback terms. The feed forward term is designed based on the steady-state mathematical model of the vehicle and relates the steering of the 4th and 3rd axles to Akerman steering geometry. In addition, an LQR controller is developed to generate the feedback term, where the 3rd and 4th axles are steered independently. Second, TVC is developed using SMC to establish a corrective moment to control the vehicle yaw rate and match a reference yaw signal. After that, the controllers' performance is evaluated and compared with the uncontrolled vehicle using DLC maneuver and a performance map is introduced proposing a new integration technique. Finally, the conclusion comes at the end to summarize the paper and state the advantages and disadvantages of each controller.

2. Vehicle Bicycle Model

A two degrees of freedom bicycle vehicle model of the 4-axle combat vehicle is shown in figure 1. This model is used to derive the mathematical equations needed in the design of the active rear-axle steering (ARS) controller using LQR and the Zero Sideslip (ZSS) controller. Furthermore, the model is used to derive the reference model for the torque vectoring controller (TVC). The model is linearized and simplified by making the following assumptions:

- The model presents the vehicle in the yaw plane.
- Constant longitudinal velocity.
- Small slip and steering angles.
- Linear tire characteristics.

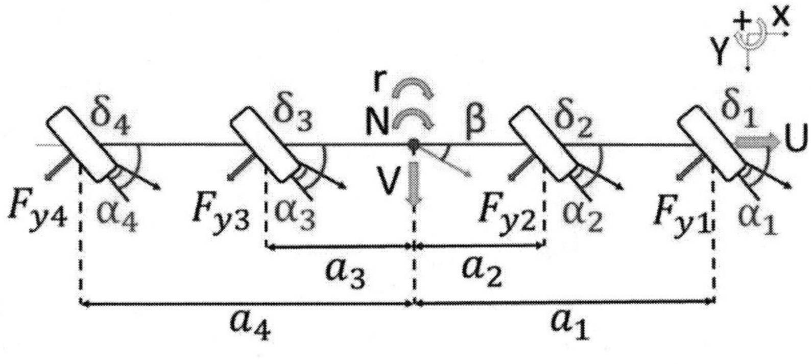

Figure 1.8x8 vehicle bicycle model.

The following equations present the vehicle mathematical model under the assumptions mentioned above, where M is the vehicle mass, U and V are the longitudinal and lateral velocities respectively, $\dot{V}$ lateral speed rate of change, r and $\dot{r}$are the vehicle yaw rate and yaw acceleration respectively, I_{zz} is the vehicle mass moment of inertia about the Z-axis,a_i is the distance from each axle i to the vehicle Center of Gravity (CG), where the axles ahead the vehicle CG have a positive sign, while the axles behind the CG takes a negative sign. N is an external yaw moment, F_{yi} is the cornering forces for each axle (combined right and left tire), $C_{\alpha i}$ is the combined cornering stiffness for each axleα_i the average slip angle, δ_i is the average steering angle, β is the vehicle sideslip, k_{21} is the ratio between the 1st and 2nd average steering angles, and k_{34} is the ratio between the 3rd and 4th average steering angles.

Lateral motion:

$$M(\dot{V} + Ur) = \sum_{i=1}^{4} F_{yi} \cos \delta_i \quad (1)$$

Vehicle sideslip:

$$\beta = \frac{V}{U} \quad (2)$$

Slip angles:

$$\alpha_i = \delta_i - tan^{-1}\left(\beta + \frac{a_i r}{U}\right) \quad (3)$$

Lateral forces under linear tire characteristics:

$$F_{yi} = C_{\alpha i}\alpha_i \quad (4)$$

Yaw motion:

$$I_{zz}\dot{r} = \sum_{i=1}^{4} a_i F_{yi} \cos \delta_i + N \quad (5)$$

Substituting equations 2,3 and 4 into equation 1 and applying small angles approximation, the general vehicle yaw model in matrix form can be expressed in the state space form as:

$$\dot{x} = Ax + Bu + w \quad (6)$$

$$y = Cx + Du \quad (7)$$

where x and $\dot{x}$ are the states and the states rate of changes vectors, A is the states matrix, u and B are the input vector and the input matrix, respectively.C is the output matrix;D is the feedback matrix, which equals to zero, and w is the disturbance. Eventually, the state-space model of the system is:

$$\begin{bmatrix}\dot{\beta}\\ \dot{r}\end{bmatrix}=\begin{bmatrix}\frac{-\sum_{i=1}^{4}C_{\alpha i}}{MU} & \frac{-\sum_{i=1}^{4}a_iC_{\alpha i}}{MU^2}-1\\ \frac{-\sum_{i=1}^{4}a_iC_{\alpha i}}{I_{zz}} & \frac{-\sum_{i=1}^{4}a_i^2C_{\alpha i}}{I_{zz}U}\end{bmatrix}\begin{bmatrix}\beta\\ r\end{bmatrix}+\begin{bmatrix}\frac{C_{\alpha 1}}{MU} & \frac{C_{\alpha 2}}{MU} & \frac{C_{\alpha 3}}{MU} & \frac{C_{\alpha 4}}{MU} & \frac{0}{MU}\\ \frac{a_1C_{\alpha 1}}{I_{zz}} & \frac{a_2C_{\alpha 2}}{I_{zz}} & \frac{a_3C_{\alpha 3}}{I_{zz}} & \frac{a_4C_{\alpha 4}}{I_{zz}} & \frac{1}{I_{zz}}\end{bmatrix}\begin{bmatrix}\delta_1\\ \delta_2\\ \delta_3\\ \delta_4\\ N\end{bmatrix}+\begin{bmatrix}0\\ 0\end{bmatrix} \tag{8}$$

$$\begin{bmatrix}\beta\\ r\end{bmatrix}=\begin{bmatrix}1 & 0\\ 0 & 1\end{bmatrix}\begin{bmatrix}\beta\\ r\end{bmatrix}+\begin{bmatrix}0 & 0 & 0 & 0 & 0\\ 0 & 0 & 0 & 0 & 0\end{bmatrix}\begin{bmatrix}\delta_1\\ \delta_2\\ \delta_3\\ \delta_4\\ N\end{bmatrix} \tag{9}$$

3. Controllers Design

This section will be dedicated to designing ARS and TVC control systems. ARS will be designed based on optimal LQR methodology and a feed forward controller, so-called ZSS controller, while TVC will be designed based on SMC. The reference of each controller will be mentioned in the design procedure as apart from it.

3.1. Active rear steering control system (ARS)

As mentioned in the literature, ARS control has the benefits of generating excessive lateral forces without the increase of the vehicle sideslip which gives better tracking performance for the designed trajectory. Therefore, the reference and the main controller objective is to minimize vehicle sideslip.

3.1.1. LQR Control design. LQR is Multi input – Multi output (MIMO) controller based on solving a quadratic optimization problem under some constraints on the system states and the control inputs. The algorithm aims to solve the Riccati equation [21]as

$$A^TP+PA-PBR^{-1}B^TP+Q=0 \tag{10}$$

where$Q$ is the weight of the state, R is the input weight, and P is a positive definite matrix. As the control optimal control input u_{LQR} equal to the following:

$$u_{LQR}=-R^{-1}(B^TP)x \tag{11}$$

where x is the states.

Before applying the LQR on the general model, it should be manipulated to express the desired ARS model. By substituting the external yaw rate N by zero, separating the vehicle front steering angle as disturbance, and relating the average steering angle of the 2nd axles to the 1st one by the ratio k_{21}, where:

$$\delta_2=k_{21}\delta_1 \tag{12}$$

The ARS state-space model will be as follows:

$$\begin{bmatrix}\dot{\beta}\\ \dot{r}\end{bmatrix}=\begin{bmatrix}\frac{-\sum_{i=1}^{4}C_{\alpha i}}{MU} & \frac{-\sum_{i=1}^{4}a_iC_{\alpha i}}{MU^2}-1\\ \frac{-\sum_{i=1}^{4}a_iC_{\alpha i}}{I_{zz}} & \frac{-\sum_{i=1}^{4}a_i^2C_{\alpha i}}{I_{zz}U}\end{bmatrix}\begin{bmatrix}\beta\\ r\end{bmatrix}+\begin{bmatrix}\frac{C_{\alpha 3}}{MU} & \frac{C_{\alpha 4}}{MU}\\ \frac{a_3C_{\alpha 3}}{I_{zz}} & \frac{a_4C_{\alpha 4}}{I_{zz}}\end{bmatrix}\begin{bmatrix}\delta_3\\ \delta_4\end{bmatrix}+\delta_1\begin{bmatrix}\frac{C_{\alpha 1}+k_{21}C_{\alpha 2}}{MU}\\ \frac{a_1C_{\alpha 1}+a_2k_{21}C_{\alpha 2}}{I_{zz}}\end{bmatrix} \tag{13}$$

$$\begin{bmatrix}\beta\\ r\end{bmatrix}=\begin{bmatrix}1 & 0\\ 0 & 1\end{bmatrix}\begin{bmatrix}\beta\\ r\end{bmatrix}+\begin{bmatrix}0 & 0\\ 0 & 0\end{bmatrix}\begin{bmatrix}\delta_3\\ \delta_4\end{bmatrix} \tag{14}$$

The states and input weight matrices Q and R are tuned based on Bryson's rule [22-27], where

$$Q=\begin{bmatrix}\frac{1}{x_{1\,max.}^2} & 0\\ 0 & \frac{1}{x_{2\,max.}^2}\end{bmatrix}\text{ and }R=\rho\begin{bmatrix}\frac{1}{u_{1\,max.}^2} & 0\\ 0 & \frac{1}{u_{2\,max.}^2}\end{bmatrix} \tag{15}$$

where ρ is a tuning positive scaler, x_{1max} and x_{2max} are the maximum allowed error of the states (β and r), which can be determined as follow [28-30]:

$$\beta_{max} = tan^{-1}(0.02\mu g) \text{ and } r_{max} = \frac{0.75\,\mu g}{U} \quad 16)$$

And u_{1max} and u_{2max} are the average permissible steering angles for the 3rd and 4th axles, respectively.

3.1.2. ZSS controller. This controller is a feed forward controller. The control signal is determined by calculating the ratio between the 4th and 1st axle steering angles as a function of the vehicle speed ranging from 0 to 120 km/h, while the vehicle sideslip is set to be zero during steady-state motion (let $\dot{\beta} = \dot{r} = 0$). In addition, the ratio between the steering angles of the 4th and 3[rd]axles, $k_{34,}$ is set based on Ackerman.

The steady-state equation of the vehicle can be written as follows:

$$\begin{bmatrix}\beta\\ r\end{bmatrix}_{ss} = -\begin{bmatrix}\frac{-\sum_{i=1}^{4}C_{\alpha i}}{MU} & \frac{-\sum_{i=1}^{4}a_iC_{\alpha i}}{MU^2}-1\\ \frac{-\sum_{i=1}^{4}a_iC_{\alpha i}}{I_{zz}} & \frac{-\sum_{i=1}^{4}a_i{}^2C_{\alpha i}}{I_{zz}U}\end{bmatrix}^{-1}\begin{bmatrix}\frac{C_{\alpha 1}+k_{21}C_{\alpha 2}}{MU} & \frac{C_{\alpha 4}+k_{21}C_{\alpha 3}}{MU}\\ \frac{a_1C_{\alpha 1}+a_2k_{21}C_{\alpha 2}}{I_{zz}} & \frac{a_4C_{\alpha 4}+a_3k_{34}C_{\alpha 3}}{I_{zz}}\end{bmatrix}\begin{bmatrix}\delta_1\\ \delta_4\end{bmatrix} \quad (17)$$

3.2. *Torque vectoring control system*

This section includes two subsections. The first section will discuss the SMC design and in the subsequent section, the reference model for the controller will be developed.

3.2.1. Sliding mode control (SMC) design. For the TVC system, a single input single output sliding mode controller is designed to ensure robust tracking performance to the reference yaw signal. As the objective of this controller is removing the error between the vehicle yaw rate and the reference one, the sliding surface is chosen as follows:

$$S = r - r_d \quad (18)$$

Substituting equation 5 in 18:

$$\dot{S} = e = \dot{r} - \dot{r_d} = \sum_{i=1}^{4}\frac{a_iF_{yi}}{I_{zz}}\cos\delta_i + \frac{N}{I_{zz}} - \dot{r_d} \quad (19)$$

To satisfy the reaching condition, the Lyapunov function is chosen as:

$$S\dot{S} \leq -K_{smc}|S| \text{where } K_{smc} > 0 \quad (20)$$

By using the law of constant reaching rate [31]:

$$\dot{S} = -K_{smc}sgn(s) \quad (21)$$

The chattering phenomena, which appears by using the previous discontinuous function can be lessened by using Quasi-Sliding mode utilizing the relay function as follows:

$$sgn(S) \approx \frac{S}{|S|+\varepsilon} \qquad \text{Where } \varepsilon \textit{ is very small and} > 0 \quad (22)$$

Finally, the controller, which is considered as the input corrective yaw moment to the vehicle *N* can be expressed as:

$$u_{SMC} = N = -\sum_{i=1}^{4}a_iF_{yi}\cos\delta_i + I_{zz}\dot{r_d} - I_{zz}K_{smc}\frac{S}{|S|+\varepsilon} \quad (23)$$

The control virtual stabilizing moment produced by the SMC is allocated using equation (24), where b is the vehicle trackwidth, and F_{xr}, F_{xl} are the summation of the driving forces on the right and left wheels respectively.

$$N = \frac{b}{2}(F_{xr} - F_{xl}) \qquad (24)$$

The driving forces have been optimized using quadratic cost function based on the constraints of the tire friction circle as in Mokhiamar et al. [32]with the additional assumption of negligible pitching motion due to vehicle acceleration and deceleration. However, the optimization procedure will not be mentioned as it is out of the scope of this paper. Furthermore, it is assumed that all the variables in the controller equations can be measured or determined. The determination of this date can be done by considering that the lateral forces of the tires can be determined by having the tire corner stiffness data as a function of the vehicle load and tire slip angle, and by assuming the ability to calculate the tire slip angle by measuring the vehicle sideslip and wheels steering angles by commercial available sensors.

3.2.2. SMC reference model. The steady-state model has been derived in Equation 18 and separating the steady-state yaw rate equation and substituting in δ_4 by zero as the rear two axles are fixed, the reference signal $r_d$is determined. However, the reference yaw rate signal has to be modified to match the vehicle response. This could be done by multiplying the signal by a Low Pass Filter (LPF) with a time constant τ such that:

$$r_{ref} = \frac{r_d}{\tau s+1} \qquad (25)$$

Noted that the reference yaw rate maximum value should be limited by the maximum desired yaw rate considered by [30] as in Equation 16.

4. Results and Discussion

The handling performance of each controller is evaluated and compared using modified Nato-DLC maneuver, where a curb is added in the beginning of the second lane to make the test more difficult. The test is performed at speed range from 30 to 80 km/h, where the surfaces COF is varied from 0.2 to 0.85. All cases will be been assessed in terms of trajectory stabilization time, yaw rate, lateral acceleration and velocity stability, and minimum lateral deviation from the designed trajectory.

A nonlinear 22 degrees of freedom 8x8 combat vehicle-model is used to evaluate the designed controllers, figure 2. The model was developed and validated in [33, 34] and used with TruckSim in previous researchers [20, 28]. Moreover, the simulation tests were performed using a TruckSim/Simulink environment and the controllers were implemented in Matlab/Simulink. Only, the results of the extreme cases at COF of 0.85 and 0.2 at vehicle speed of 80km/h will be introduced, while the results of all tests will be introduced at the end of this section.

Figure 2.8x8 vehicle configuration and the corresponding simulation model [35]

A DLC maneuver has been conducted on a dry surface (μ=0.85) at vehicle speed of 80 km/h. The simulation results are introduced in figure 3 and figure 4. Figure 3(a) shows the vehicle trajectory, which indicates that all cases completed the maneuver successively. However, the lateral deviation from the desired path is illustrated at figure 3(b) where the minimum error is produced by the conventional vehicle followed by ARS, while TVC performs maximum error where the Rote Mean Square Error (RMSE) for all cases are 0.186, 0.216, and 0.259, respectively. The corresponding average front steering angle is indicated in figure 3(c). The minimum steering angle is produced by TVC, which gives the driver more controllability than other cases.

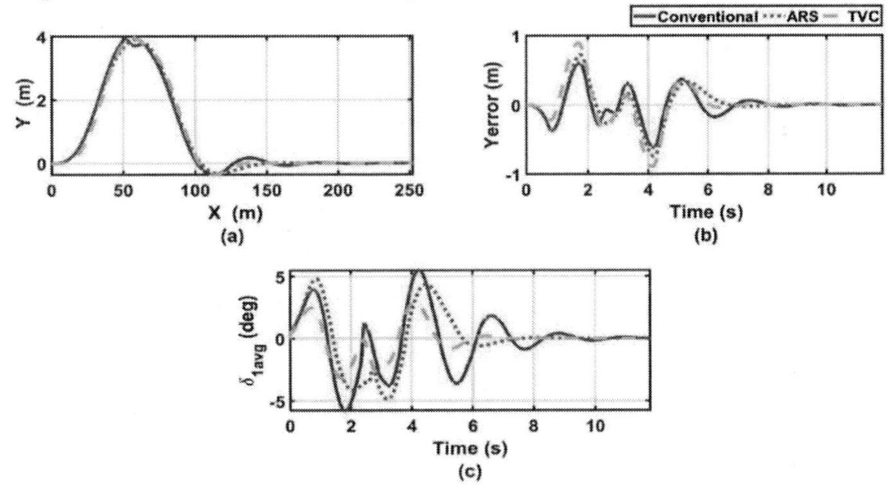

Figure 3. DLC maneuver at speed of 80km/h and μ= 0.85 a) Vehicle trajectory b) lateral error c) the corresponding average steering angle of the front axle.

Figure 4(a) shows the longitudinal velocity variation with time, where ARS produces the least variation, while TVC produces the most. The vehicle lateral acceleration and yaw rate response can be noticed in figure 4(b) and (c) respectively. The controlled vehicle is stabilized before the conventional one, where the TVC stabilizes the vehicle before the ARS by meantime 0.36 second. The vehicle side slip response is illustrated in figure 4(d), in which ARS produces the minimum vehicle slip andstabilize faster than TVC by 0.2 sec. Despite the preference of TVC in safety control systems on dry surfaces, it is recommended to utilize ARS in the case of autonomous applications due to its low lateral error, while it still gives an acceptable behavior in comparison with TVC.

Simulation results for repeating the test on a slippery surface with a coefficient of friction of 0.2 at speed of 80 km/h can be seen in figure 5 and figure 6. The vehicle trajectory is presented in figure 5(a), which shows that the conventional vehicle has been failed to complete the maneuver, while the controlled one did. However, TVC takes about 18.6 seconds to stabilize the vehicle trajectory, which is more than ARS by 9 seconds. Besides that, the RMSE for the lateral deviation from the desired trajectory in the case of ARS is 0.492, while in the case of TVC it reaches 2.037, which can be concluded from figure 5(b). figure 5(c) shows the corresponding average front steering angle, where ARS produces the minimum steering angle with maximum driver controllability. On the contrary, the conventional vehicle shows a saturation in the steering angle, which leads to loss of controllability and stability as well.

figure 6(a) shows the variation in the longitudinal velocity, where the least variation resulting from ARS. By observing the lateral acceleration and yaw rate response from figure 5(b) and (c), it can be noticed that TVC stabilizes the vehicle in about 20 seconds, which is almost double the time ARS takes to stabilize the vehicle. In figure 5(d), the vehicle side slip response with time indicates the same behavior as the yaw rate and lateral acceleration response. Furthermore, ARS produces the least side slip angle. It is concluded that the TVC behavior is dramatically deteriorating at very low COF, while

ARS shows higher reliability on slippery surfaces. Table 1and table 2summarize the vehicle response (settling time and lateral error) under each controller for DLC on a dry road and slippery road surfaces respectively.

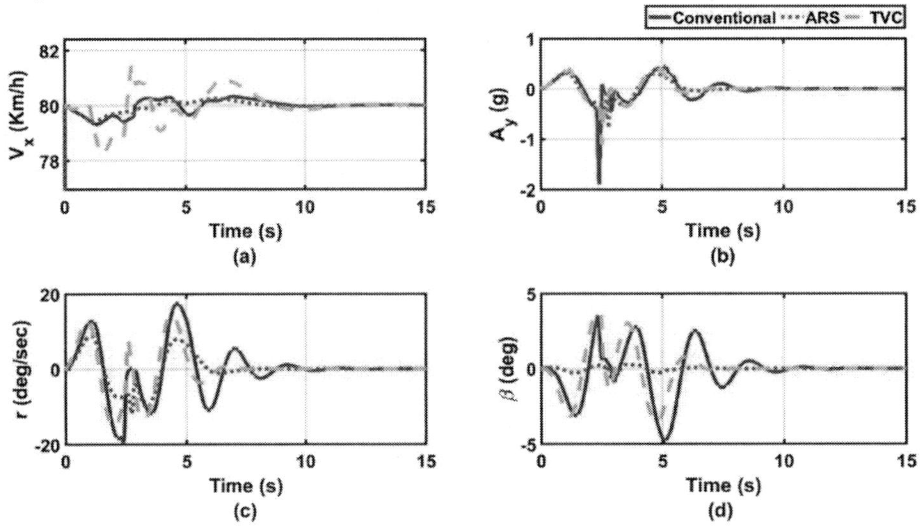

Figure 4. DLC maneuver at speed of 80km/h and $\mu = 0.85$ a) vehicle longitudinal velocity b) lateral acceleration response c) yaw rate response d) vehicle sideslip.

Table 1. Lateral RMSE and Response for DLC at $\mu = 0.85$ and speed of 80km/h.

Criteria	Conventional			ARS			TVC		
Lateral RMSE (m)	0.0542	0.0862	0.1862	0.0552	0.1065	0.2155	0.166	0.2015	0.259
Trajectory Settling time (sec.)	13.344	9.532	8.85	13.295	9.184	7.753	12.782	8.641	6.883
Lateral acceleration Settling time (sec.)	13.507	9.797	8.723	13.657	9.757	7.275	13.451	9.157	6.597
Yaw rate Settling time (sec.)	14.03	9.602	10.53	13.831	9.715	7.407	13.283	8.769	7.164
Side slip Settling time (sec.)	15.358	10.001	10.099	15.121	9.376	7.376	14.954	9.266	7.547

IOP Conf. Series: Materials Science and Engineering **973** (2020) 012009 doi:10.1088/1757-899X/973/1/012009

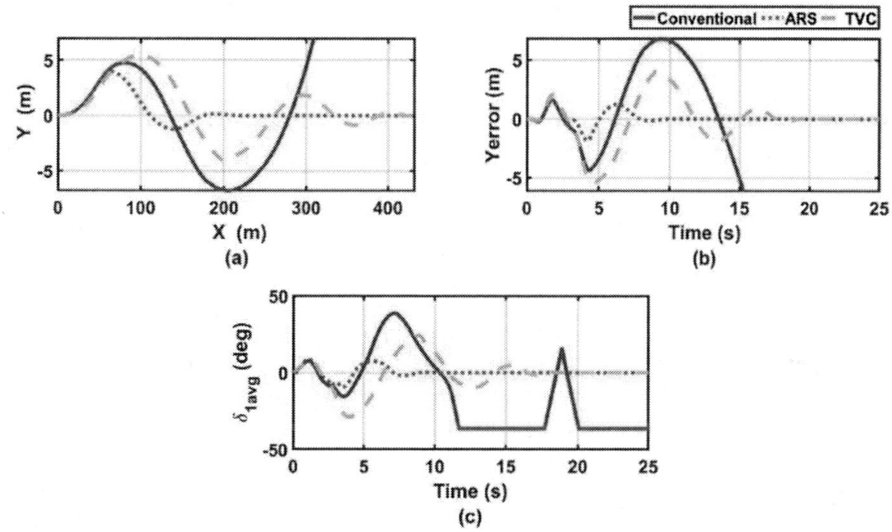

Figure 5. DLC maneuver at speed of 80km/h and $\mu = 0.2$ a)Vehicle trajectory b) lateral error c) the corresponding average steering angle of the front axle.

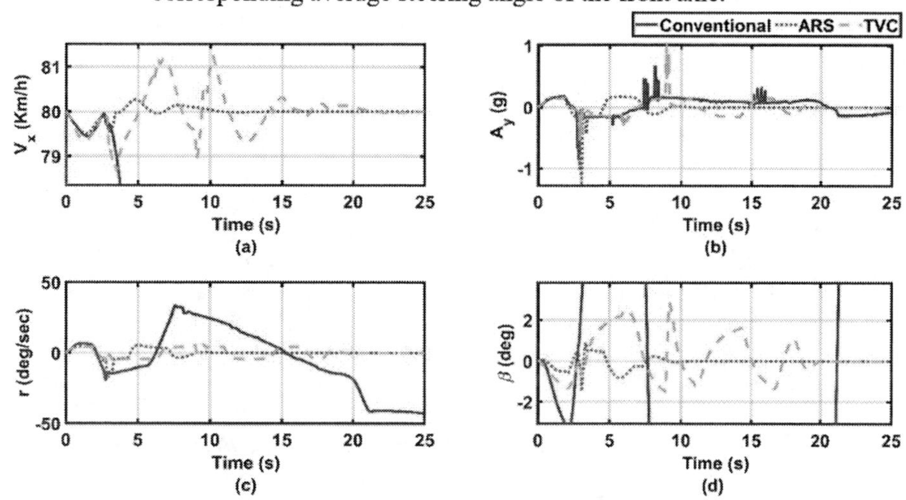

Figure 6. DLC maneuver at speed of 80km/h and $\mu = 0.2$ a) vehicle longitudinal velocity b) lateral acceleration response c) yaw rate response d) vehicle sideslip

Table 2. Lateral RMSE and Response for DLC of $\mu = 0.2$ and speed of 80km/h

Criteria	Conventional			ARS			TVC		
Lateral RMSE (m)	0.038	3.18		0.0452	0.284	0.494	0.1686	1.341	2.037
Trajectory Settling time (sec.)	14.10	19.81		13.134	11.003	9.474	13.679	17.401	18.635
Lateral acceleration Settling time (sec.)	13.97	18.99	Fail	13.512	10.541	8.956	14.078	18.229	19.545
Yaw rate Settling time (sec.)	14.14	19.82		13.63	10.579	10.108	13.699	18.858	20.262
Side slip Settling time (sec.)	15.16	19.78		15.098	10.032	9.829	14.953	19.294	20.37

IOP Conf. Series: Materials Science and Engineering **973** (2020) 012009 doi:10.1088/1757-899X/973/1/012009

To give a better insight into the advantages and preferences of each controller, the DLC maneuver is conducted at different surfaces with different COF at various speeds. The vehicle longitudinal velocity was varied from 30 to 80 km/h, while the COF changed from 0.2 to 0.85. The mean value of the time (settling time) for stabilizing the vehicle trajectory, lateral acceleration, yaw rate, and side slip is calculated. In figure 7(a) and (b), the inverse of the settling time, which denoted as Z-axis, plotted against the vehicle speed and road COF for ARS and TVC respectively, where the higher point on Z-axis indicates better stability and less settling time. It can be seen that at low COF and higher speed ARS gives much better performance than TVC, while at high COF and low-speed TVC is preferred over ARS. This preference decreases with the increase of speed at a high coefficient of friction road surface, where the difference is negligible.

Combining the two surfaces as in figure 8(a) and project the Line of Intersection (LOI) of the two surfaces on the X-Y plane (COF-Speed) can separate it into two areas in which one of the two controllers is favored. The difference between the settling time of each controller is calculated and merged with the LOI of the two controllers in figure 8(b). The driving condition that each controller has supremacy is illustrated, where TVC is preferred in the region above the LOI, while ARS is preferred below it. However, the difference between the settling time shows that there is a huge difference in behavior between the two controllers at low friction and high-speed region, which is decreased with the increase of the vehicle speed and the road COF and becomes less than 1 second above COF of 0.5. This figure clarifies that implementing ARS has the highest demand at extreme driving condition, at low COF and high speed. Moreover, if the small difference between both controllers is neglected, ARS could be implemented alone without the integration with TVC. Furthermore, if the two controllers are combined, figure 8(b) can be used for coordination with giving the percentage of contribution of each controller based on the settling time difference.

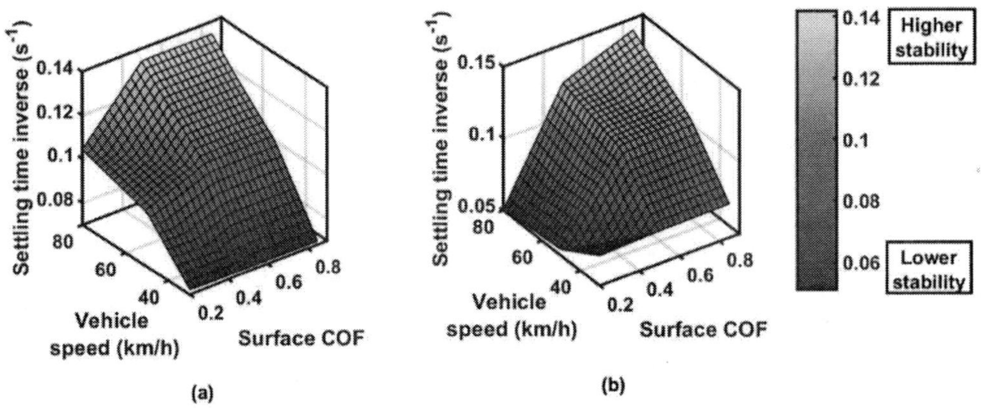

Figure 7. Controllers stability contour at various driving conditions a)ARS system b)TVC system.

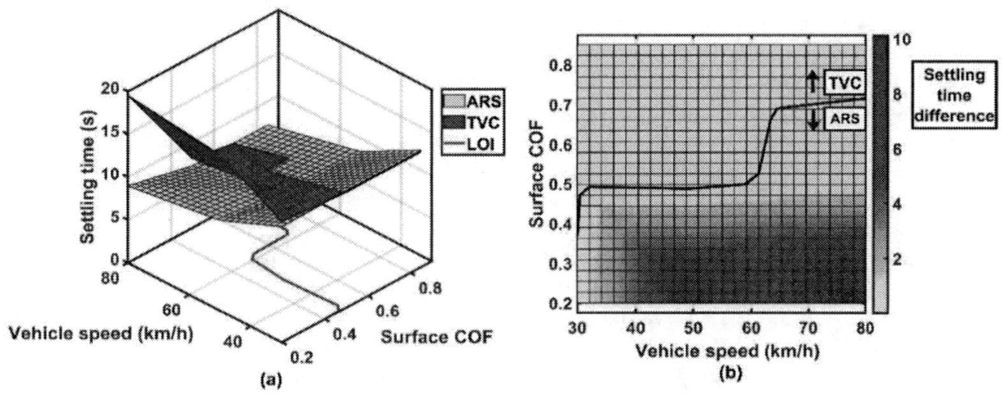

Figure 8.(a) Intersection of stability contours at various driving conditions (b) Region of preference.

5. Conclusions

In this paper, an 8x8 combat vehicle was used to implement and compare two chassis controllers. An ARS system has been developed using a combined feed forward controller and optimal Linear Quadratic Regulator (LQR) feedback controller and compared with Torque Vectoring Control (TVC) system which is designed using Sliding Mode Control (SMC) technique. Both controllers were separately evaluated and compared with conventional front steering vehicle. Double Lane Change (DLC)maneuver was used for evaluating and comparing the controllers at various vehicle speeds and road coefficient of friction. The results show that both controllers can enhance the handling performance and stability of the vehicle at all driving conditions. However, by comparing both controllers on a dry road surface, TVC offers more controllability and stability at high speed than ARS but with higher lateral error in comparison with the desired trajectory, which makes ARS recommended in the case of path follow maneuvers. On the contrary, ARS has notable superiority at extreme driving conditions such as driving on a very low coefficient of friction road surface and high speed, where there are low tire-ground traction forces. While the TVC is preferred at very low speeds. Moreover, the advantages and disadvantage of both controllers are summarized as follows:

- From the stabilization time perspective, TVC is slightly better than ARS while driving on a dry surface. In addition, TVC is recommended because of the higher degree of controllability it offers for the driver at high speed.
- ARS has superiority over TVC in case of driving on high slippery surfaces with high speed.
- ARS is also preferred due to the absence of the complications inthe reference model, which is essential for TVC, and the less need for parameter estimation and measurements than TVC.
- For path-following applications, ARS is recommended due to its lower lateral deviation from the desired trajectory.
- Economically, implementing the required component for TVC for all axles is relatively more expensive than implementing ARS for the rear axles only.

A novel method for coordinating both controllers is introduced for particular applications. This coordination is based on the road coefficient of friction, the vehicle speed, and the response time of each controller.

Finally, the integration of both controllers using the proposed coordination method can be a subject of future work. In addition, the differential braking should be examined in detail because of its simplicity and lower cost in comparison to the TVC.

References

[1] Asiabar AN and Kazemi R 2019 A direct yaw moment controller for a four in-wheel motor drive electric vehicle using adaptive sliding mode control.*P. I. Mech. Eng. K-J Mul.* p. 549-67.

[2] Ma X, et al. 2019 Cornering stability control for vehicles with active front steering system using TS fuzzy based sliding mode control strategy.*Mech. Syst. Signal Pr.***125**: p. 347-64.
[3] Park K, Joa E and Yi K 2019 Rear-Wheel Steering Control for Enhanced Maneuverability of Vehicles, SAE International.
[4] Zhao W, Qin X and Wang C 2018 Yaw and lateral stability control for four-wheel steer-by-wire system.*IEEE/ASME Trans. Mechatronics.* **23**(6): p. 2628-37.
[5] Yokoya Y, et al. 1990 Integrated control system between active control suspension and four wheel steering for the 1989 Celica, SAE Technical Paper.
[6] Kawakami H, et al. 1992 Development of integrated system between active control suspension, active 4WS, TRC and ABS, SAE Technical Paper.
[7] Selmanaj D, et al. 2013 Advantages of rear steer in LTI and LPV vehicle stability control. in *52nd IEEE Conference on Decision and Control.* IEEE. p. 3523-28.
[8] Wang R, Zhang H and Wang J 2013 Linear parameter-varying controller design for four-wheel independently actuated electric ground vehicles with active steering systems.*IEEE Trans. Control Syst. Technol.***22**(4): p. 1281-96.
[9] Jing H, et al. 2018 Robust H∞ dynamic output-feedback control for four-wheel independently actuated electric ground vehicles through integrated AFS/DYC.*Journal of the Franklin Institute.* **355**(18): p. 9321-50.
[10] Jalali M, et al. 2017 Model predictive control of vehicle stability using coordinated active steering and differential brakes.*Mechatronics.* **48**: p. 30-41.
[11] Mirzaeinejad H, Mirzaei M and Rafatnia S 2018 A novel technique for optimal integration of active steering and differential braking with estimation to improve vehicle directional stability.*ISA transactions.* **80**: p. 513-27.
[12] Nah J and Yim S 2019 Optimization of control allocation with ESC, AFS, ARS and TVD in integrated chassis control.*J. Mech. Sci. Technol.*: p. 1-8.
[13] Zhang J and Li J 2019 Integrated vehicle chassis control for active front steering and direct yaw moment control based on hierarchical structure.*T. I. Meas. Control.* **41**(9): p. 2428-40.
[14] Song J 2018 Integrated vehicle dynamic controls using active rear wheel steering and four wheel braking.*IJ VSMT.* **13**(1): p. 26-43.
[15] Rahimi S and Naraghi M 2018 Design of an integrated control system to enhance vehicle roll and lateral dynamics.*T. I. Meas. Control.* **40**(5): p. 1435-46.
[16] Ahangarnejad AH, Melzi S and Ahmadian M 2019 Integrated Vehicle Dynamics System through Coordinating Active Aerodynamics Control, Active Rear Steering, Torque Vectoring and Hydraulically Interconnected Suspension.*Int. J. Auto. Tech-Kor.* **20**(5): p. 903-15.
[17] Termous H, et al. 2019 Coordinated control strategies for active steering, differential braking and active suspension for vehicle stability, handling and safety improvement.*Vehicle Syst. Dyn.***57**(10): p. 1494-529.
[18] Kim W, Yi K and Lee J 2011 Drive control algorithm for an independent 8 in-wheel motor drive vehicle.*J. Mech. Sci. Technol.***25**(6): p. 1573.
[19] Kim W, Yi K and Lee J 2012 An optimal traction, braking, and steering coordination strategy for stability and manoeuvrability of a six-wheel drive and six-wheel steer vehicle.*P. I. Mech. Eng. D-J Aut.***226**(1): p. 3-22.
[20] D'Urso P 2016 Development of H∞ control strategy for a multi-wheeled combat vehicle.
[21] Shinners SM 1998 *Modern control system theory and design.* John Wiley & Sons.
[22] Alves SBD 2012 Helicopter flight modeling and robust autonomous control with uncertain dynamics, Universidade da Beira Interior.
[23] Murray RM 2009 Optimization-based control.*California Institute of Technology, CA.* p. 111-28.
[24] Kalbat A 2013 Linear quadratic gaussian (lqg) control of wind turbines. in *2013 3rd International Conference on Electric Power and Energy Conversion Systems.* IEEE. p. 1-5.

IOP Conf. Series: Materials Science and Engineering **973** (2020) 012009 doi:10.1088/1757-899X/973/1/012009

[25] Manchester IR 2013 AMME4500: Guidance, Navigation, and Control Course Notes University of Sydney.

[26] Wang R, et al. 2016 Composite nonlinear feedback control for path following of four-wheel independently actuated autonomous ground vehicles.*IEEE Trans. Intell. Transp. Syst.*17(7): p. 2063-74.

[27] Hespanha JP 2018 *Linear systems theory.* Princeton university press.

[28] Russell B 2018 Development and analysis of active rear axle steering for 8x8 combat vehicle.

[29] Odrigo A 2017 Development of multi-wheel drivetrain control system for future electric combat vehicle.

[30] Rajamani R 2011 *Vehicle dynamics and control.* Springer Science & Business Media.

[31] Liu J and Wang X 2012 *Advanced sliding mode control for mechanical systems.* Springer.

[32] Mokhiamar O and Abe M 2004 Simultaneous optimal distribution of lateral and longitudinal tire forces for the model following control.*J. Dyn. Sys., Meas., Control.* **126**(4): p. 753-63.

[33] Hillegass MJ, et al. 2004 Validating the directional performance of multi-wheeled combat vehicle computer simulation models. in *ASME 2004 International Mechanical Engineering Congress and Exposition.* American Society of Mechanical Engineers. p. 781-89.

[34] Hillegass MJ, et al. 2005 Validating the Vertical Dynamic Performance of a Multi-Wheeled Combat Vehicle Computer Simulation Model. in *ASME 2005 International Mechanical Engineering Congress and Exposition.* American Society of Mechanical Engineers. p. 31-40.

[35] Stryker November 2019 *https://en.wikipedia.org/wiki/Stryker.*

Evaluating the effect of shape on energy absorbing response of structures used in armored vehicles floors

HishamKamel
Automotive department, Military Technical College, Cairo, 11787, Egypt
hisham.kamel@mtc.edu.eg

Abstract. Improvised explosive devices (IEDs) are a major threat to the lives and wellbeing of soldiers transported in armored vehicles. The blast of an IED under vehicle body can cause severe injuries especially to the lower extremities of vehicle occupants. Blast-mitigating structures are developed to absorb as much energy as possible to protect occupants from the effect of blast. These structures are placed on vehicle floor to soften the impact on occupants' lower extremities thus minimizing injuries. Shape plays an important role in determining the amount of energy absorbed in these structures. This paper presents a finite element study where nonlinear finite element (FE) models were used to simulate the motion of vehicle floor due to the blast of an IED. A benchmark structure was developed to mimic Skydex that is a trademark structure. Different shapes were then used and compared with Skydex. The study shows the potential of optimizing shape to maximize the performance of blast-mitigating structures.
Keywords: improvised explosive device, vehicle structure, land mine, blast, protection.

1. Introduction

A typical improvised explosive device (IED) is an explosive material connected with a detonator, buried beneath earth's surface on the route of vehicles. It is preset to explode either remotely or through a trigger mechanism. IEDs are easy to assemble and require no high technological skills, yet they can inflict large human loss and injuries as well as property damage. During the war in Iraq and Afghanistan, more than 50% of all vehicles losses [1] 61.3% of all deaths and 65.5%:77.7% of all injuries [2, 3] were due to IEDs. The majority of injuries were to the lower extremities [4, 5]. IEDs even pose psychological deterrent by the thought that an area may be infested with IEDs; moreover, the sight of injured soldiers affects the morale of their mates [6-9].

Modern armored vehicles have been fitted with different counter IEDs mechanisms; such as V shape protection plates [10], anti-blast seats [11-15] and energy absorbing structures in the form of mats laid on vehicle floors [16-19]. This paper focuses on the latter mechanism.

Skydex is the trade name of an advanced patented structure that was originally developed reduce impact in running shoes but is currently used by the US army to protect army vehicles against IEDs explosions [20-22]. It consists of a twin-hemispheres arranged periodically made of thermoplastic polyurethane. Through its simple and lightweight structure, it absorbs large amount of impact energy at a constant stress. This mechanism reduces loads transferred to the feet of soldiers sitting inside an armored vehicle during an IED explosion [23]. The behavior of Skydex has been studied [24] and modeled [25]. Its design has been parametrized [26] and, it has been shown that filling the hollow microstructure with sand [27] or water [28] improves its performance. However, despite its success in absorbing impact energy and minimizing peak loads transmitted to vehicle floor, it has been shown that at floor speeds approaching 12 m/s loads can exceed injury threshold (5.4 kN) of the tibia [29]. The performance of energy absorbing floor mats is still far from ideal [30]. The advances in additive manufacturing techniques have been recently shown to produce complex shapes capable of absorbing more energy under compression loads [31].

Despite the importance of shape in affecting the performance of energy-absorbing floor mats, up to the author's knowledge, no previous research has addressed this aspect. In this paper, we attempt to investigate the effect of shape on the performance of energy-absorbing floor mats. Skydex has been used as a base model for comparing the performance of different shapes under the loading effect of under belly explosion of an improvised explosive device.

2. Model development

2.1 CAD development

The computer aided design CAD model shown in figure 1 was developed to represent a single cell of the Skydex structure according to the published data from Ref. [1]. The thickness of its surfaces is 1 mm. The radius of the hemisphere was assumed 16 mm.

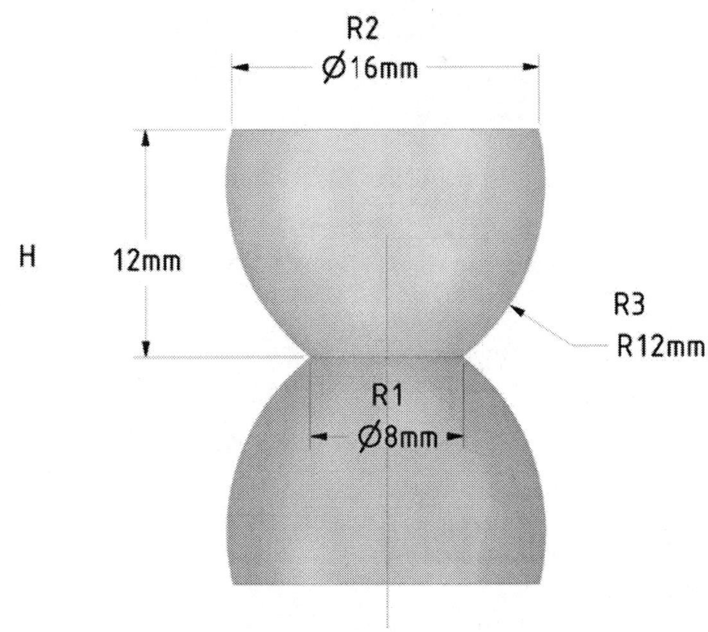

Figure 1. View of the cad model.

2.2 Load determination

For load calculation, the worst case was assumed when an occupant hits the floor with his full body weight landing on his feet. This occurs when a blast is large enough to propel the vehicle upwards after which it slams down to the ground under gravity effect. This scenario involves the effect of the occupant's feet hitting the rigid floor at a 10.8 m/s. This specific speed is within range of values that had been recorded in field [2, 3]. It is also enough to inflict a 10% risk of foot/ankle fracture, which accounts for an abbreviated injury scale of two (AIS 2+) as level recommended by the NATO/RTO HFM-090/TG-25 group in their publicly released report [4].

Considering a mid-size male weighing 78 kg [5] and measuring the surface area of the military boot of the author ~300 cm2. The load on one cell is then 0.616 kg. The FE model was developed in ANSYS, which is an efficient finite element package. As shown in figure 2, a rigid steel plate that

IOP Conf. Series: Materials Science and Engineering **973** (2020) 012010 doi:10.1088/1757-899X/973/1/012010

weighs 0.616 kg hits the cell from downwards at 10.8 m/s. The time duration was selected as 1.5 ms.This time was calculated to allow the cell to deform to more than half (16 mm) its height moving at full speed of the impactor (10.8 m/s). In reality, the speed decreases due to the resistance of the cell to movement, thus this time was enough to capture the physical phenomenon as evident in the deformed shapes in figure 4.

Figure 2. View of the FE model.

2.3 FE model development

A finite element model was developed using the CAD model in ANSYS. The SHELL163 element type was selected due to its accuracy and efficiency. To determine the optimum element size, a mesh sensitivity analysis was conducted and it converged at an element size of 0.5 mm. For material, Skydex is made of thermoplastic polyurethane according to Ref. [6]. The Cowper-Symonds material model [7] was selected to model its strain rate sensitivity in ANSYS. Material properties are given in Table 1.

Table 1. Material properties of Skydex.

Property	Value
Young's modulus of elasticity [MPa]	60
Tangent modulus [MPa]	50
Yield stress (A) [MPa]	8.6
Strain hardening coefficient (B)	50
Strain hardening exponent (n)	1
Strain rate hardening coefficients (D and q)	15 and 3
Poisson's ratio	0.49
Density [kg/m^3]	1100

3. Cellular structures

Different shapes were modeled and simulated under the aforementioned loading conditions. The model was set as shown in figure 3. The cellular structures were developed such that each structure fits inside the same cell surface area of the original Skydex structure. The thickness of each structure is adjusted to achieve the same mass as the original Skydex structure. This is accomplished in order to isolate the effect of mass and area and include only the effect of cellular shape. However, thickness also influences the response and it will be the focus of future work.

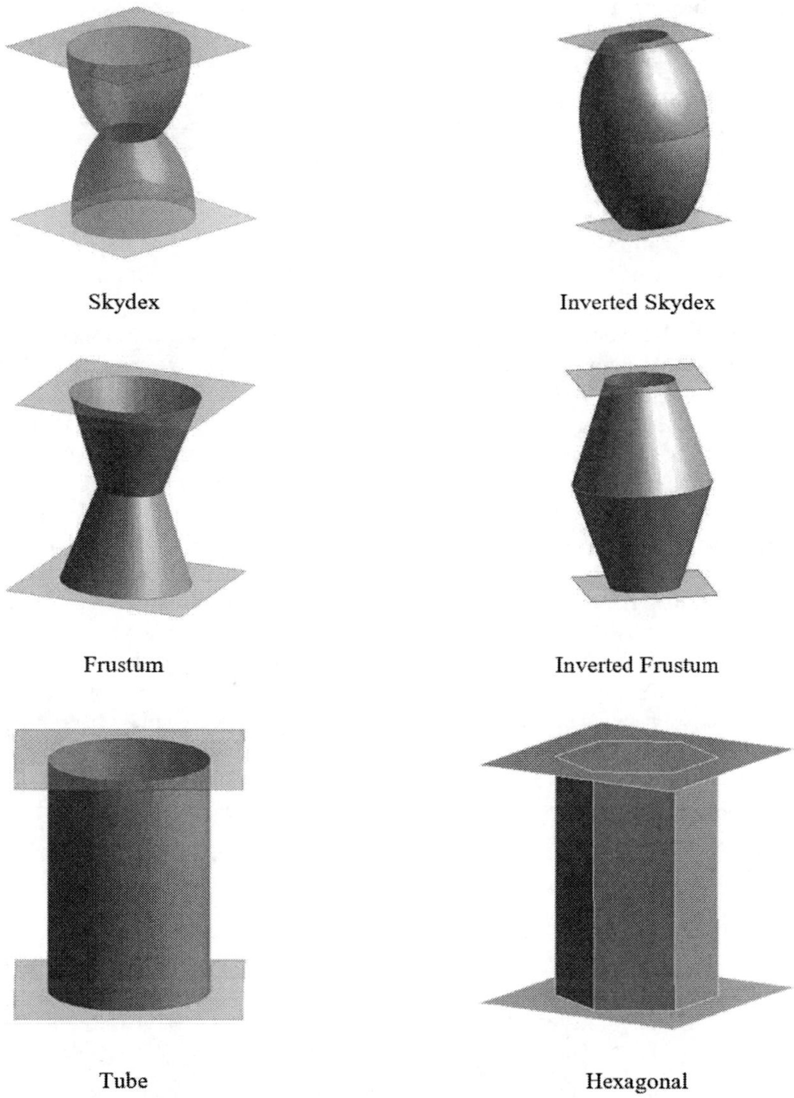

Figure 3. Views of the different CAD geometries.

4. Results and discussions

Finite element simulation is conducted for each cellular structure and the modes of deformation under impact is recorded and are shown in figure 4. The amount of energy absorbed (EA) in each structure during impact and the average transferred load (ATL) are recorded which and are summarized in Table 2Error! Reference source not found.. It should be noted here that these are the two important metrics for the design of efficient floor mats. A good design must ensure the maximum amount of EA at minimum ATL. This means that the structure protects the occupant's feet by absorbing as much impact energy as possible while transmitting as little energy as possible. Figure 5 shows the general behavior of a structure under impact. It shows the amount of EA as it increases as the structure deforms with time (figure 5-a) and the variation of transmitted load with time as shown in figure 5-

IOP Conf. Series: Materials Science and Engineering **973** (2020) 012010 doi:10.1088/1757-899X/973/1/012010

b.The average values of the transmitted load are calculated for each structure and are summarized in Table 2.

It is clear from figure 4 that in general shape has a profound effect on the modes of deformation. For example, inverting the shape of Skydex alters the mode of deformation and the structure starts to crumple from the tighter ends. The frustum shape has sides with no curvatures as in Skydex. Although it appears to deform in the same manner as Skydex, however, a closer look shows that necking in the outer walls of the frustum are developed as shown at onems in figure 4. This raises the transferred reaction force;30.3 N for frustum versus 26.6 N for Skydex.

For better visualization of results, the values of EA and ATL are normalized with respect to the ones for the original structure –Skydex- and are given in figure 6. Skydex is located at the origin of the plot where other cellular structures are located to its right upper side; this means that other cellular structures absorb higher energies but at the expense of higher transmitted load. We can then conclude that there is a trade-off between EA and ATL as shown in the increasing trend between both values. However, more insight in figure 6 reveals that the trend is not monotonically increasing since an increase in EA does not necessarily mean that ATL will increase at the same percentage. This is obvious if we look at locations of frustum and inverted frustum; they achieve approximately the same values of EA but the frustum transmits 21% less load than the inverted one. This can be further explained if we look at the deformation modes in figure 4 that shows the inefficient and irregular deformation mode of the inverted frustum structure. One can also deduce from Figure 6 that the tube structure is superior to the hexagonal one since it absorbs higher (14%) EA at less (12%) ATL than it. This is evident because the tube is further to the right and below the hexagonal structure, i.e. it absorbs more energy at lower transferred load. Finally, comparing between Skydex and the frustum structure reveals that the curvature of the outer wall play a significant role in the performance of the structure during impact. The frustum structure is simpler than Skydex since there are no curvature in the sidewalls, therefore, it will be easier and cheaper to produce. This observation can be used for developing structures of the simpler and cheaper frustum structure that achieves better 45% increase in energy absorption compared to Skydex. The 34% increase in ATL in case of the frustum structure must then be looked in more details. For example, the structure can be optimized to minimize it.

Table 2. Energy absorbed (EA) and average transferred load to each cellular structure.

	EA (mJ)	ATL (N)
Skydex	398.1	26.6
Inverted Skydex	521.2	30.3
Frustum	605.9	39.6
Inverted Frustum	574.0	41.1
Tube	466.9	33.6
Hexagonal	577.6	35.7

5. Conclusions

It can be concluded that the shape of the cellular structure plays a major role in the performance of energy absorbing structure for the application of vehicle floor mats. Simple shape changes such as the curvature in the sidewalls will significantly alter the performance of the structure. It was revealed that there is a strong trade-off between the amount of energy absorbed and the transmitted load, which are the two most important performance measures. In addition, we found that some structures perform more efficiently than others do. This means that optimization can be applied to find the shape that achieves the best/optimum balance between energy absorption and transmitted force. It is recommended that this point is suitable for further future research. Finally, this present research focused only on the effect of shape of cellular structure using limited number of shapes while the same material was used for all structures.

IOP Conf. Series: Materials Science and Engineering **973** (2020) 012010 doi:10.1088/1757-899X/973/1/012010

Based on this work, it is recommended that future study must include: initial floor speed, cell thickness and type of material. In addition, experimental testing must be included to help in verifying FE results.

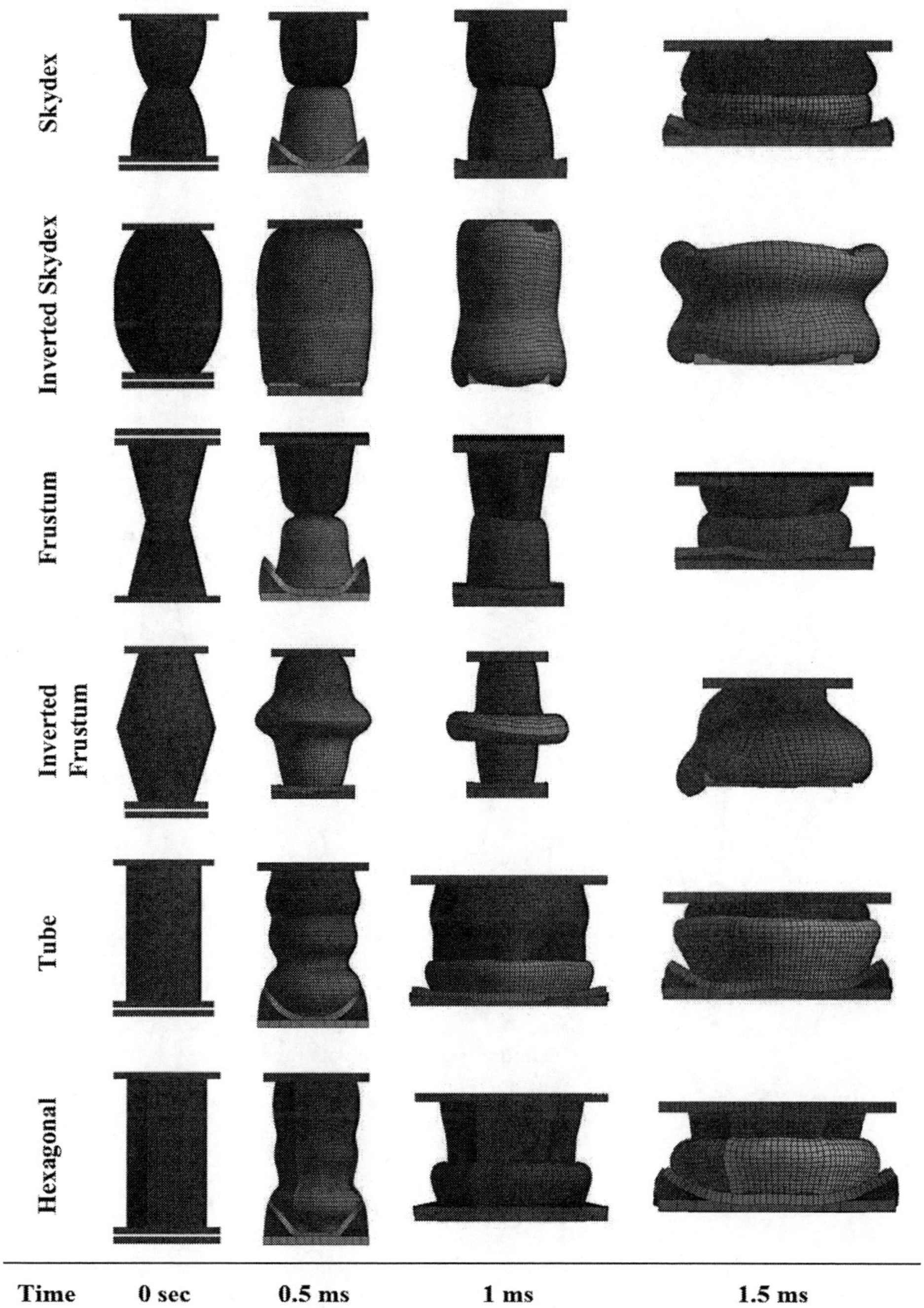

Figure 4. Modes of deformation under impact load.

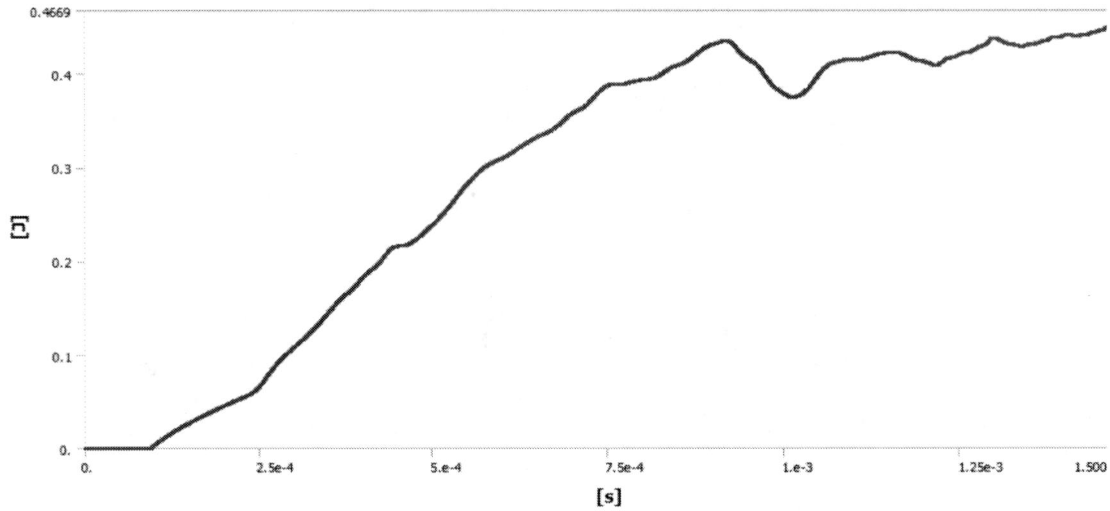

(a) Change of energy absorbed with time

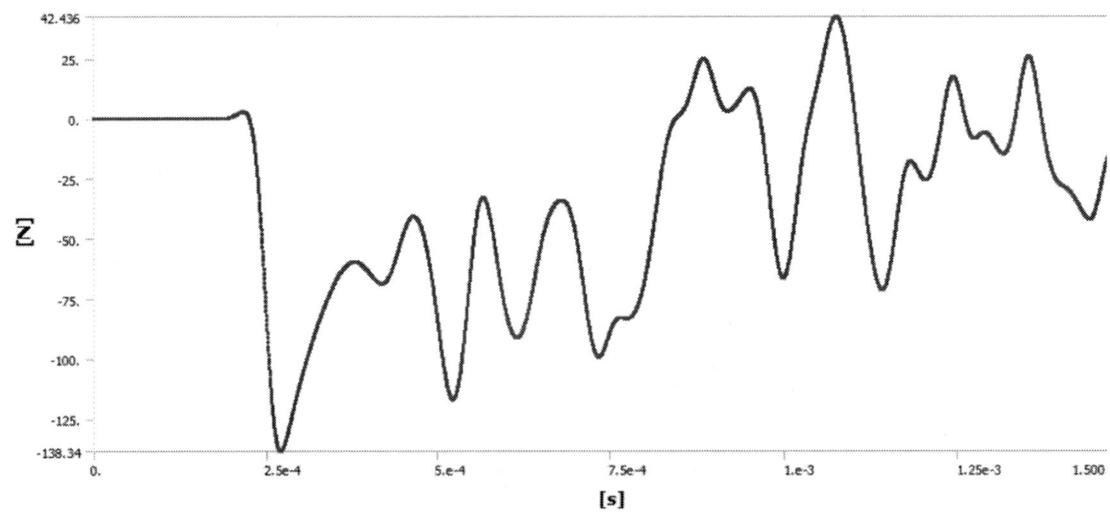

(b) Change of transmitted load with time

Figure 5.A sample of EA and transmitted load during time for Skydex structure.

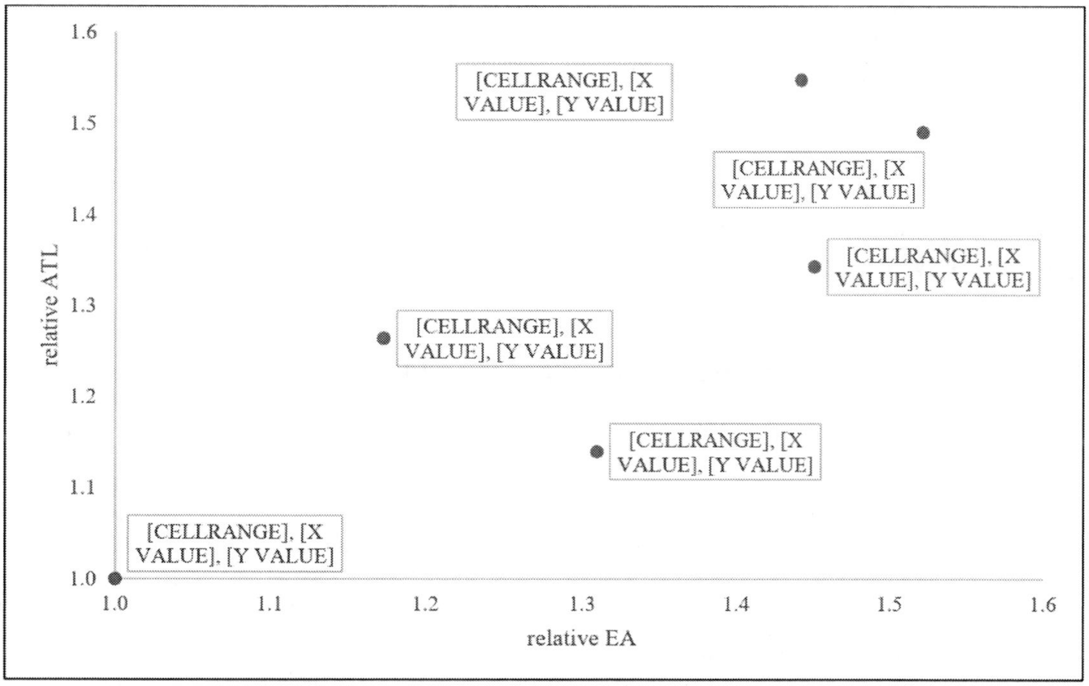

Figure 6. Representation of relative EA and ATL values for the different cellular structures.

References

[1] Zhu F et al 2014 Parameterized optimal design of a novel cellular energy absorber. *International Journal of Mechanical Sciences* **86** pp 60-68.
[2] Wang J and Bird R 2000 Improved seating to protect occupants of vehicles from landmines *Proceedings of the Land Warfare Conference*
[3] Bird R,Swinton Rand Krstic A 2001Protection of lower limbs against floor impact in army vehicles experiencing landmine explosion *Journal of battlefield technology* **4** pp 8-12
[4] HFM-090 Task Group 25 2007 Test Methodology for Protection of Vehicle Occupants against Anti-Vehicular Landmine Effects
[5] Herman IP 2016 Physics of the human body
[6] Zhu F et al 2013 Crushing behavior of SKYDEX® material in *Key Engineering MaterialsTrans Tech Publ*
[7] Cowper G and Symonds P 1958 Strain hardening and strain rate effects in the impact loading of cantilever beams *Brown Univ. App. Math. Report* p. **28**
[8] Winter DA 2009 *Biomechanics and motor control of human movement (Hoboken, N.J.: Wiley)*
[9] Reed MP 2006 Whole-body center of mass location in seated posture
[10] McKay BJ 2010 Development of lower extremity injury criteria and biomechanical surrogate to evaluate military vehicle occupant injury during an explosive blast event *Wayne State University: Ann Arbor* p. **216**

IOP Conf. Series: Materials Science and Engineering **973** (2020) 012011 doi:10.1088/1757-899X/973/1/012011

Test rig design for upper-limb socket prosthetics at transradial amputation level

M M Bondok[1], M A El-Sheikh[1] and M A El-Hadek[1]
[1] Production and Mechanical Design Department, Faculty of Engineering, Port-Said University
mogeeb500yom@gmail.com

Abstract. Statistics show that the transradial level is the most common level of upper-limb amputation seen by the prosthetist. Amputee needs an intermediate part to wear and fit the prosthetic device with the residual limb firmly. The socket is the only channel between residual limb and prosthetic components and is the crucial part of the prosthesis influencing the amputee's acceptance considerably. The purpose of this design is making a test rig that can assess the socket while simulating arm movements that are common during activities of daily living such as:(Abduction, Adduction, Flexion, Extension, Supination and Pronation) in Euler angles: Yaw, Pitch and Roll. This test rig allows to predict when dislocation between the residual limb and the socket will be happened with any weight and at any plane.

1. Introduction

The study published in 1996 of the medical records of Korean amputees showed that the ratio of upper amputation to lower amputation was approximately 1:2.2. Upper amputees have an enormous sense of frustration and problems in the rehabilitation process because of technical difficulties in reproducing delicate and complex movements and the tactile and proprioceptive sensory functions [1].

Based on information from National Center for Health Statistics in USA that showed 70% of all persons with upper limb amputations have amputations distal to the elbow. Transradial amputation level occurs in the forearm, from the elbow to the wrist [2]. Amputee needs intermediate part to wear and fit the prosthetic device firmly with the residual limb. So the socket is the only channel between residual limb and prosthetic components and is the critical part of the prosthesis affecting the amputee's acceptance considerably [3].

Assessment of transradial socket prosthetic is very important for measurement how amputees are satisfied about prosthetic interface. A survey showed that 20% of upper-limb amputees had abandoned prosthesis use. The critical factor is the comfort and functionality [4].The comfort-related problems are such as weight, fit, and heat [5].

Wayne K. Daly presented system that is adaptable to many amputation levels that maintains contact of electrode for patients who have volume changes or suffers from difficulty with the traditional suspension methods. Electrodes in roll-on sleeves have been tested to provide improved suspension. It showed that Roll-on sleeve is an excellent way to accomplish superior suspension and greater range of motion [6].

MatthewWernke used the motion capture model and Slip Detection Sensor. The 3D motion capture model allows for measurement of the movement that occurs at interface of the prosthetic socket. The Slip Detection Sensor measures the amount of socket slip which occurs between the socket wall and surface of the residual limb skin. Kinematic data was collected for each participant during the

IOP Conf. Series: Materials Science and Engineering **973** (2020) 012011 doi:10.1088/1757-899X/973/1/012011

reiteration of a series of range of motion included shoulder flexion/extension, shoulder rotation, and elbow flexion, shoulder abduction/adduction, and activities of daily living included a modified box and blocks test, bilateral and unilateral lifting task at various weight increments, walk and carry a gallon jug of water, and folding a towel. The effect of donning the prosthesis on the range of motion of participant and the amount of socket movement during the activities of daily living tasks was analyzed[7].

On the contrary of the previous studies, Michael D. Paskett designed a modular upper-limb bypass socket allows to perform tests of prosthetic systems from the prosthetic user's perception. Bypass socket accesses to forearm musculature and the hand that are necessary to provide substituted sensory feedback and allows a sufficient range of motion to accomplish tasks in the frontal working area. Tests are performed by non-amputee participants [8].

Liz Haverkate also depended on able-bodied subjects in his assessment of body-powered upper limb prostheses to assess and compare the functional performance three body-powered upper limb terminal devices which commonly used in this work, two functional tests; the Nine Hole Peg Test and the Box and Blocks Test are performed by able-bodied subjects [9].

Upper-Limb Socket Prosthetic needs different tests to be assessed such as Pull-off force test, Practical Activities Test, Live-lift test, Axial-load test, and positioning control test [6][8][10].

The goal of this study is to objectively designing and fabricating a Test Rig to assess the upper - limb socket prosthetic in Euler angles: pitch, yaw and roll while simulating the arm movements during activities of daily living such as: Abduction and Adduction, Flexion and Extension, Supination and Pronation. Also, it allows predicting when dislocation between residual limb and the socket will happen with any weight and at any plane.

2. Methods

In this work, theSolidWorks CAD software is used to build the model of the test rig. This design allows testing the upper-limb prosthetic socket in Euler angles: Yaw, Pitch and Roll as shown in figure 1.It is used to find out when the socket will be dislocated from the residual limb by the effect of different weights in different plane.The authors introduce the test rig that has ability to move universally at any plane.

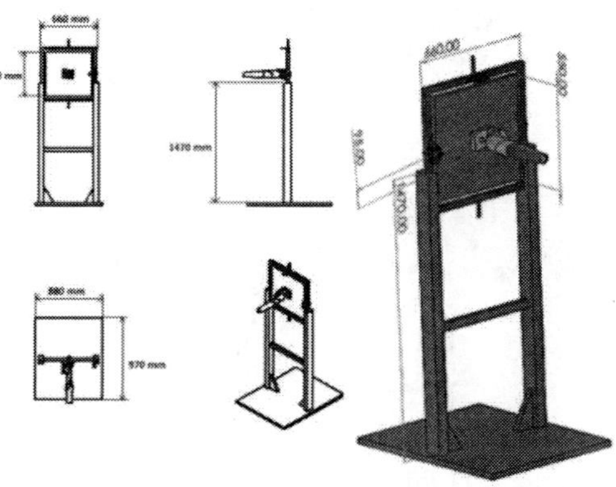

Figure 1. Test rig design on SolidWorks CAD software (Dimensions in mm).

This study is based on assessment of below - elbow socket prosthetic. The socket has been assessed was for a patient that has abandoned because of changes that happened in the volume of his residual limb. SolidWorks CAD software also was used to draw the socket with real dimensions as shown in figure 2.

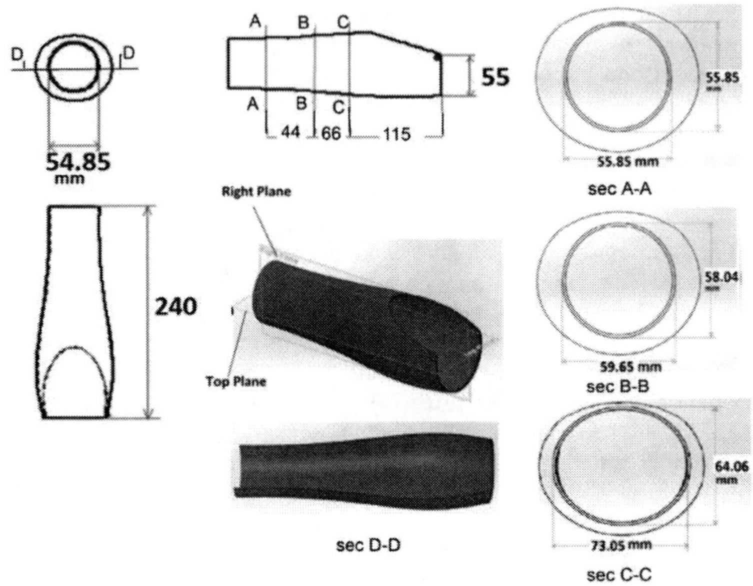

Figure 2. The Socket CAD model with sections at (C-C), (B-B), (X-X), and (D-D).

Figure 2 illustrates sections of the socket at different positions. Sections X-X and B-B have an elliptical cross section. The elliptical cross section decreases gradually to become a circular cross section at section C-C finally.

2.1. Design and manufacturing process for the test rig

Wood is the material used to build the test rig in this work. The design consists of : square plate 55 cm * 55 cm and square frame 66 cm * 66 cm, both of them are 3 cm in thickness, plat form with 1470 mm in height (Item no.3), connection part between plate and arm (Item no. 9), four wooden pins (Item no.4), two protractors (Item no.5), and two Indicators (Item no.6) as shown in figure 3 and table 1.

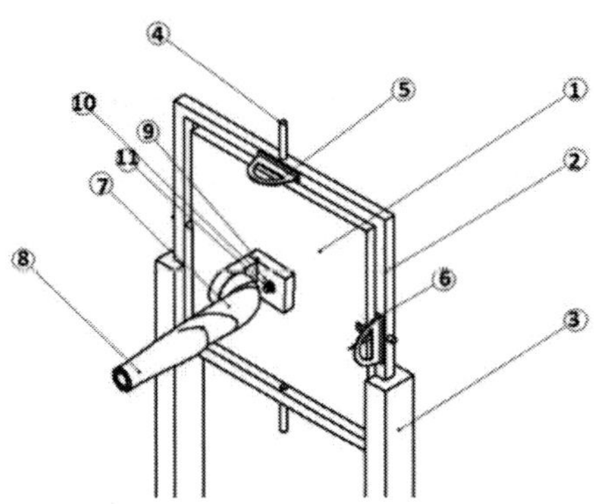

Figure 3. Parts of the test rig.

IOP Conf. Series: Materials Science and Engineering **973** (2020) 012011 doi:10.1088/1757-899X/973/1/012011

Table 1. Parts of test rig.

Name of parts	Number of parts
1 plate	1
2 Frame	1
3 Plate form	1
4 Wooden pin	4
5 Protractor	2
6 Indicator	2
7 Arm	1
8 Socket	1
9 Connection part	1
10 Nut	1
11 Pin stud	1

The purpose of this design is building a test rig that can assess the socket while simulating the common arm movements during the daily activities such as: Abduction and Adduction, Flexion and Extension, Supination and Pronation. All of these movements can be represented in Euler angles: pitch, yaw and roll as shown infigure4.

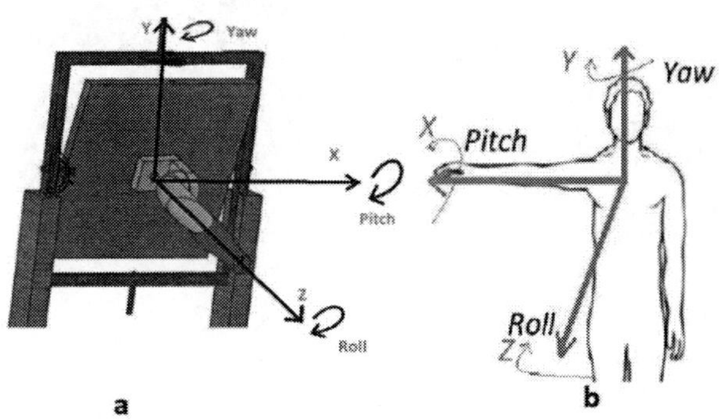

Figure 4(a) Representation of (Pitch, Yaw and Roll angles) on the Test rig. (b) Representation of (Pitch, Yaw and Roll angles) on the body human.

The test righ as been designed to simulate the movements of Euler angles by using four wooden pins and connection part between plate and arm. The four pins make the socket to rotate by alternating in X and Y axis to simulate pitch and yaw movement and connection part between plate and arm make the socket to simulate roll movement as shown in figure 5.

Abduction and Adduction are used to describe the movements that are towards or away from the midline of the body. Horizontal abduction: at the start position, arm is lifted in front of the body. The action occurs when arm moves out to the side. Horizontal adduction: at the start position, when arm is lifted out to the side of the body. The actions occur when arm moves in front of the body as shown in figure 6(a).

Flexion and Extension movements are used to describe increasing and decreasing the angle between two body parts and occurred in the sagittal plane. Flexion occurs when the angle is decreased between two body parts. Extension occurs when the angle is increased between two body parts as shown in figure 6(b).

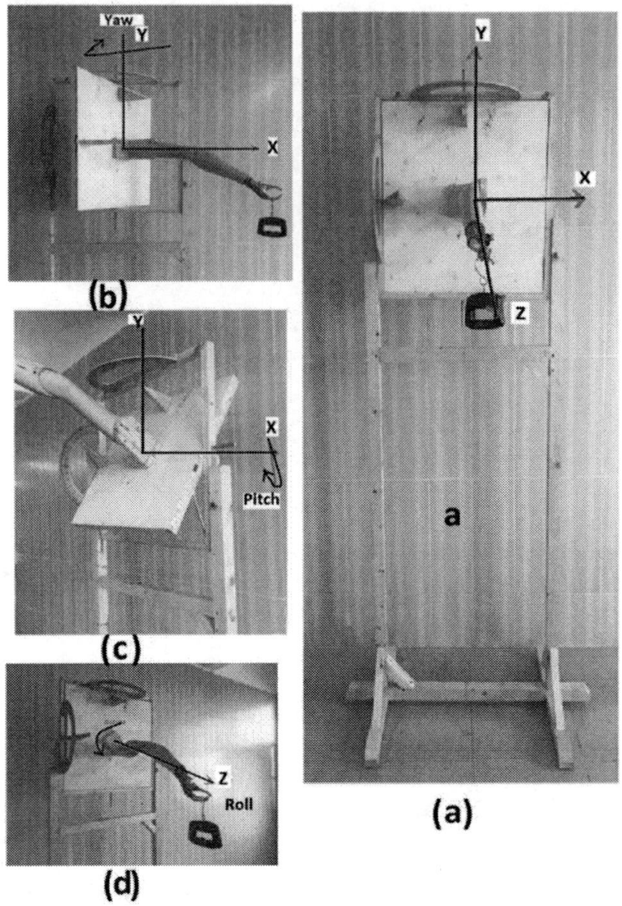

Figure 5.(a) Design and fabricating (The Test rig), (b) pitch angle by rotation about the Y axis Y-axis, (c) Yaw angle by rotation about the X-axis, and (d) Roll angle by rotation about the Z-axis.

Pronation and Supination is a pair of movements that occurs at the proximal radioulnar joint. The two unique movements allow the human body to flip the palm. Supination is occurred when palm is facing upwards and Pronation when palm is facing downwards as shown in figure 6(c).

2.2. Experimental work
Two tests were done by this test rig, Live-Lift Test, and Axial Load Test in Euler angles simulated common arm movements such as: Abduction and Adduction, Flexion and Extension, Supination and Pronation.

2.2.1. Experiment 1 (Live-lift test). Live-lift test measures the amount of vertical downward force applied to the socket that amputee can resist without occurring dislocation between the residual limb and the socket while maintaining his arm at 90°as shown in figure 7(a)[10].

2.2.2. Experiment 2 (Axial load test). Axial load test measures the amount of vertical downward force applied to the socket that amputee can resist without occurring dislocation between the residual limb and the socket while maintaining his arm in an extended position as shown in figure 7(b)[10].

IOP Conf. Series: Materials Science and Engineering **973** (2020) 012011 doi:10.1088/1757-899X/973/1/012011

Digital luggage scale is used to measure the maximum force the socket can endure as shown in figure7 (c). It has 50kg capacity and accuracy of 0.05 Kg as shown in figure 7 (c).

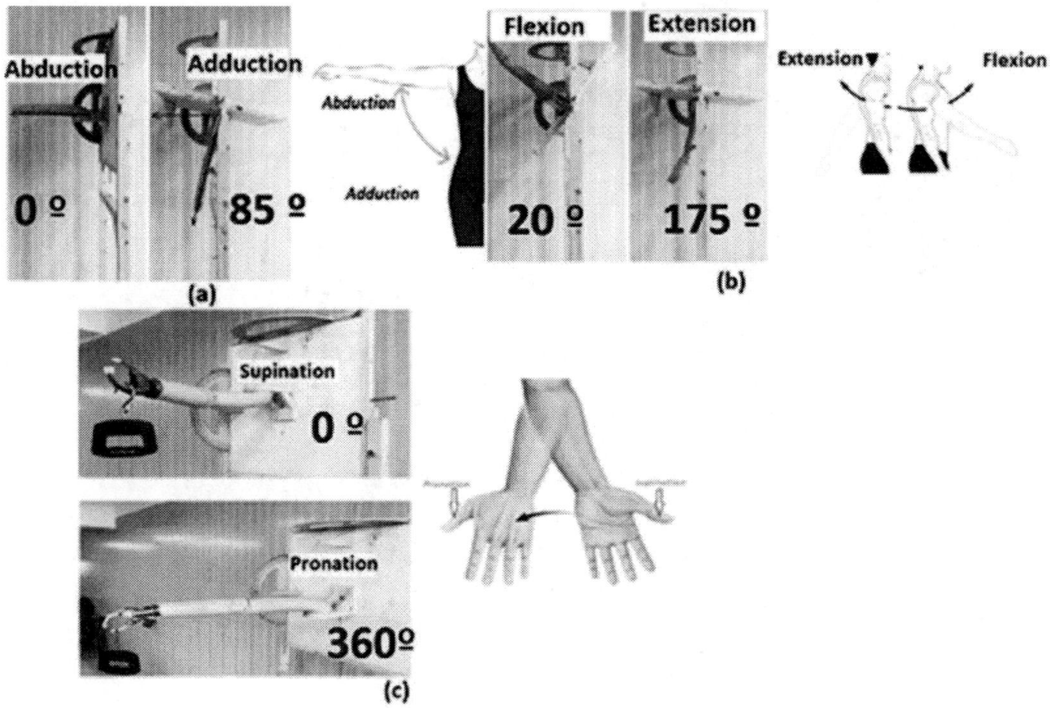

Figure 6.(a)Abduction and Adduction movements, (b) Flexionand Extension movements, and (c) Supination Pronation movements by test rig design and body human.

2.3. Special consideration

Female mannequin' arm is used to simulate arm's anatomy of human body. Female mannequin residual limb model is fitted firmly with the socket and both were suspended from a wooden connection part as shown in figure 7(c).

Revision should be considered that the socket must be donned firmly with a good level of accuracy. A mark on the stump of the arm was made where the maximum fitted position is reached it and the socket to be a guide for every time the position of the arm was made as shown in figure 7(d).

3. Results

In table 2, shows the live-lift force is larger than axial-load force at abduction case. In adduction case with an angle of 85°, live-lift force is larger than axial-load force.

In table 3, shows the live- lift force is larger than the axial-load force at flexion case with an angle of 20°. In Extension case when angle was 175 °, live- lift force was larger than axial-load force.

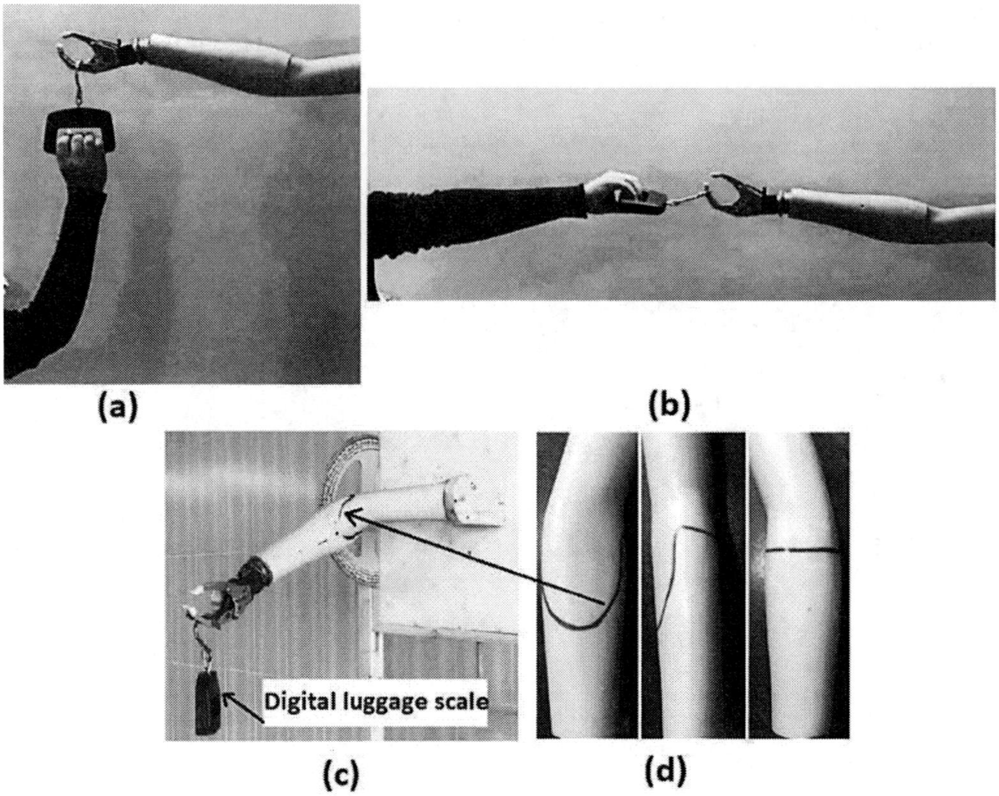

Figure 7.(a)Live-Lift Test, (b) Axial-Load Test, (c)Female Mannequin' arm simulating arm's anatomy of body human, and (d) Mark on the female mannequin' arm.

Table 2. Live-Lift and Axial-Load test Pronation and Supination movements.

	Abduction	Adduction
Angle (degree)	0 °	85 °
Live-Lift force (g)	6325	2330
Axial-Load force (g)	2330	523

Table 3. Live-Lift and Axial-Load test for Flexion and Extension movements.

	Flexion	Extension
Angle (degree)	20 °	175 °
Live-Lift force (g)	6577	875
Axial-Load force (g)	3620	523

In table 4, in pronation case, live-lift force was larger than axial-load force. In supination case, live-lift force was larger than axial-load force.

Table 4. Live-Lift and Axial-Load test for Flexion and Extension movements.

	Pronation	Supination
Angle (degree)	0 °	360 °
Live-Lift force (g)	6325	5425
Axial-Load force (g)	3620	2752

4. Discussion

The test rig was designed to simulate the human arm movement. All the parts of this test rig are made from wood. It is supplied with special parts to allow movements in Euler planes or angles. Four pins are used to allow movements in pitch and yaw angle. When two pins are connected with plate on the Y-axis and be free with X-axis, plate can rotate about the Y-axis (making yaw angle). Opposite to this, plate rotates about X-axis (making pitch angle) when pins are connected only with plate on the X-axis. Wooden connection part is responsible for rotation the arm about Z-axis (making roll angle).

The purpose of this test rig is the aid for assessment on the used commercial socket experimentally. The shape of the socket and cross sections at different locations are elliptical with different diameters. The study implemented two setup tests for the socket (Live-Lift Test and Axial-Load Test). Two tests were done at different positions as the human arm movements (Abduction, Adduction, Flexion, Extension, Supination and Pronation). The socket was suspended on Female mannequin arm without considering material used to simulate the soft tissues of the human body.

Referring to the experimental work, live-lift force is larger than axial-load force in all cases (Abduction, Adduction, Flexion, Extension, Supination and Pronation). In pronation and supination case, the variation of the measured forces was affected by the ending shape of the socket. In adduction, abduction, flexion, and extension cases, the experimental tests restricted by the edge of plate form of the test rig. Therefore, the tested socket didn't reach to 90 ° and 180 °.

5. Conclusion

The study introduced a test rig design to test the commercial socket at different positions and planes that simulate the arm movements. The experimental work covers two main tests to assess the dislocation of the socket for Guarantee the safety for amputee. In the future, the study could use special materials to resemble the soft tissues of the human body.

References

[1] Jang C, Yang H, Lee S, Kwon J, Yun B, Choi J, Kim S and Jeong H 2011 A survey on activities ofdaily living and occupations of upper extremity amputees *Annals of rehabilitation medicine* **35(6)** pp 907-921

[2] Sang Y, Li X and Luo Y January 2016 Biomechanical design considerations for transradialprosthetic interface: *A review Journal of Engineering in Medicine* **230**(3) pp 239–250

[3] Bertels T and Kettwig T 2011 Breathable liner for transradial prostheses *Proceedings of the Myoelectric Symposium*

[4] Sang Y, Li X, Gan Y, Su D and Luo Y 2014 Anovel socket design for upper-limb prosthesis *International Journal of Applied Electromagnetics and Mechanics* **45**(1-4) pp 881-886

[5] Sang Y, Li X and Luo Y March 2016 Characteristics of a volume-adjustable compression chamber for transradial prosthetic interface Journal of Engineering in Medicine **230** (7) pp 650–660

[6] Daly W 2002 Electrodes installed in roll-on suspnsion sleevesProceedings *UNB Myoelectric Controls/Powered Prosthetics Symposium* pp 45-47.

[7] Wernke M 2014 Quantification of transhumeral prosthetic socket residual limb interface movement using motion capture and a slip detection sensor *Journal of Public Transportation*

IOP Conf. Series: Materials Science and Engineering **973** (2020) 012011 doi:10.1088/1757-899X/973/1/012011

[8] Paskett M, Olsen N, George J, Kluger D, Brinton M, Davis T, Duncan C and Clark G 2019 A modular transradial bypass socket for surface myoelectric prosthetic control in non-amputees *IEEE Transactions on Neural Systems and Rehabilitation Engineering* **27**(10) pp 2070-2076.

[9] Haverkate L, Smit G, Plettenburg D 2014 Assessment of body-powered upper limb prostheses by able-bodied subjects, using the box and blocks test and the nine hole peg test *Prosthetics and orthotics international* 40(1) pp 109-116

[10] Fishman S, Kay H 1964 The munster-type below-elbow socket, an evaluation *Artificial Limbs* **8** pp 4-14

IOP Conf. Series: Materials Science and Engineering **973** (2020) 012012 doi:10.1088/1757-899X/973/1/012012

Design and manufacturing of a test rig for measuring the torque required in soil drilling operations

M A Abdeldayem[1], M EAbo-Elnor[1] and M H Mabrouk[1]
[1] Mechanical Equipment Department, Military Technical College
mabdeldayem@mtc.edu.eg

Abstract. Soil drilling operation has become one of the most important interests to researchers due to its many applications in engineering systems. Such as construction industry, soil samples for geological sciences and space sampling. The dominant factor in determining drilling parameters based on drilling operations experience or in some times based on proposed modelling techniques. As a result, soil drilling process using auger drilling is studied to obtain drilling parameters and then optimize these parameters to improve drilling performance which enables proper selection of machine for a required job. One of the main challenges that faces researchers during using modelling techniques to define the soil drilling problem is the complex nonlinear behaviour of the drilled medium itself due to its discontinuity and heterogeneous formation. This paper presents a developed apparatus that has been designed and manufactured to be used in measuring and recording the total torque required during soil drilling operation. A simplified auger drilling machine is built in soil-tool interaction laboratory, Military Technical College, to obtain experimental results that can be used to verify the presented models. Data acquisition measuring system is established to analyse experimental results using a. The Labview® program enables recording and displaying the output data collected mainly from sensors planted in the test rig. Results of both analytical and numerical models are then compared to experimental results to aid in developing the presented parametric study that can be used to define the working parameters during drilling operations in different types of soils.

1. Introduction

Auger drilling technique is used on a wide range in many flight twist drill into the ground to required profundity [1]. Continuous flight augur(CFA) piling is a broadly utilized heaping strategy offering both specialized and business favourable circumstances, when utilized suitably, in the correct ground conditions and task conditions [2]. Additional advantages of CFA are higher generation rates, flexible auger tallness, and movable twist drill width inengineering applications such as geological sampling, pile foundation engineering and space sampling. Also CFA fined to be appropriate for some kinds of soil, for example, mud, sand, residue and moderate rocks [3].. Auger drilling has many advantages such as it is considered a dry drilling method, continuous transportation of soil and the efficient construction [4, 5]. Soil is considered as a heterogeneous material, which prompts its complex behaviour[6]. For the ideal structure of soil interacting machines the genuine conduct of soil should be taken into consideration as it seriously affects large portion of parameters that portrays the hardware execution amid activity for instance weight on burden arms, pressure driven powers, footing on tire and conveyance of intensity...etc[7]. If the drilling rig couldn't satisfy the required feeding rate and auger rotating motion, the twist drills can "mine" the overlying frail soil to the surface and

IOP Conf. Series: Materials Science and Engineering **973** (2020) 012012 doi:10.1088/1757-899X/973/1/012012

causesubsidence [8]. Drilling is generally utilized in heap establishment designing, geographical testing for sandy soil, alsoit is more convenient than other boring methodologies for planetarydrilling due to its high productivity and continuity in cuttings transportation [1, 4].
The key point of this paper, the construction of the test rig, which has been built to proceed the experimental work, is represented. The test rig mainly consists of the auger-drilling system, monitoring system and control unit. The experimental test rig is installed in the laboratory of the Mech. Equip.Department, Military Technical College, Cairo, Egypt.

The testing system shown in figure 1 will analyze and calculate the torque for the cutting and drilling of loose sand to a depth of 0.5 m. A continuous auger system consists of an auger powered by an AC motor, fitted with an speed reducer of a worm form, is applied to the soil. For control the AC motor speed an AC inverter is used.The AC engine is mounted on a wagon.It is acceptable to move vertically through four linear movement systems as shown in the figure. The vertical feed used to induce penetration is generated by a linear actuator. The penetration speed is about 5mm per second.Between the auger and the output shaft from the speed reducer, a torque transducer is attached.The steel frame structure is designed to contain all the components of the proposed system; the frame is attached to the main frame of the soil bin as shown in figure 1.

The speed of the three-phase AC motor is controlled by avariable frequency drive model AS 2 - 115 is used to control the speed of the three-phase AC motor. Alsoit is used to reverse the direction of rotation.A designed system using data acquisition is developed to allow the torque measurement. The transducer "T4A" uses a 4-wire link technology is inserted between the auger and the driving shaft.The T4A output signal is then forwarded to NI 9219 via a USB connection to the device via a compact acquisition system (DAQ).

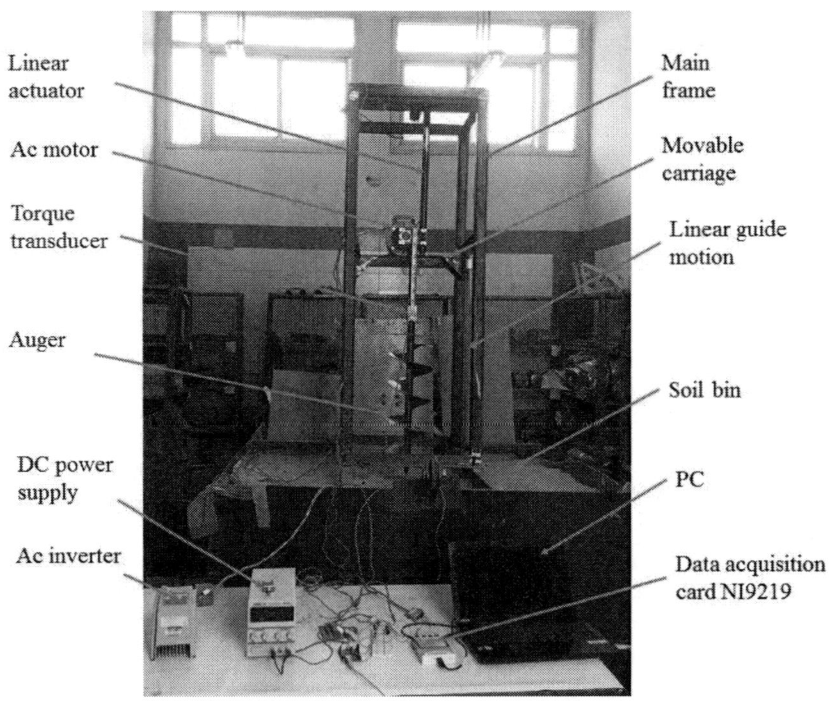

Figure 1.Specific configuration of the test platform [9].

IOP Conf. Series: Materials Science and Engineering **973** (2020) 012012 doi:10.1088/1757-899X/973/1/012012

2. Test rig components details

This section introduces the components that have been used in the manufacturing process of the test rig along with their technical specifications and data.

2.1. Soil bin

Soil bin is a lab facility that is used to contain the tested soil for different experimental work. This soil bin is established in soil lab, mechanical equipment department, Military Technical College. The main dimensions of the soil bin are (3990mm,1584mm,1300mm) (length, width, height) with soil depth up to 700mm as shown in figure 2.

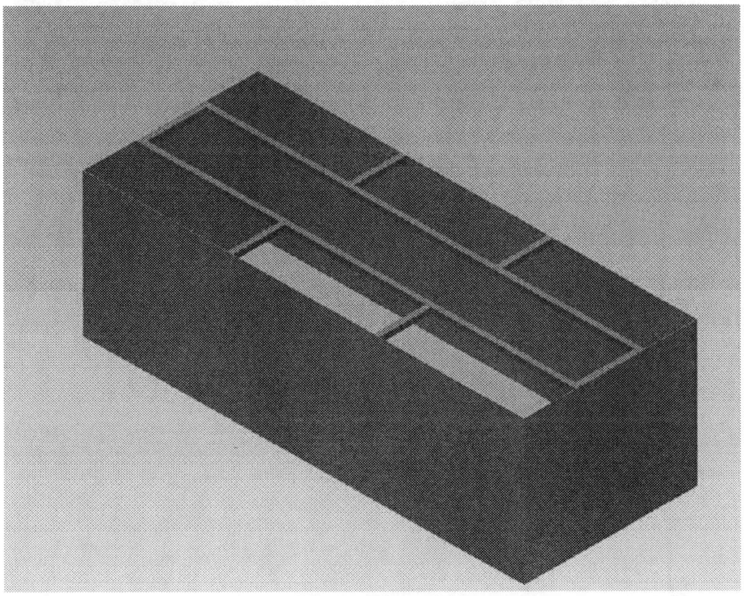

Figure 2. Soil bin.

2.2. Fixed frame

The fixed frame is considered as a skeleton of the test rig. It carries all the components that comprise the test rig with height up to1700 mm. It is equipped with an adjustable frame at its top to carry linear actuators. A hollow bar, with 5 mm thickness rectangular cross section, is the main component used in manufacturing the fixed frame as shown in figure 3.The fixed frame is attached to the main soil bin and is ready for assembly.

2.3. Movable frame

The movable frame is a simple frame used to carry the auger mechanism, which comprises of an AC motor and the auger. A mechanical linear actuator is attached to the movable frame to generate the up down motion as shown in figure 4. Four linear guide motions are used to keep a straight pass for the whole system during drilling operation.

2.4. Auger

The auger is the main working tool in the test rig. The tool consists of a 70 cm long hollow stemmed tube sharpened at its end. A continuous helix of "flights" is located all along the tube.The pile diameter and the hole diameter both equal to horizontal space between flight edges (20cm) and The vertical flight distance is the pitch(15cm) as shown in figure 5.

Figure 3. Fixed frame.

Figure 4. Movable frame.

Figure 5. Auger.

2.5. Linear guide motion system

To assure the vertical motion of the movable frame, which carries the auger drilling system, a four linear guide motion system is used as shown in figure 6. They are fitted to the fixed frame by using (16-end) support. The movable frame is supported by eight bearings, which allow it to move along the linear guide motion. Bearings are installed as two bearings for each linear guide rod.

2.5.1. Linear bearing. Figure .7 shows a SCS16UU closed type linear ball bearing with aluminium block mounted. Its inner diameter is 16 mm to mount on 16mm diameter linear shaft, which is a common choice for linear motion applications. The linear bearing consists of two main parts: self-aligning aluminium block and linear bearing. The self-aligning aluminium block permits the compensation of any misalignment of inaccuracy in the final assembly. The bearing cage is made from special polymers allowing smooth and noiseless operation. It is convenient and low cost solution for machinery that require smooth motion with minimum friction and at the same time high rigidity and less vibration with affordable cost and easy to assemble operation

IOP Conf. Series: Materials Science and Engineering **973** (2020) 012012 doi:10.1088/1757-899X/973/1/012012

Figure 6. The linear guide motion system.

2.5.2. Linear guide rod. A chromed steel bar is utilized in states of high pressure and severe conditions. In like manner, the nature of the base metal and resulting surface treatment are of principal significance. Wear obstruction, surface smoothness, high return quality and uniform quality are basic properties of chrome plated bars. Chromed plated linear rods are delivered from steel bars joining high quality with good machinability and weld capacity. Regular applications are found in direct movement, programmed apparatus, CNC machines, and injection moulding machine, farming machinery and production equipment as shown in figure8.

2.5.3. End shaft support. The Aluminium end support block can be utilized for end or irregular shaft support. The anodized aluminium end support is lightweight and solid, and can be utilized with all rod types as shown in figure9.

Figure 7. Linear bearing

Figure 8. Linear guide rod

Figure 9. End shaft support.

IOP Conf. Series: Materials Science and Engineering **973** (2020) 012012 doi:10.1088/1757-899X/973/1/012012

3. Actuating motors

Actuating motors are responsible for two types of motion, linear motion, which can be interpreted for feeding rate and rotary motion. Electric motors are used to establish the test rig, and represent the advantage of low cost and easiness to be controlled

3.1. Linear actuator

Two super-jack regular actuators mounted on the top of the fixed frame are employed to produce the feeding action. The end of the extended cylinders are attached to the movable frame as shown in figure 10 Super-jack actuator is used as the source of the linear motion or feeding. This type is characterized by low noise and high performance with a stroke up to 0.61 meter. The operating voltage is up to 36 volt and Load capability is 136 kilogram. Rated, (250-kilogram dynamic/454 kilogram. static).The two linear actuator are powered by a DC power supply with maximum volt equal 30 volt.

Figure 10 Super lack linear actuator.

3.2. A three phase AC motor

A three-phase AC motor, shown in figure 11, is responsible for developing the rotary motion. The motor base is fitted to the movable frame with four bolts. A worm gear set is attached to the motor output shaft giving the advantage of self-lock with a reduction ratio 1:20.

An AC inverter is used to control the motor output speed which by its role controls the auger rotating speed. The motor power is 0.75 hp and the output shaft speed is up to 1500 rpm.

3.3. AC inverter

To control the 3-phase AC motor an AC inverter is used as shown in figure 13. This inverter is responsible for reversing the direction of rotation of the AC motor from clockwise to counterclockwise and vice versa. An AC inverter model AS 2 – 115 is used to control the 3-phase motor by changing the inverter frequency. The input voltage is 220 volt and the output voltage obtained from the inverter is 360 volt.

Figure 11..A three phase AC motor.

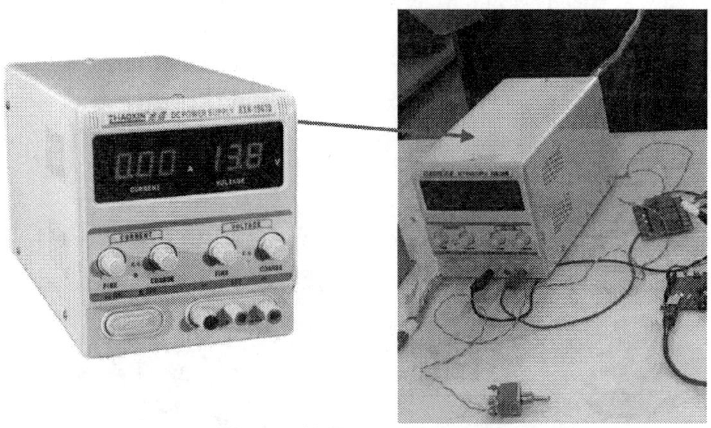

Figure 12.DC power supply.

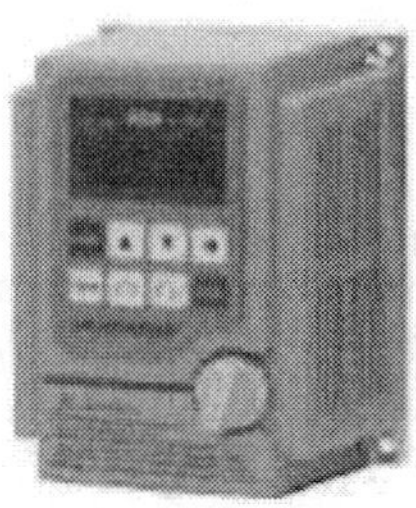

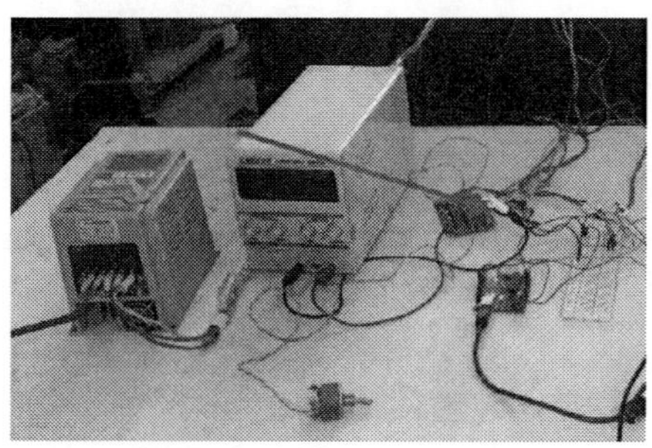

Figure 13. The variable frequency drive AS 2– 115.

3.4. Torque transducer

Torque can be identified as a rotational force that represents the rate of change of angular momentum of an object[10]. The requirement for torque estimations has prompted a few strategies for gaining trusted data from items moving. A torque sensor or transducer changes over torque into an electrical signal.

The most well-known transducer is a strain gauge that changes over torque into a change in electrical resistance. The strain gauge is glued to a bar or basic part that misshapes when a torque or power is connected. Deflection encourages a stress that variates its resistance. A Wheatstone bridge changes over the resistance change into an electrical signal. The structure of a response torque cell looks to take out side loading and is sensitive just to torque loading.

Other generally acknowledged strategies for measuring torque depend on mounting transducers in the machine train or on the turning shaft, these strategies use strain gauges. Every strategy has preferences and confinements, a lot of which are application dependent. The best arrangement requires a top to bottom comprehension of the application.

Two regular approaches to get torque measurements are by strain-gauging the shaft and by utilizing in-line torque transducer. Both have two technical obstacles: getting capacity to the measures over the stationary turning hole and recovering the signal. The strategies to overcome any issues are either contact or non-contact.

The used transducer in this research is a screw torque transducer model (T4A) as shown in figure 14. This model characterized by the ability of measuring torque in any direction also any angle of rotation and any rotational speed. Nominal torques 5, 10, 20, 50, 100, 200, 500 and 1000 N.m. The transducer "T4A" uses a 4-wire link technology is inserted between the auger and the driving shaft. The T4A output signal is then forwarded to NI 9219 via a USB connection to the device via a compact acquisition system (DAQ).

Figure 14. The T4A torque transducer.

3.5. Limit switches

In order to detect the feeding speed, two limit switches are used as shown in figure 15. The output signal from limit switches are sent to Arduino toolkit. The distance between the two limit switches equal to the proposed hole depth (0.5) meter. The feeding speed can be calculated by dividing the distance between the two limit switches by the time calculated from the limit switches output

3.6. Arduino

The ArduinoUno rev.3 board show in figure 16It is an ATmega328-based microcontroller board. The system consists of 14 digital pins, 6 of which can be used for PWM outputs, 6 Analog inputs, the 16 MHz oscillate, a USB interface, an ICSP Header and a reset button. The device also has a USB connection. It contains all the microcontroller needs; simply connect it to a USB cable device or power it to continue using an AC-to-DC converter or a charger.

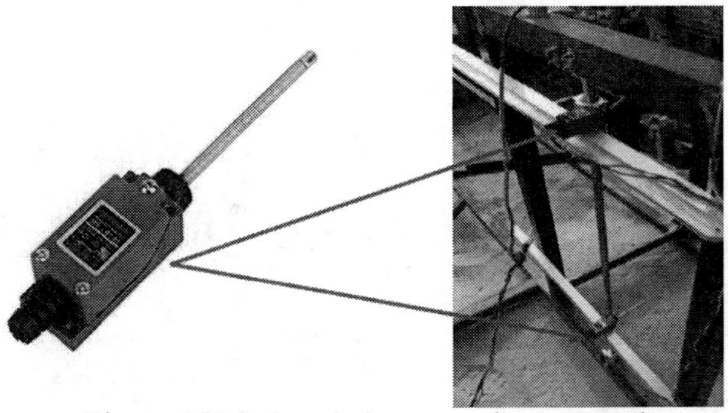

Figure 15. Limit switches mounting on test rig.

Figure 16. Arduino Uno.

3.7. Relay

RK4 is a basic, inexpensive module of a relay motor. It supports eight output relays, which control electrical AC and DC devices including lights, engines and keys, as shown in the Figure 17, for the control of the path of the linear actuator. The 5V, 12V or 24V DC supply RK4 can be operated.

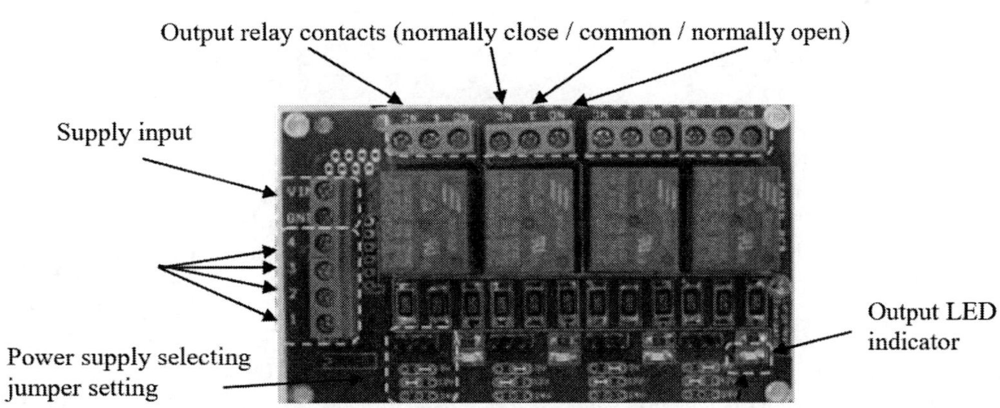

Figure 17. The RK4 layout.

IOP Conf. Series: Materials Science and Engineering **973** (2020) 012012 doi:10.1088/1757-899X/973/1/012012

3.8. Computer based data acquisition (DAQ)

Figure 18 shows the NI 9219 Compact DAQ, which is a universal C series module. Signal from torque transducer can be measured easily with a simple Labview® program as shown in figure 19. The range of measurements for each method of measurement is special, with voltage up to ±60V and current up to ±25 mA.

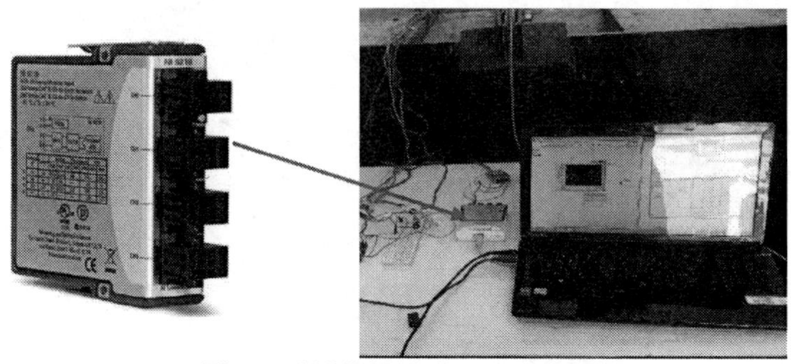

Figure 18.NI 9219 Compact DAQ.

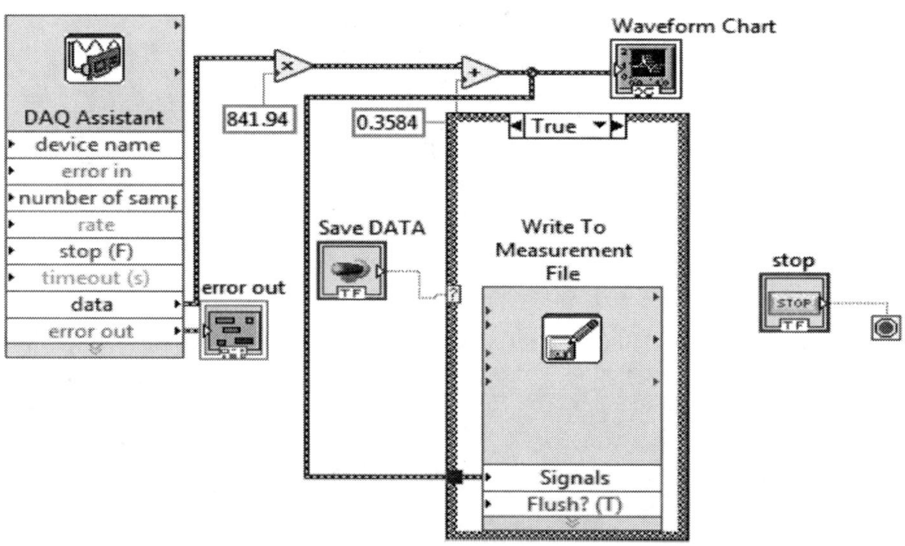

Figure 19. Labview program.

4. Torque transducer calibration process

In order to increase the accuracy of drilling torque measurements, torque transducer is calibrated and sensor calibration curve is deduced. The calibration process is conducted by using different known weights, which are hanged at a certain distance to produce torque action as shown in figure 20. (a). The Labview ® Software Package is used to design a data acquisition system to allow the viewing and recording of the output data directly from the transducer. The output readings from torque transducer is then exported to excel. The loads generate torque, which is calculated by multiplying the hanged load weight by the orthogonal distance between the axis of symmetry of the hanged load and torque transducer. Finally, both torque value obtained from torque transducer output signal and calculated by multiplying hanged weights by distance are recorded and plotted to obtain the torque transducer calibration curve as shown in figure 20. (b).

(a)

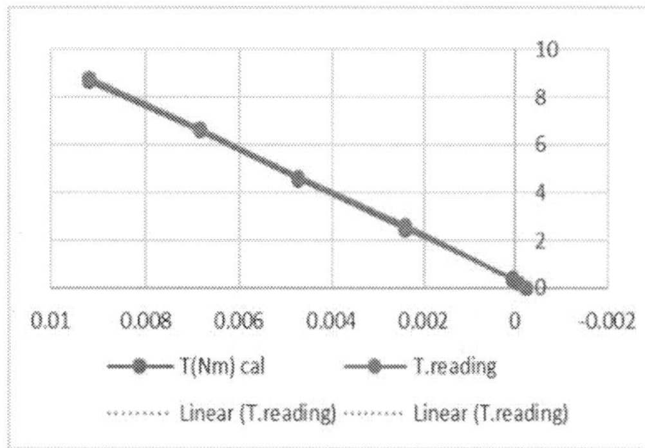

(b)

Figure 20. Torque transducer calibration process.

5. Data processing flow chart

A simple program is created by using Labview® program to enable recording and viewing the collected data. Figure 21 shows the general configuration of the flow map of the drilling test rig[11].

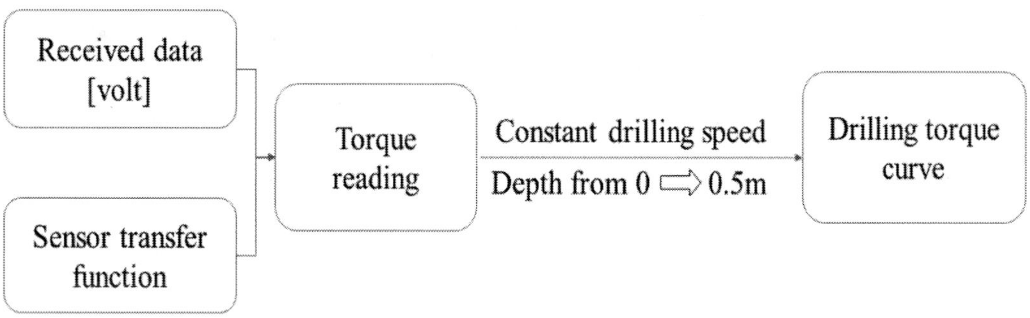

Figure 21. General configuration of the flow map of the drilling test rig.

6. Uncertainty calculations and results

The main cause for measuring uncertainty is to know the measurement range. The uncertainty of the measurement operation has been studied thoroughly in separate paper [11]. Type A uncertainty, type B uncertainty and combined uncertainty are calculated. Expanded uncertainty is calculated by multiplying a coverage factor to the combined uncertainty.

The required torque for drilling in sandy soil is calculated analytically and measured experimentally for 0.5 meter depth and plotted as shown in figure 22. The auger speed is 50 rpm and the feeding rate is 5 mm/sec. To achieve 95% confidence level the standard uncertainty has been multiplied by a coverage factor = 2, then error bars have been added to the measured values of the required drilling torque as shown in figure 23[11].

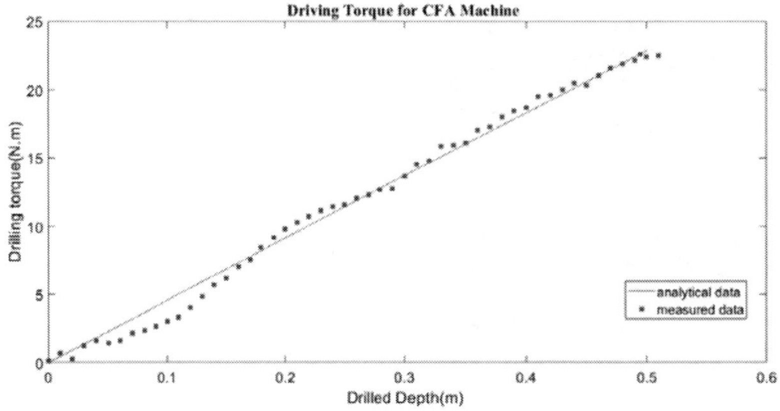

Figure 22. Drilling torque required in CFA machine.

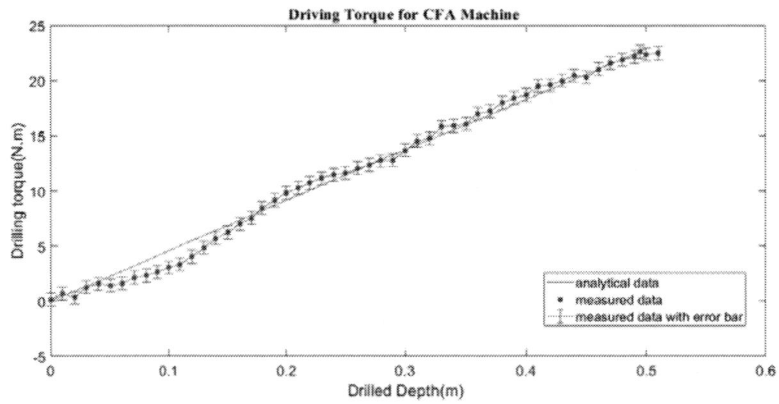

Figure 23. Drilling torque required in CFA machinewith error bar.

7. Conclusion

A test rig is set up to calculate the torque of the auger during soil digging, and uses a variety of mechanical systems, control circuit and many sensors. Using the Labview ® software package to view and monitor the calculated data directly via the transducers, a simple data acquisition program is developed. To have the validity of measurements and the sensor sensitivity scale, the torque transducer is calibrated. To order to calculate performance, the uncertainties of type A and type B are calculated. Both forms are mixed for determining the disparity in readings from different sources. The overall device precision of the drilling torque is dependent on the calculation of calculated torque uncertainties. An error bar with 95 percent conviction is given for each reading showing the potential variance. Comprehensive measurement of the driving torque during the drilling is determined according to the directly calculated values on a continuous flight auger unit.

IOP Conf. Series: Materials Science and Engineering **973** (2020) 012012 doi:10.1088/1757-899X/973/1/012012

References

[1] Hosny H, Ibrahim A and Fraig R 2016 Cost analysis of continuous flight auger piles construction in Egypt*Alexandria Engineering Journal* **55** pp 2709-20
[2] Zacny K, Bar-Cohen Y, Brennan M, Briggs G, Cooper G , Davis K, et al 2008Drilling systems for extraterrestrial subsurface exploration *Astrobiology* **8** pp 665-706
[3] Brown D, Dapp S, Thompson W and Lazarte C 2007 Design and construction of continuous flight auger piles*Geotechnical engineering circular* **8**
[4] Tan S, Duan L, Tan S Shi H 2011 Study on critical drilling parameters for auger drilling," in *Advanced Materials Research* pp 3331-40.
[5] F. O. P. Specialists 2014 CFA Piling: Preventing ground & riginstability through over-flighting
[6] Peurifoy R and Ledbetter W 1985 *Construction planning, equipment, and methods*
[7] E. team 2017 Optimizing Heavy Equipment Design for Handling Bulk Materials *www.edemsimulation.com*, ed: www.edemsimulation.com.
[8] Fayek A, Dissanayake M and Campero O 2004 Developing a standard methodology for measuring and classifying construction field rework *Canadian Journal of Civil Engineering,* **31** pp 1077-89
[9] Abdeldayem M, Mabrouk M and Abo-Elnor M 2019 Analytical and numerical modeling of soil cutting and transportation during auger drilling operation *ASME International Mechanical Engineering Congress and Exposition*
[10] Serway R and Jewett J 2003 *JR. Physics for Scientists and Engineersn (Thomson Brooks/kole)* p. 1296
[11] Abdeldayem M, Mabrouk M and Abo-Elnor M 2019 Estimating uncertainties for the driving torque in continuous flight Augur machine during space sampling drilling operation*IOP Conference Series: Materials Science and Engineering* p. 012024.

IOP Conf. Series: Materials Science and Engineering **973** (2020) 012013 doi:10.1088/1757-899X/973/1/012013

Assessment of steel wire's fatigue life using finite elements modelling and experimental testing

H M Lotfy[1,2], A B El-Shabasy[1,3], T A Attia[1,4] and HA Hassan[1,3]
[1]Faculty of Eng., Ain Shams University, Cairo, Egypt
[2] M.Sc. Student
[3]Professor
[4]Associate Professor
hanaamohamedlotfy@yahoo.com

Abstract. Various applications consist of wire rope as one of their most important construction parts. Wire ropes formed of three parts as basis; wires, strands and core. to get a deep understanding of wire ropes and their mechanical properties, it is essential to study their main part; the wire. For this reason, steel wire has been studied in this work to get its fatigue life as the number of cycles to failure using finite element modeling. FEM results have been verified using experiments; Flex tester machine and fatigue apparatus which is created and structured for this study. Outcomes have proved that using of bigger bending diameters resulted in increasing the fatigue life of the steel wire and vice versa. It is noticed that the FEM results have small and reasonable value of deviation comparing to the experimental results. Other mechanical properties of the steel wire have been obtained from FEM such as; equivalent Von-Mises stress, equivalent elastic strain, equivalent plastic strain and equivalent total strain.

1. Introduction

Wire ropes are considered as an essential element for many applications such as elevators, bridges, cranes and mining [1-3]. Multiple researches have been undertaken to examine the wire ropes and their mechanical specifications, performance, drawbacks and development of them during various conditions and using different forms of tests. Mechanical properties of the wire rope under tension have been studied by Juan Wu [4] using FEM. The results have been verified using theoretical analysis and experiments. Gerdemeli, et.al. [5], used FEM to estimate the fatigue life of one simple strand that has been projected to axial force, and several parameters such as, force, helix angle and length of the strand have been studied. While, Finite Element Modeling has been used by Sung-Yun Kim and Phill-Seung Lee [6], to get the mechanical properties of the wire rope using axial and transverse forces. Wire ropes have three fundamental components; wires, strands and core, Figure 1. Wires are the major parts of the wire ropes. These wires spirally convoluted to form the strands. Strands spirally convoluted around the core which is the interior part to form the final wire rope geometry [6]. It is significant to study the wires as they are the basic construction parts.

IOP Conf. Series: Materials Science and Engineering **973** (2020) 012013 doi:10.1088/1757-899X/973/1/012013

Figure 1. Wire rope construction [6].

Several types of tests such as; tension and fatigue have been conducted to evaluate the performance of the wire ropes [7-8]. Fatigue life and other mechanical characteristics could be estimated using finite element modeling. FEM basically depends on splitting the part concerned with the study into many small divisions each of them has to be studied individually. Applied forces, displacements and other factors that can affect the mechanical properties have to be studied as well [9-12]. The intended properties can be acquired by solving the stiffness matrix (1).

$$\{F\} = [K] * \{U\}. \quad (1)$$

Where, {F} is the applied forces array, [K] is the stiffness matrix, and{U} is the displacement array.

2. Experimental work

There are two types of experiments have been conducted to assess the fatigue life of steel wire similar to that used in elevator wire ropes. Fatigue life of the steel wire has been detected as the number of cycles that the wire can withstand till failure occurred. Tests have been done using three test specimens at least for each bending diameter for both types of tests. The two types of tests have been conducted using the Flex tester machine, figure 2 and the fatigue apparatus, figure 3 that has been designed and constructed for this study.

2.1 Flex tester machine

This machine is located at the Faculty of engineering, Ain Shams University. Flex tester machine consists of mandrel, supporting pulleys, counter, switch button, and two clips. It depends on bending the tested wire on the mandrel of two cylinders with different bending diameters (2, 4, 6, 8, 10, 12.7, 19.4 & 25) mm that have been used to evaluate the fatigue behavior of the steel wire with different strain values. The two ends of the steel wire have been fixed using two clips. The steel wire tested has 0.8 mm diameter

Figure 2. Flex tester machine.

2.2 Fatigue Test Setup:

This fatigue test setup has been designed and constructed for this study. Fatigue setup consists of drive and driven pulleys, two fixed indicators, movable indicator, counter, steel spring, motor, and power supply. It relies on the notion of the elevator wire ropes that expose to tension and bending over drums. Mainly, the steel wire has to be bent on the two pulleys with center distance between them equals to 272 mm, two ends of the wire have to be grasped using the movable indicator which has two bolts to make sure that the wire is properly fixed and tied. The drive pulley with the diameter of 133 mm gets its rotational motion from the motor connected to it. Due to the friction between the pulleys and the wire, the rotational force transmits to the wire which consequently rotates around the pulleys. The two fixed indicators refer to the start and the end of the stroke and send signals when the movable indicator presses on one of them to reverse the direction of motion.

Different diameters (35, 45 & 75) mm of the driven pulleys have been used to evaluate the fatigue behavior of the steel wire with different strain values.

To guarantee that the wire will be under tension, a steel spring with calculated force has been used. The spring force has been calculated using different weights with different spring elongations, and then substituted in the equation (2) to get the spring constant. Once the spring constant is calculated, the spring force in return can be calculated using spring constant with the actual elongation of the spring when it has been installed on the machine and by substitution in the same equation.

$$F = k \times \Delta x. \qquad (2)$$

Where, F is the spring force (N), k is the spring constant (N/mm), Δx is the spring elongation (mm).

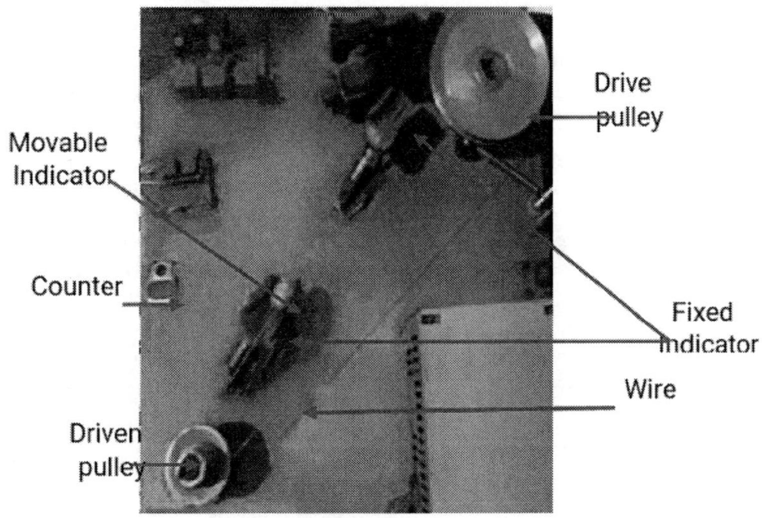

(a)

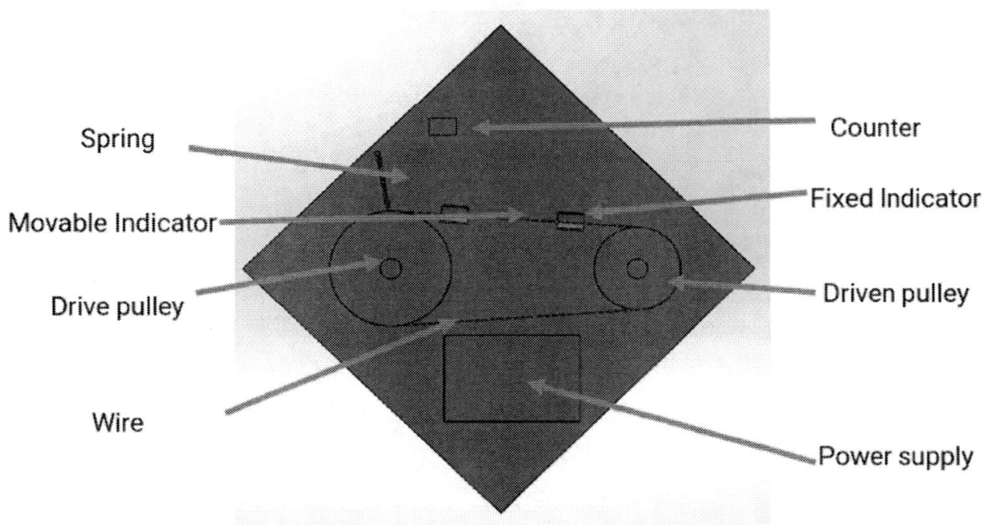

(b)

Figure 3. (a)The fatigue test setup. (b) Schematic drawing of the fatigue setup.

3. Finite element analysis

SOLID WORKS has been used for drawing of the models geometry and ANSYS WB program has been used for analyzing of finite element models. Two models have been set, the Flex tester machine model and the fatigue apparatus model. The model material's specifications have been obtained from tension test that has been done on the steel wire. Tensile yield strength = 2.5×10^{8} Pa, Young's modulus = 2×10^{11} Pa, Strength coefficient = 9.2×10^{8} Pa, Strength exponent = -0.106, Ductility coefficient = 0.213 and Ductility exponent = -0.47.

IOP Conf. Series: Materials Science and Engineering **973** (2020) 012013 doi:10.1088/1757-899X/973/1/012013

3.1 Flex tester machine

Only three diameters have been used to obtain the fatigue behavior using finite element modeling due to time concern. These diameters were 4, 6 and 8 mm which resulted in strain of 0.2, 0.133 and 0 1 respectively. Model joints have been set using two types, fixed joint for the machine base and rotational joints for all pulleys.

Contacts of the model have been adjusted to be frictionless between the wire and the mandrel, rough between the wire and the pulleys 1 & 2 and friction between the wire and the pulleys 3 & 4 with the symmetric behavior as the pure penalty formulation and normal nodal from contact as the detection method.

Meshing of the model has been determined to be 0.2 mm element sizing and inflation as summarized in Table 1 for the wire and the mandrel areas of contact, 2.5 mm element sizing for the wire and the pulleys 1 & 2 areas of contact, 0.5 mm element sizing for the wire and the pulleys 3 & 4 areas of contact and 1.5 mm face sizing for the wire. Number of nodes and elements are summarized in Table 2.

Table 1.Inflation details.

Mesh type	Geometry	Boundary	Inflation Option	Transition Ratio	Maximum Layers	Growth Rate
Inflation	Wire	1 Face (Symmetry Surface)	Smooth Transition	Default (0.272)	5	1.2

Table 2.Number of nodes and elements used in FE modeling for flex tester machine.

No.	Part name	Number of nodes Bending diameter = 4 mm	Number of nodes Bending diameter = 6 mm	Number of nodes Bending diameter = 8 mm	Number of elements Bending diameter = 4 mm	Number of elements Bending diameter = 6 mm	Number of elements Bending diameter = 8 mm
1	Wire	11570	11748	11996	5675	5728	5822
2	mandrel	7602	13990	21799	4184	7899	12610
3	Pulley 1	5206	5206	5206	858	858	858
4	Pulley 2	2200	2200	2200	400	400	400
5	Pulley 3	16546	16546	16546	3240	3240	3240
6	Pulley 4	8812	8812	8812	1719	1719	1719
7	Base plate	962	962	962	117	117	117
8	Total	52898	59464	67521	16193	19961	24766

Motion of the model parts have been restricted as shown in figure 4. The wire has been limited to move only in X-Y direction because not to move away from the pulleys. The mandrel has been limited to move only in Y direction. Force of 0.1 N has been applied on the two terminals surfaces of the wire.

IOP Conf. Series: Materials Science and Engineering **973** (2020) 012013 doi:10.1088/1757-899X/973/1/012013

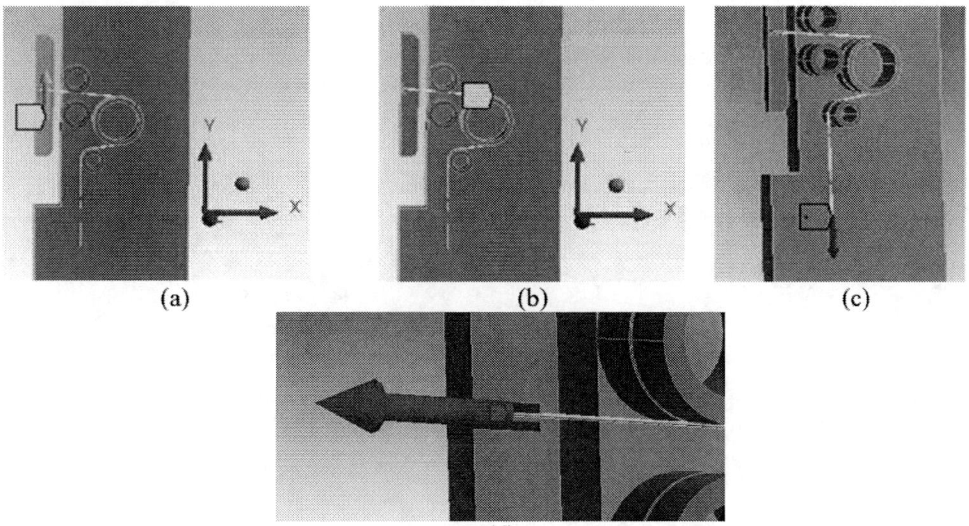

(d)

Figure 4. Setting of the model constrains. (a) Wire constrain to move in X-Y direction only. (b) Movable indicator constrains to move in X-direction only. (c) Force direction. (d) Rotational direction of the drive pulley.

Fatigue details of the model have been chosen to be 0.88 as the Fatigue strength factor (k_f), Fully reversed loading type (i.e. stress ratio (R) is equal to -1), Strain life as the analysis type, Morrow as the Mean stress theory and Max shear as the stress component.

3.2 Fatigue Test Setup

Only two diameters (35 & 45) mm have been used to obtain the fatigue behavior using finite element modeling due to time constrain. Model joints have been set using three types, fixed joint for the machine base, translational joint for the movable indicator and rotational joints for both drive and driven pulleys. Contacts of the model have been adjusted to be bonded between the wire and the movable indicator and rough between the wire and the pulleys with the symmetric behavior as the pure penalty formulation and normal nodal from contact as the detection method. Meshing of the model has been determined to be 2.5 mm contact sizing for the wire and the pulleys' areas of contact, and 2.5 mm sweep method for the wire. Number of nodes and elements are summarized in Table 3.

Table 3.Number of nodes and elements used in FE modeling.

No.	Part name	Number of nodes Bending diameter 35 mm	Number of nodes Bending diameter 45 mm	Number of elements Bending diameter 35 mm	Number of elements Bending diameter 45 mm
1	Wire	2156	2312	179	192
2	Drive pulley	23524	23733	4992	5034
3	Driven pulley	5378	12922	1044	2676
3	Movable indicator	2280	2280	360	360
4	Base plate	960	960	121	121
5	Total	34298	42207	6696	8383

Motion of the model's parts have been restricted as the following, Figure 5. The wire has been limited to move only in X-Y direction to avoid wire slipping. The movable indicator has been limited to move only in X direction. To ensure that the wire is under tension all the time, a force of 28 N has been applied. The drive pulley has been set to rotate counter clockwise.

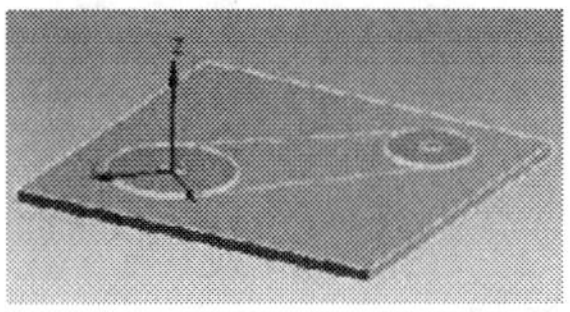

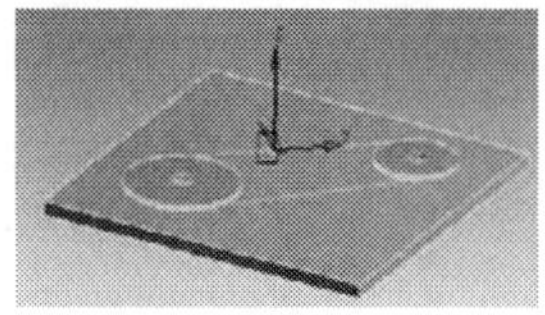

(a) (b)

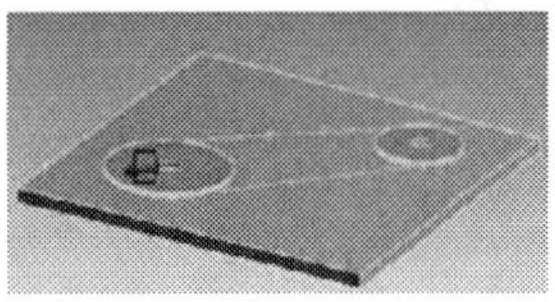

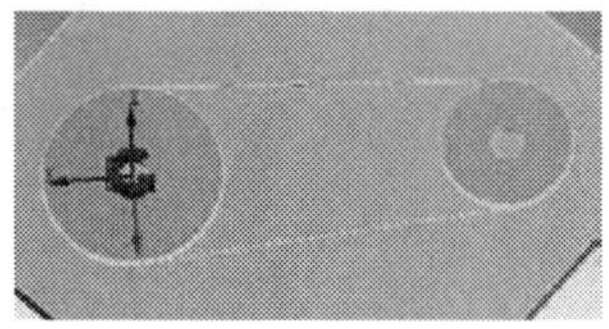

(c) (d)

Figure 5. Setting of the model constrains. (a) Wire constrain to move in X-Y direction only. (b) Movable indicator constrains to move in X-direction only. (c) Force direction. (d) Rotational direction of the drive pulley.

Fatigue details of the model have been chosen to be 0.78 as the Fatigue strength factor (kf), the strain ratio varies from 0.03 to 0.06 respectively according to bending diameter, Strain life as the analysis type, Morrow as the Mean stress theory and Max shear as the Stress component.

4. Results

4.1. Experimental Fatigue test

The resulted strain from both types of fatigue tests have been calculated using equation (3).

$$\varepsilon \text{ (Strain)} = d \text{ (wire's diameter in mm)}/D \text{ (bending diameter in mm)}. \quad (3)$$

4.1.1 Flex tester machine. Table 4, summarizes the fatigue life obtained from FE Modeling for Flex tester machine of different bending diameters. Figure 6 represents the relation between strain and fatigue life for tested wire using different pulley diameters.

Table 4. Fatigue life of the tested wire in cycles for flex tester machine.

Bending Diameter, (mm)	Strain = (ε)	N = fatigue life, cycles
2	0.4	1805
4	0.2	2014
6	0.13	2250
8	0.1	2484
10	0.08	2700
12.7	0.063	10612
19.4	0.04	12505
25	0.032	16726

IOP Conf. Series: Materials Science and Engineering **973** (2020) 012013 doi:10.1088/1757-899X/973/1/012013

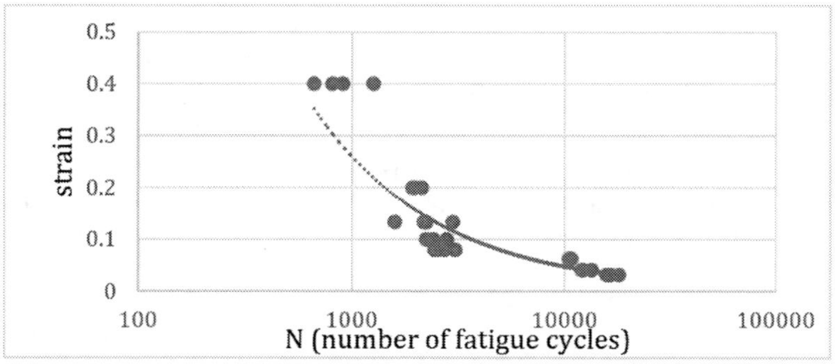

Figure 6. Relationship between fatigue life and strain for the flex tester machine.

From the previous results; it is noticed that using bending diameter of 2 mm which is the smallest bending diameter among the other values (2.5 times the wire diameter of 0.8 mm) has given the minimum fatigue life of the all results due to the higher value of strain and hence the stress applied on tested wire. Fatigue life values increased gradually with small values by increasing the bending diameter till reach the bending diameter of 10 mm which is 12.5 times the wire diameter. At this point the fatigue life increased to be 1.5 times of fatigue life resulted from using of 2 mm bending diameter. Then, the fatigue life increased relatively with high values by using bigger bending diameters from 12.7 mm which is near to 16 times of the wire diameter till 25 mm which is 31 times the wire diameter. Using of the biggest bending diameter which is 16 times of the wire diameter and 12.5 times the smallest one has given fatigue life of 9.3 times the value that resulted from using of the smallest bending diameter which is 2.5 times the wire diameter.

4.1.2 Fatigue Test Setup. Table 5 summarizes the fatigue life obtained from FE Modeling for fatigue apparatus of 4, 6 and 8 bending diameters. Figure 7 represents the relation between strain and fatigue life for tested wire using different pulley diameters.

Table 5.Fatigue life of the tested wire in cycles for fatigue test setup.

Bending diameter, (mm)	Strain (ε)	N = fatigue life, cycles
75	0.0107	15010
45	0.018	10591
35	0.0229	4399

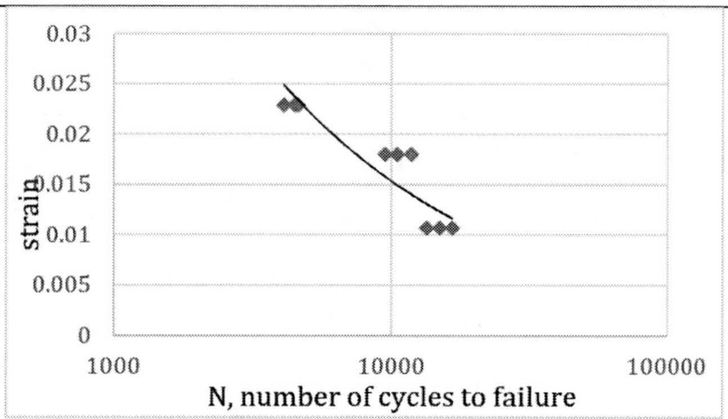

Figure 7. Relationship between fatigue life and strain for the fatigue test setup.

From the previous results; it is noticed that using of 35 mm as a bending diameter which is the smallest bending diameter among the other values (near to 44 times the wire diameter of 0.8 mm) has given the minimum fatigue life of the all results. Fatigue life values increased gradually by increasing the bending diameter till reach to 75 mm which is the biggest bending diameter that is near to 94 times the wire diameter and 2.1 times the smallest bending diameter which recorded the maximum fatigue life of the all results. The fatigue life resulted from using of the biggest bending diameter reached to be 3.4 times of using the smallest one.

Finally; it can be concluded that decreasing the bending diameter has a great effect on fatigue life, as it decreased sequentially due to the higher value of strain and hence the stress applied on tested wire.

4.2 Finite element analysis results

4.2.1 Fatigue life. Table 6 and 7 summarize the fatigue life obtained from FE Modeling for Flex tester machine and for fatigue test setup respectively for different bending diameters in mm.

Table 6. Fatigue life obtained from the finite element model of the flex tester machine.

Diameter (mm)	Fatigue life (cycle)
4	1967
6	2052
8	2206

Table 7. Fatigue life obtained from the finite element model.

Diameter (mm)	Fatigue life (cycle)
45	10759
35	4232

From the previous results; it is noticed that the resulted fatigue life from both finite element models increased by increasing the bending diameter. Which is in a good agreement with experimental results.

4.2.2 Von-Mises parameters. Tables 8 and 9, summarize Von-Mises parameters, such as equivalent stress, equivalent strain, equivalent plastic strain, and total strain obtained from FE modeling for the flex tester machine and fatigue apparatus respectively for different bending diameters mentioned before.

Table 8.Von-Mises parameters results.

Bending Diameter (mm)	4 mm		6 mm		8 mm	
	Minimum	Maximum	Minimum	Maximum	Minimum	Maximum
Average Equivalent (Von-Mises) Stress, MPa	0.17895	562.52	0.17839	557.04	0.17798	547.88
Equivalent (Von-Mises) Elastic Strain, mm/mm	8.9935×10^{-7}	2.8133×10^{-3}	8.9433×10^{-7}	2.7862×10^{-3}	8.9405×10^{-7}	2.7435×10^{-3}
Equivalent Plastic Strain, mm	0	0.10357	0	9.5594×10^{-2}	0	7.9814×10^{-2}
Equivalent Total Strain, mm	8.9935×10^{-7}	0.10557	8.9433×10^{-7}	9.7542×10^{-2}	8.9405×10^{-7}	8.165610^{-2}

Table 9. Von-Mises parameters results.

Bending Diameter (mm)	35 mm		45 mm	
	Minimum	Maximum	Minimum	Maximum
Average Equivalent Stress, MPa	0.47047	417.71	0.99736	337.9
Equivalent Elastic Strain	6.8891×10^{-5}	2.0885×10^{-3}	9.0784×10^{-5}	2.0557×10^{-3}
Equivalent Plastic Strain, mm	0	6.7661×10^{-2}	0	6.0595×10^{-2}
Equivalent Total Strain, mm	6.8891×10^{-5}	6.9414×10^{-2}	9.0784×10^{-5}	6.2288×10^{-2}

The results of Von-Mises parameters for the flex tester machine model are shown in figures 8 and 9.

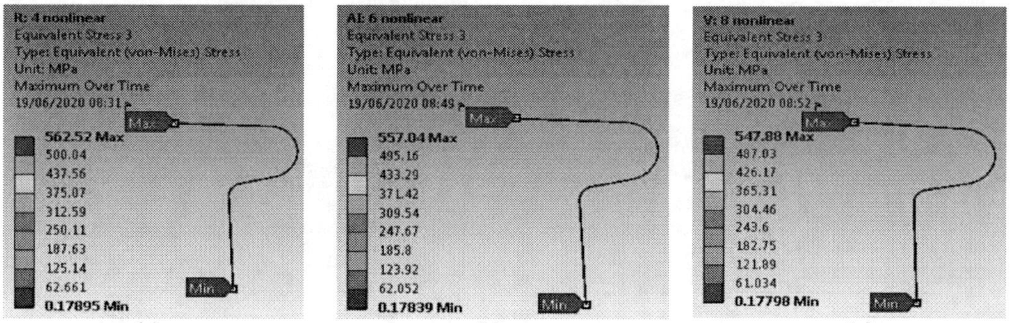

(a) (b) (c)

Figure 8. Equivalent (Von-Mises) stresses (a) Bending diamster 3mm (b) bending diameter 6 mm (c) bending diameter 8 mm

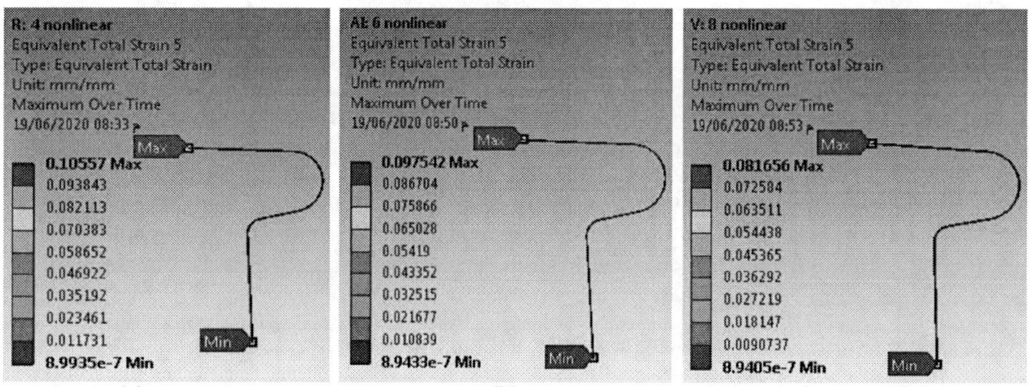

(a) (b) (c)

Figure 9. Equivalent (Von-Mises) totalstrainstresses (a) Bending diamster 3mm (b) bending diameter 6 mm (c) bending diameter 8 mm

The results of Von-Mises parameters for the fatigue test setup model are shown in figures 10 and 11.

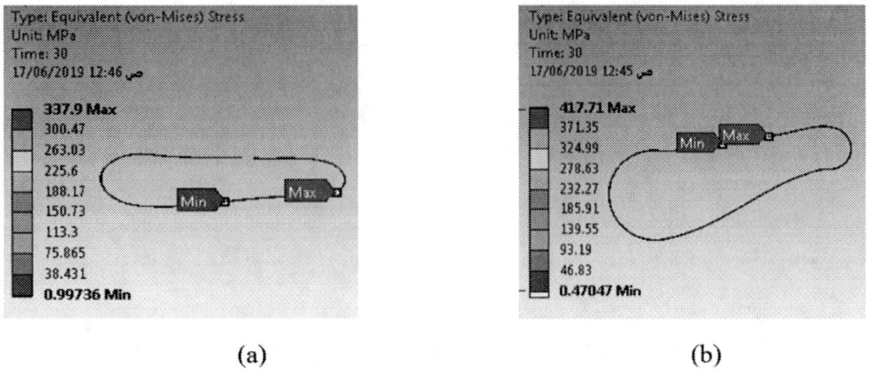

Figure10. Equivalent (Von-Mises) Stress. (a) Bending diameter 35 mm. (b) Diameter 45 mm.

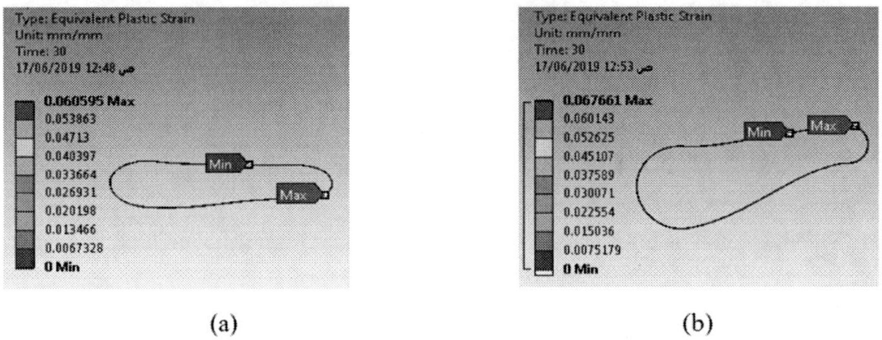

Figure 11.Equivalent plastic strain.(a) Bending diameter 35 mm. (b) Bending diameter 45 mm.

From the above results; it can be seen that the resulted Von-Mises parameters decreased by increasing the bending diameter for both finite element models.

Table 10 summarizes the comparison between the fatigue life obtained from experimental fatigue tests and from the FEM for the flex tester machine. While table 11summarizes the comparison between the fatigue life obtained from experimental fatigue tests and from the FEM for the fatigue test setup.

Table 10. Comparison of experiment and FEM fatigue life results for flex tester machine.

Diamete r, (mm)	Average Fatigue Life (Cycle) Experiment	Fatigue Life (Cycle) FEM	Differenc e	Deviatio n %
4	2014	1967	47	2.33
6	2250	2052	198	8.8
8	2484	2206	278	11.19

Table 11. Comparison of experiment and FEM fatigue life results for fatigue test setup.

Diameter ,(mm)	Average Fatigue Life (Cycle) Experiment	Fatigue Life (Cycle) FEM	Difference	Deviation %
45	10591	10322	269	2.5415
35	4399	4232	167	3.7936

From Tables 10 and 11, it can be concluded that the results obtained from the finite element models are fair and have reasonable drift percentage compared to the results obtained experimentally from conducting fatigue tests using both testing devices, flex tester machine and the fatigue apparatus.

5. Conclusion

Fatigue life of steel wire has been studied using FEM by ANSYS program as well as conducting fatigue tests using flex tester machine and fatigue apparatus. Fatigue test setup has been created and structured in order to conduct experimental fatigue tests for steel wire. Flex tester machine has been used successfully to obtain the fatigue life of steel wire using different bending diameters of 2, 4, 6, 8, 10, 12.7, 19.4 and 25 mm. Fatigue apparatus has been used successfully to obtain the fatigue life of steel wire using different pulley diameter of 35, 45, and 75 mm.
From the results obtained from this work, the following conclusions can be drawn.

- Decreasing the bending diameter of Flex tester as well as the Fatigue Test Setup has a great effect on fatigue life, as it decreased sequentially due to the higher value of strain and hence the stress applied on tested wire.
- Using of biggest bending diameter that is 16 times the wire diameter increased the fatigue life by 9.3 times compared to using of the smallest diameter, which is 2.5 times the wire diameter.
- Using of biggest bending diameter of Fatigue Test Setup that is 94 times the wire diameter increased the fatigue life by 3.4 compared to using of the smallest one, which is 44 times the wire diameter.
- Using smaller bending diameter achieved a higher values of Von-Mises equivalent stress, equivalent strain, equivalent plastic strain, and total strain.

- Using smaller bending diameter resulted in reducing the fatigue life due to the increase of the bending strain and bending stress.
- The results have been obtained from Finite Element model were in a good agreement with the experimental results with an acceptable percentage of drift.

References

[1] Jay C Martin 2014 Strands That Stand: Using Wire Rope to Date and Identify Archaeological Sites *The International Journal of Nautical Archaeology* **43**.1 pp151–16
[2] Yusuf A O and Erdemİmrak C 2012 Experimental and Theoretical Investigation of Bending Over Sheave Fatigue Life of Stranded Steel Wire Rope *Indian Journal of Engineering & Materials Sciences,* **Volume. 19** pp189-195.
[3] Beltrána JF and De Vico E 2015 Assessment of Static Rope Behavior with Asymmetric Damage Distribution *Engineering Structures* pp 84-98.
[4] Wu J 2014 The finite element modeling of spiral ropes *Int. J Coal Sci. Technol.*, **1**(3) pp 346–355.
[5] Gerdemeli I, Kurt S and Anıl AS 2012 *Analysis with Finite Element Method of Wire Rope (Turkey : Istanbul Technical University Faculty of Mechanical Engineering)*
[6] Kimab S Yand Lee P S 2017 Modeling of Helically Stranded Cables Using Multiple Beam Finite Elements and Its Application to Torque Balance Design *Construction and Building Materials* pp 591-606.
[7] Kim S H, Bae R H and Kwon J D 2107 Bending Fatigue Characteristics of Wire Rope *Journal of Mechanical Science and Technology* ***26*** pp 2012 -2110.

[8] Basana R et al 2010A Novel Approach to the Estimation of Strain Life Fatigue Parameters *Procedia Engineering* pp 417–426.

[9] Hussain N, Baharom M and Maharun M 2016 Analysis of Optimum Wire Rope Configuration for Equal Unidirectional Torsional Stiffness for Flexible Steering Shaft *proceeding of MATEC Web of Conferences* **74**.

[10] Hui-ling Z, Xin-yinZ and Yu-lin W 2014 Numerical Analysis of Sliding Friction Coefficient and The Normal Pressure between the Cable Wire *International Journal on smart sensing and intelligent systems*, **volume. 7**, NO. 4.

[11] Fraldi M, Perrella G , Ciervo M, Bosia F and MPugno N 2017 A hybrid deterministic-probabilistic approach to model the mechanical response of helically arranged hierarchical strands *Journal of Mechanics and physics of solids* pp 338-352.

[12] Madenci E and Guven I 2015 The Finite Element Method and Applications in Engineering Using ANSYS Second Edition, *Springer International Publishing*.

IOP Conf. Series: Materials Science and Engineering **973** (2020) 012014 doi:10.1088/1757-899X/973/1/012014

Interactions between microorganisms and composite material with marble & granite filler

H Ahmed[1] and O Darwesh[2]
[1] Dr. Eng. consultant in materials sciences & materials applications, Egypt
[2] Agricultural Microbiology Department, National Research Centre, Egypt
hebatalrahman11@yahoo.com

Abstract. It is promising to develop new composite materials which have antimicrobial activity for many applications; this is our goal in this study. All experiments were done to achieve this an important objective by studying the interaction between the new composites and some pathogenic reference microorganisms. The produced composite was manufactured from industrial wastes. Recently, many research groups in all fields have focusing work on removing the wastes and recycling it as useful materials. In this study, the preparing new composite from marble and granite wastes with thermoplastic matrix PMMA was investigated. This composite is considered as new type of micro- nano composite because it has fillers in the nano and micro sizes. Pathogenic microorganisms, used in this study, obtained from the American type culture collection (ATCC; Rockville, MD, USA), yeast and one pathogenic fungus were applied. The evaluation of analysis was done by spectrophotometer SEM and EDX techniques. The suitable applications of new composite material were mentioned based upon the interaction recorded in the study. The study ends with recommendation for future modifications in the new composite materials to improve its resistances to microorganism.

1. Introduction

Exposure to microorganisms, such as fungi, bacteria, viruses and their biological by-products, may cause a lot of defects in materials. The type of defects takes different forms; it may be lead to change in materials micro-structure, change in chemical composition, change in dimension which leads to harmful and cracks in the structure [1,2]. Contact of materials and machine parts with pathogenic microorganisms and their by-products inside or outside parts usually occurs by interaction between moisture, lubricant and different substances which contain microorganisms during manufacturing and operation [3,4]. In order to achieve the acceptable level of interaction between materials and microbes in different application like buildings, machinery and daily life applications, the contacting technology with fungi and other pathogens should be minimized or eliminated [5]. The pathogens distribute and develop daily to resist used antibiotics and antimicrobial materials [6]. For that, researchers work hard to find solution for this problem. Some constructions made to face the problem of pathogens distribution in different sectors like production of new biomaterials [7], development of antimicrobial nanomaterials [8] and converting the wastes to useful substances [9]. Prevention and protection of the material must be related to the factors causing interaction between materials and microorganisms which can be summarized in three main actions;

1. There must be a reservoir (i.e., location where an unusually high concentration of microorganisms is present).
2. The microorganisms must be released into the air.

3. The microorganisms must be allowed to reproduce favorable conditions are needed for reproduction to occur [10]. For example, fungal growth is usually optimized when moisture levels are high. Figure 1 shows the micro-organisms types and effects [11].

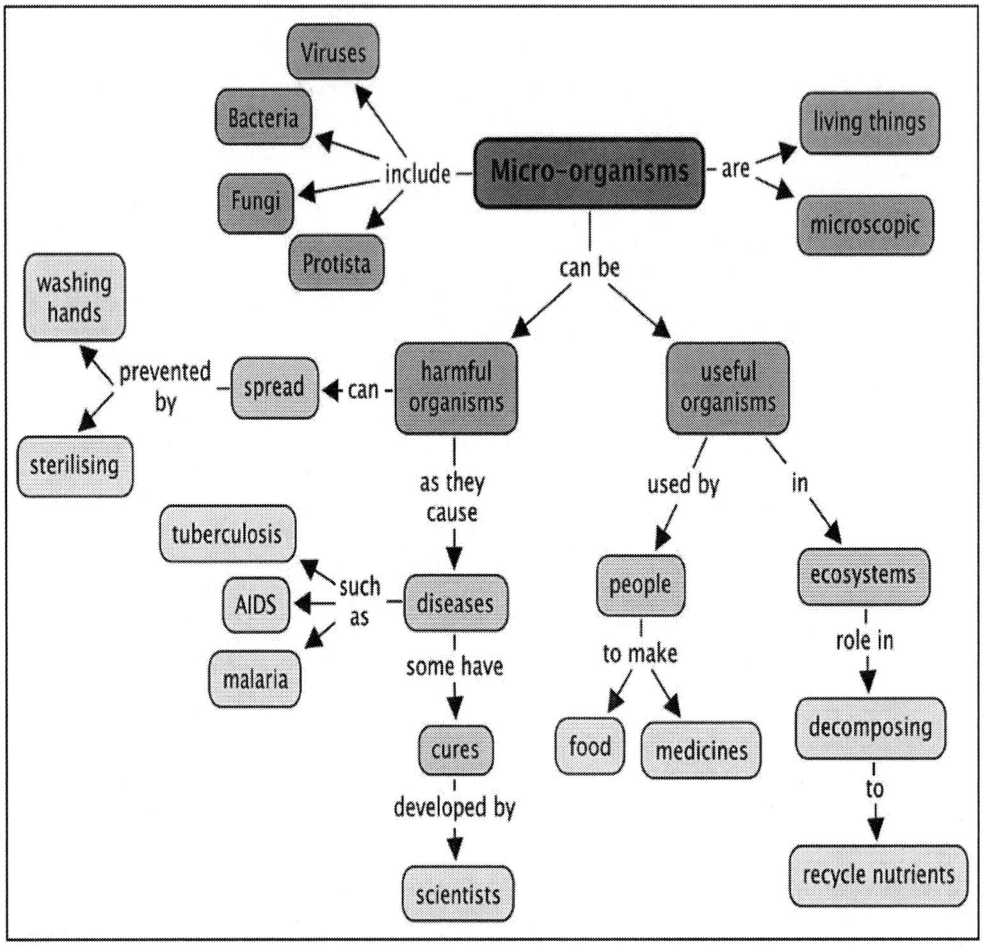

Figure 1. The micro-organisms types and effects [11].

Fault diagnosis, qualitative and quantitative analysis are the promising techniques to evaluate the interaction between microorganisms and materials from the material point of view, re-maintenance and preventive maintenance program should be implemented that provides regular inspections [12].
Inspection techniques must be followed by treatment methods before cracks initiation and propagation to the critical size. Management requirements for the materials exposed to contamination leads to failure by microorganisms must include commissioning, operation, maintenance, repair, replacement, and expansion of new and existing material [13]. Different techniques such as coating, surface alloying, painting, lubrication and material replacement by new materials may be used [14]. The selection of best technique to eliminate and/or reduce the interaction results between material and microorganism depend on different factors; some of them related to materials and their operation environment conditions and others related to the type and amount of microorganism, the factors may be correlated and interrupt together [15]. In other words, we can say that as the interaction between microorganism and material are complicated process the protection techniques are also need so many precautions to be considered [16]. In this trend, the present study focused on converting the granite

and marble wastes to useful materials having antimicrobial activity to face the problems causing by interaction between materials and pathogenic microbes. The main focus of the paper is to investigate the interaction between microorganisms and the developed composite material.

2. Experimental Work

2.1 Materials
The matrix used in this study is poly methyl methacrylate (PMMA). It is a thermoplastic agent prepared from monomer (methyl methacrylate) by additional polymerization process; the form of polymer granule was in less than 5mm diameter.

2.2 Filler
Marble & granite refuses consists of wet refuse (Sahla). It is a type of the wet refuse where its percentage of water reaches about 70%. It is sticky and it is produced from the processes of leveling and polishing. Commercially it is known as Al Sahla. The sticky refuse is heavy in weight and big in size. Lime 28-32%, Silica 3-30%, Magnesium oxide 20 to 25%, Iron oxides 1-3%, and losses 20-45%. After drying, the particles size ranges from 50 to 300 μm for marble wastes, while granite wastes particle size ranged from 150 to 500 μm.

2.3 Preparation of wet refuse
The mixture is automatically pushed to the drying room. The heaters were adjusted at 120°C for period from three to four h. For guaranteeing the drying process, wet refuse (Sahla) is got out of the dryer in the form of fragile masses which are transferred by a conveyor to the automatic turning over room and made in the form of powder suitable for mixing [17-18].

2.4 Manufacturing of the composites
Plastic powders and treated marble and granite wastes were mixed together in the solid state with 30% volume fraction. The mixture was heated at the required temperature about 300 -400° C, according to its components. The heating rate depends on the type of the joining material; compression 160-200 Kg/cm^2 and cooling were done for heated mixture. Nano and micro -composite material was fabricated and the new composite material was distinguished with the light weight and the ability to endure scratch and damping capacity comparing with the natural marble and /or granite. It was also labled for the easy formation in different colors and shapes [19].

2.5 Preparation of pathogenic microbe's cultures and making interaction with produced composites
Pathogenic microorganisms, used in this study, were obtained from the American type culture collection (ATCC; Rockville, MD, USA). Two gram-positive bacteria, two gram-negative bacteria, yeast and pathogenic fungus were applied [20]. The details of pathogenic microorganisms subjected for this study were represented in table 1.

Table 1. Pathogenic microorganisms used to study their interaction with produced compounds.

Type	Microorganisms
Gram +ve Bacteria	Bacillus cereus ATCC- 12228
	Staphylococcus aureus ATCC-47077
Gram -ve Bacteria	E. Coli ATCC- 25922
	Salmonella typhi ATCC 15566
Fungi	Candida albicans ATCC- 10231
	Aspergillus niger ATCC- 16888

A loopful of bacterial growth from culture slope was inoculated into Minerals salt medium (MSM) broth medium and incubated at 30°C under shaking conditions (100 rpm) for 48 h. After incubation period, 100 µl of growing cultures were inoculated into flasks (250 ml) containing 100 ml of MSM broth medium and the treatments (surface mechanical attrition treated sheets at times and frequencies, also etched as treated surfaces) were added to flasks [21]. The inoculated flasks were incubated at 37°C for 7 days under shaking conditions with speed (120 rpm). In case of fungal strain, 100 µl of spore suspensions were inoculated into flasks (250 ml) containing 100 ml of MSM broth medium (pH 5.5) and the sheets were added. The flasks were incubated under shaking conditions for 7 days at 30°C. The microorganism growth was recorded by two methods: -

1. Weighting the cell biomass daily by digital electrical balance of fungal strain and measuring the changing in cultures optical density using spectro-photometer (Jenway UV/Visible- 2605 spectro-photometer, England) at wavelength 550 nm [22].
2. Measurement of the changing in pH using digital Orion pH meter (model 420A).

Composite material exposed to microorganism in disc shape with 10mm length and 16mm diameter.

2.6 Scanning Electron Microscope (SEM)

The specimens were examined by scanning electron microscope (SEM) operating at a nominal accelerating voltage of 30kv. Specimen preparation is very simply accomplished by cutting a thin slice of the specimen containing the surface of interest, the samples were inserting into the specimen chamber for direct examination of the effects microorganisms on the structure [23].

2.7 Energy dispersive X-ray "EDX"

The quantitative method of elemental analysis of the samples has been examined by SEM JSM-T200 at 25KV acceleration voltage, 20mm working distance, and magnification 200x, (1peak omitted 0.02 KeV). Each value is at least an average of 2 readings.

3. Results and discussion

Pathogenic microorganisms had distribution in various environments; agriculture, industrial, medical and municipal waste water. These pathogens cause many problems toward people, animals and plants [47, 48]. The global researchers have continuously research about new technology for pathogens disinfection such as using plant extracts nano-materials and new formulations. In this study, we used the new formulation (micro-nano composite) from marble and granite wastes to obtain composite materials with behavior of antimicrobial activity.

Six microbial pathogens (St. aureus, B. cereus, E. Coli, S. typhi., C. Albicans, A. niger) were developed in specialized and liquid environments together for a week under optimal growth conditions and microbial growth was monitored during the growth period. Samples were extracted and washed. The microorganisms in the study range were divided into three main categories G^+ bacteria include (St. aureus, B. cereus), G^- bacteria include (E. Coli, S. typhi) and fungi which include (A. niger, C. Albicans). the interaction between each category and the new thermoplastic based composite with marble and granite filler were studied [24]. The influence of pH on the relative importance of the decomposer in thermoplastic base micro and nano composite. The behaviour of composite was investigated along a continuous pH gradient. This experimental location provides a uniform pH gradient, ranging from pH 7 or 5.5, incubated at 30 °C under shaking conditions (100 rpm) for 48 h, it has been continuously grown for more than 7 days as shown in figures (2), (3) and (4), respectively, which represent the rate of change in basicity for bacteria G+, bacteria G^- and fungi, respectively. The changing in microbial growth with time give indication about the interaction between microorganisms and composite material figures (5), (6) and (7) show these variations of weight for all microorganisms in the study range. The bacteria G+ specially, the composite material control sample basicity remains constant at normal basicity PH 7 for 4th days and in 5th day, it dropped down to PH 6.8, the decrease in basicity is continue with time and reach PH 6.5. The rate of basicity decrease with B. cereus, it is decreased and just reach B cereus PH 6.8 after 7 days while St. aureus have the same behaviour of the control sample as shown in figure (2). The bacteria G^- such as E. Coli basicity is dropped down faster after only three days, the rate of decrease continue at the same rate and reach PH 6.7 at the end of the 7 days, S. typhi has the behaviour but the rate of decrease is less, it reaches 6.9 PH at the end of the 7 days while rate of decrease began after 4 days figure (3). The fungi basicity as shown in figure (4) yeast (C. Albicans) have the same behaviour of the control sample while the A. niger is dropped down after only 2 days and reach to PH 2.2 at the end of the 7th day. The fungal growth was recorded by weighting the cell biomass daily. The weight changes with microbial growth almost have the same behavior for all microorganism under study. The results in figures (5), (6) and (7) showed that when water was absorbed by capillary action to the materials, fungal and bacterial growth started fast and was abundant in the composite materials with time above 20 %. Such a limit value could not be defined for fungal contamination in the composite material because of complex behavior of moisture covering. The condensation under the varying RH and temperature conditions caused only restrained fungal growth in the materials [25-27].

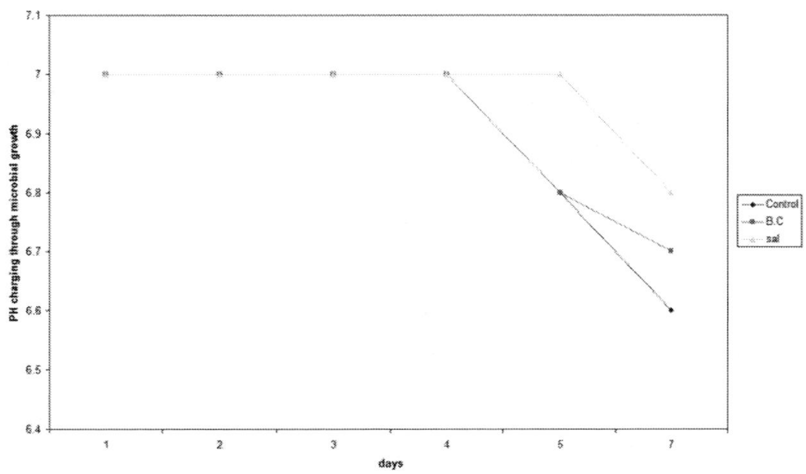

Figure 2. PH (basicity) change during microbial growth for G^+ bacteria.

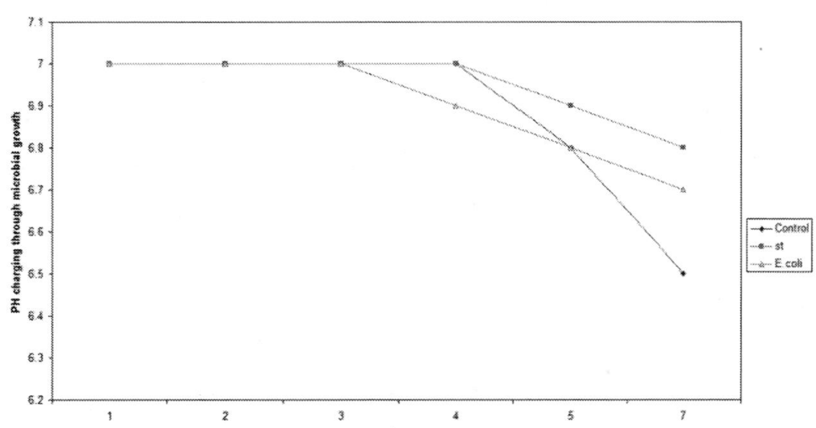

Figure 3. PH (basicity) change during microbial growth for G^- bacteria.

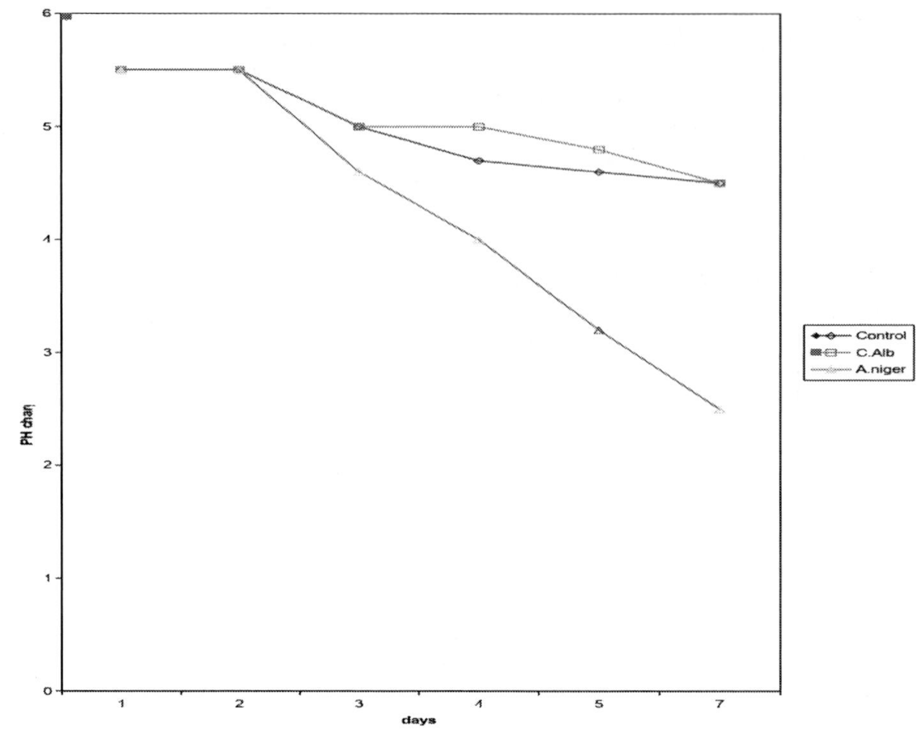

Figure 4. PH (basicity) change during microbial growth for fungai.

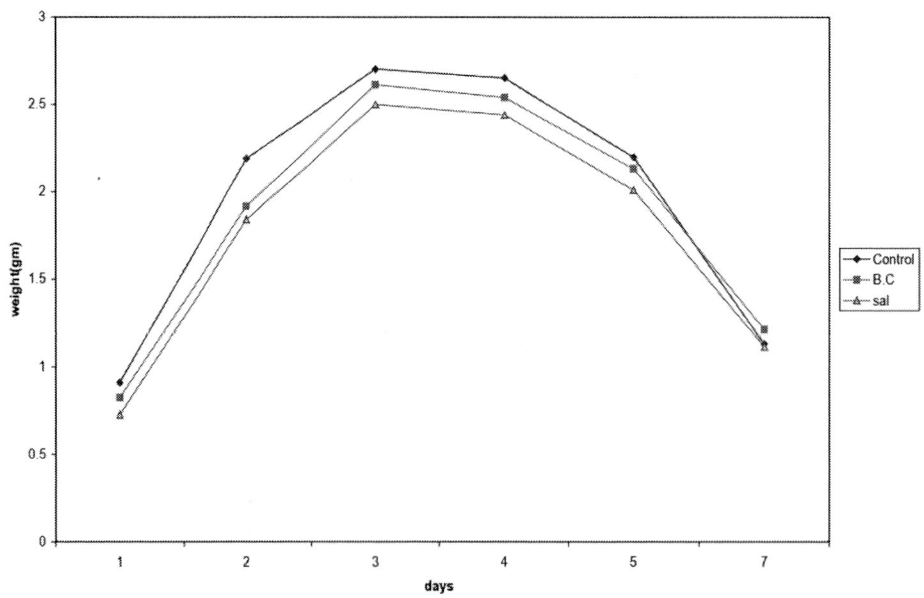

Figure 5. Weight change during microbial growth for G^+ bacteria.

IOP Conf. Series: Materials Science and Engineering **973** (2020) 012014 doi:10.1088/1757-899X/973/1/012014

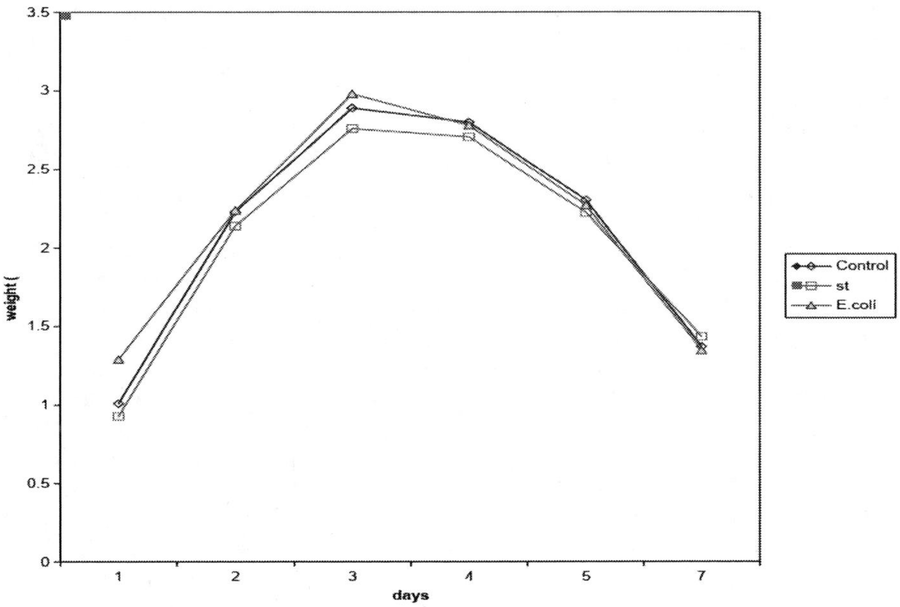

Figure 6. Weight change during microbial growth for G⁻ bacteria.

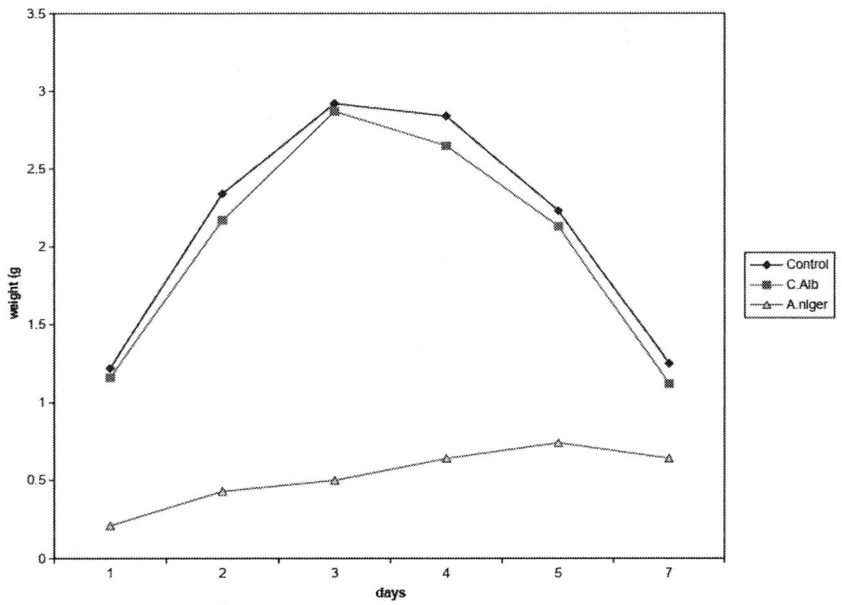

Figure 7. Weight change during microbial growth for fungai.

The shapes of the growth curves were different depending on the nutrient medium used and the species tested. Microbes can interact in the environment with materials/surfaces in so many ways, the interactions lead to change in micro structure and chemical composition of the composite material. The discs of the produced composite were incubated with the tested pathogenic microbes for 7 days and then examined for their surface changing using scanning electron microscopy (SEM). The pictures produced from SEM were illustrated in figure 8. No clearly changing was detected in the internal structure of composites, just few changing resulting from microbial growth or metabolities adhesion onto discs surface were detected. The elemental analysis by EDX as represented in figure (9), changing in elemental profile of composite surface was noted.

Certain species of fungi is the engines of the process of decomposition. Fungi is the major organisms decomposing of organic substances. Decomposition can be a problem on a large and small scale. Organic describes any material made up of molecules containing carbon and hydrogen atoms. The matrix of the composite material under consideration is PMMA is considered organic. Decomposition is a complex process. Organic matter is broken down into carbon dioxide and the mineral forms of nutrients like nitrogen. It is also converted into fungi and bacteria through these organisms feeding on the organic material and reproducing. Scientists call the organisms that decompose organic matter decomposers, saprobes or saprotrophs. Decomposition is made up of a number of sub -processes. Consider the decomposition of matrix which is the binding element in the structure, it breaks the matter into smaller pieces in a process called fragmentation. This is an important step because smaller fragments have more surface area to support the growth of fungi. Fungi growth is especially affected by fragment size, since fungi can penetrate substances more easily than bacteria or other microorganisms.

The fast drying (RH 30%) seemed to decrease the viability of fungi but along with the experiment fungal flora was modified to tolerate fluctuating conditions and the drying at RH 50% had only a slight effect on the viability of fungi. In the case of yeast there was a decrease in the amount of organic matter, which means that Yeast interact with the matrix and swelling of matrix was done by yeast. Strong precautions must be considered in using Composite material in applications have Yeast (Candida albicans ATCC- 10231), which has bad reaction with the organic matrix. Staphylococcus aureus ATCC-47077 is spread in hospitals and it is recommended to use composite material with marble filler samples under study in hospital floors and walls. The adverse effect appears clearly for Staphylococcus aureus ATCC-47077 with the composite material with granite filler, the matrix erosion appears clear in SEM photos and decrease in oxygen and hydrogen percentages appear in EDX analysis, so using of these materials in hospitals is prevented. Salmonella typhi ATCC 15566 did not have any obvious effect on the composite materials with marble and /or granite filler used in poultry raising wards and poultry processing places. Environmental factors influence rate of bacterial growth such as acidity (pH), temperature, water activity, macro and micro nutrients, oxygen levels, and toxins. Conditions tend to be relatively consistent between bacteria with the exception of extremophiles

IOP Conf. Series: Materials Science and Engineering **973** (2020) 012014 doi:10.1088/1757-899X/973/1/012014

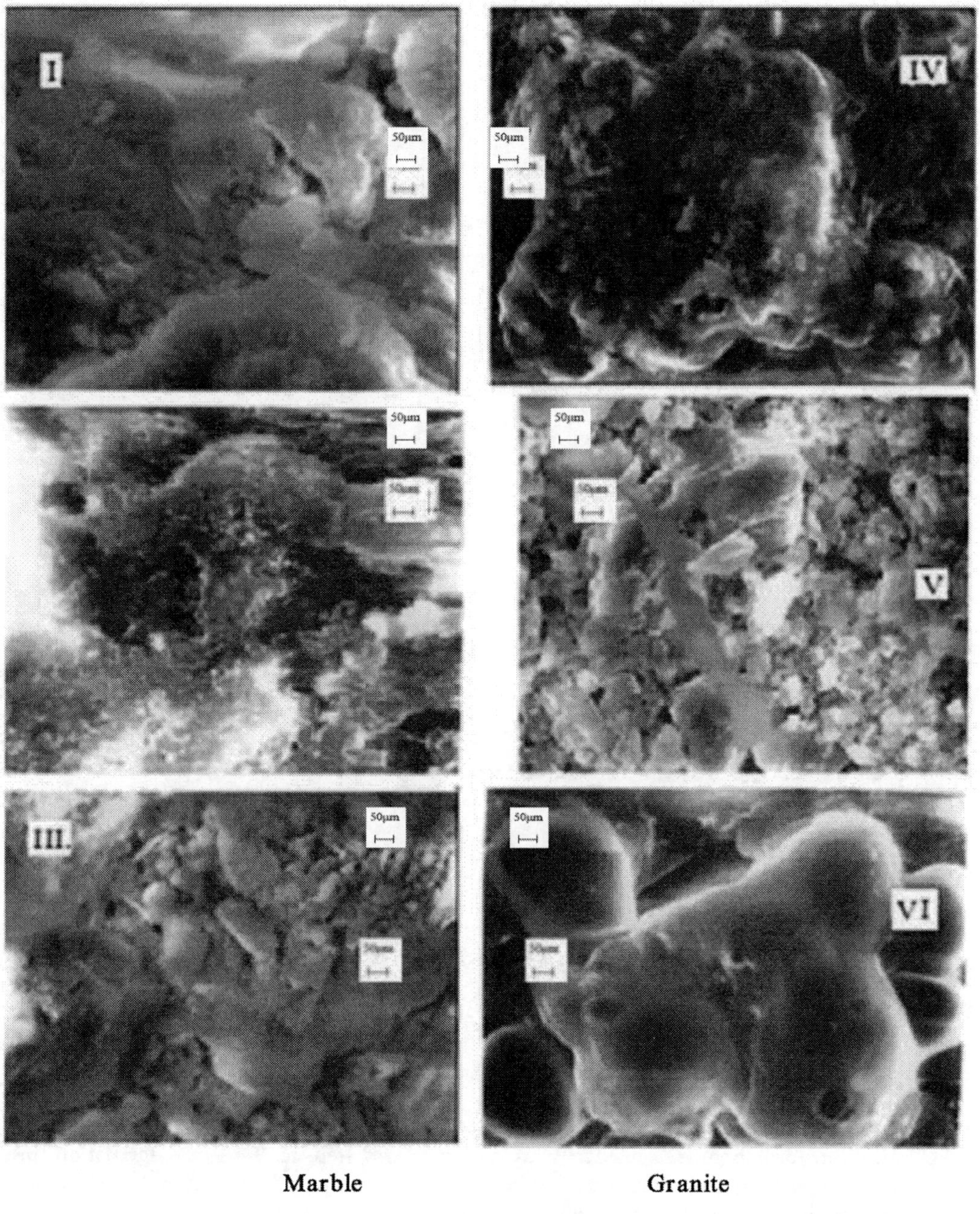

I. Control marble	II. Staureus Marble	III. Albicans Marble
IV. Control Granite	V. Staureus Granite	VI. Albicans Granite

Figure 8. Scanning electron microscopy image for interaction between microorganisms and composites.

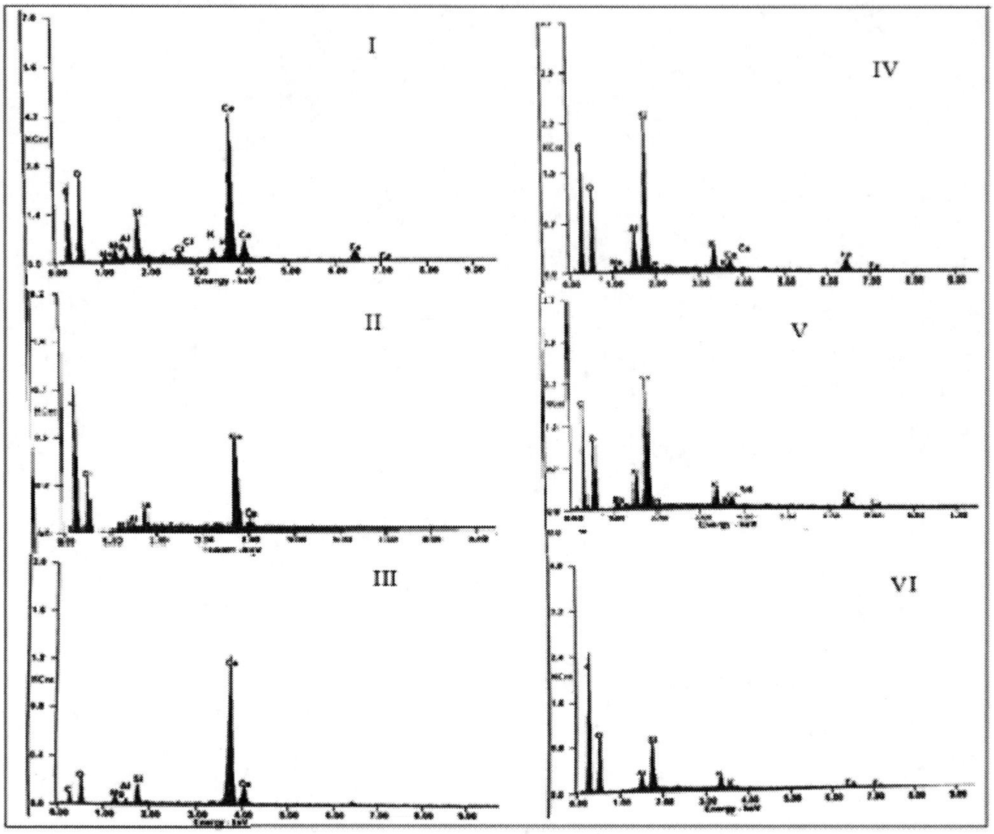

Marble **Granite**

I. Control Marble II. Staureus Marble III. Albicans Marble
IV. Control Granite V. Staureus Granite VI. Albicans Granite

Figure 9. EDX images for interaction between microorganisms and composites.

4. Conclusions

- The interaction between microbes and materials is function of time and type of microorganism, weight loss and change in basicity PH can be recorded significantly during the interaction of microorganism with composite material, the behavior depends mainly on the composition of composite materials and type of microorganism.
- Biofilm formation is considered as the primary step to start a degradation process, microorganism become in close contact with the surface and creating a micro environment that can be totally different from the bulk with distinct properties
- Six pathogenic microbes were developed in specialized and liquid environments
- For yeast there was a decrease in the amount of organic matter, which means that Yeast interact with the matrix and swelling of matrix was done by yeast. Strong precautions must be considered in using Composite material in applications have Yeast (Candida albicans ATCC-10231), which has bad reaction with the organic matrix.
- Staphylococcus aureus ATCC-47077 is spread in hospitals and it is recommended to use composite material with marble filler samples under study in hospital floors and walls.

- The adverse effect appears clearly for Staphylococcus aureus ATCC-47077 with the composite material with granite filler, the matrix erosion appear clear in SEM photos and decrease in oxygen and hydrogen percentages appear in EDX analysis, so using of these materials in hospitals is prevented.
- Salmonella typhi ATCC 15566 did not have any obvious effect on the composite materials with marble and /or granite filler used in poultry raising wards and poultry processing places. Environmental factors influence rate of bacterial growth such as acidity (pH), temperature, water activity, macro and micro nutrients, oxygen levels, and toxins. Conditions tend to be relatively consistent between bacteria.

References

[1] Annual Review of Nursing Research 2015 *Traumatic Brain Injury* (New York USA: Springer publishing company) Vol **33**.
[2] Mohamed LA and et al 2015 Yeast cell-based analysis of human lactate dehydrogenase isoforms. *J Biochem* **158** (6) pp 467-476
[3] https://www.sciencemag.org/news 2015 05 silver-turns-bacteria-deadly-zombies.
[4] https://www.silverhealthinstitute.com 2015 microscope-view-of-silver-killing-bacteria
[5] https://www.flickr.com photos 121935927@N06/13537347284.
[6] http://www.ucmp.berkeley.edu/fungi/fungi.html
[7] Samuels GJ 2006 Trichoderma: systematics, the sexual state and ecology *Phytopathol.* **96** (2) pp195–206.
[8] Zisova LG 2009 Malassezia species and seborrheic dermatitis *Folia Medica.* **51** (1) pp 23–33.
[9] Gadd GM 2007 Geomycology: biogeochemical transformations of rocks, minerals, metals and radionuclides by fungi bioweathering and bioremediation *Mycological Research.* **111** (Pt 1) pp3–49.
[10] Lindahl BD and et al 2007 Spatial separation of litter decomposition and my corrhizal nitrogen uptake in a boreal forest *The New Phytologist.* **173** (3) pp 611–620.
[11] Desjardin DE and et al 2010 Luminescent Mycena: new and noteworthy species *Mycologia* **102** (2) pp 459–477.
[12] Chandler PJ 2010 *A Dipterist's Handbook* (U.K The Amateur Entomologists' Society 2nd Edition) pp 1–525.
[13] Christian V and et al 2005 Degradation of xenobiotic compounds by lignin-degrading white-rot fungi: enzymology and mechanisms involved Indian *J Experimental Biol.* **43** (4) pp301–12.
[14] Chandler PJ 2010 *A Dipterist's Handbook* (U.K The Amateur Entomologists' Society 2nd Edition) pp 1–525.
[15] http://www.kew.org/science-conservation/plants-fungi/fungi/about
[16] Soll DR 2012 Signal Transduction Pathways Regulating Switching Mating and Biofilm Formation in Candida Albicans and Related Species. In: G. Witzany (ed) Biocommunication of Fungi Springer pp85-102.
[17] Erdogan A and Rao SS 2015 Small intestinal fungal overgrowth *Curr. Gastroenterol Rep* **17** (4) pp 16.
[18] Martins N and et al 2014 Candidiasis: predisposing factors, prevention, diagnosis and alternative treatment *Mycopathologia* **177 (5-6)** pp223–240.
[19] Tortora and Gerald J 2010 *Mibrobiology: an Introduction* (San Francisco CA: Pearson Benjamin Cummings) pp 758.
[20] Hebatalrahman 2012 Study of The Manufacturing Parameters Affect The Fabrication of Nano and Micro Composites Researcher vol 4(1) pp15-23 (ISSN: 1553-9865)., http://www.sciencepub.net/researcher/research0401
[21] Hebatalrahman 2011 Risks Concerning Work in Building Materials Industries Case Study of Marble and Granite Mining Region in Tora Researcher vol 3(12) pp46-56 ISSN: 1553-9865 http://www.sciencepub.net/researcher.,http://www.sciencepub.net/researcher/research0312/009

7555breseach0312_46_56.pdf.,
[22] Hebatalrahman 2010 A new Method for Fabrication and Laser Treatment of Nano-Composites *Journal of American Science* **6** (7) pp149-154 ISSN: 1545-1003 www.americanscience.org/journals/am-sci/am0607/am0607.pdf.
[23] Hygiene 2017 World Health Organization (WHO).
[24] Hebatalrahman and T. M. Meaz 2014 Marble Wastes as Filler in Heterogenious Dielectric Composites *1st International Conference on Innovative Building Materials.*
[25] Hebatalrahman 2017 Characteristics of Micro & Nano Composite from Marble& Granite Wastes *Sustainability in Envrionment* **2** No 3
[26] Mukherjee P K and et al 2015 Mycobiota in gastrointestinal diseases *Nat. Rev. Gastroenterol Hepatol* **12** (2) pp 77–87.
[27] https://en.wikipedia.org/wiki/Candida_albicans

IOP Conf. Series: Materials Science and Engineering **973** (2020) 012015 doi:10.1088/1757-899X/973/1/012015

Optical non linear characteristics of particles from natural resources

H Ahmed[1] A Salah[2] S Hasabelnabi[2] and A Khalil
[1] Consultant in materials sciences & materials applications, EGYPT
[2] National Institute of laser enhanced Science, NILES, CU, EGYPT
hebatalrahman11@yahoo.com

Abstract. In these work some nano and micro particles such as granite, zeolite, diopside ferrous and sand glass which considered as waste or byproduct from mining industries. The micro and nano particles from natural resources were prepared by sieving and mixing with acidic solution at certain concentration, the particles are soaked and immerged in the acidic solution for long time to achieve reasonable hemogenity, sterring is done directly before measurements. The non linear characteristics of the prepared solution were measured by z-scan open aperture technique. The results explained in terms of the composition and structure of the micro and nano particles. Granite particles does not introduce any emission due to compact structure, while zeolite, diopside ferrous and sand glass have significant non linear spectrum related to their structure. The research introduces new applications of the ceramics materials micro and nano particles in non linear optical devices. The research ends with conclusions and recommendation for future work in this new field of applications. The new techniques and the new materials selected in the current study have good non linear optical characteristics; these are promising for practical application. It can be used for various optical applications such as optical modulators, sensors and other optical protection applications.

1. Introduction

Nanoparticles often possess unexpected optical properties as they are small enough to confine their electrons and produce quantum effects[1],[2]. For example gold nanoparticles appear deep-red to black in solution. Nanoparticles of yellow gold and grey silicon are red in color. Gold nanoparticles melt at much lower temperatures (~300 °C for 2.5 nm size) than the gold slabs (1064 °C);.[3] Absorption of solar radiation is much higher in materials composed of nanoparticles than it is radiation in thin films of continuous sheets of material. In both solar PV and solar thermal applications, controlling the size, shape, and material of the particles, it is possible to control solar absorption [3],[4]. other size- dependent property changes include quantum confinement in semiconductor particles, surface plasmon resonance in some metal particles and super para magnetism in magnetic materials, which would appear the changes in physical properties[5],[6]. Ferromagnetic materials smaller than 10 nm can switch their magnetisation direction using room temperature thermal energy, thus making them unsuitable for memory storage [7],[8]. Nonlinear effects in electromagnetism had been observed as early as the late nineteenth century by Kerr, Röntgen, Kundt, and Pockels, and later by Raman in 1927 when he discovered spontaneous scattering of light into new wavelengths is passing through a transparent medium[9],[10]. the first observation of coherent nonlinear optical effects was made by Franken et al. (1961), who demonstrated second-harmonic generation of light in the crystal of quartz. This discovery propelled the field of modern nonlinear optics and initiated intensive research in materials science and crystal technology, figure 1 shows different applications of Non linear optical elements [11]. Non linear optical (NLO materials) has distinct crystal structure which is anisotropic with respect to electromagnetic radiation. It is the branch of optics that describes the behavior of light in nonlinear media, as a field of electromagnetic wave propagation, Media in which the dielectric polarization P responds nonlinearly to the electric field E of the light[12],[13]. The nonlinearity is typically observed only at very high light intensities (values of the electric field comparable to interatomic electric fields typically 108 V/m) such as those provided by lasers The fundamental aspects of nonlinear optical (NLO) materials and their role in modern communication were presented. NLO materials has distinct crystal structure which is anisotropic with respect to electromagnetic radiation [14],[15].

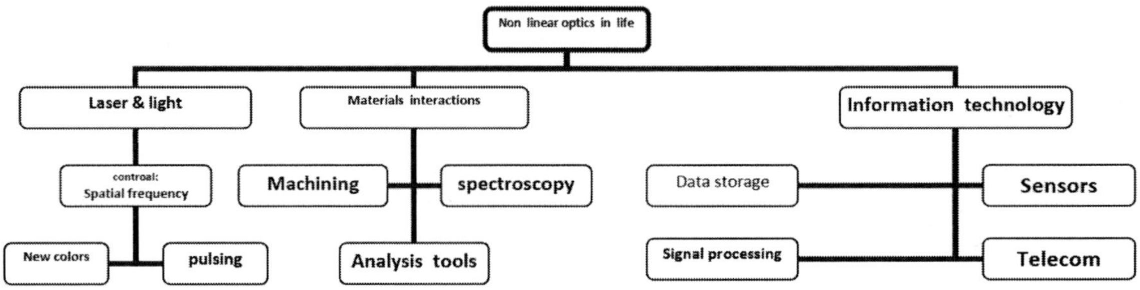

Figure 1. Different applications of Non linear optical elemets.

2. Experimental

2.1 Materials and methods

Preparation of nano and micro Ceramics Material (natural materials)
The micro- nano particles from natural resources (Granite, Zeolite, glass sand, diopside ferrous) were soaked for 100 hour in standard solution from distilled water with 30 % Sulphuric acid $H_2 SO_4$. Table 1 shows the relation between crystal structure and chemical composition for the miro and nano particles of ceramics materials used in the current study. The micro and nano particles were found as waste or by products during mining process. The particle size of waste mateial in this study was less than 0.125 mm. Figure 2 shows the apparatus use in mixing and preparation of samples.

Figure 2. The experimental apparatus for samples preparation.

IOP Conf. Series: Materials Science and Engineering **973** (2020) 012015 doi:10.1088/1757-899X/973/1/012015

Table 1. Chemical composition and structure of ceramics materials,

	Commercial name	Composition	Structure
1	**Granite**	SiO2 ,Al_2O_3,K_2O.Na_2O_3, CaO,FeO,Fe_2O_3,MgO,Ti O_2,P_2O_5,MnO	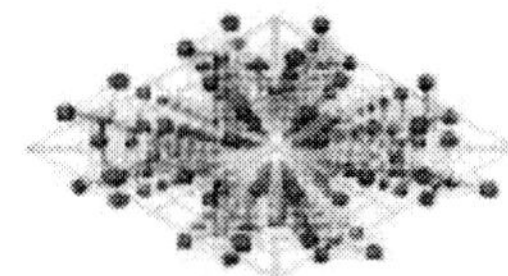
2	**Diopside ferrous**	SiO2 Ca.92Fe.08 Al.14 Fe.33 Mg.53Si $_2O_6$	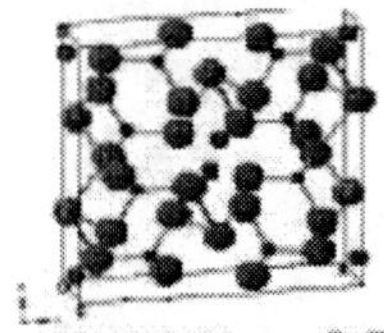
3	**Zeolite**	SiO2 CaO, TiO_2,, Na_2O, Al_2O_3, K_2O, Fe_2O_3, P_2O_5,MnO,SO_3	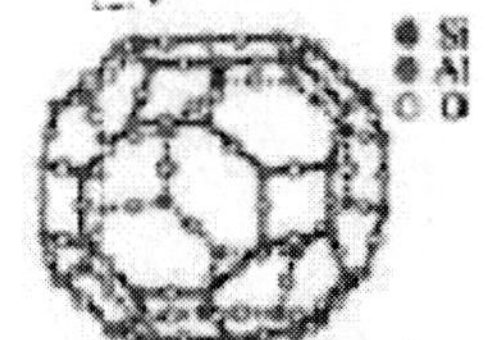
4	**Sand glass**	SiO2 ,Al_2O_3, Fe_2O_3. TiO_2, MgO,,K_2O.Na_2O_3,CaO, Cr_2O_3	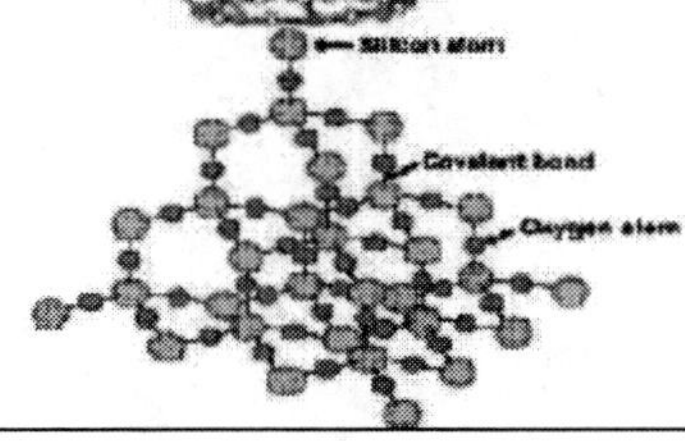

2.2 Z-scanning method

Z-scan is a single beam technique developed by Sheik Bahae to measure the magnitude of nonlinear absorption as well as the sign and magnitude of nonlinear, refraction in these measurements in the current study using the standard single beam technique. The signs and values of nonlinear absorption coefficient and nonlinear refractive index were determined by the open aperture z-scan technique. In our measurements, a Q-switched 532 nm (2.33 eV) frequency doubled Nd:YAG laser (Continuum) characterized by pulse duration of 6 ns was employed. The pulse repetition rate was set at 1 Hz to reduce the possible thermal accumulative effect. The laser energy was remain constant. below the optical breakdown threshold for the samples investigated. Care must be taken to prevent material optical breakdown and irreversible changes produced by laser radiation. the beam passing through the sample is inblocked (S=1) and all the transmitted light is recorded as a function of z., measurements were performed by placing the sample (liquid solution contain particles) in a quartz cuvette of thickness 3mm . The samples were placed at the focus of the lens with the focal length 15 cm.

Figure 3 shows the experimental set up used in this work. The Z-scanning method of the recording device and the test process are not complicated, but this method provides very rich information about the tested object [14],[15].

IOP Conf. Series: Materials Science and Engineering **973** (2020) 012015 doi:10.1088/1757-899X/973/1/012015

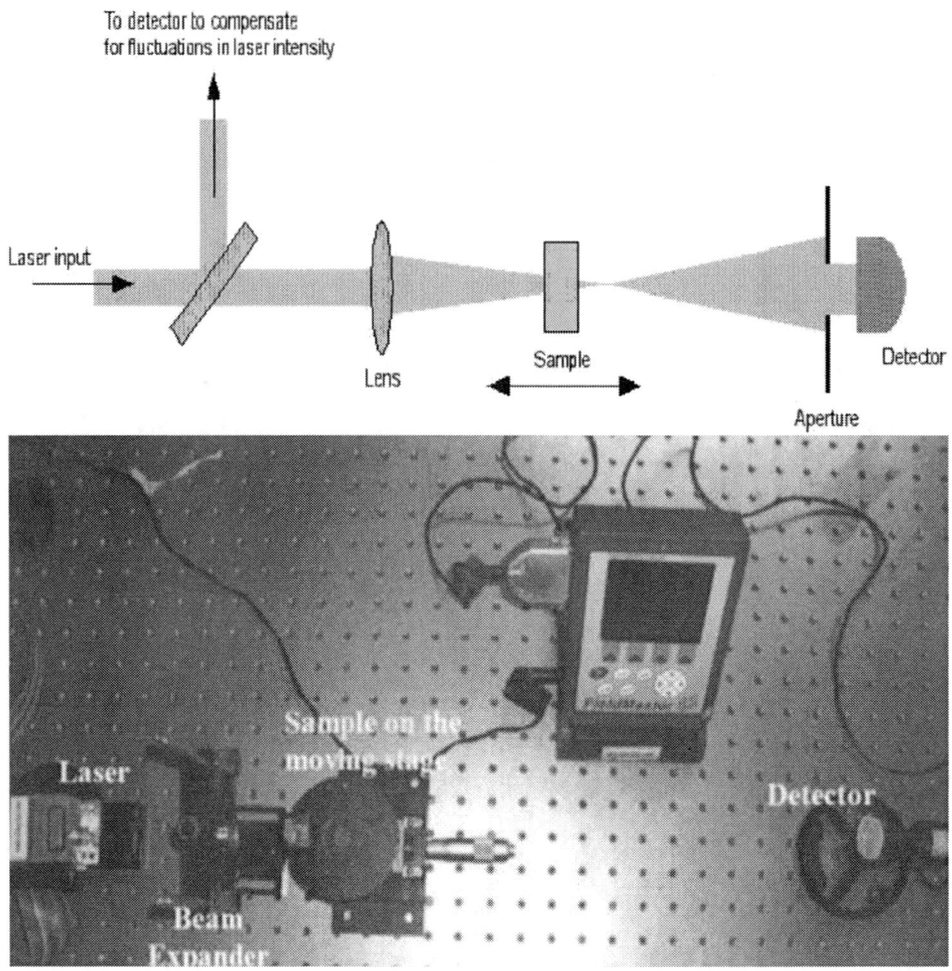

Figure 3. The z- sczn technique used in this work

2.3 Theory explain Z- scan aperture

Open aperture z-scsm and reverse saturable absorption (RSA) positive nonlinear absorption

β , $\beta>0$, ESA>GSA

Saturable absorption (SA), Negative Non linear absorption

β, $\beta<0$, ESA<GSA

$$T(z,S=1)=\sum_{m=0}^{\infty}\frac{\left|-\beta I_0 Leff\right|^m}{(m+1)^{3/2}(1+\frac{z^2}{z_0^2})^m} \quad (1)$$

$$L_{eff}=\frac{1-e^{-\alpha_0 L}}{\alpha_0} \quad (2)$$

Non linear refraction (NLR) $n=n_0+n_2I$ (3)

Non linear absorption (NLA) $\alpha(I)=\alpha_0+\beta I$ (4)

$$\beta=\Sigma\beta_i \quad (5)$$

Positive nonlinear refractive index (V-P), n2>0

IOP Conf. Series: Materials Science and Engineering **973** (2020) 012015 doi:10.1088/1757-899X/973/1/012015

$$T(z,\Delta\phi_0) \approx 1 - \frac{4\Delta\phi_0 x}{(x^2+9)(x^2+1)} \quad (6)$$

Negative nonlinear refractive index(P-V), n2< 0

$$\Delta T_{p-v} \approx 0.406(1-S)^{0.25}|\Delta\phi_0| \quad (7)$$

$$\Delta\phi_0 = kI_0 n_2 L_{eff} \quad (8)$$

α	The low intensity absorption coefficient
n_2	refractive index (The third -order nonlinear refractive index)
n_0	The linear index of refraction
n	The refractive index in the presence of nonlinearity
β	The two -photon absorption coefficientt
I	Intensity of incident laser radiation
ϕ	the on-axis phase shift
Leff	the effective interaction length
S	the aperture linear transmittance
χ	the constant of proportionality known as the linear susceptibility.
L	interaction length
E	electric field
P	induced polarization
ε	Dielectric constant

3. Results and discussion

3.1 The new topic in this work can be summerized in two main points:-

First using insulated ceramics material in the powder form as non linear materials almost all previous work does not study this type of insulated materials.

Second preparation of the material as hinged particles in acidic solution at specific concentration solution phase measurement is often used to get information on microscopic nonlinearities. In the fluid phase, anisotropic molecules also exhibit another kind of nonlinearity, not observed in the case of inorganic semiconductors or salts, that is achieve flexibility for optimization of nonlinear optical responseand necessary for studying non linear characteristics of material under fabrication.

3.2 Nonlinear properties of the samples

They were confirmed using open aperture Z-scan technique. Dependence of nonlinear optical properties on laser excitation energy is detected for different chemical composition and crystal structure Table 2 shows the z- scan results for different ceramics materials prepared in acidic solution at the same volume percent and the same concentration of acid 30%.(battery water).

Table 2. The z- scan of different ceramics materials.

	Solution type	Composition	Additive volume %	Effect
1	Battery water	0	0	null
2	Battery water	Granite	%۲۰	null
3	Battery water	Diopside ferrous	%۲۰	positive
4	Battery water	Zeolite	%۲۰	positive
5	Battery water	Sand glass	%۲۰	Weak

From the above table, the physical origins of NLO phenomena can be categorized as either structural or compositional. Here structural refers to light-induced structure changes, There is no change in light-induced chemical reactions, there is no significant difference in chemical composition between the selected ceramics material under study all of the have silicate origion SiO_2 besides some other oxides[5],[9]. the significant difference between the selected ceramics materials under study were in the crystal structure which leads to change of electronic density, average interatomic distances, molecular orientation, phase transition, the difference in the crystal structure as shown in the table (1) leads to change in the non linear optical characteristics of the materials as shown in the following curves.

Granite samples immerged in the acidic solution, the compact structure without any significant interatomic distances in the structure prevent any transmission of light so all the incident light was absorbed, granite cover prepared by the technique developed in this work can be used as optical filter or optical protection to protect the sensitive detectors from the action of high power Lasers. So the development of optical devices protecting eyes from laser radiation optical filters work at all wavelenghts as result of high attenuation of the transmission in the nonlinear regime and complete beam decay happen, the new technique marked by high thermal and chemical stabilities beside low cost The available protecting devices are optical filters, which don't transmit radiation at specific laser wavelengths. The new devices made from granite particles immerged in acidic solution can overcome the main disadvantages in the common filter which can be mentioned in the some points. The filters may become transparent at high input intensity because of saturation of absorption. The optical damage phenomena also may be occurs.

The open aperture z- scan of the acidic solution H_2SO_4 with 30% was shown in figure (4). There is Gaussian beam distribution appear with maxmium value at 50mm and minimmum values at 0-100mm, the solution does not introduce any nonlinear characteristics so it is ideal as media for preparation of new particles from ceramics materials, it allows the determination of nonlinear behaviour of the immerged particles without any confusion.

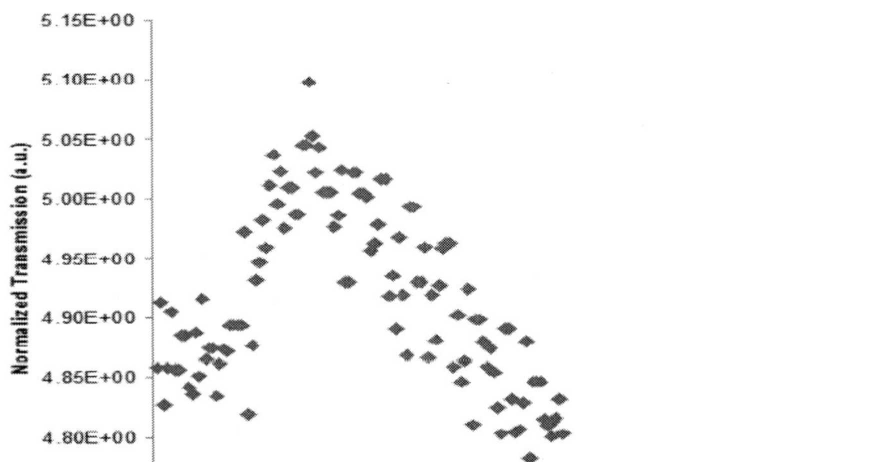

Figure 4. Open aperture z-scan for acidic solution (H_2SO_4 30% concentration).

Saturable absorption (SA) can be observed clearly in figure (5) where the open aperture z-scan for zeolite particles imerged in the standard solution used in this work (H_2SO_4,with 30% concentration) the amount of particles were determined to be 20% volume percent [14]. Saturable absorption (SA)

IOP Conf. Series: Materials Science and Engineering **973** (2020) 012015 doi:10.1088/1757-899X/973/1/012015

where the absorption coefficient decreases with increasing intensity, saturable absorption can occur when the cross-section of excited state less than the ground state cross section. Zeolite solution on glass covett measured by Z-scan technique, it is recorded saturable absorption beak at 38mm and other beak can be recorded at 55mm but as less value of normalized transmission.

The second-order nonlinear optical interactions can occur only in non centrosymmetric crystals that is happeded in crystals that do not display inversion symmetry. Since liquids, gases, amorphous solids (such as glass), and even many crystals display inversion symmetry. The solution prepared in our work has inversion symmetry and suitable for second -order nonlinear optical interaction. Third-order nonlinear optical interactions can occur for both centrosymmetric and noncentrosymmetric media [13] so it is can occur in the solution prepared in our study range. In other words the new technique developed in this work is suitable for second and third order non linear optics.

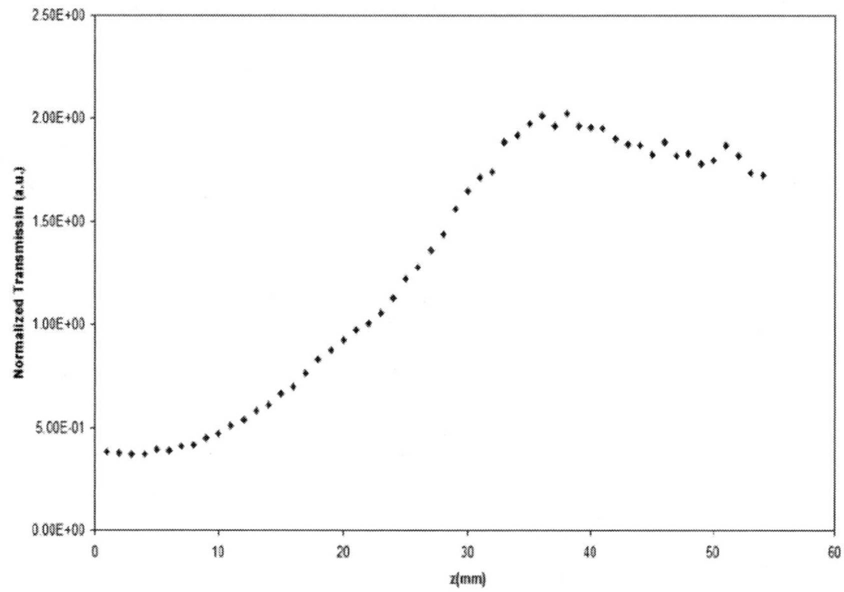

Figure 5. Open aperture (z-scan) for zeolite nano and micro particles at volume fraction 20%.

In figure (6) The measured nonlinear refractive index and absorption coefficient of the diopsite ferrous solution on glass covett by Z-scan technique was recorded. It is observed saturable absorption (SA) at 40mm and a transition from saturable absorption to reverse saturable absorption at 50mm and 60mm with decrease in normalized transmission. reverse saturable absorption (RSA), where the absorption coefficient increases with increasing intensity. RSA can occur when the absorption cross-section of excited states exceeds the ground state cross section. The behaviour is re-reverse to saturable absorption(SA) at 70mm, The complex behaviour of the diopside ferrous is related to the existance of some compounds mixed with the original structure such as Illite-2M1 (NR), Actinolite,Hematite and Montmorillonite[1],[3]. The electronic structure change is the distribution of electrons at different electronic states. The nonlinearity can be enhanced by several orders of magnitude by sacrificing response time and absorption loss.

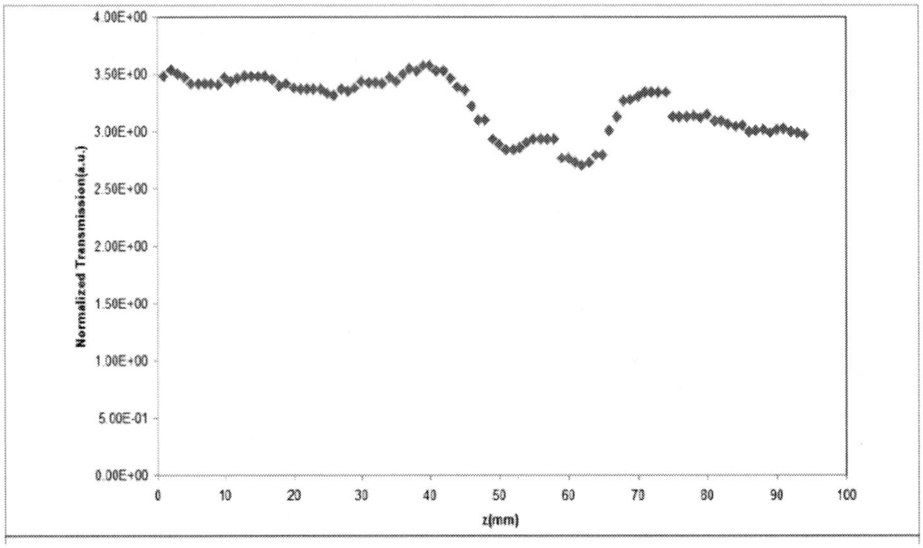

Figure 6. Open aperture (z-scan) for diopside ferrous micro and nano particles with volume fraction 20%.

Reverse saturable absorption (RSA) can be observed clearly in figure (7), the nonlinear behaviour of sand glass emmerged in standard acidic solution (H_2SO_4 30%),with volume fraction 20% .

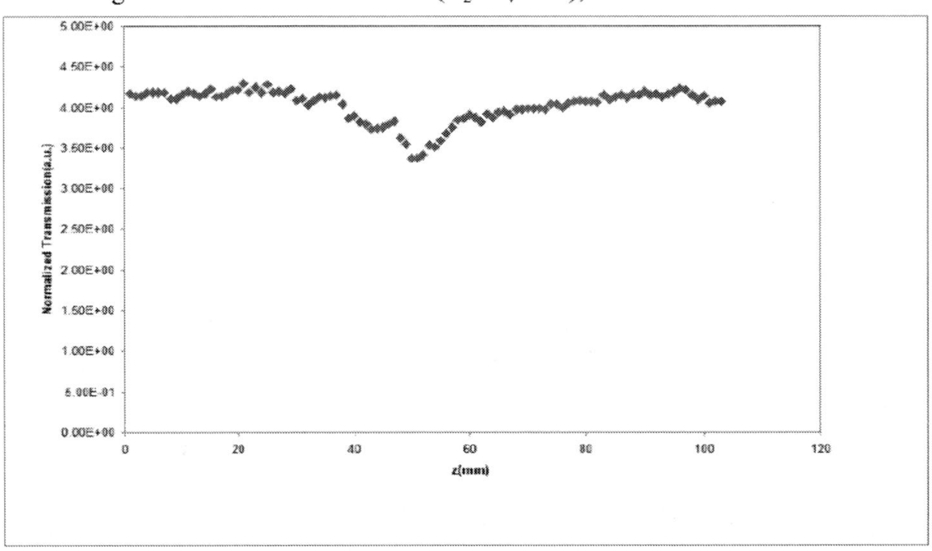

Figure 7. Open aperture z-scan for sand glass micro and nano particles at 20% volume fraction.

When the absorption cross-section from excited state is larger than that from the ground state, the transmission of the system will be less under intense laser fields. The normalized transmission exhibits a small one valley with respect to the focus at (z = 50mm), implying that the RSA plays a primary role for the nonlinear properties of material. Sand glass introduce special pure and simple behaviour when

IOP Conf. Series: Materials Science and Engineering **973** (2020) 012015 doi:10.1088/1757-899X/973/1/012015

comparting this with the behaviour introduce by other ceramics materials under study. the behaviour of sand glass can be explained in terms of the purity and transparency of the crystal [8].

3.3. Disadvantages of the new technique
The new technique has only one disadvantage that the concentration of solution is inhomogenious due to precipation of particles under their weights. The future reseach will be developed to overcome this disadvantage

3.4. Advantages of the new technique
1) Abundant and low cost achieve economic version of non linear optical material
2) Avoid effects occurs in resonant NLO interactions due to the absorption of light and the generation of phonons.
3) Avoid thermal effect
4) Suitable for different applications
5) Suitable for second order and third order non linear phenomenas

4. Conclusions
1. Development of new type of materials from natural resources suitable for optical non linear characteristics
2. The new techniques and the new materials selected in the current study have good non linear optical characteristics ; these are promising for practical application. It can be used for various optical applications such as optical modulators, sensors and other optical protection applications
3. Granite particle prepared by the technique mentioned in the current work can be used as optical protect and/or optical filter which can overcome the disadvantages in the common filters or protection devices.
4. Open the way to the study of new nonlinear optical effects and the introduction of new concepts, and find and develop materials presenting large nonlinearities and satisfying at the same time all the technological.

References
[1] Kyle J R and Ketcham R A 2015 Application of high resolution X-ray computed tomography to mineral deposit origin, evaluation, and processing *Ore Geology Reviews* **65** pp821–839
[2] Gu Y, Schouwstra R. P. and Rule C 2014 The value of automated mineralogy. Miner. Eng. **58** pp 100– 103.
[3] Pooler R. and Dold B 2017 Optimization and Quality Control of Automated Quantitative Mineralogy Analysis for Acid Rock Drainage Prediction *Minerals* 7(1), 12.
[4] Berry R, Walters,S. G. and McMahon C 2008 Automated mineral identification by optical microscopy *Australasian Institute of Mining and Metallurgy Publication Series* pp 91–94.
[5] Pirrie D. et al. 2009 Automated SEM-EDS (QEMSCAN®) mineral analysis in forensic soil investigations: testing instrumental reproducibility Criminal and Environmental Soil Forensics ed Ritz K , Dawson L and Miller D) (Netherlands: Springer,) pp 411–430 .
[6] Sylvester P J 2006 *Nanomaterials Handbook* (Taylor & Francis Group, LLC)
[7] Yang GW 2007 Laser ablation in liquids: Applications in the synthesis of nanocrystals *Progress in Materials Science* **52** (4) pp 648-98.
[8] Zeng H, Du X W Singh S C, Kulinich SA, Yang S, He J and Cai W 2012 Nanomaterials via Laser Ablation/Irradiation in Liquid: A Review *Advanced Functional Material* **22**(7) pp 1333-1353.

IOP Conf. Series: Materials Science and Engineering **973** (2020) 012015 doi:10.1088/1757-899X/973/1/012015

[9] Bäuerle D 1996 Thermal, Photophysical, and Photochemical Processes *Laser Processing and Chemistry* (Springer Berlin Heidelberg) pp 13-38.
[10] Houle, M. A. et al. 2017 Rapid 3D chemical-specific imaging of minerals using Stimulated Raman Scattering (SRS) microscopy J. *Raman Spectrosc* **48** (5) pp 726–735.
[11] Song W. et al. 2016 Fully integrated reflection-mode photoacoustic, two-photon, and second harmonic generation microscopy in vivo. *Scientific reports* **6** 32240.
[12] Fan H, Wang X et al. 2010 Investigation of the nonlinear absorption and optical limiting properties of two [Q][Cu(CS)] compounds *Opt. Laser Technol.* **42**(5) pp 732–736 .
[13] TH. Stimulated Optical Radiation in Ruby 1960 *Nature* **187** 493-94.
[14] Shirk M D, Molian P A 1998 A review of ultrashort pulsed laser ablation of materials *Journal of Laser Applications* **10** (1) 18-28.

IOP Conf. Series: Materials Science and Engineering **973** (2020) 012016 doi:10.1088/1757-899X/973/1/012016

Improving water desalination efficiency by using carbon nanotubes as pre-treatment

A AbdAllah[1], W El-Sallamy[2], B Azzam[3] and T Osman[4]
[1] Production Engineering, Education Development Fund, Cabinet of Ministers, Cairo,
[2] Production Engineering and Printing Technology Department, Akhbar Elyom Academy, Giza
[3] Mechanical Design and Production Department, Taibah University, KSA
[4] Mechanical Design and Production Department, Cairo University
wsallamy@yahoo.com

Abstract. Filters based on carbon nanotubes (CNTs) have been highlighted as an emerging technology for water desalination. In this paper, the weight percentage (wt.%) of CNTs required for desalination is discussed. The idea is based upon addition of CNTs to simple filters that is employed as adsorbent for dissolved solids in water (sodium chloride removal). Different CNTs filters (using CNTs Multi walled tube type) specimens were fabricated. Experimental work was done, through building a test rig, to show the effect of percentage of weight of CNTs on desalination of (Red Sea and Brackish) water. The effect of multiple filtration stages has been also shown. It was shown that the filters containing CNTs would be effective if used at the intake of modern RO (Reverse Osmosis) water desalination plants. Using filters with CNTs as pre-treatment of intake water for the RO units will reduce feeding pressure, specific energy and power consumption.

1. Introduction

Desalination means that removal of dissolved mineral salts from water (brackish water or seawater). About 71% of the earth surface is covered by water in form of the oceans, seas and the ices in the poles. However, only about 3% of water is fresh and suitable for use. Whereas, the water of the oceans and seas have high salt concentration. Therefore, some special processes are needed for desalination to overcome the water shortage [1–5]. Conventional methods for water desalination are being used to remove dissolved minerals and ions including oxidation, ion exchange reduction, membrane filtration and adsorption [6,7]. Yunhui Wang et al. [8] have used Nano-porous graphene for water desalination under external pressure, the graphene membrane has rejected the salt while transmitting water with good performance.

CNTs, a filter in carbon family, are relatively adsorbents that have been proven for removing many kinds of pollutants such as dioxin from air [9] and dissolved minerals from water [10,11]. CNTs powder is mixed through ultrasonic techniques with salty water for desalination. This technique has shown to be unaccepted due to its high cost and the difficulty of removing powder of CNTs from salty water.

Therefore, CNTs filters were synthesized as a practical adsorbent for economical desalination, without any leakage of CNTs into water. The objective of the present work is to study the effects of

CNTs on water desalination. Effects of percentage of weight (wt.%) of CNT in filters on initial salt concentration for RO units were also investigated.

1.1. Carbon Nanotubes

Several techniques have been developed to produce carbon nanotubes in sizeable quantities, including arc discharge, laser ablation, high-pressure carbon monoxide disproportionation (HiPco), and chemical vapor deposition (CVD). Most of these processes take place in vacuum or with process gases. CVD growth of CNTs can occur in vacuum or at atmospheric pressure. Large quantities of nanotubes can be synthesized by these methods; advances in catalysis and continuous growth processes are making CNTs more commercially viable.

1.1.1. CNTs Synthesis The type of CNTs is Multi walled tube. CNTs were synthesized by the electric arc discharge. The arc is generated between two electrodes (size φ 6 x 100 mm) using Distilled water. The cathode and the anode are from graphite (99.9% pure), and was performed under AC current, 75 A and 75 V. Figure 1 shows a view and a real photograph of the fully automatic machine for producing CNTs, this was designed and produced in Akhbar El-Youm Academy CNTs laboratory (through which the CNTs were brought and used in this work) . Each part is pointed out as follows: (1) Base frame, (2) lower tank, (3) fans for cooling, (4) valve, (5) upper tank, (6) carbon holder, (7) control unit, (8) electrical unit, (9) steeper motor, (10) nuts housing, (11) deionized water, (12) power screw, (13) bearing, (14) cover, (15) pure graphite, (16) cover hand [12].

Figure 1. Automatic Machine for Producing CNTs *(through which the CNTs were brought and used in this work)* [12].

1.1.2. CNTs Specifications Produced CNTs using arc discharge were characterized with scanning electron microscope (SEM: BPI-T) and high-resolution transmission electron microscope (HRTEM: JEM-2100). Figure 2 shows the SEM image of the synthesized CNTs sample. Through transmission electron microscope (TEM) measurements, it can be observed CNTs are typically of 3-10 nm in diameter and on the order of 0.5–1 mm long, together with some amorphous carbon and graphite flakes, as shown in figure 3 and figure 4.

IOP Conf. Series: Materials Science and Engineering **973** (2020) 012016 doi:10.1088/1757-899X/973/1/012016

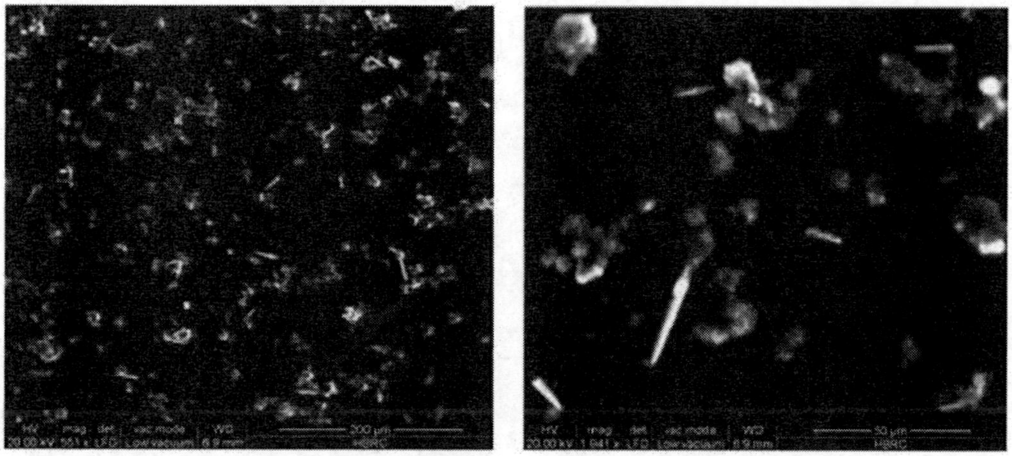

Figure 2. (SEM) images of the CNTs which were synthesized [12].

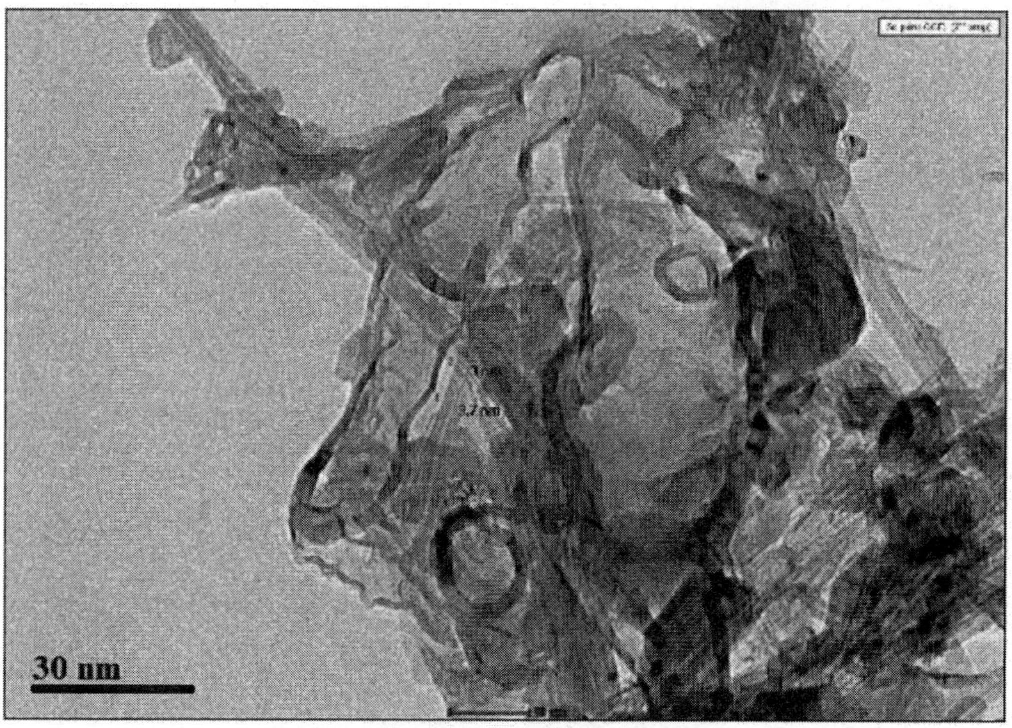

Figure 3. HRTEM image of the sample which were synthesized [12].

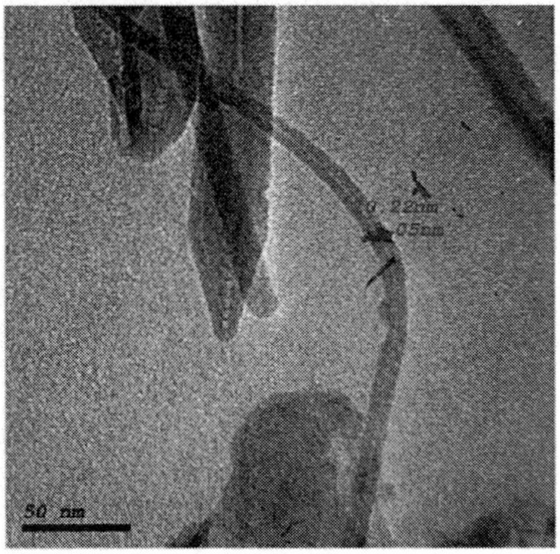

Figure 4. HRTEM image of a Multi Walled CNTs bundle [12].

2. Experimental Procedures

The main idea of CNTs filter desalination is based on the batch adsorption, where the dissolved salts are adsorbed inside the CNTs. The next part shows the experimental work which has been accomplished, presenting a detailed description of test rig specifications, synthesis of CNTs filters and experimental procedure.

2.1. Test Rig

A test rig was built to test the effect of CNTs filters with different wt.%, on different salty water types by using several stages. The test rig is a filtration unit which provides water desalination without phase change. This unit consists of inlet tank that is directly connected to the pump, the pump delivers the water to CNTs filter (specimen to be tested) finally after soaking of water inside the filter, water is transferred to the outlet tank, as shown in figure 5.

The rig specifications are as follows:

a) Tank capacity: 1 Liter,
b) Pump type: Diaphragm electric motor driven pump with 125 PSI maximum pressure, 1.2 Liter/min open flow rate and maximum power of 28 Watt.

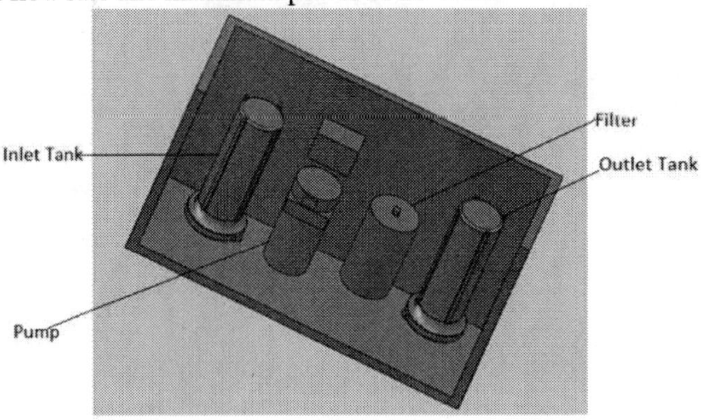

Figure 5. Test Rig used.

2.2. Synthesis of the CNTs filters

The CNTs filter consists of; Granular Activated Carbon (GAC), CNTs (Multi walled tube-type), epoxy (OXSIED 50 (KI) and EUXIT 50 (KII)), as shown in Table 1. GAC is the matrix material for the CNTs filter, it is a favored water treatment material as it removes organics and residual disinfectants contaminants from water. GAC is mostly made from raw materials such as nutshells, wood, coal and petroleum. The two principal mechanisms by which GAC removes contaminants from water are adsorption and catalytic reduction. The epoxy OXSIED 50 (KI) is a type of epoxy used for attraction between GAC particles. EUXIT 50 (KII) is a hardener for the epoxy. Epoxy resin is also a durable product which can be used with various materials, including wood, fabric, glass, or metal. It's important to note, however; epoxy resin is not considered to be water resistant [14].

Table 1. CNTs Filters Specifications.

Unit (gram)	Specimen					
	Filter No.1: 0%	Filter No.2: 0.5%	Filter No.3: 1 %	Filter No.4: 1.5%	Filter No.5: 2%	Filter No.6: 3%
KI (1/16 GAC)	10.000	10.050	10.100	10.150	10.200	10.300
KII (1/3 KI)	3.333	3.350	3.367	3.383	3.400	3.433
GAC	160.00	160.000	160.000	160.000	160.000	160.00
CNT (%GAC)	0.000	0.800	1.600	2.400	3.200	4.800

One of the challenges in the manufacturing process of CNTs is the well dispersion of CNTs into the matrix material. Only well separated and homogeneously distributed nanotubes can lead to unique properties of the material. To reach this goal; a well proven Ultrasonic dispersion technique was used. Typical procedures and same parameters used by Bittmann et al. for CNTs dispersion in epoxy resin was typically repeated [13]. An ultrasonic horn was used to disperse CNTs powder into epoxy resin. The fabrication of Nano phased epoxy was carried out in the following steps:

a) The CNTs are ultrasonically mixed with OXSIED 50 (KI) of epoxy resin.
b) CNTs were added by 0.5, 1, 1.5, 2 and 3 wt. %, respectively by weight of the resin.
c) The mixing process was carried out in an ultrasonic processor up to 200 sec, with the following parameters:
   i. Power: 200 watts.
   ii. Frequency: 24 kHz.
   iii. Amplitude: 20-100%.
   iv. Pulse: 0-100%.
d) The mixing is carried out at 55% of the amplitude and for about 20 min for 0.5, 1, 1.5 wt.% and 90 min for 2 and 3 wt. % of the resin.
e) In order to avoid rise in temperature during sonication, the mixing beaker is placed in a cooling mixture of ice and water.
f) The CNTs powder is gradually added into the mixing beaker, mixing process took 20 to 90 minutes depending on the composition. Figure 6 shows the mixing process.
g) Hardener EUXIT 50 (KII) is added to the mixture at a ratio of 1:3. The mixture is then poured into the mold, figure 7 shows the molding process.
h) The tube is left to cure at room temperature for 5 days to completely solidified.

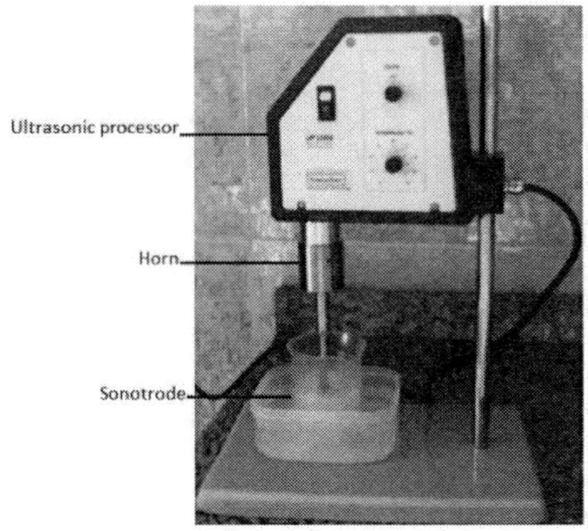

Figure 6. Ultrasonic mixing process.

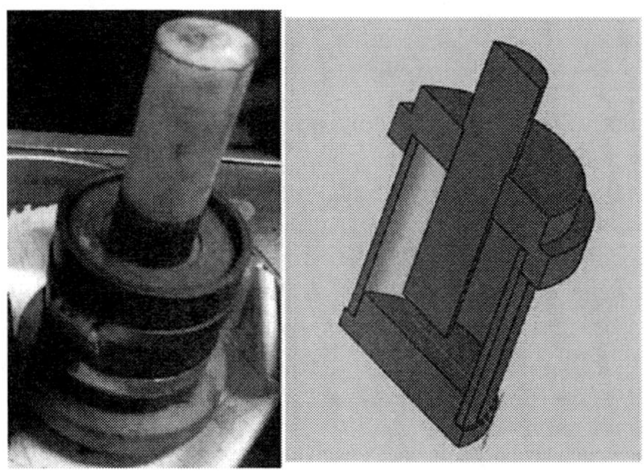

Figure 7. Moulding process

Six different filters with the following dimensions were produced:

a) Outer diameter 65 mm.
b) Inner diameter 30 mm.
c) Height 75 mm.

The filters finally appeared, as shown in figure 8. Table 1 shows the specifications of produced filters and their compositions percentages.

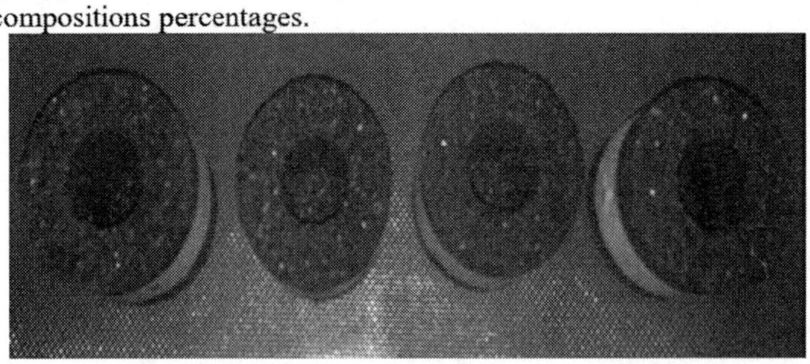

Figure 8. Final CNTs filters.

IOP Conf. Series: Materials Science and Engineering **973** (2020) 012016 doi:10.1088/1757-899X/973/1/012016

2.3. Measurement procedures

The efficiency of desalination of the proposed filters were examined using salty water samples from Red Sea and brackish. Brackish water sample was brought from a deep well (water depth; 21 m) at Wadi El Natroun area 100 Km North Cairo. Several experiments have been performed via soaking of Red Sea and brackish water through the test rig. All the water samples have been sent for examination (dissolved salts concentration) at Governmental certified labs of the Ministry of Agriculture in Egypt. The effect of CNTs filters desalination was examined for each of the 6 filters (0, 0.5, 1, 1.5, 2 and 3 wt.% of CNTs) for Sea and brackish water at room temperature. The filter with highest performance has been examined to see the effect of multiple desalination stages on its performance.

3. Results and Discussion

The effect of weight percentages of CNTs and number of filtration stages on desalination were investigated. Figure 9 and figure 10 show the effect of water desalination when using different percentages of CNTs filters. It was shown that filter without CNTs (0 wt.%) almost has no effect on desalination (dissolved salts were reduced by 9.2 %, from 39,963 of the raw water to 36,983). Increasing the weight percentages of CNTs has increased the water desalination effect. The laboratory analysis results show that the Red Sea water dissolved salts were reduced by 27.3% (from 39,963 PPM of the raw water to 29,057 PPM) when using CNTs of 3% wt. The filter with best performance was selected to show the effect of using series of this filter (multiple filtration stages). Figure 11 and figure 12 show the effect of number of filtration stages when using 3% wt. CNTs filter on dissolved salts of Red Sea water sample. It was shown that the Red Sea water dissolved salts concentration has been reduced by a ratio of 53.8% (from 39,872 PPM of the raw water to 21,440 PPM) when using 5 stages of CNTs filter (3% wt.).

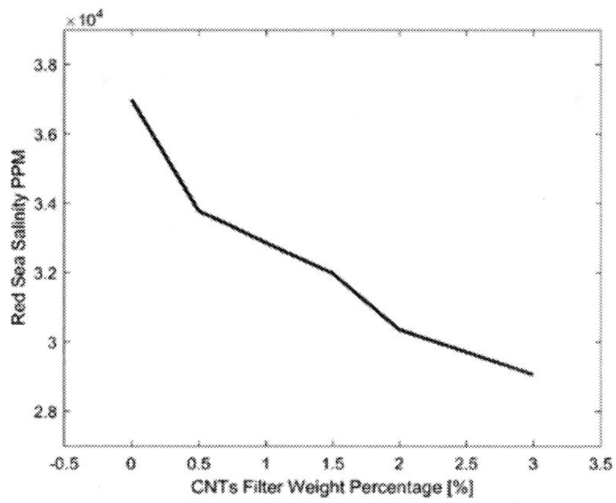

Figure 9. Effect of CNT wt.% on Dissolved Salts of Red Sea Water.

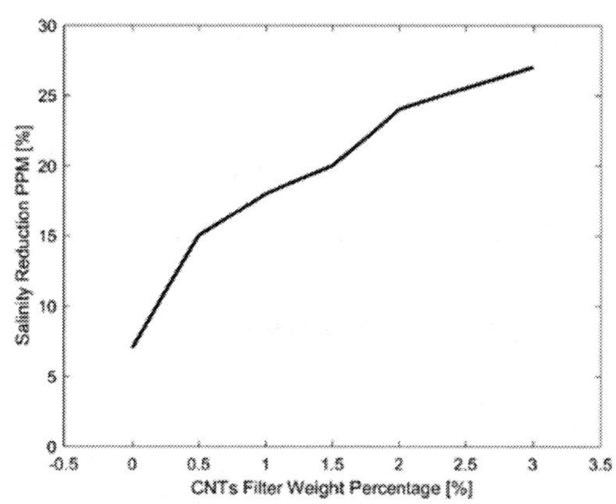

Figure 10. Effect of CNTs Weight on Reduction Percentage of Dissolved Salts for Red Sea Water PPM%.

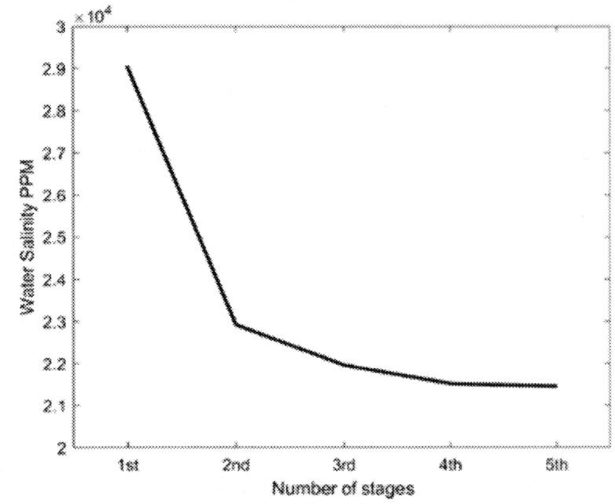

Figure 11. Effect of Filtration stages at 3 wt.% filter on Dissolved Salts of Red Sea Water.

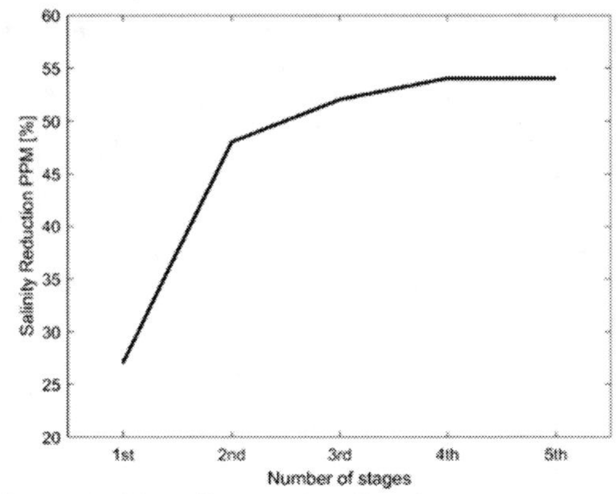

Figure 12. Effect of Filtration Stages on Reduction Percentage of Dissolved Salts for Red Sea Water PPM%.

Same experimental procedures were repeated for brackish water. Figure 13 and figure 14 show the effect of percentages of CNTs on brackish water dissolved salts. The laboratory analysis results show that the brackish water dissolved salts were reduced by 44.6 % (from 7,168 PPM of the raw water to 3,968 PPM) when using CNTs filter of 3% wt. CNTs filter of 3% wt. is shown to have the best performance. Figure 15 and figure 16 show the effect of number of filtration stages on salinity. It was shown that the brackish water dissolved salts concentration has been reduced by about 60% (from 3,968 PPM to 2,835 PPM) when using 3 stages of CNTs filter (3% wt.). Using more than 3 stages shows that there is almost no effect on desalination.

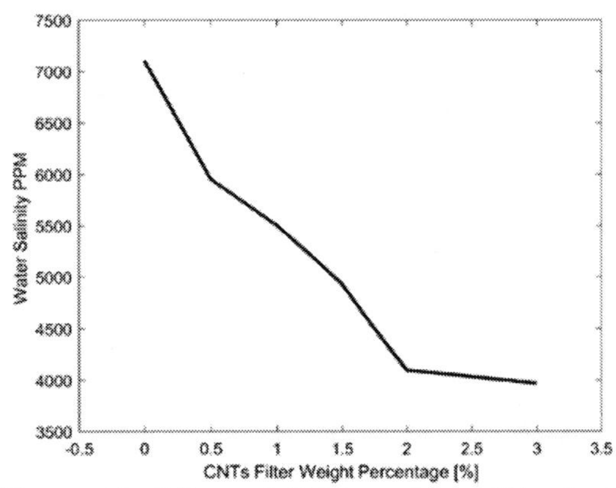

Figure 13. Effect of CNTs wt.% on Dissolved Salts of Brackish Water.

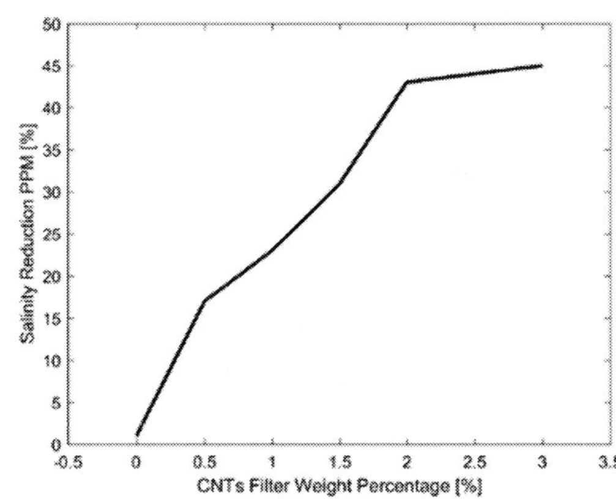

Figure 14. Effect of CNTs wt.% on Reduction Percentage of Dissolved Salts for Brackish Water.

IOP Conf. Series: Materials Science and Engineering **973** (2020) 012016 doi:10.1088/1757-899X/973/1/012016

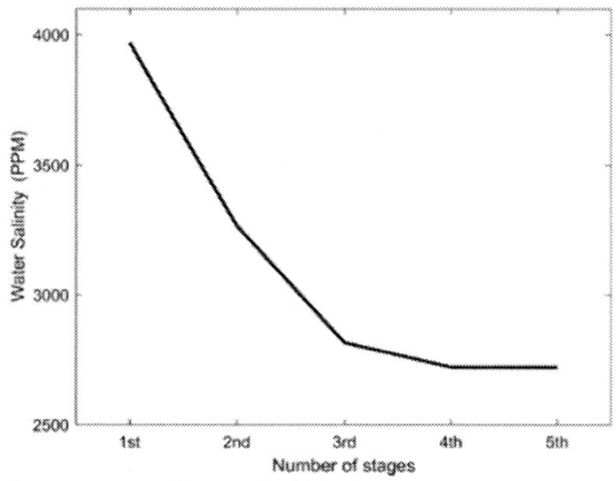

Figure 15. Effect of Filtration Stages at 3% wt. filter on Dissolved Salts for Brackish Water.

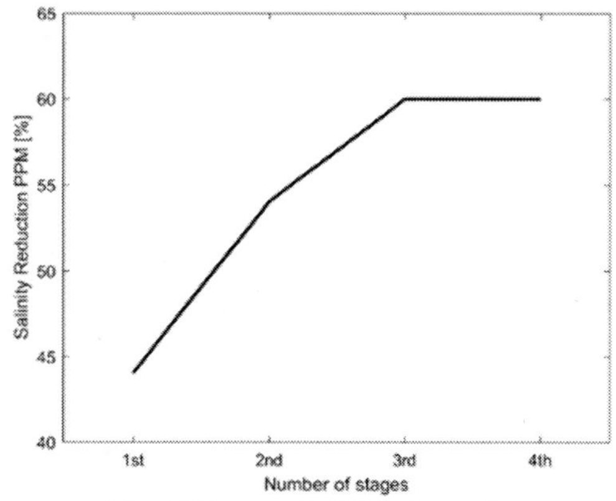

Figure 16. Effect of Filtration Stages on Reduction Percentage of Dissolved Salts for Brackish Water PPM%.

3.1. Combining carbon Nanotubes filters with RO desalination units

Reverse Osmosis technique is widely used now for water desalination processes. Cost of water desalination using RO filters is mainly depending on power consumed for raising water inlet pressure, this is directly related to inlet water salinity. The target is to show the effect of reduction in pressure and power consumption when using CNTs coupled to the RO desalination plant to simulate its effect on energy reduction. The effect of using CNTs filter attached with RO plant; for Red Sea and brackish water desalination was studied through well proven software modeling. Reverse Osmosis System Analysis (ROSA) software is used for modeling, it is one of the leading RO system design tools used to design RO plants. The same parameters for the samples with and without CNTs filter effect was stated; the raw water flow rate, inlet pressure, chemical dosage, total membrane active area. The filter with best performance was simulated when coupled with RO plant through several filtration stages for Red Sea and brackish water.

Simulation results show a considerable reduction in the plant's power, pressure and energy consumption. Figure 17 shows a reduction of 24.3% in power, pressure and energy consumption when using 3% wt. CNTs filter. Figure 18 shows the effect of using 3% wt. CNTs filter on several stages. For Red Sea sample analysis, it was shown that a reduction percentage of 44.5% in power, pressure and energy consumption was achieved after 3 stages (the values of the reduction in power is the same as the reduction in Pressure), while almost no reduction was shown during 4th and 5th stages. Brackish water was simulated; figure 19 shows the reduction percentage achieved for different weight percentages of CNTs filters. A Maximum reduction in energy of 24% was shown when using a filter of 3% wt. Figure 20 shows a maximum reduction of 46.1% for energy when using the 3% wt. filter for 3 filtration stages.

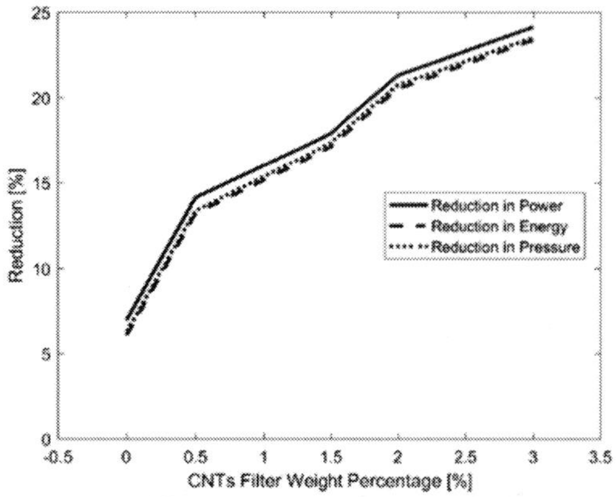

Figure 17. Effect of CNTs wt.% on Percentage of Reduction in Power, Energy and Pressure for Red Sea Water.

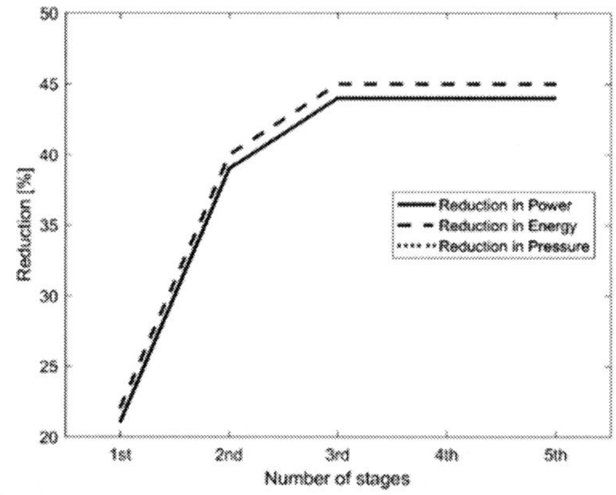

Figure 18. Effect of Filtration Stages at 3 wt.% filter on Percentage of Reduction in Power, Energy and Pressure for Red Sea Water.

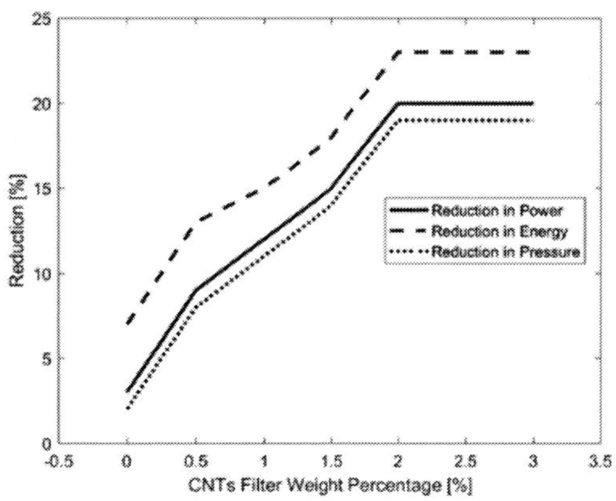

Figure 19. Effect of CNTs wt.% on Percentage of Reduction in Power, Energy and Pressure for Brackish Water.

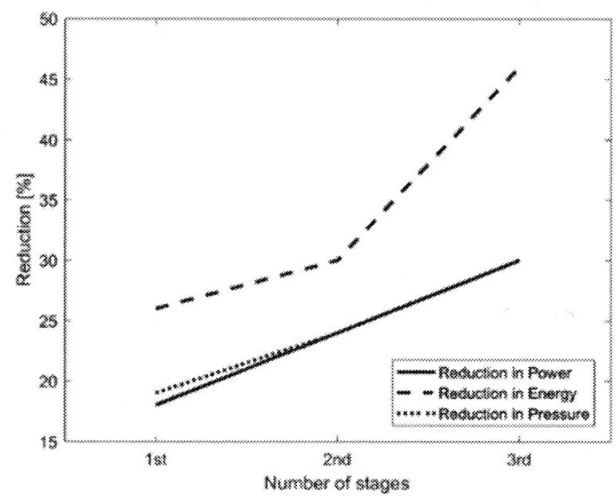

Figure 20. Effect of Filtration Stages at 3 wt.% filter on Percentage of Reduction in Power, Energy and Pressure for Brackish Water.

4. Conclusion

CNTs filters as an emerging technology is shown to be a promising solution for water desalination. A test rig was built, different CNTs filter specimens were fabricated for testing the effect of CNTs on desalination. It was found that addition of CNTs (Multi-tube type) to epoxy resin with granular activated carbon (GAC) can increase the rate of desalination of water. Experimental work was done to show the effect of weight percentage of CNTs on water desalination. Water samples used was from Red Sea and brackish (a deep well of water depth; 21 m). Several experiments were performed via soaking of Red Sea and brackish water through the test rig. Water samples were sent for examination of dissolved salts concentration at Governmental certified labs of the Ministry of Agriculture in Egypt. Several experiments were done without and with CNTs; where CNTs content reached 0.5, 1, 1.5, 2 and 3 wt. %. The effect of multiple filtration stages were also shown. Filters without CNTs (0 wt. %) were shown to have no effect on salinity. This can be used for a variety of agricultural crops withstanding medium salinity such as soya beans, mango, grapes, etc.

IOP Conf. Series: Materials Science and Engineering **973** (2020) 012016 doi:10.1088/1757-899X/973/1/012016

The effect of using CNTs filter (as pre filtration) attached with RO plant; for Red Sea and brackish water desalination was simulated. Well proven software modeling; Reverse Osmosis System Analysis (ROSA) software was used for modeling. CNTs filters output were simulated, results showed a considerable reduction in the plant's power, pressure and energy consumption. Software simulation for RO water desalination units show that using filters with 3 wt.% after 3 stages can reduce significantly feeding pressure and specific energy. Hence, it was shown that the consumed power will be reduced by 45% for Red Sea desalination, whereas, for brackish water the consumed power could be reduced by 24%. This would impact highly the production cost for desalinated water as it is well known that more than 50% of the production cost is due to energy consumption.
As a future work more analysis must be done to study the interaction between water and filter and expected degradation rate of filter and the effect of filter on pollutant water and its effect on the water quality and health in case of water direct using.

References

[1] Sadrzadeh M and Mohammadi T 2008 Sea water desalination using electrodialysis *Desalination* **221** pp 440–447. doi:https://doi.org/10.1016/j.desal.2007.01.103.
[2] Mohammadi T and Kaviani A 2003 Water shortage and seawater desalination by electrodialysis *Desalination* **158** pp 267–270. doi:https://doi.org/10.1016/S0011-9164(03)00462-4.
[3] Belessiotis Vand Delyannis E 2001 Water shortage and renewable energies (RE) desalination-possible technological applications *Desalination* **139** 133–138. doi:https://doi.org/10.1016/S0011-9164(01)00302-2.
[4] Hadadin N, Qaqish M, Akawwi E and Bdour A 2010 Water shortage in Jordan — Sustainable solutions *Desalination* **250** pp 197–202. doi:https://doi.org/10.1016/j.desal.2009.01.026.
[5] Kaldellis J K and Kondili E M 2007 The water shortage problem in the Aegean archipelago islands: cost-effective desalination prospects *Desalination* **216** pp 123–138. doi:https://doi.org/10.1016/j.desal.2007.01.004.
[6] Rao G, Lu C, and Su F 2007 Sorption of divalent metal ions from aqueous solution by carbon nanotubes: A review *Sep. Purif. Technol* **58** pp 224–231. doi:https://doi.org/10.1016/j.seppur.2006.12.006.
[7] Das R, Ali E, Abd Hamid S B, Ramakrishna S, Chowdhury Z 2014 Carbon nanotube membranes for water purification: A bright future in water desalination, *Desalination* pp 336 , 97–109. doi:https://doi.org/10.1016/j.desal.2013.12.026.
[8] Wang Y, He Z, Gupta K, Shi Q and Ruifeng L 2017 Molecular dynamics study on water desalination through functionalized nanoporous graphene *Carbon N. Y.* **116** pp 120–127. doi:https://doi.org/10.1016/j.carbon.2017.01.099.
[9] Long R Q and Yang R T 2001 Carbon Nanotubes as Superior Sorbent for Dioxin Removal *J. Am. Chem. Soc.* **123** pp 2058–2059. doi:10.1021/ja003830l.
[10] Y.-H. Li, S. Wang, J. Wei, X. Zhang, C. Xu, Z. Luan, D. Wu, B. Wei, Lead adsorption on carbon nanotubes *Chem. Phys. Lett.* **357** (2002) pp 263–266. doi:https://doi.org/10.1016/S0009-2614(02)00502-X.
[11] Lior N 2017 Sustainability as the quantitative norm for water desalination impacts, *Desalination* **401** pp 99–111. doi:https://doi.org/10.1016/j.desal.2016.08.008.
[12] Yousef S, Khattab A, Osman T and Zaki M 2013 Effects of Increasing Electrodes on CNTs Yield Synthesized by Using Arc-Discharge Technique *J. Nanomater pp* 1–9. doi:http://dx.doi.org/10.1155/2013/392126.
[13] Bittmann B, Haupert F and Schlarb A K 2009 Ultrasonic dispersion of inorganic

IOP Conf. Series: Materials Science and Engineering **973** (2020) 012016 doi:10.1088/1757-899X/973/1/012016

nanotubes in epoxy resin *Ultrason. Sonochem.* **16** pp 622–628. doi:https://doi.org/10.1016/j.ultsonch.2009.01.006.
[14] Geoffrey Pritchard 1999 *Reinforced Plastics Durability* (Woodhead Publishing Ltd and CRC Press LLC ©, Woodhead Publishing Ltd).

IOP Conf. Series: Materials Science and Engineering **973** (2020) 012017 doi:10.1088/1757-899X/973/1/012017

Optimization of friction stir welding parameters using response surface methodology

I Sabry[1], N Gadallah[1] and M Abu-Okail[1]
[1] Manufacturing Engineering Department, Modern Academy for Engineering and Technology, Cairo, Egypt
ibrahim_sabry@f-eng.tanta.edu.eg

Abstract. The aim of this article is to create a new technique for predicting discontinuity formation, its place and magnitude during aluminium alloy (AA6061) friction stir welding (FSW). The effectiveness of the technique was demonstrated using visual inspection, hardness and tensile test of the friction stir welded joints. The measured current was analysed through power calculations. In each of the FSW stages, the energy consumption is significantly varied, clearly distinguishing the penetration of the tool, its revolution, its traverse movement and its metal removal rate. The findings of tracking the energy consumption indicate that using power consumption means the significance of weld quality. FSW has been carried out based on two factors - two levels. Response surface methodology (RSM) is employed to develop a mathematical model. Analysis of variance (ANOVA) technique checks the adequacy of the developed mathematical model, which is used effectively at 95% confidence level. In contrast, tensile and hardness tests also showed that welds at high power usage failed continuously within the welding area, due to reduced welding temperature and absence of penetration in the welding zone.

1. Introduction

Friction stir welding, a comparatively fresh joining method invented at the Welding Institute in 1991, is a method that falls into a solid-state joining one that allows a plastic deformation and physical intermixing of two or more materials [1]. A friction stir welding process scheme is shown in figure1. A non-consumable rotating tool composed of a shoulder and a pin plunging into the workpieces during friction stir welding so that the shoulder and the workpieces are in contact.

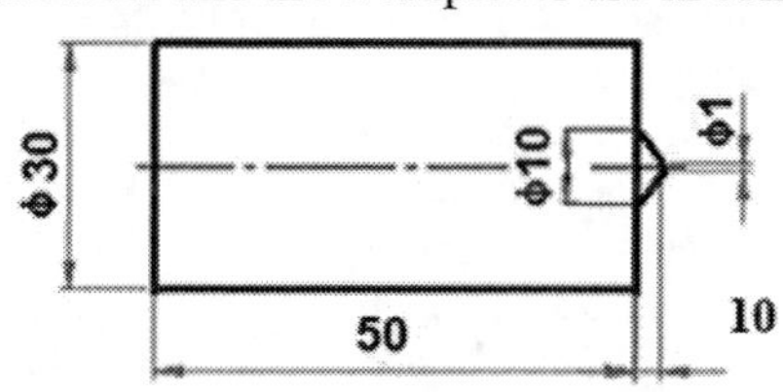

Figure 1. Conical friction stir welding tool

Due to an elevated heat field arising from the mixture of high tool-workpiece interface and friction forces, a plastic deformation under the rotating tool is produced. The instrument passes along the joint interface and joins the neighbouring workpieces as the instrument leaves the processing area. One of the main approaches to define critical surveillance and forecast characteristics has concentrated on

IOP Conf. Series: Materials Science and Engineering **973** (2020) 012017 doi:10.1088/1757-899X/973/1/012017

frequency analysis of measured output signals [2]. The main reason behind this is that discontinuity formation results from the uneven material flow that has been plasticized. Evidence indicates that the plasticized material around the tool probe flows into a cavity created at the probe's trailing edge as the tool rotates and moves. The plasticized material flow is extremely regular in relation to the tool rotation frequency under correct welding conditions [3].

In case of defective welds, the cavity is partially filled at the tool's rotational frequency, and the material flow is not purely periodic. The ability to detect and monitor the formation of voids is to analyse the output signals. These measured signals include acoustic emission, dipping force, vertical and traverse forces, spindle or feed engine current, and surface pictures. With a mixture of discrete wavelet conversion, Chen et al.[4] used acoustic emission signals. Based on the research conducted by Chen et al, Sundararajan et al. [5] diverse welding parameters (welding velocity, spindle speed and dip depth) are used to explore the impact of process parameters on changes in acoustic emission signals. The option of automatically identifying gap-induced failures provided by Fleming et al. [6], with a Fast Fourier Transform, the plunge forces were gathered and processed. The statistical analysis showed that the frequency spectrum of measured dip force could be a main characteristic in identifying defective welds by discovering discriminating features.

Kumar and Kumari et al. [7] evaluated surface defect correlation with abrupt modifications in plunge force. In order to locate the faulty surface area, statistical characteristics such as variance and square of detail coefficient mistakes obtained from constant and discrete wavelet transformations were introduced. Das et al. [8] obtained spindle, feed motor currents, and extracted decomposed characteristics from the signals using wavelet conversion. The characteristics along with process parameters (welding velocity, spindle speed and the tool's shoulder diameter) were supplied to artificial neural network algorithms to predict ultimate tensile strength of the workpieces and weld strength. Attempts to classify welds as defect-free or faulty using different statistical techniques and approaches to machine learning. Instant frequency and stage data were obtained from the measured plunge force through the established technique. Sectioning the welds at random places verified the inner voids. The writers indicated that both characteristics (instant frequency and phase) could describe the existence of defects in the welded samples effectively.

Boldsikhan et al. [9] explored the use of a discrete Fourier transformation and a method of machine learning to separate faulty welds from defect-free welds. For classification, the decomposed reduced frequency sections of the vertical and traverse force signals frequency spectrum were fed into the neural network. The data set tested showed a classification precision of roughly 95%..Shrivastava et al. [10] introduced computerized tomography (i.e. 3D X-ray imagery) to evaluate the complete volumetric size of sub-surface voids generated in the friction stir welded specimens. Specimen with variable error sizes (formed by variable input parameters such as spindle velocity and weld velocity) were evaluated using the frequency content (via quick Fourier Transform) of the force signal measured in the weld travel direction and using a statistical method. Their works showed a nice correlation between the welding force frequency content and the occurrence and voids size.

The power usage of 6061-T6 aluminium friction stir welding and gas metal arc welding was earlier compared [11] using a single parameter situation (welding velocity: 400 mm/min, spindle velocity: 1100 rpm and steel backing plate) where general power consumption in friction stir welding was discovered to be 40% lower than gas arc welding in this specific welding situation. Hence, the FSW process parameters, rotation speed and travels speed effect the power consumption, current, Cooling rate, ultimate tensile strength, yield strength, and percentage of elongation in FSW joints. There is a gap between predicting power consumption, current, Cooling rate, ultimate tensile strength, yield strength, and percentage of elongation for different parameters and for various materials. This gap could be filled by developing an RSM based model for the FSW process. In this work, an attempt was made to develop a straightforward RSM for the prediction of weld strength and hardness, within the range of the process parameters, which could provide a perfect joint through utilize FSW.

2. Experimental work

2.1 Material and Experimental Setup:

The AA6061 aluminum alloy plates (workpieces) were milled to nominal dimensions of 10 x 75 x 300 mm.

Table 1.chemical composition (wt%) of AA 6061 (ASTM D3039).

Alloy	6061
Si	0.6
Fe	0.7
Cu	0.28
Mn	0.15
Mg	0.2
Ti	0.15
Cr	0.2
Zn	0.25

Table 2. Mechanical Properties of AA 6061

Description	6061
σy MPa	85
σu1MPa	122
E1 %	16
VHD	62

The plates are mounted in a rigidly developed fixture. During the welding process, the tool is fed with 2mm penetration depth. The plate were welded using friction stir welding equipment Egypt (EG) FSW. The spindle current consumed for the welding process is acquired for every weld. The rotational speed and feed rate were fixed at 1000, 1400 and 1800 rpm and 10, 16 and 31.5 mm/s respectively after several welding trials. The tool tilt was 2°. The geometrical dimensions of the conical friction stir welding tool that is used for the experimentation are shown in figure 1. Table 3 shows the parameters used for the welding of AA 6061 plate. The Scheme of the overall approach used in monitoring and controlling friction stir welding is shown in figure 2 (a) and The Scheme of the overall approach used in monitoring and controlling friction stir welding is shown in figure 2 (b). The scheme of the definitions of cooling rate and temperature gradient is shown in figure 3.

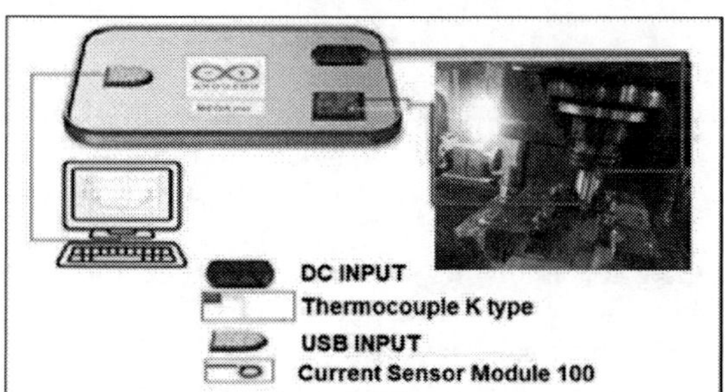

Figure2. Scheme of friction weld overall approach used in monitoring and controlling friction stir welding

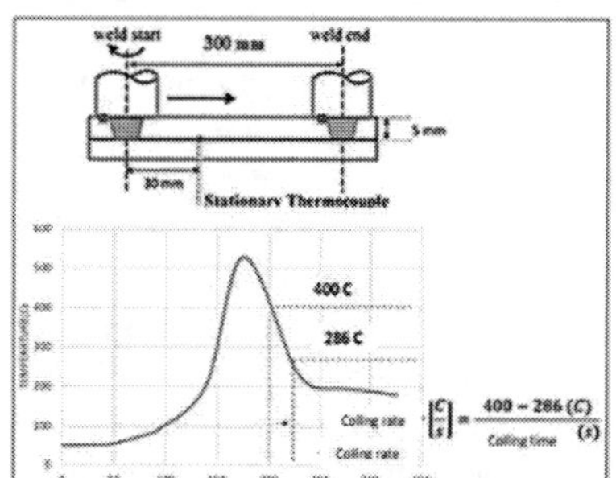

Figure3. Definitions of cooling rate and temperature gradient.

2.2 FSW process variables

Based on preliminary trials and previous studies, the independent process parameters affecting the power consumption, current, Cooling rate, temperature, ultimate tensile strength, and percentage of elongation were identified as rotation speed (N), and travel speed (S). Table 3 shows the friction stir welding parameters.

Table 3. Parameters for friction stir welding of Al 6061

Parameter	Value
Tool rotation [rpm]	1000,1400,1800
Tool feed rate [mm/min]	10,16,31.5

Trial runs proceeded to find the upper and lower limit of process parameters for AA 6061 by changing one parameter only at each time. A parameter range was tuned in such a way that the final welded joint has no defects recognized by visual screening. The upper limit of a factor was coded as 1 and lower limit as −1. The intermediate coded values were calculated using equation(1).

$$X_i = 2X - \frac{X_{max} + X_{min}}{X_{max} - X_{min}} \quad (1)$$

where X_i, X,X_{max} and X_{min} are the wished-for coded value, the variable value, the lower limit of the variable and the upper limit of the variable respectively [14]. Table 4 specify the considered process parameters with their limits, units and notations.

Table 4. Parameters Process and Their Levels in FSW

Process Parameters	Unit	Levels		
		-1	0	1
Tool rotation[rpm]	N	1000	1400	1800
Tool feed rate[mm/min]	S	10	16	31.5

Specimens of required size were cut from the welded plate to execute out metallurgical studies as explained in figure 4.a. Figure 4.b represents the sample of FSW welded at 10 mm/min and 1000 rpm.

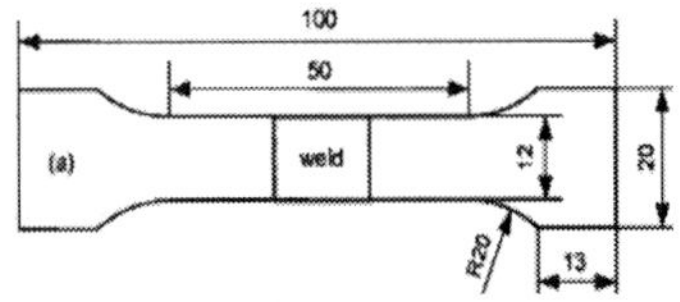

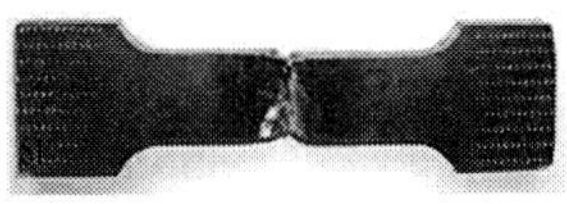

Figure4(a). Tensile test sample geometry.**Figure4(b).** A sample of FSW welded.

3. Results and discussion

The tensile test samples were according to ASTM D3039. The ultimate tensile strength of the FSW joints were measured through computerized universal testing machine. For each welded joint, four specimens were prepared and tested. Table 5 presents the average values of the results acquired from FSW experimental values.

The experiment was based on two factors with three levels of full factorial experimental design. As prescribed in the Experimental design matrix night joints were carried out by considering three levels of a process parameter, namely tool rotational speed and travels speed as given in the Table 2. It is to be further noted that the experiments were conducted with a constant tool rotational speed and travels speed.

IOP Conf. Series: Materials Science and Engineering **973** (2020) 012017 doi:10.1088/1757-899X/973/1/012017

Table 5. Experimental design matrix and results

Trial no.	Coded value		Real value		Experimental work					
	N	S	Tool speed (rpm)	Travels speed (mm/min)	Temperature	Cooling rate	Tensile strength	Power process	Elongation	Current consumption
1	1	1	1000	10	480	65	235	2500	0.68	4.8
2	1	2	1000	16	471	29	220	1500	0.60	4.5
3	1	3	1000	31.5	450	11	210	1450	0.55	4.1
4	2	1	1400	10	500	55	245	2700	0.65	6.5
5	2	2	1400	16	485	25	230	1700	0.612	6.2
6	2	3	1400	31.5	465	15	212	1500	0.6	6
7	3	1	1800	10	550	70	250	3000	0.70	8
8	3	2	1800	16	510	22	237	1900	0.6256	7.8
9	3	3	1800	31.5	500	19	215	1600	0.63	7.3

3.1 Mathematical Model:

Ultimate tensile strength, hardness and surfaces roughness of the FSW joints is function of rotation speed, travel speed and axial force, and it can be expressed as in Eq. 2.

$$Y = f(N, S) \tag{2}$$

where Y is the response N is the rotation speed and S is the travel speed. For the three factors, the chosen polynomial could be expressed as in equation (3).

$$Y = \beta_0 + \beta_1 N + \beta_2 S + \beta_{11} N^2 + \beta_{22} S^2 + \beta_{12} NS \tag{3}$$

where β_0 is the free term of the regression equation β_1, and β_2 are the linear terms the coefficients β_{11}, and β_{22} are quadratic terms the coefficients β_{12} are the interaction terms. The values of the coefficient of the polynomial are calculated by regression analysis with the help of following equations [3]:

$$\beta_0 = 0.1663 \sum(Y) - 0.0568 \sum\sum(X_{ii}Y) \tag{4}$$

$$\beta_j = 0.0732 (X_i Y) \tag{5}$$

$$\beta_j = 0.0625 \sum(X_{ii}Y) + 0.00689 \sum\sum(X_{ii}Y) - 0.0568 \sum(Y) \tag{6}$$

$$\beta_{ij} = 0.1250 \sum(X_{ij}Y) \tag{7}$$

where i ; j = 1, 2, 3 and i< j

DESIGN EXPERT 8.0.4 software package has been used to estimate the values of those coefficients for different responses and the results are presented in Table 6.

Table 6 Calculated regression coefficients of mathematical model

Regression factor	Temperature	Cooling rate	Tensile strength	Power	Elongation	current
β_0	568.6	225.8	224.2	4781.	0.9980	-0.4065
β_1	-0.09721	-0.08063	0.04845	0.3893	-0.0001976	0.006697
β_2	-5.473	-13.72	-3.056	-337.8	-0.02725	-0.06733
β_{11}	6.354E-005	2.708E-005	-7.292E-006	0.0001563	6.417E-008	-9.375E-007
β_{22}	0.1144	0.2636	0.06110	7.460	0.0004395	0.0009669
β_{12}	-0.0007541	0.0003808	-0.0006584	-0.02031	3.668E-006	-1.608E-006

IOP Conf. Series: Materials Science and Engineering **973** (2020) 012017 doi:10.1088/1757-899X/973/1/012017

3.2 Checking the adequacy of the developed models using ANOVA

The sufficiency of the advanced models employ the analysis of variance technique (ANOVA). The contribution percentage (p) signalizes the linear square and interaction process parameters with the response functions. These (p) values are employed to identify the significant parameters on the response functions [12]. Table 6 demonstrates the results for tensile strength model of the ANOVA. The model F-value of 17.25, 19.98, 99.63, 2011.3, 9.58 and 370.33 for temperature, cooling rate, tensile strength power process, elongation%, current consumptions implies the model is significant. There is only a 0.01% chance that a model F-value could occur due to noise. Value NS is significant model terms. Values better than 0 1 indicate that the model terms are not significant. The coefficient of determination R2 values gives the goodness of fitness of the model. The determined values of the developed model are presented in Table 7.

Table 7. Coefficient of determination values for temperature, cooling rate, tensile strength, elongation, current and power process

	F-value	R^2	*Adusted* R^2	*Predicted* R^2	*Adequate Precision*
Temperature model	17.25	0.9664	0.9104	0.6648	12.58
Cooling rate model	19.98	0.9709	0.9223	0.7023	10.35
Tensile strength model	99.63	0.9940	0.9840	0.9274	28.10
Elongation model	9.58	0.9409	0.8425	0.3107	8.882
Current model	370.33	0.9984	0.9957	0.9753	50.40
Power processes model	2011.3	0.9997	0.9992	0.9981	113.1

3.3 Effect of FSW process parameters

The tensile strength, elongation%, current consumption, power, cooling rate and temperature of FSW welded aluminium alloy 6061 were predicted by the mathematical models using the experimental observations presented in figure 5, showing the general trends between predicted and experimental results. The results of the ANOVA are given in table 6. The validity of regression models developed is further tested by drawing scatter diagrams presented in Fig.5 (a-f). The observed values and predicted values of the responses are scattered close to the 45° line, indicating an almost perfect fit of the developed empirical models [13].

Surface plot for temperature due to interaction effect of the travels and rotational speed is shown in figure 6 (a). The increase in temperature is significant at larger rotation speed while insignificant increase in temperature at low travels speed is observed. Surface plot for cooling rate due to interaction effect of the rotational and travels speed is shown in figure 5 (b). Figure 5 (c) shows the interaction effect of rotational and travels speed on the tensile strength. Figure 5 (c) shows surface plot for tensile strength due to interaction effect of the travels and rotational speed. The increase in tensile strength is significant at larger rotational speed while insignificant increase in tensile strength at low travels speed is observed. Surface plot for elongation percentage due to interaction effect of rotational and travels speed is shown in figure 6 (e). Surface plot for power process due to the interaction effect of the rotational and travels speed is shown in figure6 (d). Figure 6 (f) shows the increasing trend of current consumption at higher rotational and travels speeds as compared to lower rotational speed. Surface plot for current consumption due to interaction effect of the rotational and travels speed is shown in figure6 (f).

IOP Conf. Series: Materials Science and Engineering **973** (2020) 012017 doi:10.1088/1757-899X/973/1/012017

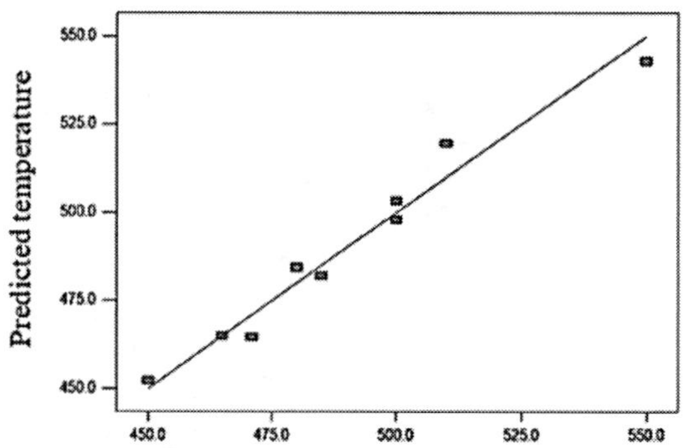

Fig.5 a) Experimental temperature

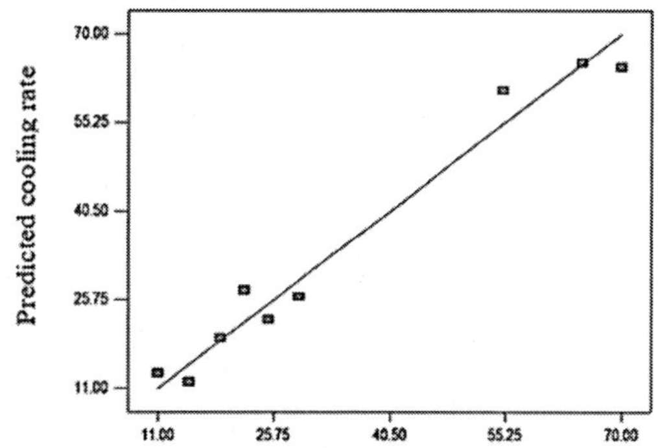

Fig.5 b) Experimental cooling rate

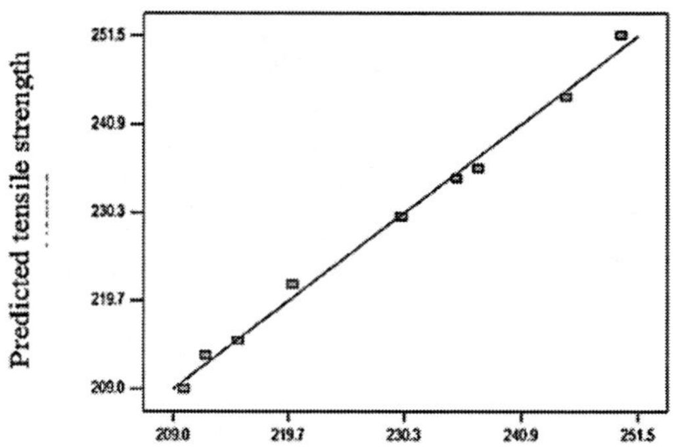

Fig.5 c) Experimental tensile strength

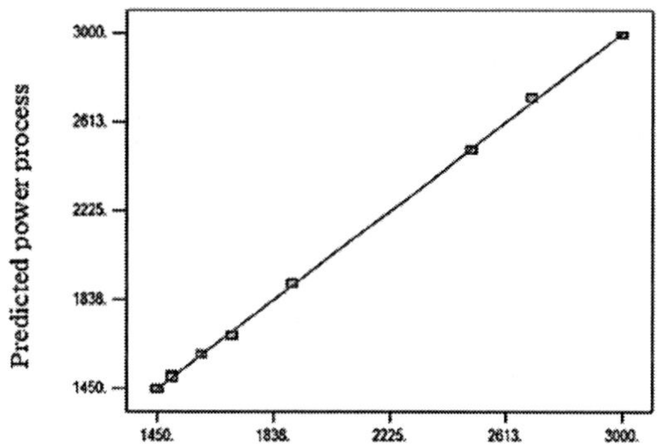

Fig.5 d) Experimental power process

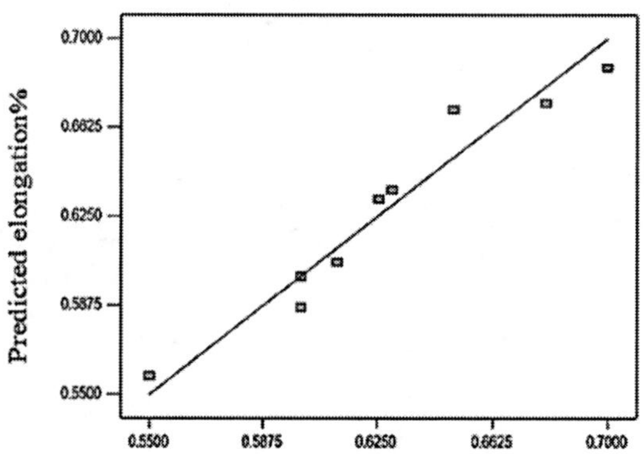

Fig.5 e) Experimental elongation%

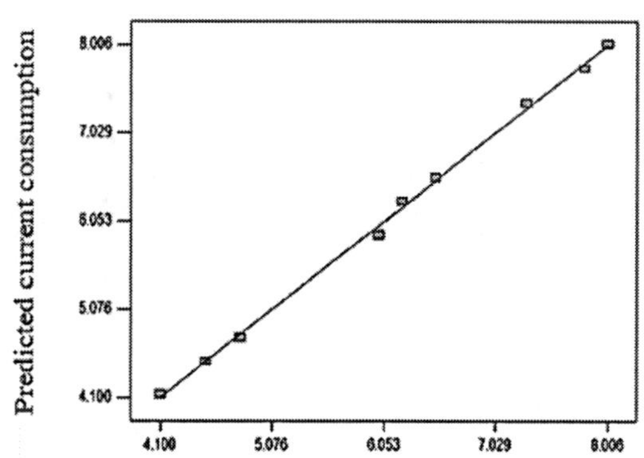

Fig.5 f) Experimental current consumption

Figure5. Scatter diagram of predicted and experimental (a)temperature (b)cooling rate (c) tensile strength (d)power process (e)tensile strength (f) current consumption.

IOP Conf. Series: Materials Science and Engineering **973** (2020) 012017 doi:10.1088/1757-899X/973/1/012017

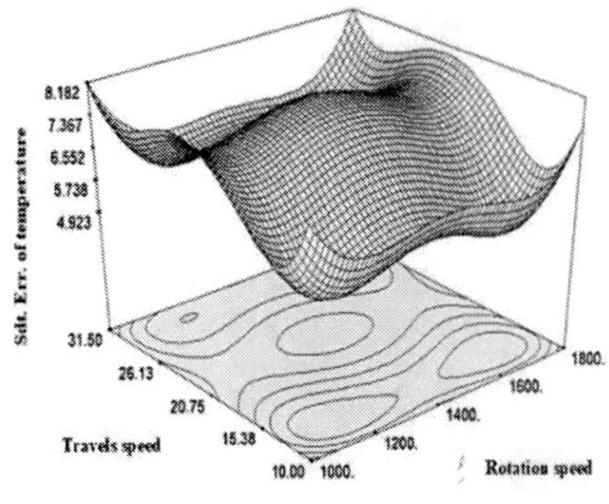

Fig. 6 a)

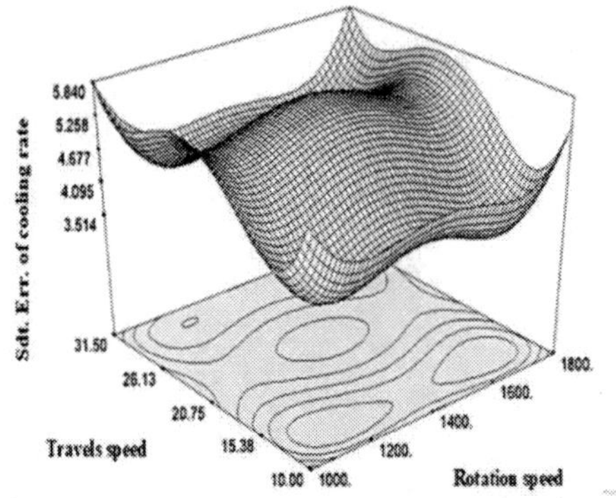

Fig. 6 b)

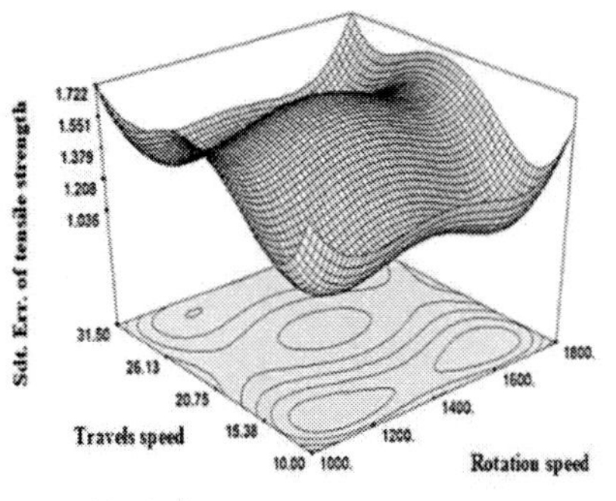

Fig. 6 c)

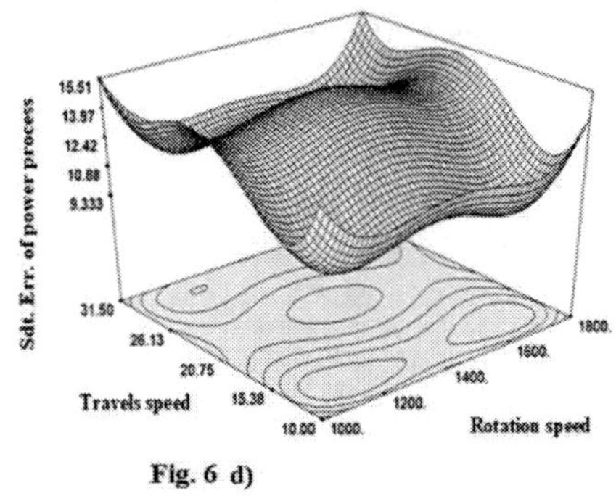

Fig. 6 d)

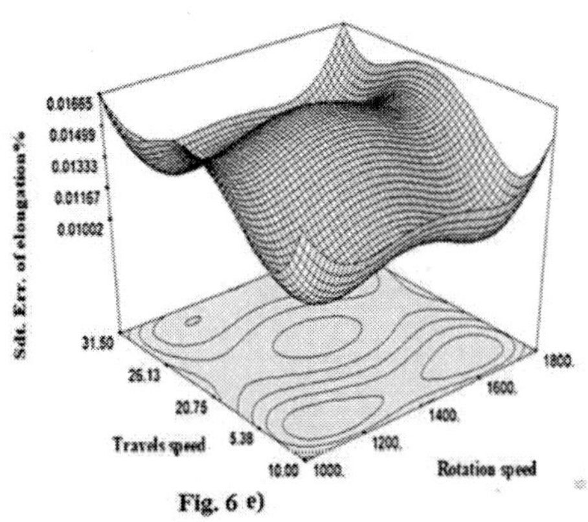

Fig. 6 e)

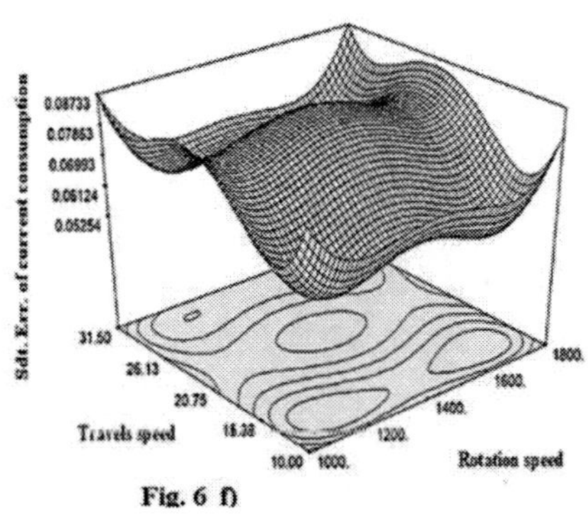

Fig. 6 f)

Figure6. Total stander error of predicted and experimental (a)temperature (b)cooling rate (c) tensile strength (d)power process (e)tensile strength (f) current consumption.

The heat production of FSW is related to the rotational speed and travels speed of the tool, yield shear stress, friction coefficient, contact pressure between the tool and the matrix, etc. the effect of different rotational speeds on the recorded welding thermal cycles. As the rotational speed increases from 1800 rpm to 1000 rpm, the peak temperature of the retreating side increases. When the speed reached 1800 rpm and travel speed 10 mm/min , the maximum temperature fell back to 55℃.For the aluminium

IOP Conf. Series: Materials Science and Engineering **973** (2020) 012017 doi:10.1088/1757-899X/973/1/012017

alloy, the tensile strength increases with increasing temperature, so the yield shear stress, contact pressure and friction coefficient will decrease with increasing temperature.

Therefore, when the rotational speed of the tool is increased and travels speed decrease, although the heat production may be increased and the temperature will rise, the yield shear stress, contact pressure and friction coefficient will decrease with increasing temperature, which will lead to a decrease in the actual heat production Because of this, FSW can keep the base metal in the solid state for welding.

The Rotational speed is directly proportional to the tensile strength of the weld. The frictional heat input increases with increase in the rotational rate of the tool. Thus the increase in the Rotational speed enhances the heat input of the process which in turn results in better material flow and increases the material to be displaced in a unit time. The maximum tensile strength, power and current consumption of the weld is also seen when the process parameter rotational speed set at a higher level in the process.

As travel speed is one of the important process parameters, the considered level doesn't affect the responses as the variation in the level average is not appreciable. This implies within the selected range of process window the levels can be fixed at any level to have the desired output show

3.4 Optimization of parameters of FSW

One of the most important aims of this investigation is to maximize tensile strength, elongation % and cooling rate and to minimize current consumption, power process, and temperature of FS welded joints of AA 6061 plate. From the mathematical model, the optimum process parameters are obtained with the desirability ranges between 0 and 1.The predicted optimal results from above technique are tensile strength, elongation percentage, current consumption, power, cooling rate and temperature that obtained using Design Expert software at the combined desirability value of 0.631.Figure7 shows the immediate influence of rotational speed on mechanical properties. It is seen that as the rotational speed increases the tensile strength, yield strength, hardness and percentage of elongation of Friction Stir welded aluminium alloy 6061 increases. In FSW as the rotational speed increases the heat input also increases and additional heat input destroys the regular flow behaviour. Figure 7 shows the immediate influence of travel speed on mechanical properties. It is evident that as traveling speed decreases from 31.5 to 10mm/min the mechanical properties of the FS welded AA 6061 increases and then decreases. At lowest traveling speed (10 mm/min) and highest traveling speed (31.5 mm/min) lower tensile strength is observed. This is due to the increased frictional heat and insufficient frictional heat generated respectively. Also higher traveling speed produce poor plasticflow of the material it causes poor consolidation of the metal interface.

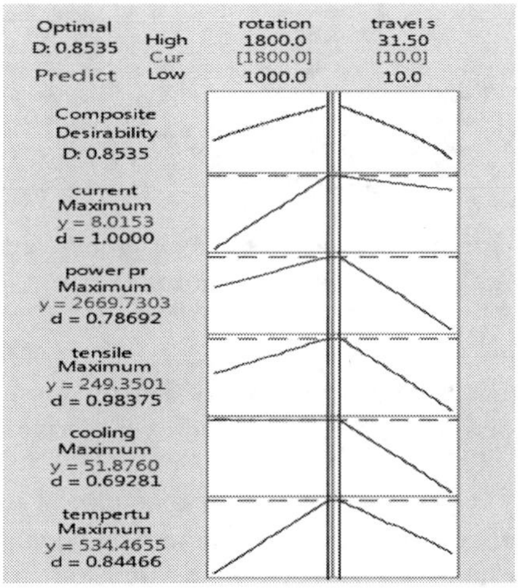

Figure7. Optimized parameters and responses using response optimizer in ANOVA

3.5 Analysis of results

The effects of the different process parameter on the mechanical properties of FS welded aluminium alloy 6061 are predicted from the mathematical models using the experimental observations are presented in figure 8 showing the general trends between cause and effect.In figure 9 shown contour Plot of current consumption, power process, tensile strength, cooling rate and temperature

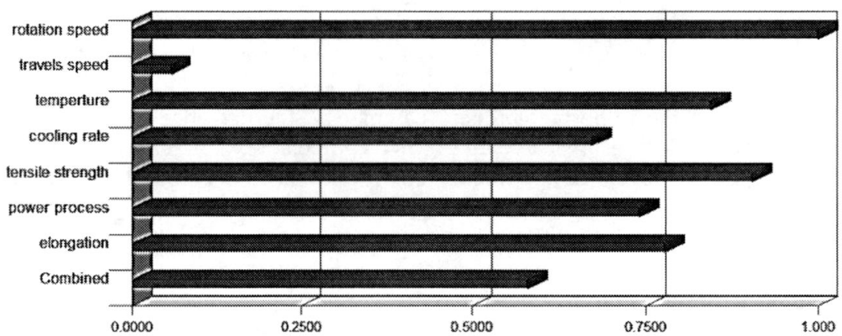

Figure8. Bar graph showing the maximum desirability for the combined desirability value of 0.631.

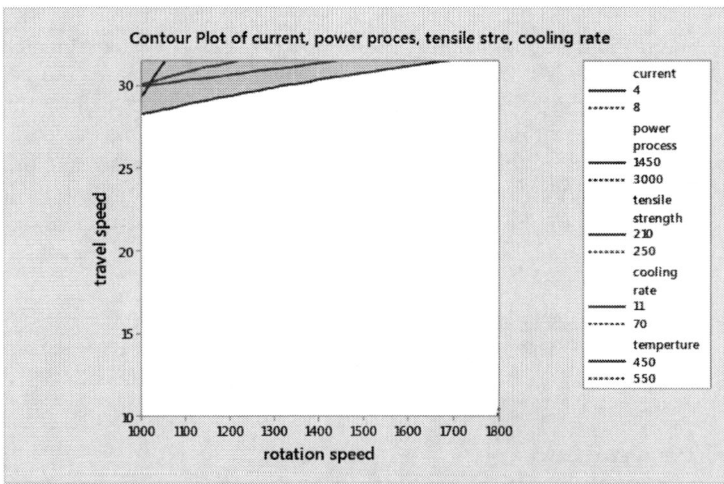

Figure9. Contour Plot of current consumption, power process, tensile strength, cooling rate and temperature.

4. Conclusions

In the present work, the minimum power and energy consumption of AA6061 butt welds created by friction stir welding were experimentally investigated over a wide range of spindle speeds from 100, 1400 and 1800 and travel speeds from 10, 16 and 31,5mm/min.

1- The consumed power is minimum with a combination of lowest spindle speed and high weld speed, while the current consumption is minimum by operating at the highest weld speed.
2- The consumed power increases with both increasing spindle speed and travel speed.
3- The results show that the measured temperature gradient at the tool-weld interface is an effective proxy for both the cooling rate trend and the resulting ultimate tensile strength.

References

[1] Mishra R Sand Ma Z Y 2005Friction stir welding and processing*Mater. Sci. Eng.R Reports* **50** pp 1–78.

[2] Fleming P, LammleinD, Wilkes D, Fleming K, Bloodworth T, Cook G, Strauss A, Lienert T and Bement M 2008 In-process gap detection in friction stir welding *Sens. Rev.* **28** pp 62–7.

[3] Arbegast W 2008A flow-partitioned deformation zone model for defect formation during friction stir welding*ScriptaMaterialia* **58** pp 372–6.

[4] Chen C, Kovacevic R and Jandgric D 2003 Wavelet transform analysis of acoustic emission in monitoring friction stir welding of 6061aluminum*Int. J. Mach. Tools Manuf.***43** pp 1383–90.

[5] SoundararajanV, Atharifar H and Kovacevic R 2006 Monitoring and processing the acoustic emission signals from the friction-stir-welding process*Proc. Inst. Mech. Eng. B J. Eng. Manuf.***220** pp 1673-85

[6] Upadhyay P and Reynolds A 2014 Effect of backing plate thermal property on friction stirwelding of 25-mm-thick AA6061*Metall. Mater. Trans.*A**45** pp 2091–100.

[7] Kumar U, Yadav I, Kumari S, Kumari K, Ranjan N, Kesharwani R, Jain R, Kumar S, Pal S, Chakravarty D and Pal S 2015 Defect identification in friction stir welding using discrete wavelet analysis*Adv. Eng. Softw.***85** pp 43–50.

[8] Das B, Pal Sand Bag S 2017 Weld quality prediction in friction stir welding usingwavelet analysis *Int. J. Adv. Manuf. Technol.***89** pp 711–25.

[9] Boldsaikhan E, Corwin E M, Logar A M and Arbegast W 2011 The use of neural network and discrete Fourier transform for real-time evaluation of friction stir welding *Appl. Soft Comput.***11** pp 4839–46.

[10] ShrivastavaA, Zinn M, Duffie N A, Ferrier N J, Smith, C B and Pfefferkorn F E 2017 Force measurement-based discontinuity detection during friction stir welding*J. Manuf. Process.***26** pp 113–21.

[11] El-Kassas A M and Sabry I 2019 A new quality monitoring system for friction stir welded joints of aluminumpipes *Int. J. Eng.Technol.* **11** pp 78–87.

[12] Shrivastava A, Krones M and Pfefferkorn F E 2015Comparison of energy consumption and environmental impact of friction stir welding and gas metal arc welding for aluminium *Proc. Inst. Mech. Eng. B J. Eng. Manuf.* **9** pp 159–68.

[13] Kumari S, Jain R, Kumar U, Yadav I, Ranjan N, Kumari K, Kesharwani R K, Kumar S, Pal S, Pal S K and Chakravarty D 2016 Defect identification in friction stir welding using continuous wavelet transform *J. Intell. Manuf.* **30** pp 483–94

[14] El-Kassas A M andSabry I 2019 Optimization of the underwater friction stir welding of pipes using hybrid RSM-fuzzy approach *Int.J. of Appl. Eng. Res.***14** pp 4562–72.

IOP Conf. Series: Materials Science and Engineering **973** (2020) 012018 doi:10.1088/1757-899X/973/1/012018

Study of peak temperatures in friction stir spot welding of AA2024-Al/polycarbonate and AA2024-Al/polypropylene systems

R K Mohammed[1], S S Mohammed[1] and T S Mahmoud[1]
[1] Mechanical Engineering Department, Faculty of Engineering at Shoubra, Benha University, Cairo, Egypt.
engreemkamal81@gmail.com

Abstract. Compared to ordinary fusion welding processes, friction stir spot welding (FSSW) have significant advantages for example: joining of conventionally non-weldable alloys, improved mechanical properties and reduced distortion of weldable alloys joints due to the pure solid-state joining of metals. This welding technique gives the online feedback control and opportunity of automation, allowing automatic adaptation to geometrical and environmental variations of the component. However, several fundamentals of the process such as the temperature inside the stirred zone of the weld and its power on mechanical properties, are not yet fully understood. In the present study, FSSW was adopted to weld AA2024 aluminium alloy sheets with both polycarbonate (PC) and polypropylene sheets using two different tools having two different pin configurations. The two tools have conical pin with 2.5 mm height, however, tool (T1) has draft angle of ≈ 16.69° and the tool (T2) has a draft angle of ≈ 6°.The effects of FSSW parameters of the process, typically, the tool rotational speeds and dwell time, on temperature variation in the tool/work piece interface was also recorded, via an infrared camera. The results revealed that, increasing the tool rotational speed and/or the dwell time increase(s) the peak temperature in FSSW of both AA2024Al/polycarbonate and AA2024Al/polypropylene welded joints. For AA2024Al/polycarbonate joints, the peak temperatures recorded for tool T1 are higher than those recorded using tool T2. The peak temperature ranges were found to vary from 54 to 150 °C and from 45 to 136 °C, for joints welded using T1 and T2 tools, respectively. While, for AA2024Al/polypropylene joints, the peak temperatures recorded for tool T2 are higher than those recorded using tool T1. The peak temperature ranges were found to vary from 40 to 115 °C and from 61 to 143 °C, for joints welded using T1 and T2 tools, respectively.

1. Introduction

Metals and plastics are extensively used in the manufacturing industry to produce automobile components, electronic devices, and aircraft. Demands for the utilization of lightweight materials such as thermoplastics and aluminum alloys have increased in recent years to reduce the weight of products without sacrificing its structural strength[1].

Different promising welding techniques for joining dissimilar materials have been developed to solve problems related to convectional joining techniques. Friction stir spot welding (FSSW) is one of these methods. In FSSW, a rotating tool (non-consumable) is plunged into the work pieces to be joined. When the selected plunge depth is reached, the rotating tool is held in that position for a predestined time sometimes referred to as dwell period (t). Afterwards, the rotating tool is retracted from the welded joint leaving behind a friction stir spot weld. During FSSW, dwell period and the tool penetration mainly determine the heat generation, weld geometry, material plasticization around the pin and hence the mechanical properties of the welded joint. The shoulder generates bulkiness of the frictional or deformational heat while; the pin assists in material flow between the work pieces [2].Besides the tool, the other parameters involved in FSSW are, tool plunge depth, the dwell time (t)and the tool rotation speed [3]. During FSSW the low temperatures avoids several of the defects usually observed in fusion welding processes such as cracks and porosities, and so presenting good mechanical properties over arc welding [4]. In addition, the reduced heat input results in lower deformation. In this process, the heat is initially derived from the friction between the welded material and the welding tool (including the shoulder and the probe), which causes the welded material to soften at a temperature less than its melting point. The softened material beneath the shoulder is subjected to extrusion by the tool rotational and transverse movements. It is expected that this process will innately produce a weld with less distortion and residual stress as compared to the fusion welding methods, since no melting of the material occurs during the welding[5].

Joining techniques of lightweight dissimilar materials, mostly metals and polymers, are becoming more and more important in the manufacturing of hybrid structures and components for engineering applications [6-8].The aim of the present study is to examine the temperature variation during the FSSW of AA2024Al-to-polycarbonate AA2024Al-to-polypropylenesheets at different working conditions via the manipulation of the rotational speeds, the dwell times and the tool design.

2. Experimental work

2.1 Materials

In the present investigation, 1.5 mm thick AA2024 aluminium alloy sheets were joined with 3 mm thick polycarbonate (PC) as well as polypropylene (PP) sheets. The AA2024 Al alloy has chemical compositions (weight %) of 3.73% Cu, 1.17% Mg, 0.33% Fe, 0.23% Si, 0.24% Cr, 0.56% Zn and Bal.-% Al. Table 1 lists the mechanical and thermal properties of the AA2024 Al alloy [9]. Table 2 lists the thermal and mechanical characteristics of the polypropylene and polycarbonate sheets [10].

Table1. The mechanical and thermal properties of AA2024 Al alloy [9].

Parameter	Value
Density	2.78 g/cm^3
Melting point	502 - 638 °C
Thermal conductivity (W/m-K)	151 W/m-K
Coefficient of thermal expansion	24.7 μm/m-°C
Tensile strength	Min 427 MPa
Specific Heat Capacity	875 J/kg.°C

Table 2. Thermal characteristics of the polypropylene and polycarbonate sheets [10].

Material	Coeff. of thermal expansion (10-6/°K)	Specific heat (J /K^{-1} kg)	Thermal conductivity (W/m.K)	Melting Temp. (°C)	Tensile strength (MPa)	Density (g/cm^3)
Polycarbonate	66-70	~1200	0.19-0.22	288-316	59	1.15-1.52
Polypropylene	100-180	1700-1900	0.1-0.22	163.8	48	0.97-1.25

2.2.Tool Design

Two tools having different pin configurations were used to perform the FSSW for the AA2024Al-to-polycarbonate and AA2024Al-to-polypropylene sheets. Figure 1 shows a schematic illustration of the FSSW used tools. The tools have conical pins. The pin for the tool (T1) has slope angle of 16.69°, while the tool (T2) has slope angle of 6°. The pin height and diameter of shoulder are kept constant at 5 mm and 15 mm, respectively, for both tools. The tools are made from H13 tool steel.

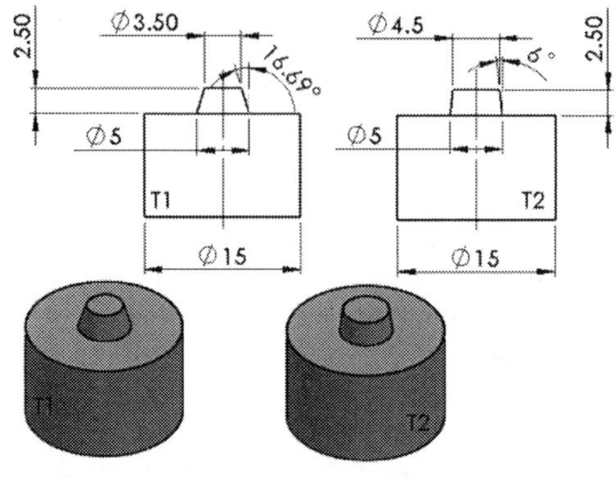

Figure 1. Schematic illustration of the tools used for FSSW (dimensions in mm).

2.3.FSSW Process

The FSSW was carried out using CNC verical milling machine. Figure 2 shows the experimental setup used for FSSW. The AA2024 Al sheets were located at the top (i.e. facing the tool), while the polycarbonate (PC) or polypropylene (PP) sheets were located at the bottom of the lap-joint configuration shown in figure 3.The FSSW was carried out using three rotational speeds of 500, 1000 and 1500 rpm and three dwell times (t) of 4, 5 and 6 sec.

Figure 2. The FSSW process.

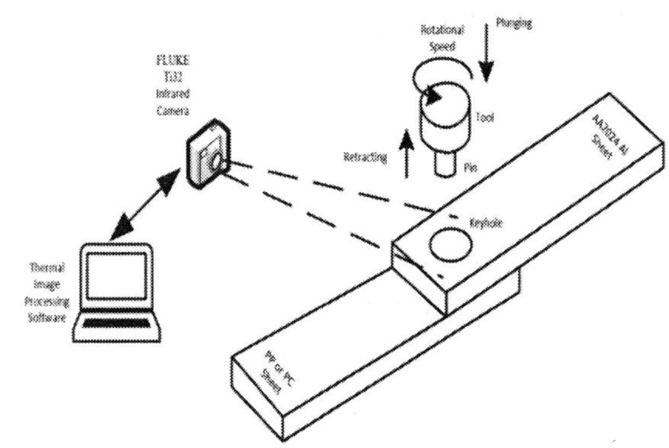

Figure 3. The Lap-joint configurations.

2.4.Temperature Measurements

The temperature meaurements during FSSW were recorded using FLUKE Ti32 infrared camera. Since it is very difficult to measure the temperature at the metal/polymer interface region using the inferared camera, the camera was aimed, to capture the temperature images, only at the top surface of the AA2024 Al sheet of the welded joint. The thermal images was immediattly captured, after the retracting of the tool, after finishing the FSSW process. It has a powerful 320×240 resolution and it has a thermal senstivity of 0.045°C. The temperature range of the camera varies from -20 to 600°C. Figure 4 shows a typical output from the infrared camera during FSSW specimen. The figure shows the temperature distribution at the contact between the tool and the workpiece.Fluke Thermographyimaging software was used to calculate the temperature distribution on the surfaces of the welded joints.

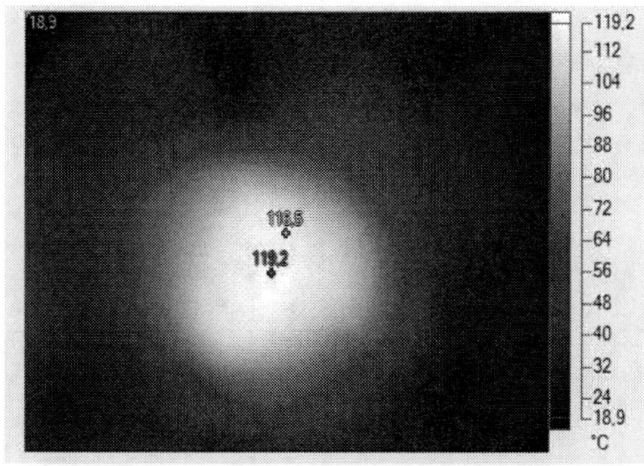

Figure 4. Typical photograph showing the temperature distribution during FSSW obtained using FLUKE Ti32 infrared camera.

3. Results and discussion

3.1.Macrostructure of the Joints

Typical cross-section of the AA2024Al/polymer system joint welded using FSSW is shown in figure 5. The figure shows the interface between the AA2024 Al sheet and the polypropylene sheet after FSSW. A nub or protrusion is also noticed due to the penetration of the pin during FSSW. Figure 6 shows macrographs of AA2024-Al/polycarbonate joint welded using tool T1 and 500 rpm and 4 sec. (see figure 6a); and 1500 rpm and 6 sec. (see figure 6b). It is clear from figure 6 (c) that the black color in the interface is the blending that happened between the AA2024-Al and the polycarbonate at the mentioned conditions. The Al particles that appeared in figure 6 (d) was due to the heat transfer that happened from the aluminum to the polycarbonate sheets due to the high temperature resulted from the high rotational speed (1500 rpm) and high dwell time (6 sec), since the high dense of aluminum rather than the polycarbonate. Therefore, these particles fall and immerse (indulge) in the mushy zone of the polycarbonate. In figure 7 it is also clear the blending regions that happened in the polypropelene and aluminum sides after tensile shear test under the mentioned conditions.

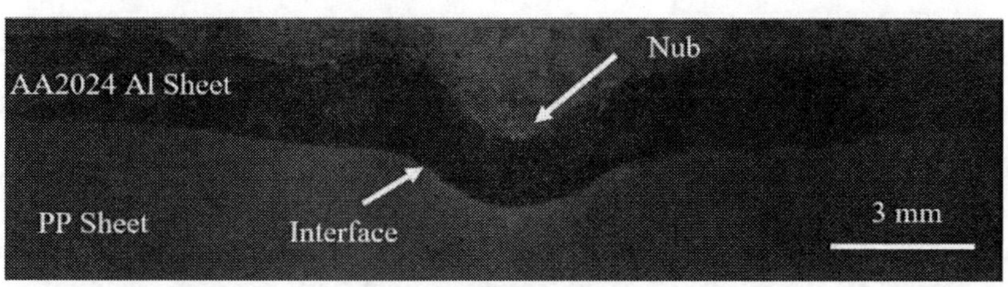

Figure 5. Typical cross-section of FSSW joint illustrating the interface between the AA2024 Al sheet and the polypropylene sheet.

At higher tool rotational speed and prolonged dwell time (t), an intensive stirring of the welded region can take place which increases the heat generated due to friction that occurs between the tool's shoulder and the Al sheet [7,8]. The heat is then transferred, from the upper layer of the Al sheet to the lower layer at which the metal/polymer interface is located. This increasing the temperature at the interface region and leads to the welding of the metal/polymer sheets. It has been observed that the interface regions are not clearly seen using optical microscopy for joints friction stir spot welded at high rotational speeds and high dwell times (t).

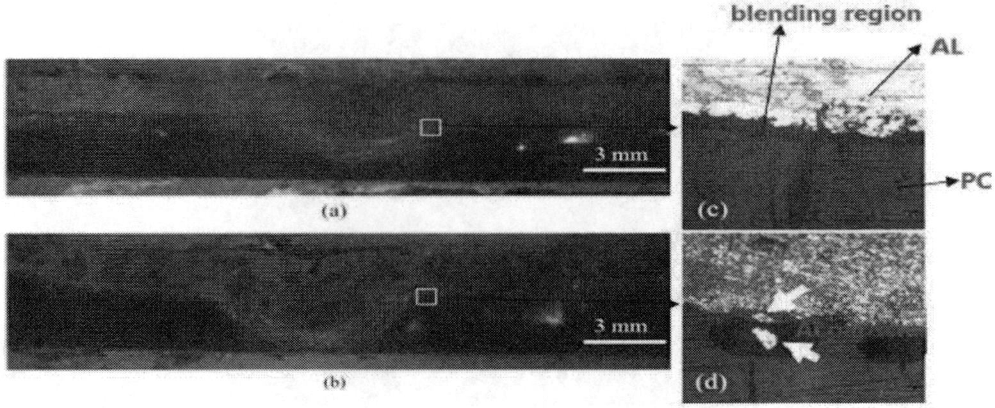

Figure 6. Macrographs of the AA2024-Al/polycarbonate joint welded using tool T1 and (a) 500 rpm and 4 sec.; and (b) 1500 rpm and 6 sec.

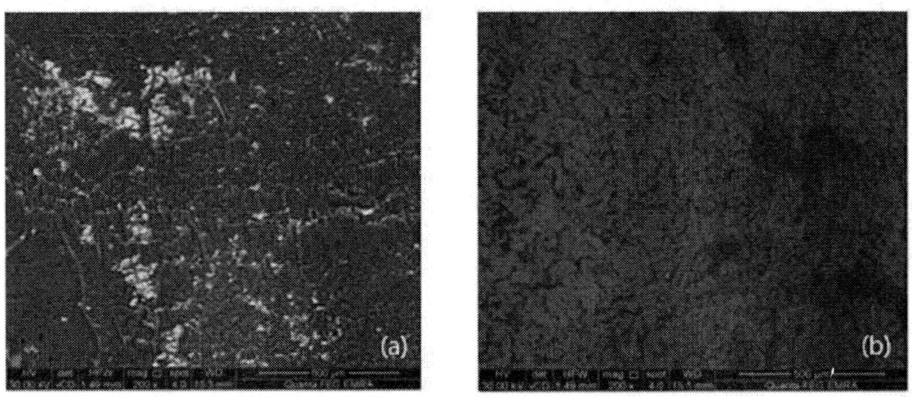

Figure 7.SEM images after tensile shear test at 1500 rpm, 4 sec, T1 (a) polypropelene side (b) aluminium side

3.2.Peak Temperature Results

Figures 8 and 9 show typical thermographs, taken immediately after retracting the tool, for AA2024Al/polycarbonate and AA2024Al/polypropylene joints, respectively. The joints were welded using the T1 and T2 tools at different tool rotational speeds and constant dwell time (t) of 6 sec. The cross-marks shown in thermographs are located on the surface of the workpieces for tracking the temperature changes during the FSSW process. The points are located in the surrounding area of the tool shoulder boundary. The results revealed that the temperatures near the center of the keyhole are maximum (represented by white and yellow colors). Getting away from the center of the keyhole, the temperatures tend to decrease (represented by red and blue colors).

Figures 10 and 11show the variation of the peak temperature in FSSW with the tool rotational speed at various dwell times for AA2024Al/polycarbonate and AA2024Al/polypropylene joints, respectively. The results revealed that, in most cases, increasing the dwell time and/or the tool rotational speed increase(s) the peak temperature during FSSW of both AA2024Al/polycarbonate and AA2024Al/polypropylene welded joints. For example, in AA2024Al/polycarbonate joints welded using tool T2 and constant dwell time of 6 sec., increasing the tool rotational speed from 500 rpm to 1500 rpm increases the peak temperature from $\approx$89°C to 136°C.

IOP Conf. Series: Materials Science and Engineering **973** (2020) 012018 doi:10.1088/1757-899X/973/1/012018

For AA2024Al/polycarbonate, the peak temperatures recorded for tool T1 are higher than those recorded using tool T2. The peak temperature ranges from 54 to 150 °C and from 45 to 136 °C, for joints welded using T1 and T2 tools, respectively. The maximum recorded peak temperature was about 150 °C, for AA2024Al/polycarbonate joints welded using tool T1 and, tool rotational speed and dwell time of 1000 rpm and 6 sec., respectively. While the maximum peak temperature was 136 °C for joints welded using T2 tool and rotational speed and dwell time of 1500 rpm and 6 sec., respectively.

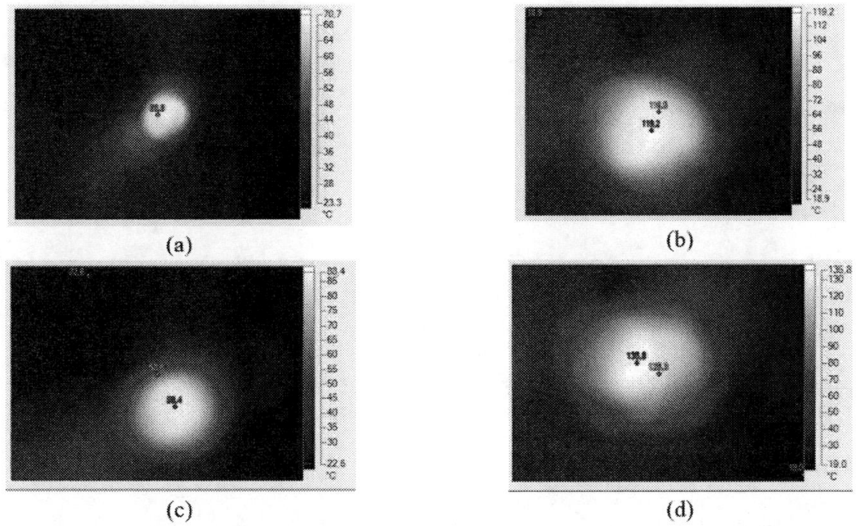

Figure 8. Thermographs of the temperature distributions on the top surfaces of AA2024Al / polycarbonate joints. The joints were welded using dwell time of 6 sec and rotational speed and tools of (a) 500 rpm and T1, (b) 1500 rpm and T1, (c) 500 rpm and T2, (d) 1500 rpm and T2.

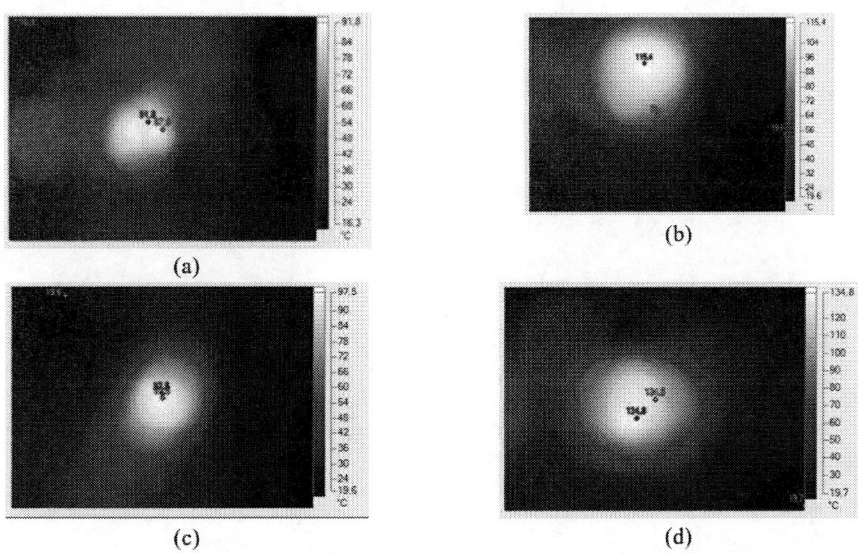

Figure 9. Thermographs of the temperature distributions on the top surfaces of AA2024Al / polypropylene joints. The joints were welded using dwell time of 6 sec and rotational speed and tools of (a) 500 rpm and T1, (b) 1500 rpm and T1, (c) 500 rpm and T2, (d) 1500 rpm and T2.

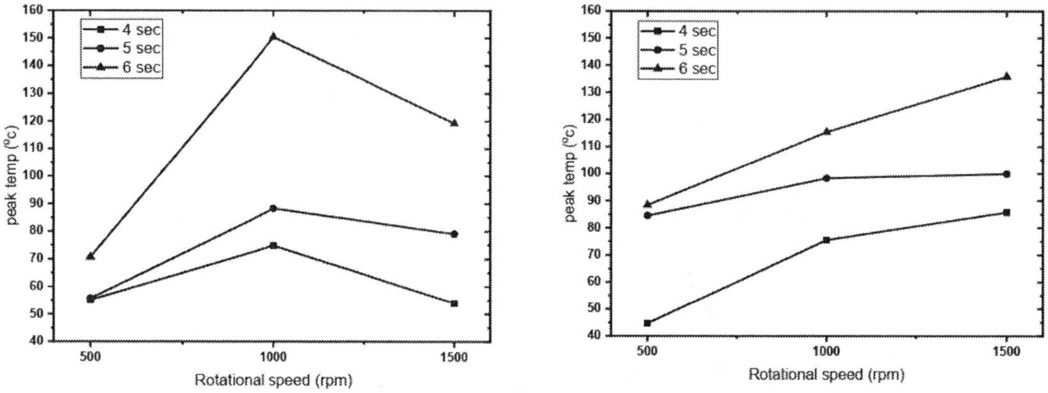

(a) (b)

Figure 10. The variation of the peak temperature with the tool rotational speed at several dwell times during FSSW of AA2024Al/polycarbonaste joints for (a) tool (T1) and (b) tool (T2).

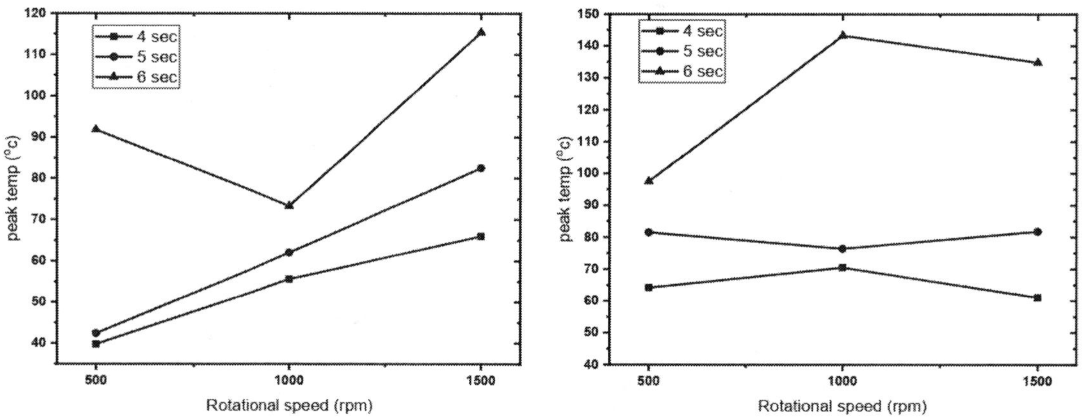

(a) (b)

Figure 11. The variation of the peak temperature with the tool rotational speed at several dwell times during FSSW of AA2024Al/polypropylene joints for (a) tool (T1) and (b) tool (T2).

In contrast, for AA2024Al/polypropylene, the peak temperatures recorded for tool T2 are higher than those recorded using tool T1. The peak temperature ranges from 40 to 115 °C and from 61 to 143 °C, for joints welded using T1 and T2 tools, respectively. The maximum peak temperature for tool T1 was 115 °C and recorded for joints welded using rotational speed and dwell time of 1500 rpm and 6 sec., respectively. While for tool T2, the maximum peak temperature was 143 °C and recorded for joints welded using rotational speed and dwell time of 1000 rpm and 6 sec., respectively.

Figures12 and 13 show the variation of the peak temperatures with the tool rotational speed and the dwell time(t) for AA2024Al/polycarbonate and AA2024Al/polypropylene joints welded using T1 and T2 tools, respectively. The red colored areas indicate the maximum peak temperature regions. While the blue colored areas represent the minimum peak temperature regions. The red areas are shifted toward the higher rotational speeds and dwell times.

There are three possible heat sources generated during FSSW process, i.e., friction work at the tool'sshoulder and top workpiece interface, friction work at the interface of top and bottom workpiece, and plastic deformation of the workpiece material [11]. It is well known that the tool's shoulder has greater influence on the heat generation process than the influence of the tool's pin [12]. So, the tool's shoulder and top workpiece interface contributes the most heat to the FSSW process. In the present work, only the temperatures distribution at the interface between the tool's shoulder and top workpiece are recorded. During FSSW process, the heat transfers from the top surface to the bottom of the joint i.e. to the interface between the metal/polymer interface as well as to the bottom polymer sheet. The heat transfer to the polymer/metal interface depends on the heat transfer coefficient of the AA2024 alloy (1000 W/m^2.k) [13].

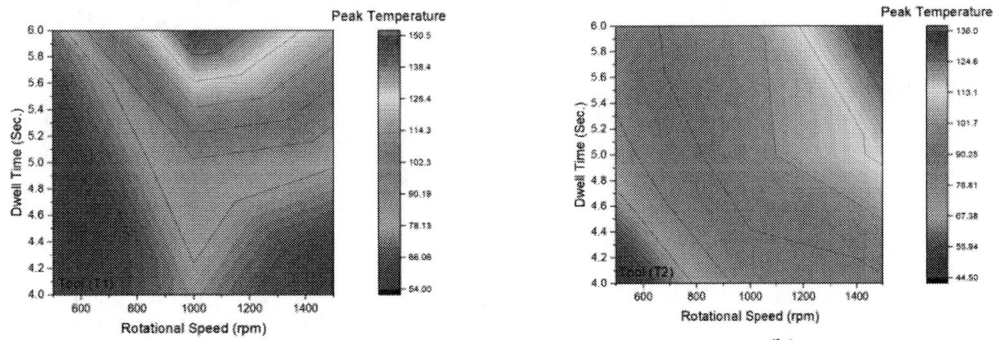

(a) (b)

Figure 12. Contour plots showing the variation of the peak temperatures with the tool rotational speed and dwell times during FSSW of AA2024Al/polycarbonate joints for (a) tool (T1) and (b) tool (T2).

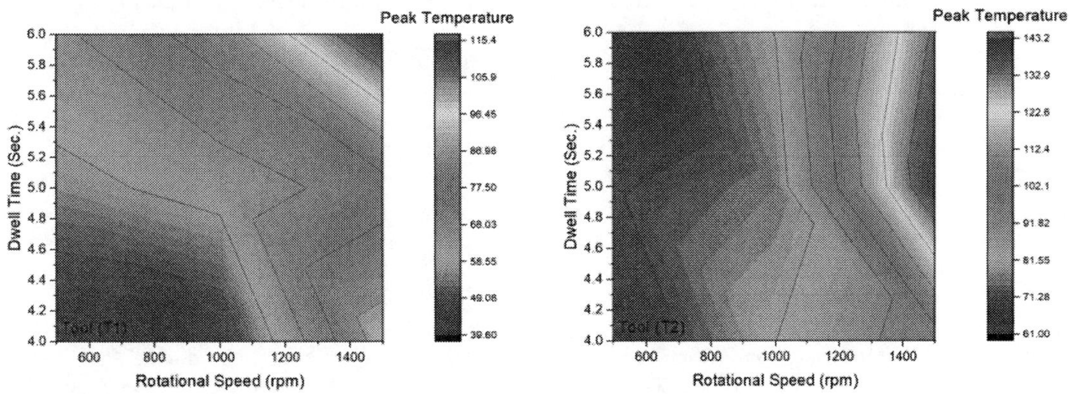

(a) (b)

Figure 13. Contour plots showing the variation of the peak temperatures with the tool rotational speed and dwell times during FSSW of AA2024Al/polypropylene joints for (a) tool (T1) and (b) tool (T2).

The heat input during FSSW can be calculated using the following equation [14]:-

$$Q = \frac{13}{12}\mu\frac{N}{K_A}\omega R \qquad \ldots(1)$$

Where: Q is the heat input in Watt, R is thestir tip radius in mm, N is the vertical compressive force in Newton, ω is the rotational speedin rad/sec and it is equal to 2πN, N is number of revolution per minute (rpm), μ is the coefficient of friction between the workpiece and tool's shoulder, $K_A$is the tool shoulder profile surface area ratio expresses by the ratio of shoulder profile contact area (A_c) to tool circular cross sectional area, (i.e. $K_A = A_c/A$). According to this equation, and as noticed in the present investigation, increasing the tool rotational speed increases the heat input. This is also reported by many investigators [10-14].

4. Conclusions

The main conclusions are:

- Increasing the dwell time and/or the tool rotational speed increase(s) the peak temperature during FSSW of both AA2024Al/polycarbonate and AA2024Al/polypropylene joints.
- For AA2024Al/polycarbonate joints, the peak temperatures recorded for tool T1 are higher than those recorded using tool T2. The peak temperature ranges vary from 54 to 150 $^{\circ}$C and from 45 to 136 $^{\circ}$C, for joints welded using T1 and T2 tools, respectively.
- For AA2024Al/polypropylene joints, the peak temperatures recorded for tool T2 are higher than those recorded using tool T1. The peak temperature ranges were found to vary from 40 to 115 $^{\circ}$C and from 61 to 143 $^{\circ}$C, for joints welded using T1 and T2 tools, respectively.

5. References

[1] Raza Moshwan, RahmatFirman, Farazila Yusof, Mohsen A Hassan and Mohd Hamdi, MohdFadzil 2015 Dissimilar friction stir welding between polycarbonate and AA 7075 aluminum alloy,*Int. J. of Materials Research*, **106(3)**, pp 258-266.

[2] Mubiayi M P and Akinlabi E T 2015 An overview on friction stir spot welding of dissimilar materials, In: Kim H., Amouzegar M., Ao S. (eds) *Transactions on Engineering Technologies. Springer*, (Dordrecht), pp 537–549.

[3] Silva A C F, Backer J De and Bolmsjö G 2017 Temperature measurements during friction stir welding, *Int. J. Adv. Manuf. Technol.*, **88**, pp 2899–2908.

[4] Zaidan A 2011 Theoretical analysis of temperature distribution in friction stir welding, **vol. 17**, no. 3.

[5] Hassan R Hassan, Ahmad Zaidan and Qasim Doos M 2011 Theoretical analysis of temperature distribution in friction stir welding, *J. of Engineering* , (Baghdad University), **17(3)**, pp 517-533.

[6] Anjal R Patel, Chirag G Dalwadi and Harikrishna G Rana 2016 A review: Dissimilar material joining of metalto polymer using friction stir welding (FSW), *Int. J. of Sci. Tech. & Engineering*, **2(10)**, pp 702-706.

[7] Bilici M K and Yukler A I 2012 Effects of welding parameters on friction stir spot welding of high-density polyethylene sheets, *Mater. Des.*, **33**, pp 545–550.

[8] Yang X W, Fu T and Li W Y 2014 Friction stir spot welding : a review on joint macro- and microstructure, property and process modelling, *Advances in Materials Science and Engineering*, http://dx.doi.org/10.1155/2014/697170, pp 1-11.

[9] http://asm.matweb.com/search/SpecificMaterial.asp?bassnum=ma2024t6.

[10] https://www.curbellplastics.com/Research-Solutions/Plastic-Properties.

[11] Mokhtar Awang and Victor H Mucino 2010 Energy generation during friction stir spotwelding (FSSW) of Al 6061-T6 plates, *Materials and Manufacturing Processes*, **25**, pp 167–174.

[12] Darko M Veljić, Marko P Rakin, Milenko M Perović, Bojan I Međo, Zoran JRadaković, Petar M. Todorović and Mirko N. Pavišić 2013 Heat generation during plunge stage in friction stir welding, *Thermal Science*, **17**, pp 489-496.

[13] Nilesh Kumar, ParthaSarathi De and Rajiv Sharan Mishra 2014 Friction stir welding and processing: *science and engineering*, Springer,.

[14] Ahmet Atak, AydınŞık and VeyselÖzdemir 2018 Thermo-mechanical modeling of friction stir spot welding and numerical solution with the finite element method, *Int. J. of Eng. and Appl. Sci. (IJEAS)*, **5(2)**, pp 70-75.

The effect of tire rubber particles on the mechanical and physical properties of polyester

K F Abo Elenien[1], N A Azab[2], G Bassioni[3] and M H Abdellatif[4]
[1] PhD student, Faculty of Engineering, Ain Shams University
[2] Associate Professor, Faculty of Engineering, Ain Shams University
[3] Professor, Chemistry Departments, Faculty of Engineering, Ain Shams University[3]
[4] Professor, Faculty of Engineering, Ain Shams University
Eng.karimfathy@gmail.com

Abstract. In this work, a novel composite of polyester with rubber particles -obtained from recycled wasted car tires -was developed. Worn tires generate a large amount of wastes that do not derogate resulting in environmental problems. This has the advantage of producing useful product that is economical and eco-friendly. The effect of the volume fraction of the rubber particles in the composites on the physical and mechanical properties of the composites was studied. Varying the volume fraction from 0-60% resulted in an increase in the impact properties. The addition of rubber particles had also the effect of reducing the density of the composites which is a desirable property

1. Introduction

1.1. Solid waste management

Solid waste management is one of the most significant issues now a days. Waste tires are one of the top issues as they do not derogate resulting in environmental problems. Consumption of rubber was 26.166 ton in 2012[1]. This material is used to produce many products mainly tires [2]. In USA 300,000,000 tires are turned to scrap tires annually, some are used as a fuel in ovens for about 40 % of the scrap tires, some are shredded and ground into particles which is about 26 % of the scrap tires, the rest are in landfills or used in some civil applications [3]. Waste tire rubber is representing a big part in waste disposal. This is considered as a huge problem in all over the world. In USA and Japan, they have a high environmental standard that lead to the need to dispose the waste tire rubber in an efficient way. Many countries now are developing and putting law to face the environmental hazards resulting from wastes and on the top of that is the waste tire rubber. USA and Japan were the leaders to put the laws in place to face these issues. That makes them now the leaders in waste tires recycling [4]. As for now the way to dispose the waste tire rubber in landfills is not accepted anymore because the landfills sites are now fully occupied [5, 6]. European union direction now is to implement a time schedule to solve this problem, so new disposal ways for waste tire rubber is an urgent need to face this huge waste problems [7].Also, the waste tire rubber is used as a fuel in cement factories, as a source of carbon black, and a source to create reefs in marine[9, 10]. In order to use it as a fuel it is technically applicable but economically expensive, as to get the carbon black from the tires, they need shredding and grinding process which is costly, even to get it from pyrolysis is more costly and produces lower quality compared to normal producing from petroleum [11, 12]. For civil engineering applications the waste tires are cut and shredded to be used in leachate landfills in dams. Also, it is used as a filler and modifier in asphalt mixture and as an additive in cement [13, 14]. Several research efforts were directed during the last decades to develop composite materials. While some of these composites had metal matrices [15,16], some others had ceramic or polymeric matrices [16-20]. The reinforcing phase varied from synthetic to natural fibers or particles. In this work, waste rubber tire particles are inserted into polymeric matrix to develop a novel composite material that can have useful engineering applications [21] Waste tire rubber

Tires are fundamental things in vehicles that forms the point of contact between wheel and the road. They provide the friction between wheel and road so that they can accelerate and brake and provide proper handling and in order to do this tire had to with stand wear, tear and impact [22]. Tires are manufactured from vulcanized rubber, natural rubber, synthetic rubber, fiber and steel wires. The two major components used in the manufacturing of tires are natural rubber and synthetic rubber [23]. By analyzing rubber, it is a high molecular weight, amorphous polymeric material. As per application the manufacturing of rubber differs according to the properties needed. Car tire is consisting of 58.6 wt.% rubber, 29.2 wt.% carbon black, 2.9 wt.% zinc, 1.8 wt.% stearic acid, 1.2 % sulfur, 5.85 % extending oil, and 0.4 wt.% accelerators [18]. Rubber can be oxidized in the amorphous regions. Attacking the tire by chemicals and microbes can results in decreasing in the mechanical properties and decreasing the working life of tire [24]. Natural rubber is sticky in nature and can be deformed under heat and can be turned to brittle when cooled under glass transition temperature. As a raw material it cannot be used to have products with needed elasticity, and this is because un-vulcanized rubber is consisting of long chains that can move independently. The vulcanization process is done by adding Sulphur, cross linking is formed and that results in the elasticity which means that by applying stress, deformation happens and when remove stress, the rubber goes back to its initial [25].

Many ways are now used for reuse, recycle and reclaim of waste tires such as retreading and can also be used as floaters [2]. In engineering applications, the ground tire rubber is used in civil engineering, asphalt, cement and composites. The waste tire rubber particle is used in composite mixtures.

1.2. Waste tire rubber in polymer blends

Waste tire rubber can be recycled by using them in polymeric composites. Polymeric composites can be divided into two types; thermoset and thermoplastic rubber composites. Waste tire rubber could be blended with the two types to obtain blends. The market use of polymers is huge, only 10 wt.% of waste tire rubber is blended with polymers, particularly thermo plastic [26]. Thermoset and thermoplastic can use these waste tire rubber as a filler in form of particles to increase the toughness and to reduce the cost of final product.

Polyethylene, Polypropylene and polyvinyl chloride are the most used in composites materials due to their competitive cost and easy blending processing. These composites are done to have thermoplastic elastomer composite to be a solution to utilize the waste tire rubber in thermoplastics. There are many factors that affect the mechanical properties as the influence of particle size of waste tire rubber. Particle size of waste tire rubber has a noticeable effect on the mechanical properties, also the properties may vary according to the processing as injection molding or pressurized molding. Ismail et al. [27] reported that particle size <500 μm has better mechanical properties compared to bigger size. Colom et al. [28] reported a decrease in tensile strength at 20 wt.% Waste tires rubber containing HDPE blend with <200 μm particle size whereas when bigger particle size (>500 μm) was used the decrease was 51%. Same results were noted by Mujal-Rosas et al. [29] in their study, they used three different particle sizes (<200, 200–500, >500 μm). Sonnier et al. [30] also used three different waste tire rubber particle sizes which were all >500 μm and did not results in significant effect of particle size on the mechanical properties. Tantayanon and Juikham [31] reported the impact strength of PP/Waste tire rubber blend comprising 420 μm, 1.2 mm and 2.4 mm sized particles and found only the smallest particle size results in an tangible improvement in impact strength (20%) whereas the other two bigger particle size composites results in only slight improvement.

Tensile properties generally decrease with addition of waste tire rubber. This was accompanied with poor adhesion between waste tire rubber and polymer matrix at interface. Poor interface results in high interfacial tension, making the waste tire rubber particles agglomerate and creating voids around waste tire rubber. Clear indication of lack of interfacial adhesion can be concluded from the easy removal of waste tire rubber particle from the polymer. Colom et al. [28] observe a decrease in tensile modulus of approximate 25%when only 20 wt.% Waste tire rubber was added to HDPE. Recycled HDPE/ Waste tire rubber blends properties were studied by Sonnier et al. [32] and they observe all tensile properties decrease considerably with increasing Waste tire rubber content. At 50 wt.% GTR content, tensile properties showed decrease of 40–80%. Study using PP/ Waste tire rubber also showed similar pattern where the tensile modulus decreased by 20%at 30 wt.% Waste tire rubber loading [27].

2. Methodology

2.1. Materials
Polyester were supplied by SUNPOL with density 1.2 gm/cm3, Recycled rubber particles were supplied by HOPPEC Company, the used average rubber particle size is2.5 mm of density 0.4 gm/cm^3.

2.2. Composite preparation
Polyester/ rubber particles composite is prepared as follows:

1. Calculate volume required for sample size
2. Divide the volume calculated according to the required volume fraction for rubber and polyester.
3. Polyester is mixed with waste tire rubber particles and stirred till homogeneity at Room Temperature.
4. Hardener is added at 2.5 wt.% to the mixture and stirred for 2 min
5. The mixture is placed in a vacuum chamber for 5 min at -0.8 bar.
6. The mixture is poured in a silicon rubber mold with required sample size

2.3. Physical testing.
Density tests were conducted on Aczet CY (0.1 mg & 0.001ct) as shown in Figure 1 according to standard ASTM D792-00 to measure the density of the waste tire rubber particles, polyester and the composite. Samples of density were prepared with the dimensions as shown in Figure 2

Figure 1 Aczet CY (0.1 mg & 0.001ct)

IOP Conf. Series: Materials Science and Engineering **973** (2020) 012019 doi:10.1088/1757-899X/973/1/012019

Figure 2 Density sample size

2.4. Mechanical testing

Tensile tests were conducted on LLOYD universal testing machine-30 ton according to ASTM D 3039/D 3039M – 00 as shown in Figure 3, with crosshead speed 5 mm/min in order to determine the ultimate and yield tensile strength. Impact tests are carried according to ASTM D 6110-04 using XJJU-5.5/50J Izod & Charpy Impact Tester as shown in Figure 4

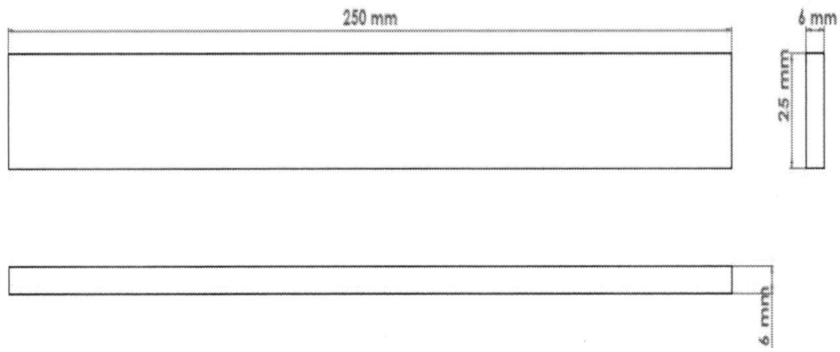

Figure 3. Tension Test sample size

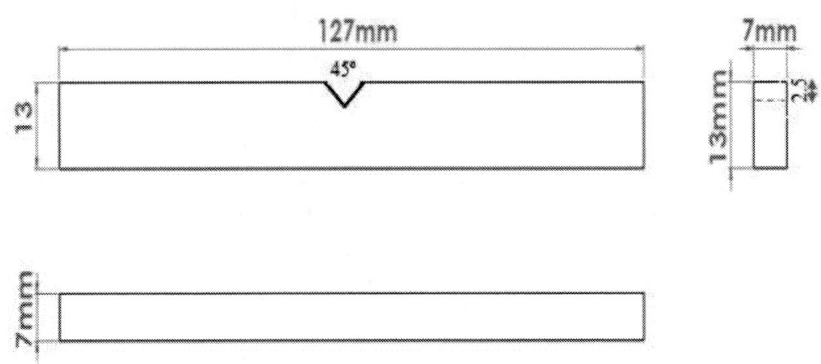

Figure 4 Impact sample size

2.5. Scanning electron microscopy

The fracture surface of the specimens that underwent impact test was studied to assess the mechanical adhesion and the interface of rubber particles with the polymer in the composite. Field emission scanning electron microscope Leo Supra 55was used.

3. Results and discussion

3.1. Physical properties

The effect of the waste tire rubber particles size on the density of the Polyester composites is shown in Figure 5. The density decreases as the vol% of waste tire rubber particles increases because of the lower density of the waste tire rubber particles than that of polyester, and this agrees with what was mentioned in Scrap Tires - Material Description [33]. The theoretical density was calculated using simple mixture rule and compared to the practical measured density. It was observed that there is an error gap between theoretical density and practical measured one, this error is increasing with the increase of waste tire rubber vol.%. The error is stabilized from and after 50 vol.% as shown in Figure 5.

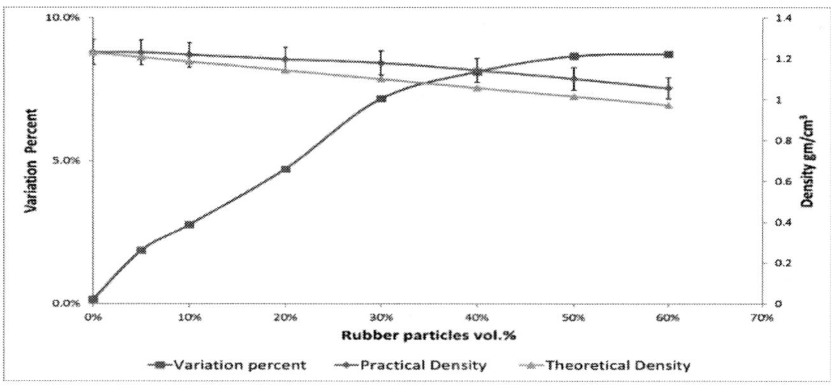

Figure 5. Theoretical density vs Practical density ofwaste tire rubber particles polyester composite.

3.2. Mechanical properties

3-2-1 Tensile Test

Some mechanical properties of composite material composite are obtained as shown in Figure 6 to Figure 11typical stress strain curve shows the decrease of strain rate up to failure with the increase of waste tire rubber vol% as shown in Figure 6.

The stress strain curve shows that the ultimate tensile strength decreases with the increase of waste tire rubber vol.%. and this is expected to be due to the lower strength of rubber particles compared to polyester.

Area under stress-strain curve which is known as the toughness was calculated. The toughness of the composite decreases with the increase of waste tire rubber vol.%. yield tensile strength was also obtained from tension test as shown in Figure 7.Yield strength decreases up to 5 vol.% of waste tire rubber particles. However, it shows an increase compared to the value of yield strength for 0vol.%at 10 vol% of waste tire rubber particles. This can be explained as up to 10 vol.%, the rubber particles act as a barrier that block the crack propagation and act as a reinforcement particle inside the matrix. Yield strength decreases with the increase of waste tire rubber particles after the 10 vol.% and up to 60 vol.% and this can be explained as the vol.% of particles increases the agglomeration of particles-so, particle to particle contact effect, is more than the effect of stopping the crack propagation that leads to decrease in yield strength and this is can be verified with SEM as shown in Figure 12 and Figure 14.

IOP Conf. Series: Materials Science and Engineering **973** (2020) 012019 doi:10.1088/1757-899X/973/1/012019

The ultimate tensile strength decreases with the increase of waste tire rubber particles vol.% as shown in Figure 8. and this can be due to the form of the rubber as ductile particles can't stand tensile stress unlike fibers.

The toughness was calculated form the area under curve of stress-strain. The toughness decreases with the increase of waste tire rubber particles vol % as shown in Figure *9* with agreement with previous researches [34]. This is due to the decrease both of ultimate tensile strength and strain to fracture.

The strain up to yield was measured and calculated form from the stress-strain curve. The strain up to yield increase with the increase of waste tire rubber particles vol.% up to 10 vol %. After 10 vol % the strain up to yield decrease with the increase of waste tire rubber particles vol.% as shown in Figure *10*. and this can be explained as in the phase of elastic deformation the matrix transfer the load to the rubber particle in which the particle has already high stain property but when the load goes beyond to the elastic phase the load transfer and cracks increase by amount more than that the rubber can absorb inside the matrix and crack propagation cannot be any more stopped by rubber particles.

3-2-2 Impact test

Impact strength was obtained from Charpy impact test and the results are shown in Figure 11. The specific impact strength is calculated by dividing impact value by density value. The impact strength decreases with the increase of waste tire rubber particles up to 10 vol %. with agreement with previous research [17],[344],[35]. then the impact strength starts to increase with the increase of waste tire rubber vol.% till reach the highest at 50 vol.%. and this may be explained by the high impact properties of waste tire rubber as this is the main property that tire rubber was manufactured for. The impact strength after 50 vol.% decreases with the increase of waste tire rubber particles vol.% as shown in Figure 11. and this can be explained from the agglomeration and particle to particle contact [36].

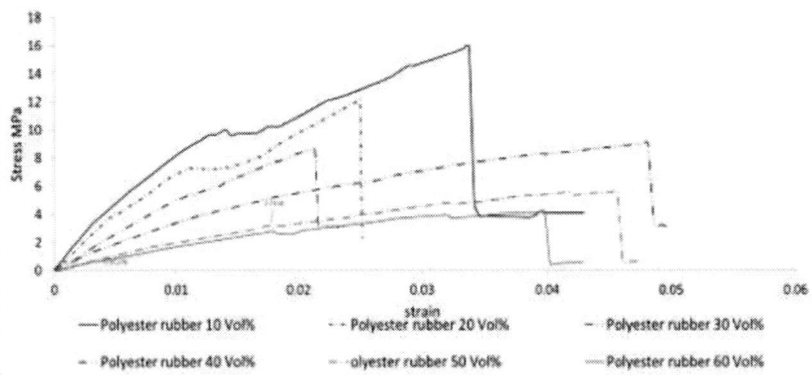

Figure 6. Typical Stress-Strain of waste tire rubber particles polyester composite.

IOP Conf. Series: Materials Science and Engineering **973** (2020) 012019 doi:10.1088/1757-899X/973/1/012019

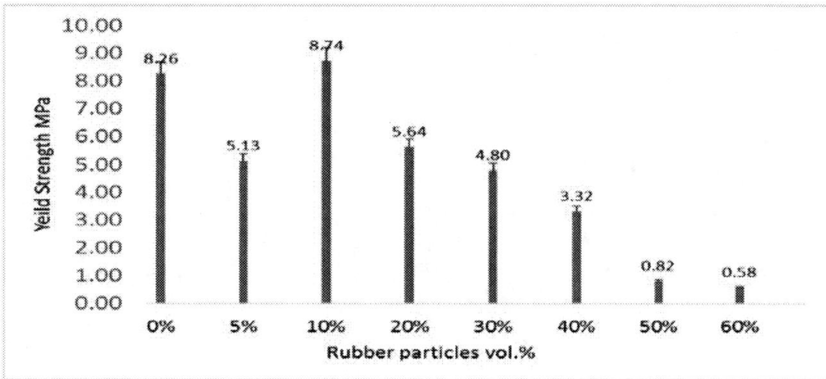

Figure 7. Yield strength vs waste tire rubber particles polyester composite vol.%.

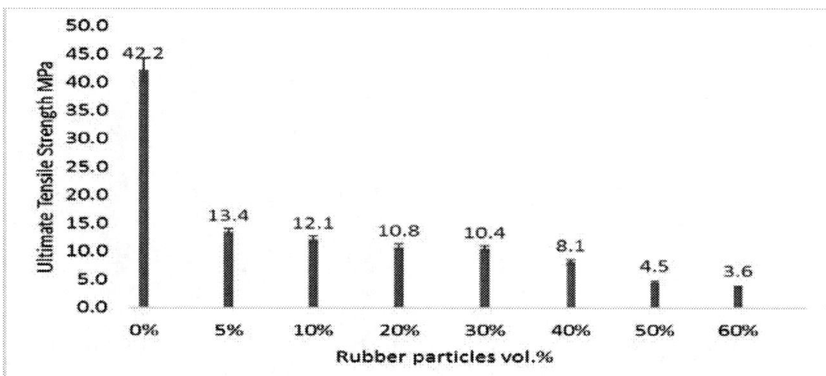

Figure 8. Ultimate tensile strength vs waste tire rubber particles polyester composite vol.%.

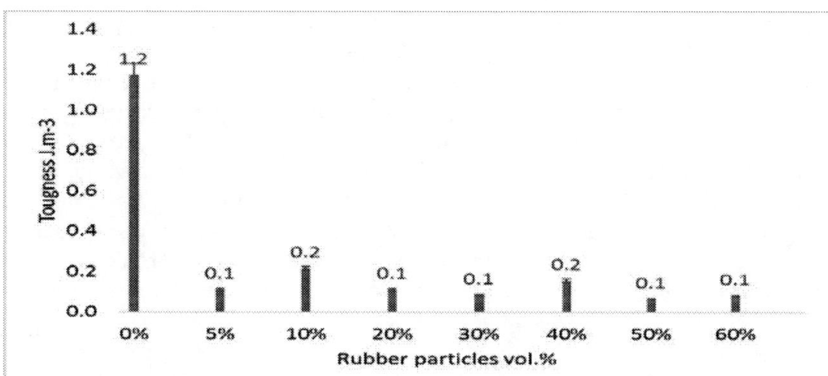

Figure 9. Toughness vs waste tire rubber particles polyester composite vol.%.

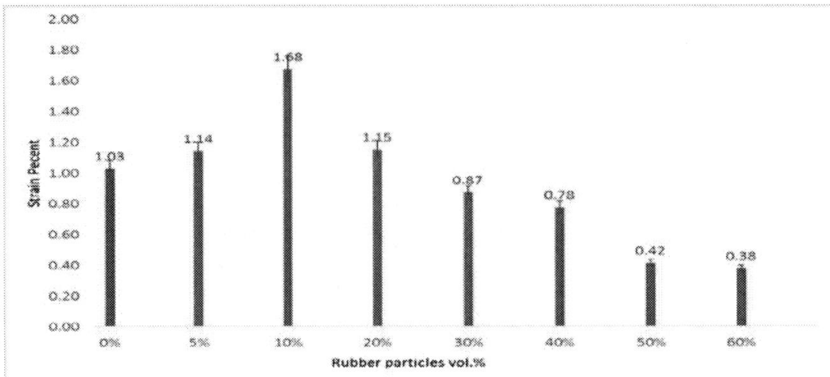

Figure 10. Strain up to Yield vs waste tire rubber particles polyester composite vol.%.

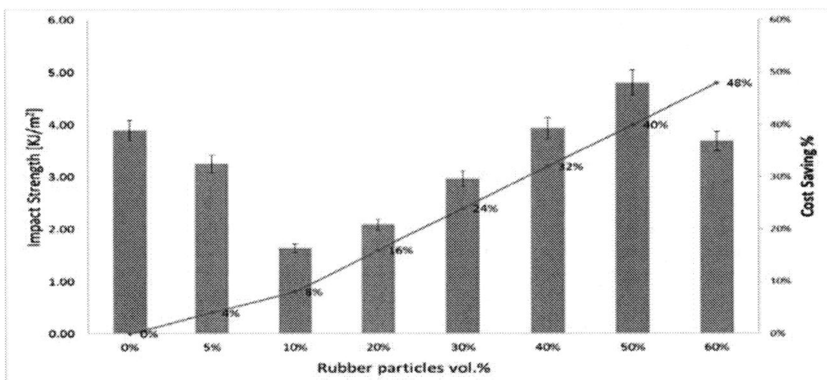

Figure 11. Specific impact strength vs waste tire rubber particles polyester composite vol.%. -cost saving % vs waste tire rubber particles polyester composite vol.%.

3.3. Morphology

SEM micrographs of impact fractured surface of polyester waste tire rubber particles composites are shown in Figure 12 to Figure 14. The surfaces of samples were gold sputtered and then examined under the SEM. The surface of waste tire rubber particles is fractured which confirms the contribution of particle in the impact fracture test as shown in Figure 12. There were cracks in matrix, debonding between matrix and particle. On the surface there are tearing in rubber particle. This morphology shows the weak adhesion between the waste tire rubber particle and the polyester resin as shown in Figure 12 with magnification 30X. and agrees with previous researcher [17].the particle surface was examined by SEM Figure 13 with magnification 422X and this shows that the particle is smooth and this may be the reason of the debonding and to increase the bonding surface of rubber need to be treated so the surface can have more roughness to bond and increase the interfacial forces between rubber and matrix. For 60 vol.% of waste tire rubber particles, a particle to particle contact is observed as shown in Figure 14.

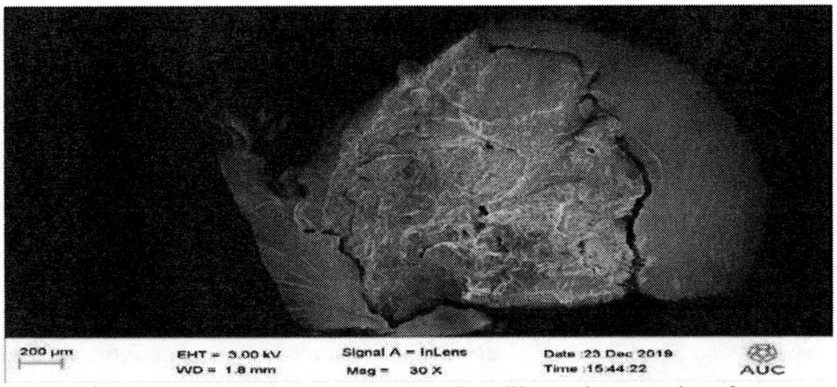

Figure 12.SEM of the fractured surface illustrating poor interface

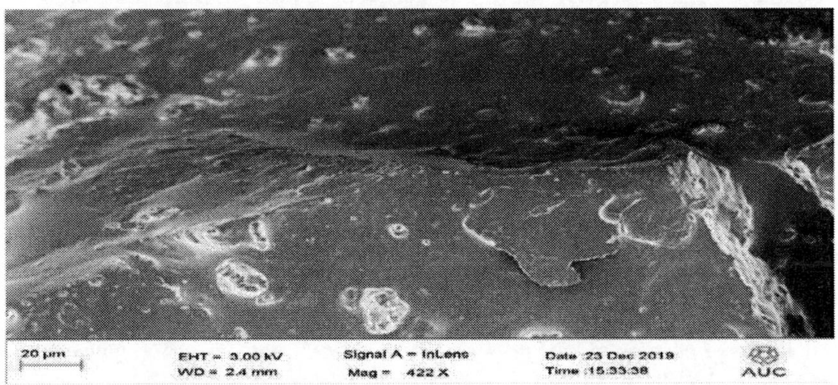

Figure 13 SEM rubber particle with magnification 422X

Figure 14 SEM particle to particle contact

4. Conclusions

In this study, the following can be concluded;

The developed rubber composite present a considerable improvement in its Mechanical Properties. The Impact strength increased by 23% at 50 vol.% of waste tire rubber particles while the yield strength and strain up to yield increased by 10% and 63 % at 10 vol.% of waste tire rubber particles respectively. Moreover, the developed composites became lighter with a density decrease with the increase of waste tire rubber particles vol.% which is a desirable property. On the other hand, the ultimate tensile strength decreased with the increase of waste tire rubber particles vol%. These results were also verified using field emission scanning electron microscope.

References

[1] International Rubber Study Group 2011. http://www.rubberstudy.com/welcome

[2] Karger-Kocsis J, Mészáros L and Bárány T 2013 Ground tyre rubber (GTR) in thermoplastics, thermosets, and rubbers *Journal of Materials Science* **48**(1) pp 1-38.

[3] *US scrap tire* management summary 2005–2009 2013 (Washington: Rubber Manufacturers Association)

[4] The recycling industry: a global view 2003. http://www.irevna.com/pdf/Industry%20report.pdf. Accessed 15 February 2014

[5] Nehdi M and Khan A 2001 Cementitious composites containing recycled tire rubber: an overview of engineering properties and potential applications. *Cement, concrete and aggregates* **23**(1) pp.3-10.

[6] Meyer, C 2009 The greening of the concrete industry *Cement and concrete composites* **31**(8), pp.601-605.

[7] Reschner K 2008 *Scrap tire recycling. A summary of prevalent disposal and recycling methods* (Berlin :Entire-Engineering).

[8] Tantala MW, Lepore JA and Zandi I 1996 Quasi-elastic behavior of rubber included concrete *12th international conference on solid waste technology and management, Philadelphia, PA*

[9] Siddique R and Naik TR 2004 Properties of concrete containing scrap-tire rubber—an overview *Waste Manag.* **24**:563–569. doi:10.1016/j.wasman.2004.01.006

[10] Ye, Z G , Kong X M , Yu J Y and Wei L Q 2003 Microstructure and properties of desulfurized crumb rubber modified bitumen *Journal of Wuhan University of Tehnology-Materials Science Ed.* **18**(1), pp.83-85.

[11] Paul J 1985 *Encyclopedia of polymer science and engineering* (Newyork: John Wiley & Sons, Inc.) 14:787–802

[12] Siddique R 2008 *Waste materials and by-products in concrete* (Berlin: Springer)

[13] Sunthonpagasit N and Duffey MR 2004 Scrap tires to crumb rubber: feasibility analysis for processing facilities *Resour Conser Recycl* 40:281–299.doi:10.1016/S0921-3449(03)000739

[14] Shu X and Huang B 2014 Recycling of waste tire rubber in asphalt and portland cement concrete: An overview *Construction and Building Materials* 67, pp.217-224.

[15] Zhu K, and Schmauder S 2003 Prediction of the failure properties of short fiber reinforced composites with metal and polymer matrix *Computational Materials Science* **28**(3-4) pp 743-748.

[16] Serena PA et al. 2019 Manufacturing of a metal matrix composite coating on a polymer matrix composite through cold gas dynamic spray technique *Journal of Materials Engineering and Performance* **28**(6) pp 3211-3219.

[17] Elenien K F A , Abdel-Wahab A , ElGamsy R and Abdellati M H 2018 Assessment of the properties of PP composite with addition of recycled tire rubber *Ain Shams Engineering Journal* **9**(4) pp.3271-3276.

[18] Awad A H and Abdellatif M H 2019 Assessment of mechanical and physical properties of LDPE reinforced with marble dust *Composites Part B: Engineering* **173** p.106948.

[19] Yousri O M ,Abdellatif M H and Bassioni G 2018 Effect of Al_2O_3 Nanoparticles on the Mechanical and Physical Properties of Epoxy Composite *Arabian Journal for science and engineering* **43**(3),pp.1511-1517

[20] Awad A H , El-gamasy R , El-Wahab A A A and Abdellatif M H 2019 Mechanical behavior of PP reinforced with marble dust *Construction and Building Materials* **228**, p.116766.

[21] Diaconescu RM, Barbuta M, Harja M 2013 Prediction of properties of polymer concrete

composite with tire rubber using neural networks *Mater Sci Eng B* **178** pp 1259–1267. doi:10.1016/j. mseb.2013.01.014

[22] Suparat T 2013 *Waste tyre management in Thailand: a materialflow analysis approach* (Thailand: Asian Institute of Technology PH.D Dissertation)

[23] Nakao Y, Yamamoto K 2002 Waste tire and its collection systems *Nippon Steel technical report* no. 86

[24] Huynh H, Raghavan D 1997 Durability of simulated shredded rubber tire in highly alkaline environments *Adv Cem Bas Mater* **6** pp 138–143. doi:10.1016/S1065-7355(97)90021-3

[25] Lo Presti D 2013 Recycled tyre rubber modified bitumens for road asphalt mixtures: a literature review *Con Build Mat* **49** pp 863–881. doi:10.1016/j.conbuildmat.2013.09.007

[26] Jana G K and Das C K 2005 Devulcanization of natural rubber vulcanizates by mechanochemical process *Polym – Plast Technol Eng* **44** pp1399–412.

[27] Ismail H, Awang M and Hazizan M 2006 Effect of waste tire dust (WTD) size on the mechanical and morphological properties of polypropylene/waste tire dust (PP/WTD) blends *Polym – Plast. Techno.l Eng.* **45** pp 463–8.

[28] Colom X, Cañavate J, Carrillo F and Suñol JJ 2009 Effect of the particle size and acid pretreatments on compatibility and properties of recycled HDPE plastic bottles filled with ground tyre powder *J. Appl. Polym. Sci.* **112** pp 1882–90.

[29] Mujal-Rosas R, Orrit-Prat J, Ramis-Juan X, Marin-Genesca M and Rahhali A 2011 Study on dielectric, thermal, and mechanical properties of the ethylene vinyl acetate reinforced with ground tire rubber *J. Reinf. Plast. Compos.* **30** pp 581–92.

[30] Sonnier R, Leroy E, Clerc L, Bergeret A, Lopez-Cuesta J 2007 Polyethylene/ground tyre rubber blends: influence of particle morphology and oxidation on mechanical properties *Polym Testing* 26 pp 274–81.

[31] Tantayanon S and Juikham S 2004 Enhanced toughening of poly(propylene) with reclaimed-tire rubber *J. Appl. Polym. Sci.* 91 pp 510–5.

[32] Sonnier R, Leroy E, Clerc L, Bergeret A and Lopez-Cuesta J M 2006 Compatibilisation of polyethylene/ground tyre rubber blends by c irradiation. *Polym. Degrad. Stab.* **91** pp 2375–9.

[33] *Scrap Tires - Material Description - User Guidelines for Waste and Byproduct Materials in Pavement Construction* 2019 (US Federal Highway Administration Research and Technology FHWA-RD-97-148 27/7/2019)

[34] Abo-El-Enin K F, Abdelwahab A, Abdellatif M H and El-Gamasy R 2016 PP Reinforced With Recycled and Treated Rubber Particles *International Journal of Mechanical and Production Engineering (IJMPE)* **4** Issue-8 pp 84-88.

[35] Mousa B H , Abdel-Wahab A, El-Gamasy R and Abdel Latif M H 2016 Mechanical Behaviour of H_2So_4 Treated Tire Rubber- HDPE Composites *International Journal of Mechanical and Production Engineering (IJMPE)* **4** Issue-9 pp. 67-70.

[36] Mousa B H and Abdel-Latif M H 2019 Mechanical behaviour of rubber hybrid composites *IOP Conference Series: Materials Science and Engineering* **610** No. 1. IOP Publishing.

Acknowledgement
This paper and the research behind it would not have been possible without the exceptional support of my supervisor, Mohamed Hazem Abdellatif, Nahid Azab, Ghada Basinoi. I would express my sincere regards to my colleague Bassant Hany Mousa.

IOP Conf. Series: Materials Science and Engineering **973** (2020) 012020 doi:10.1088/1757-899X/973/1/012020

Effect of modified bouc-wen model parameters on the dynamic hysteresis of magnetorheological dampers

M Abdelhamed[1], W G Ata[2], A M Salem[3], A Z Khafagy[4]
[1]MScStudent, Tanks Dept., Military Technical College
[2]Lecturer, Tanks Dept., Military Technical College
[3] Associate Professor,Tanks Dept, Military Technical College
[4]Chief of Mechanical Engineering Branch, Military Technical College
wael.kotb@mtc.edu.eg

Abstract. Magnetorheological (MR) dampers are a semi active control device that becoming a promising actuators in vibration mitigation. The unique properties of MR fluids are used to control the damping force without consuming large power source. Despite MR dampers are fail-safe, efficient, and robustness devices, they poses significant nonlinear characteristics. A well-known Bouc-Wen models and their extension modified Bouc-Wen are the most accurate models that predict the hysteresis of the MR dampers. However, these models suffer from some complexity and limitations and they miss a good fitting to the measured data. In this paper, the existing modified Bouc-Wen (MBW) models are introduced and their dynamic characteristics are simulated using MATLAB Simulink environment. The robustness of the model is judged by a comparison between the experimental tests of the MR damper (on and off state) on the MTS damper test machine in MTC and the simulated characteristics. To better understand the damper nonlinearity, an investigation into the effect of the model parameters on the dynamic characteristics of the damper is studied. A parametric study is proposed to identify the optimal model parameters that best fits the experimental data. The response of the model with the obtained parameters is validated across the measured MR damper characteristics under different sinusoidal excitations and command current. The results show a good agreement between the simulated and measured responses of the MR damper.

1. Introduction

Recently, the magnetorheological (MR) damper has regarded as the most promising actuators in vibration control and structure isolations. Suspension system differs according to their construction to passive, active and semi active suspension. Passive suspension has linear and constant characteristics that limit the suspension performance. However, in the active suspension the system generates damping force based on the terrain excitations. This ability to change the damping force provides the opportunity to increase the rid comfort and vehicle handling at the expense of high power demand and cost [1] On the other hand, semi-active suspension overcomes the drawbacks of passive and active suspension systems.MR fluid consists of polarized particles affected by magnetic field in micron size suspended in a carrying medium like water or mineral oil. This type of fluid is not sensitive for impurities. Moreover, the yield strength of this fluid could be altered by changing the magnetic flux intensity in millisecond[2]. The magnitude of damping force changed with respect to electric power applied on damper terminals. When damper is not energized it acts as passive dampers (a fail-safe) actuators[3].

MR dampers possess high nonlinear characteristics which make the dynamic modelling of such dampers a challenge issue [4]. Despite the nonlinearity, MR fluid dampers have a lot of advantage as a mechanical simplicity, high dynamic range, low power requirement, large force capacity and robustness. Moreover, MR fluid could be changed from Newtonian fluid to semi solid(non-Newtonian) in millisecond[5].Several models such as Bingham, Bouc-wen and modified Bouc-Wen models have been developed to characterize the intrinsic dynamic of the MR damper[6]. These models classified as parametric models that have many parameters. These parameters are identified by comparing the simulated data with the experimental responses[7]. The difference between the parametric models and nonparametric models also introduced. The parametric models employ analytical expression based on testing data result and device principle for working[8].The nonparametric models can predict damper behaviour with no need to system physical characteristics[9]

In this article, a comparison between the experimental results obtained from testing of MR damper (RD-8040-1)of Lord Corporation and simulated results obtained from modified Bouc-Wen models. A parametric study held also to the modified Bouc-Wen model with 14 parameters to recognize the effect of each parameter on the damper hysteresis. A Simulink model constructed in Matlab/Simulink for the governing equations of the modified Bouc-Wen model. The model parameters varied to study their influence on the hysteresis performance. Finally, a parametric study proposed to identify the optimal parameters values that best fit the simulated data with the experimental responses.

2. Magnetorheological damper model

The MR damper is a smart device that has intrinsically nonlinear characteristics. These nonlinear dynamics make the modelling and design of control techniques of such device a challenging issue. To obtain the full advantages of the MR damper nonlinearity, an accurate and precise model has tobe identified. The ModifiedBouc-Wen model shown in Figure 1 is an extension of the Bouc-Wen model proposed by Spencer[8, 10]and it is considered as an accurate model that best tracks the MR damper dynamics. The damping force of the model expressed as follows:

$$F = c_1(\dot{y}) + k_1(x - x_0) \tag{1}$$

where the evolutionary variables given by:

$$\dot{z} = A(\dot{x} - \dot{y}) \; - \beta(\dot{x} - \dot{y}) \, |z|^n - \gamma z|\dot{x} - \dot{y}||z|^{n-1} \tag{2}$$

$$\dot{y} = (c_0 - c_1)^{-1}(c_0\dot{x} + k_0(x - y) + \alpha z) \tag{3}$$

The following equations introduced for the model tobe taken into account for the applied voltage and/ or current dependent parameters.

$$\alpha = \alpha(u) = \alpha_a + \alpha_b u \tag{4}$$

$$c_1 = c_1(u) = c_{1a} + c_{1b}u \tag{5}$$

$$c_0 = c_0(u) = c_{oa} + c_{ob}u \tag{6}$$

$$\dot{u} = -\eta(u - v) \tag{7}$$

In the proposed model the gas stiffness is represented by a spring with constant, k1, the damping at high velocities is denoted by, co. A dashpot of damping c1, is presented in the model to be taken into account for the hysteresis characteristics at low frequencies. However, a spring, with constant, ko is introduced into the model for controlling stiffness at large velocities. It is supposed that the initial displacement due to the accumulator stiffness is, xo. A Simulink model that represents the modified Bouc-Wen model equation developed in Matlab environment as illustrated in Figure 2.

In the simulation, the damping force of the MR damper considered as the output of the model while the displacement, velocity and command voltage are the excitation inputs.

IOP Conf. Series: Materials Science and Engineering **973** (2020) 012020 doi:10.1088/1757-899X/973/1/012020

In modified Bouc-Wen model, there are 14 parameters describing the MR damper hysteresis these parameters have been determined in [10, 12] and are shown in Table 1. These two references denoted as MBW1 and MBW2 respectively. In order to judge these published parameters and their degree of fit to the experimental data, various experiments have been conducted on the pre-denoted short stroke MR damper.

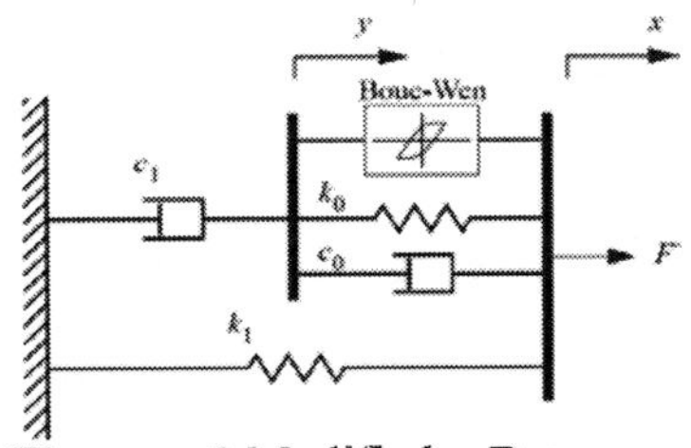

Figure 1. Modified Bouc-wen model [11].

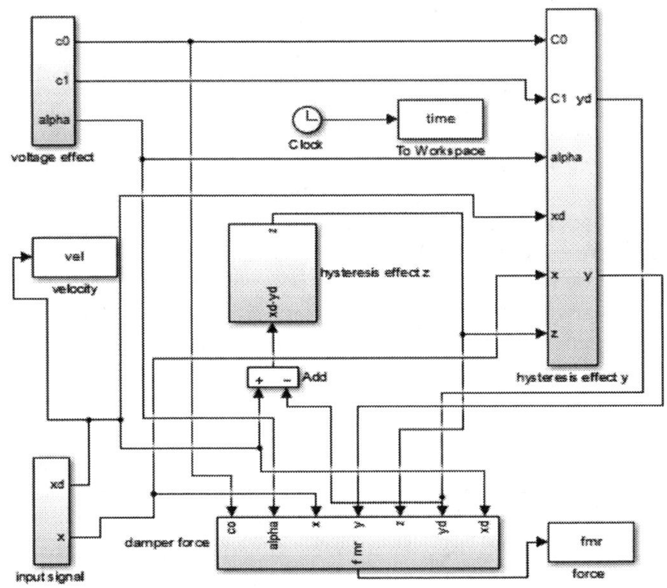

Figure 2. Simulink block diagram for modified Bouc-Wen model.

Determining the values of system parameter is necessary to minimize the deviation of the data obtained from the parametric model from the experimental results. A number of optimization methods used to select the 14- parameter values of modified Bouc-Wen model. [10, 12] have used two sets of parameter values that are supposed to represent the damper hysteresis accurately.

Table 1. Modified Bouc-Wen parameters

Parameters	MBW1, Ref [10]	MBW2,Ref [12]	Unit
A	301	58	-
β	3630000	2059020	m^{-2}
n	2	2	m^{-2}
η	190	196	-
γ	3630000	136320	s^{-1}
c_{oa}	2100	784	N.s/m
c_{ob}	350	1803	N.s/m.V
c_{1a}	28300	14649	N.s/m
c_{1b}	295	34622	N.s/m.V
α_a	14000	12441	N/m
α_b	69500	38430	N/m.V
k_o	4690	3610	N/m
k_1	500	840	N/m
x_o	0.134	0.0245	m

3. Experimental testing of MR damper

The magneto-rheological damper used for obtaining the experimental result is a fixed orifice damper filled with MR fluid developed by Lord Corporation. The damper installed on the MTS Damper Test Machine as shown in Figure 3. The MTS machine consists of an upper crosshead frame with (5)KN force transducer and the actuator with linear variable differential transducer (LVDT) for measuring displacement. A 12V DC power supply used to provide the necessary command voltage and current for MR damper. The MR damper is excited with frequencies from 1 to 2.5 Hz with increment of 0.5 Hz and a dynamic amplitude of 5 mm at 0 and 0.5 A. The measured force and velocity filtered and smoothed in Matlab environment to avoid unwanted frequencies. The force-displacement and force-velocity histories presented for the measured and simulated data for comparison.

Figure 4 and Figure 5 show a comparison between the data obtained from two models denoted as MBW1, MBW2 and experimental results obtained at different frequencies with amplitude of 5 mm and an input current of zero and 0.5 A, respectively.

Finally, the comparison carried out between the model results and experimental results prove that modified Bouc-Wen parameter not valid for all MR dampers. Consequently, a parametric study has to be done in order to investigate the effect of each parameter on the dynamic hysteresis of the damper.

4. Influence of model parameters on MR hysteresis

A comprehensive analysis of the effect of modified Bouc-Wen [10] parameters on the dynamic behavior of the damper is addressed. To better understand the influence of variation of the model parameters on the damper output and identification of the optimal parameters, the parametric study is introduced. The method based on changing one parameter out of 14 parameters in a certain range and keeping the other parameters unchanged. The parameters are changed from the reference value (MBW1[10]) to ± 50% with interval 10%. Then, the simulation of the MBW1 model executed in Matlab for all values of the first parameter and keeping the rest of parameters unchanged. After that the simulations is repeated for all values of the second one parameter and further till the last parameter. The root mean square (rms) of the output damping force calculated each time for each run as shown in Table 2. For the sake of brevity, the force-velocity and force-displacement graphs from simulations are presented for only the reference and extreme values (-50% and +50%).

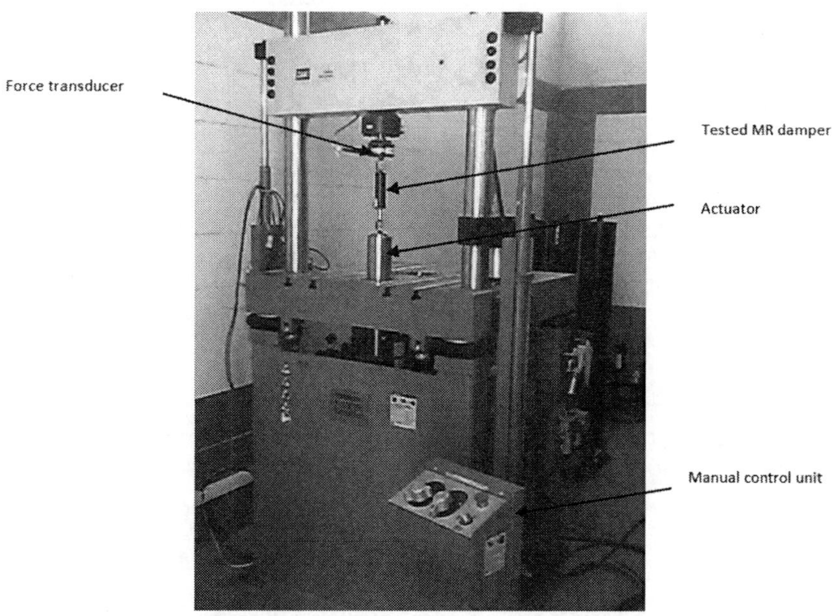

Figure 3. Damper test system used to identify MR damper characteristics.

IOP Conf. Series: Materials Science and Engineering **973** (2020) 012020 doi:10.1088/1757-899X/973/1/012020

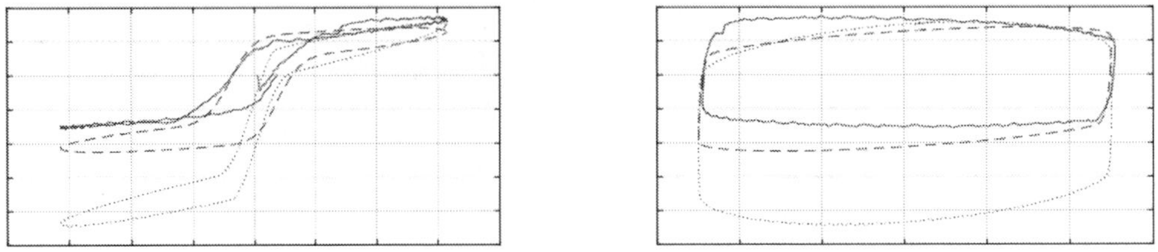

(a) At 1 Hz

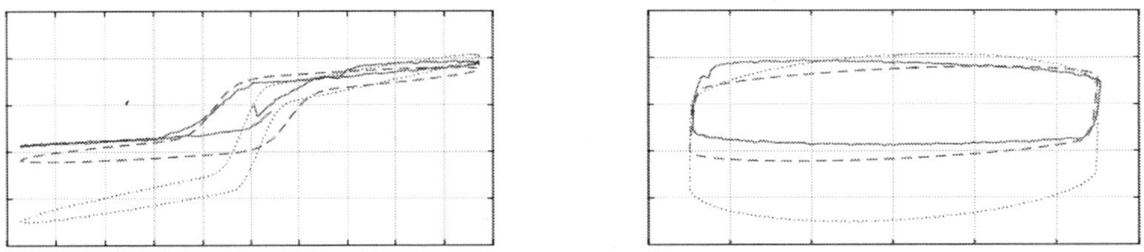

(b) At 1.5 Hz

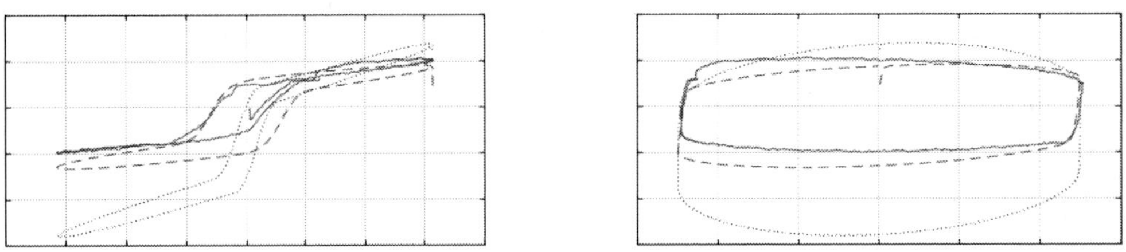

(c)At 2 Hz

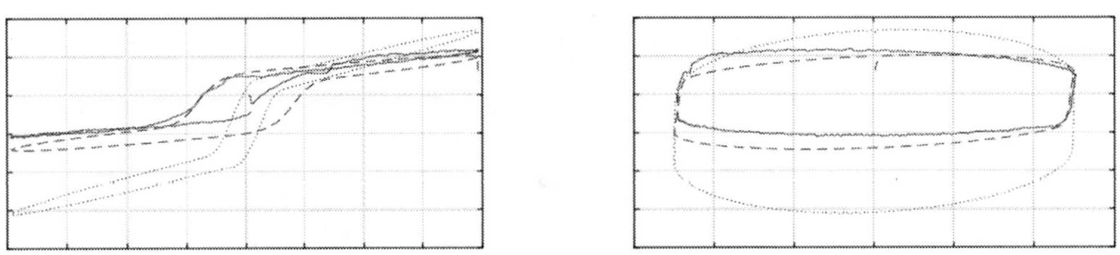

(d) At 2.5 Hz

Figure 4. Comparison between simulated and experimental data at zero current, amplitude of 5 mm and different frequencies; (solid, experimental; dot, MBW1; dash, MBW2)

IOP Conf. Series: Materials Science and Engineering **973** (2020) 012020 doi:10.1088/1757-899X/973/1/012020

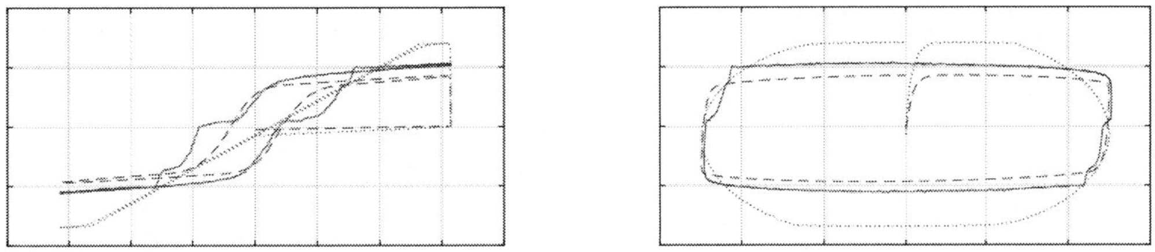

(a)At 1 Hz

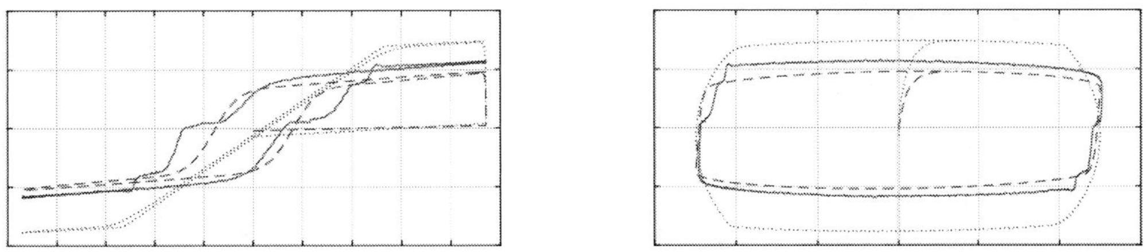

(b) At 1.5 Hz

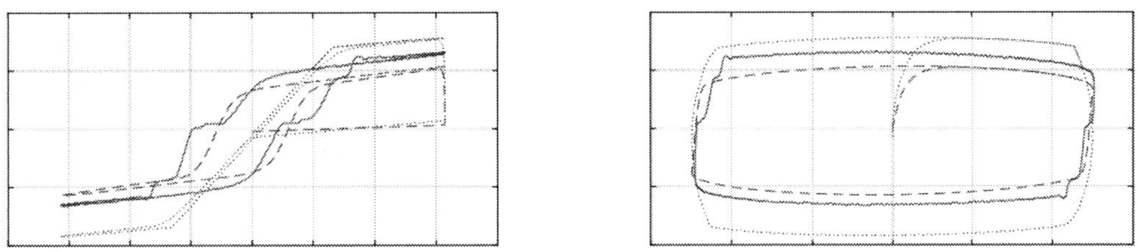

(c)At 2 Hz

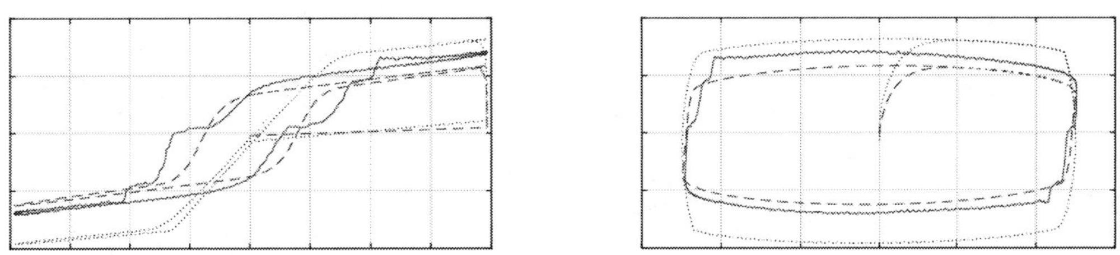

(d) At 2.5 Hz

Figure 5. Comparison between simulated and experimental data at current of 0.5 A, amplitude of 5 mm and different frequencies; (solid, experimental; dot, MBW1; dash, MBW2)

IOP Conf. Series: Materials Science and Engineering **973** (2020) 012020 doi:10.1088/1757-899X/973/1/012020

Table 2. Rms values of damping force for MBW1 parameters (15 mm, 2.5Hz,1.5 volt)

Parameters	-50%	-40%	-30%	-20%	-10%	Ref	10%	20%	30%	40%	50%
A	870	914	955	993	1029	1063	1095	1125	1155	1183	1210
α_a	1024	1032	1093	1047	1055	1063	1071	1078	1086	1094	1102
α_b	772	830	888	947	1005	1063	1121	1178	1236	1293	1350
β	1164	1140	1118	1098	1080	1063	1047	1032	1018	1006	993
c_{1a}	972	1000	1022	1039	1052	1063	1072	1080	1086	1092	1097
c_{1b}	1062	1062	1062	1062	1062	1062	1062	1063	1063	1063	1063
c_{oa}	940	965	990	1015	1939	1063	1087	1110	1133	1156	1179
c_{ob}	1033	1039	1045	1051	1057	1063	1069	1075	1081	1086	1092
γ	1164	1140	1118	1998	1080	1063	1047	1032	1018	1005	993
k_1	1061	1061	1061	1062	1062	1063	1063	1063	1064	1064	1064
k_o	1062	1062	1062	1063	1063	1063	1063	1063	1063	1064	1064
n	439	666	972	1332	1696	1063	1915	1915	1916	1916	1917
η	1062	1062	1063	1063	1064	1063	1064	1064	1064	1064	1064
x_o	1061	1061	1061	1062	1062	1063	1063	1063	1064	1065	1066

Figure 6 depicts the force-displacement and force-velocity responses with respect to three values; a maximum level (+50%), a minimum level (-50%) and a reference value for the parametersβ, A, αa and αb. It is clearly seen that the force range decreases by increasing the value of β. But when increasing the parameters A, αa andαb, the force is increased. In addition, the ascending of parameter,a contributes to the narrowing of the enclosed area of the force-velocity loop.

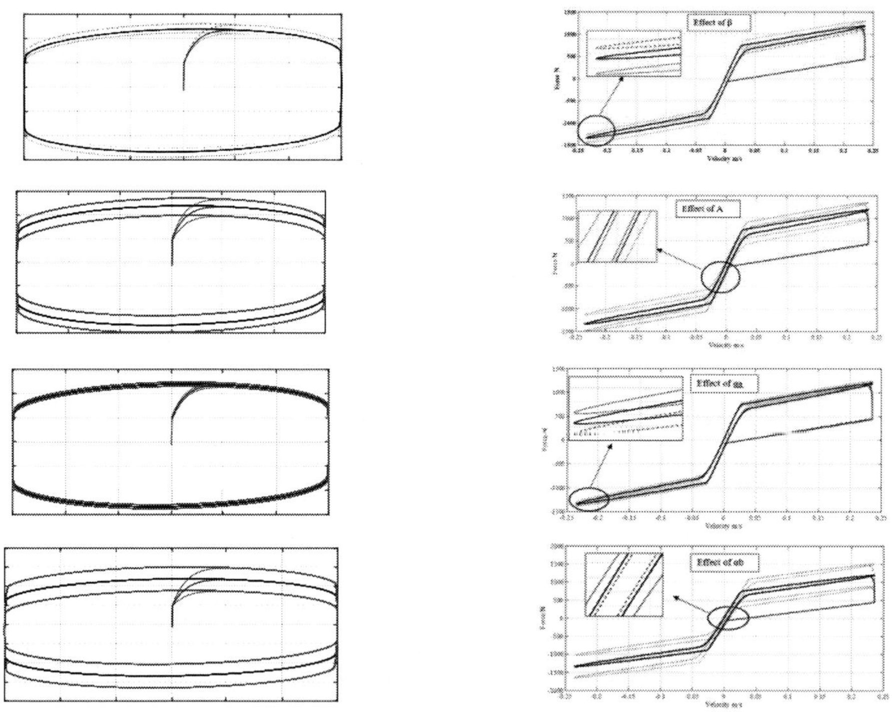

Figure 6. Force-displacement and force-velocity simulated responses with different values of parameters β, A, αa and αb

IOP Conf. Series: Materials Science and Engineering **973** (2020) 012020 doi:10.1088/1757-899X/973/1/012020

Figure 7 shows the consequent hysteretic responses of the model by varying parameters; C1a, C1b, Coa, Cob, andγ. It is obvious that parameter C1b, has very little influence on the slop of the loop (effective stiffness). However, the increase of the parameters C1a, Coa and Cob result in the increase of the dynamic force of the model. Moreover, the ascending of parameter, γ, results in descending of the effective stiffness.

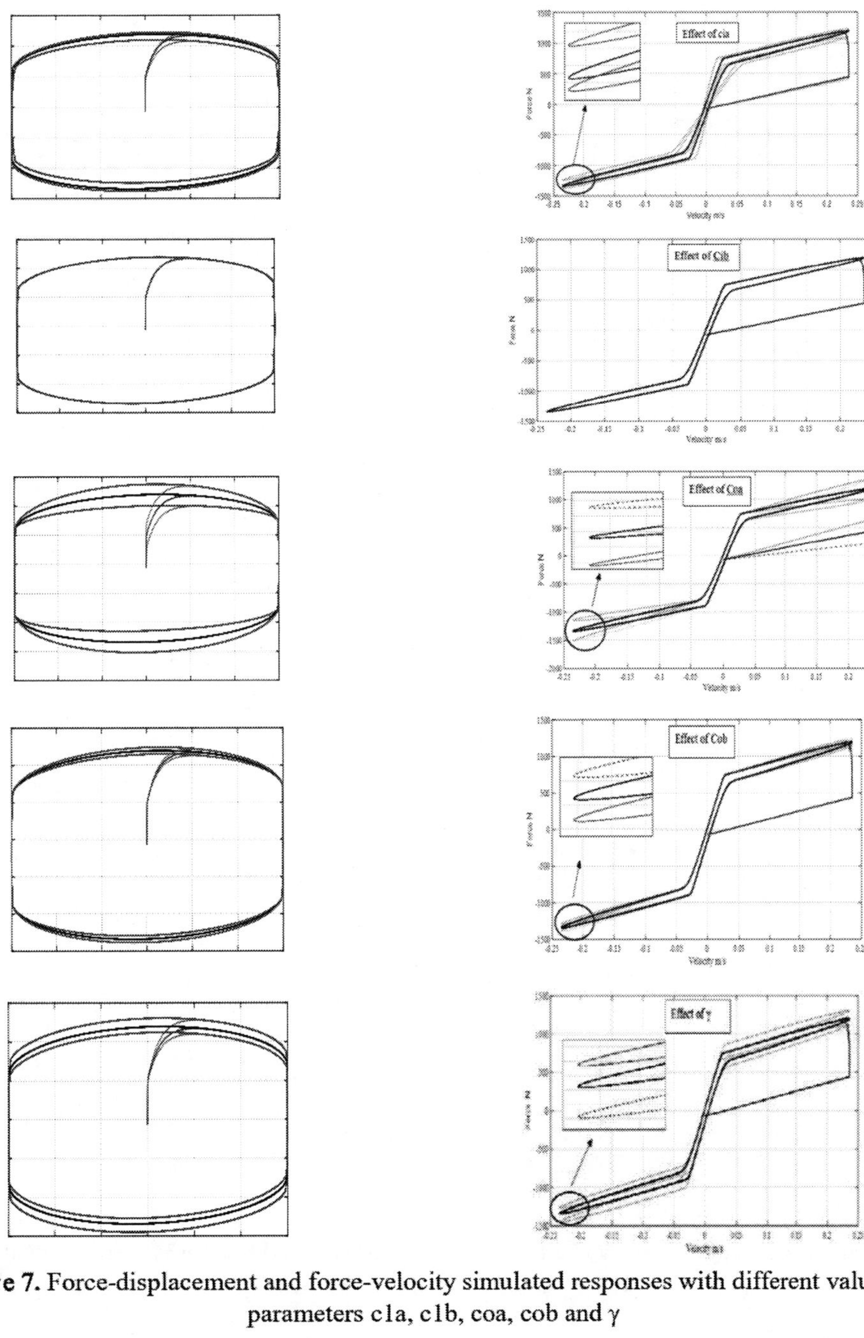

Figure 7. Force-displacement and force-velocity simulated responses with different values of parameters c1a, c1b, coa, cob and γ

IOP Conf. Series: Materials Science and Engineering **973** (2020) 012020 doi:10.1088/1757-899X/973/1/012020

Figure 8 describes the damper hysteresis with the variation corresponding to the parameters, k1, ko, xo, n and η. Different from the aforementioned parameters, there is no remarkable effect noticed for the parameters ko and η. It is noticed that the dynamic range increases by the increase in value of parameter k1. The variation of parameter xo results in shifting the hysteres is profile up and down from the equilibrium position. Exponent n is remarked as the crucial factor that controls the shape of the hysteresis loop. It is shown from the figure that the degree of nonlinearity becomes more obvious when the exponent, n is decreased. On the other hand, the hysteresis tends to be almost linear when the exponent, n is increased.

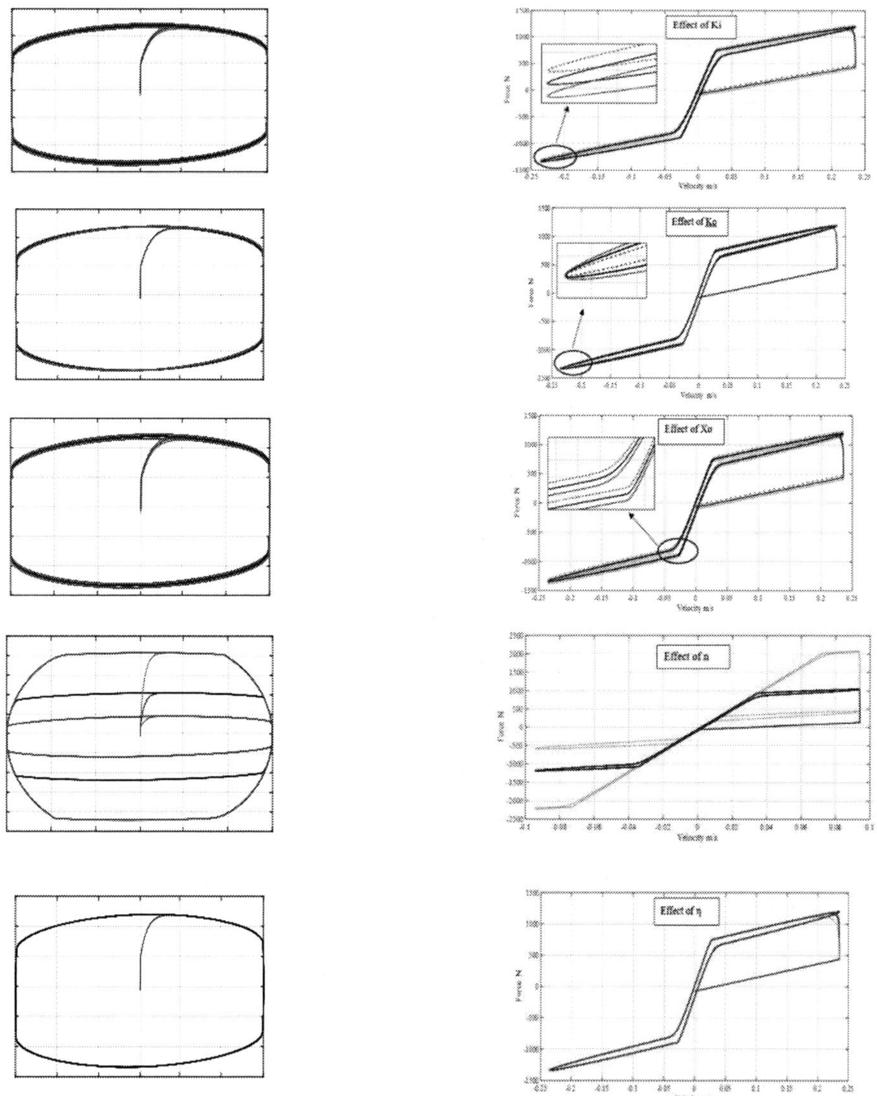

Figure 8. Force-displacement and force-velocity simulated responses with different values of parameters k1, ko,xo,n and η

5. Identification of optimal model parameters

The accuracy of the MR damper model is determined by the difference between the measured and simulated characteristics. When the simulated response of the model agrees closely with the measured data, the model can be used in simulation to represent the MR damper dynamics. Consequently, it is of rather importance to identify accurately the MR damper parameters that fit the experimental results. The accurate identification of the parameter values can be done using systematic search procedures with the objective to make the simulated characteristics similar to the measured response.

In order to demonstrate the influence of the identified parameters on the model response, and there, to obtain better estimates of the damper parameters of the modified Bouc-Wen model, a systematic method is employed. By this manner, all values of the reference parameters are varied systematically in order to predict characteristics that are similar to the measured data. The rms values of the damping force between the varied and reference parameters are used to identify the optimal parameters that best fit the damper output. The parameters A, β, γ, n. ko, k1, xo control the shape of the damper hysteresis with zero applied voltage. In addition, the parameters Coa,Cob,C1a,C1b, αa,αb control the dynamic force range.

Figure 9 and Figure 10 show the comparison between the force-displacement and force-velocity dependence of the simulated characteristics using the optimized parameters shown in Table 3and the measured response of the MR damper under sinusoidal excitation of 1 and 2 Hz and 5 mm amplitude at zero and 0.5 Amp respectively. It can be seen from the figures that the simulated response shows good agreement with the measured data.Also, the rms values and error percentage of the simulated and experiment responses have been calculated in order to judge the accuracy of the optimized parameters as shown in Table 4. The results show the rms values of the simulated damping force is closely match the rms values from experiment which prove the good match between the model and the experimental tests.

Table 3. Optimal parameters for fitting Modified Bouc-Wen model to experimental data

Parameter	Modified parameters	Unit
A	63	-
β	1963086	m^{-2}
n	2	m^{-2}
γ	206320	-
η	194	s^{-1}
c_{oa}	783	N.s/m
c_{ob}	2603	N.s/m.V
c_{1a}	15649	N.s/m
c_{1b}	45072	N.s/m.V
α_a	10441	N/m
α_b	45430	N/m.V
k_o	1750	N/m
k_1	317	N/m
x_o	0.0032	m

Table 4. Rms values and percentage of error

Parameters	Model (optimized parameters)	Practical	Percentage of error
1Hz, 0 A, 5mm	68.27	67.72	-0.81%
1Hz, 0.5 A, 5mm	475.82	463.36	-2.68%
2Hz, 0 A, 5mm	84.31	83.55	-0.9%
2Hz, 0.5 A, 5mm	572.14	561.67	-1.86%

The negative sign in error percentage means the model force is larger than practical force.

IOP Conf. Series: Materials Science and Engineering **973** (2020) 012020 doi:10.1088/1757-899X/973/1/012020

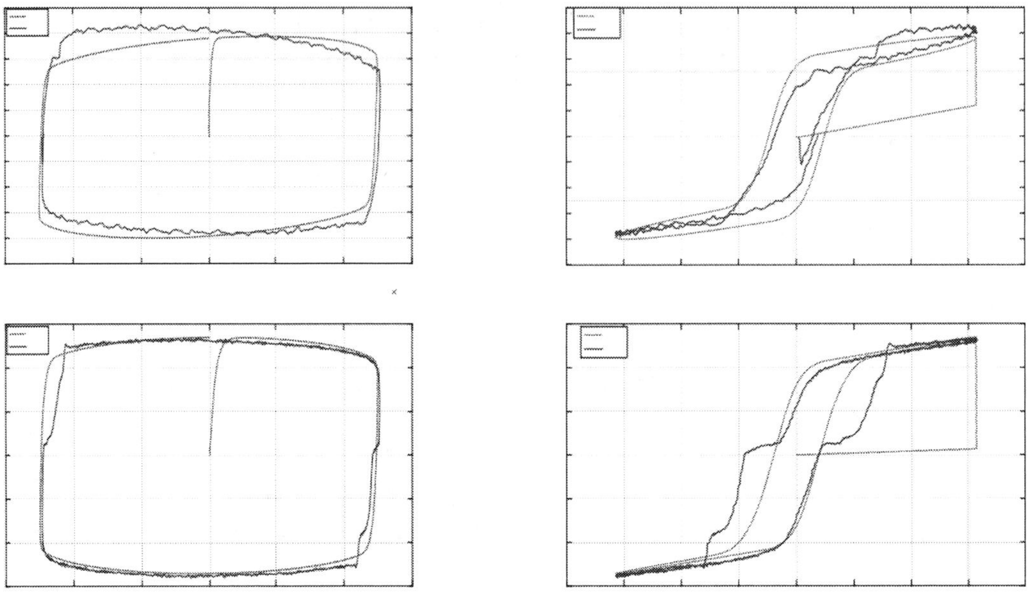

Figure 9. Comparison between force-displacement and force-velocity responses predicted by modified Bouc-Wen using optimal parameters and the measured data under excitation of 1Hz and 5 mm amplitude at zero and 0.5 Amp

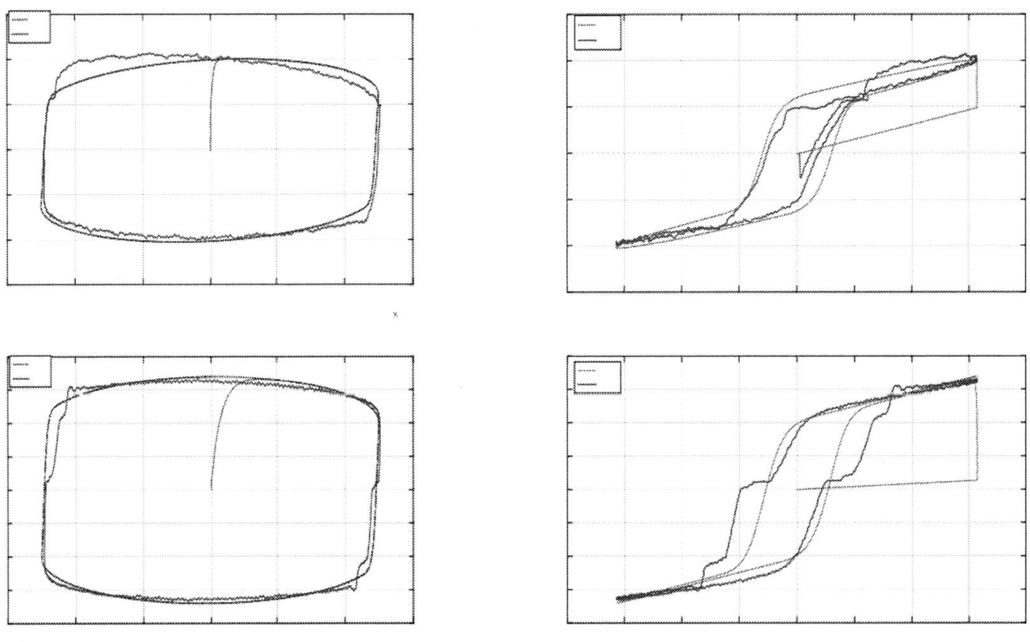

Figure 10. Comparison between force-displacement and force-velocity responses predicted by modified Bouc-Wen using optimal parameters and the measured data under excitation of 2Hz and 5 mm amplitude at zero and 0.5 Amp.

6. Conclusion

This paper presents an investigation on the effect of the modified Bouc-Wen on the dynamic characteristics parameters of a short stroke MR damper. The Experimental tests on the existing MR damper show the lack of the existing Bouc-Wen models (MBW1 and MBW2) to closely fit the measured damper characteristics. To improve the model accuracy, a parametric study was held to investigate the effect of each parameter on the hysteresis shape of the MR damper. The results show that the MR damper model characteristics are highly nonlinear and its parameters have a significant effect on the damper hysteresis. A parametric study is introduced to find the optimal parameters values that increase the degree of fitting between the simulated and measured data. Finally, the optimized parameters are introduced in the model and validated with the experimental responses which show good agreement between the simulated and measured results.

References

[1] Bai X-X, Chen P and Qian L-J2015 Principle and validation of modified hysteretic models for magnetorheological dampers*Smart Materials and Structures* **24** (8)

[2] Domínguez-González A, Stiharu I, and Sedaghati R2014 Practical hysteresis model for magnetorheological dampers*Journal of Intelligent Material Systems and Structures* **25** (8) pp 967-79

[3] Balamurugan L, Jancirani J and Eltantawie M 2014 Generalized magnetorheological (MR) damper model and its application in semi-active control of vehicle suspension system *International Journal of Automotive Technology* **15**(3) p p419-27

[4] Zhao X, Qin W, Wu S and Pan H 2018 A dynamic hysteresis model for MR dampers based on particle swarm optimization*8th Annual International Conference on CYBER Technology in Automation, Control, and Intelligent Systems*

[5] Yu J, Dong X and Zhang Z 2017 A novel model of magnetorheological damper with hysteresis division*Smart Materials and Structures* **26** (10)

[6] Rossi A, Orsini F, Scorza A, Botta F, Belfiore N and Sciuto S 2018 A review on parametric dynamic models of magnetorheological dampers and their characterization methods*Actuators* **7** (2)

[7] Peng Y and Yang J, Li J2018 Parameter identification of modified Bouc–Wen model and analysis of size effect of magnetorheological dampers*Journal of Intelligent Material Systems and Structures* **29**(7) pp 1464-80

[8] Zamani A-A, Tavakoli S, Etedali S and Sadeghi J2019 Modeling of a magneto-rheological damper: An improved multi-state-dependent parameter estimation approach*Journal of Intelligent Material Systems and Structures* **30** (8) pp 1178-88

[9] Metered H2012 Application of nonparametric magnetorheological damper model in vehicle semi-active suspension system*SAE International Journal of Passenger Cars-Mechanical Systems* **5**715-26

[10] Spencer Jr B, Dyke S, Sain M and Carlson J 1997 Phenomenological model for magnetorheological dampers*Journal of engineering mechanics* **123**(3) pp 230-38

[11] Şahin İ, Engin T, and Çeşmeci Ş2010 Comparison of some existing parametric models for magnetorheological fluid dampers*Smart materials and structures* **19** (3)

[12] Lai C and Liao W2002 Vibration control of a suspension system via a magnetorheological fluid damper. *Modal Analysis* **8**(4) pp 527-47

IOP Conf. Series: Materials Science and Engineering **973** (2020) 012020 doi:10.1088/1757-899X/973/1/012020

Nomenclature

symbol	Meaning	Symbol	Meaning
v	Voltage applied to the current driver	α	Represent rheological effect
f	The resulting force from MR damper	z	Hysteresis variable
y	Internal dynamic variable	$\beta, A, \gamma, n, \eta$	Characterization parameter for MR damper nonlinearity
k_1	Accumulator stiffness	k_o	Control stiffness at large velocities
c_1	Represent the dashpot at low velocities	c_0	Viscous damping at large velocities
x_o	The initial displacement of spring k_1		

IOP Conf. Series: Materials Science and Engineering **973** (2020) 012021 doi:10.1088/1757-899X/973/1/012021

Experimental and computational dynamic structural analysis of free-flight rockets

M A Habaka[1,2], **SS Farag**[1], **MKhalil**[1], **M Y Ahmed**[1]
[1]Aerospace Engineering Department, Military Technical College, Cairo, Egypt
Email: [2]mohamed.habakah@hotmail.com

Abstract. Aerospace vehicles' designers often strive to approach the lightweight structures in order to improve flight performance without compromising vehicles durability. However, this goal may be achieved on the expense of structural rigidity of the vehicle. A missile with high slenderness ratio is a typical vehicle that should be considered as flexible body if precise dynamic behaviour analysis is sought. This analysis is crucial to identify the flight performance of such missile with high accuracy. The present paper discusses the results of a side-by-side experimental and numerical analysis of the vibration characteristics of a full-scale free-flight missile having 70mm caliber. The experimental modal analysis is conducted on the missile with empty motor to represent its unpowered flight regime. Experimental modal analysis setup involved accelerometer sensors while vibration excitation is achieved using an impact hammer. Modal analysis is also conducted numerically using a well-used high-fidelity commercial tool. Results of experimental modal analysis have shown that the first four modes as bending modes with frequencies of 134.4, 400.6, 819.4 and 1173.6 Hz, respectively. The results also demonstrate the close agreement between numerical and experimental approaches.

1. Introduction

Dispersion is one of the key issues related to unguided missiles that can degrade their accuracy and lethality and increase their undesired lateral impact. Enhancing dispersion is thus an essential practice for missile flight experts. One main source of dispersion is aeroelasticity and structural dynamic behavior of missiles during their flight which is more pronounced for those with high fineness (slenderness) ratios; understanding and predicting such missile features is one of the designers' goals. Aeroelasticity is concerned with the coupled dependency between loads and deformation and involves three disciplines namely, elasticity, dynamics, and aerodynamics. Structural dynamic characteristics are fundamental aspects of aeroelasticity. This interaction of elastic and inertial forces that effects on the structure body, leads to mechanical vibrations. All structural bodies that have mass and elasticity are capable of free vibration in the absence of an external excitation. To better investigate structural vibration problems, the natural frequencies and mode shapes need to be studied. Moreover, exploring structural dynamics helps finding new useful ways to solve aeroelasticity and, hence, dispersion problems. Unguided air-to-surface missiles that are fired against stationary ground targets are vulnerable to dispersion. In addition, they are commonly of a high fineness ratio and unpowered flight dominates their trajectory. As a result of increasing slenderness ratio, the structure flexibility of a missile is increased and the vibration longitudinal and transverse natural frequencies are decreased. Accordingly, it is of a high importance to study the dynamic behavior of such "flexible" missiles to ensure the safety of flight and accurate arrival at the target. Modal analysis is a more and more paramount engineering tool that is used for representing structural dynamic behavior.

IOP Conf. Series: Materials Science and Engineering **973** (2020) 012021 doi:10.1088/1757-899X/973/1/012021

Modal analysis can be performed in many approaches namely, theoretically, experimentally, or numerically. The numerical (finite element) method is widely used in industry for producing a true engineering structure with an excellent representation. However, results of finite element method need to be validated against experimental counterparts. The coupling of both finite element method and experimental modal testing is widely used for determining dynamic characteristics of complicated engineering structure with maximum confidence.

In the literature, a considerable body of researches investigated the structural characteristics of missiles using the different approaches. For instance, Vernon et al. [1] studied the natural frequencies and mode shapes of the multistage launch vehicle. The first three natural frequencies and mode shapes of a full-scale four-stage launch vehicle were measured experimentally and calculated numerically. The results of the experimental and numerical analysis of the vibration characteristics of the launch vehicle were compared. Larry et al.[2]verified experimentally the numerically calculated vibration characteristics and natural and bending frequencies of complex launch vehicle structure. Similarly, Basant et al.[3]studied the vibration characteristics of a two-stage launch vehicle experimentally and numerically. The first three natural frequencies and mode shapes were calculated and compared with the experimental results. Robert and Theodore[4]studied experimentally the vibration characteristics of a launch vehicle. The natural frequencies, mode shapes and damping characteristics of this launch vehicle were measured in different conditions of flight. Platus [5]investigated numerically the effect of structural damping on dynamic behavior of hypersonic launch vehicle. The natural frequencies and mode shapes were calculated for the launch vehicle. A similar missile category was examined by Yun et al. [6], who investigated numerically dynamic behavior, natural frequencies, and mode shapes of hypersonic launch vehicle. Abbas et al. [7]calculated theoretically the natural vibration characteristics of a two-stage launch vehicle under variable axial load. The natural frequencies and mode shapes were calculated and compared with flight test results and numerical simulations [1]. Launch vehicles were also examined numerically in [8, 9]. The first five natural frequencies and bending mode shapes were deduced numerically by Li et al.[8]whereas Chen et al.[9]computed the first six natural frequencies and mode shapes theoretically and numerically.

Numerical finite element approach was adopted in a number of studies. Natural frequencies and mode shapes of a flexible launch vehicle models were examined by Rajan and Narasimhan[10]and Başkut and Akgül [11]whereas Tsushima et al.[12]studied the dynamic bending vibrations of such vehicles; the results were compared with experimental results in [4].Recently, Li and Ye[13]investigated numerically the dynamic behavior of a flexible spinning missile.

In the present paper, modal analysis has been implemented on 70 mm missile. An experiment was carried out on the missile's structure without presence of propellant and explosive for safety to confirm the numerical calculation that were performed using ANSYS Workbench.

2. Case-study

The case study in concern is an unguided, fin-stabilized, free-flight air-to-surface missile with caliber 70 mm as shown in figure 1. The dummy empty missile has a total mass of 4.208 kg and 1.445 m total length (fineness ratio is 20.6). As illustrated in table 1, the missile is represented as a non-uniform beam consisting of five parts. Material properties for each part are listed in table 2.

Table 1. Dimensions, masses, and materials of missile body sections.

Part	Length [cm]	Mass [kg]	Material
Nose tip	8.5	0.189	AL-1050
Ogival warhead fairing	8.9	0.886	Steel Alloy
Dummy warhead	27.7	0.717	AL-1050
Motor case	86.8	1.102	AL-1050
Nozzle	12.6	1.314	Steel Alloy

Table 2. Density and mechanical properties of missile body material.

Material	Density [kg/m^3]	Young's modulus [GPa]	Poisson's ratio
AL-1050	2710	71	0.33
Steel Alloy	7850	210	0.30

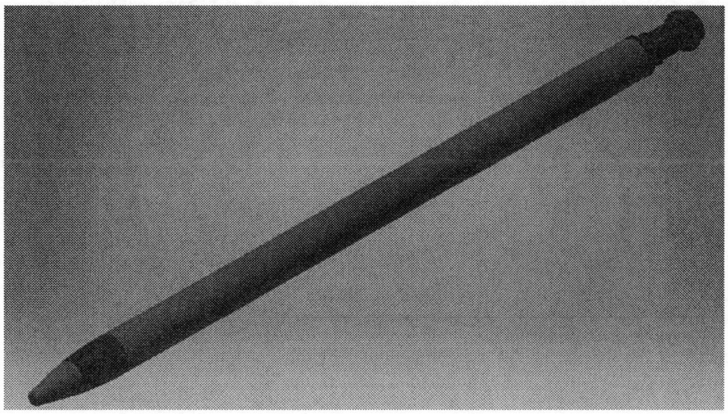

Figure 1. Configuration of the case study model.

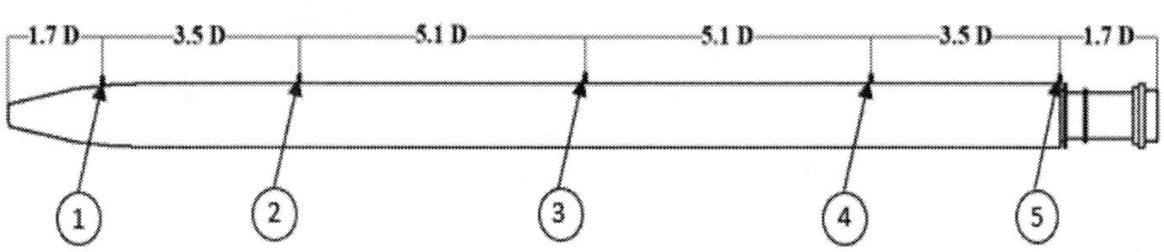

Figure 2. Locations of accelerometer sensors along the missile.

3. Methodology

3.1. Experiment setup and instruments

For experimental analysis, the missile body is horizontally hanged by using polyester ropes that allow one degree of freedom around its hang point. Experimental modal analysis of the missile is observed using accelerometers, impact hammer slightly excites the body vibration. Accelerometer sensors measure the local accelerations due to vibration and convert this mechanical motion into electrical signal. Vibration is excited by applying an impact for infinitely short duration to generate constant amplitude in the frequency domain yielding all modes of vibration. Five accelerometer sensors are fitted on the missile airframe at the positions as shown in figure 2.

Three accelerometer sensors (type PCB352C33) with voltage sensitivity (±10%): 100 mV/g [14] and two sensors (type PCB352C03) with voltage sensitivity (±10%): 10mV/g [15]are used. Accelerometers of the first type are fitted at positions 1, 3 and 5 as shown in figure 2. Frequency range for both types is (±10%): 0.3 to 15000 Hz. The used impact hammer is (type PCB086C03) with voltage sensitivity (±15%): 2.25 mV/N, constant current excitation: 2 to 20 mA, and Quartz sensing

IOP Conf. Series: Materials Science and Engineering **973** (2020) 012021 doi:10.1088/1757-899X/973/1/012021

element with Epoxy sealing [16]. Data acquisition is performed using NI (National Instruments) PXIe-4499 hardware along with NI Signal Express control software [17].

Firstly, one accelerometer sensor was used in the experiment. It was located in the middle of the missile. The NI Signal Express program was established for measuring ten times in one experiment. Then, the impact hammer was used to excite the missile for ten times at the same point in the nose part. After every strike, the experiment was held until the NI Signal Express reading data. Then, NI Signal Express was stopped automatically after the last one. Then, the output data were exported in an excel sheet from NI Signal Express. This experiment with the same settings was repeated five times. For every time data were exported to ensure the validity of the results of the experiment.The experimental work of the modal shape analysis and sample data is shown in figure 3.

Finally, the experiment was done but by using five accelerometer sensors. These five accelerometer sensors were placed at certain positions on the missile as shown in figure 2. This experiment had been done similarly. However, in this experiment, the output data were exported in Excel sheet for each accelerometer.

3.2. Numerical modal analysis

Finite element simulation approach is applied to the missile using a commercial software (ANSYS). An unstructured mesh is constructed on the structure with high smoothing. A grid sensitivity check is performed for five grid resolutions to enhance solution accuracy. In the sensitivity check, a free-free boundary condition is represented and with option twenty max modes to find because the first six calculated modes in ANSYS were taken as a rigid body. The results of the grid sensitivity check are shown in table 3. As a result, grid 4 is adopted in the present study.

Table 3. Results of the grid sensitivity check.

Grid trail	No. of mesh elements	1st bending mode [Hz]		Change [%]
		Y-direction	Z-direction	
1	30687	147.16	148.26	-----
2	71791	144.57	145.37	1.76
3	89555	142.53	142.53	1.41
4	280906	142.50	142.50	.02
5	1667890	142.47	142.47	.02

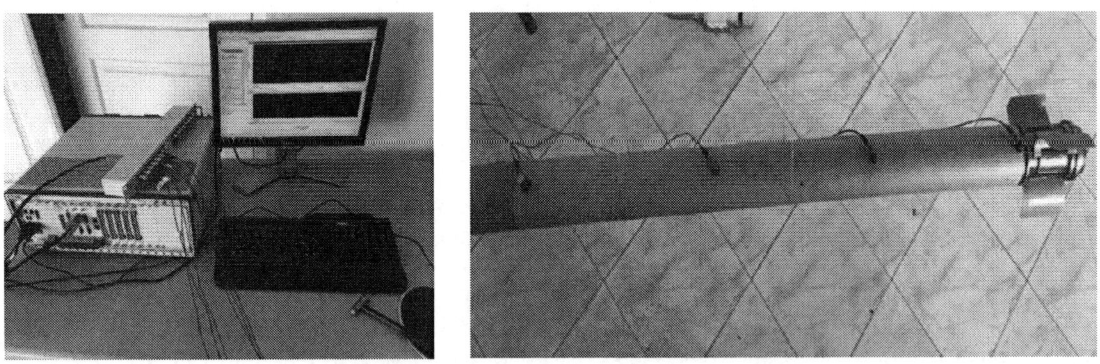

Figure 3. Missile modal analysis experiment setup.

4. Results

The first four natural frequencies and mode shapes were obtained for the missile with 70mm caliber numerically by using finite element model (ANSYS Workbench). Also, the natural frequencies of the same missile were obtained experimentally. The modal analysis was performed as a free-free body.

4.1 Experiment results

Through experiment procedure illustrated in subsection 3.1, the first four natural frequencies of bending mode are obtained. The experiment has been repeated five times, where corresponding results are listed in table 4.

4.2 Numerical results

The obtained first four natural frequencies of bending mode for the missile using ANSYS Workbench is shown in Table 5. The obtained mode shapes for a free-free boundary conditions of the same missile are shown in Figures 4, 5, 6 and 7. A comparison has been implemented between bending natural frequencies obtained with the experimental results as illustrated in table 5.

Table 4. The first four bending natural frequencies resultant from the five experiments.

	Mode 1	Mode 2	Mode 3	Mode 4
EXP 1	133	397	815	1174
EXP 2	133	399	817	1172
EXP 3	134	408	829	1174
EXP 4	134	398	818	1174
EXP 5	138	401	818	1174
Mean	134.4	400.6	819.4	1173.6

Table 5. Comparison between numerical and experimental results for bending modes.

	Numerical approach [Hz]	Experimental work [Hz]	Error
Mode 1	142.47	134.4	6%
Mode 2	440.41	400.6	9%
Mode 3	909.72	819.4	11%
Mode 4	1441.4	1173.6	22%

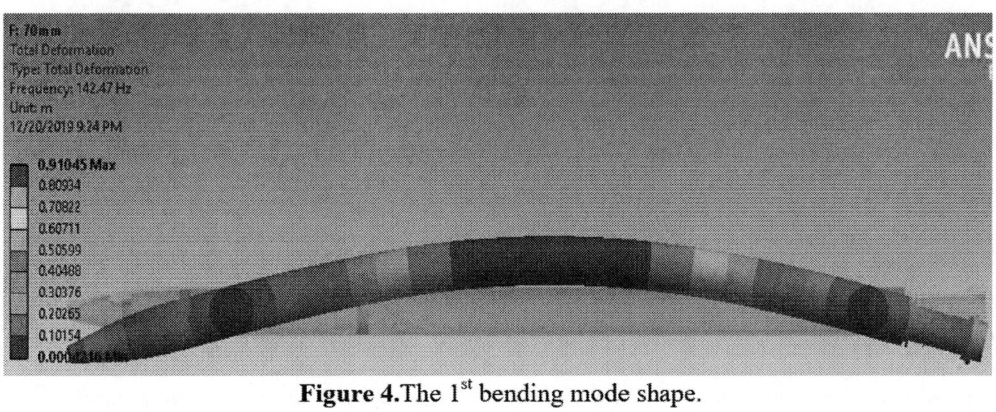

Figure 4. The 1st bending mode shape.

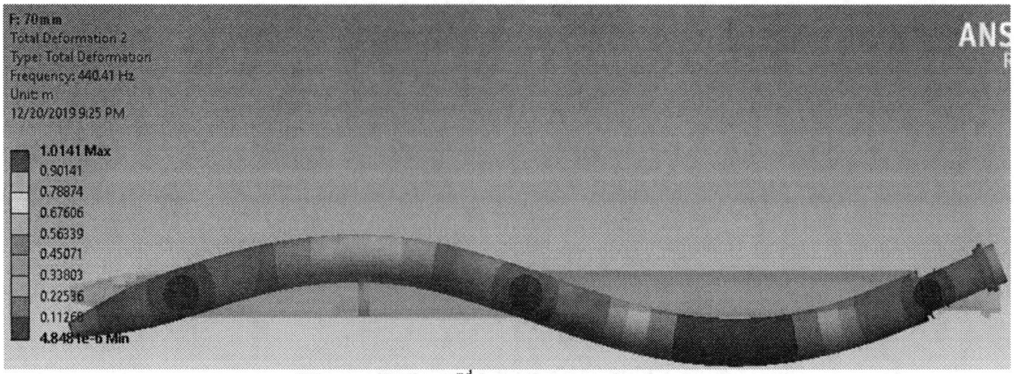

Figure 5. The 2nd bending mode shape.

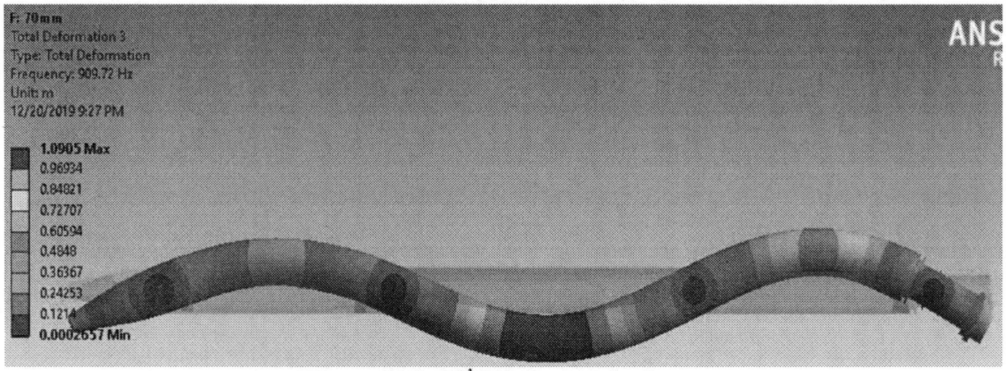

Figure 6. The 3rd bending mode shape.

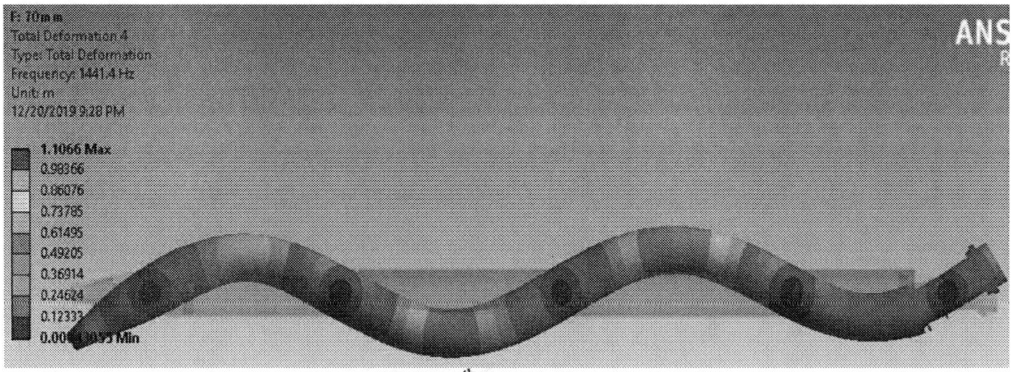

Figure 7. The 4th bending mode shape.

5. Conclusion

In present work, numerical and experimental modal analysis of the missile with 70mm caliber were performed in free-free boundary conditions. In the experiment modal analysis, it was not possible to determine the higher modes in a free-free boundary conditions of the missile using the impact hammer. The numerical and experimental results for the missile were found to have a highly good correlation as showed in table 5. Some errors were introduced between the results of the numerical and experimental modal analysis of the missile. After analyzing the results of overall modal analysis, it can be concluded that the results are all acceptable within the affordable error margin. The presented techniques of modal analysis in this paper are extremely helpful in dynamic analysis, optimizing and developing of designing complex engineering structures and components.

References

[1] Alley V and Leadbetter S 1963 Prediction and measurement of natural vibrations of multistage launch vehicles *AIAA Journal* 1 374-9

[2] Mixon L C, Schaeffel Jr J A, Green P L, Salas R L and Garner R 1988 Modal Analysis and Testing of Missile Systems. *Army missile command redstone arsenal al structures directorate*

[3] Kumar B, Rao N, Gupta R and Mohan J R 2013 Prediction and Measurement of Free Vibration Response of Multistage launch vehicles

[4] Miller R and Gerus T 1969 Experimental lateral bending dynamics of the Atlas-Centaur-Surveyor launch vehicle

[5] Platus D 1992 Aeroelastic stability of slender, spinning missiles. In: *16th Atmospheric Flight Mechanics Conference,* p 3393

[6] Xu Y, Xie C and Yang C 2008 Effect of thrust on aeroelastic stability of a slender missile. In: *49th AIAA/ASME/ASCE/AHS/ASC Structures, Structural Dynamics, and Materials Conference, 16th AIAA/ASME/AHS Adaptive Structures Conference, 10th AIAA Non-Deterministic Approaches Conference, 9th AIAA Gossamer Spacecraft Forum, 4th AIAA Multidisciplinary Design Optimization Specialists Conference,* p 2022

[7] Abbas L K, Min-Jiao L and Xiao-Ting R 2013 Transfer matrix method for the determination of the natural vibration characteristics of realistic thrusting launch vehicle—part I *Mathematical Problems in Engineering*

[8] Li M, Rui X and Abbas L K 2015 Elastic dynamic effects on the trajectory of a flexible launch vehicle *Journal of Spacecraft and Rockets* **52** pp1586-1602

[9] Chen G, Rui X, Yang F and Zhang J 2016 Study on the natural vibration characteristics of flexible missile with thrust by using Riccati transfer matrix method *Journal of Applied Mechanics* **83** 031006

[10] Rajan K and Narasimhan K 2005 Design and structural analysis of a thrust chamber for a spinning supersonic rocket–a case study *The Aeronautical Journal* **109** I-VI

[11] Başkut E and Akgül A 2011 Development of a coupling procedure for static aeroelastic analyses *Scientific Technical Review* **61** pp 39-48

[12] Tsushima N, Su W, Wolf M G, Griffin E D, Whittaker J T and Dumoulin M P 2017 Bending Vibration Monitoring for Flexible Rockets through a Reference Strain Structure. In: *58th AIAA/ASCE/AHS/ASC Structures, Structural Dynamics, and Materials Conference,* p 1126

[13] Li H and Ye Z 2019 Effects of rotational motion on dynamic aeroelasticity of flexible spinning missile with large slenderness ratio *Aerospace Science and Technology* **94** 105384

[14] PCB-Piezotronics 2009 Piezo-electric ICP Accelerometer 352c33 18292 15

[15] PCB-Piezotronics 2009 Piezo-electric ICP Accelerometer 352c03 18292 15

[16] PCB-Piezotronics 2007 Piezo-electric ICP Impact Hammer 086c03 19198 15

[17] National-Instruments 2012 NI 449x Specifications 37212F 18

IOP Conf. Series: Materials Science and Engineering **973** (2020) 012022 doi:10.1088/1757-899X/973/1/012022

Numerical investigations of bistable piezoelectric cantilever for enhanced energy harvesting

M A Abdelnaby[1], M S Shiba[2]and SAli[3]
[1]Production Engineering and Printing Technology Department, Akhbar El Yom Academy, Egypt
[2]Automotive Engineering Department, Higher Technology Institute, 10th of Ramadan 6th of October, Egypt
[3]Department of Mechatronics, Canadian International College, Cairo, Egypt
E-mail: muhmmad.aly@akhbaracademy.edu.eg

Abstract. Piezoelectric energy harvesting applications have been highly impacted research area recently. Numerical investigations for bistable piezoelectric cantilevers have great importance to understand its dynamic performance. This paper aims to develop comprehensive numerical models to simulate the dynamic response of aunimorph piezoelectric bender under bistability effect for enhanced energy harvesting. To this end, the equivalent stiffness, damping, electromagnetic coupling and capacitance coefficients of a bistable piezoelectric cantilever have been numerically developed. Besides, finite element models are developed to determine the bistability repulsion force between magnets. The resulting simulations are then implemented in an electromechanical model to predict the output voltage and power over a range of load resistance.

1. Introduction

Over the last decade, vibration-based energy harvesting has been discussed as a promising field to convert ambient waste energy into useful electrical energy[1-5].Furthermore, the vibration energy harvesting approach has been developed to power wireless sensor nodes in the structural health monitoring field[6-10]. One of the most promising solutions in this context is using the magnetic bistability piezo electric systems, which transition from one stable state to another and accordingly enhance the output power over a wider frequency range [11-15].Some of the developments on bistability piezoelectric energy harvesting have been presentedin[16-23].

Previous researches about using piezoelectric cantilever in energy harvesting applications indicate two main theoretical challenges. First, estimating the magnetic repulsion force between a couple of magnets, which often experimentally developed by mounting one magnet on an electronic scale while placing another magnet on an arm that is linearly guided on a movable frame whose displacement is measured by a Vernier caliper [20], [23], after that can be expanding the vector of repulsion force as a function of horizontal gap and vertical elevation difference between magnets. Second, theoretical relations for understanding electromechanical performance of piezoelectric beam are very complicated, especially for multi-layer piezoelectric beams as in [24], [25], so in this paper, the magnetic repulsion force model has been developed to simulate the behavior of repelling magnets at different horizontal gaps and vertical elevations. Also benefited from previous research and linked it in a distinct way have been done to create theoretical relations, which used for estimating equivalent

stiffness and electromechanical coupling coefficients for unimorph piezoelectric beams at different design parameters. All numerical models have been examined by comparing its results with actual parameters of commercial products to ensure the power of this comprehensive study.
The remainder of this paper is structured into five sections. Section 2 presents the design and electromechanical model of the proposed energy harvester. Sections 3 presents the comparative numerical analysis of the proposed design and Section 4 presents an electromechanical response for the proposed harvester. Finally, Section 5 is dedicated to conclusions.

2. Harvester design and electromechanical model

Figure 1shows a schematic illustration of the proposed bistable piezoelectric energy harvester. The harvester consists of unimorph piezoelectric cantilever with tip magnet, which repels against another one in front of it. Whole system subjected to vertical sinusoidal excitation as shown in figure 1. The effects of base excitation and repulsion force between magnets, causes a strain in the piezoelectric layer, which generates output voltage across a load resistance.

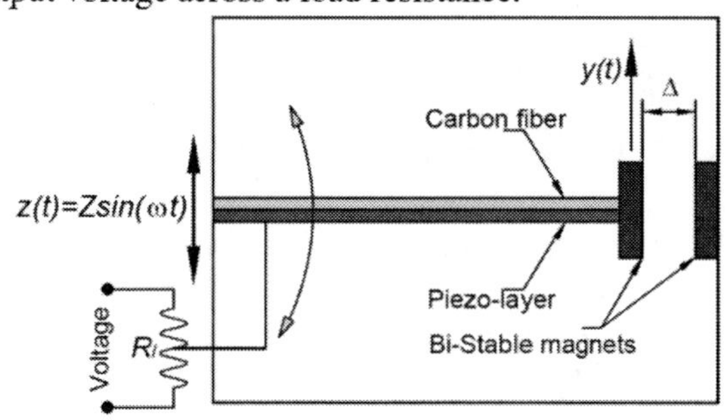

Figure 1. Schematic illustration of the proposed harvester.

Electromechanical governing equations for the unimorph piezoelectric cantilever subjected to a repulsion magnetic force as in[23].

$$\begin{bmatrix} m & 0 \\ 0 & 0 \end{bmatrix}\begin{Bmatrix} \ddot{y} \\ \ddot{v} \end{Bmatrix} + \begin{bmatrix} c & 0 \\ \theta & c_p \end{bmatrix}\begin{Bmatrix} \dot{y} \\ \dot{v} \end{Bmatrix} + \begin{bmatrix} k & -\theta \\ 0 & 1/R_l \end{bmatrix}\begin{Bmatrix} y \\ v \end{Bmatrix} = \begin{Bmatrix} F_m \\ 0 \end{Bmatrix} \tag{1}$$

Where:m[kg], k[N/m], and c[Ns/m] are the equivalent mass for piezoelectric cantilever with tip magnet, equivalent stiffness, and equivalent damping, respectively, also θ [N/V] is the equivalent electromechanical coupling, c_p [F] is the equivalent capacitance of piezo layer, R_l [Ω] is the load resistance, y [mm] tip displacement, v [V] is the output voltage and F_m[N] magnetic repulsion force. In order to solve numerically the electromechanical equations of the model in figure 1state-space representation will be as follows [23]:

state variables,$u_1 = y, u_2 = \dot{y} = \dot{u_1}$ and $u_3 = v$, so equation (1)can write in the form:

$$\begin{gathered} \dot{u_1} = u_2 \\ \dot{u}_2 = -\frac{c}{m}u_2 - \frac{k}{m}u_1 + \frac{\theta}{m}u_3 + \frac{1}{m}F_m \\ \dot{u}_3 = -\frac{1}{c_p R_l}u_3 - \frac{\theta}{c_p}u_2 \end{gathered} \tag{2}$$

In 2004, the relationship between displacement and charge or force and voltage of the unimorph piezoelectric cantilever beam has been investigated by Park and Moon [24], as follows:

$$\begin{bmatrix} y \\ q \end{bmatrix} = \begin{bmatrix} b_{11} & b_{12} \\ b_{21} & b_{22} \end{bmatrix}\begin{bmatrix} F_m \\ v \end{bmatrix} \tag{3}$$

IOP Conf. Series: Materials Science and Engineering **973** (2020) 012022 doi:10.1088/1757-899X/973/1/012022

Where, q[C] is the charge, b_{11}[m/N] is a coefficient relating between tip displacement and force, b_{12}= b_{21}[m/V] are the coefficients relating stiffness and electromechanical coupling and b_{22} [F] equals to the equivalent capacitance (c_p). Detailed equations for b_{11}, b_{12}=b_{21} and b_{22}were found in [24]. By comparing equations (1 and 3)we can determine equivalent stiffness, capacitance and electromechanical coupling for the unimorph piezoelectric cantilever beam as following:

$$b_{11} = 1/k\,,\ b_{12} = b_{21} = \frac{\theta}{k} and b_{22} = c_p \tag{4}$$

3. Comparative analysis

In order to achieve the main contribution of this study, which is developing a comprehensive numerical analysis to examine the electromechanical behavior of the proposed harvester in figure 1, a commercial piezoelectric cantilever (American Piezo Stripe Actuator 40-2010[26]) and nickel-plated NdFeB of Grade 42, F200-10 (first4magnets[27]) have been selected in this work. Accordingly, the analysis will successively show in four steps.

3.1. Determination of electromechanical parameters

Table 1, demonstrate all required parameters for the selected piezoelectric cantilever, to determine coefficients b_{11}, b_{12}=b_{21} and b_{22}in equation (3)which used after that to calculate equivalent stiffness, capacitance and electromechanical coupling from equation (4).

Table1. American piezostripe actuator 40-2010 parameters.

Parameters	Values
Piezoelectric constant	-320×10^{-12}
Total beam length	60 mm
Piezoelectric layer length	60 mm
Length ratio (piezo layer/total beam)	1
Compliance ratio (piezo layer/carbon fiber)	4
Thickness ratio (carbon fiber/ piezo layer)	0.67
Carbon fiber layer thickness	0.16 mm
Piezoelectric layer thickness	0.24 mm
Cantilever beam width	20 mm
Dielectric constant	3800-1700
Electromechanical coupling factor	1.7×10^{-12}

According to previous conditions in this work, the calculated electromechanical parameters for equation (4)will be as follows:

$$b_{11} = 0.0030375 \text{ m/N},\ k = 329 \text{ N/m},\ b_{12} = -8.8437\times10^{-6} \text{ [m/V]},\ \theta = -2.9\times10^{-3} \text{ [N/V]},\ and\ b_{22} = c_p = 19{,}000 \text{ pF}. \tag{5}$$

3.2. Electrostatic model

At this point, the electrostatic module of a finite element package (COMSOL Multi-physics) has been used to analyze and examine the electromechanical parameters, which calculated in section 3.1. By using this model, we can estimate the tip displacement of the piezoelectric beam at certain electric potential volt input.

Figure 2a shows the y-displacement along a piezoelectric beam with subjected to 100 V (electric potential volt), while figure 2bshows the electric potential across the piezoelectric layer (0-100 V) and figure 2cshows the stresses distribution along the piezoelectric beam. By comparing the value of y-displacement from the electrostatic model with its calculated value from equations (3-5) finding at 100 V (electric potential), tip displacement is 0.92 mm (electrostatic model) and 0.8844 mm (results from equations (3-5)), which are very close to each other.

IOP Conf. Series: Materials Science and Engineering **973** (2020) 012022 doi:10.1088/1757-899X/973/1/012022

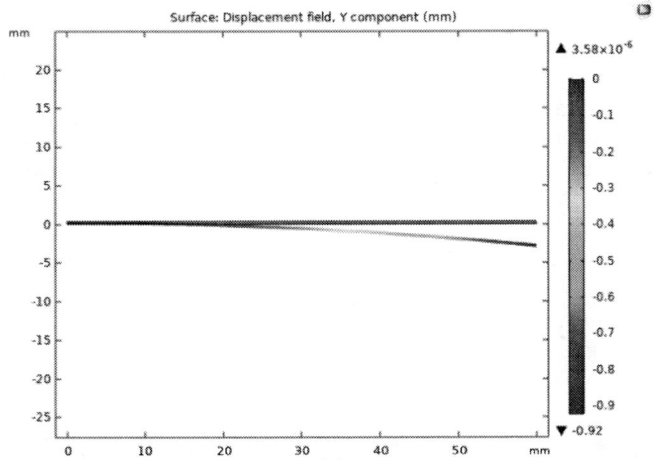

(a)

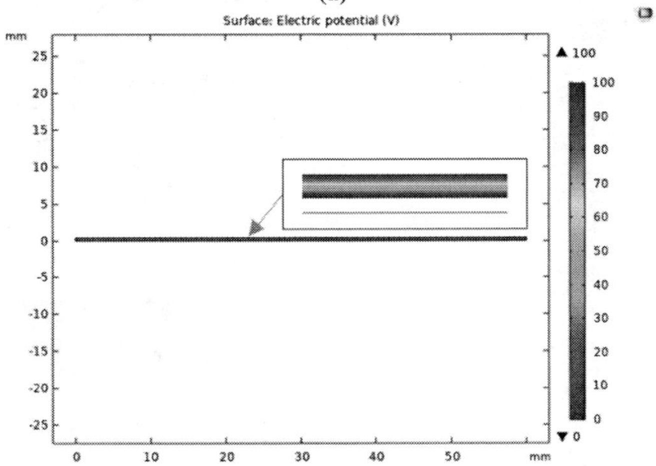

(b)

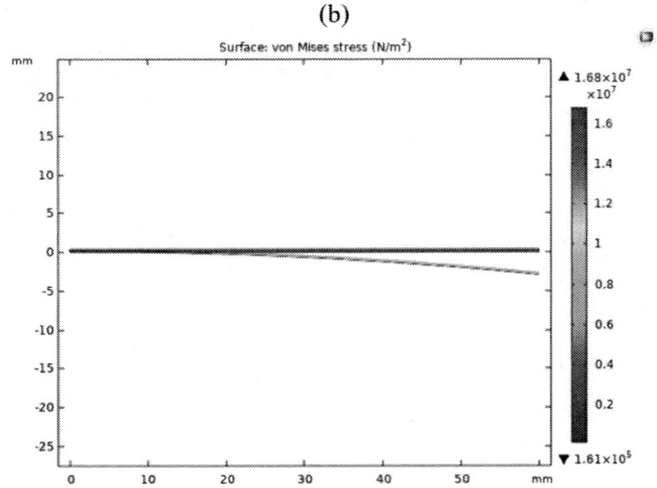

(c)

Figure 2. Electrostatic finite element model.

3.3. Magnetic repulsion force determination

In this section, we discuss the simulation of magnetic repulsion force between two magnets at a defined gap,Δremnant flux density, and vertical elevation difference. Therefore, a finite element model has been developed by using COMSOL Multi-physics to numerically determine the magnetic repulsion force, for a couple of commercial magnets [27] at a gap, Δ = 0.0001 mm from -40 to 40 mm vertical elevation difference.

Figure 3ashows the two magnets atΔ = 0.0001 mm, with a refining mesh design that can be achieved by a small fillet radius at magnets corner. Figure 3b and figure 3cshows the magnetic flux density distribution and close view in figure 3dto see the maximum magnetic flux density zone. The physical study in this model has been a magnetic field (no current).The characteristics of the magnet (material, magnetic flux conservation, and magnetic scalar potential) have been imported to the model from the magnets datasheet.

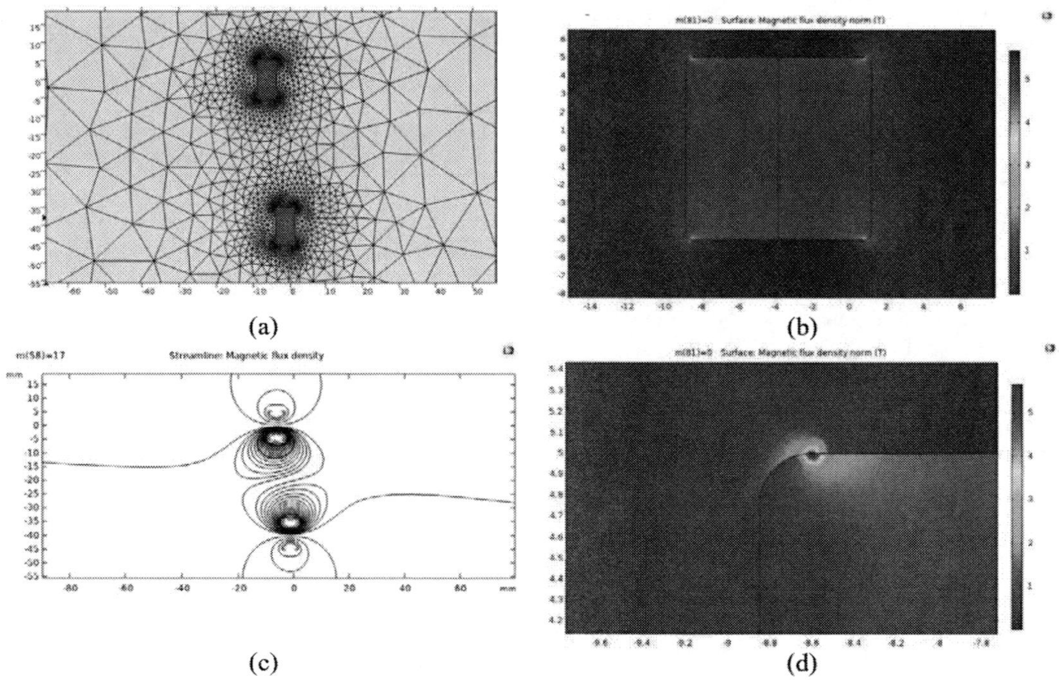

Figure 3. Magnetic repulsion force model.

A parametric sweep analysis has been attached to the magnetic field physical study, in order to estimate horizontal and vertical repulsion force between the magnets at different elevation differences, from -40 to 40 mm with step size equals to 1 mm.

Figure 4shows the magnetic repulsion force in *X* and *Y* directions with relative vertical distance between the two magnets. It is clear from figure 4 and table2,that the numerical results compared favorably with the commercial datasheet of the selected magnets.

IOP Conf. Series: Materials Science and Engineering **973** (2020) 012022 doi:10.1088/1757-899X/973/1/012022

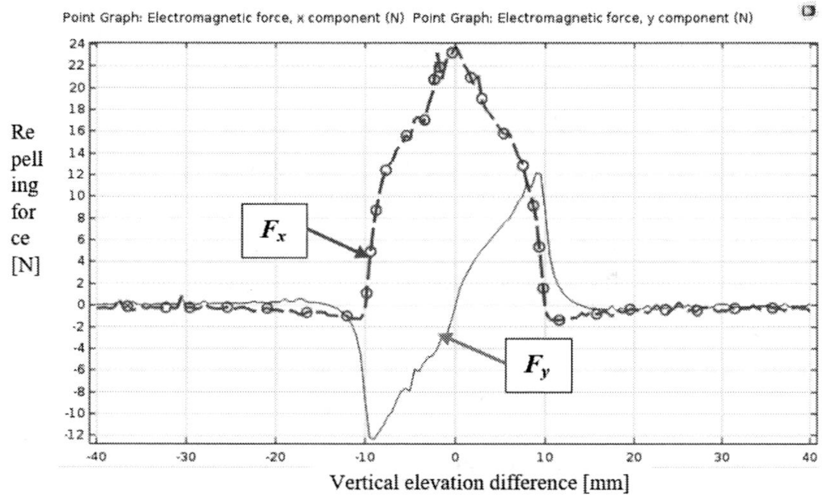

Figure 4. Repulsion force versus elevation difference.

Table 2. Repulsion force comparisons.

Repulsion force	Data Sheet (KJ-N42)	Finite Element Model	Error %
X direction [N]	22	24 at zero relative distance	≈8
Y direction [N]	11	12.1 at 9.5 mm relative distance	≈9

3.4. Mechanical characteristics of the piezo beam

In order to calculate natural frequencies and equivalent damping coefficient of the piezoelectric beam with tip magnet (= 5 g), the numerical models by COMSOL multi-physics have been developed, namely Eigen-frequency and base excitation analysis. Pre-described displacement with 2 mm amplitude and 27 Hz($\approx\omega_{n1}$) frequency, subjected to the beam root, then measure tip displacements as shown in figure 5 and from the dynamic relations in the case of a base exciting beam, can estimating the equivalent damping coefficient which equals0.025 N.s/m.

Table 3, shows the first three mode shapes of the piezoelectric cantilever with the tip magnet.

Table 3. Eigen-frequency analysis.

First mode shape	Second mode shape	Third mode shape

ω_{n1}=28 Hz	ω_{n2}=381 Hz	ω_{n3} =1220 Hz

IOP Conf. Series: Materials Science and Engineering **973** (2020) 012022 doi:10.1088/1757-899X/973/1/012022

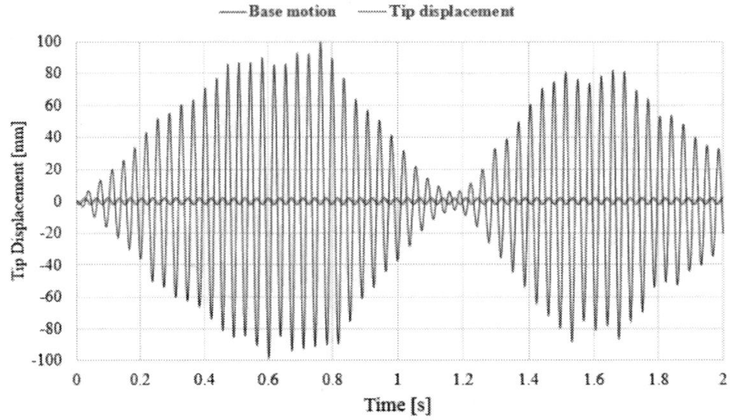

Figure 5. Tip displacement response (beating phenomenon).

As a sample figure 6a and figure 6bshow the displacement field in *Y*-direction along the piezoelectric beam at time = 0.82and 1.046 s, respectively.

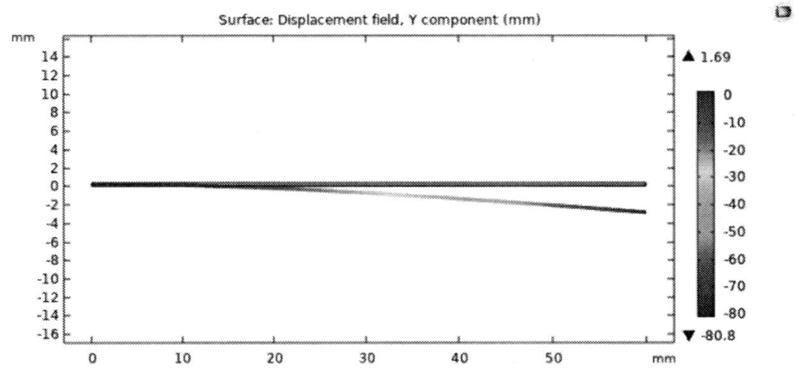

(a) t = 0.82 s

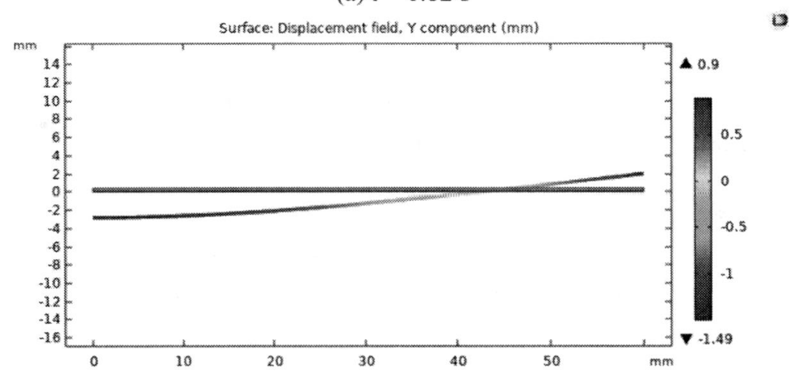

(b) t = 1.046 s

Figure 6. Displacement field distribution (base excited beam analysis).

IOP Conf. Series: Materials Science and Engineering **973** (2020) 012022 doi:10.1088/1757-899X/973/1/012022

4. Electromechanical response

Afterward, all required parameters to solve equation (1) have been numerically developed and examined by using commercial products. So, the governing ordinary differential equations were numerically solved (*ode15s* command on MATLAB is used in this study) at different load resistances. Figure 7 and figure 8 show the displacement and velocity responses, respectively, at 8 Hz excitation frequency (for low frequency applications), excitation amplitude equals 5 mm, load resistance = 100 kΩ and gap (Δ = 20 mm).

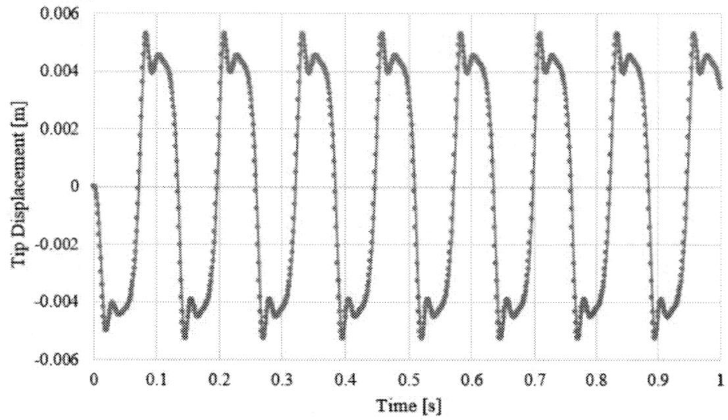

Figure 7. Displacement time history.

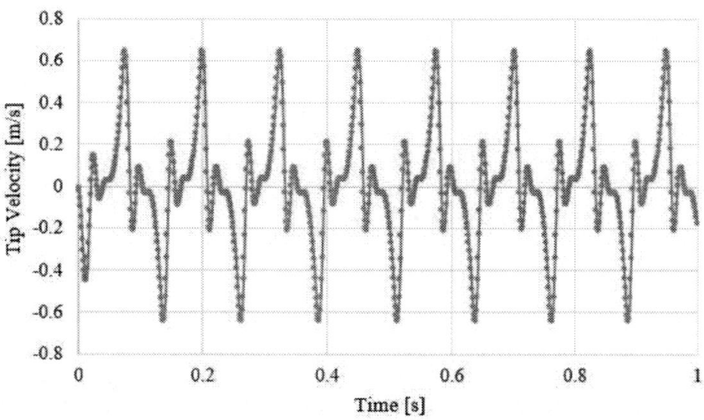

Figure 8. Velocity time history.

Figure 9 and figure 10 shows a sample of the time history plot for the output voltage and the output voltage versus load resistance, respectively, at the same conditions for figure 7 and figure 8.

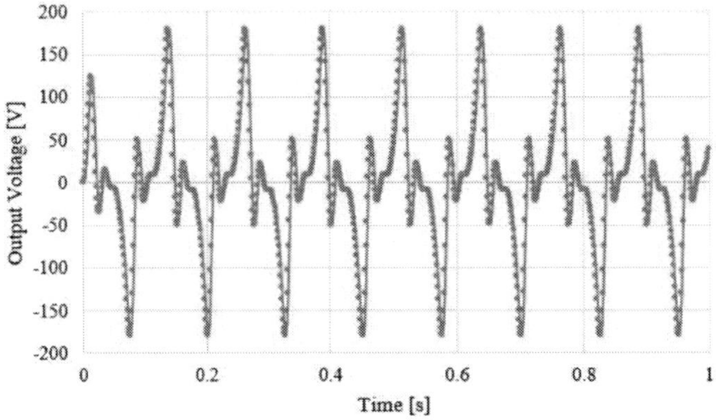

Figure 9. Output voltage time history.

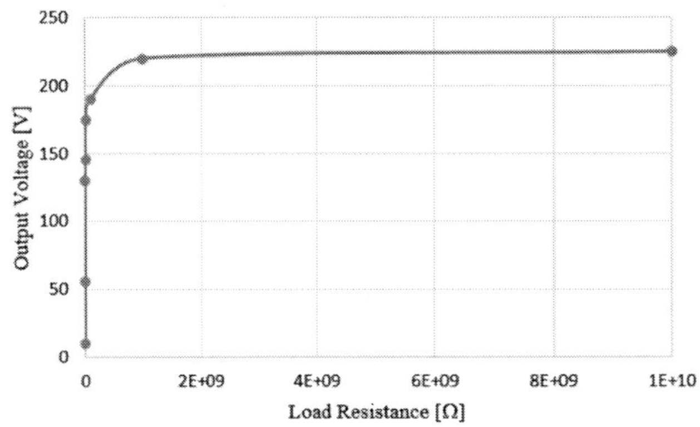

Figure 10. Output voltage versus load resistance.

5. Conclusions

A bistable piezoelectric energy harvester was comprehensively simulated in this paper. First, two previous studies were linked together in order to obtain the equivalent values for stiffness, electromechanical coupling and capacitance coefficients of the piezoelectric cantilever beam. Second, numerical models were developed to simulate the electromechanical, electrostatic, magnetic repulsion force, Eigen-frequency analysis and dynamic time responses for the proposed harvester. In order to examine each numerical module used in this work, a commercial unimorph piezoelectric cantilever beam and nickel-plated NdFeB couple of magnets have been selected. Moreover, a comparative analysis has been discussed to compare finite elements results with other values from previous research or commercial datasheet for piezoelectric beam and repelling magnets. Comparing data shows good convergence between numerical models results and previous studies. Finally, depending on finite element models results, electromechanical ordinary differential equations of the proposed bistable piezoelectric harvester have been numerically solved at a wide range of load resistance.

References

[1] Abdelnaby M A and Arafa M 2016 *J. Intell. Mater. Syst. Struct.***27(19)**.
[2] Liu H, Zhang S, Kathiresan R, Kobayashi T and Lee C 2012*Appl. Phys. Lett.***100(22)** pp 2012-2015.
[3] Dinulovic D, Brooks M, Haug M and Petrovic T 2015*Phys. Procedia* **75** pp 1244–1251.
[4] Leland E S and Wright P K 2006*Smart Mater. Struct.***15(5)** pp 1413–1420,.

IOP Conf. Series: Materials Science and Engineering **973** (2020) 012022 doi:10.1088/1757-899X/973/1/012022

[5] Masana R and Daqaq M F 2011*J. Vib. Acoust. Trans. ASME***133(1)**.
[6] Sazonov E, Pillay P, Li H, and Curry D2009*IEEE Sens. J.***9 (11)** pp 1422–1429.
[7] Rhimi M and Lajnef N2012 *J. Intell. Mater. Syst. Struct.***23(15)** pp 1759–1770.
[8] Chew Z J, Ruan T and Zhu M 2016*Procedia Eng.***168** pp 1717–1720.
[9] Bonessio N, Zappi P, Benzoni G, Rosing T S, and Lomiento G 2016 *Open Constr. Build. Technol. J.***10 (1)** pp 136–149.
[10] Cao S and Li J 2017*Adv. Mech. Eng.* **9(4)** pp 1–14.
[11] Harne R L, Thota M, and Wang K W2013 *Smart Mater. Struct.***22(12)**.
[12] Harne R L and Wang K W, 2013*Smart Mater. Struct.***22 (2)**.
[13] Pellegrini S P, Tolou N, Schenk M, and Herder J L 2013 *J. Intell. Mater. Syst. Struct.***24(11)** pp 1303–1312.
[14] Andò B, Baglio S, Maiorca F, Trigonaa C 2012*Procedia Eng.***47(0)** pp 1061–1064.
[15] Abdelnaby M A and Arafa M 2016*Active and Passive Smart Structures and Integrated Systems Proc. of SPIE* **9799, 97990M.**
[16] Stanton S C, McGehee C C, and Mann B P 2010*Phys. D Nonlinear Phenom.***239 (10)**640–653.
[17] Zhu D and Beeby S P 2013 *J. Phys. Conf. Ser.***476(1)**.
[18] Singh K A, Kumar R, and Weber R J 2015*IEEE Trans. Power Electron.***30(12)** pp 6763–6774.
[19] Harris P, Arafa M, Litak G, Bowen C R, and Iwaniec J 2017*Eur. Phys. J. B***90(1)**.
[20] Zhou S, Cao J, Erturk A, and Lin J 2013*Appl. Phys. Lett.* **102(17)** pp 3–6.
[21] Stanton S C, McGehee C C, and Mann B P 2009*Appl. Phys. Lett.***95(17)**.
[22] Upadrashta D and Yang Y 2016*journal of applied physic***120** 054504-1–054504-9.
[23] Erturk A and Inman D J 2011 *J. Sound Vib.* **330 (10)** pp 2339–2353.
[24] Park J K and Moon W K 2005 *Sensors Actuators, A Phys.***117(1)** **pp** 159–167.
[25] Peng S, Zheng X and Peng Ke 2012*journal of micromechanics and microengineerin***22(6)**DOI. 10.1088/0960-1317/22/6/065005.
[26] https://www.americanpiezo.com/
[27] https://www.first4magnets.com/

IOP Conf. Series: Materials Science and Engineering **973** (2020) 012023 doi:10.1088/1757-899X/973/1/012023

Simulating ultrasonic vibration enhanced FSW process of AA6082 using finite element method

OM Mabrouk[1], M S El-Wazery[2] andM A EL-Desouky[3]
[1,2]Production Engineering &Mechanical Design Department, Faculty of Engineering, Minoufia University, Shebin El-Kom, Egypt
[3]Civil EngineeringDepartment,NileHigh Institute of Engineering and Technology - Mansoura, Egypt
eng_mahmoudsamir@yahoo.com

Abstract. Several modifications have been made on friction stir welding process to overcome some certainlimitations which have been reported. One of these modifications is to use ultrasonic energy as an assistance tool in FSW process. In the present paper, a mathematical model is developed to express the heat generation during ultrasonic vibration enhanced friction stir welding process (UVeFSW). A finite element model is built to perform a transient thermal heat transfer analysis using ANSYS mechanical APDL software package. The temperature contours, temperature distribution and the thermal cycles were predicted using the moving heat source technique. To validate the model, k-type thermocouples were used to measure the temperature and thermal cycles at five locations. The results showed a good agreement between the simulated and experimentally measured results.

1. Introduction

Friction stir welding process (FSW) is considered to be one of the most advances in welding technology. As a solid-state welding technique, FSW has many advantages over fusion welding processes,as it consumes less energy, no fumes or gases are emitted during the process, no fillers or fluxes are needed making the process, energy efficient and environmentally friendly, in addition to low distortions and residual stresses.FSW process has been used successfully to weld a wide range of soft materials such as Al and Mg alloys. It has achieved acceptable level of weld quality and joint properties. Now days, many attempts are made to widen the range of materials that can be welded using FSW to include harder aluminium alloys, high strength, and high melting point materials. Such hard materials necessitate more amount of frictional heat to achieve sufficient degree of material plasticization at the weld zone. This requires low welding speed, high downward force and high spindle torque. Such requirements represent some limitations in friction stir welding of hard materials.

To overcome these limitations, the researchers have proposed to integrate FSW process with an auxiliary energy source to provide additional energy. Thermal and mechanical sources were used to obtain this required energy. The thermal means include laser assisted FSW[1], arc assisted FSW[2] and electrically assisted FSW[3]. On the other hand, the mechanical source includes only ultrasonic vibration assisted FSW.Compared to other auxiliary energy sources, ultrasonic energy has many advantages as it helps in enhancing the material flow as it produces finer grains size, improving the mechanical properties of the joint, consumes less energy, simple and more save than other processes[4]. Ultrasonic vibration can be integrated with FSW by two techniques, the first is to apply the waves to the tool which transmit the vibration to the weld area and it is known as ultrasonic vibration assisted FSW (UAFSW)[5]. The second is to apply the waves directly to the workpiece and it is known as ultrasonic vibration enhanced FSW (UVeFSW) as invented by Lai et al. in 2015[6]. The mechanism of FSW process and its related modifications depends basically on the heat generated during the process. many properties of the resulting joints are affected by the thermal behaviour of the material. one of the most useful approaches to study this behaviour isthe finite element modelling. Several models have been developed to simulate the heat transfer in conventional FSW. Choetal. analysed the heat transfer in FSW using computational fluid dynamics approach[7]. Heetal, predicted the temperature distribution in FSW by developing a thermo-mechanical model[8]. A finite element model was developed by buffa et al. to investigate the influence of process parameters on temperature distribution[9]. On the other hand, there are few researches that investigated the heat transfer in ultrasonic vibration assisted FSW using finite element method. The additive inertial force exerted by ultrasonic vibration of the tool in UAFSW process was considered in the model developed by Lai et al. to analyze the effect of ultrasonic vibration on thermal process[10].A mathematical model was developed by Shi et al. to investigate the effect of ultrasonic vibration on the heat generated during UVeFSW process[11].

There is still a shortage in developing a finite element model to simulate UVeFSW process. The objective of the present work is to develop a clear mathematical model to calculate the heat generation rate and heat flux as well as developing a finite element model to predict the temperature distribution and thermal cycles resulting from UVeFSW process.

2. Mathematical model

The heat generated during FSW process has a significant influence on the success of the process as it affects the weld quality, microstructure and the amount of residual stresses and distortions. In general, there are two main sources for heat generation in FSW process including the frictional heating at the contacting interfaces of the tool and the workpiece in addition to the heat generated due to plastic deformation of the material at the stirring zone.

An expression was introduced by Schmidt et al.to estimate the heat generation rate at the tool/workpiece contact interface, equation (1)[12]:

$$q(r) = [\eta(1-\delta)\tau_c + \delta\mu P_N](\omega r^2) \tag{1}$$

Where, $q(r)$ is heat generation rate, δ is the dimension slip rate, η is the mechanical efficiency, τ_c is the contact shear stress, μ is coefficient of friction, P_N is the normal pressure of the tool over the workpiece surface, ω is the rotational speed of the tool, and r is the distance between the elemental area and the tool axis.

The heat generated at the tool shoulder/workpiece contact surface is obtained by integrating equation(1) over the surface of the tool shoulder as follows in equation(2):

$$Q = \int_{R_P}^{R_s} \int_0^{2\pi} [\eta(1-\delta)\tau_c + \delta\mu P_N](\omega r^2) d\theta dr \tag{2}$$

whereR_s and R_P are the radius of the tool shoulder and the tool pin respectively.Based on von-Mises criteria, the contact shear stress during the UVeFSW process can be determined from the following equations(3), (4), (5):

$$\tau_c = \frac{\sigma_{yo}}{\sqrt{3}} \tag{3}$$

Where, $\boldsymbol{\sigma_{yo}}$ is the yield stress of the material without the influence of ultrasonic vibration. To consider the effect of the ultrasonic vibration on the yield stress of the material a model was proposed by Kelly et.al[14]:

$$\boldsymbol{\sigma_{yo} = K\varepsilon_p^n} \quad (4)$$

$$\sigma_y = \xi K \varepsilon_p^n \quad (5)$$

Where, $\boldsymbol{\xi}$ is the percentage of acoustic softening, the value of $\boldsymbol{\xi}$ ranges from zero to one. At $\boldsymbol{\xi = 1}$ there is no acoustic softening and the material will deform normally. At $\boldsymbol{\xi = 0}$ the acoustic softening is sufficient to reduce the yield stress of the material to zero (a theoretical case).The percentage of acoustic softening can be predicted depending on the ultrasonic vibration parameters through the following equation (6):[14]:

$$\xi = 0.177 + \frac{1.74}{1 + \exp(0.109 + 171\Lambda - 47.0\Lambda P_s)} \quad (6)$$

Where, $\boldsymbol{P_s}$ is the dimensionless pressure and $\boldsymbol{\Lambda}$ is the dimensionless amplitude and is given byequation (7):

$$\Lambda = \frac{\lambda}{t} \quad (7)$$

Where, $\boldsymbol{\lambda}$ is the ultrasonic vibration amplitude, and t is the workpiece thickness. The dimensionless pressure ($\boldsymbol{P_s}$) is given byequation (8):

$$P_s = \frac{P_h}{\sigma_{yo}} = \frac{F_h}{\pi R_h^2 \sigma_{yo}} \quad (8)$$

Where, P_h is the pressure of the sonotrode on the workpiece, F_h is the clamping force of the sonotrode, R_h is the radius of the sonotrode.By substituting in equation (2), the final equation to estimate the heat generation due to friction between the tool shoulder and the plastic deformation of the material can be expressed as followingequation (9):

$$Q = \frac{2}{3}\pi\left[\eta(1-\delta)\frac{\xi\sigma_{yo}}{\sqrt{3}} + \delta\mu P_N\right](R_S^3 - R_P^3)\omega \quad (9)$$

3. Finite element modelling

The heat generated from the tool surface is transferred into the workpiece according to the Fourier's law of heat conduction [15]. The heat transfer equation is as the followingequation (10):

$$\boldsymbol{\rho C_p \frac{\partial T}{\partial t} = \frac{\partial}{\partial x}\left(K_x \frac{\partial T}{\partial x}\right) + \frac{\partial}{\partial y}\left(K_y \frac{\partial T}{\partial y}\right) + \frac{\partial}{\partial z}\left(K_z \frac{\partial T}{\partial z}\right)} \quad (10)$$

Where $\boldsymbol{\rho}$ is the material density, T is the absolute temperature, $\boldsymbol{C_p}$ is the specific heat, $\boldsymbol{K_x, K_y, and\ K_z}$are the heat conductivitiesthat vary with temperature. The transient thermal analysis model is based on a moving heat source which is generated by the frictional heating and the plastic deformation of the material. The material used in this simulation is AA6082-T6. One half of the parts to be welded is modelled assuming symmetry about the weld line to reduce simulation time. The part dimensions are $\boldsymbol{160 \times 100 \times 3\ mm}$. Defining the material properties properly plays a dominant role in obtaining more accurate results. The temperature dependent properties of

IOP Conf. Series: Materials Science and Engineering **973** (2020) 012023 doi:10.1088/1757-899X/973/1/012023

AA6082-T6 are listed in table (1) [16]. Since the material density is little affected by the temperature variation, a constant value of $\mathbf{2710\ kg/m^3}$ is defined.

Table 1. *The material temperature dependent properties of AA6082.*

Temperature (°C)	Conductivity (w/m °C)	Specific heat (j/kg °C)
20	179.1	870
100	194.6	934
200	205.2	967
300	209.8	1002
400	211.6	1066
500	225.3	1167

3.1. Meshing and Boundary conditions

The element type used in the present model to mesh the workpiece is a brick element called SOLID 70. The element has 8-nodes with temperature as a single degree of freedom at each node.The boundary conditions applied for the present thermal model are the conductive and radiative heat losses from all free workpiece surfaces to the surrounding except the bottom surface were conductive heat loss is only presented. The ambient temperature is 25°C. A convection coefficient of 30 W/m^2°C is applied for top and side surfaces and 300 W/m^2°C[15] is applied to the bottom surface of the workpiece as shown in figure 1.

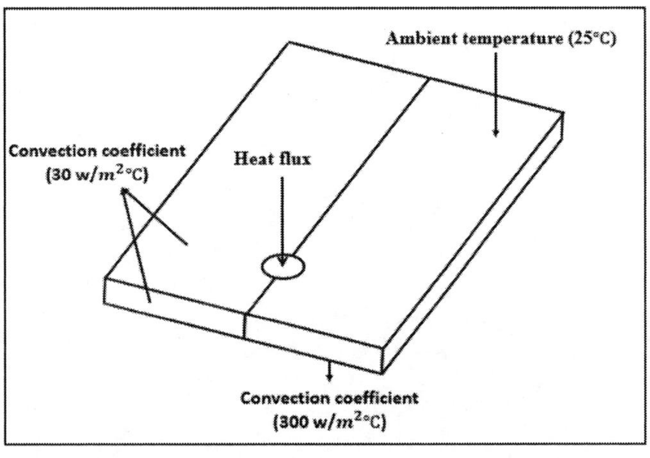

Figure 1. The boundary conditions defined in the model.

3.2. Heat flux

In the present thermal model, the heat generated during the UVeFSW process is simulated as a moving thermal heat flux, which is applied as a surface load. The heat flux is calculated as the following:

$$heat\ flux = \frac{heat\ input(Q)}{Area} \tag{11}$$

The mechanical efficiency ($\boldsymbol{\eta}$) is taken equal to 0.9 according to Awang[17], the dimension-less slip rate ($\boldsymbol{\delta}$) calculations is discussed in details in the study made by Hamilton [18]. the yield stress of the material approximately approaches $\mathbf{60\ MPa}$at$\mathbf{400}$ °C[19].The coefficient of friction $(\mu)$is taken to be 0.4.

$\boldsymbol{axial\ force\ (F) = 3000\ N}$

$\boldsymbol{P_N = \frac{F}{A}}$where: A is the area

$\boldsymbol{R_s = 7\ mm}$and$\boldsymbol{R_p = 2.5\ mm}$

$\boldsymbol{A = \pi(R_s^2 - R_P^2)}$

$\boldsymbol{\omega = \frac{2\pi N}{60}}$ N: is the tool rotational speed $(\boldsymbol{rpm})$

$heat\ flux = 3.2 \times 10^6$

4. Experimental work

4.1. Material and welding procedure

The material to be welded is AA6082-T6 which is a promising alloy as it has the highest strength value of all 6000 series aluminium alloys. The chemical composition as provided by the supplier is shown in table (2). The specimen dimensions are $(160 \times 100 \times 3\ mm)$. A conventional milling machine was used to perform the welding operation. The process parameters utilized were a rotational speed of 800 rpm, a welding speed of 80 mm/min, tilt angle of 2° and plunge depth of 0.15 mm. an FSW tool of 14 mm shoulder diameter, a pin diameter of 5 mm and a pin length of 2.7 mm.

An ultrasonic processor (model up 400S) manufactured by (Hielscherultrasonics GmbH) was used to obtain the required ultrasonic vibration. Ultrasonic vibration waves of a frequency of 20 KHz, a power of 85 watt and amplitude of 40 μmwere employed. The ultrasonic processor was attached to the head of the milling machine via a suitable attachment. so that, the sonotrode can move along the welding line ahead of the FSW tool by a distance of 30 mm and with inclination angle between the axis of the sonotrode and the workpiece surface of 60°to keep it away from the tool. Thereby, the ultrasonic vibration waves can be transmitted directly into the localized area of the workpiece without any loss in the transmitted energy compared to other methods for transmitting the ultrasonic vibrations as shown in figure 2.

Figure 2. Experimental setups of ultrasonic vibration friction stir welding process.

Table 2. Chemical composition of AA6082

Si	Fe	Cu	Mn	Mg	Cr	Ni	Ti	Al
0.7-1.3	0.5	0.1	0.4-1.0	0.6-1.2	0.25	0.2	0.1	REM

4.2. Temperature measurement

The temperature was measured during the welding process on the advancing side using K-type thermocouple at five locations (P1, P2, P3, P4. and P5) at a depth of 2 mm from the top surface as shown in figure (3). In order to monitor the temperature with time an Arduino circuit consisting of

IOP Conf. Series: Materials Science and Engineering **973** (2020) 012023 doi:10.1088/1757-899X/973/1/012023

Arduino board mega 2560, module MAX 6675 to amplify the signal and the temperature sensor was utilized.

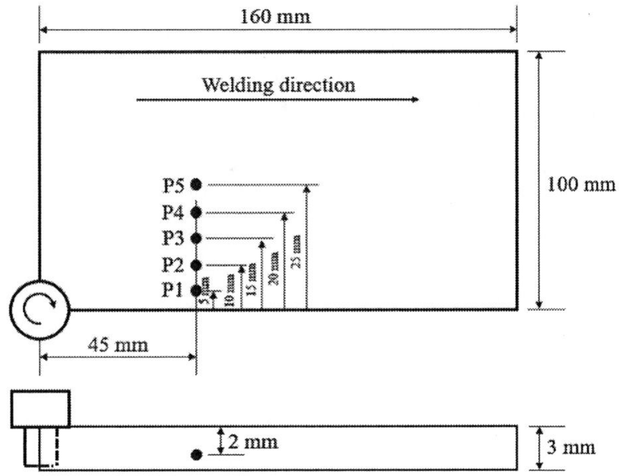

Figure 3. Locations of thermocouples in the workpiece.

5. Results and discussion

As building the finite element model and applying various loads such as calculated heat flux and convection coefficient has been accomplished, the transient heat transfer analysis is performed. After the solution of the model has been completed, three types of data can be obtained which are necessary from point of view.

5.1. Temperature contours

The moving heat source is simulated successfully in the present model as shown in figure 4. The red area represents the location of the moving heat source (location of the FSW tool). The temperature distribution over the welded workpiece is represented. At the first step a peak temperature of 388 C$^{\circ}$ is obtained which is increased to 396 C$^{\circ}$ in the second step. This increase may be attributed to the pre-heating effect of the former step on the later step.

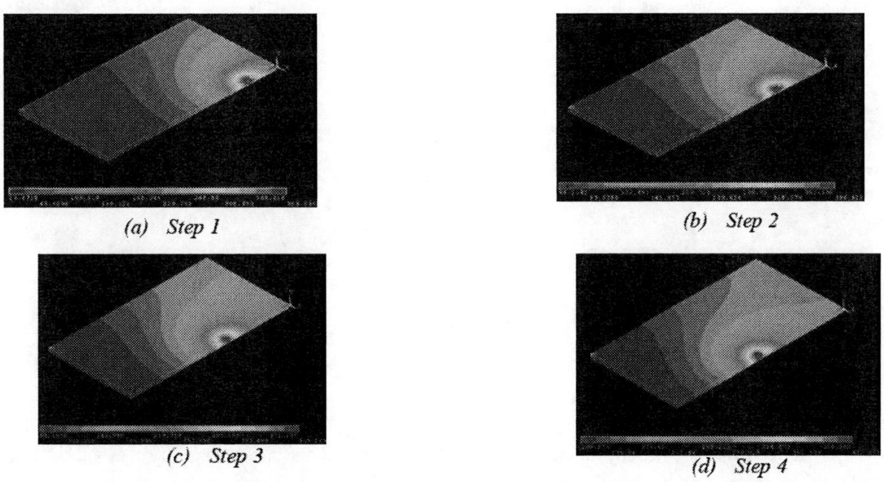

(a) Step 1 *(b) Step 2* *(c) Step 3* *(d) Step 4*

Figure 4. Temperature contours at various steps.

5.2. Peak temperatures at different locations

Simulated and experimentally measured values of the peak temperatures at five different locations from the weld center line (P1=5, P2=10, P3=15, P4=20 and P5=25 mm) are represented at figure 5 to compare the results and hence determining the model accuracy and percentage error (error %) between simulated and experimentally measured data. As shown from the figure, the error% at P1 is about 3.27% and reaches the maximum value at P3 where error% is 6.14%. It is obvious from the results that the predicted results are in good agreement with the experimentally measured ones.

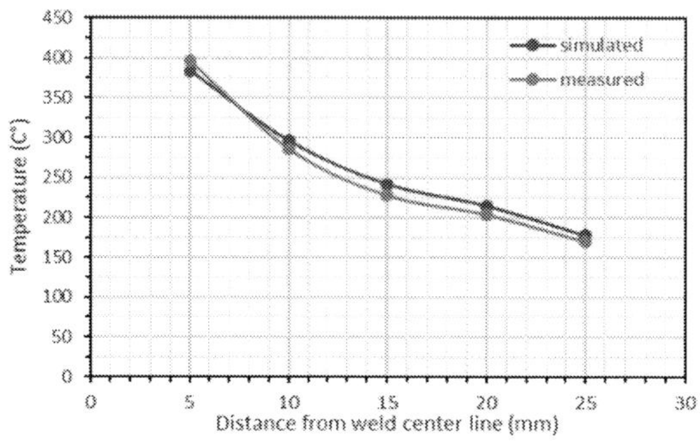

Figure 5. *Comparison between simulated and measured peak temperature at each location*

5.3. Thermal cycles

The variation of temperature with time (thermal cycles) has an important effect on the microstructure as well as the mechanical properties of the welded joints thus the thermal cycles at locations P1 and P2 were determined experimentally and predicted by the model. Also, the comparison between them at each location is shown in figure (6a,b). As the seen in figure 6a,b, it was observed that the variations between predicted and measured results are in the allowed limits.

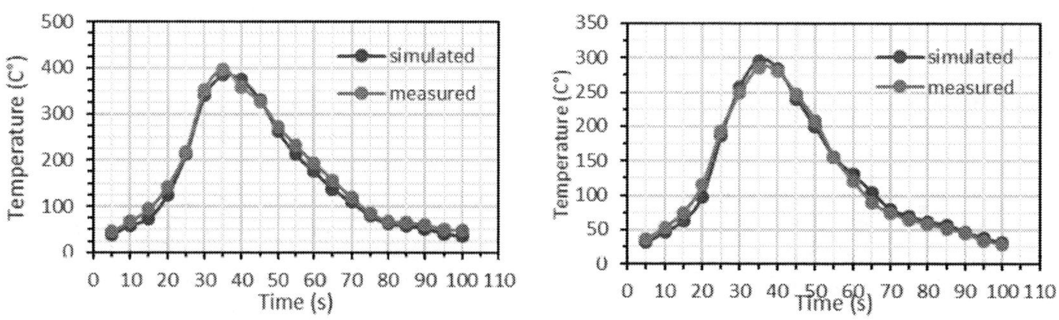

Figure 6.The thermal cycle at locations A) location P1, B) Location P2.

6. Conclusions

In the present study a transient heat transfer analysis was performed using finite element method. The moving heat source technique was applied to simulate the welding process. Temperature contours, temperature distribution and thermal cycles was predicted using ANSYS mechanical APDL software. The model was validated by measuring the temperature and thermal cycles using K-type thermocouple. The comparison between simulated and measured results showed acceptable accuracy of the model.

IOP Conf. Series: Materials Science and Engineering **973** (2020) 012023 doi:10.1088/1757-899X/973/1/012023

References

[1] Campanelli S, Casalino G, Casavola C and Moramarco V 2013 Analysis and comparison of friction stir welding and laser assisted friction stir welding of aluminum alloy. *J. Materials.* **6(12)**pp 5923-5941

[2] Bang H,Bang H, Jeon G, Oh I and Ro C 2012Gas tungsten arc welding assisted hybrid friction stir welding of dissimilar materials Al6061-T6 aluminum alloy and STS304 stainless steel *J. Materials &Design.* **37**pp 48-55

[3] Telmo GS, Rosa MM and Pedro V2014Friction stir welding assisted by electrical joul effect to overcome lack of penteration in alumimum alloys. *Key Engineering Materials***611-612,** pp 763-772

[4] Padhy GK, CS Wu, Gao S 2015 Auxiliary energy assisted friction stir welding–status review.*J. Science and technology of welding and joining***20(8)** pp 631-649

[5] Park K, Kim GY and Ni J 2009Design and analysis of ultrasonic assisted friction stir welding.*ASME International Mechanical Engineering Congress and Exposition.* American Society of Mechanical Engineers Digital Collection IMECE2007-44007 pp 731-737

[6] Liu XC, CS Wu, Rethmeier M and Pittner A2013 Mechanical properties of 2024-T4 aluminium alloy joints in ultrasonic vibration enhanced friction stir welding. *China welding* **22(4)**pp 8-13

[7] Cho HH , Hong ST, JH, Roh HS Choi and Kang SH 2013 Three-dimensional numerical and experimental investigation on friction stir welding processes of ferritic stainless steel. *J. Acta Materialia.* **61(7)** pp 2649-2661

[8] He X, F Gu and AJPiMS Ball 2014 A review of numerical analysis of friction stir welding.*J. Progress in Materials Science.* **65** pp 1-66

[9] Buffa G, Fratini L, Schneider M and Merklein M 2013 Micro and macro mechanical characterization of friction stir welded Ti–6Al–4V lap joints through experiments and numerical simulation. *J. Materials Processing and Technology.* **213(12)** pp 2312-2322

[10] Ruilin L, Diqiu H, Luocheng L and Shaoyong Y 2014 A study of the temperature field during ultrasonic-assisted friction-stir welding.*I. J. of Advaned Manufacuring Technology***73(1-4)** pp 321-327

[11] Shi L, Wu CS and Liu XC 2015 Modeling the effects of ultrasonic vibration on friction stir welding.*J. Materials Processing and Technology***222** pp 91-102

[12] Schmidt H, HattelJ and Wert J 2003 An analytical model for the heat generation in friction stir welding. *J. Modelling and Simulation in Materials Science and Engineering.* **12(1)** pp 143-150

[13] Shi L , CS Wu,Padhy GK and Gao S 2016 Numerical simulation of ultrasonic field and its acoustoplastic influence on friction stir welding. *J. Materials &Design.* **104** pp 102-115.

[14] Kelly, Gregory S Advani, Suresh G Gillespie Jr, John W Bogetti and Travis A 2013 A model to characterize acoustic softening during ultrasonic consolidation. *J. Materials Processing Technology.* **213(11)**pp 1835-1845

[15] Kıral B, Tabanoğlu M, and Serindağ H 2013 Finite element modeling of friction stir welding in aluminum alloys joint. *J. Matehematical and Computational Applications.* **18(2)** pp 122-131

[16] Peel M. J,Steuwer A,Withers P J,Dickerson T, ShiQ andShercliff H 2006 Dissimilar friction stir welds in AA5083-AA6082. Part I: process parameter effects on thermal history and weld properties. *J. Metallurgical and Materials Transactions A***37(7)** pp 2183-2193

[17] Awang M, Mucino VH,Feng Z andDavid SA 2005 Thermo-mechanical modeling of friction stir spot welding (FSSW) process: use of an explicit adaptive meshing scheme *SAE World Congress Exhibtion*01-1251

[18] Hamilton C,Dymek S, Sommers A 2008 A thermal model of friction stir welding in aluminum alloys. *I. J. Machine Tools and Manufacture* **48(10)** pp 1120-1130

IOP Conf. Series: Materials Science and Engineering **973** (2020) 012023 doi:10.1088/1757-899X/973/1/012023

[19] SummersPT,Matulich RD, Scott WC, Lattimer B, 2012 Post-fire mechanical properties and hardness of 5083 and 6082 aluminium alloys. *ASME International Mechanical Engineering Congress and Exposition. American Society of Mechanical Engineers Digital Collection* IMECE2012-88175 pp 1251-1259

IOP Conf. Series: Materials Science and Engineering **973** (2020) 012024 doi:10.1088/1757-899X/973/1/012024

Passive vibration attenuation: a comparison study

A A Abdelghany[1], M M Hegazy[2] and A Badawy[3]
[1] Egyptian Armed Forces, Cairo, Egypt
[2] Arab Academy for Science, Technology & Maritime Transport, Cairo, Egypt
[3] October University for Modern Sciences and Arts, Cairo, MSA
83ahmedali83@gmail.com

Abstract. A comparison study between two different approaches for passive vibration attenuation of solar array had been presented. A powerful finite element software "Ansys" was used to perform this study. Finite element models of a cantilever rectangular aluminum plate are created in order to model the satellite solar array. Stiffeners and circular patches (passive masses) are used separately in order to attenuate the vibration of the plate. The output results of the mode shapes and the FRF of the plate are investigated in order to figure out the best and adequate vibration attenuation technique for satellite solar array, the results showed that using stiffeners have a better effect on vibration attenuation more than using circular patches. A further investigation for vibration attenuation using stiffener is conducted to study the effect of changing the aspect ratio of stiffener.

1. Introduction

The control of Noise and vibration has a significant importance in nearly all fields and branches of engineering, especially in the aerospace applications the vibration control is regarded as a basic design issue. In space applications many sources of vibration "during launch phase and in-orbit operations such as the maneuver of the satellite, and the adjustment / movement of payload" affects all parts of satellite [1]. One of the main goals and objectives that must be achieved during the process of satellite structure design is to reduce the mass of the satellite structure as maximum as possible in order to increase the payload fraction and reduce the satellite launching cost, and hence some satellite components such as the solar array has a large size and a relatively low weight / stiffness so they are subjected to several sources of vibration [2]. For these issues the understanding of vibration behavior in satellite solar array and the best ways to govern, regulate or deal with all sources of vibration are paid a great attention in order to reduce or eliminate it by altering the design and /or designing a suitable control mechanism.

In general there are different types or methods for vibration attenuation of space structure; as passive, active, and hybrid vibration control techniques. The energy of vibration in passive damping technique is dissipated and / or controlled by a damping element (mass for example) but without feedback capability, while in active damping technique an equal force to the external excitation force is applied in the opposite direction by an actuating element as a piezoelectric actuator, finally the hybrid damping technique combines both passive and active elements for vibration attenuation [3].

The passive vibration attenuation technique is the most suitable technique for space application because it is simple, requires no power and also can be used to suppress a wide range of mechanical vibrations. Taking into consideration that the two main designing issues in passive damping techniques is to maintain the weight of the damping element as minimum as possible and maximizing the vibration damping as maximum as possible [4].

M. Ansari [5], studied the effect of the shape of the passive masses (patches) that are used to attenuate the vibration of light weight structures. It was found that the optimal shape of added masses that gives higher damping is circular. Circular patches gave 10% more damping ratio when compared with square ones.

E. Askar [1], studied the effect of using circular patches as passive masses to attenuate a plate vibration. By optimizing the locations of the circular patches using Ansys workbench to find the optimal locations that give maximum vibration attenuation, it was found that the amplitude of the plate frequency response function (FRF) was minimized by 65% and 81% and the resonant frequencies were maximized by 19% and 7% for the first and second bending modes, respectively.

S. Jafarpour [6], used both conventional and super finite element methods to study the vibration of stiffened plates, both the Mindlin plate and Timoshenko beam theories were used in order to formulate the plate and stiffeners equation of motion respectively. The proposed method helps to model any configuration of plate and stiffeners. It was proven that the fundamental natural frequency value increases as the stiffeners number is increased but at a certain extent the rate of increase in natural frequency decreases with adding more stiffeners because of the effect of added mass, and it was also found that the maximum fundamental frequency is obtained when the orientation angle of stiffener is 80°.

Amar N. [7], studied the free vibration characteristics of stiffened plates. The study concluded that the square plate (a/b = 1) has the maximum fundamental frequency followed by the rectangular long/narrow plates (a/b = 1.5 and 2) also the clamped boundary condition shows its superiority to the simply supported one. It was proven that the increase in the number of stiffeners to a certain value leads to the increase in the fundamental frequency, also the study figured out that when the value of stiffener depth to plate thickness is increased the fundamental frequency increases.

C. KUMAR [8], studied the forced vibration characteristics of a stiffened plate with a various arrangement of stiffeners and different boundary condition, it was proven that the natural frequency value is maximum as the concentrated load position coincides with the location of the stiffener, also the Natural frequencies in the case of biaxial loading condition is higher compared to axial loading condition and The frequency parameters increase by increasing the restraint at the edges according to the boundary condition.

2. Finite element model creation and verification

Using ANSYS workbench package version R15.0, The solar array panel is modeled by an isotropic aluminum rectangular plate with the same characteristics and boundary conditions of the FE model introduced in [1] as shown in both table 1 and figure 1.

Table 1. FE model Specification.

	Property	Value
Mechanical properties of the plate	Material	Aluminum 5056
	Density (kg/m^3)	2660
	Young's modulus E (GPa)	70
	Poisson's ratio	0.33
Plate dimensions	Length (mm)	500
	Width (mm)	200
	Thickness (mm)	1.2
Number of model elements and nodes	Number of elements	4036
	Number of nodes	28962

IOP Conf. Series: Materials Science and Engineering **973** (2020) 012024 doi:10.1088/1757-899X/973/1/012024

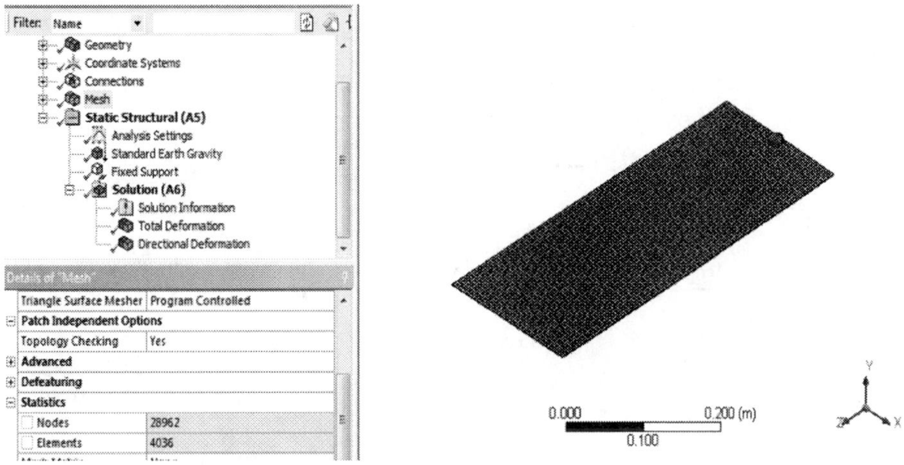

Figure 1. Ansys FE model for the aluminum plate.

A transverse excitation force of 1 N with sine sweep signal varies from 0 to 200 Hz is applied at the plate structure In order to study the plate FRF. The created FE model is verified by comparing the output results from Static structural analysis, Modal analysis and Harmonic response with the published results introduced in [1]. The comparison showed that the resulting values are almost the same as shown in table 2 and table 3.

Table 2. Maximum deformation comparison.

Max. deformation (mm)	[1]	Obtained result
	25.781	25.781

Table 3. First six mode shapes comparison.

Mode shape number	Mode shape type	Natural frequency (Hz)	
		[1]	Obtained result
1st	Bending	3.8	3.7681
2nd	Torsional	21.1	21.091
3rd	Bending	23.9	23.927
4th	Torsional	67.5	67.529
5th	Bending	67.7	67.729
6th	Torsional	125.8	125.79

3. Results and discussion

By using the verified FE model, the vibration attenuation is conducted by adding the aluminum circular patches with optimal diameter and locations as introduced in [1], then instead of patches an aluminum rectangular cross section stiffeners are used to attenuate the vibration. Four different FE models for plate with different stiffeners number and cross sections are created in order to select the best stiffeners arrangement that produces best vibration attenuation. In order to have a reasonable comparison all created

FE models are made from the same material and have almost the same weight. The arrangement of plate models with circular patches and stiffeners is illustrated in the following tables and figures.

Table 4. Plate with aluminum patches/stiffeners characteristics.

Property	Plate with patches	Plate with stiffeners			
		1 stiffener	2 stiffeners	3 stiffeners	3 stiffeners
Mass (kg)	0.36312	0.36326	0.36355	0.36724	0.36355
Material	Aluminum (with the same properties as given in table 1)				

Table 5. Optimal patches diameters and locations.

Patch number	Diameter (mm)	Center location (mm)	
		X	Y
1	44	43.887	150
2	44	64.01	100
3	44	81.397	50
4	44	184.47	150
5	44	144.14	100
6	44	200.8	50

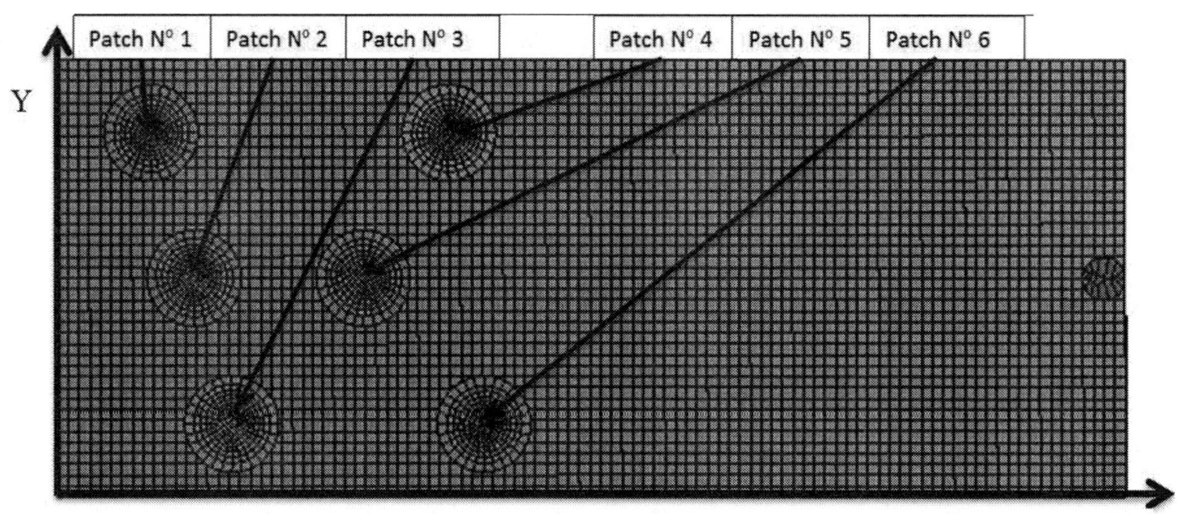

Figure 2. Plate with optimal patch location.

IOP Conf. Series: Materials Science and Engineering **973** (2020) 012024 doi:10.1088/1757-899X/973/1/012024

Table 6. Plate with stiffener arrangement.

N° of stiffeners	Stiffener center location Y [mm]	Stiffener cross section [mm]	Length [mm]
1	100	4 x 5.5	500
2	50 150	1.5 x 2	500
3	50 100 150	1.5 x 1.5	500
3	50 100 150	1 x 2	500

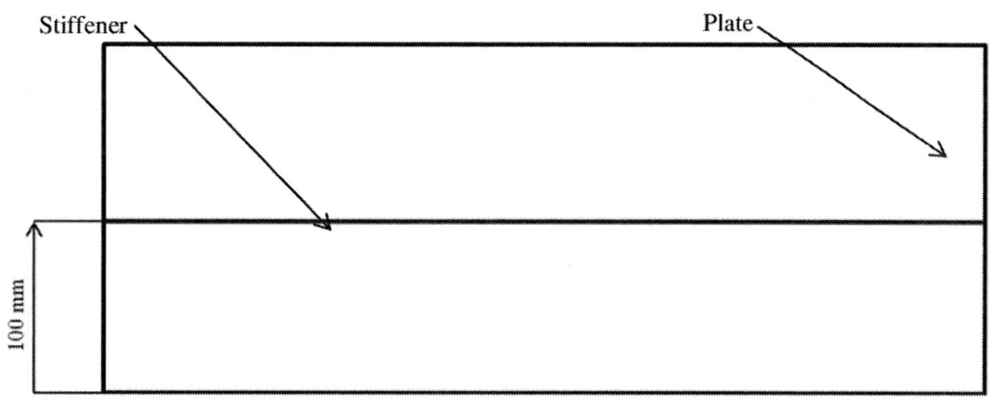

Figure 3. Plate with one stiffener arrangement.

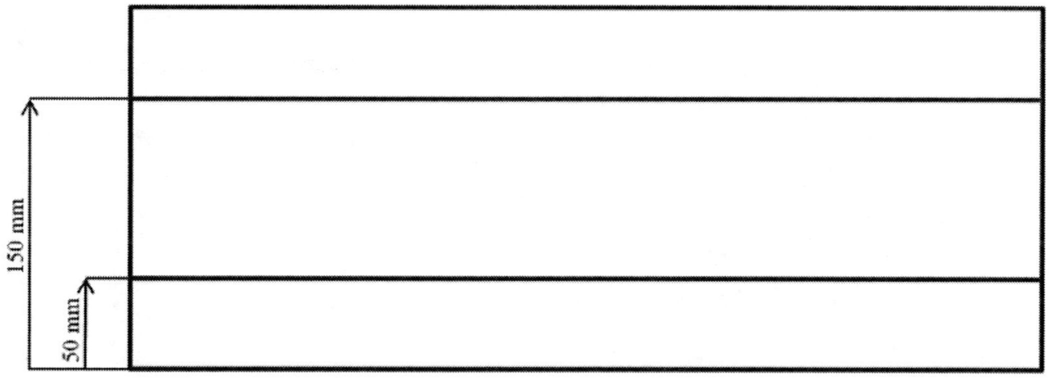

Figure 4. Plate with two stiffeners arrangement.

IOP Conf. Series: Materials Science and Engineering **973** (2020) 012024 doi:10.1088/1757-899X/973/1/012024

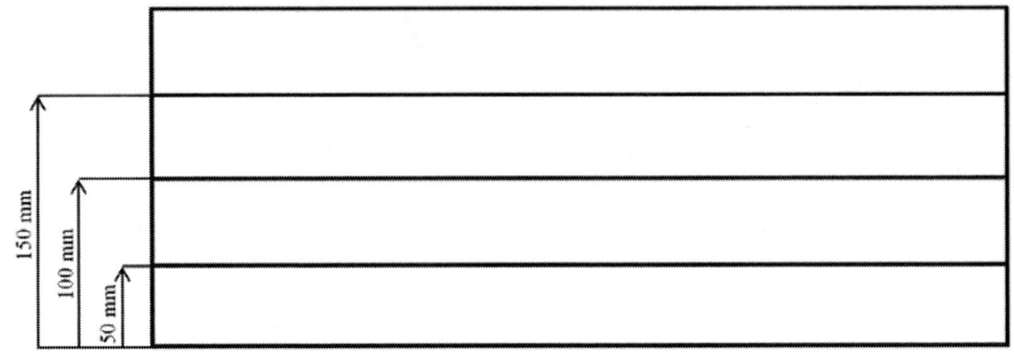

Figure 5. Plate with three stiffeners arrangement.

3.1 analyzing the output results of FE models of plate with stiffeners

By comparing the first six natural frequencies and also the amplitude FRF of the plate at different resonant frequencies, it was found that the best vibration attenuation (minimizing the amplitude FRF and maximizing the resonant frequency of the plate vibration) occurred when using one stiffener as follow.

Table 7. Bare plate / plate with stiffeners first six natural frequencies.

Modes of Vibration	Natural frequency (Hz)				
	Bare plate	1 stiffener	2 stiffeners	3 stiffeners	3 stiffeners
1	3.7681	9.568	5.6833	5.3004	5.699
2	21.091	27.324	22.602	22.649	22.042
3	23.927	59.827	35.874	33.415	35.918
4	67.529	83.496	74.612	73.543	72.359
5	67.729	135.97	99.501	92.925	99.442
6	125.79	146.45	146.1	140.13	140.18

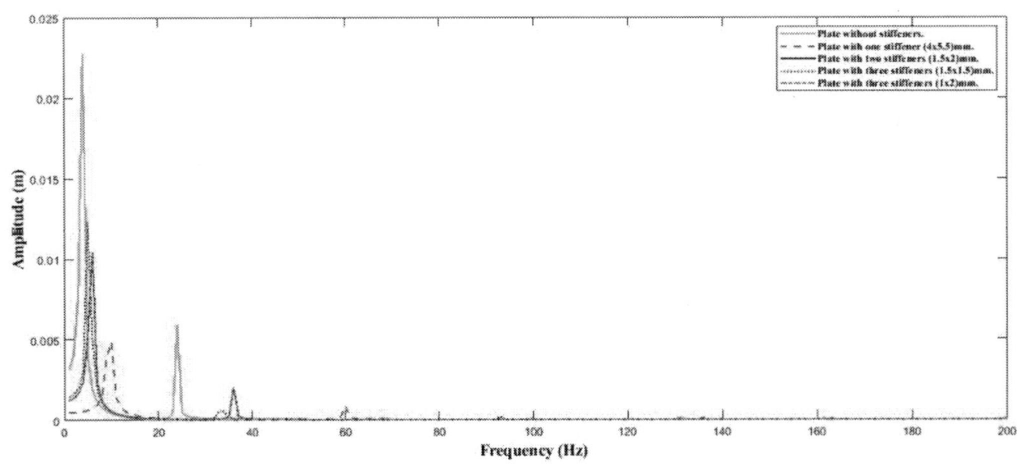

Figure 6. Bare plate / plate with different stiffeners arrangement FRF comparison.

3.2 Comparing the effect of using stiffeners and patches on Vibration Attenuation

Vibration attenuation of mechanical structures at low frequencies is a challenging issue, and it is clear that by analyzing the output results of vibration attenuation using circular patches and stiffener with nearly same volume and mass. the first and third resonant frequencies of stiffened plate is more than the double of that for plate with patches and also the maximum FRF amplitude occurs at the first mode in both cases and its value for stiffened plate is approximately less than the half of the FRF amplitude of the plate with patches.

Table 8. Plate with patches/stiffeners first six natural frequencies.

Modes of Vibration	Natural frequency (Hz)		
	Bare Plate	Plate with patches	Plate with stiffener
1	3.7681	4.3854	9.568
2	21.091	24.512	27.324
3	23.927	24.757	59.827
4	67.529	69.457	83.496
5	67.729	70.667	135.97
6	125.79	131.97	146.45

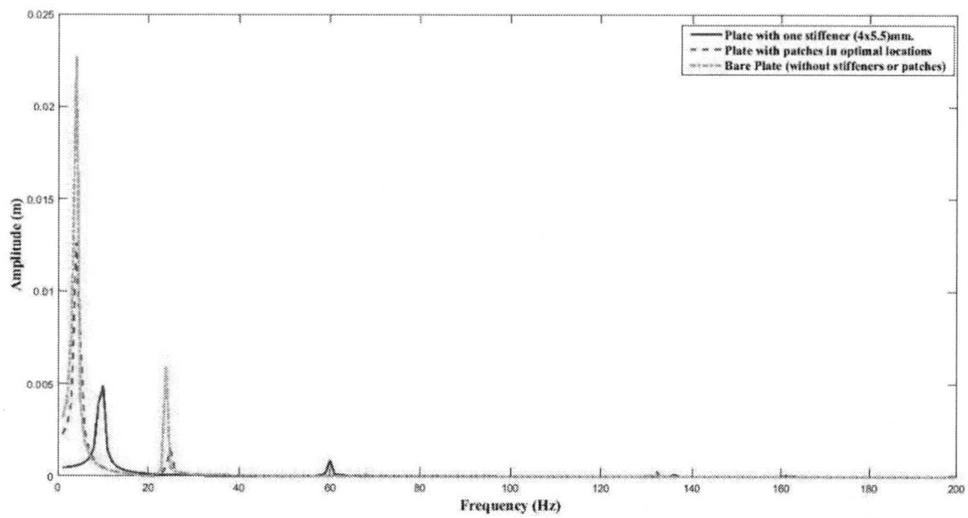

Figure 7. Plate with patches \ stiffener FRF comparison.

3.3 Effect of Stiffener aspect ratio on vibration attenuation

The effect of changing the ratio (a/b) of the stiffener's cross section area is studied by carrying up seven different trials using an aluminum plate with three stiffeners as seen in figure (8), for a reasonable comparison the variation in mass and volume of stiffeners is very small with changing the values of "a" and "b" as illustrated in table (9).

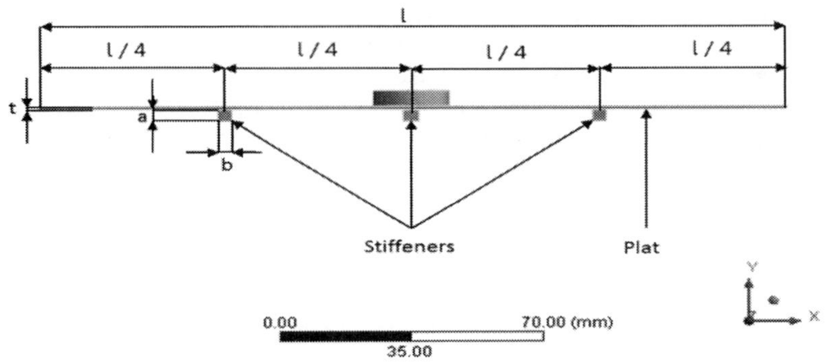

Figure 8. Aluminum plate with three stiffeners arrangement.

3.3.1 Static structural analysis.

Table 9. Maximum total deformation of plate with different aspect ratio stiffeners.

Model N°	Aspect ratio a/b	Plate max. deformation (mm)
1	Bare plate	25.781
2	1/7	16.589
3	2/6	9.0584
4	3/5	6.0329
5	4/4	4.7277
6	5/3	1.589
7	6/2	3.8785
8	7 / 1	4.1298

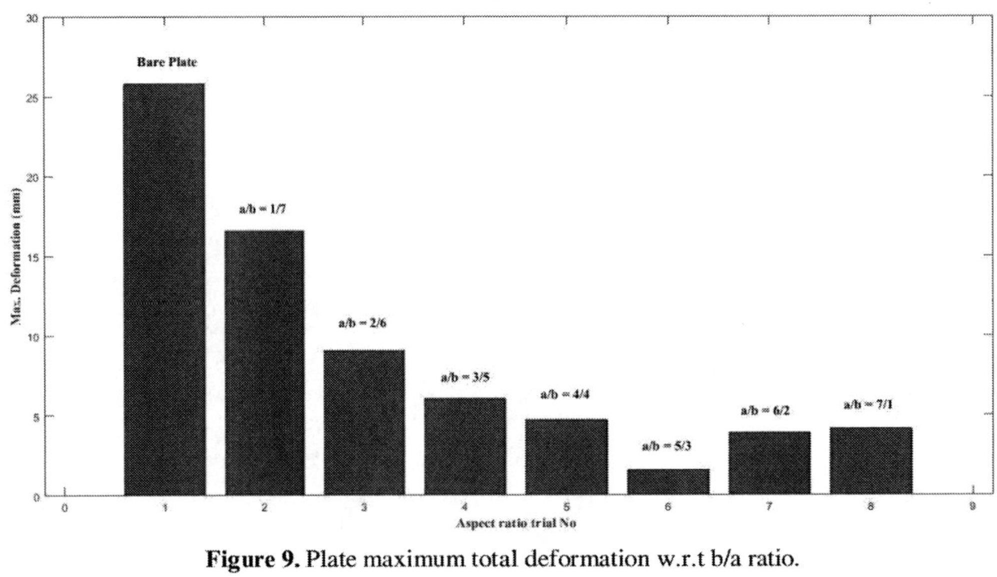

Figure 9. Plate maximum total deformation w.r.t b/a ratio.

By analyzing the values of the plate total maximum deformation at different aspect ratios (a/b), it was found that the value of plate total maximum deformation starts to decrease by increasing the aspect ratio, the rate of total deformation decrease is changing inversely with the increase in the aspect ratio. The best value was achieved when the aspect ratio was (a/b = 5/3) and then the maximum deformation starts to increase by increasing the aspect ratio.

3.3.2 Natural frequencies Comparison

Table 10. Plate natural frequencies at different mode shapes.

Model N°	Aspect ratio a/b	Natural frequency (Hz)					
		Mode 1	Mode 2	Mode 3	Mode 4	Mode 5	Mode 6
1	Bare Plate	3.7681	21.091	23.927	67.529	67.729	125.79
2	1/7	4.7052	24.585	29.794	78.296	83.574	144.73
3	2/6	6.372	28.958	40.754	92.539	113.52	171.35
4	3/5	7.8063	31.788	51.782	102.72	142.65	169.23
5	4/4	8.8126	31.99	61.681	106.39	158.59	181.07
6	5/3	14.742	56.354	87.079	126.69	162.3	213.19
7	6/2	9.7129	25.503	73.425	94.834	160.47	197.95
8	7/1	9.4075	22.785	68.881	85.712	158.46	189.29

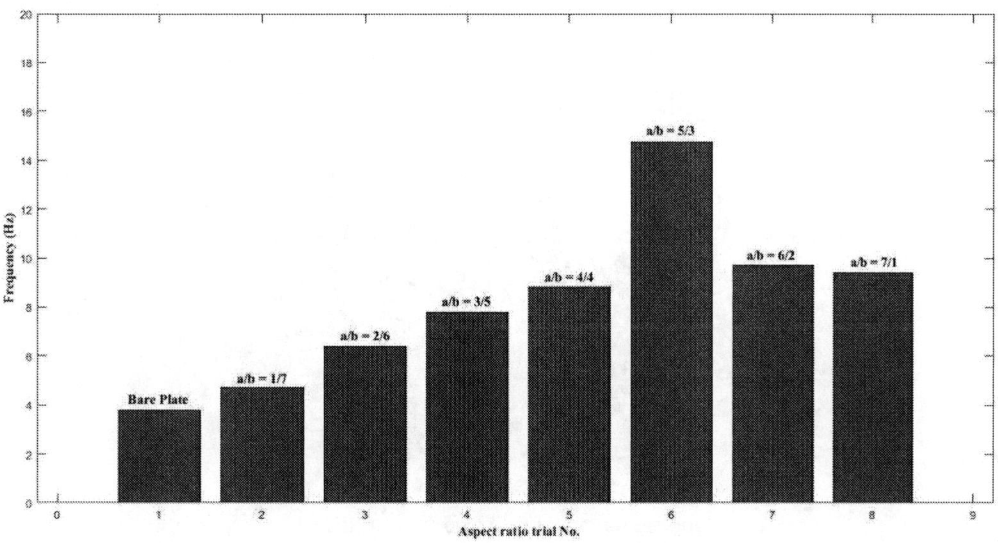

Figure 10. 1st mode shape natural frequency at different aspect ratios.

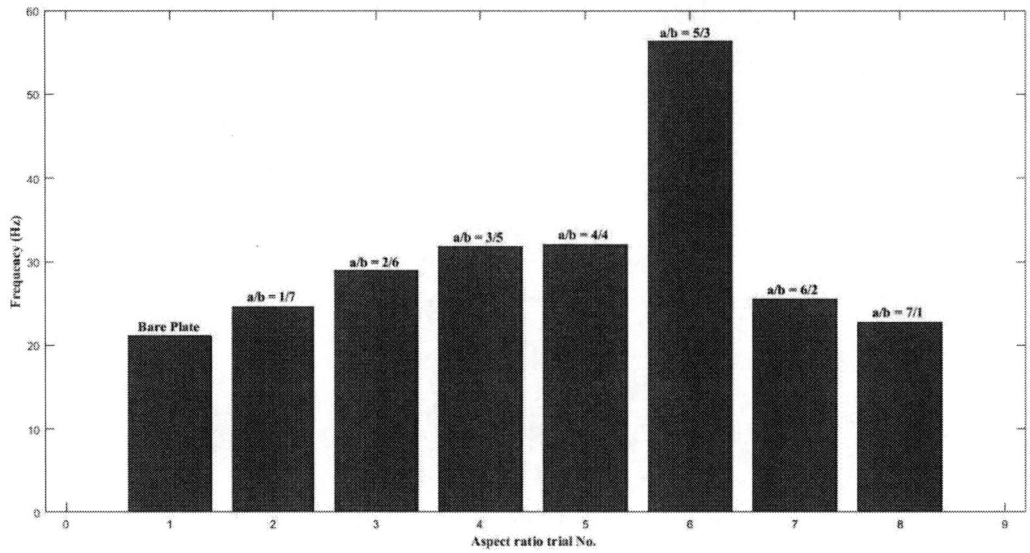

Figure 11. 2nd mode shape natural frequency at different aspect ratios.

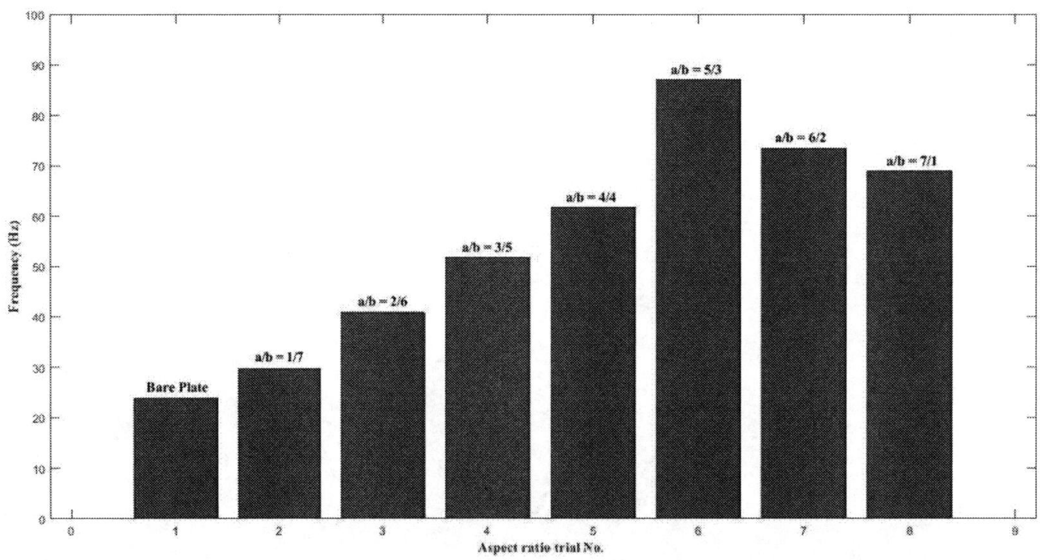

Figure 12. 3rd mode shape natural frequency at different aspect ratios.

By analyzing the values of the plate natural frequencies at the first three modes, it was found that the values of natural frequencies starts to increase by increasing the aspect ratio (a/b), the highest mode natural frequencies are achieved at aspect ratio (a/b = 5/3). The difference between the values of the resonant frequencies with respect to aspect ratio decreases at higher mode shapes. Using aspect ratios (a/b <1) is inefficient for vibration attenuation.

3.3.3 FRF Comparison.

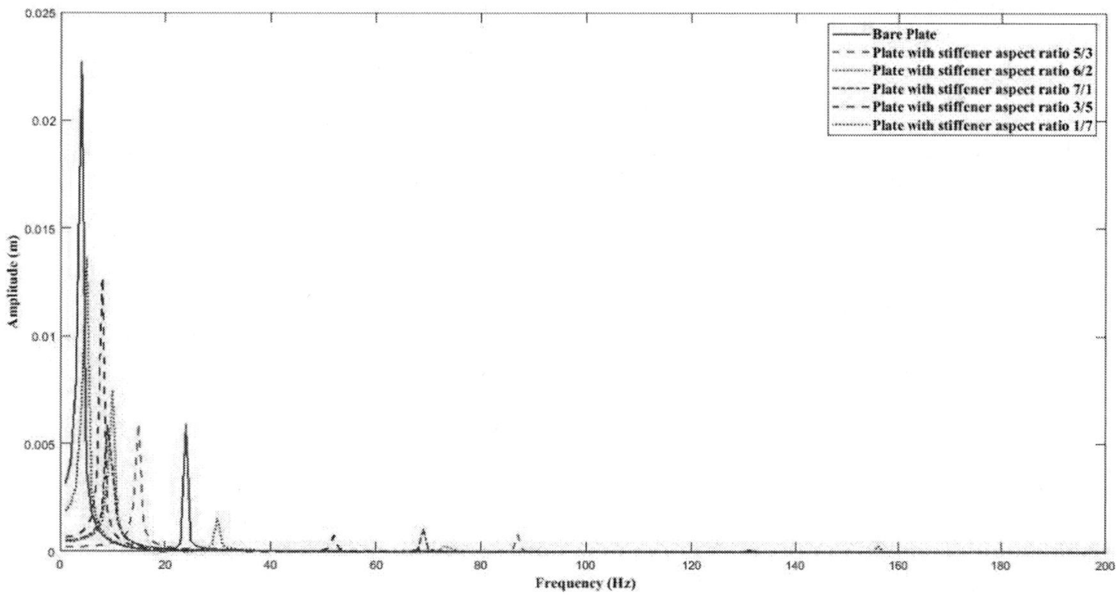

Figure 13. FRF of plate with different b/a ratios.

By analyzing the values of the plate FRF, it was found that the values of maximum FRF amplitude is very high when the aspect ratio (a/b < 1), the smallest FRF amplitude occurs when the aspect ratio (a/b = 5/3).

From the previous results, it is obvious that the vibration characteristics of the plate are improved by increasing the ratio (a/b) of the stiffener until a certain extent although keeping the stiffener's.

4. Conclusion

It can be concluded from simulations that using stiffeners is the best solution for passive vibration attenuation when compared with circular patches, the first and third resonant frequencies of stiffened plate is more than the double of that for plate with patches, also the FRF amplitude of the stiffened plate is approximately less than the half of the FRF amplitude of the plate with patches. The stiffener aspect ratio (a/b) has a great effect on vibration attenuation. The highest natural frequencies and smallest FRF amplitudes were achieved at aspect ratio (a/b = 5/3) and it is obvious that for aspect ratios (a/b <1) the vibration attenuation is not efficient.

References

[1] Emad A 2017 *Passive Vibration Attenuation Analysis and Optimization of Satellite Solar Cell Panels.* (Cairo Egypt: Military Technical College PH. D. Dissertation)

[2] Venkata T and Seetha P 2013 Embedded computer based active vibration control system for vibration reduction of flexible structures. *Journal of Computer Science* **9**

[3] Hongli J, Jinhao Q and Pinqi X 2010 *Semi-active Vibration Control Based on Switched Piezoelectric Transducers*, ed Mickaël L (Nanjing University of Aeronautics and Astronautics, China)

[4] Aglietti1 G Aglietti G, Langley R, Rogers E and Gabriel S 2004 Model Building and Verification for Active Control of Microvibrations with Probabilistic Assessment of the Effects of Uncertainties. *J. Mechanical Engineering Science* **218** part c

[5] Masoud A 2013 Optimal Vibration Control in Structures using Level set Technique. *A thesis of Doctor of Philosophy, Waterloo, Ontario, Canada*

[6] Shahed H, Mohammed K, and Saeed A 2012 Vibration analysis of stiffened plates using finite element method. *Latin American Journal of Solids and Structures* **9**

[7] Amar N, Laren S and Prasant T 2018 Free vibration characteristics of stiffened plates. *International Journal of Advanced Structural Engineering* **10**, pp 153–167

[8] Chetak K 2018 Free Vibration Analysis of Rectangular Plate with Partial Stiffener Subjected to Static Uniform Loading. *A thesis of Master of technology, NIT, Jamshedpur*

IOP Conf. Series: Materials Science and Engineering **973** (2020) 012025 doi:10.1088/1757-899X/973/1/012025

Phase stability of mechanically alloyed Ti-Fe-Al alloys

S A Ibrahim[1], M. M Seleman[1], H M Ahmed[1] and A E Hannora[2]

1 Metallurgical and Materials Engineering Department, Faculty of Petroleum and Mining Engineering, Suez University, 43521 Suez, Egypt
2 Department of Science and Mathematical Engineering, Faculty of Petroleum and Mining Engineering, Suez University, Suez 43521, Egypt
E-mail: ahmed.hannora@suezuniv.edu.eg

Abstract Iron alloying element addition to pure titanium can be used to enhance strengthening as well as β-phase stabilizer. However, the outcome depends on the processing route to some extent. Mechanical alloy technique is adapted in this work to produce titanium alloys with 7wt% iron content and 1wt% Aluminum. The obtained nano-size particles have two phases α and β structures and the formation of the second β-phase is mainly enhanced by the milling force. Heating the produced grain compacted powder to a temperature above the beta phase transus temperature followed by quenching to ambient temperature resulted in α-martensite partial transformation. This result can indicate that the obtained β-phase for the present alloys is not fully stabilized by the iron additions.

1. Introduction

The Ti-6Al-4V (in wt.%) is one of the most widely used alloys for aircraft high performance automobiles and biomedical prosthesis. This alloy has a dual phase microstructure α and β at room temperature [1]. The aluminum addition to this Ti-base alloy is intended to stabilize the hcp α-phase and increase the strength whiles the vanadium to introduce a second β-phase to the microstructure and enhance both ductility and toughness. However, for medical applications vanadium is cytotoxic and it's also an expensive element [2-6]. Also, titanium alloys such as Ti-6Al-7Nb, Ti-5Al-2.5Fe and Ti-6Al-4V still contained Al which has been suggested to cause osteolysis and neural disorders such as Alzheimer's disease [7, 8].

Recently Ti-based alloys with up to 7wt.% Al and a small addition of Fe that doesn't exceed the 2wt.%have been developed with the aim of reducing the cost and maintaining mechanical properties and biocompatibility to a level close to Ti-6Al-4V alloy[9]. Substitution of vanadium in Ti-6Al-4V with iron and niobium did not significantly affect the electrochemical behavior[10]. The amount of formed β-phase in the microstructure relative to α phase depends not only on alloy composition but also on the processing conditions and subsequent treatment. Furthermore, the volume fraction of β phase has been reported to increase with strain during superplastic deformation, where stress induced transformation take place[11]. The deformation behavior of β-Ti alloys is shown to be strongly dependent on the degree of the β-phase stability [12]. The deformation mechanism changes from slip deformation to deformation twinning and/or martensitic transformation with decreasing the β phase stability. The stress-induced martensitic transformation $\beta \rightarrow \alpha''$(orthorhombic) is not stabilizing alloying elements dependent but also stress level can have a determining role [13].

2. Material and techniques

Ti-Fe-Al alloys were prepared from the constituent elements Ti, Fe, and Al with purity grades higher than 99.9%. In the mechanical alloying process, the required mixture and hardened stainless steel balls were loaded together into a water-cooled stainless steel pot under argon atmosphere at speed of 500 rpm with 1:20 powder to ball ratio. The process was stopped almost each one hour to keep the milling temperature close to room temperature. The green compacted mechanically alloyed powder particles have been solution treated at 950 and 1100°C under argon atmosphere and followed by water quenching to examine structure thermal stability and sintering tendency.

XRD analysis was performed using a Siemens D5000 powder diffractometer equipped with Cu Ka radiation (wavelength = 0.15406 nm) with a nickel filter at 40 kV and 30 mA. Thermal analysis was performed in the temperature range from 25 to 1100°C using a AT instrument SDT Q600, with heating rate 10 K/min. The microstructure features of the samples have been investigated using a Field-Emission Scanning Electron Microscopy Quanta FEG 250and Transmission Electron Microscopy JEOL TEM-2100. Sample composition was analyzed by energy dispersive X-ray spectroscopy (EDX).

3. Results and discussion

Mechanical alloying is a solid-state powder processing technique involving repeated welding, fracturing, and rewelding of powder particles. In this process, the mixture of microcrystalline pure Ti, Fe, and Al elements, with alloy composition of Ti-7Fe-1Al weight percent were milled as a function of mechanical alloying time up to 12 hours. The typical XRD pattern of the mixture before and after milling time is shown in figure 1. The original shape lines of all elements gradually became broader and their intensity decreases with milling time. The partial formation of Fe and Al solid solution in the α-Ti matrix can be deduced from such decreases in intensity of X-ray peaks for both Fe and Al. Since the powder particles are subjected to severe plastic deformation, high density of defects generated and subsequently a distortion in the formed solid solution crystal structure are to be expected.

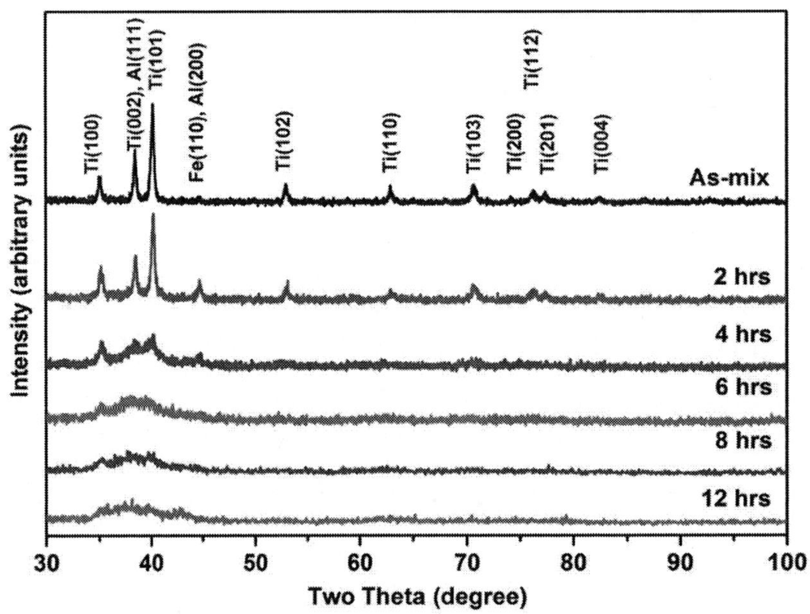

Figure 1. XRD patterns of Ti-7Fe-1Al up to 12 hours of ball milling.

The original XRD patterns contains mainly the α-Ti phase which completely matched with the JCPDS-ICDD standard cards number (44-1294), in additions to two small in intensity peaks for both

Al (65-2869) and/or Fe (01-1267). After 4 hours of milling, these two extra lines from Fe and Al disappear which can be attributed to the deformation process, the refining of the microstructure and/or to the possible formation of a supersaturated solid solution in the α-Ti matrix. Peak broadening continues with milling time representing the refinement in the particles structure and an increase of the internal strain. The line (101) of alpha phase has been shifted to lower angle values with milling time and resulted in differences in $\Delta d = 0.0442$nm decrease after 4h milling. This might suggest an interface orientation relation between α and β-phase mainly α-(101) and β-(110).

During milling process, both iron and aluminum penetrate in Ti simultaneously leading to increases in Fe/Al concentration ratio in newly formed phase as suggested by the different elements mapping for their distribution, figure 2. The Al-element appears almost homogeneously distributed in the whole area, figure 2b, however iron distribution seems to be high concentrated in the lighter zone of the iron map, figure 2d, the higher iron concentration zone corresponding to the β-phase and a rough estimation of the volume fraction of this phase particles to be around ≈ 10% at this stage of milling. Therefore, such observation confirms the β-phase presence in addition to α-phase matrix. This also suggested that the nano-particles β-phase originated from the supersaturated Al/Fe solid solution in titanium.

The degree of crystallinity to amorphization formed in the structure can only be conformed using TEM investigation. TEM analysis of mechanically alloyed powder indicated the presences of nano-crystalline phase. As shown in figure 3, the diffraction pattern indicates the presence of randomly oriented nano-crystal. The pattern also consists of rings sampling all possible planes. Further, the micrographs confirmed the nano-particles microstructure and the spot arranged in circles suggests the presence of high angle-boundaries in ultra-fine structure. The diffuse characteristic of the spots is a strong indication of the lattice distortion associated with milling stress. The destabilization of the crystalline phase is thought to occur during mechanical alloying by increasing the free energy due to the accumulation of structural defects such as vacancies, dislocations, and grain boundaries. The continuous refinement in grain size is therefore increasing the boundaries area and lattice expansion could also contribute to increasing the free energy of the deformed phase.

A differential thermal analysis DTA test was carried out for Ti-7Fe-1Al mechanically alloyed powder sample in a range of temperatures up to 1100°C, as shown in figure 4. Two exothermic peaks are observed at 120 and 325°C. Both peaks could be attributed to annealing the high concentration of defects, associated with the milling process. Further heating to a relatively higher temperature resulted in the formation of an endothermic peak at ≈ 786°C which can be associated with the possible allotropic transformation of α→β phase change. Such a transformation is generally known to occur at 882°C for pure Ti and the decrease of this transition temperature is mainly due to the presence of alloying elements, Fe and Al and additional reduction could be due to the small powder particle size used for the test compared to the bulk compacted specimen. Since the starting powder for the thermal test is mainly composed of α+β phases, therefore the amount of β is expected to increase with heating and a fully single β phase would be reached passing the 786°C temperature.

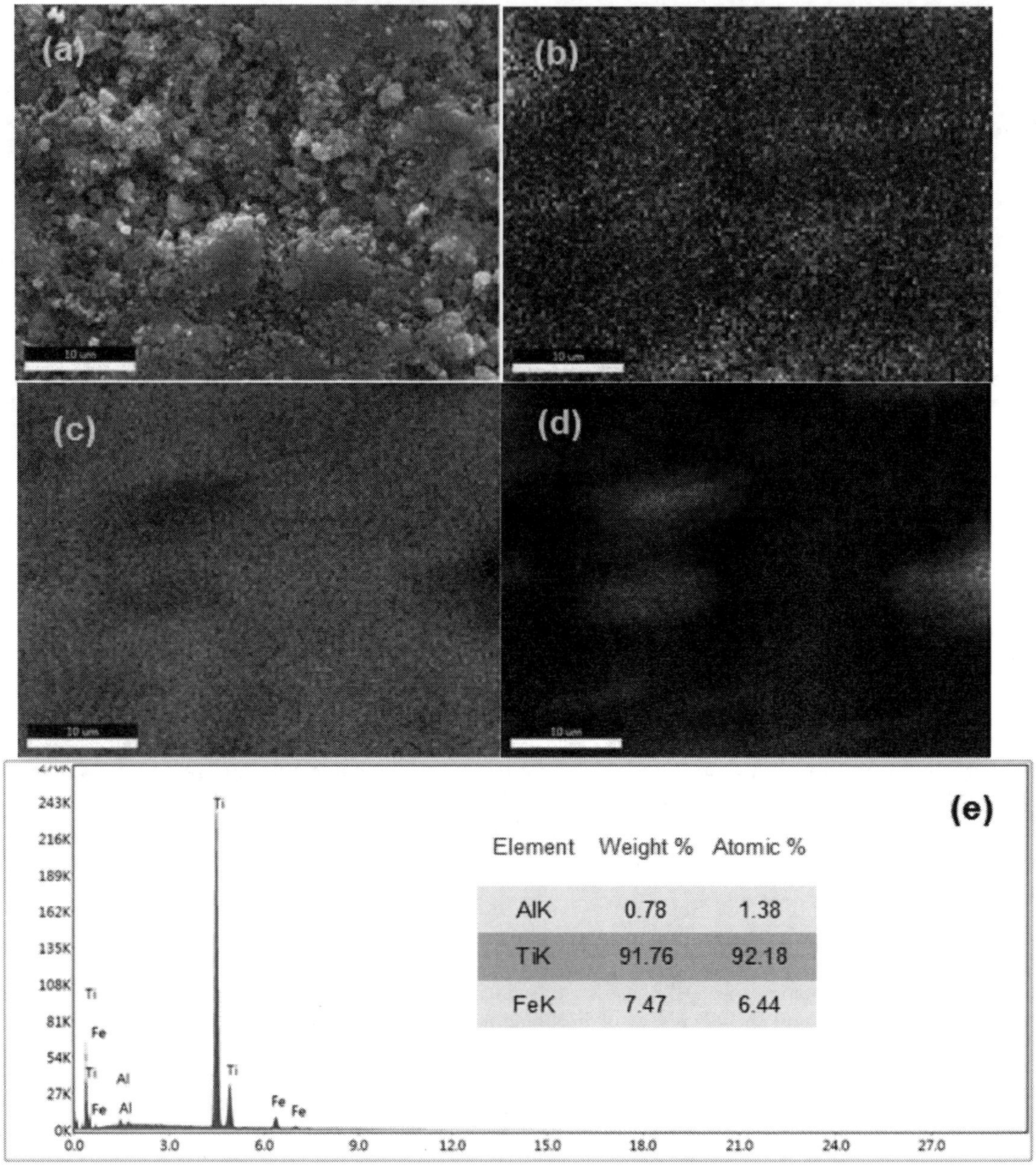

Figure 2. (a) SEM image of the as-milled sample Ti–7Fe–1Al alloy and its X–ray element distribution maps: (b) Al, (c) Ti, (d) Fe and (e) EDX

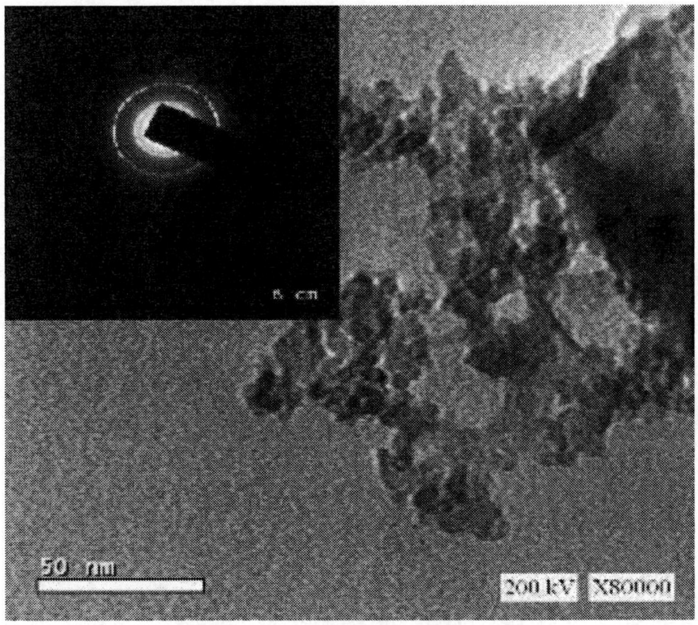

Figure 3. TEM micrographs of Ti-7Fe-1Al after 6 hours of ball milling.

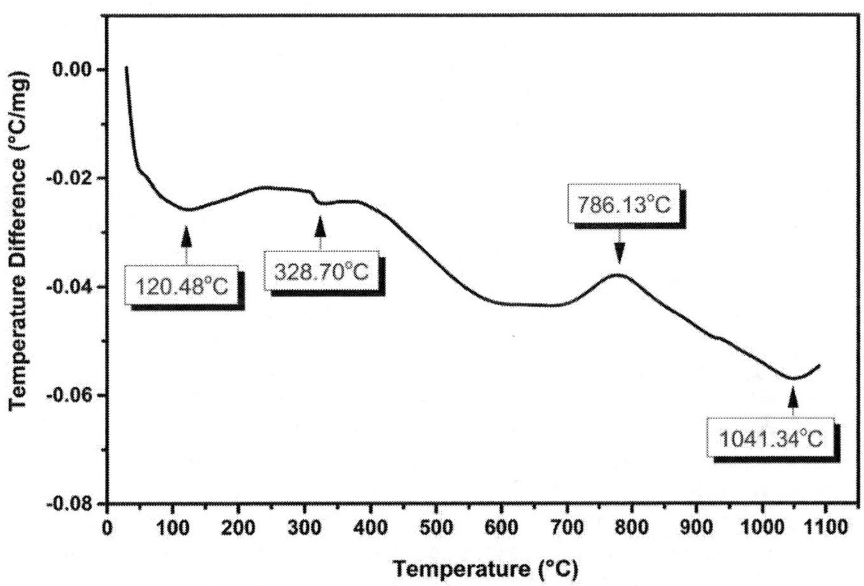

Figure 4. DTA curve for Ti-7Fe-1Al mechanically alloyed powder sample.

In the binary equilibrium diagram for Ti-Fe [14, 15], a temperature of ≈ 826°C for fully β-phase with 7wt.%Fe, which is 40°C more than the measured one, indicates a possible enhancement of α→β transformation by stored energy during milling. In that case, the Fe lean α phase loses its iron content to further stabilize the high temperature β-phase. High temperature binary Ti-Fe system shows a eutectic reaction at 1085°C in the Ti side of the diagram, similar exothermic presuming due to the eutectic reaction but occurs at slightly lower temperature of 1041°C. On the other hand, there is another eutectoid reaction at 595°C of β→α+TiFe, but no trace of such reaction is observed on the thermal diagram. The suppression of the eutectoid reaction and consequently the formation of the intermetallic TiFe phase has been reported by several works assuming kinetic reasoning for such changes [16].

The α-Ti phase is known to have low symmetry (hcp) structure which lack sufficient independent slip system when compared to Ti-β phase with (bcc) symmetry. The amount of newly formed β-phase in the present ternary system at early stage of milling will continue to generate a high rate of dislocation during milling. The formed dislocations are not expected to pass through the α-phase and accumulate at the interface between α/β phases, creating a large stress concentration at the interface resulting in a stress gradient from the interface to α-grain interior. Such condition can enhance the diffusion flow mainly for the fast diffusion element Fe. The iron migration from the interface to the interior of α phase, will enrich α phase with iron atoms with the subsequent effect of increasing the structure instability and compel α phase to transform to β-phase. This transformation starts at the interface and gradually extends to grain interior with strain. Further, the high stress concentration formed at the interface will significantly increase the Gibb's free energy in addition to the enrichment by Fe atoms will eventually cause a high order of lattice instability of the α phase near the interface. Transformation from α to β phase to reduce the rise in free energy will occur gradually and depend on both Fe diffusivity and the amount of strain produced at the interface. Similar trend of result has been observed with Ti-Al-Fe alloy but at lower iron content [11, 17]. Thermal stability of the pre-milled alloys was examined at of 950 and 1100°C for different holding times. X-ray diffraction patterns indicated that, the 950°C quenched samples transformed to α'-Ti phase. The intensity of the peaks corresponding to α' phase decreases for the 4hrs milled sample while 6 hrs milled sample quenched from 950 shows α' and α''-Ti phases formation. Also, the 1100°C quenched sample indicate that, the α''-martensitic phase is the dominate structure.

The microstructures of the Ti–7Fe-1Al without milling shows α'-martensite phases formation. According to XRD results, increasing milling time leads to the significant increasing in the amount of α''-phase after quenching, hence β phase mostly not found. The quantity of α' phase is decreased with milling time. The length of the longest martensitic plates was about 30.6 μm. Only primary and secondary martensite plates are visible in figure 6. Figures 7 and 8 show SEM image of the quenched samples from 950 and 1100°C, a nano size particle appeared to dominate the microstructure in addition to porosity. SEM observation indicated that the times for isothermal holding up to two hours at both selected temperatures were not enough for the green compacted particles to rich a complete densification microstructure. The porosities in the obtained microstructures are observed to decreases with both times and temperatures of the thermal treatment and they are mainly located along the particles boundaries and some appears interconnected with irregular shape as shown in figure 6. It therefore possible to suggests that the thermal energy during this early stage of particles sintering is mainly spent in the alloying elements diffusion for α + β → β transformation, rather the compact densification. The diffusivity and solubility limit of alloying elements in the high temperature β-phase which dictate the kinetic of densification process. As previously indicated the iron has a high diffusivity in titanium and mainly in β-phase which will enhance the alloys sinterability. On the other hand, this high diffusivity of iron can be responsible for the formation of Kirkendall porosities the titanium matrix.

IOP Conf. Series: Materials Science and Engineering **973** (2020) 012025 doi:10.1088/1757-899X/973/1/012025

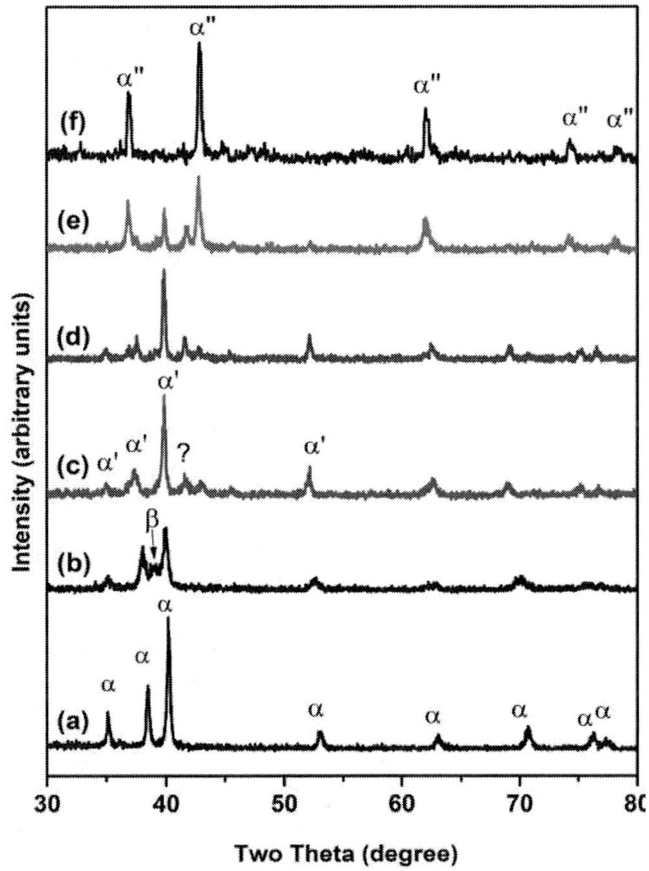

Figure 5. XRD patterns of as-quenched Ti-7Fe-1Al from 950°C (b) without milling, (c) 2 hrs, (d) 4 hrs and (e) 6hrs, and from 1100°C (f) 6hrs, compared with initial α-Ti (a)

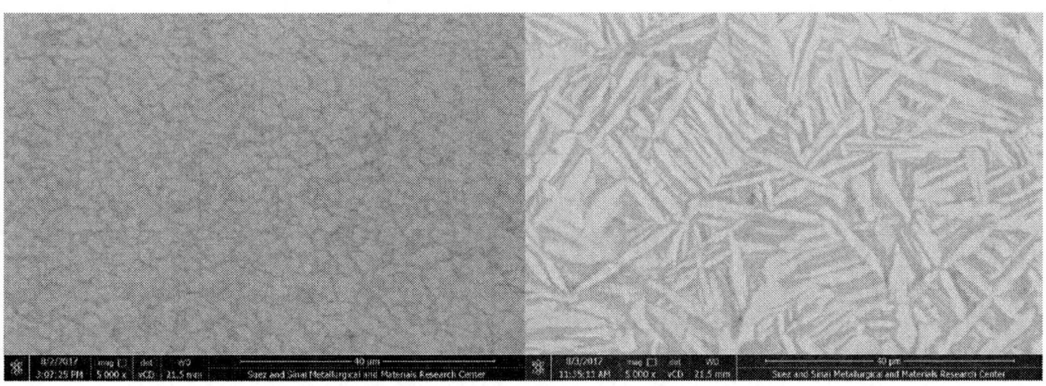

Figure 6. SEM of Ti-7Fe-1Al without milling quenched from 950°C.

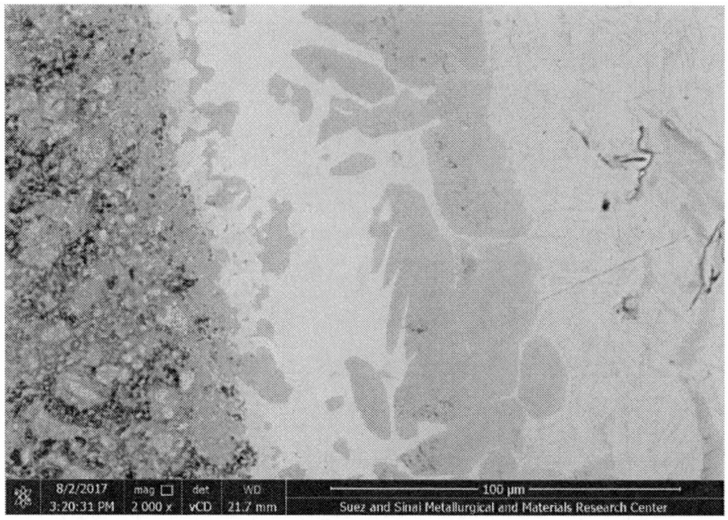

Figure 7. SEM of 4hrs milled sample Ti-7Fe-1Al quenched from 950°C.

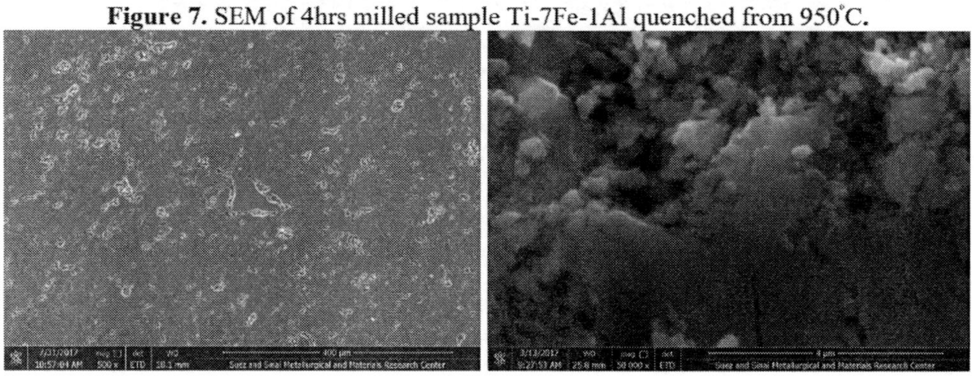

Figure 8. SEM of 6 hours milled Ti-7Fe-1Al quenched from 1100°C.

A volume contraction is observed following the solution treatment indicating the start of densification process however it is not completed due to the relatively short isothermal holding times; between 1 and 2 hours before quenching. In this process of the solid solution above β transition temperature the primary effect will be to enrich the β-matrix with β-stabilizing elements, mainly Fe. The high residual strain produced by the previous milling process is expected to enhance densification for nano particles of the compacted powder in addition to accelerating any expected diffusion process of particle growth. Moreover, at the higher solution treatment temperature of 1100°C, the presence of Fe will lead to the formation of a eutectic reaction at 1085°C leading to an increase in the alloy sinterability due to the transient liquid phase formation. Furthermore, the diffusivity of Fe in β-Ti is known to be higher than that in α-Ti and can lead to a fast decomposition of α-phase and the formation of the single β-phase above T_{β}. The presence of Al in the alloy composition is not likely to contribute to improvement of sintered density since it has been suggested that a partial Ti-Al diffusion can lead to a residual pore formation. On the other hand, Al-addition has been shown to act as β-stabilizing elements, but only with single β-Ti phase [18].

XRD results of different solution treated samples at 1100°C showed that the quenching to room temperature produced a $\beta \rightarrow \alpha''$ martensitic transformation. However, the intensities of α″-lines are less strong at higher Fe content alloys with a few lines of β and even α-phase. This could be an indication that the higher Fe-content is more stable than the lower one and therefore, the metastable β-structure with lower Fe-content is expected to give more martensitic transformation. Similar results were

IOP Conf. Series: Materials Science and Engineering **973** (2020) 012025 doi:10.1088/1757-899X/973/1/012025

obtained after specimens treated at 950°C, however it appears that a different quenching rate has a considerable effect on martensitic volume fractions. In that respect, the change in quenching rate is expected to create a change in the degree of solute partitioning during cooling.

The phase stability indicators parameter calculated for present alloy composition and shown in table1 predicted the formation of the β-stable and/or near β-phase to occur during processing. Since the fully β-stability is associated with retarding the displasive phases at room temperature by quenching. The absence of such results is considered as discrepancies to experimental and calculated parameters. Such discrepancies can be attributed with a possible shift of the boundary between the phases with ternary content of alloying elements. Moreover, the additive effect of each alloying element on titanium may need to be reconsidered and replaced by interactive form between trio-elements electronic structures. It has been generally indicated that the finer the parent phase grain size, martensite will form easier and would be thus in a more stable condition. Also, refinement of grain size could lead to an increase in the transformation temperatures and in a more general term, the effect of the stored energy due to milling process will reduce the temperature for recovery, recrystallization and grain growth.

4. Conclusions

The milling of the titanium base alloys showed a nanosize double phase α and β with particles size been iron content dependent. The accumulated stored energy during milling is partially enhancing the second β-phase formation in the α-matrix. TEM observation of milled powder confirm the presence of a nanocrystalline microstructure with high angle boundaries. The thermal stability of the pre-milled particles in a grain compacted form undergoes a martensite type of transformation α-phase with orthorhombic structure following the quenching from a temperature above the beta-transus temperature. Such transformation is considered as an indication that a full β-stable phase is not reached with the present iron content in titanium base alloy but rather a β metastable phase is obtained.

References

[1] Wang Q, Liu Z, Wang B, Hassan Mohsan AU. 2017. *Materials Science and Engineering: A* 690: 32-6
[2] Wang K. 1996. *Materials Science and Engineering: A* **213** pp 134-7
[3] He G, Eckert J, Dai QL, Sui ML, Löser W, et al. 2003. *Biomaterials* **24** pp 5115-20
[4] Xie KY, Wang Y, Zhao Y, Chang L, Wang G, et al. 2013. *Materials Science and Engineering: C* **33** pp 3530-6
[5] Song Y, Xu DS, Yang R, Li D, Wu WT, Guo ZX. 1999. *Materials Science and Engineering A* 260: 269-74
[6] Yang YF, Luo SD, Schaffer GB, Qian M. 2012. *Metallurgical and Materials Transactions A* **43** pp 4896-906
[7] Perl DP. 1985. *Environmental health perspectives* **63** pp 149-53
[8] Kiss T. 1995. *Archives of Gerontology and Geriatrics* **21 pp** 99-112
[9] Fujii H. 1998. *Materials Science and Engineering: A* **243** pp 103-8
[10] Choubey A, Basu B, Balasubramaniam R. 2005. *Trends Biomater. Artif. Organs, Vol 18 (2), January* **18** pp 64-72
[11] Koike J, Shimoyama Y, Ohnuma I, Okamura T, Kainuma R, et al. 2000. *Acta Materialia* **48** pp 2059-69
[12] Xu W, Kim KB, Das J, Calin M, Eckert J. 2006. *Scripta Materialia* **54** pp 1943-8
[13] Jackson M, Ahmed W. 2007. In *Surface Engineered Surgical Tools and Medical Devices*, pp 533-76(Boston, MA: Springer US)
[14] Ozaki T, Matsumoto H, Watanabe S, Hanada S. 2004. *Materials Transactions* **45** pp 2776-9
[15] Neelakantan S, Rivera-Díaz-del-Castillo PEJ, van der Zwaag S. 2009. *Scripta Materialia* **60** pp 611-4
[16] Moiseev VN. 1969. *Metal Science and Heat Treatment* **11** pp 335-9

[17] Kilmametov A, Ivanisenko Y, Straumal B, Mazilkin AA, Gornakova AS, et al. 2017. *Scripta Materialia* **136** pp 46-9
[18] Abdel-Hady M, Hinoshita K, Morinaga M. 2006. *Scripta Materialia* **55** pp 477-80

IOP Conf. Series: Materials Science and Engineering **973** (2020) 012026 doi:10.1088/1757-899X/973/1/012026

Fracture toughness assessment of axial partially through cracks in X65 and X70 steel pipes

A E Said[1], H A Hassan[1], T G Abu-El-Yazied[1] and A M Makady[2]
[1]Design and Production Eng. Dept., Faculty of Engineering, Ain Shams University, Egypt.
[2]Department of Manufacturing Technology, College of Technological Studies, PAAET, Kuwait.
a.e.said@eng.asu.edu.eg

Abstract This paper presents analysis of fracture assessment methods of axial partially-through crack in X65 and X70 steel pipes with internal pressure. Two analytical methods Folin–Ciocalteu method (FC method) and Gauss–Seidel method (GS method)) are used to make assessment for two steel pipes (steel X65 and X70). Finite Element model of X65 and X70 steel full-scale pipes with axial part-through crack was established. In this work, a comparison is made between results obtained from FC and GS methods and finite element model with previous experimental results. The GS is more conservative assessment method as it provides smaller crack depth (a) corresponding to (J_{cr}). Finite Element model in case of steel X70 is more conservative than the analytical methods and its results close to the experimental values.

1. Introduction

The defects in the wall thickness of thin-wall pipelines are very serious. During service, these defects are sources for crack initiation and propagation until the pipe failure [1-4]. In some cases, catastrophic failure with human casualties occurs especially if these fluids/gases are flammable.

Investigation of the crack propagation in pipes is very important to avoid catastrophic failure of pipes. The pipes material undergo in elastic-plastic fracture mechanics methods [5]. J-integral [6] and crack tip opening displacement (CTOD) [7] are two fracture assessment methods used to describe the behavior of elastic-plastic fracture material [8].

Fracture assessment methods of defected pipes are exposed to internal pressure got a great consideration in lots of previous works [9-16]. In most cases, the pipe got axial crack (parallel to pipe axis), circumferential crack (on the pipe circumference) or inclined to the pipe axis.

Two analytical methods (i.e. FC method [17] and GS method [18]), for crack propagation, were used to get critical crack depth (a_{cr}) from each method, beyond this value unstable crack growth occurs. The objective of this work is to establish a finite element model for full scale pipe with axial partially-through crack and compare the results of finite element model and analytical methods with experimental results. This study is considered very useful for maintenance planes of gas pipelines.

2. Material

This work focused on two types of steel pipes steel X65 and X70. Mechanical properties in a circumference direction were determined previously by Ľ.Gajdoš [9]. The fracture toughness of the material (J_{cr}) was determined using J-R curve from compact tension specimen (CT) [9]. Table 1 summarizes the mechanical properties and fracture toughness of these pipes.

Table 1. Mechanical properties and fracture toughness of steel X65 and X70 [9].

Steel grade	Yield strength in (MPa)	Ultimate strength in (MPa)	Ramberg–Osgood parameters		J_{cr} N/mm
			α	m	
Steel X65	496	582	5.34	8.45	432
Steel X70	536	643	5.92	9.62	439

Analytical fracture assessment methods (FC and GS methods) and Finite Element method were applied to these two materials. Table 2 summarizes both the fracture pressure and crack depth at fracture obtained from previous work [9, 15].

Table 2. Fracture pressure and critical crack depth for tested pipes [9].

Steel grade	Outer diameter (D) mm	Thickness (t) mm	Half crack length (c) mm	critical crack depth (a) mm	Fracture pressure (MPa)	Plastic constraint factor C
Steel X65	820	10.6	100	7.0	9.86	2.30
Steel X70	1018	11.7	127	6.7	9.86	2.07

3. Analytical Methods

3.1. The Folin–Ciocalteu method (FC method)

It is an analytical method for estimating J-Integral for non-linear material Equation (1). This method depends on Ramberg-Osgood parameters for the material. This method was described in addendum A16 of the French nuclear code RCC-MR.[17]

$$J = \frac{K^2}{E`}\left[\left(1+\alpha\left(\frac{\sigma}{\sigma_o}\right)^m\right)+\frac{0.5\left(\frac{\sigma}{\sigma_o}\right)^m}{\left(1+\alpha\left(\frac{\sigma}{\sigma_o}\right)^m\right)}\right] \quad (1)\ [17]$$

where

K: stress intensity factor in case of pipe with axial semi-elliptical partially through crack

σ: nominal stress $\sigma = \frac{\sigma_h}{1-\frac{\pi a c}{2t(t+2c)}}$ [9]

σ_h: hoop stress $\sigma_h = \frac{PD}{2t}$

c: half crack length

a: crack depth, as shown in Figure 1

$E` = E$ for the plane stress

$= \frac{E}{1-\upsilon^2}$ for the plane strain, where υ: poisson' ratio

σ_0 is the yield strength and $\varepsilon_o = \frac{\sigma_o}{E}$

α, m : material constants (Ramberg-Osgood parameters) [8]

$$\frac{\varepsilon}{\varepsilon_o} = \frac{\sigma}{\sigma_o} + \alpha\left(\frac{\sigma}{\sigma_o}\right)^m$$

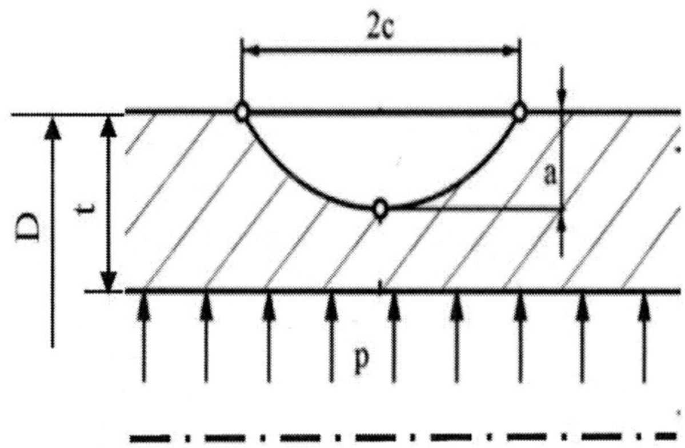

Figure 1. Semi-elliptical partially through crack in pipe.[9]

3.2. The Gauss–Seidel method (GS Method)

It is another analytical method for determining J-Integral particularly for semi-circular crack, Equation(2).[18]

$$J = \frac{K^2}{E`}\left[1 + \frac{2\alpha\, m\left(\frac{\sigma}{\sigma_o}\right)^{m-1}}{m+1}\right] \quad (2)\ [18]$$

4. Finite Element Method

ANSYS program is used to simulate semi-elliptical axial partially-through crack in a full-scale pipe similar to the pipes tested previously [9, 15]. Comparison is made between the values of fracture pressure and crack depth obtained from finite element and experimental values obtained by Gajdoš et. al. [9, 15].

The steel X65 and X70 properties were identified to Ansys program via the material yield strength and the tangent modulus. From figure 2, the tangent moadulus of steel X70 was calculated to identify the material to ANSYS program, figure 3.[21]

The proper mesh for the crack configuration was determined to facilitate the steps of establishing the Finite Element model. Many trials were conducted to get the appropriate meshing. In this work, tetrahedral mesh was selected because it is more proper and suitable for this kind of crack.

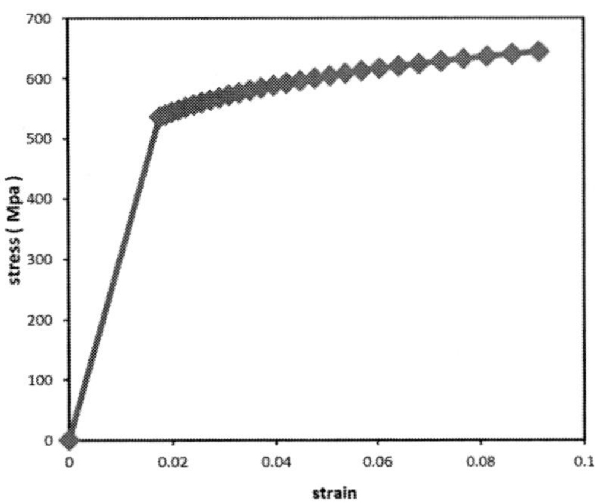

Figure 2. The behavior of steel X70 according to Ramberg-Osgood parameters.

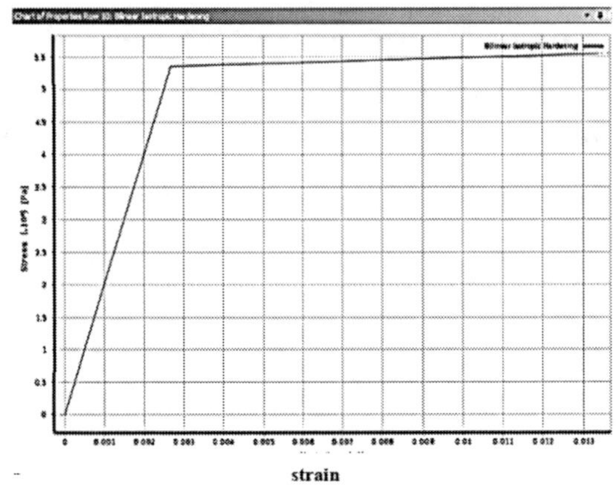

Figure 3. The behavior of bilinear material steel X70 from ANSYS program.

5. Results and Discussion

The J-integral was determined using ANSYS under same fracture pressure obtained experimentally for each pipe at different crack depth to establish relationship between J and a using finite element to get the critical crack depth (a_{cr}) corresponding to J_{cr}. Figure 4 shows the two analytical methods (FC and GS methods) compared to the Finite Element method.

Generally, the Finite Element method is in good agreement with the analytical methods (FC and GS methods). This is obvious hence, a_{cr} for steel X70 according to Finite Element method = 7.169 mm while according GS method a_{cr} = 7.218 mm, Table 3. The Finite Element method is more conservative than FC and GS methods.

Table 3. The critical crack depth (a_{cr}) for steel X65 and X70 using FC, GS and Finite Element methods.

Steel grade	critical crack depth (a_{cr}) mm		
	The FC method	The GS method	The Finite Element method
Steel X65	7.109	7.065	7.511
Steel X70	7.348	7.218	7.169

The critical crack depths (a_{cr}) corresponding to (J_{cr}) obtained at the fracture pressure by two analytical methods and Finite Element are summarized in Table 3.

Figure 5 shows the value of J-integral at the critical crack depth (a_{cr}) which obtained from finite element method for steel X65 and X70 at the same fracture pressure obtained experimentally [9].

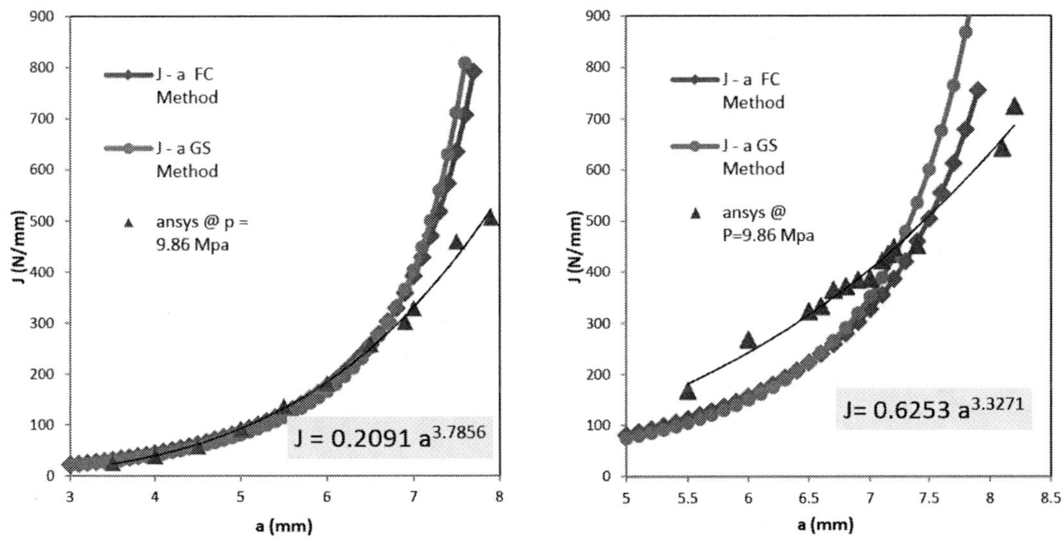

(a) Steel X65 (b) Steel X70

Figure 4. Comparison between the FC, GS and Finite Element methods for: (a) steel X65 and (b) steelX70.

Table 4 refer to the deviation between critical crack depth (a_{cr}) determined by different methods and the one obtained experimentally [9, 17].

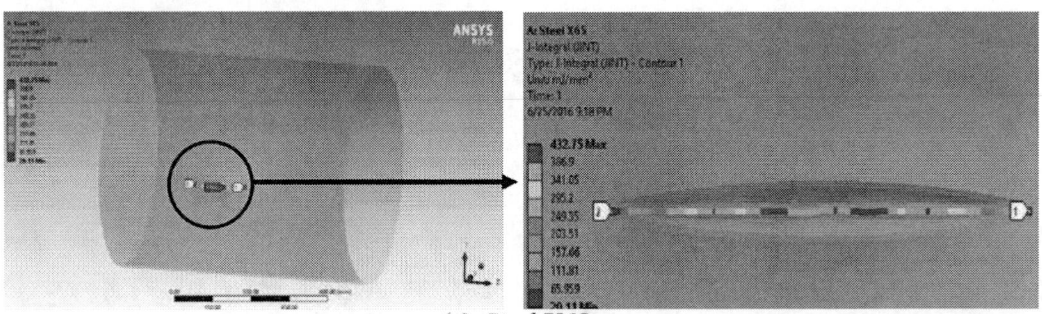

(a) Steel X65

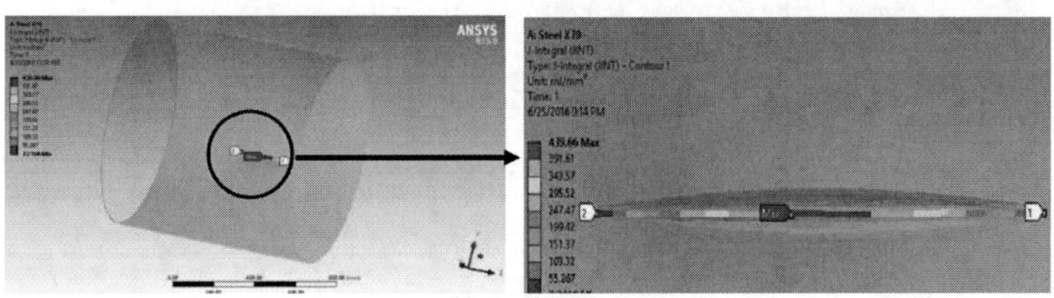

(b) Steel X70

Figure 5. The J-integral at fracture pressure for (a) Steel X65 and (b) Steel X70.

IOP Conf. Series: Materials Science and Engineering **973** (2020) 012026 doi:10.1088/1757-899X/973/1/012026

Table 4. The deviation in crack depth of each method relative to the experimental value.

Steel grade	Deviation from experimental value		
	The FC method	The GS method	The Finite Element method
Steel X65	1.56%	0.93%	7.30%
Steel X70	9.67%	7.73%	7.00%

6. Conclusions

The FC and GS analytical methods for fracture assessments worked well in determining fracture parameters. GS method is more conservative to be used in maintenance plans. This means it gives the minimum crack depth at fracture under same internal pressures.

The Finite Element method is successfully simulated fracture assessment of natural gas pipelines. In addition, Finite Element method is very simple tool to get the critical crack depth (or fracture parameters) in case of semi elliptical partially-through crack. Finite Element method is more conservative compared to analytical methods.

References

[1] Shalaby H, Riad W and Behbehani M 2006 Failure analysis of fuel supply pipeline *Engineering Failure Analysis* **13** pp789–96

[2] Azevedo C R 2007 *Engineering Failure Analysis* **14** pp978–94

[3] Hasan F, Iqbal J and Ahmed F 2007 Stress corrosion failure of high-pressure gas pipeline *Engineering Failure Analysis* **14** pp 801–9

[4] Majid Z, Mohsin R, Yaacob Z and Hassan Z 2010 *Engineering Failure Analysis* Failure analysis of natural gas pipes **17** 818–37

[5] API Specifications 5L 2012 *Specifications for Line Pipe* 43rd Edition (Washington D.C: American Petroleum Institute)

[6] Rice J R A 1968 Path independent integral and the approximate analysis of strain concentration by notches and cracks *J. of Applied Mechanics* **35** pp 379–86

[7] Wells A 1961 Unstable crack propagation in metals: cleavage and fast fracture *Proc. of the crack propagation Symp.* vol 1 (Cranfield, UK)

[8] Anderson T 2005 *Fracture mechanics: fundamentals and applications* (Boca Raton: CRC)

[9] Gajdoš Ľ and Šper M 2011 Application of a Fracture-Mechanics Approach to Gas Pipelines *Proc. of World Academy of Science* **73** 480–7

[10] El-Sayed M, El Domiaty A and Mourad A 2015 Assessment of Axial Crack in Steel Pipe under Internal Pressure *Procedia Engineering Fracture* **130** pp 1273–87

[11] Sharma R, Shrivastava S and Basu S 2006 *International Journal of fracture* **140** 269–75

[12] Hopkins P 2002 *Comprehensive Structural Integrity* (Berlin: Elsevier)

[13] Hernandez-Rodriguez M and Rodriguez J 2007 Corrosive wear failure analysis in a natural gas pipeline *Wear* **263** 5 pp 67–71

[14] Cravero S and Ruggieri C 2006 Structural integrity analysis of axially cracked pipelines using conventional and constraint-modified failure assessment diagrams *Int. J. of Pressure Vessels and Piping* **83** pp 607–17

[15] Gajdoš Ľ and Šperl M 2012 Evaluating the Integrity of Pressure Pipelines by Fracture Mechanics *Applied Fracture Mechanics* ed Alexander Belov (IntechOpen) chapter 10 pp 283–310

[16] Adib-Ramezani H, Jeong J and Pluvinage G 2006 Structural integrity evaluation of X52 gas pipes subjected to external corrosion defects using the SINTAP procedure *Int. J. of Pressure Vessels and Piping* **83** pp 420–32

IOP Conf. Series: Materials Science and Engineering **973** (2020) 012026 doi:10.1088/1757-899X/973/1/012026

[17] D. RCC-MR 1985 *Construction Rules for Mechanical Components of FBR Nuclear Islands* vol. 1 Section I Subsection B
[18] Gajdoš Ľ and Srnec M 1994 An approximate method for J integral determination *Acta Technica CSAV* **39** pp 151–64
[19] Newman J 1976 *NASA-TN* Fracture Analysis of Surface and Through Cracks in Cylindrical Pressure Vessels D-8325
[20] Folias E 1970 On the theory of fracture of curved sheets **2** *Engineering Fracture Mechanics* pp 151–64
[21] Tian C and Gao Y 2005 Role of elasticity in elastic–plastic fracture *Int. J. of Solids and Structures* **42** pp 951–70
[22] Shih C, Moran B and Nakamura T 1986 Energy release rate along a three-dimensional crack front in a thermally stressed body *Int. J. of Fracture* **30** pp 79–102

IOP Conf. Series: Materials Science and Engineering **973** (2020) 012027 doi:10.1088/1757-899X/973/1/012027

Assessment of St. St. 316L joints welded by laser beam and metal active gas MAGprocesses

AM Magdy[1,3], AB El-Shabasy[2,3] andHA Hassan[2,3]
[1] Assistant Lecturer,
[2]Professor
[3]Ain Shams University, Faculty of Eng., Design & Prod. Eng. Dept.
ayman_hussien@eng.asu.edu.eg

AbstractStudy the welding of Stainless steel 316L sheet was carried out using two welding techniques, Laser beam and Metal Active Gas MIG. Strips of 1.5 mm thick were cut from the Stainless-steel sheet and then welded using nitrogen gas for shielding. Metallographic examination using optical microscopy and SEM was conducted to study the microstructure of the welded joints. The tensile stress for joints welded by Laser beam and MAG were ranged from 386 to 632 MPa and from 126.5 to 599 MPa respectively according to the parameters of each welding techniques. The average of microhardness values measured at successive distances for joints welded by Laser beam and MAG were 265 HV and 261 HV, respectively. The obtained results were analysed and discussed based on the recent research.

1.Introduction

Austenitic stainless steels are a group of the materials that have the highest service performance in industry, besides their high weldability percent and economical nature. They have a high corrosion resistance due to the high chromium percentage that ranges from 16 to 26% and the presence of Nickel that reaches 20%[1]. 316L grade is one of the most common types that is characterized by its high immunity from sensitization, pitting and crevice corrosions resistance in warm chloride environments, and to stress corrosion cracking above 60 °C. It has a good oxidation resistance up to about 900 °C in continuous work.

Na Zhang Liu and Walcott [2] made a series of experiments to investigate correlations between welding parameters and the weld pool geometry. He found a significant nonlinearity in the diode laser welding process and it was found the model takes the reciprocal of the welding speed as the input and the top side surface width of the weld pool as the output.Seang, David and Ragneau[3] studied the effect of Nd:YAG laser welding parameters on the hardness of lap joint. They observed the interaction between the welding parameters and it was reported that the hardness along the depth of the weld bead is affected strongly by the parameters, but they have no influence on the hardness inside the softening zone. Wang et. al. [4] investigated the ductility of laser welded joint by applying three-point bending test. It was clear from their experiments thatthe decrease of heat input by the laser beam welding will increase the ductility of the welded joint. The presence of fine and irregular grains leads to better crack propagation resistance and higher fracture strains.

Sun et. al. [5] showed various advantages such as HAZ was narrow and the grain growth degree was small. Moreover, it could well satisfy motor vehicle exhaust pipe production.Khan et. al.[6]stated that the weld penetration and width mainly increase at higher laser power values because this leads to higher energy deposition on the weld area and the deposited energy will have longer time to diffuse in the material.Xueet. al.[7]reported that high penetration rate can be achieved when the range of laser power reached 60% and full penetration at laser power greater than 60%. However, microstructure showed that penetration depth is directly proportional with the welding speed. They explained that the temperature of the material increased with increasing welding speed while the cooling time decreased, hence the ability of the laser beam to penetrate the material is increased at higher welding speeds. El-Batahgyet. al. [8] showed that low welding speeds increased the fusion zone dimensions incredibly rather than higher speeds. El-Batahgy[9], Y.Feng et. al. [10] concluded that welding speed had an opposite effect on the penetration depth and bead width, when welding dissimilar joint of austenitic stainless steel joint.Bolut Kong et. al. [11] reported after their trials to weld 6 mm thick stainless steel by Yb-fibre continuous wave laser that full penetration is achieved below 4m/min when the laser power is 5kW, and 1 to 2 m/ min when the laser power is greater or equal to 4kW, and 0.5 m/ min at minimum power of 2.5kW.

Mitsuhiro Okayasu et. al. [12] confirmed that the tensile strength of the welded samples were higher than the base metal when he welded commercial cold rolled low carbon steel by MIG welding. It was also observed that the endurance limit of some samples is higher than the base metal and any other sample.Sin Lin et. al. [13] studied the mechanical properties of MIG welded samples from extruded Al-Mg-Si and it was observed that the ultimate tensile strength of the welded metal is 76.5% from the base metal and the fracture appeared at the least hardened location in the HAZ. The microstructure examination showed the presence of equiaxed coarser grains in the centre zone of the welded samples than the top and side zones.Sudhir Kumar et. al. [14] investigated the effect of preheating, current and voltage on welded steel joint by MIG. The results showed that the efficiency of the joint reached 94.37% at 275 °C preheating temperature, 120 Ampere, and 35volt. It is also confirmed that the ultimate strength and percent elongation are affected greatly by preheating temperature and the current.

2.Experimental Procedure

Stainless steel sheets of 1.5mm thickness were cut to 30mm width and 200 mm length by wire EDM to obtain high precise cuts. MAG electrode wire of 1 mm was chosen for this experiment.The chemical compositionof the used stainless steel sheets used in this work and the welding wire electrodeobtained from spectroscopy analyser isshown in table 1 and table 2 respectively. The stainless steel sheet is specified according to ASTM A240-12 and the welding electrode is typical to the specifications of AWS 5.9.

Table 1. Chemical composition of 316L stainless steel sheet.

C	Mn	P	S	Si	Cr	Ni	Mo	N
0.03	0.97	0.072	0.006	0.49	16.86	9.73	2.0	0.1

Table 2. Chemical composition of MIG Welding electrode.

C	Mn	P	S	Si	Cr	Ni	Mo	N
0.02	1.73	0.032	0.37	0.37	18.87	12.37	2.3	0.13

Figure 1 shows the laser welding robotic arm used in this work. It is a solid-state Nd: YAG laser pump with maximum power of 2.2kW and focal point of 500 μm, which used for welding processes. The laser beam is delivered through a fibre optic cable which is connected to the focusing lens that is fixed in a robotic arm

IOP Conf. Series: Materials Science and Engineering **973** (2020) 012027 doi:10.1088/1757-899X/973/1/012027

Figure 1. Laser Welding robotic arm.

MAG weldingWas carried out by using a designed CNC 3 axis machine and Gedik welding machine of Maximum current of 600 Amp as shown in figure 2.

The specimens are cleaned with ethyl alcohol to remove any impurities that may affect the weld bead properties and are held in a custom fixture designed for the experiment and fixed by toggle clamps as shown in figure 3 to prevent the distortion of welded samples by residual stresses. The gap between 2 parts are adjusted to be zero.The ranges of Laser and MIG welding parameters were selected and arranged by designing a factorial experiment as shown in table 3 andtable 4. Two replications were added in the factorial experiments of both MIG and Laser welding. The number of samples of laser were 18 and the number of samples of MIG were 54.Nitrogen is used as shield gas. The flow rate of the shield gas was kept constant at 30 l/min during the whole experiment.

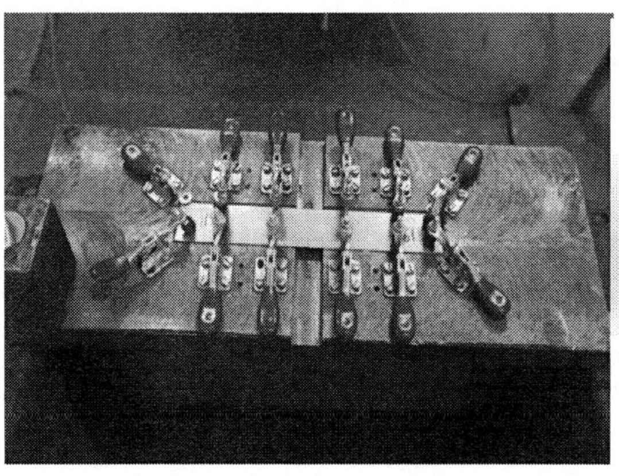

Figure 2. Specimen fixture used during welding processes

Figure 3. MAG welding CNC

Table 3. Laser Welding parameters ranges.

Laser Power (W)	Welding Speed (m/min)
800	0.5
1000	1
1200	1.5

Table 4.MAG welding parameters ranges.

Current (Amp)	Welding Speed (m/min)	Wire Feed (rpm)
50	0.5	25
60	1	35
70	1.5	60

The values of the laser power and speed are programmed in the machine. The track is positioned such that the weld line is straight. The height is adjusted to achievezero focal point and in the middle weld line to equally distribute the power density and to prevent any probability offormation of amisaligned welded joint.
The MIG power and wire feed speed were adjusted in the MIG head and the welding speed, weld position and torch height were adjusted in the CNC controller.

To make sure that the welding process is stable, the process is started and ended 5 mm away from the samples. After the end of the welding process, the sample is kept for a short period to cool down and then the clamps are released to fix another sample. Visual inspection was carried out on all the welded sheet to ensure the welding quality. Then the defect free welded joints were subjected to next step of specimen's preparation for the further tests and examination.

EDM wire cut method has been used to get high precision dimensions' specimens used for the mechanical testing as well as macroscopic and microscopic examination. The tension samples were cut according to ASTM E8M-04 standard as shown in figure 4 .The strain rate of the test was adjusted at 4.5 mm/min.

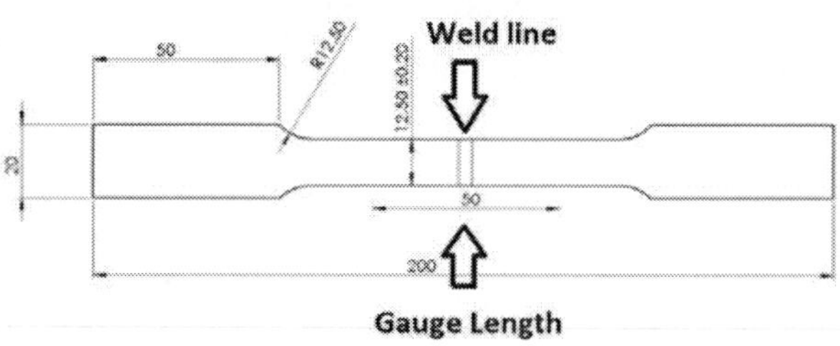

Figure 4. Tensile test specimen standard.

Moreover, other samples were cut and prepared for macroscopic examination of the weld bead. The samples were electro etched by 10% oxalic acid to reveal the welded zone of the sample. Analysis has been performed using metallographic macroscope with image analysis software.
Microhardness test is applied at 100 gram-force and dwell time 20 seconds and the distance between each indentation is adjusted to be 100 μm.
Tensile specimens fracture surfaces have been examined using SEM to reveal the failure type and feature of the fracture surfaces.

3.Results and discussion

3.1 Visual Inspection

The Visual inspection of the Laser showed no sign of imperfections or any air voids as shown in figure5 and this is a prove of thermal stability of laser during the operation.on the other hand,the inspection ofMAG welding specimens has shown some imperfections in some samples and this means that some parameters values did not provide stable heat input during welding. The results are shown in figure 6.

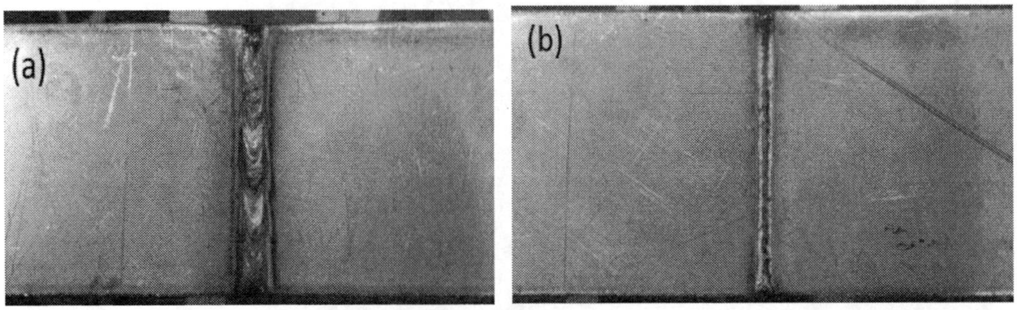

Figure 5. Laser welded specimens at (a) 1000 W, 0.5 m/min. (b) 1000 W, 1 m/min

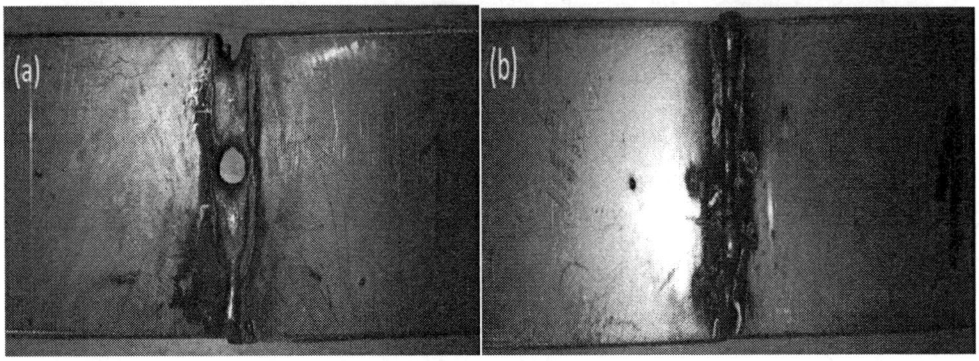

Figure 6. Discontinuities in MIG welded samples at (a) 70 Amp, 0.5 m/min, 60 rpm. (b) 50 Amp, 0.5 m/min, 35 rpm.

3.2.Mechanical properties

Table 5shows that the Average ultimate stress and max strain of laser welded samples of 12 samples.It is noted that the stress increases with increasing laser power at the same speed and that the stress decreases at an increase in the speed. Due to the small surface area subjected to the load, the material was fractured at the weld area. Tension test was applied on 26 accepted samples and the average results of tension test forMAG specimensare shown in table 6. It is noted that the ultimate tensile strength increases by the increase of MAG welding current and decreases by the increase of welding speed or wire feed rate. In addition, the welding speed and wire feed rate are mutual to each other, such that the change of one of them without the change of the other will affect the quality of the weld, which are illustrated in figures 7 and 8.

Table 5. Average maximum stress at different welding power and speed values.

Power (W)	Welding speed (m/min)	Ultimate Strength (MPa)	Elongation %
800	0.5	580.53	1.58
1000	0.5	631.85	2.55
1000	1	385.77	1.72
1200	0.5	605.21	3.92
1200	1	474.09	2.18
1200	1.5	521.52	1.23

Table 6. Average MAG tension test resultsat different welding parameters.

Current	Welding Speed (m/min)	Wire Feed (rpm)	Ultimate Strength (MPa)	Elongation %
60	0.5	25	368.7	14.1
70	0.5	25	445	34.5
60	0.5	35	508.7	30.5
70	0.5	35	597.65	72.5
60	1	25	456.6	3.7
70	1	25	521.65	47
50	1	35	260.15	4.7
60	1	35	298.6	9.2
70	1	35	480.4	30.3
70	1	60	387.8	17.9
60	1.5	35	126.4	1.4
70	1.5	35	369.3	11.4
60	1.5	60	306.1	6.8
70	1.5	60	303.5	5.9

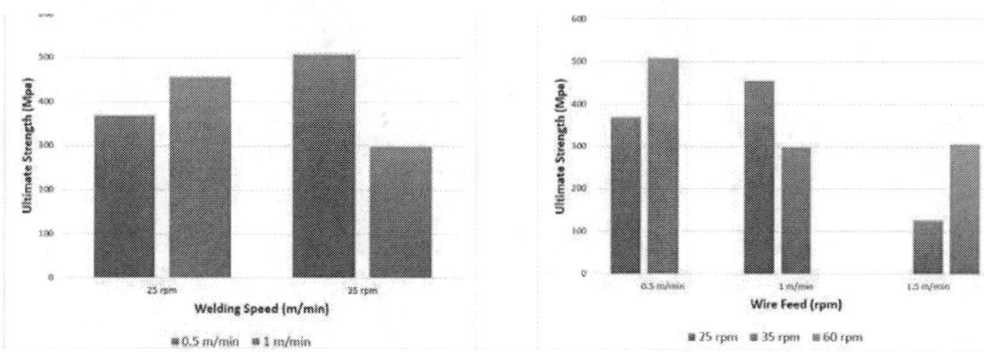

Figure 7. The influence of welding speed and wire feed rate at 60 amp

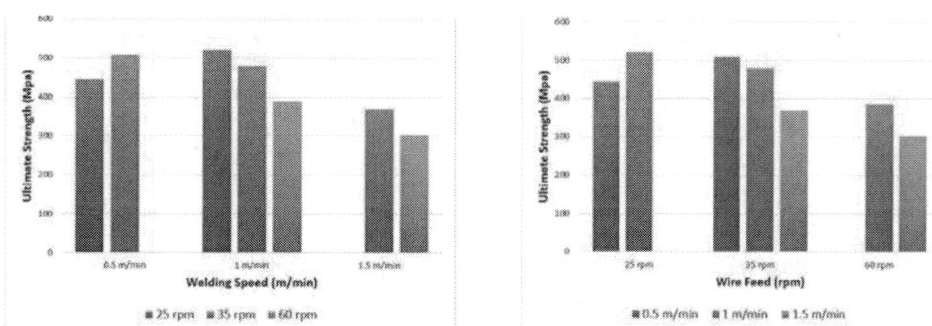

Figure 8. The influence of welding speed and wire feed rate at 70 amp

The Laser welding joints and MAG welding joints fractured surfaces from specimens tested under tension test have undergone SEM examination. The analysis has proved ductile fracture for both techniques and the results are shown figure9 and figure 10, respectively.

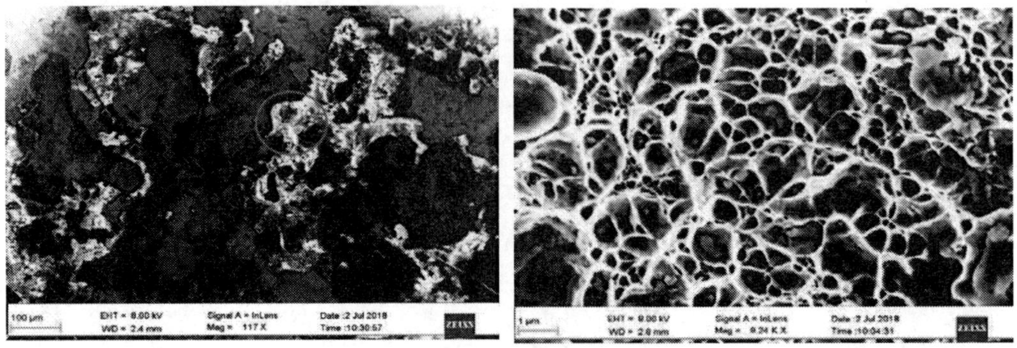

Figure 9. Fracture surface of Laser welded joint at 1200 W, 0.5 m/min using SEM.

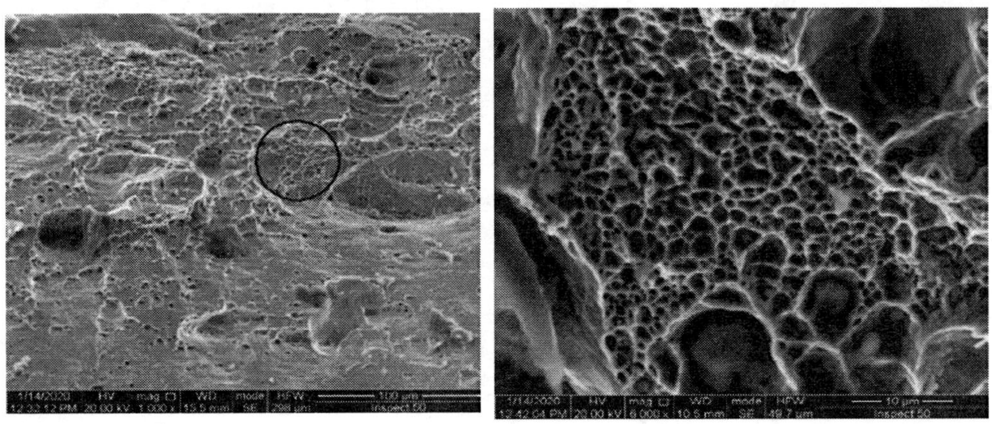

Figure 10. Fracture surface of MAG welded joint at 60-amp, 1 m/min, 25 rpm using SEM.

3.3 Microstructure Analysis

The microstructure of the weld is the main impact in controlling the mechanical properties and service efficiency of the joint. The percentage of the elements in the material and the cooling rate are domaining the nature of solidification and solid-state transformation that dominate the microstructure of the fusion zone. The mode of solidification determines the microstructure shape of the welded zone. EDX analysis was conducted at the middle of the weld zone. The chemical composition of the welded joint is measured by EDX analysis as shown in presented inTable 7, and table8.

According to D.J.Kotecki and T.A.Siewert[15] the equations of $Cr_{eq}$and Ni_{eq} are

Cr_{eq} = %Cr + %Mo + 0.7 %Nb

Ni_{eq} = Ni + 35 %C + 20 %N + 0.25 %Cu

Bsy substituting values in the equations of Creq and Nieq, the results of Laser will be:

Cr_{eq}= 17.97 %

Ni_{eq}= 10.68 %

While the results of MAG will be:

Cr_{eq}=20.49%

Ni_{eq}= 8.65 %

Table 7. EDX Analysis of Laser welded samples.

N	Si	Cr	Mn	Ni	Mo
0	0.317	15.46	0.25	10.68	1.58

Table 8. EDX Analysis of MAG welded samples.

N	Si	Cr	Mn	Ni	Mo
0	0.15	18.12	0	8.65	2.37

TheCr_{eq}/Ni_{eq}ratio of Laser is 1.68, while that of MAG is 2.36. From the pseudo-binary phase diagram (figure 11), the solidification mode of laser is expected to be at Ferrite-Austenite(FA) zone, and the expected ferrite number is 6.8, which indicates that the weld will contain about 6.8% ferrite phase. While the ferrite number for MAG is 18, hence the expected solidification mode of MAG is at Ferrite (F) zone.

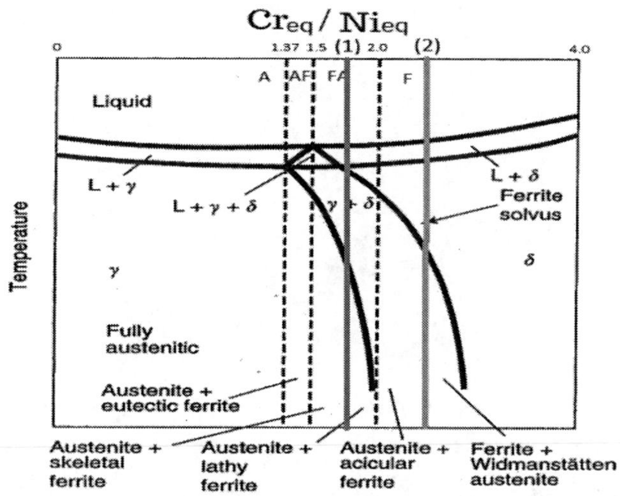

Figure 11. Predicted Solidification mode of (1) Laser, (2) MIG.

The microstructure of Laser welded joint showed different results from expected calculations. With the aid of J.C.Lippold[16],vermicular ferrite was observed which refers to Austenite Ferrite(AF)solidification mode as shown in figure 12(a).It is also observed the presence of skeletal and lathy ferrite phases with some regions of Widmanstatten austenite and intercellular austenite which confirms the primary FAsolidification mode as shown infigure 12(b)

MAG welded joint microstructure showed coarse grains of skeletal ferrite in most regions of the weld, which confirms the occurrence of F solidification mode as predicted in pseudo-binary phase diagramand this is shown in figure 13. The MAG specimens also showed the presence of HAZ that contains precipitates of Ferrite at the grain boundaries, while Laser Samples did not show any HAZ and this is shown in the figures 14 (a) and (b).

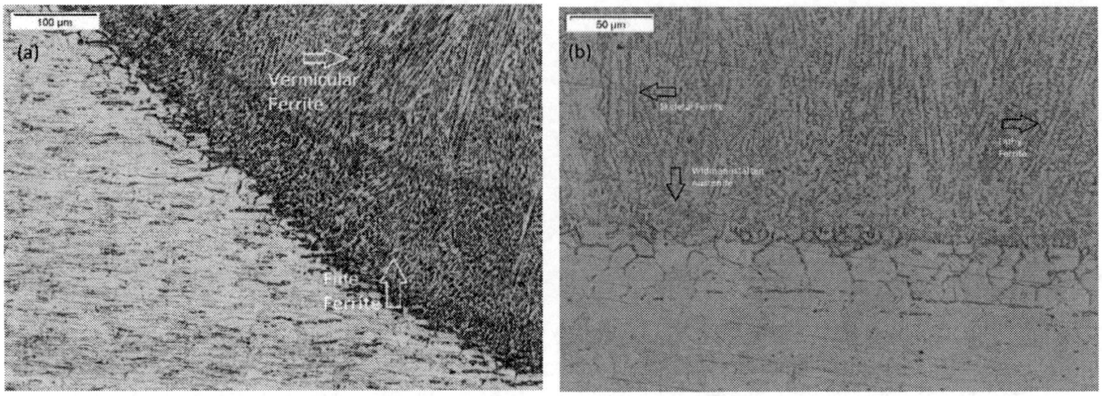

Figure 12. Laser sample microstructure (a) vermicular and fine ferrite microstructure (b)skeletal ferrite, lathy ferrite and Widmannstatten austenite.

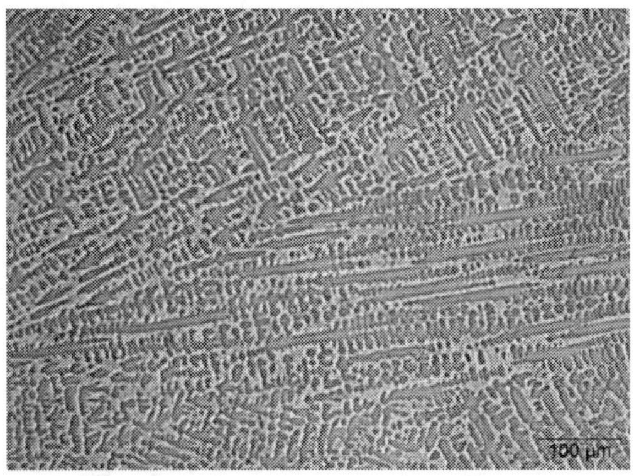

Figure 13. Microstructure of MAG welded joint showing Skeletal Ferrite phase.

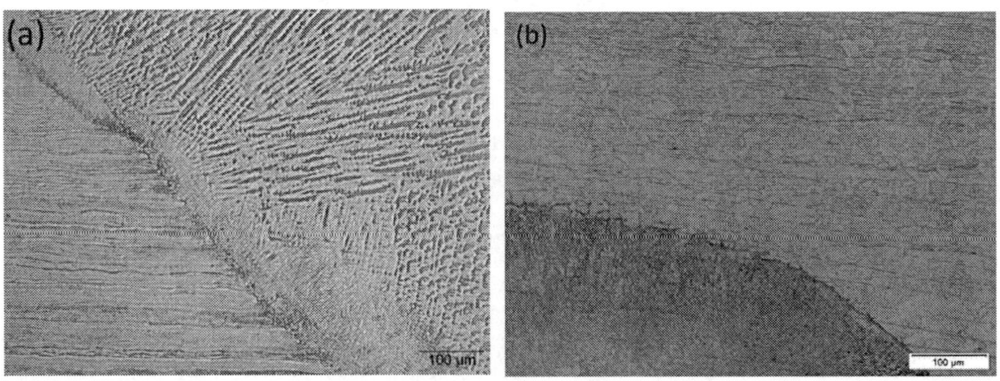

Figure 14. Welded zone boundaries with the base metal in (a) MAG joint and (b) Laser joint

3.4. Microhardness test

Microhardness test is applied on welded specimens to find out whether hardness is the same in different conditions or if there is any change in the hardness value. It is noted that the hardness number of the welds are nearly the same andtheir values are more than that of base metal as shown in figure15andfigure16. The laser weld has non-uniform distribution of material contained in the weld zone which changes the behaviour of the weld.This leads to a change in the hardness of the material. The appeared peak in the hardness values is due to the presenceof sigma phase in the laser weld bead, which contains 17.42 wt.%of chromium and 2.25 wt.% molybdenum, and this is from the EDX analysis shown in figure 17. While the MAG welded joint has shown that the fluctuations of in the hardness numbers were lower than that of Laser, and this is due to the homogeneity of phase distribution in the major region of the MAG weld bead, hence the change of hardness values is minimised.The average value reported bySeang[3] was 400 HV, while value of 240 HV was achieved by Khan et. al. [17].

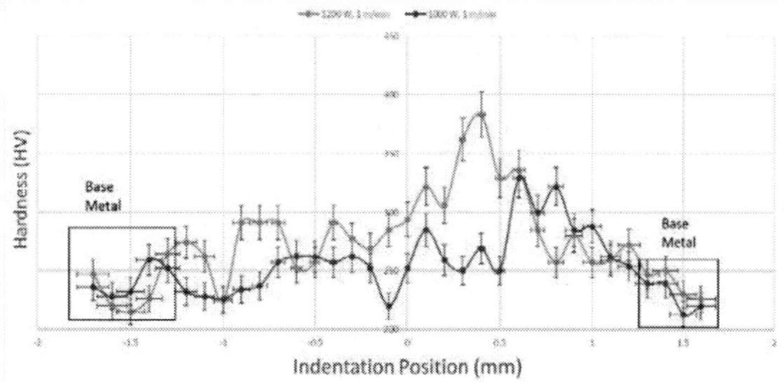

Figure 15. Hardness values across two Laser welded samples at 1200 W and 1 m/min, and 1000 W and 1 m/min.

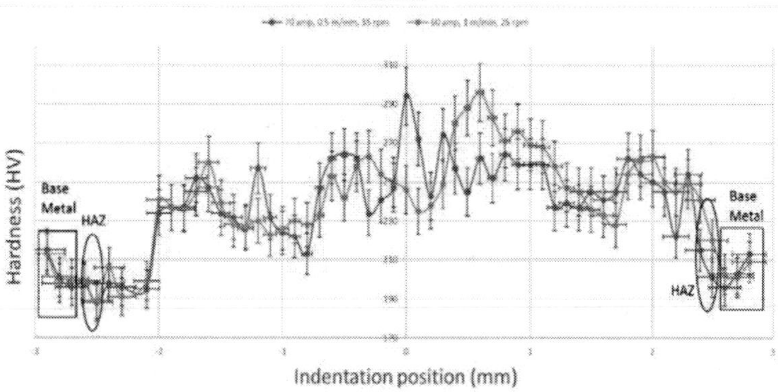

Figure 16. Hardness values of MAG joints at 70-amp, 0.5 m/min, 35 rpm, and 60-amp, 1 m/min, 25 rpm.

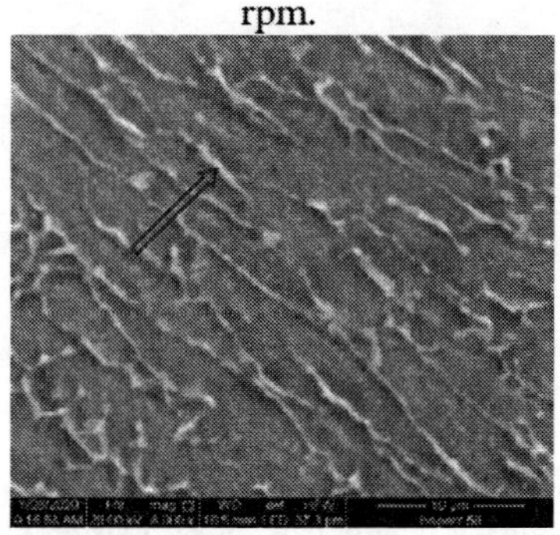

Figure 17.EDX analysis location of sigma phase.

4.Conclusions

Stainless steel 316L sheet has been welded using two welding techniques, Laser beam and Metal Active gas MAG. Based on several results have been obtained from the experimental work, the following conclusions could be drawn:

- The tensile stress for joints welded by Laser beam and MAG were ranged from 386 to 632 MPa and from 126.5 to 599 MPa, respectively.
- The maximum tensile stresses of the Laser specimens were increased by increasing laser power. On the other hand, the increase of the maximum tensile stress values of MAG specimens are independent from increasing current, but they depend on optimising the values of current, welding speed and wire feed.
- The average of Microhardness values measured at successive distances for joints welded by Laser beam and MAG were 265 HV and 261 HV, respectively.
- The microstructure examination revealed that, the laser welded joints have shown dendritic non uniform microstructure which may affect the properties of the microstructure along the weld line. While, the MAG welded joints have shown uniform microstructure, which means that the mechanical properties are distributed equally along the weld line.
- The SEM analysis has shown that the fracture of Laser and MAG welded specimens are ductile, which means that the weld did not have a major effect in the properties of the material.

5.References

[1] Folkhard E 1988 *Welding Metallurgy of Stainless Steels.*

[2] Na X, Zhang Y, Liu Y and Walcott B 2010 *Nonlinear identification of laser welding process* (IEEE Trans. Control System Technology vol. 18 no. 4) pp 927–934.

[3] Seang C, David A K and Ragneau E 2013 *Effect of Nd:YAG laser welding parameters on the hardness of lap joint: Experimental and numerical approach* (Physics Procedia vol 41) pp 38–40.

[4] Wang G, Wu A, Zou G, Zhao Y, Chen Q and Ren J 2009 *Bending Properties and Fracture Behavior of Ti-23Al-17Nb Alloy Laser Beam Welding Joints* (Tsinghua Science and Technology vol. 14 no. 3) p 293–299.

[5] Sun A, Liu J and Liu W 2011 Proc. Int. Conf. Electronic Mechanical Engineering Information Technology EMEIT vol. 8 p 3931–3933.

[6] Khan M M A, Romoli L and Dini G 2013 *Laser beam welding of dissimilar ferritic/martensitic stainless steels in a butt joint configuration* (Opt. Laser Technology vol 49) p 125–136.

[7] Xue X, Pereira A, Amorim J and Liao J 2017 *Effects of Pulsed Nd:YAG Laser Welding Parameters on Penetration and Microstructure Characterization of a DP1000 Steel Butt Joint* (Metals Basel vol 7 no 8) p 292.

[8] El-Batahgy A M, Khourshid A F and Sharef T 2011 *Effect of Laser Beam Welding Parameters on Microstructure and Properties of Duplex Stainless Steel* (Material Science Applications vol 2 no 10) p 1443–1451.

[9] El-Batahgy A M 2012 *Laser Beam Welding of Austenitic Stainless Steels–Similar Butt and Dissimilar Lap Joints* (Welding Processes) p 93–116.

[10] Feng Y, Luo Z, Liu Z, Li Y, Luo Y and Huang Y 2015 *Keyhole gas tungsten arc welding of AISI 316L stainless steel* (Material Design vol 85 no 11) pp. 24–31.

[11] Bolut M, Kong C Y, Blackburn J, Cashell K A and Hobson P R 2016 *Yb-fibre laser welding of 6 mm duplex stainless steel 2205* (Physics Procedia vol 83) pp 417–425.

[12] Okayasu M, Ohkura M, Sakamoto T, Takeuchi S, Ohfuji H and Shiraishi T 2013 *Mechanical properties of SPCC low carbon steel joints prepared by metal inert gas welding* (Material

IOP Conf. Series: Materials Science and Engineering **973** (2020) 012027 doi:10.1088/1757-899X/973/1/012027

Science Engineering A vol 560) pp 643–652.
[13] Lin S *et al 2019Microstructures and fatigue behavior of metal-inert-gas-welded joints for extruded Al-Mg-Si alloy* (Material Science Engineering A, vol 745) pp 63–73.
[14] Kumar S and Singh R 2019 *Optimization of process parameters of metal inert gas welding with preheating on AISI 1018 mild steel using grey based Taguchi method* (Measurement vol 148) pp 106924.
[15] Kotecki D J and Siewert T A 1992 *WRC-1992 Constitution Diagram for Stainless Steel Weld Metals□: A Modification of the WRC-1988 Diagram* (AWS Annual Meeting) pp 171–178.
[16] Kotecki D J and John Lippold C 2005 *welding metallurgy and weldability of stainless steels.*
[17] Khan M M A, Romoli L, Fiaschi M, Dini G and Sarri F 2012 *Laser beam welding of dissimilar stainless steels in a fillet joint configuration* (Material Process Technology vol 212 no 4) pp 856–867.

IOP Conf. Series: Materials Science and Engineering **973** (2020) 012028 doi:10.1088/1757-899X/973/1/012028

Pre-Cracking techniques of polymeric materials: an overview

A Kamal[1], AH Elsheikh[1] and EShowaib[1]
[1]Production Engineering and Mechanical Design Department, Tanta University, Tanta, Egypt.
ammar_elsheikh@f-eng.tanta.edu.eg

Abstract. In fracture toughness tests, a number of notched specimens with identical artificial pre-cracks are essential to obtain accurate fracture parameters. The test results are critically depending on the initiation stage quality. Any slight variance in pre-cracks front shape, length and orientation could significantly affects the test results; therefore, producing identical pre-cracks is a critical issue to obtain accurate results. The pre-cracking technique should be selected carefully to fulfil controllably and repeatability requirements of the standard pre-cracks for a certain material while preserving the induced residual stresses at the crack tip at a minimal value. The notching and pre-cracking standards for metallic material have been well specified in ASTM E399. However, the case is more cumbersome for polymeric materials due their viscoelastic nature. ASTM D5045, ISO 13586:2000 and ASTM D6068 specified different procedures to prepare a sharp pre-crack for polymeric materials. Many pre-cracking techniques have been proposed in literature. The present work introduces an overview of the pre-cracking techniques for polymeric materials.

1. Introduction

Recent days, polymers are widely used in the industry due to their good properties against metals[1-3]. The usage of polymeric materials in critical applications makes the determination of accurate fracture toughness parameters an important issue especially when the structural integrity of the component is critical. For the fracture toughness tests, a sharp pre-crack must be introduced firstly. The pre-cracking of solid materials is investigated fairly and well specified in ASTM E399[4]. But polymeric materials still have some problems associated with obtaining a sharp pre-crack without developing residual stresses a head the crack tip especially for ductile polymers and sudden failure of brittle polymers. According to the standards, ASTM D5045 and ASTM D6068, a natural pre-crack can be initiatedin Compact Tension ***(CT)*** specimens or Single Edge Notch Bending ***(SENB)*** specimens using a razor blade by tapping in the centre of the notch with a new razor blade or sliding a new razor blade in either a sawing motion (side-to-side) or slicing repeatedly in a single direction. The standards warn against pressing a razor blade for ductile resins. ASTM D6068 provides fatigue pre-cracking standards requirements. Earlier problems associated with specimen preparation appeared since the labs use different techniques that lead to different results. Brown et al. [5]have investigated the effect of pre-cracking technique on the lifetime of polyethylene specimens and shows pressing a razor blade at a rate of 0.05 mm/min produced the fastest growing cracks. Dapp and Rimnac[6] investigated the effect of different pre-cracking techniqueson the results of J-R curve determination for Ultra High Molecular Weight Polyethylene resin, also indicating that the pressing of a razor blade at slow feed is the most accepted method since it introduced minimum residual stresses at the crack tip. Many investigations studied the different pre-cracking techniques of different polymeric material such as (PE- EPBC-PETG)[7-11] and it was reported that the fracture toughness results depend on the pre-cracking

methodology. New technique was introduced by Tamura [12];toinduce pre-crack controllable length in brittle polymeric materials by applying tension and compression load. Then, the technique was improved and optimized in[8]. Recently femto-laser is used to pre-crack polymeric materials. Many researches [13-15]investigate femto-laser as a promising pre-cracking technique against conventional techniques since femto-laser ablation causes minimum residual stresses ahead the crack tip .
This paper introduces an overview of pre-cracking techniques of polymeric materials, discusses the advantages and disadvantages of each technique, and highlights the recent developed techniques.

2. Notching

Prior to pre-cracking, a small machined notch is introduced firstly in a CT or SENB specimen according to standards as shown in Figure 1 (a). The notch can be produced either by conventional machining process such as hand saw, sawing machine, or rotary cutting machine. The effect of notch geometry (square or angular) ona single edge notched bending (SENB) specimen made of commercial Poly(Methyl Methacrylate)PMMA resin was investigated by Souza et al. [7]. Square and angular notches were produced by a sewing machine and rotary cutting device, respectively. The used rotary cutting device is supplied with cutting disk of 45 tip angle as shown in Figure 1 (b). The results of photoelasticity tests showeda slight perturbation of the patterns ahead the notch tip for the notch produced by rotary cutting device as shown inFigure 1 (c).The results indicated that the notch geometry has a very slight effect on the fracture toughness values; however, the angular sharp notch shape is preferred because it allows better control to direct the razor blade into a single point, thus insuring reproducibility of pre-cracks.

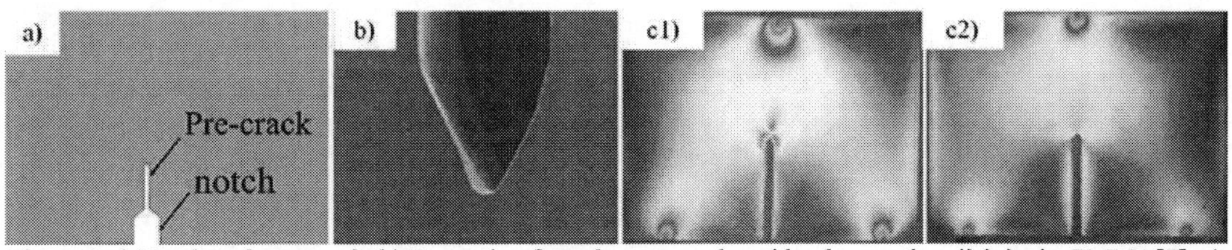

Figure 1 a) Notch and pre-crack; b) Example of notch root, produced by the rotating disk in the PMMA[7]; c) Examples of Moiré patterns for notch produced by (1) rotary cutting device, (2) sawing[7]

3. Pre-cracking Techniques

It is known that fatigue tests of ductile polymers are long term testing [16-18]; as it could take days or weeks at room temperature. Many attempts to obtain stable slow crack propagation at room temperature for such polymers in short time testing were induced using different pre-cracking techniques. The results showed that the pre-cracking technique has a critical role for obtaining stable propagation in a short time testing. Also, the fracture toughness test results of polymeric materials are sensitive to pre-cracking method since pre-cracking introduce large residual stresses at the crack tip for ductile polymers. And pre-cracks need very little energy to propagate for brittle polymers. The pre-cracking techniques can be classified into two categories conventional techniques and un conventional techniques as shown in Figure *2*.**Error! Reference source not found.**represents a summary of the pre-cracking techniques.

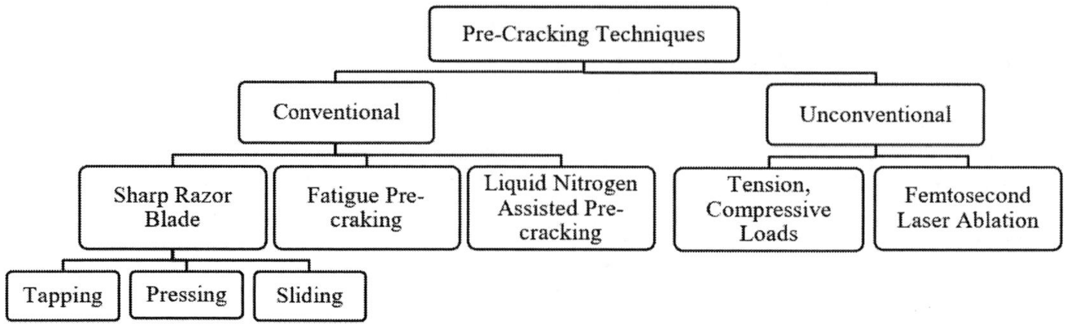

Figure 2. Classification of thePre-Cracking Techniques.

Table 1. Summary of Pre-Cracking Techniques.

Pre-Cracking Techniques			***machine***	***tool***	***International Standard***	***Standard size of Specimen***	***Material recommended***
Conventional	*Using Sharp Razor Blade*	*Pressing*	*Special pressing machine*	*Razor Blade*	*ASTM D5045 and ASTM D6068*	*Compact Tension **CT** or Single Edge Notch Bending **SENB***	-
		Tapping	*Manual*				*Ductile polymers*
		Sliding	*Sawing machine or manual*				-
	Fatigue Pre-Cracking		*Universal Testing Machine*	-	*ASTM D6068*		-
	Liquid Nitrogen Assisted Pre-Cracking		-	*Razor Blade*	-		-
Unconventional	*Tension – Compression Loads Application*		*Universal Testing Machine*	*Clamps*	-		*Brittle polymers*
	Femto-Second Laser Ablation		*Laser Machine*	-	-		*Both Brittle and Ductile Polymers*

3.1 Pre-cracking via Fresh Razor Blade

Pre-cracking via fresh razor blade is one of the most popular techniques according to ASTM D6068 and ASTM D5045. In this technique sharp cracks are produced either by tapping in the center of the notch with a fresh razor blade, pressing a fresh razor blade perpendicularly into specimen at very slow feed, sliding a new razor blade in a sawing motion (side-to-side) or slicing repeatedly in a single direction. Although razor blades are used in a wide range of applications, it has many problems. It is very difficult to obtain identical pre-cracks via razor blades, beside safety concerns.

3.1.1.Pressing a Fresh Razor Blade. Sharp pre-cracks are produced by pressing a fresh razor blade perpendicularly into a specimen at very slow feed to avoid deformation and minimizing the residual stresses at the crack tip using a suitable pressing machine. This technique is firstly developed by Brown et al. [19-23] to prepare fracture toughness specimens. The influence of pre-cracking by pressing a fresh razor blade at feed of (0.05 mm/min) on the test results for a tough polyethylene is discussed in [5]. The results showed an appropriate failure time. Although this technique provides high

IOP Conf. Series: Materials Science and Engineering **973** (2020) 012028 doi:10.1088/1757-899X/973/1/012028

controllable pre-cracks lengths and orientations, the standards warn against pressing a razor blade for ductile resins since it introduces a residual stress at the crack tip as shown in Figure 2(a). Souza et al. [7] have studied the effect of inserting a fresh razor blade on PMMA as a pre-cracking technique; and concluded that this technique produces non-planar propagation of the unstable crack started from the center of the pre-cracksince the resultant propagation show concentric lines initiated from the center as shown in figure 2 (b). The photo elasticity tests show a perturbed stress state at the crack tip as shown in Figure 2(c) leading to high K_{Ic} values.

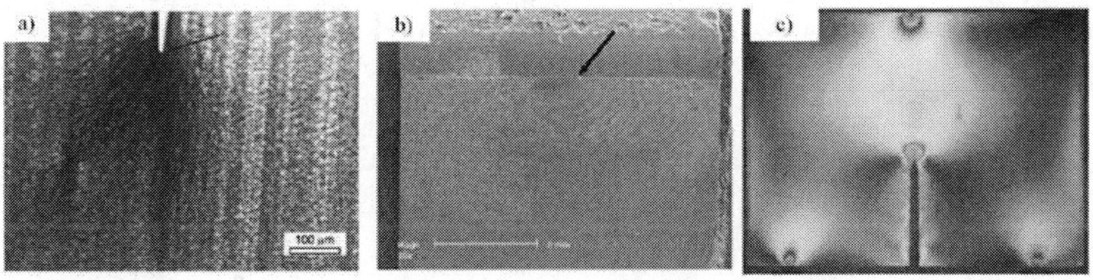

Figure 3. a) Residual stresses developed by inserting a razor blade[14]; b) Morphology for the case in which the pre-crack was produced by inserting a razor blade[7]; c) Moiré patterns produced in samples pre-cracked by inserting a razor blade[7]

3.1.2.Tapping a Fresh Razor Blade. Tapping a fresh razor blade manually in a center of the notch to produce pre-cracks as shown in Figure 3 (a) is acceptable technique according to the standards(ASTM D5045 and ASTM D6068). Souza et al. [7]have investigated the effect of tapping a fresh razor blade on PMMA; and concluded that this technique is very acceptable compared to inserting a fresh blade technique for PMMA since it produces planar propagation of the crack started from the root line of the pre-crack as show in Figure 3 (b).The frac to graphic patternshow less disturbed stresses at the crack tip as shown in figure 3 (c) which leads to lower K_{Ic} values. But this technique is very poor since it is depending on the operator skills. Positioning and orientation of razor blade, tapping force and pre-cracks lengths are very sensitive to the operator so there producibility of pre-cracks is impossible[9]. Some of the problems associated with the operator are highly eliminated by the pre-crack rig that introduced by Kuppusamy and Tomlinson [8]. The pre-crack rig is designed to control the length and the orientation of the produced pre-cracks by guide springs as shown in Figure 3 (d).Although many problems of this technique are eliminated by using the pre-crack rig, but the tapping force is still sensitive to operator skills and specimen material, blade usage is very risky, and tapping is not suitable for brittle polymers.

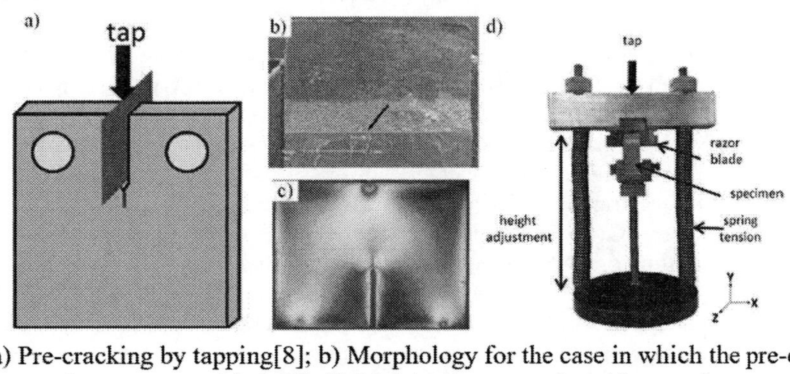

Figure 4. a) Pre-cracking by tapping[8]; b) Morphology for the case in which the pre-crack was produced by tapping on a razor blade[7]; C) Moiré patterns produced in samples pre-cracked by tapping on a razor blade[7]; d)pre-cracking rig[8].

3.1.3.Sliding a Fresh Razor Blade. According to the standard(ASTM D5045 and ASTM D6068), pre-cracks can be obtained either by sliding a new razor blade in a sawing motion (side-to-side) or slicing repeatedly in a single direction manually or using a suitable machine. Although this technique is allowed by standards, it is uncommon technique and is rarely used.

3.2.Fatigue Pre-Cracking

Fatigue pre-cracking is a common used technique especially for metals pre-cracking using a fatigue testing machine but it is restricted for polymers since it has the following limitations: it is very time-consuming technique, producing unstable crack growth, and it is very difficult to crack the brittle material without failure [8]. Moreover, the loading frequency must be lower than 4 Hz to minimize the residual stresses at the crack tip that introduced by the hydrostatic heating [24]. An attempt was performed to grow a pre-crack by fatigue in polyethylene [5]; the results showed the same trend as that obtained by standard razor blade[6].Moreover, it was reported that the damage developed ahead the crack tip is more than that developed by pressing razor blade leading to high fracture resistance.

3.3.Liquid Nitrogen Assisted Pre-Cracking

A very sharp pre-cracks can be obtained via pressing or tapping on razor blade after immersing the polymer in liquid nitrogen for a period of time. As known that the freezing process embrittle the polymers. Brown et al. [5]have investigated producing pre-cracks for tough polyethylene using liquid nitrogen by two methods. First method, initial notch is produced prior immersing the specimen into liquid nitrogen then additional notch is produced. Second method, after producing the initial notch the specimen is loaded during cooling process. After pre-cracking the specimens are tested at 80 °cunder maximum stress of 2.4 MPa. The results indicated that both methods have a slight effect on the failure time. Another attempt was conducted by [11]to study the influence of pre-crack technique on K_{Ic} for medium density polyethylene (MDPE) with cryogenic conditions. K_{Ic} test is conducted after pre-cracking and cooling the specimen in liquid nitrogen (-196 °c). The pre-cracking was produced by two techniques (inserting and tapping) fresh razor blade. The results showed that freezing the polymer in liquid nitrogen is a good method to obtain proper K_{Ic} values.

3.4.Unconventional Procedures

3.4.1.Tension and Compression loads. Tamura et al. [12] have developed a technique to grow a pre-crack in brittle polymeric materials. This technique depends on applying load in two directions as shown in Figure 4 (a). A tension load is applied in the direction of crack opening causing crack growth. And compression load is applied in the transverse direction, ahead of the notch causing crack arrest at the required crack length. Although this technique introduced a safe manner to produce pre-cracks, it has some problems to obtain repeatable pre-cracks front shape. Pre-cracks front shape problem that that induced by Tamura technique have been treated in Kuppusamy and Tomlinson [8]. In that investigation, the effect of contact area type between the specimen surfaces and clamping that apply the compressive load on pre-crack front shape for Arcan-type specimens [25] made from an untoughened epoxy resin used in aerospace composites has been explored as shown in Figure 4 (b)The results showed that the rectangular contact provided by (T clamp) type generate a very straight crack front whereas ball contact (G clamp) generate a curved crack front. Kuppusamy and Tomlinson [8]. have determined the optimum compressive load required to produce a pre-crack of a known length for untoughened epoxy modified Arcan specimen using a slipper torque wrench on (T clamp). If the compressive load is insufficient the pre-crack will be larger than the desired length. if the compressive load is large the pre-crack will travel along the axis then grow perpendicular to the pre-crack axis. The results show that the optimum compressive and tensile loads required to grow a pre-crack for a 4mm thickness specimen are 300 N. This technique shows high repeatable, controllable pre-crack lengths and shapes with minimum residual stresses at the crack tip. As shown in Figure 4-d-f, T-clamp method did not introduce any residual stresses at the crack tip.

IOP Conf. Series: Materials Science and Engineering **973** (2020) 012028 doi:10.1088/1757-899X/973/1/012028

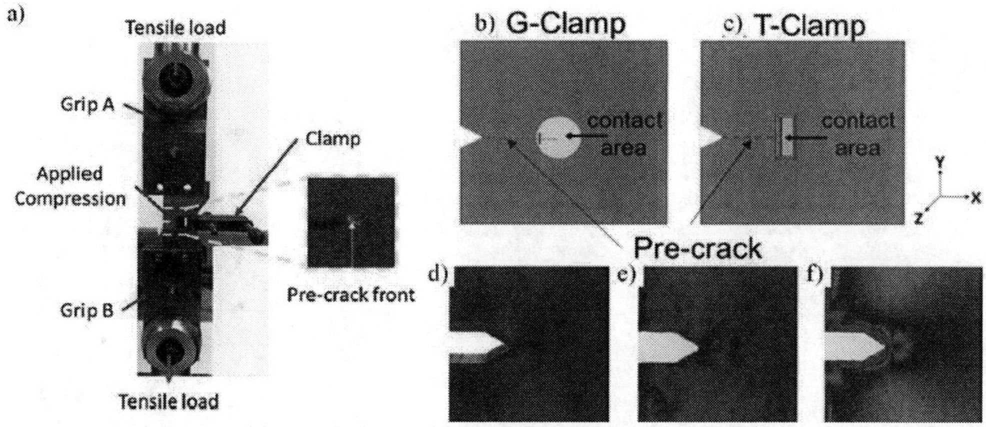

Figure 5. a) Pre-cracking technique without using a blade; b)Compressive load contact area (G clamp); c) Compressive load contact area (T clamp);(d-f)Showing residual strain CNC-machined specimens with a crack introduced (d) using the T-clamp method; (e) using razor tapping; and (f) a specimen with a notch only [8].

3.4.2.Noncontact Technique (Femtosecond Laser Ablation). Femtosecond laser pulses [26] are used to ablate polymers since 1994 [27]. Many researchers studied the effect of femtosecond laser pulses on polymers [28]and polymer based composites [29]and show minimum thermal damage around the ablation area. Recently Femto-laser technique is used to prepare specimens of fracture toughness tests since it produces very sharp pre-cracks with a very small crack tip radius as shown in Figure 6 (a)and minimum residual stresses a head the crack tip leading to accurate determination of K_{Ic} and G_{Ic}[13]and [14].Thermoplastic polymers such as (semi-crystalline multi-phase ethylene–propylene block copolymer and an amorphous polycarbonate) are better to pre-cracked by femto-laser techniques since very small crack tip can be achieved resulting in proper fracture toughness values that can be explained that the plastic deformation developed at the crack tip is very small compared to the damage resulted from the conventional techniques [10]. Also, femto-laser technique produces a well-defined pre-crack that are independent on the dimensions of specimens, on the contrary of the conventional techniques in which the higher specimen width the more damage introduced at the crack tip [15].. The influence of different pre-cracking techniques (contact conventional techniques such as razor blade tapping, sliding, pressing, broaching opposite recent non-contact techniques such as femto-laser ablation) on the fracture parameters have been analyzed for SENB specimen made of poly (ethylene terephthalate) modified glycol (PETG)by Martínez et al.[30]. As shown in figure 6 (b),sharp pre-cracks with 0.5 μm tip radii can be obtained via the non-contact femto-laser technique, leading to the low values of fracture toughness. But the conventional tapping on razor blade technique generated large crack tip resulted in higher values of fracture toughness. Broaching and pressing razor blade techniques produced smaller crack tip radii compared to pre-cracks produced by razor tapping, and hence leading to intermediate values of fracture toughness.

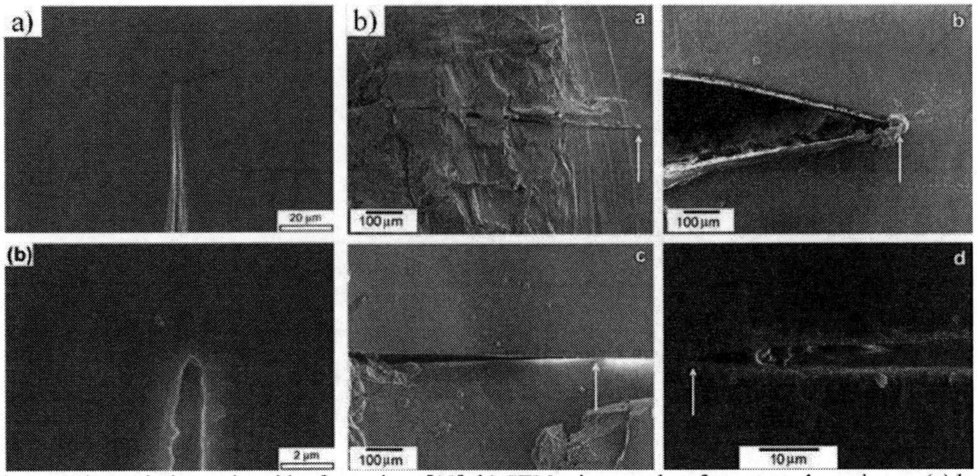

Figure 6 a) Crack tip produced by femto-laser[10]; b) SEM micrographs of non-tested specimens (a) by Tapping, (b) by sharp tool (c) by Pressing, (d) by femto-laser. The arrow points out the end[9].

4. Conclusion

Fracture toughness and initiation stage of the long-term fatigue tests results of polymeric materials are very sensitive to the pre-cracking methodology. Conventional techniques (tapping on, pressing, slicing by) a fresh razor blade that specified in the standards procedures are insufficient and unsuitable for all polymers and need more study since resulted large damage at the pre-crack tip is observed leading to variance in test results. Pre-cracking procedures must be specified independently for ductile and brittle polymers to facilitate the selection of proper pre-cracking technique. New promising pre-cracking technique is developed via femto-laser. Femto-laser produces very small crack tip radii with minimum damage independent on the tested specimen dimensions leading to accurate fracture toughness values.

References

[1] Katunin A, Krukiewicz K, Turczyn R, Sul P, Łasica A and Bilewicz M 2017 Synthesis and characterization of the electrically conductive polymeric composite for lightning strike protection of aircraft structures *Composite Structures* **159** pp 773-83

[2] Elsheikh A H, Deng W and Showaib E A 2019 Improving laser cutting quality of polymethylmethacrylate sheet: experimental investigation and optimization *Journal of Materials Research and Technology*

[3] Showaib E A and Elsheikh A H 2020 Effect of surface preparation on the strength of vibration welded butt joint made from PBT composite *Polymer Testing* **83** 106319

[4] E399-90 A 1997 Standard Test Method for Plane-Strain Fracture Toughness of Metallic Materials. ed A International: ASTM International)

[5] Lu X, Qian R and Brown N 1991 Notchology-the effect of the notching method on the slow crack growth failure in a tough polyethylene *Journal of Materials Science* **26** pp 881-8

[6] Dapp E K and Rimnac C M 1998 Effect of Precracking Method on the Static (J-integral) Fracture Resistance of Ultrahigh Molecular Weight Polyethylene *ASME-PUBLICATIONS-BED* **39** 349-50

[7] de Souza J M, Yoshimura H N, Peres F M and Schön C G 2012 Effect of sample pre-cracking method and notch geometry in plane strain fracture toughness tests as applied to a PMMA resin *Polymer Testing* **31** pp 834-40

[8] Kuppusamy N and Tomlinson R A 2016 Repeatable pre-cracking preparation for fracture testing of polymeric materials *Engineering Fracture Mechanics* **152** pp 81-7

[9] Martínez A B, León N, Arencón D, Rodríguez J and Salazar A 2013 On the effect of the different notching techniques on the fracture toughness of PETG *Polymer Testing* **32** pp 1244-52

[10] Salazar A, Rodríguez J and Martínez A B 2013 The role of notch sharpening on the J-fracture toughness of thermoplastic polymers *Engineering Fracture Mechanics* **101** pp 10-22

[11] Peres F M, Schön C G and Tarpani J R 2010 Effect of precracking method on KIc results for medium-density polyethylene tested under cryogenic condition *Polymer Testing* **29** pp 667-73

[12] Tamura K, Tamura S and Hashimoto A 2012 Precrack Introducing Method in CT-Specimens for Measuring K Values of Brittle Materials. In: *Porto, 15th International Conference on Experimental Mechanics,*

[13] Salazar A, Rodríguez J, Segovia A and Martínez A B 2010 Influence of the notch sharpening technique on the fracture toughness of bulk ethylene–propylene block copolymers *Polymer Testing* **29** pp 49-59

[14] Salazar A, Rodríguez J, Segovia A and Martínez A B 2010 Relevance of the femtolaser notch sharpening to the fracture of ethylene–propylene block copolymers *European Polymer Journal* **46** pp 1896-907

[15] Salazar A, Segovia A, Martínez A B and Rodríguez J 2010 The role of notching damage on the fracture parameters of ethylene-propylene block copolymers *Polymer Testing* **29** pp 824-31

[16] Chudnovsky A and Shulkin Y 1999 Application of the crack layer theory to modeling of slow crack growth in polyethylene *International Journal of Fracture* **97** pp 83-102

[17] Hamouda H B H, Simoes-betbeder M, Grillon F, Blouet P, Billon N and Piques R 2001 Creep damage mechanisms in polyethylene gas pipes *Polymer* **42** pp5425-37

[18] Brown N 2007 Intrinsic lifetime of polyethylene pipelines *Polymer Engineering & Science* **47** pp 477-80

[19] Bhattacharya S K and Brown N 1984 Micromechanisms of crack initiation in thin films and thick sections of polyethylene *Journal of Materials Science* **19** pp 2519-32

[20] Lu X and Brown N 1986 The relationship of the initiation stage to the rate of slow crack growth in linear polyethylene *Journal of Materials Science* **21** pp 2423-9

[21] Brown N and Wang X-q 1988 Direct measurements of the strain on the boundary of crazes in polyethylene *Polymer* **29** pp 463-6

[22] Huang Y L and Brown N 1988 The effect of molecular weight on slow crack growth in linear polyethylene homopolymers *Journal of Materials Science* **23** pp 3648-55

[23] Lu X, Wang X and Brown N 1988 Slow fracture in a homopolymer and copolymer of polyethylene *Journal of Materials Science* **23** pp 643-8

[24] Moore D R, Williams J and Pavan A 2001 *Fracture mechanics testing methods for polymers, adhesives and composites* vol 28: Elsevier)

[25] Banks-Sills L, Arcan M and Bortman Y 1984 A mixed mode fracture specimen for mode II dominant deformation *Engineering Fracture Mechanics* **20** pp 145-57

[26] Chichkov B N, Momma C, Nolte S, von Alvensleben F and Tünnermann A 1996 Femtosecond, picosecond and nanosecond laser ablation of solids *Applied Physics A* **63** pp 109-15

[27] Kumagai H, Midorikawa K, Toyoda K, Nakamura S, Okamoto T and Obara M 1994 Ablation of polymer films by a femtosecond high‐peak‐power Ti:sapphire laser at 798 nm *Applied Physics Letters* **65** pp 1850-2

[28] Baudach S, Bonse J, Krüger J and Kautek W 2000 Ultrashort pulse laser ablation of polycarbonate and polymethylmethacrylate *Applied Surface Science* **154-155** pp 555-60

[29] Moreno P, Méndez C, García A, Arias I and Roso L 2006 Femtosecond laser ablation of carbon reinforced polymers *Applied Surface Science* **252** pp 4110-9

IOP Conf. Series: Materials Science and Engineering **973** (2020) 012028 doi:10.1088/1757-899X/973/1/012028

[30] Martínez A B, Segovia A, Gamez-Perez J and Maspoch M L 2009 Influence of femtolaser notch sharpening technique in the determination of essential work of fracture (EWF) parameters *Engineering Fracture Mechanics* **76** pp 1247-54

Abrasive water jet machining of CFRPs: single response optimization using taguchi method optimization

I I Edriys[1], M Fattouh[2] and R Masoud[2]

[1]Kaha for chemical industries (270 army factor), Egypt
[2]Faculty of Engineering, Minufiya University, Shebin El Kom, Egypt
E-mail: Islam_edriys@yahoo.com

Abstract. Taguchi method was applied to assess and optimize machining parameters and their effect on kerf characteristics during abrasive water jet machining (AWJM) of carbon fiber reinforced polymeric composites (CFRPCs). The main responses selected for these analyses are kerf width, kerf taper, metal removal rate, and surface roughness, the consistent machining parameters focussed for this study are abrasive flow rate, pressure, traverse rate, thickness of the workpiece and standoff distance, each parameter has three levels. Twenty-seven experiments were conducted on a typical CFRP composite workpiece materials based on Taguchi L27 design. The response curves and response tables were used to assess the data obtained to control the major significant process factors statistically affecting the kerf characteristics. The optimal settings of process parameters for each response are set up. From the analysis, it was detected that the percentage contribution of the control factors affecting the kerf width is standoff distance, workpiece thickness, abrasive flow rate, traverse rate, and pressure correspondingly. The results exposed that the thickness, feed rate, and standoff distance were the most significant factors affecting the kerf taper, metal removal rate, and surface roughness respectively.

Introduction

Abrasive water jet machining (AWJM) is one of the most newly developed non-traditional cutting processes. It uses a fine jet of ultrahigh-pressure water and abrasive slurry to cut the board material by means of corroding. AWJ cutting is being progressively used to machine a wide-ranging of metals and non-metals, mainly 'difficult-to-cut' materials such as ceramics, marble, and fiber-reinforced polymeric composites, due to its various different advantages over other technologies such as no thermal distortion, high machining adaptability, ability to contour and small cutting force [1-2]. Since the introduction of the AWJ cutting technology for commercial use, a large amount of researches and progress has been made to discover its applications and related science [4-24]. However, this technology is still below progress and there are many features of the technology that persist to be fully understood [8]. The work presented in this paper describes experimental work that has been undertaken with the objective of improving the current absence of understanding in the AWJM of CFRP composites. The experimental results can be used to provide approvals for the selection of cutting parameters for AWJM applications. Definitely, the objective of the research described in this study is as follows: To achieve a detailed experimental study of the kerf characteristics when AWJM of CFRPCs to gain a complete understanding of the effects of several major process variables and to give optimum cut quality using Taguchi method.

IOP Conf. Series: Materials Science and Engineering **973** (2020) 012029 doi:10.1088/1757-899X/973/1/012029

Experimentation

The experimental work was accompanied on abrasive water jet cutting machine in which the tests were made is an OMAX 5555 jet machining center (figure 1). This model cuts multifarious flat parts out of most materials directly from a CAD drawing or DXF file. It includes a completely sealed and safe ball screw drive system, providing strength and dependability while offering high correctness. mixing tube of 0.762mm in diameter and 76.2mm in length were used to produce the AWJ. Garnet abrasives of 80 mesh (0.18 mm average diameter and 4.1 g/cm3 density) were applied. The CFRP composite material specimens used in this study were fabricated by stacking prepare, which is composed of one-way carbon fiber and epoxy resin bidirectional (0–90) and specimens were stacked to a total of 12, 24 and 36 plies respectively. The specimens were fabricated by compressing the material at a curing temperature of 125 °C using a heater located inside a cavity, with a curing time of 180 min and a forming pressure of 5 kg/cm2 as revealed in figure 2. The mechanical properties of the carbon-fiber prepares are publicized in Table 1. The specimen was 50x50 mm in vertical and horizontal direction figure 3.

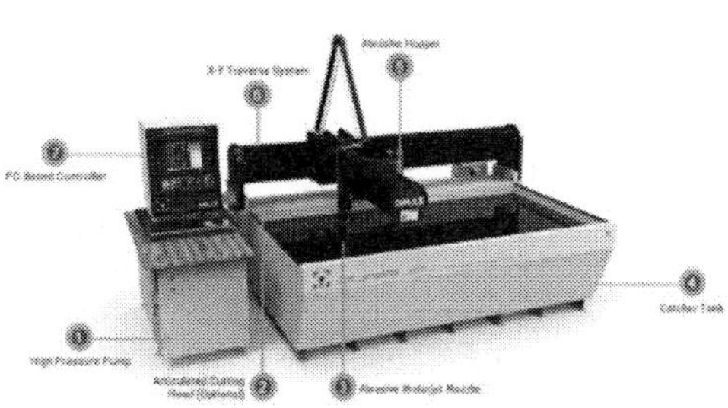

Figure 1. OMAX 5555 Jet Machining Centre

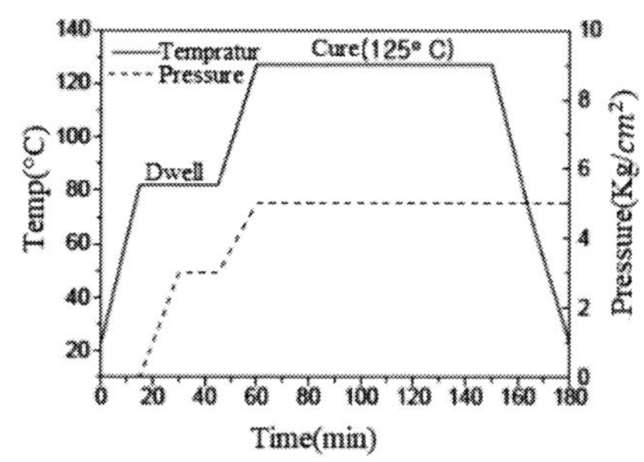

Figure 2. Stacking process for CFRPCs

The specimen was 50x50 mm in vertical and horizontal centring. Different workpiece geometries are planned to study the kerf characteristics in the straight cutting and profile cutting or contouring as presented in figure 3. The first geometry design is a straight cutting (kerf) as illustrated by figure 3. This design allows to measure the surface roughness on the depth of cut and to measure the distance between each cut side (kerf width) in the upper and lowest positions while providing the capability to picture the angle formed (kerf taper angle). MRR can be calculated. The second geometries design are square opining, hole opening, and external curving. The two holes characterized in the center of the plate were made for the resolve of holding it to the stand in the coordinate machine. In this paper, the experimental study is restricted to straight cutting under a range of AWJ process parameters.

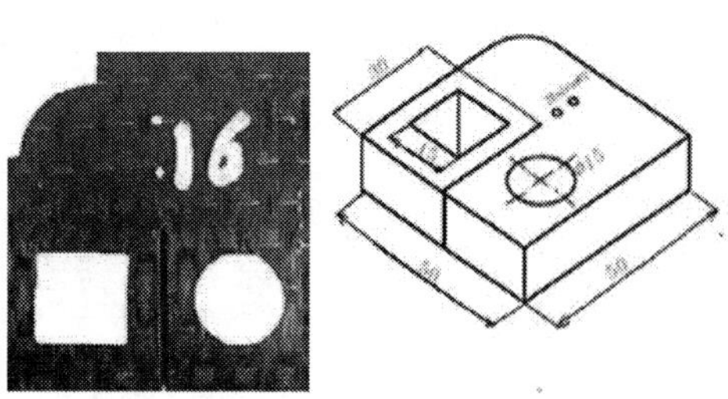

Figure 3. Geometry of the Workpiece.

Table 1. Mechanical properties of the Carbon-Fibre Sample

Properties	Value
Tensile strength (MPa)	5490
Tensile modulus (MPa)	294
Elongation (%)	1.9
Density (g/cm^3)	1.81

IOP Conf. Series: Materials Science and Engineering **973** (2020) 012029 doi:10.1088/1757-899X/973/1/012029

1.1. Experimental Design Methodology

A Taguchi orthogonal array is used in this investigation as an experimental plan. For ease and avoiding lengthy statement, the concept of Taguchi was offered and employed in Ross [3]. Table 2 illustrations the cutting parameters and their levels reflected for the experimentation on this paper. The parameters and levels were selected agreeing to the review of some papers [18-23] that has been recognized on AWJM on CFRPCs. Rendering to the Taguchi method, an L27 orthogonal array was working for the experimentation. Based on this, a total number of 27 experiments were done, each having a unlike combination of cutting parameters as shown in table 3. The responses were noted for each experimental run.

Table 2. Control factors (AWJM Parameters) and their levels.

Parameters	symbol	Level 1	Level 2	Level 3
Material Thickness (mm)	T	4	8	12
Water Pressure (MPa)	P	100	200	300
Traverse speed (mm/s)	V	1	3	5
Abrasive flow rate (g/min)	AFR	100	200	300
Standoff distance (mm)	SOD	2	4	6

Table 3. Orthogonal array.

Exp. No.	T	P	V	A.F.R	S.O.D	Exp. No	T	P	V	A.F.R	S.O.D
1	4	100	1	100	2	15	8	200	5	200	6
2	4	100	3	200	4	16	8	300	1	100	4
3	4	100	5	300	6	17	8	300	3	200	6
4	4	200	1	200	4	18	8	300	5	300	2
5	4	200	3	300	6	19	12	100	1	300	4
6	4	200	5	100	2	20	12	100	3	100	6
7	4	300	1	300	6	21	12	100	5	200	2
8	4	300	3	100	2	22	12	200	1	100	6
9	4	300	5	200	4	23	12	200	3	200	2
10	8	100	1	200	6	24	12	200	5	300	4
11	8	100	3	300	2	25	12	300	1	200	2
12	8	100	5	100	4	26	12	300	3	300	4
13	8	200	1	300	2	27	12	300	5	100	6
14	8	200	3	100	4						

1.2. Data Acquisition of Kerf Characteristics

Kerf width: A Portable 600x 3.6 MP digital microscope was used to measure the kerf width in upper and bottom positions. The measurements are occupied in three points: near to the upper edge, in the middle, near to lower edge of the Kerf and the usual value of measurements was used and preserved as the result of a single experiment for analysis (Figure 4).

Kerf taper: It was expected that in AWJ cutting the two kerf walls might not be regular due to the jet tailback effect. Figure 5. displays the kerf geometry.

The kerf taper can be acquired by measuring the upper and bottom kerf width and changing to taper kerf trend by the following relation:

$$\text{Kerf Taper Angle} \quad \theta = \tan^{-1}(W_t - W_b)/2T \qquad (1)$$

Where, W_t , W_b , T are the upper kerf width, bottom kerf width, and workpiece thickness correspondingly.

Metal Removal Rate (MRR): MRR for each experiment was calculated using the next formula:

$$MRR = 0.5\,(W_t + W_b).T.V \qquad (2)$$

Where V is cutting speed and unit for MRR is mm^3/sec.

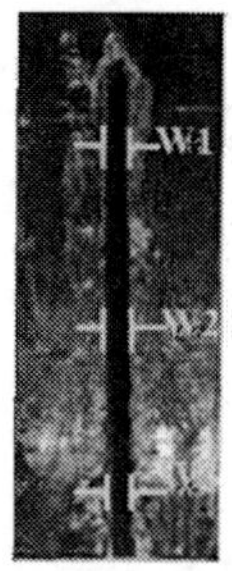

Figure 4. Schematic Illustration of kerf image.

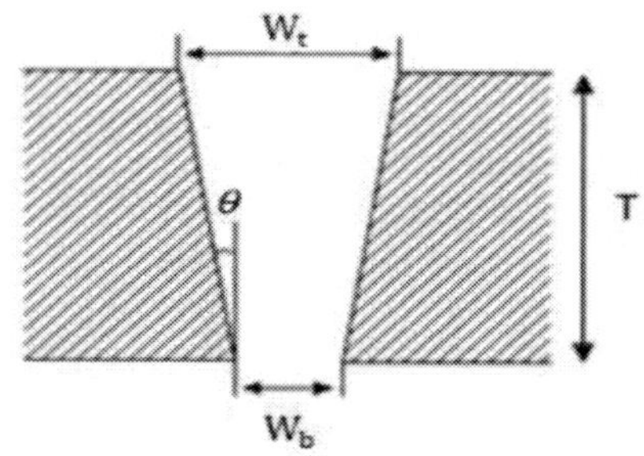

Figure 5. Schematic Illustration of Kerf Geometry.

Surface Roughness (R_a): The surface texture parameter for the Kerf throughout this investigated was the arithmetic mean roughness (R_a).

The measurements were taken at the direction of the cut in three areas: near to the upper edge, in the middle, near to lower edge in the center of the Kerf and the average value of measurements was used and treated as the result of a single experiment for analysis, figure 6. Figure 7 illustrations the kerf image.

The results of the four cut quality characteristics of kerf specifically, upper and bottom kerf width, taper angle, MRR and Ra for each of 27 trials are listed in Table 4. conferring to the performed experiment design.

Table 4. Results of experimental plan.

Exp. No.	Kerf (mm)		Taper Angle θ	MMR (mm^3/s)	SR (μm) Ra	Exp. No.	Kerf (mm)		Taper Angle θ	MMR (mm^3/s)	SR (μm)
	W_t	Wb					W_t	Wb			Ra
1	1.04	0.32	5.14	2.72	3.58	15	1.58	0.86	2.57	49.1	4.36
2	1.17	0.33	6.05	9.05	3	16	1.36	0.89	1.68	9.04	3.01
3	1.43	0.38	7.47	18.1	3.96	17	1.55	1.05	1.77	31.26	3.26
4	1.48	0.97	3.64	4.9	2.77	18	1.16	0.71	1.61	37.73	3.74
5	1.62	1.01	4.34	15.82	4.35	19	1.76	1.16	1.42	17.53	3.51
6	1.05	0.60	3.20	16.62	2.86	20	1.61	1.01	1.45	47.29	4.36
7	1.77	1.07	4.98	5.69	2.86	21	1.23	0.88	0.82	63.67	3.58
8	1.21	0.83	2.75	12.3	2.87	22	1.71	1.29	0.99	18.06	4.07
9	1.30	0.86	3.11	21.65	3.72	23	1.39	0.94	1.07	42.21	4.01
10	1.56	1	2.00	10.24	4.74	24	1.49	1.02	1.10	75.45	4.63
11	1.22	0.56	2.35	21.34	4.10	25	1.45	1.22	0.55	16.06	3.37
12	1.14	0.31	2.98	29.1	5.09	26	1.54	1.17	0.87	48.82	3.66
13	1.50	0.93	2.05	9.77	2.70	27	1.51	0.87	1.52	71.55	4.13
14	1.19	0.86	1.18	24.78	4.41						

IOP Conf. Series: Materials Science and Engineering **973** (2020) 012029 doi:10.1088/1757-899X/973/1/012029

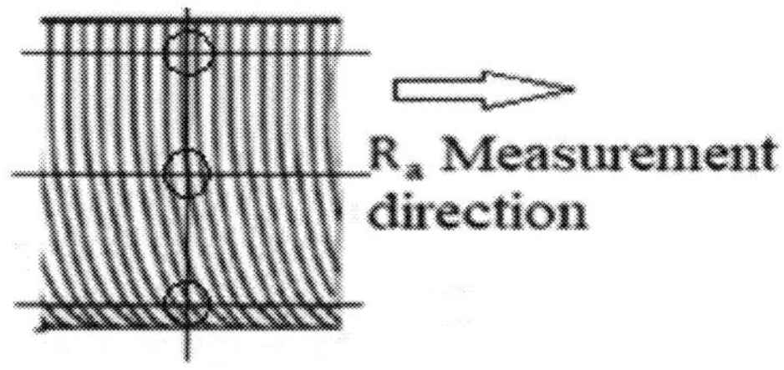

Figure 6. The Direction of Ra Measurement

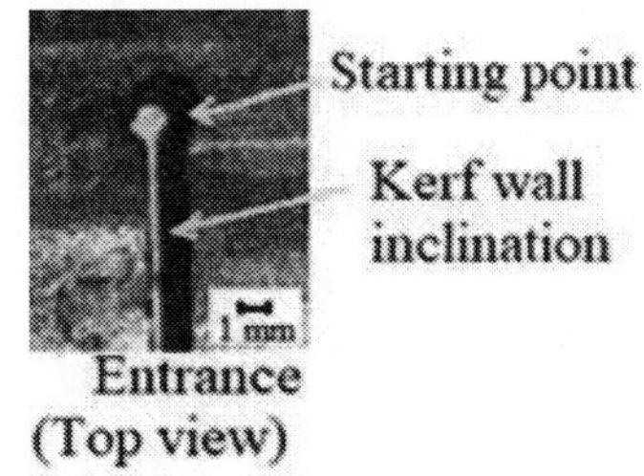

Figure 7. Image of Kerf Geometry

3. Result and Discussion

1.3. Effect of Process Variables on Kerf Top Width (W_t)

The average values of W_t for each parameter at levels 1, 2, and 3 for the raw data are schemed in figure 8. It shows that the W_t increases with the increase of T, AFR, and SOD where the standoff distance adopts the area of cutting which increases or decreases the impact area (whereas deviation of jet takes place with increase in stand-off distance), besides this, the effect of stray abrasive particles is prominent with high standoff distance with the rise of abrasive mass flow rate, cutting ability of jet that rises kerf top width and damage the surface. During increasing the abrasive particles (AFR), they own more energy to cut the material and they constantly lose the energy during decreasing its quantity. By the fact of increasing T, the time of piercing rise causes wider top kerf. Also, W_t decreases with increase in V due to less abrasive impingement at high traverse rate and results in less overlying of machining action, which reduces the kerf top width whereas the effect of water pressure is not important on W_t . It is also obvious that W_t is minimum at first level of SOD and maximum at third level of SOD.

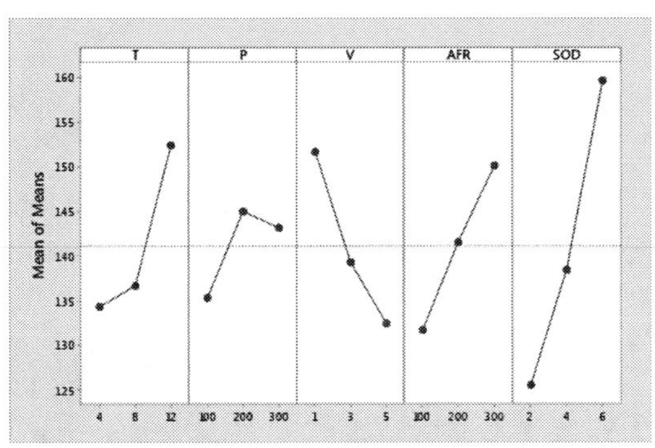

Figure 8. Effects of Process Parameters on W_t (Raw Data).

The response table 5 shows the average of W_t (S/N data) and its proportion contribution. The table contains ranks based on delta statistics, which equate the relative magnitude of effects. The delta statistic is the highest minus the deepest average for each factor. Minitab allocates ranks based on delta values; rank 1 to the highest delta value, rank 2 to the second-highest, and thus on. The ranks designate the relative importance of each factor to the response. The ranks and the delta values show that SOD has the highest effect on W_t and is tailed by T, AFR, V, and P in that order. From the analysis, it was saw that the percentage contribution of the control factors affecting the W_t is SOD (33.96%), T (18.81 %), AFR (18.78%), V (18.35%), and P (10.08%) one-to-one. As W_t is the "lower the better" type quality characteristic, it can be seen from figure 7 that the first level of T, first level of P, third level of V, first level of AFR, and first level of SOD offer minimum value of W_t. The S/N data analysis (Table 5) also advises the same levels of the variables (T = 4 mm, P =100 MPa, V= 5 mm/sec, AFR = 100 g/min and SOD 2 mm) as the best levels for minimum W_t. Because this combination parameters are

IOP Conf. Series: Materials Science and Engineering **973** (2020) 012029 doi:10.1088/1757-899X/973/1/012029

selected from the response table (table 5) conferring to Taguchi analysis, and this combination parameters are not found in the orthogonal array (table 3), the confirmation test analysis for W_t is processed. The result of confirmation test analysis is associated with the initial condition of setting parameters. It is clear from table 4, that experiment No.7 has the poorest W_t value (1.77 mm) compared to the other experiments. therefore, it can be decided that experiment No. 7 possesses initial setting parameters. Table 6 displays the comparative results of the near-optimum setting parameters (T= 4 mm, P=100 MPa, V= 5 mm/sec, AFR= 100 g/min and SOD = 2mm) and initial setting parameters (T = 4 mm, P = 300 MPa, V= 1 mm/sec, AFR = 300 g/min and SOD = 6 mm). For the single performance characteristic, the W_t is reduced from 1.77 mm to 1.03 mm. The forecast (calculated) of W_t using the optimal level setting parameters can be calculated from Minitab 17. The conforming improvement in W_t is 58 %.

Table 5. Response for Signal to noise ratios of kerf width.

Level	T	P	V	A.F.R	S.O.D
1	-2.434*	-2.505*	-3.521	-2.267*	-1.906*
2	-2.641	-3.140	-2.807	-2.976	-2.742
3	-3.618	-3.047	-2.365*	-3.450	-4.045
Delta	1.185	0.635	1.156	1.183	2.139
Rank	2	5	4	3	1
contribution	18.81%	10.08%	18.35%	18.78%	33.96%

Table 6. Results of confirmatory experiment of kerf width

Worst value (Exp. No.7 Table.4.5)	W_t= 1.772 mm
Near optimum combination	T=4 mm, P =100 MPa, V= 5 mm/sec, AFR = 100 g/min and SOD 2mm
Predicted value (TAGUCHI)	W_t = 0.948 mm
Experimental value	W_t = 1.03 mm
Improvement %	1.029/1.772 = 58%

Figure 9 displays that there is very weak interaction between the process parameters in affecting the W_t since the responses at different levels of process parameters for a given level of parameter value are nearly parallel. It can be realized from figure 10 that the data follow a roughly straight line in normal probability plot display that the data are normally scattered.

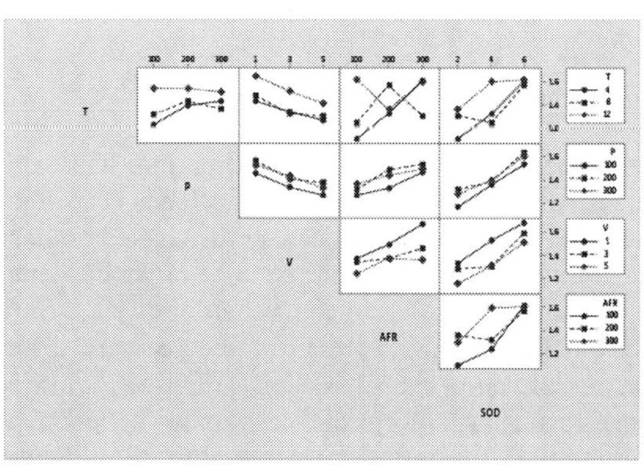

Figure 9. Effects of Process Parameters Interactions on W_t (Raw Data).

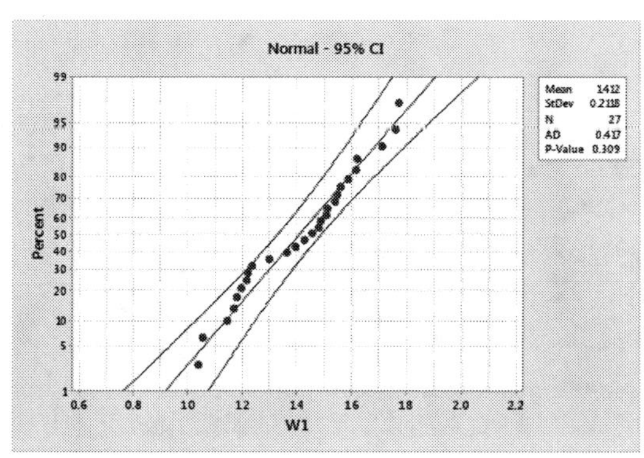

Figure 10. Normal Probability Plot for Kerf Top Width.

1.4. Effect of Process Variables on Kerf Taper (K_t)

Figure 11 illustrations that the K_t drops with the increase of T and P but increases with rise in AFR and SOD which the effect of V is frailly on K_t . During the jet penetrates into the material, the jet loses its kinetic energy when moving from upper surface to bottom surface, which reduced cutting ability constantly cases the kerf taper. The parameters similar jet pressure and feed rate decide the kinetic energy and the cutting time for the process and that control K_t. But the standoff distance agrees the area of cutting which rises or decreases the impact area cases high K_t at high level of SOD, with increase T the angle of cut by the jet penetration reduced cases little K_t.

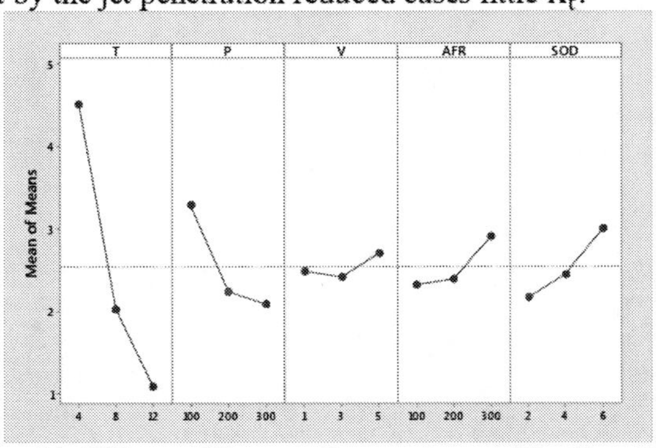

Figure 11. Effects of Process Parameters on $\mathbf{K_t}$ (Raw Data).

Table 7 displays the average of K_t (S/N data) and the optimal level setting parameters and its ratio contribution. Table 8 shows the forecast (calculated) value of K_t and the conforming improvement.

Table 7. Response for signal to noise ratios of kerf taper.

Level	T	P	V	A.F.R	S.O.D
1	-12.6725	-8.3357	-6.010	-6.1594	-4.98*
2	-5.8401	-5.7167	-5.975*	-5.5393 *	-6.1183
3	-0.3958*	-4.8559*	-6.922	-7.2097	-7.8039
Delta	12.2767	3.4798	0.947	1.6704	2.8178
Rank	1	2	5	4	3
contribution	57.93%	16.42%	4.4%	7.88%	13.29%

Table 8. Results of confirmatory experiment of kerf taper

Worst value (Exp. No.3 Table.4)	K_t = 7.477°
Near optimum combination	T=12mm, P =300MPa, V= 3mm/sec, AFR = 200 g/min and SOD 2 mm
Predicted value (TAGUCHI)	K_t=0.01°
Experimental value	K_t= 0.803°
Improvement %	0.803563/7.477 = 10.73%

Figure 12 displays that there is very frail interaction between the process parameters in touching the Kt since the responses at different levels of process parameters for a given level of parameter value are virtually parallel. It can be understood from figure 13 that the data follow a roughly straight line in normal probability plot indicates that the data are normally scattered.

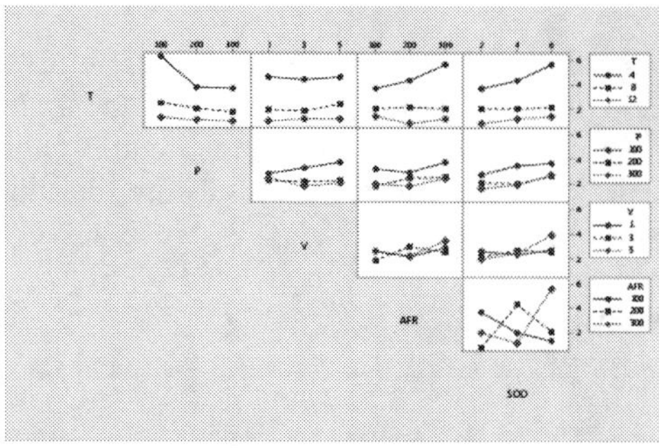

Figure 12. Effects of Process Parameters Interactions on K_t (Raw Data).

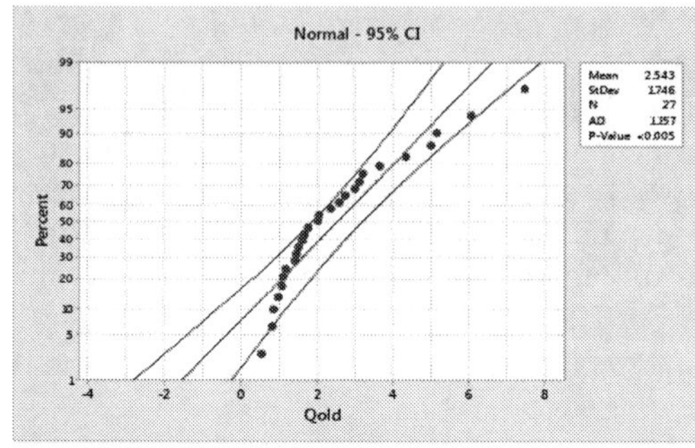

Figure 13. Normal probability Plot for Kerf Taper.

1.5. Effect of Process Variables on Metal Removal Rate (MRR)

Figure 14 displays that the MRR rises with the rise of T and V, and the effects of both P, AFR and SOD is frailly on MRR. By the fact of growing kinetic energy due to an increase in V, T, and P that give higher MRR which rises in feed rate and pressure the abrasive particle becomes less time to cut the higher material thickness and new particles arrive in cutting region. Also, aids to remove more volume of material. Also, the effect of SOD is frail because the deviation in jet and low kinetic energy of the abrasive particles due to more distance between the jet and the workpiece beside the sharp cutting of the material is not possible, the cutting ability reduced during traveling owing to distance travelled that reduces its capability of material removal but SOD increase the cutting region and remove higher MRR.

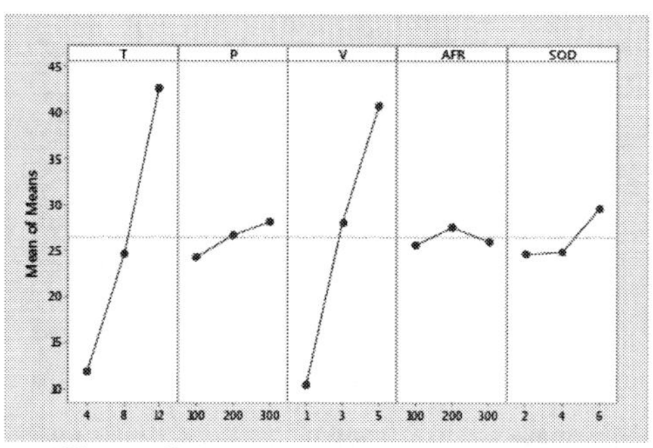

Figure14. Effects of process parameters on MRR (Raw Data).

Table 9 displays the average of MRR (S/N data) and the optimal setting parameters and its percentage contribution. Table 10 shows the forecast (calculated) value of MRR and the corresponding progress. The near optimum setting parameters (T = 12 mm, P =300 MPa , V= 5 mm/sec , AFR = 300 g/min and SOD = 6 mm).

Table 9. Response for Signal to noise ratios of MRR.

Level	T	P	V	A.F.R	S.O.D
1	19.87	24.83	18.98	25.21	25.06
2	26.46	26.54	27.67	26.25	25.79
3	31.63*	26.57 *	31.29*	26.49 *	27.10*
Delta	11.76	1.74	12.31	1.27	2.04
Rank	2	4	1	5	3
contribution	40.38	5.97%	42.27%	4.36%	7.00%

Table 10. Results of confirmatory experiment of MRR.

Worst value (Exp. No.1 Table4)	MRR = 2.72 mm
Near optimum combination	P =300 MPa, V= 5 mm/sec, AFR = 300g/min and SOD = 6 mm
Predicted value (TAGUCHI)	76.32 mm^3/sec
Experimental value	MRR= 75.69 mm^3/sec
Improvement %	2.72/75.69 = 3.59%

Figure 15 displays that there is very frail interaction between the process parameters in affecting the MRR since the responses at different levels of process parameters for a given level of parameter value are nearly parallel. It can be understood from figure 16 that the data follow an approximately traditional line in normal probability plot designates that the data are normally distributed.

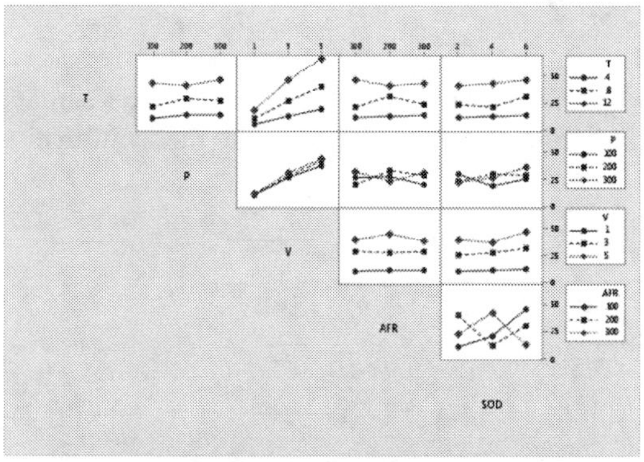

Figure 15. Effects of Process Parameters Interactions on MRR

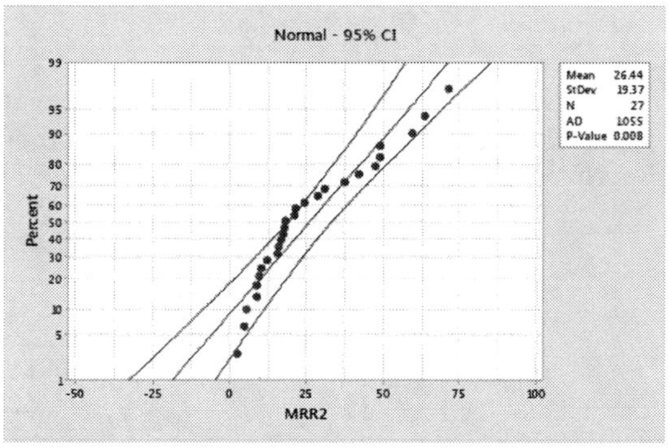

Figure 16. Normal Probability Plot for MRR

1.6. Effect of Process Variables on Surface Roughness (R_a)

Figure 17 displays that the R_a rises with the increase of T, V, and SOD, and decreases with an increase in P where lower pressure deteriorated the finish on the cut surface by creating lays and flaws with observed strong scratches and grooves, resulting in a poor finish. Also, with increasing the area of cut surface by increasing T there is more lays and flaws and the surface waviness produced a large difference between the peak and valley from the mean line, resulting in a poor finish. The excessive abrasives penetrate into the layers of material which result in abrasive embedment. Abrasive embedment is mainly observed at high AFR and low SOD. At low SOD, abrasives cannot accelerate with high-speed water jet which causes abrasives to impinge on material with low kinetic energy. These abrasives penetrate into the layers and machined surface cases rough surface. There are small effects of AFR. which the jet lag at higher V resulted because of insufficient time for cutting the CFRP. Thus, the fibers that poked out from the cut surface got forced up with the stylus probe during measurement, increasing the roughness. With the overlapping of machining action and also reduced number of abrasive particles to impinge on surface. Also, at low pressure, the surface waviness produced a large difference between the peak and valley from the mean line, resulting in a poor finish. However, the increased energy at high pressure improved the cutting efficiency and produced a smooth surface. Figure 18 displays the defects and damage in the cut surface.

IOP Conf. Series: Materials Science and Engineering **973** (2020) 012029 doi:10.1088/1757-899X/973/1/012029

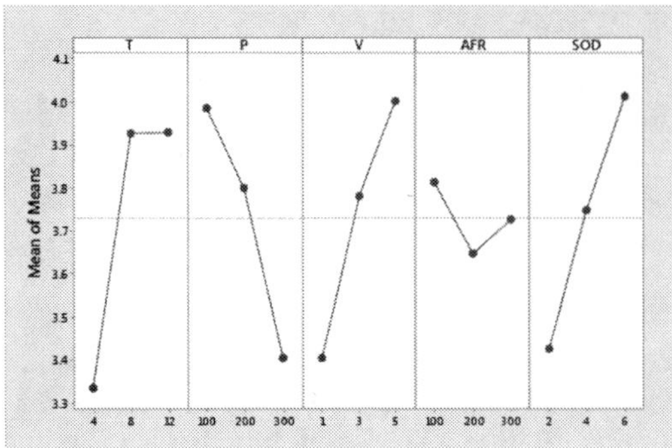

Figure 17. Effects of Process Parameters on R_a (Raw Data).

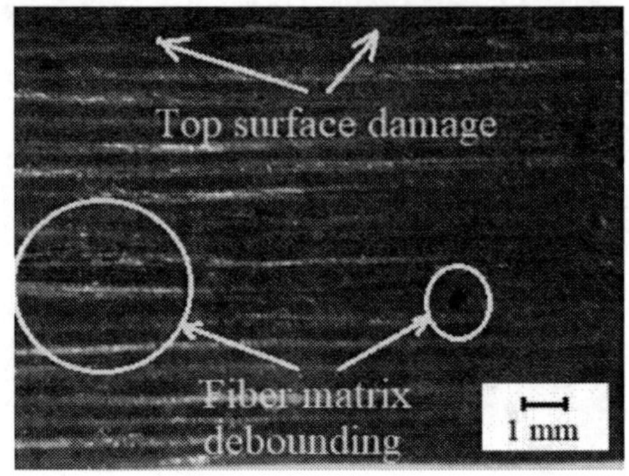

Figure 18. Stacking process for CFRPCs.

Table 11 displays the average of R_a (S/N data) and the optimal setting parameters and their percentage contribution. Which table 12 shows the predicted (calculated) value of R_a and the conforming improvement.

Table 11. Response for signal to noise ratios of surface roughness.

Level	T	P	V	A.F.R	S.O.D
1	-10.35*	-11.91	-10.50*	-11.46	-10.61*
2	-11.71	-11.41	-11.45	-11.12*	-11.31
3	-11.84	-10.58*	-11.94	-11.31	-11.98
Delta	1.49	1.33	1.44	0.34	1.37
Rank	1	4	2	5	3
contribution	24.95%	22.27%	24.12%	5.69%	22.94%

Table 12. Results of confirmatory experiment of surface roughness.

Worst value (Exp. No.12 Table.4.5)	R_a = 5.09 μm
Near optimum combination	T=4 mm, P =300 MPa, V= 1mm/sec, AFR = 200 g/min and SOD 2 mm
Predicted value (TAGUCHI)	R_a = 2.301 μm
Experimental value	R_a = 2.498 μm
Improvement %	2.698 /5.09 = 53.96%

Figure 19 shows that there is very weak interaction between the process parameters in affecting the R_a since the responses at different levels of process parameters for a given level of parameter value are greatest parallel. It can be seen from figure 20 that the data follow an approximately traditional line in normal probability plot indicates that the data are normally scattered.

IOP Conf. Series: Materials Science and Engineering **973** (2020) 012029 doi:10.1088/1757-899X/973/1/012029

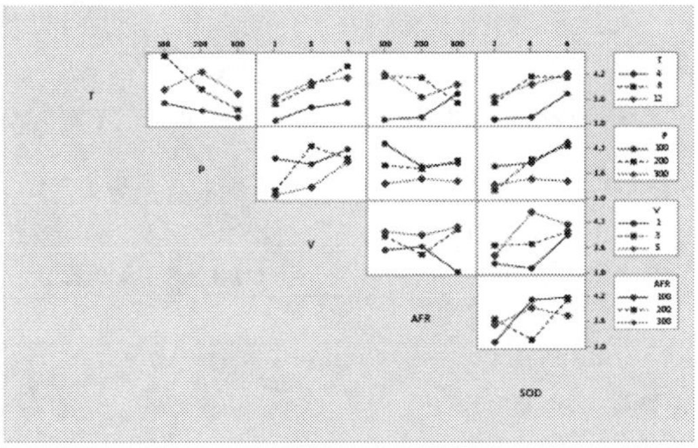

Figure 19. Effects of Process Parameters Interactions on Surface Roughness

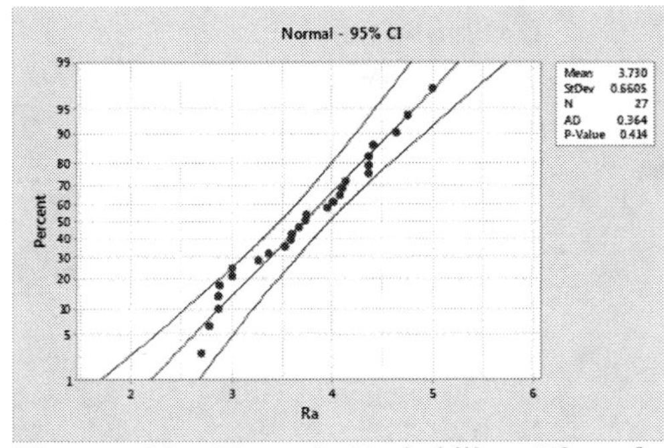

Figure 20. Normal Probability Plot for Surface Roughness

Conclusions

In cutting CFRPs using AWJM, numerous process parameters have effect on the performance measures. The effect of process variables on the characteristics of straight cutting was discussed. The optimal process parameters are found for various performance measures using the Taguchi design of experiment methodology. (single response optimization). The following assumptions can be drawn from the results of the present work:

1) SOD has the utmost effect on W_t and is trailed by T, AFR, V, and P in that order. From the analysis, it was detected that the percentage contribution of the control factors affecting the W_t is SOD (33.96%), T (18.81 %), AFR (18.78%), V (18.35%), and P (10.08%) respectively.
2) W_t rises with the increase of T, AFR, and SOD, also, W_t decreases with rise in V.
3) K_t decreases with the rise of T and P, also, Kt increases with rise in AFR and SOD. V has a very frail influence.
4) There is a very frail interaction by the process parameters on affecting the $\mathbf{K_t}$
5) MRR increases with the increase of T and V, and the outcome of P, AFR and SOD are frailly on MRR.
6) From the analysis, it was noticed that the percentage contribution of the control factors affecting the MRR is V (42.27%), T (40.38 %), SOD (7.00%), P (5.97%), and AFR (4.36%) respectively.
7) R_a rises with the increase of T, V, and SOD, and decreases with an increase in P.

References

[1] Luttervelt C A v 1989 On the selection of manufacturing method illustrated by an overview separation technique for sheet materials *Annals of the CIRP* **38**(2) pp 587-607.

[2] Ramulu M, and D Arola 1994 The influence of abrasive waterjet cutting conditions on the surface quality of graphite/epoxy laminates *International Journal of Machine Tools and Manufacture* **34** pp 295-313.

[3] Ross P J 1996 *Taguchi Technique for Quality Engineering second edition* (McGraw-Hill, NewYork).

[4] Arola D and M Ramulu 1996 A study of kerf characteristics in abrasive waterjet machining of graphite/epoxy composite *J. of engineering materials and technology* **118** pp 256-265.

[5] Hocheng H, Tasi H Y, Shiue J J and Wang B 1997 Feasibility study of Abrasive water jet milling of fiber-reinforced plastics *Journal of manufacturing science and engineering* **119** pp 133-142.

[6] Arola D and Ramulu M 1997 Material removal in abrasive waterjet machining of metals Surface integrity and Texture *Wear* **210** pp 50-58.

[7] Jun W 1999 A machinability study of polymer matrix composites using abrasive waterjet cutting technology *Journal of Materials Processing Technology* **94/1** pp 30-35.

[8] H Liu 2004 A study of the cutting performance in abrasive water jet contouring of alumina ceramics and associated jet dynamic characteristics (Queensland university of technology Ph

D Dissertation).
[9] Azmir M A and Ahsan A K 2009 A study of abrasive water jet machining process on glass/epoxy composite laminate *Journal of Materials Processing Technology* **209** pp 6168–6173.
[10] Tauseef U S, Mukul S and Pankaj B T 2009 Comparative investigation of abrasive water cut kerf quality characteristics for aramid, glass, and carbon fiber-reinforced composites used in transport aircraft applications *American WJTA Conference and Expo* August 18-20 (Houston, Texas).
[11] Ramulu M, I Hwang and V Isvilanonda 2009 Quality Issues Associated with Abrasive Water jet Cutting and Drilling of Advanced Composites *American WJTA Conference and Expo* (Hou., TX).
[12] Shanmugam D K and Masood S H 2009 An investigation on kerf characteristics in abrasive water jet cutting of layered composites *Journal of Materials Processing Technology* **209** pp 3887–3893.
[13] Alberta A, Suáreza A, Artazaa T, Escobar G A, Ridgwayb K 2013 Composite Cutting with Abrasive Water Jet *the Manufacturing Engineering Society International Conference* (MESIC) pp 421-429.
[14] Badgujar P P and Rathi M G 2014 Abrasive waterjet machining-A state of Art IOSR *Journal of Mechanical and Civil Engineering* **11** Issue 3 pp 59-64.
[15] Prasad D U, M D Gayakwad, N G Patil, R S Pawade, D G Thakur and P K Brahmankar 2015 Experimental Investigations into Abrasive Waterjet Machining of Carbon Fiber Reinforced Plastic *Journal of Composites* vol 2015 Article ID 971596 pp 9.
[16] Jignesh K P and Abdulhafiz A S 2015 The Influence of Abrasive Water Jet Machining Parameters on Various Response-A review *Int. J. Mech. Eng. & Rob. Res* **4** No 1 pp 383-403.
[17] Vishal S, R S J and Sakshi G 2015 A Review paper on most significant and subsignificant parameters of abrasive water jet machining *International Journal of Advance Research in Science and Engineering* **4** Iss 1 pp 249-256.
[18] Neelambari K S, Mangesh Y K and Vilas B S 2016 A Review on Parameters Optimization in Abrasive Water Jet Cutting *International Journal of Innovative and Emerging Research in Engineering* **3** Iss 2 pp 11-14.
[19] Rajyalakshmi M and P Suresh B 2016 Abrasive Water Jet Machining - A Review on Current Development *International Jour of Science Techn. & Engineering* **2**, Iss 12 pp 428-434.
[20] Rushabh B J, Dr A A Shaikh and Brijesh K G 2016 A Review on Abrasive Water Jet Machining On Different Material *International Journal of Engineering Development and Research* **4** Iss 2 pp 860-866.
[21] Ajit D, Shailendra K and R V Karmakar 2016 Abrasive Water Jet Machining of Carbon Epoxy Composite *Defence Science Journal* **66** pp 522-528.
[22] Thirumalai S K, Tae J K and M Uthayakumar 2017 Prediction of surface roughness in abrasive water jet machining of CFRP composites using regression analysis *Journal of Alloys and Compounds* **724** Iss 1 pp 249-256.
[23] Kumaran S T, Tae J K, R Kurniawan, C Li and M Uthayakumar 2017 ANFIS modeling of surface roughness in abrasive waterjet machining of carbon fiber reinforced plastics *Journal of Mechanical Science and Technology* **31**(8) pp 3949-3954.
[24] Irina W M M, Azwan I A, Lee C C and Ahmad F M 2017 Experimental study and empirical analyses of abrasive waterjet machining for hybrid carbon/glass fiber-reinforced composites for improved surface quality *The International Journal of Advanced Manufacturing Technology*.

Acknowledgments
Authors would like to thank Eng. Mohamed Daw director of Daw design Company at Cairo. Egypt, for providing the opportunity to accomplish the experimental work on OMAX 5555 Jet Machining Centre.

IOP Conf. Series: Materials Science and Engineering **973** (2020) 012030 doi:10.1088/1757-899X/973/1/012030

An experimental study on mechanical and ballistic characteristics of different HTPB composite propellant formulations

A F NourEldin[1], WMAdel[1], Y AAttai[2] and M A Ismail[2]
[1]Egyptian Armed Forces
[2] Mechanical Power Engineering Dept.,Faculty of Engineering,Helwan University
ahmednoureldin1212@gmail.com

Abstract. The main goal of this paper is to investigate the changing in the tensile behavior and the burning rate characteristics of hydroxyl-terminated polybutadiene (HTPB) propellantunder the variations of the crosslinking density, which was predominantly determined by the equivalent ratio ofdiisocyanate to total hydroxyl (NCO/OH ratio), which known as the curing ratio. Four variousbatches with different curing ratios (NCO/OH) percentage were produced in which the production processes were fixed. Uniaxial tensile tests were conducted at different temperatures (-40, 20 and 55°C), and different strain rates (0.000656 1/s, 0.0328 1/s) using a Zwick universal test machine. In order to measure the burning rate, cured solid propellant strands were tested using the acoustic wave emission method under different pressure ranging from 4 to 10 MPa. The experimental results indicate that the tensile behavior of HTPB propellant is remarkably influenced by curing ratio, strain rate, and temperature. It was observed that a great change on stress-strain curves affected various curing ratios and temperatures on the mechanical behavior of propellant composition. The results showed that high curing ratio leads to increase theultimatestress and decrease the strainat maximum stress, but higher temperatures lead to decrease theultimate stress and the strain at maximum stress.The curing ratios (NCO/OH) have an intense impact on mechanical characteristics, but slightly impact on ballistic characteristics for propellant. Furthermore, careful measurements of these parameters are important to control the production quality and to provide a reliable comparison between different propellant batches.

1. Introduction

Composite solid propellants have wide applications in strategic, tactical missiles, as it is responsible for delivering the required thrust to the missiles [1]. Additionally, it is characterized with its simplicity of design, simplicity of operation, safety, dependability, high specific impulse, stretched of life time and low cost [2,3]. Composite solid propellant is a heterogeneous mixture of solid particles including oxidizer and metallic fuel which are scattered in a polymeric matrix [4]. Hydroxyl terminated poly-butadiene (HTPB) constructed composite solid propellants are the most common classification of composite propellants due to their developed performance and conventional processing technology. Polymeric binder is answerable for stock the solid components into the matrix and providing the composite propellant with its viscoelastic nature [5].

Composite propellants are unprotected to thermal and mechanical loads during manufacturing, transportation, storage and operation, as a result composite formulation must endure throughout these numerous times [6]. Mechanical disappointment means a decrease in life time or it is a complete system failure [7]. The mechanical properties and life time of the composite propellant are mostly evaluation by applying uniaxial tensile [8,9]. Crafting design of composite propellant with perfect performance parameters, ballistic characteristics and mechanical properties are critical features for propellant formulation selection [10,11]. The composite solid rocket propellant (CSRP) burning rate has been reported to be an essential ballistic parameter that must be predicted and controlled throughout the combustion process [12-14]. The result show that, a smallchange in operating pressure would result in a dramatic increase in the burning rate, which could lead to a catastrophic ignition process resulting in an explosion of the missile motor. The ballistic characteristics could be evaluated using small scale test motor, strand burners include (i.e. Crawford burner, Acoustic wave emission technique, etc.) [14 -16]. However, small scale test motor provides actual ballistic parameters, strand burners are characterized with its easiness, low price, high reliability although sample with minor amount of propellant are existing [16].

On the other hand, the mechanical behavior of solid propellants is mainly influenced by the polymeric nature of binder, curing agent and interaction between binder and solid elements [17]. The curing ratio (NCO/OH) plays a critical role because it has a significant impact on the crosslinking kinetics and on the mechanical properties of the solid propellant composition [18]. In this study, the impact of the curing ratio (NCO/OH) on the mechanical and ballistic characteristic of composite propellants was successfully estimated using uniaxial tensile test and acoustic wave emission technique respectively [19]. The tensile tests were performed at different operating temperatures (-40, 20 and 55°C) and different strain rates (0.000656 1/s, 0.0328 1/s) using Zwick universal test machine. For burning rate measurement, the acoustic emission system was used at different pressures (40, 50, 68, and 90 bar) as a superior tool for measuring statistical effects in small preparation changes for quality control work and it is used as a regular technique of propellant strands in full-scale motors. Such technique was mentioned previously by some researchers [20, 21 and 22].

2. Experimental Work

2.1. Mechanical Characteristics Evaluation

2.1.1 Propellant Formulations and Specimens Preparation. In this work, four different propellant formulations containing 87% solids and 13% HTPB based binder with different curing ratio (NCO/OH) as listed in lable 1 were produced. The main solid additives include ammonium perchlorate (AP) as an oxidizer, aluminum powder (Al) as metal fuel. The employed binder composed of hydroxyl-terminated polybutadiene (HTPB) as pre-polymer, dioctylsebacate DOS as plasticizer, NG321 as a bonding agent based on aziridine groups, and hexamethylenediisocyanate (HMDI) as a curing agent. These gradients are mixed under vacuum at specified temperatures and times, so the newly prepared fresh viscous slurry is a vacuum discharge brick to produce bulk molds. Then the molds are placed in a large curing oven with temperature controlled at 55 °C, for a total curing time of 168 hours. After curing the molds are cut into sheets with a uniform thickness. Then the test samples are produced using special pneumatic cut press according to Joint Army- Navy-NASA-Air Force Propulsion Committee (JANNAF) standard, and the dimensions of the test specimens are illustrated in figure1 [14].

To ensure the results of the mechanical tests, the produced samples were checked by a non-destructive method for voids like air bubbles or micro cracks by X-Ray as a quality control step. After that, the accepted specimens were stored in desiccators at ambient temperature and relative humidity RH $\leq$ 30%.

Table 1. Composite Solid Propellant Formulations

	Binder HTPB (%)	Oxidizer AP (%)	Fuel AL (%)	NCO/OH
BM 10	13	70	17	1.0
BM 11	13	70	17	0.975
BM 12	13	70	17	0.94
BM 13	13	70	17	0.9

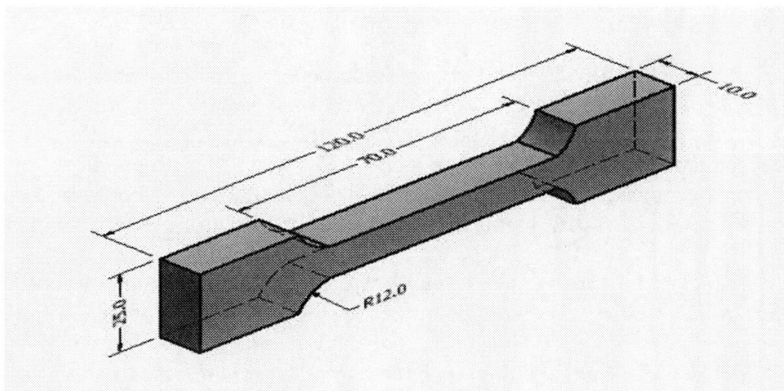

Figure 1. Main dimensions of JANNAF specimen.

*2.1.2 Tensile Tests.*All the tensile tests were conducted using a computer controlled universal test machine (Zwick /Roell) Z050. The machine was provided by a thermal chamber to set and keep the temperature constant at the desired temperature during the test. Before the test, the thickness and width of each specimen were measured in its gauge length region accurately. The effective gauge length (L_O) is 70 mm, and the crosshead speed of the machine was maintained constant during the test at the desired value according to the producerinstruction. Every experimental test was carried out on fivespecimens and the average value of the measurements was used for the analysis. Also, for every test the specimens were pre-conditioned for two hours in an external conditioning chamber. The specimens were tested at three different temperatures (20, 55, and -40 °C), and at two different crosshead speed (2, 100 mm/min).

The load-displacement data was used to generate stress-strain curves at different loading conditions, from which the ultimate stress, strain at maximum stress, were obtained for all solid propellant batches used in this investigation.

2.2. Ballistic Characteristics Evaluation

*2.2.1 Strands Preparation.*Ballistic characteristic of the applied formulations was evaluated applying acoustic wave emission technique where the specimens were cut into the specified position and direction of the block by using pneumatic grain cutter. Specimens must be in accordance with standard dimensions as (5 x 5 x 100) mm as shown in figure2. The strands were visually inspected and rejected if shape irregularities, cracks, or pores were observed.

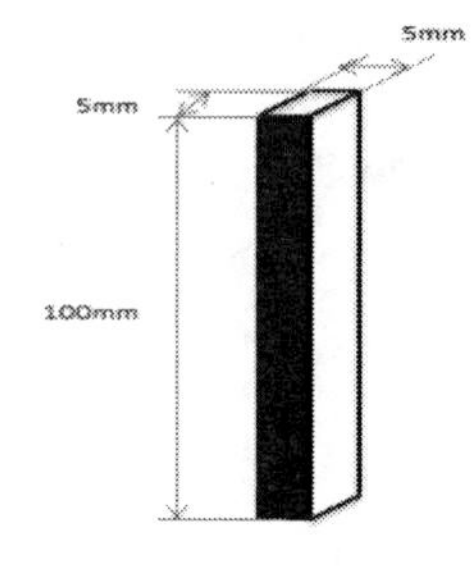

Figure 2. The dimension of the strand burning specimen.

2.2.2 Test Methodology. The propellant strand is mounted on a panel board with the help of Ni-chrome ignition wire and kept inside the bomb partially filled with water. The bomb is pressurized with nitrogen gas at the required pressure (4, 5, 6.8 and 10 MPa). The strand is ignited electrically by the firing unit. The acoustic emission (AE) wave generated due to the propellant burning in water is sensed by the acoustic emission transducer. The electrical signal is sequentially passed through the preamplifier and the post amplifier to the oscilloscope and the burning rate is calculated from the recorded graph.

This method has several advantages including low cost, a more accurate pressure value, and more reliable and accurate results. It avoids complex wiring, inhibition of the strands and sample preparation, and thus helps to improve the precision and accuracy of the measurement. An acoustic emission may be defined as a transient wave generated by the rapid release of energy within a material as sensed by a piezoelectric transducer. The signal passes through amplifier and recorded on an oscilloscope. A schematic diagram of an acoustic emission system is presented in figure 3.

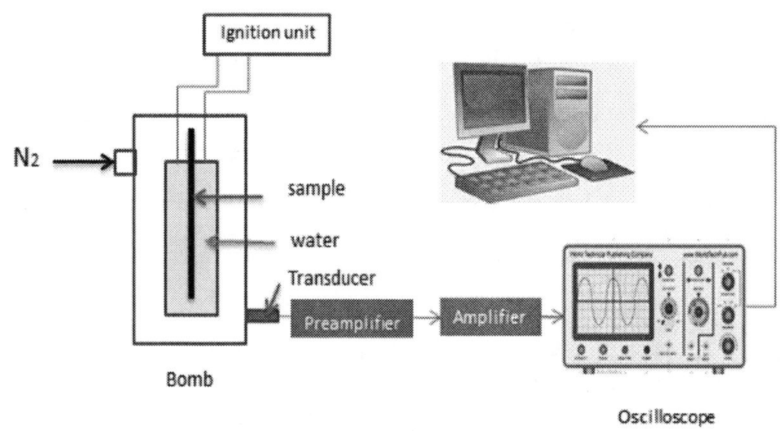

Figure 3. Acoustic Emission System (AES).

2.2.3 Burning Rate Calculation. The burning rate of propellants plays a vital role among the parameters controlling the operation of solid rocket motors; therefore, it is essential to surely measure the burning rate in the successful design of a solid missile motor. The burning rate of a solid propellant as a function of pressure is widely measured by an acoustic strand burner. The burning speed is calculated as per the following formula by equation (1), as mentioned in references[23,24].

$$r = \frac{L}{\Delta t} \tag{1}$$

where: L is the sample exact length (mm) and Δt is the burning time (s). Additionally, the pressure exponents as the main ballistic parameter could be calculated using the empirical vielle equation represented by equation (2),

$$r = aP^n \qquad (2)$$

where: r is the burning rate (mm/s), a is the burning rate constant, P is the operating pressure (bar) and n is the pressure exponent. Figure4 demonstrate how to obtain the burning rate from equation(2), then the calculated coefficient with the burning rate values were listed in table 2.

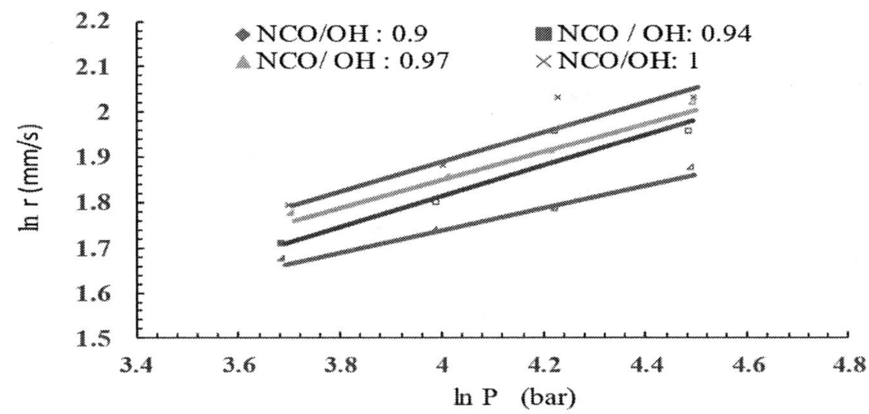

Figure 4.Chart of determination the ballistic characteristics.

Table 2.Main ballistic parameters of prepared formulations.

	0.9	**0.94**	**0.975**	**1.0**
a	1.366	1.136	1.127	1.583
n	0.392	0.440	0.444	0.365
r at (40 bar)	5.8	5.75	5.797	7.41
r at (50 bar)	6.33	6.35	6.4	6.6
r at (68.65 bar)	7.168	7.30	7.37	7.41
r at (90 bar)	7.97	8.227	8.31	8.18

3. Results and Discussion

During the tensile test, the physical nature of the propellant increases the sample size, due to the presence of vacuum holes around some crystalline fillings due to the breakage of glue bonds between the binder and oxidizer, resulting in the formation of voids. Two characteristic mechanical properties have to be determined in each single test, the ultimate stress (σ_m), and the strain at maximum stress (ε_m). The values of these parameters are changing with each propellant type, and also with the temperature, and strain rate. As seen in figures 5, 6, and 7 at all test temperatures and strain rates, after a rather straight initial part, the shape of stress-strain curves shows a sharp bend and then the stress continues to increase gradually over a prolonged strain range before rupture finally occurs.

IOP Conf. Series: Materials Science and Engineering **973** (2020) 012030 doi:10.1088/1757-899X/973/1/012030

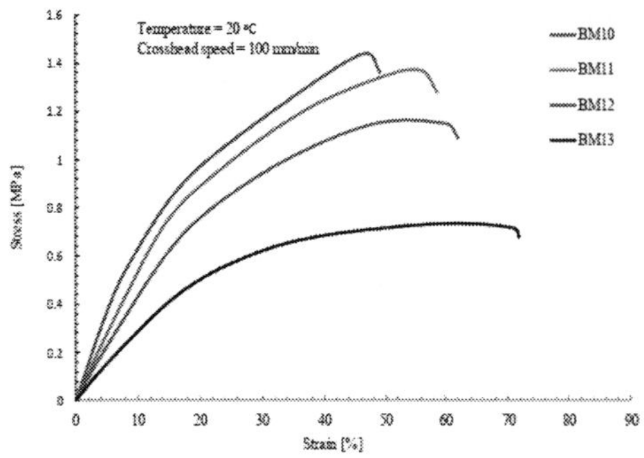

Figure 5. Stress-strain curves for different propellant batches at temperature (20°C).

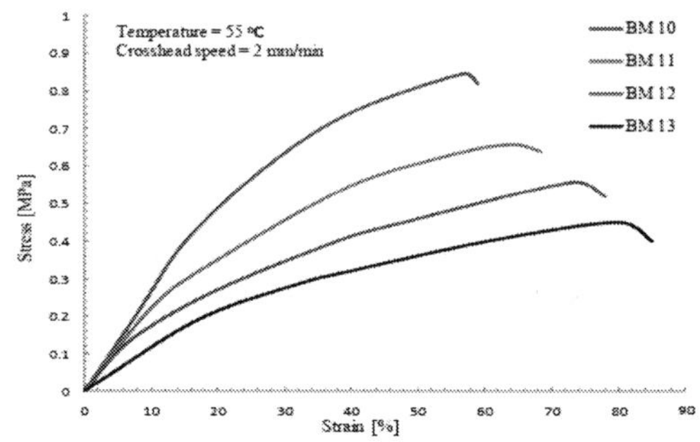

Figure 6. Stress-strain curves for different propellant batches at temperature (55°C).

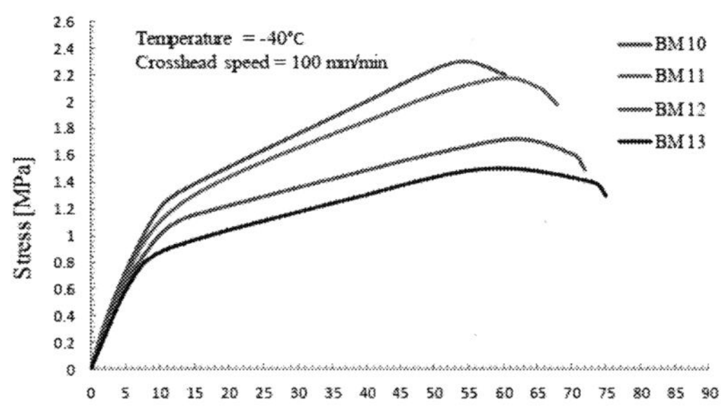

Figure 7. Stress-strain curves for different propellant batches at temperature (-40°C).

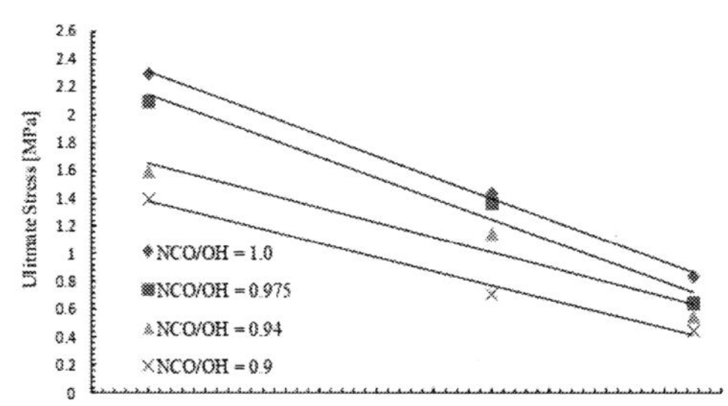

Figure 8. The ultimatestress at different test temperatures.

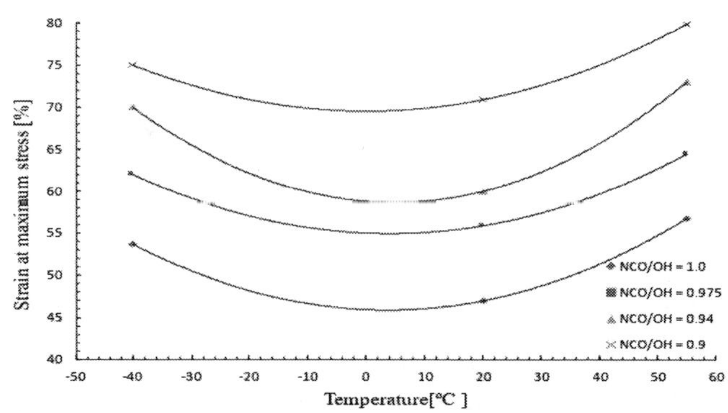

Figure 9. Thestrain at maximum stress at different test temperatures.

In figures 8 and 9 we can show clearly the effect of the temperature on the stress and strain respectively, the stress increased when the temperature decreased, but the strain increased when the temperature decreased from (20°C to -40°C), and also when the temperature increased from (20°C to 55°C) due to the nature of the bonding agent.For more details description, when the temperature decreased from (55°C to 20°C), and curing ratio (NCO/OH) is 0.9, the stress increased by about 58.89% and also the elongation decreased by about 11.28%. Also, when the temperature decreased from (20°Cto -40°C), the stress increased by about (95.8 %) and also the elongation increased by about 5.78%. When the temperature decreased from (55°C to 20°C), curing ratio (NCO/OH) is 0.94 , the stress increased by about 106.4 % and also the elongation decreased by about 17.8 %. Also, when the temperature decreased from (20°C to-40°C), the stress increased by about 39.13 % and also the elongation increased by about 16.67 %. When the temperature decreased from (55°C to 20°C), curing ratio (NCO/OH) is 0.975 , the stress increased by of about 108.2 % and also the elongation decreased by about 13.27 %. Also, when the temperature decreased from (20°C to -40°C), the stress increased by about 58.4 % and also the elongation increased by about 9.84 %.When the temperature decreased from (55°Cto 20°C), and curing ratio (NCO/OH) is 1, the stress increased by about 70.21 % and also the elongation decreased by about 17.21 %. Also, when the temperature decreased from (20°C to -40°C), the stress increased by about 60.42%, and also the elongation increased by about 10.91 %.

Whereas curing ratio (NCO/OH) increased from (0.9 to 1) at a temperature (20°C) the ultimate stress increases by (50.34 %), and maximum straindecreases by (33.71 %). Figures 10 and 11 demonstrate the effect of curing ratio (NCO/OH) on the mechanical properties of the propellant. The curing ratio is defined as the ratio between isocyanate (NCO) group and hydroxyl (OH) group to form the polyurethane (—O—CO—NH—), which is practically formatted within the curing period. By increasing the curing ratio the number of (NCO) groups increased and by default the density of crosslinking inside the polyurethane matrix increased, also the stress increased as seen in figure 10, maximum strain decreased at the ultimate stress as seen in figure 11, and finally the burning rate increased as seen in figure 12, due to the opportunity to form a large number of hard segments.

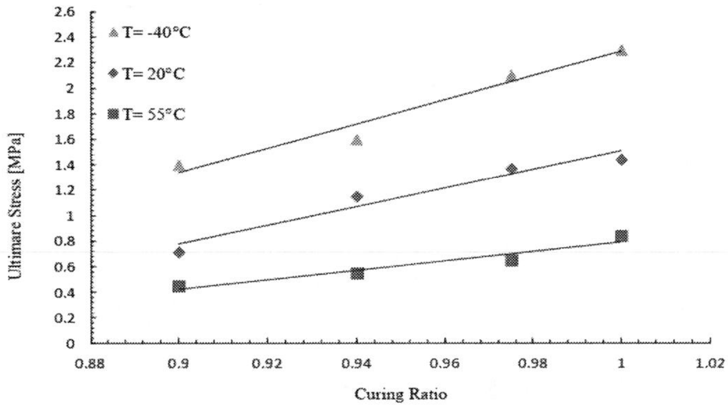

Figure 10.The ultimatstress at different curing ratio.

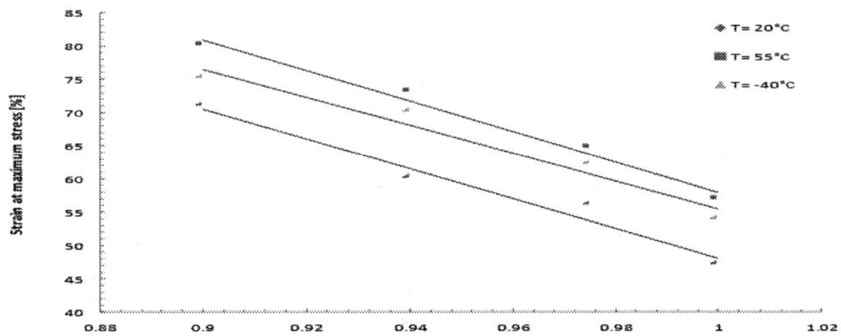

Figure 11.The strainat maximum stressat different curing ratio.

The tensile characteristics of a composite propellant depend on the properties of the polymeric binder matrix due to the differences in density of the crosslink, which is intensely affected by thecuring ratio NCO/OH . The differences in the mechanical characteristics of propellants with the curing ratioNCO/OH at the end of the curing timewere shown in figures 10 and 11 for different temperatures values for the mechanical characteristics of fivesamples .A careful check up of the results given in Table 3 specifies that the stress and strain at maximum stress , first diminution with the curing ratioNCO/OH , reaching a minimum at an curing ratio NCO/OH of 0.9,and then start to rise for the propellants with increase curing ratioNCO/OH.

Whereas for propellant at 55 °C with increase curing ratioNCO/OH from 0.9 to 0.94 the stress increase by 23.8% and strain at maximum stress decrease by 8.7%,whereas increase curing ratioNCO/OH from 0.94 to 0.975 the stress increase by 18.13% and strain at maximum stress decrease by 11.71% whereas increasecuring ratio NCO/OH from 0.975 to 1 the stress increase by 28.6% and strain at maximum stress decrease by 11.92 %.

Whereas for propellant at 20 °C with increasecuring ratio NCO/OH from 0.9 to 0.94 the stress increase by 45.5% and strain at maximum stress decrease by 15.4% ,whereas increasecuring ratio NCO/OH from 0.94 to 0.975 the stress increase by 19.13% and strain at maximum stress decrease by 6.8% whereas increasecuring ratio NCO/OH from 0.975 to 1 the stress increase by 5.11% and strain at maximum stress decrease by 15.9 %.

Whereas for propellant at -40 °C with increasecuring ratio NCO/OH from 0.9 to 0.94 the stress increase by 14.3% and strain at maximum stress decrease by 6.7%,whereas increasecuring ratio NCO/OH from 0.94 to 0.975 the stress increase by 35.6% and strain at maximum stress decrease by 11.4% whereas increase curing ratioNCO/OH from 0.975 to 1 the stress increase by 6.5% and strain at maximum stress decrease by 13.4 %.

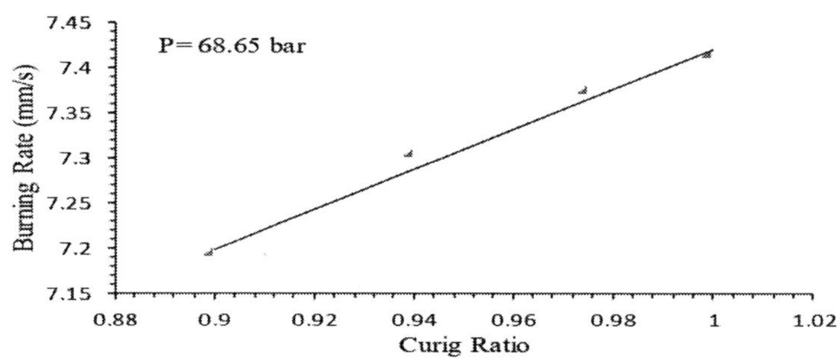

Figure 12. Burning rate VS. Curing ratio.

In figure 12, shows that, at the reference pressure, the burning rate increases slightly with increasing the curing ratio (NCO/OH), such slight increase was recorded to be around 3 %. For different batches, the values of the ultimate stresses and strains at maximum stress were listed in Table 3 for all temperature tests.

Table 3.Mechanical properties for propellants at different temperature and curing ratio.

NCO/OH) Ratio	55 °C		20 °C		-40 °C	
	σ_m(MPa)	□$_m$%	σ_m(MPa)	□$_m$%	σ_m(MPa)	□$_m$%
0.90	0.45	79.91	0.715	70.9	1.40	75
0.94	0.557	73	1.15	60	1.60	70
0.975	0.658	64.45	1.37	55.9	2.17	62
1.0	0.846	56.77	1.44	47	2.31	53.7

4. Conclusions

In this investigation, the mechanical and ballistic characteristics of different chemical compositions for HTPB solid propellant,depending on the curing ratio percentage, were studied at different temperatures, strain rates, and pressures. The results presented in this work provided the following conclusions:

1- The change of curing ratios by small percentage has a great effect and an intense impact on mechanical characteristics of the HTPB propellant but slightly impact on ballistic characteristics for propellant.
2- The stress-strain curves of HTPB propellant are remarkably influenced by temperature Theultimate stress, improves significantly with reduction of temperature. However, the strain shows different trend, when the temperature decreased from 20 °Cto -40°C the strain increased, but when the temperature increased from 20°C to 55°C the strain also increased.
3- The propellants with an NCO/OH ratio of 0.9 have minimum ultimate stress, with maximum strain capability, whereas the propellants with an NCO/OH ratio of 1 show just the opposite behavior.
4- The HTPB propellant is capable for large deformation even at low temperatures and high strain rates, also displays nonlinear material behavior, which is usually associated with the occurrence of the damage in the propellant during the tensile deformation.
5- All the curing ratios percentage used in this workgive us a satisfactory mechanical properties with a good ballistic performance, and they can be applied in real solid rocket motors, but the optimum one will depend on the application of the rocket motor, grain geometry, and grain structure analysis under a wide range of mechanical loads, which are imposed on it during storage and operational periods.
6- The knowledge of the curing ratio percentage effect on the propellant formulation is useful tool for adjusting the crosslinking density and then for controlling the physical and mechanical properties of HTPB based propellants.

References

[1] RaoNP, SolankeC , BihariBK , SinghP Pand BhattacharyaB2016 Evaluation of Mechanical Properties of Solid Propellants in Rocket Motors by Indentation Technique*Propellants, Explos. Pyrotech.* vol **41** no 2 pp 281–285
[2] KuentzmannP 2002 Introduction to Solid Rocket Propulsion*deRecherches Aérospatiale*vol**2** no5 pp 27–31
[3] Giuseppe T, Ludwig E, Günter M, Manfred and Bohn A 2016 Non-destructive Ageing State Determination of Solid Rocket Motors Charges*8th European Workshop On Structural Health Monitoring*vol**3** no 12 pp 5–8
[4] Shalini C , Pragnesh and Dave N 2015 Solid propellants AP / HTPB composite propellants*Arab. J. Chem*vol**15**no6 pp 201-209
[5] Adel WM and Guo-zhu L 2017 Analysis of Mechanical Properties for AP / HTPB Solid Propellant under Different Loading Conditions*in International Journal of Aerospace and Mechanical Engineering*vol**11** no 12 pp 1915–1919
[6] Himanshu S 2012 Studies On Stress-Strain Curves Of Aged Composite Solid Rocket Propellants*Def. Sci. J* vol **62** no 2 pp90–94
[7] Ramshorst G L, Duvalois W and Hooijmeijer H 2016 Investigation Of The Failure Mechanism Of HTPB/Ap/Al Propellant By In-Situ Uniaxial Tensile Experimentation In Sem*Propellants, Explos. Pyrotech* . vol **41** no 4 pp 700–708
[8] Nikola , Gligorijević , Vesna , Rodić Z , Saša ,Živković Z ,Bojan ,Pavković M ,Momčilo N , Stevan ,Kozomara , Sredoje and SubotićD 2015 Mechanical Characterization Of Composite Solid Rocket Propellant Based On Hydroxyl-Terminated Polybutadiene*Hemijska industrija*vol**47**no 1 pp 581–594
[9] Adel W, Kamal H and El-Soualey D 2019 Experimental Determination of Some Design

Properties of Viscoelastic Solid Propellant Using Uniaxial Tensile Test*Int. Conf. Aerosp. Sci. Aviat. Technol*vol**14** no7 pp 1–11

[10] Hunley J D 1999 The History Of Solid-Propellant Rocket*8th European Workshop On Structural Health Monitoring*vol**5**no25 pp68-79

[11] Fry R S, Columbia MD, DeLuca L , Politecnicoand Frederick R 2015 Evaluation Of Methods For Solid Propellant Burning Rate Measurement*central european Journal of Energetic Materials*vol **298**no 0704

[12] Lengellé G, Duterque Jand Trubert J F 2002 Combustion of Solid Propellantsvol**16** no 13 pp 27–31

[13] Aziz A , Mamat R and Ali W K W 2013 Development Of Strand Burner For Solid Propellant Burning Rate Studies*IOP Conf. Ser. Mater. Sci. Eng* vol**50**no 1

[14] Deluca L T, Maggi F, Hessler R O and Glick R L 2003 Burning Rates from Small-Scale Test Motors*Conference Proceedings*vol**52** no3

[15] Gupta G , Jawale L , Mehilal and Bhattacharya B 2015 Various Methods For The Determination Of The Burning Rates Of Solid Propellants - An Overview*Cent. Eur. J. Energ. Mater* vol**12** no 3pp593–620

[16] Deluca L T , Maggi F, Annovazzi A, Hessler R O and Glick R L 2003 Burning Rates From Small-Scale Test Motors Burning Rate From Small-Sczle Test Motor*Conference Proceedings*vol**14** no2

[17] Hakobu B,Naminosuke and Kubota 2000 Effect Of Binders On The Burning Rate Of Ap Composite Propellants vol **316** no 4pp 312–316

[18] Mohammed M M , Lateef A A A , Coll A andBiochem E 2008 Effect Of Nco / Oh Nco / Oh On The Mechanical Properties OfPolyurethane Elastomers*the 1st regional conference of eng.sci. nucej spatial issue*vol **7** no 3 pp 485–493

[19] Abusaidi H , Ghaieni , Hamid R and Ghorbani M 2017 Influences Of Nco / Oh And Triol / Diol Ratios On The Mechanical Properties Of Nitro-HTPB Based Polyurethane Elastomers*iranian journal of chemistry & chemical engineering-international english edition*vol **36**no 5 pp 55–63

[20] Hocaog R, Lay O Pekel F and Saim O 2002 Fine-Tuning the Mechanical Properties of Hydroxyl- Terminated Polybutadiene / Ammonium Perchlorate-Based Composite Solid Propellants by Varying the NCO / OH and Triol / Diol Ratios*Applied Polymer Science*vol**12**no 3pp 2072–2079

[21] Robert N 2006 An Extension Of The Time-Temperature Superposition Principle To Non-Linear Viscoelastic Solids*Int. J. Solids Struct*vol **43** no 17pp 5295–5306

[22] Nikola I , GligorijevićV Ž , Rodić S , Živković Z , Bojan P M , Momčilo M ,Stevan K and Mand SS D 2016 Mechanical Characterization Of Composite Solid Rocket Propellant Based On Hydroxyl-Terminated Polybutadiene2016*Hem. Ind.* vol **70** no 5 pp 581–594

[23] Aziz A , Mamat R and Wan A 2013 Development Of Strand Burner For Solid Propellant Burning Rate Studies*IOP conference*vol**35**no3

[24] Hayri Y, Veli Ç and ErcanD 2014 Experimental Investigation Of The Factors Affecting The Burning Rate Of Solid Rocket Propellants*fuel* vol **115**no 8 pp 794–803

IOP Conf. Series: Materials Science and Engineering **973** (2020) 012031 doi:10.1088/1757-899X/973/1/012031

Penetration of long rods into semi-infinite metallic targets over wide range of impact velocities

M A Abdel-Kader[1], A M Riad[1,2]andA Z Ibrahim[1]
[1]Weapons and Amm.Dept., Mechanical Engineering Branch, Military Technical College, Kobri El-Kobba, Cairo, Egypt.
[2] Corresponding author, Email: ahmed.riad@mtc.edu.eg.

Abstract. In this paper, the analytical model that describes the penetration process of a long rod into a semi-infinite metallic target over wide range of impact velocities has been presented. In this model, the target strength factor $R_t$is not considered as a constant but it is a function of penetration velocity whereas, the rod strength factor Y_P is represented by Hugoniot Elastic Limit (HEL) of its material. The penetration process of the present model consists mainly of primary phase during which the rod front is subjected to erosion and secondary (after-flow) phase. The main equations of the used model are presented, arranged and compiled into a computer program. The program is capable of predicting the time histories of the parameters associated with the penetration process. Autodyn hydrocode package is used herein to simulate the penetration processes of long rods into targets examined by the present analytical model. The hydrocode is fed with the same data used in the model for rods and targets materials. Both the model predictions and the obtained numerical results of Auto dynhydro code, respectively, are compared with the corresponding experimental measurements of other investigators. Good agreement is generally obtained. In addition, representative samples of the model predictions and their corresponding simulation results of Autodyn hydro code are presented with relevant analyses and discussions. The obtained results prove the predictive capabilities of both the present model and Autodyn hydro code, respectively, where each of them could be used as a quick tool for determining the main parameters associated with the studied penetration process.

1. Introduction

Long rod penetrators with large aspect ratios (length to diameter) have been widely used with the modern kinetic energy projectiles. These penetrators have high specific impact energies and great penetration capabilities to the modern armors. Very dense materials are used for manufacturing these penetrators such as tungsten alloys and depleted uranium. The penetration of high speed long rods into metallic targets has been studied experimentally, analytically and numerically since long time ago.

The developed one-dimensional hydrodynamic theories of penetration by Alekseevski [1] and Tate [2] have been used as reference models for describing the penetration of high-speed long rods into semi-infinite metallic targets. Both models contained two strength factors for target and rod materials; they denoted byR_t and Y_p, respectively. Both factors were determined experimentally. Tate used the Hugoniot Elastic Limit (HEL) of the rod material for Y_p and a value of 3.5 x HEL of the target material for R_t. Their models are capable of predicting the penetration depth and the rod deceleration into target.

IOP Conf. Series: Materials Science and Engineering **973** (2020) 012031 doi:10.1088/1757-899X/973/1/012031

Jones et al. [3] modified Tate's model of Ref. [2] by incorporating the mushroom-type deformation at the impact end of the rod and the deceleration of rod rigid end. The predicted penetration depths using Tate's model were inconsistent with those predicted by Jones et al. model for the same material strength factors. They attributed this difference to the lack of mushroom at the rod front in Tate's model. Wilson et al. [4] conducted an experimental program to check the validity of Jones et al. predictions.They got a quite good agreement between their measurements and the predictions of Ref. [3].They also reported that it was possible to get better agreement when a velocity-dependent mushroom strain was considered in theirmodel.

He and Wen [5] analyzed the long rod penetration process into semi-infinite metallic targets based on energy balance. They considered that a part of kinetic energy of the long rod was remained in the ejected rod debris.Thisdebris may continue to penetrate the target causing secondary penetration especially when a dense rod impacted into a semi-infinite metallic target with high velocities. Their model predictions are in good agreement with test results of Refs. [6-7].

Several analytical models that describe the penetration of high speed long rod into semi-infinite and finite thickness metallic targets have been reviewed by Anderson [8]. Upon review, he deduced the followingmain points: i) for the target strength factor of the modified hydrodynamic models; however,it depended on material properties, Anderson et al. [9] showed that it also depended on the impact velocity, ii) the use of compressible cylindrical cavity expansion model gave a good agreement with experimental measurements at impact velocity below 2 km/s as noted by Chocron et al. [10] when values of material flow stresses consistent with the behavior of materials were used, iii) the analytical models based on conservation of momentum such as Walker-Anderson model [11] tended to be applicable to a wide range of materials and impact velocities, and iv) the area concerned with the penetration of multi-material layered targets by high speed long rod was neededfor further works.

Jiao and Chen [12] developed approximate solutions of the main equations representing Tate's model of Ref. [2]. In "theoretical solution", these equations were solved numerically to predict the parameters associated with the rod penetration into a semi-infinite target implicitly as function of time. They also derived another two sets of explicit approximate algebraic solutions based on the implicit solution deduced from the equations of Tate's model. In their first approximate solution, they considered that the tail and penetration velocities varied logarithmically with time while the other parameters were deduced from heir integrals. In their second approximate solution, they assumed that the tail and penetration velocities were constant and the other parameters varied linearly with time.They also derived a special case of first approximate solution named "first-order perturbation solution". The different solutions have been examined using the data for rods and targets of Refs. [13-14]. They found that the results of their first approximate solution were close to the theoretical predictions of Tate's model than other derived solutions. In addition, they strongly recommended this solution in engineering applications.

In the following, the analytical model of Ref. [15] that describes the penetration of a long rod into a semi-infinite metallic target is presented. The main phases and the considered assumptions in constructing the present model are introduced. In addition, the governing equations representing each penetration phase are presented. These equations are arranged and compiled into a computer program. The input data to the program are easily determined. Representative samples of the model predictions are presented and discussed. Autodyn 2D hydrocode is also used to assess the obtained analytical results.

2. Analytical model

In the present model, the projectile is idealized as a cylindrical rod with a hemispherical nose havinginitialshank length L_o and initial diameter D_o. The rod is assumed to strike the target normally with a high velocity. At such a velocity, Lan and Wen [15] reported that the ratio of shear stress to the pressure at rod-target interface can be neglected based on the numerical simulation results of Ref. [16]. Moreover, the behavior of the materials near the rod-target interface can be treated as a fluid similar to the one-dimensional model developed by Tate [2].Therefore, new response regions due to long rod penetration intoa semi-infinite metallic target can be seen in figure 1 [15]. These regions along the penetration centerline are named by flow, plastic and elastic regions in the target; whereas the regions in the penetrating rod are named by flow and rigid tail regions. The points A, O and B on figure1 connect the penetration centerline with rigid flow interface in the rod, rod-target interface, flow plastic interface (FP) in the target, respectively. In the flow regions of rod and target, their materials are treated as non-viscous incompressible fluids. In addition, particle velocity and stress at the interface FP are continued in the normal directions. The penetration velocity is denoted by u;whereas thevelocity of rod rigid tail is denoted by v.

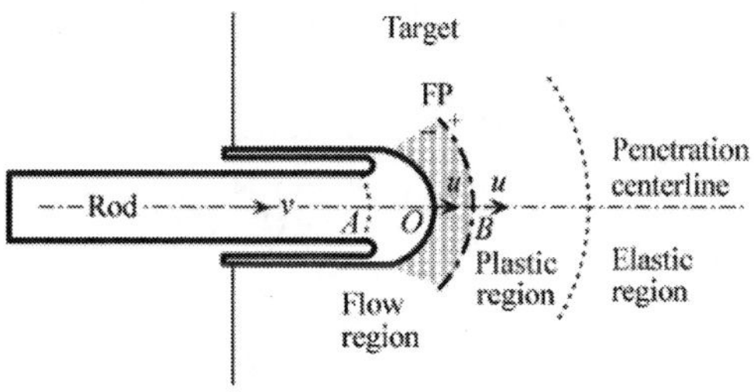

Figure 1. Response regions along the centerline at high impact velocity [15].

The present model identifies two different phases associated with the penetration of along rod into a semi-infinite metallic target; these are primary penetration phase and after-flow phase. At a high impact velocity, the primary penetration phase is dominated and the rod front is continuously erodedeven at low penetration velocity due to lateral confinement of the eroded rod. The velocity of rod rigid mass at which no flow region in the target V_C and the interface defeat velocity V_{ID} at which the penetration velocity u is equal to zero will be introduced, respectively.

For the primary penetration phase, a system of equations has been presented. The penetration time is taken as independent variable. This system is solved numerically to determine the time-histories of parameters associated with the penetration process. The current value of penetration velocity u is determined using its respective interface equation. In the following, the main assumptions considered during the analysis are introduced. In addition, the main equations and end conditions of the primary penetration phase are presented.

2.1.Main assumptions

- There exists a critical penetration velocity U_{Fo}. When the penetration velocity $u \geq U_{Fo}$, a flow region appears in the target; i.e. a thin layer appears in the target near the rod-target interface within which the material can be treated as "fluid" due to loss of its shear strength, otherwise there are only plastic and elastic regions.
- At the critical state ($u = U_{Fo}$), no particle passes through the interface FP, and U_{Fo} is equal to the velocity of the particle at position B; i.e. $\delta(u) = U_{Fo}$ where $\delta(u)$ is a function of u with defined field $[U_{Fo}, +\infty]$.
- When the penetration velocity u exceeds the critical penetration velocity U_{Fo}, the velocity of interface FP along the penetration direction is always equal to the penetration velocity u. This meansthat the width of the flow region remains constant during penetration so that the Bernoulli's equation can be used which is applicable only in steady flow field.
- The cavity expansion model of Ref. [17] is applicable for predicting the displacement and velocity field of plastic and elastic regions.
- The function $\delta(u)$ decreases with increasing the penetration velocity u and tends to zero when u becomes infinity.
- As the rod decelerates, the pressure at the position O decreases with decreasing the penetration velocity u.

IOP Conf. Series: Materials Science and Engineering **973** (2020) 012031 doi:10.1088/1757-899X/973/1/012031

2.2. Governing equations

2.2.1. Primary penetration phase. When u $\geq U_{Fo}$, the flow field between points A and B can be treated as steady flow relative to position O. The particle velocity at position B is [u-δ(u)] relative to position O. When u $<U_{Fo}$, there is no flow region in the target. The governing equations of the current phase are:

a) The pressure at position O of rod-target interface P_o [15]:

For u $\geq U_{Fo}$:

$$P_o = R_t + \frac{1}{2}\rho_t u^2 = Y_P + \frac{1}{2}\rho_P (v-u)^2 \qquad (1)a$$

where

$$R_t = 2\rho_t \delta(u)^2 + S - \rho_t u \delta(u) \qquad (1)b$$

and

$$\delta(u) = U_{F0}. exp\left[-\left(\frac{u-U_{F0}}{nU_{F0}}\right)^2\right]. \qquad (1)c$$

For u $<U_{Fo}$:

$$P_o = S + C\rho_t u^2 = Y_P + \frac{1}{2}\rho_P(v-u)^2 \qquad (1)d$$

R_t is the target strength factor, u is the penetration velocity, ρ_t and ρ_p are the densities of rod and target materials, respectively, δ(u) is the particlevelocity at position B, S is the static target resistance, C is the constant of dynamic resistive pressure and is equal to 3/2 for incompressible materials, v is the velocity of rod rigid part, Y_P is the rodstrength factor, U_{Fo} is the critical penetration velocity and n is constant which is equal to 2.45.

The rod strength factor Y_P is equivalent to Hugoniot Elastic limit $(HEL)_P$of its material as reported in Ref. [18] whereas; the critical penetration velocity U_{Fo} is equivalent to $\sqrt{(HEL)_t/\rho_t}$. The subscripts P and t refer to rod and target, respectively.

The static target resistance is determined by [19]:

$$S = \frac{2Y_t}{3}\left[1+\left(1-\frac{E_{2t}}{E_{1t}}\right)ln\left(\frac{2E_{1t}}{3Y_t}\right)\right] + \frac{2}{27}\pi^2 E_{2t} \qquad (2)$$

where Y_t,E_{1t} and $E_{2t}$are yield strength, elastic and plastic modulii of target material, respectively.

From Eqn. (1)a, the current penetration velocity can be calculated as a function of velocity of rod rigid part (v) when the target flow region exists. Equation (1)d determines the current penetration velocity as function of v when plastic and elastic regions are dominated in the target.

b) From the equation of motion of rod rigid part, the deceleration of its rigid mass during the penetration process is represented by:

$$\frac{dv}{dt} = -\frac{Y_P}{\rho_P.l}, \qquad (3)$$

wherel is the current length of rod rigid mass.

c) The decreasing rate of length of rod rigid part is:

$$\frac{dl}{dt} = -(v-u). \qquad (4)$$

d) The time rate of change of primary penetration depth of rodinto target is:

$$\frac{dZ_P}{dt} = u \qquad (5)$$

where$Z_p$is the rod penetration depth in the primary penetration phase.

e) The current diameter of crater D_C which is represented by [20]:

IOP Conf. Series: Materials Science and Engineering **973** (2020) 012031 doi:10.1088/1757-899X/973/1/012031

$$\frac{D_C}{D_o} = \sqrt{\frac{Y_P + \rho_P(\phi+1)(v-u)^2}{S}} \tag{6}$$

where

$$\phi = \sqrt{1 - \frac{2\,Y_P}{\rho_P\,(v-u)^2}}\;. \tag{6)a$$

f) The velocity of rod rigid mass $V_C$is represented by:

$$V_C = U_{FO} + \sqrt{\frac{2(S + C\,\rho_t\,U_{FO}^2 - Y_P)}{\rho_P}}. \tag{7}$$

and

g) The interface defeat velocity $V_{ID}$is represented by:

$$V_{ID} = \sqrt{2(S - Y_P)/\rho_P}\,. \tag{8}$$

For $u \geq U_{FO}$, equations (3), (4), (5) and (6) in addition to Eqn. (1)arepresents a system of equationsthat solved numerically.The initial conditions for solving this system are:

$$at\ t = 0.0,\ v = v_i,\ u = u_i,\ l = l_o + \frac{2}{3}r_o,\ Z_p = 0.0\ and\ D = D_o$$

When $v < V_C$, the same system of equations is used after replacing equation (1)a by (1)d. This is meant that there is no flow region in the target; i.e. the plastic and elastic regions are only exist. This system is also used when the impact velocity v_i< V_C. The penetration process is continued until the velocity of rod rigid part v reaches a velocity named interface defeat velocity, V_{ID} (c.f. Eqn. (8)).

Based on the impact velocity of rod into target; the condition terminating the primary penetration phase is: (i) when the length of rod is totally consumed or (ii) when the penetration velocity u reaches zero.

2.2.2.After-flow penetration phase. The Tate's empirical equations [21] are employed to calculatethe depth of penetration which occurs in the after-flow phase. Thedepth of penetration $Z_{a\text{-}f}$is evaluated by:

$$Z_{a-f} = \frac{D_{cp}}{4}\left[\left(1 + \frac{3\rho_P u_i^2}{2R_{spt}}\right)^{1/3} - 1\right], \tag{9}$$

Where D_{Cp} is the diameter of the crater at the end of primary penetration phase, u_i is the penetration velocity corresponding to the initial impact velocity v_i, and R_{spt} is evaluated by the following equation:

$$R_{spt} = \frac{2Y_t}{3}\left[1 + ln\left(\frac{2E_{1t}}{3Y_t}\right)\right]. \tag{10}$$

The total depth of penetration of rod into target Z at the end of penetration process is the sum of penetration depth at the end of primary penetration phase Z_{pt} and that of the after-flow phase $Z_{a\text{-}f}$; i.e.

$$Z = Z_{pt} + Z_{a-f}.$$

The governing equations representing the penetration process due to the impact of high-speed long rods into semi-infinite metallic targets are converted into a computer program. The input data to the program are diameter of hemispherical-cylindrical rod and its length as well as its mechanical properties (density, yield strength, modulus of elasticity and Poisson's ratio), and impact velocity in addition to the mechanical properties of the target material (density, yield strength, moduliof elasticity and plasticity and Poisson's ratio).

The present model is capable of predicting time-histories of the velocity of rod rigid mass, the penetration velocity, the rod penetration depth into target, the length of rod rigid mass and the crater diameter in the target. Moreover, the model is also capable of predicting the total depth and total time of penetration process, respectively. Experimental measurements of Ref. [22]have been used to validate the model predictions. In addition, samples of the model predictionswill be presented with relevant analyses and discussions.

3. Numerical simulation

In the following, Autodyn-2D hydrocode simulates the penetration process of long rods into semi-infinite metallic targets at different impact velocities using the data of tested rods and targets of Ref. [22]. The tested rods, with aspect ratios of 10, were made of tungsten alloy having different mechanical properties. In addition, the tested semi-infinite metallic targets were made of steel with different mechanical properties. For groups 1,the rod is designated by D17 whereas; the target is designated by St-52. For group 2,the rod is designated by D17.6whereas; the target is designated by HzBA. For each group, the rod strikes the target normally with wide range of impact velocities.

3.1. Material description

3.1.1. Rod material. For each group, the input data of the rod material to the hydrocode are listed in Table 1. The equation of state of each rod material was selected to be shock; the erosion model was selected to be incremental geometrical strain. Both the strength and failure models of the rod material were selected to be Johnson-Cook model [23].

3.1.2. Target material. For each group, the input data of the target material to the hydrocode are listed in Table 1.For each target material, linear equation of state and erosion model of incremental geometric strain was selected. Johnson-Cook model was used as strength and failure models, respectively.

The main dimensions of rod and target (x and y) for each interaction are converted into sub-grids with vertical and horizontal lines (grid lines) as shown in figure 2. The main procedures representing the interaction problem using Autodyn-2D hydrocode are reported in Ref. [23]. The numerical results of Aurodyn-2D hydrocode will be validated with the available experimental measurements of Ref. [22] and they are used to assess the predictions of the present analytical model.

3.2. Mesh sensitivity

It is well known that the density of mesh affects the simulation results. In general, simulation with fine meshes produces appropriate solution; but it takes a longer time than that needed with coarse meshes. Therefore, the mesh density was taken to be fine along the penetration depth into target while the coarse meshes were taken away of the penetration direction. The mesh sensitivity study for the penetration of a long rod into a semi-infinite metallic target was done for each group of rod-target interaction at three different impact velocities. It was predicted that the mesh size at the penetration zone is 0.25 x 0.25 mm for the rod and target, respectively. This size gave a close penetration depth to the corresponding measured value of Ref. [22] at each tested impact velocity.

Table 1. Input data of tungsten alloy rods and steel targets to the code [22].

Parameter	**Value**			
	Group 1		Group 2	
	Rod	Target	Rod	Target
Material designation	D17	St-52	D17.6	HzBA
Reference Density, (g/cm^3)	17	7.85	17.6	7.85
Reference Temperature, (K)		300		
Rod length, (mm)	60	--	58	--
Rod diameter, (mm)	6	--	5.8	--
Specific heat (C.V.), (J/ kg K)	134	477	134	477
Shear modulus, (kPa)	160×10^6	73×10^6	160×10^6	73×10^6

IOP Conf. Series: Materials Science and Engineering **973** (2020) 012031 doi:10.1088/1757-899X/973/1/012031

Yield stress, (kPa)	985	500	1360	849
Hugoniot Elastic Limit, HEL (MPa)	1665	---	2312	---
Melting temperature, (K)	1723	1573	1723	1573
Erosion strain,	1.0	1.25	1.0	1.25

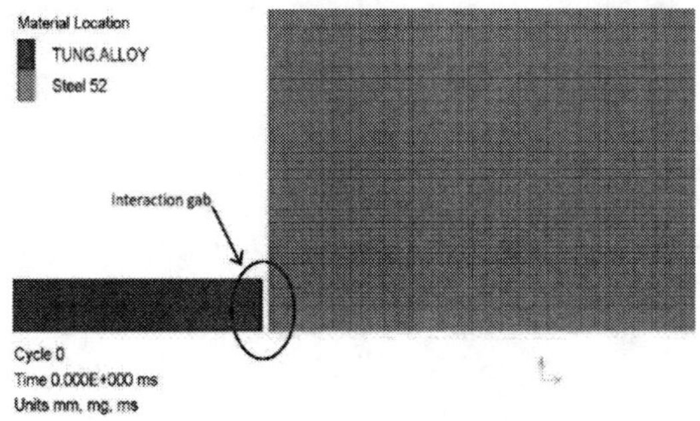

Figure 2. Representative D17 rod and St-52 target into Autodyn-2D hydrocode.

4. Results and discussions

In the following, the results of the analytical model are classified into: (i) validation and (ii) predictions. For each group of rod-target interaction, the validation is concerned with comparing the predicted penetration depths and crater radii obtained by the analytical model and Autodyn-2D hydrocode, respectively, at different impact velocities with the corresponding experimental measurements of Ref. [22]. In predictions, samples of the time-histories of the parameters associated with the penetration process and the corresponding simulation results of Autodyn-2D hydrocode are presented and discussed.

4.1. Validation

The predicted results of the analytical model and Autodyn-2D hydrocode are validated, respectively, with the corresponding experimental measurements of Anderson et al. [22]. They listed the measured penetration depths and surface crater diameters due to the impact of D17 tungsten alloy rod into St-52 steel semi-infinite target over a wide range of impact velocities. The input data of rod and target materials, respectively, to the model and Autodyn-2D hydrocode are listed in Table 1. The predicted change of penetration depths obtained by the model and hydrocode with impact velocity are depicted into figure 3a. The corresponding experimental measurements of Ref. [22] are also depicted in the same figure. Moreover, figure 3b plots the predicted change of surface crater diameters obtained analytically and numerically as well as the corresponding measurements of Ref. [22] with impact velocity.

Good agreement is generally obtained between the predictions of the analytical model and simulation

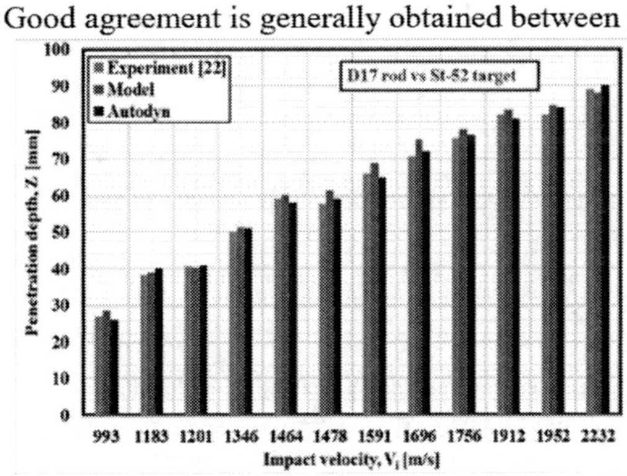

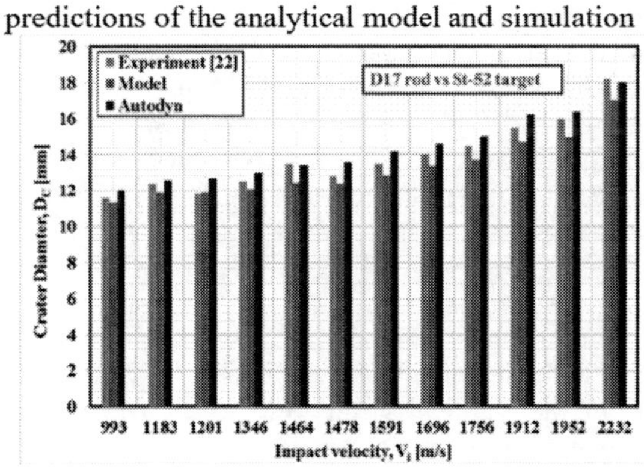

Figure 3a. Predicted change of penetration depth with rod impact velocity obtained by analytical model and Autodyn-2D hydrocode, respectively; measured depths of Ref. [22].

Figure 3b. Predicted change of crater diameter with rod impact velocity obtained by analytical model and Autodyn-2D hydrocode, respectively; measured diamters of Ref. [22].

results of Autodyn-2D hydrocode, respectively, and the corresponding experimental measurements of Ref. [22].The maximum absolute errors between the measured depth and the corresponding predicted one obtained by the model and Autodyn hydrocode, respectively, are found to be 8.2% at v_i = 1005 m/s and2.5% at v_i = 1632 m/s. In addition, the maximum absolute errors between the predicted surface diameters obtained analytically and numerically with the corresponding measurements are found to be 8.4% at v_i = 1982 m/s and 2.9% at v_i = 1909 m/s, respectively.

Another validation of the present model and Autodyn-2D hydrocode, respectively, is performed. Their predictive capabilities are tested using another measured data of Ref. [22] due to the impact of 17.6 tungsten alloy rods into HzBA steel targets at different impact velocities. The input data of rod and target materials to the model and hydrocode are listed in table 1. Figure 4a plots the predicted change of rod penetration depths obtained analytically, numerically and experimentally [22] with impact velocity. Similarly, figure 4b plots the predicted change of surface crater diameters with impact velocity.

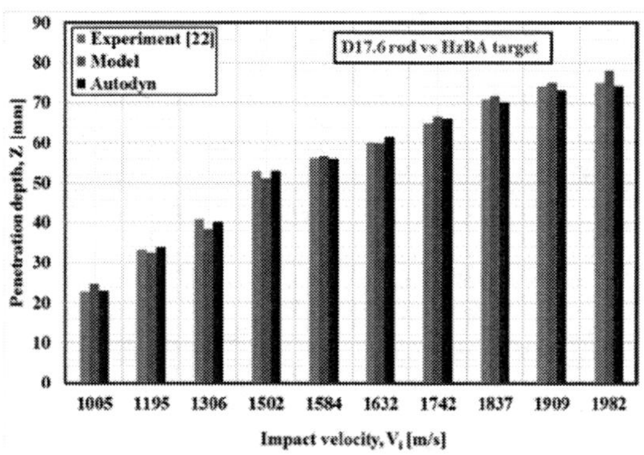

Figure 4a. Predicted change of rod penetration depth with impact velocity obtained by the analytical model and Autodyn-2D hydrocode, respectively; measured depths of Ref. [22].

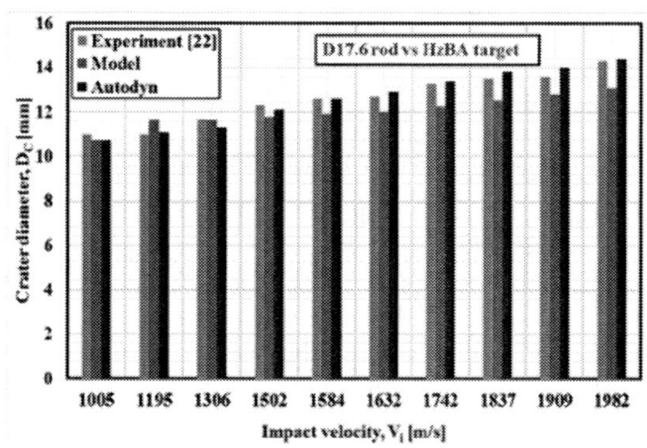

Figure 4b. Predicted change of crater diameter with rod impact velocity obtained by the analytical model and Autodyn-2D hydrocode, respectively; measured diamters of Ref. [22].

The obtained analytical and numerical results are consistent with the corresponding experimental measurements of Ref. [22]. The obtained results prove the predictive capabilities of the analytical model and Autodyn-2D hydrocode. The maximum absolute errors between the predicted penetration depths obtained analytically and numerically with the corresponding measurements are found to be 6.4% at v_i = 1696 m/s and 3.7% at v_i = 993 m/s, respectively. In addition, the maximum absolute errors between the predicted surface diameters obtained analytically and numerically with the corresponding measurements are found to be 6.38% at v_i = 1952 m/s and 7.63% at v_i = 1201 m/s, respectively.

4.2. Predictions

Figure 5a plots the predicted time-histories of the velocity of rod rigid part and the penetration velocity due to the impact of D17 rod into St-52 target at different velocities. At each impact velocity, it is seen from the figure that the velocity of rod rigid part, v, is slightly decreased at the beginning of penetration process however, the decreasing rate of this velocity increases gradually until the end of the penetration process due to the increase of rod erosion. The rod front is subjected to erosion during the whole penetration process, the target flow region (FP) in front of eroded rod is replaced with plastic region when v < V_C. Similarly, the penetration velocity of the rod front, u, follows the same trend as the velocity of rod rigid part and the target penetration stops when u = 0.0; i.e. when the corresponding velocity v is equal to the velocity V_{ID} (=464 m/s). At the highest impact velocity, the penetration process is terminated due to the total consumption of the rod. It is also seen from this figure that the total time of penetration increases with the decrease of impact velocity. The predicted total penetration times are 80.2, 96 and 112 μs at v_i = 1952, 1346 and 993 m/s, respectively. The trends of the model predictions are similar to that presented in Refs. [10,13].

Figure 5b plots the time-histories of the velocity of rod rigid part due to the impact of D17 rod into St-52 target at different velocities predicted by the analytical model and Autodyn-2D hydrocode, respectively. At each impact velocity, it is seen from the figure that the predicted deceleration of the rod into target obtained by the model is consistent with the corresponding deceleration obtained by Autodyn hydrocode; except at the end of the penetration process where the deceleration of the rod obtained by Autodyn is slightly greater than that predicted by the model. This may be attributed to the difference in erosion rate of the rod length (v-u) and its influence on the predicted deceleration of rigid rod part. In addition, the difference on remaining lengths of rod rigid parth as a small influence on the total rod penetration depths into target. The predicted total penetration times obtained by the analytical model are 80.2, 96 and 112 μs and by Autodyn hydrocode are 84, 100 and 114 μs at v_i = 1952, 1346 and 993 m/s, respectively.

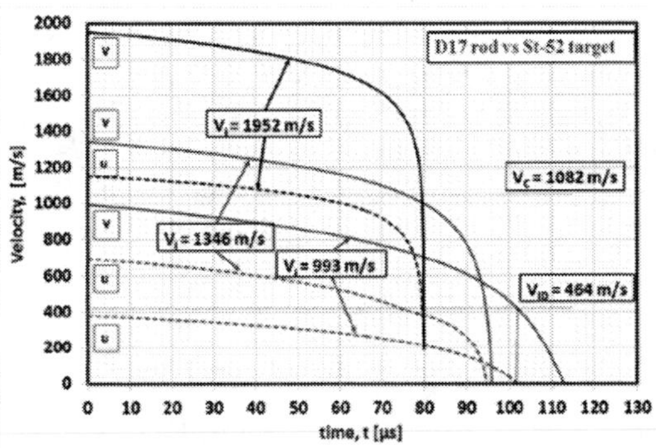

Figure 5a. Predicted time histories of velocity of rod rigid part and penetration velocity due to the impact of D17 rod into St-52 target at different velocities.

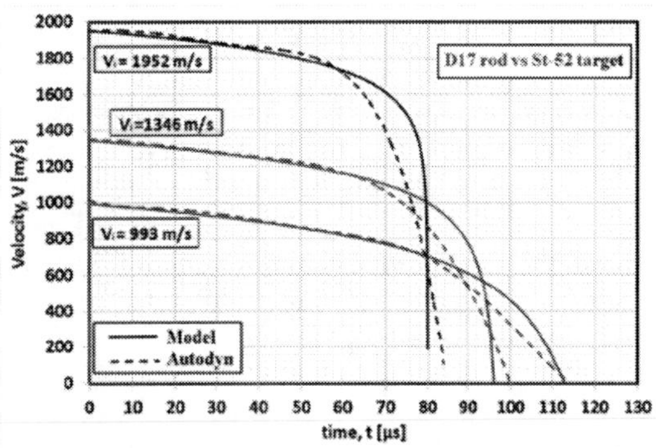

Figure 5b. Predicted time histories of velocity of rod rigid part and the corresponding predictions of Autodyn-2D due to the impact of D17 rod into St-52 target at different velocities.

Figure 6a plots the time histories of the velocity of rod rigid part, v, and the penetration velocity, u, due to the impact of D17.6 rod into HzBA target at different velocities. At each impact velocity, similar trends as that presented for the velocity, v, and the penetration velocity, u, in figure 5a are presented. It is also predicted that the rod is completely consumed at v_i = 1982 m/s. For the other two impact velocities, small rigid rod parts are remained at the end of penetration process. The predicted total penetration times obtained by the analytical model are 76.8, 91 and 107 μs at v_i = 1982, 1306 and 1005 m/s, respectively.

In addition, figure 6b depicts the time-histories of the velocity of D17.6 rod rigid part that predicted by the model and Autodyn hydrocode, respectively, at different velocities. Similar to that presented in figure 5b, good agreement is obtained between the velocity-time histories obtained by the model and Autodyn hydrocode except at the end of penetration process when v_i = 1982 m/s. The predicted total times of penetration predicted by the model are compared with that of Autodyn hydrocode and the maximum absolute error between them is found to be 5.49% at v_i = 1306 m/s.

IOP Conf. Series: Materials Science and Engineering **973** (2020) 012031 doi:10.1088/1757-899X/973/1/012031

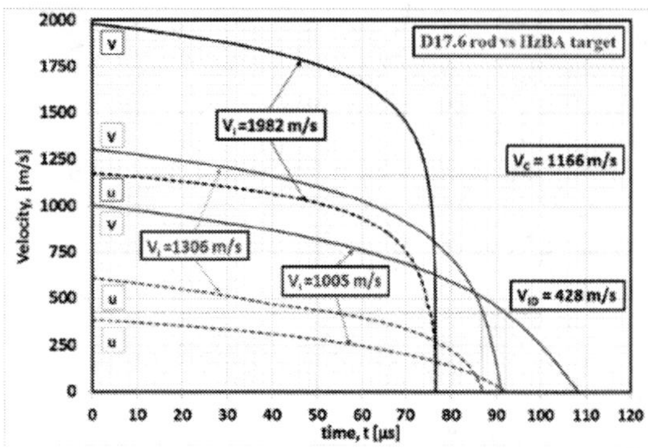

Figure 6a. Predicted time histories of velocity of rod rigid part and penetration velocity due to the impact of D17.6 rod into HzBA target at different velocities.

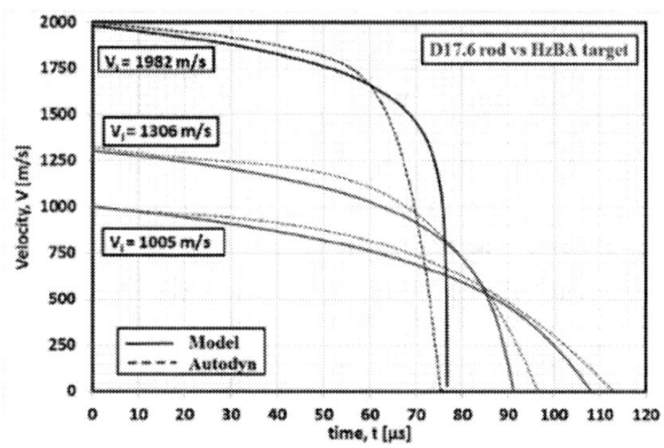

Figure 6b. Predicted time histories of velocity of rod rigid part and the corresponding predictions of Autodyn-2D due to the impact of D17.6 rod into HzBA target at different velocities.

Figure 7a plots the time-histories of the penetration depth due to the impact of D17 rod into St-52 target at different velocities. The corresponding predicted time-histories obtained by Autodyn hydrocode are depicted on the same figure. It is seen from the figure that the depth of penetration increases with the impact velocity; this is due to the increase of the penetration velocity with impact velocity. At each impact velocity, it is also seen from the figure that the predicted depth-time history of the analytical model is consistent with that obtained by Autodyn hydrocode. The maximum absolute errors between the model prediction and that of Autodyn hydrocode are 2.48%, 2.95% and 1.24% at v_i = 993, 1346 and 1952 m/s, respectively. Moreover, figure 7b plots the depth-time histories due to the impact of D17.6 into HzBA target at different velocities. Similar trends for the depth-time histories as that presented in figure 7a are plotted. In addition, the maximum absolute errors between the model prediction and that of Autodyn hydrocode are 5.7%, 3.9% and 0.39% at v_i = 1005, 1306 and 1982 m/s, respectively.

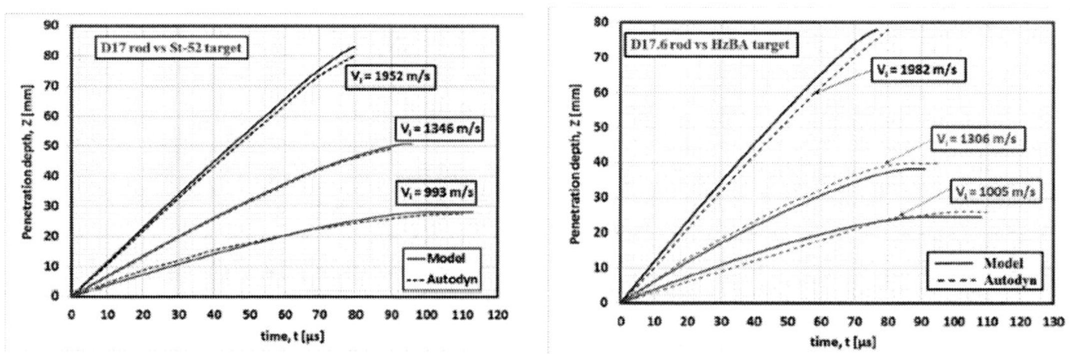

(a) (b)

Figure 7. Predicted time histories of penetration depth and the corresponding predictions of Autodyn-2D due to the impact of: a) D17 rod into St-52 target and b) D17.6 rod into HzBA target at different velocities.

The present analytical model is used to predict the penetration depth during after-flow phase due to the impact of D17 rod into St-52 target over wide range of velocities. The obtained penetration depth represents 2.8% and 3.45% from the total penetration depth at impact velocities of 993 and 2232 m/s, respectively. This percent could be increased with impact velocity. Moreover, the depth of penetration of after flow phase due to the impact of D17.6 into HzBA target represents 2.18% and 2.95% from the

IOP Conf. Series: Materials Science and Engineering **973** (2020) 012031 doi:10.1088/1757-899X/973/1/012031

total depths at v_i = 1005 and 1982 m/s, respectively. The obtained results are similar to that predicted by Lan and Wen [15].

Figure 8a depicts the time-histories of rigid rod length due to the impact of D17 rod into St-52 target at different velocities. The corresponding time-histories obtained by Autodyn hydrocode are plotted on the same figure. At each impact velocity, it is seen from the figure that the length of rod rigid part decreases due to continuous erosion of its front during penetration. At the highest impact velocity, the rod is completely consumed during penetration whereas; tines of rigid rod lengths are remained at the end of penetration process for the other two impact velocities. This is attributed to the high erosion rate (v-u) at the highest impact velocity compared with the other two velocities which leads to the complete erosion of the rod length at the end of penetration process. It is also seen from the figure that the predicted rigid rod length-time history by the analytical model is consistent with that obtained by Autodyn hydrocode at each impact velocity. The predicted remaining rigid rod lengths at the end of penetration process of D17 rod into St-52 target obtained by the analytical model are 2.25, 0.45 and 0.0 mm whereas; the corresponding predictions by Autodyn hydrocode are 3.0, 1.0 and 0.0 at v_i = 993, 1346 and 1952 m/s, respectively.

Moreover, figure 8b plots the rod rigid length-time histories due to the impact of D17.6 into HzBA target at different velocities. Similar trends for the rod rigid length-time histories as that presented in figure 8a are presented. In addition, the predicted remaining lengths of the rod by the model are 4.5, 1.35 and 0.0 whereas; the corresponding predictions by Autodyn hydrocode are 4.5, 1.8, and 1.0 at v_i = 1005, 1306 and 1982 m/s, respectively.

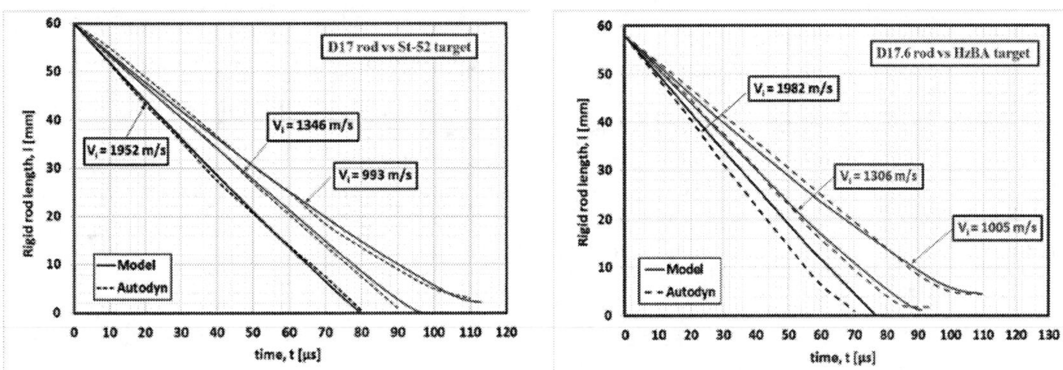

(a) (b)

Figure 8. Predicted time histories of rigid rod length and the corresponding predictions of Autodyn-2D due to the impact of: a) D17 rod into St-52 target and b) D17.6 rod into HzBA target at different velocities.

5. Conclusions

The main conclusions of the present work are:

- The analytical model of Lan and Wen [15] has been used herein to describe the penetration of semi-infinite metallic targets by long rods over a wide range of impact velocities. The target strength factor is not considered as a constant during penetration but; it is considered as a function of penetration velocity. The predictions of the model are compared with the available experimental measurements of Ref. [22], good agreement is generally obtained.
- The depth of penetration during the after flow phase is shallow at low impact velocity; and it has a considerable value when the impact velocity exceeds 2 km/s.
- Autodyn-2D hydrocode simulates the penetration of long rods into semi-infinite metallic targets using the rod and target data that fed into analytical model. The numerical results are compared with the available experimental measurements of Ref. [22] and the model predictions; good

agreement is generally obtained. In addition, the time-histories of penetration parameters that predicted by the analytical model and Autodyn-2D hydrocode are in good consistent.

- The obtained analytical results of the analytical model and numerical results of Autodyn-2D hydrocode reflect their predictive capabilities; each of them could be used as a quick tool for predicting the main parameters associated with the penetration of a long rod into a semi-infinite target over wide range of impact velocities.

References

[1] Alekseevskii VP1966Penetration of a rod into a target at high velocity*Combustion, Explosion and Shock Waves* **2** pp 63-6.
[2] Tate A1967 A theory for the deceleration of long rods after impact*J. Mech. Phys. Solids* **15** pp 387–99.
[3] Jones SE, Gills PP and FosterJ C1987On the penetration of semi-infinite targets by long rods*J. Mech. Phys. Solids* **35** pp 121-31.
[4] Wilson LL, Foster JC, Jones JC, Jr and Gillis PP1989Experimetal rod impact results*Int. J. Impact Engng.* **81** pp 5-25.
[5] He Yand Wen H M2013 Predicting the Penetration of Long Rods into Semi-Infinite Metallic Targets *Sci. CHINA* **56** pp 2814–20.
[6] Silsby G F1984Penetration of semi-infinite steel targets by tungsten rods at 1.3 to 4.5 km/s *Proc. of 8th Int. Symp. on Ballistics* (Orlando Florida **TB**/31-5.)
[7] Cullis I G and Lynch N J1994Hydrocode and experimental analysis of scale size jacketed ke projectiles*Proc.14th Int. Symp. on Ballistics* (Quebec, Canada **TB-7** 271-80).
[8] Anderson Jr. CE2017Analytical models for penetration mechanics: A review", *Int. J. Impact Engng.* **108** pp 3-26.
[9] Anderson Jr. CE, Littlefield DL and Walker JD1993Long rod penetration, target resistance and hypervelocity impact*Int. J. Impact Engng.* **14** pp 1-14.
[10] Walker JD and Anderson Jr CE1995A time dependent model for long rod penetration*Int. J. Impact Engng.* **16** pp 19-48.
[11] Chocron S, Anderson Jr CE and Walker JD1998Long rod penetration: Cylindrical vs spherical cavity expansion for the extent of plastic flow", *Proc. 17th Symp. on Ballistics*, South African Ballistics Organization pp319-26.
[12] Jiao WJ and Chen XW2018Approximate solutions for Alekseevskii-Tate model of long-rod penetration *Acta Mech. Sin.* **34** **pp** 334-48.
[13] Anderson Jr. CE and Walker JD1991An examination of long rod penetration*Int. J. Impact Engng.* **11** pp 481-501.
[14] Walter W, Williams Cand Normandia M2006An explicit soluion of Alekseevskii-Tate penetration equations*Int. J. Impact Engng.* **33** pp 837-46.
[15] Lan Band Wen HM2010 Alekseevskii-Tate revisited: An extension to the modified hydrodynamic theory of long rod penetration *Sci. CHINA* **53** pp 1364 373.
[16] Lan Band Wen H M2008Numerical simulation and analysis of the penetration of tungesten alloy long rods into semi-infinite steel targets *Chinease J. High Pressure Physics* **22** 245-52.
[17] Forrestal M J and Luk V K 1988Dynamic cavity-expansion in a compressible elastic-plastic solid *J. Appl. Mech.* **55** pp 275-79.
[18] Lan Band Wen H M2009Numerical simulation of the penetratin of a spherical-nosed 4340 steel long rod into semi-infinite 6061-t6511 aluminium targets", *J. Eng. Mech.* **26** 183-190.
[19] Lu Z C and Wen H M2017On the penetration of high strength steel rods into semi-infinite aluminium alloy targets *Int. J. Impact Eng.* **111** pp 1–10.
[20] Wen H M, He Yand Lan B2010Analytical model for cratering of semi-infiniye metal targets by long rod penetration *Sci. ChinaTech Sci.* **53** pp 3189-196.
[21] Tate A 1986A long rod penetration models-part ii. extensions to the hydrodynamic theory of penetration *Int. J. Mech Sci.* **28** pp 599-612.

IOP Conf. Series: Materials Science and Engineering **973** (2020) 012031 doi:10.1088/1757-899X/973/1/012031

[22] Anderson Jr CE, Morris BL and Littlefield D1992A penetration mechanics database *Report No. 3593/C01* San Antonio, Texas, U.S.A. Southwest Research Institute.
[23] AUTODYN 2016Interactive non-linear dynamics analysis software *Century Dynamics Inc.*

IOP Conf. Series: Materials Science and Engineering **973** (2020) 012032 doi:10.1088/1757-899X/973/1/012032

Transient heat transfer for GPMG's barrel 7.62x51mm

M Ghanem[1] and O Abdelsalam[1]
[1] Mechanical branch, MTC, Cairo, Egypt
m.ghanem.mtc@gmail.com

Abstract. In this paper, a modified analytical solution for time-dependent one dimension heat diffusion equation in cylindrical coordinate had been mathematically solved to calculate the temperature distribution through thick-walled weapon barrel subjected to successive shots. The analytical solution has been compared with experimental measurements for the temperature at the outer surface of the barrel at two different positions. The **G**eneral **P**urpose **M**achine **G**un (GPMG) 7.62X51 mm was considered as the case of study. The results show a good agreement between the analytical model and the experimental work with a difference less than 3 %. The main difference between them mainly as a result of the very short time of the heating phenomena as it takes about 6 (ms) only, although the used Data Acquisition System was able to take a high sampling rate and a high sensitive thermocouple type k with only 0.5 mm thickness had been used.

1. Introduction

Not only the weapon's barrel receives bulky amount of heat due to successive shots but also it has a very short time to cool down which leads to heat accumulation on the barrel wall thickness. Consequently, the effect of the thermal pulses with rapid heating and cooling to the barrel wall leads to a dramatic change in temperature distribution through the gun barrel wall. A time-dependent heat diffusion equation should be mathematically solved in order to estimate the temperature variation through the barrel wall thickness. The chromium layer at the inner surface of the barrel is very brittle, it is responsible to resist the cracks and crack extension under the action due to repeated of the fire. The repeated fire makes the barrel bore geometry changes and this leads to decrease the projectile muzzle velocity, projectile stability, the firing accuracy and the tactical technical performance of the machine gun declining, [1].The solution of time dependant heat diffusion equation with nonhomogeneous boundary conditions is solved numerically [2,3], or algebraically solved [4-7] and using advanced software like MATLAB and Mathematica in one dimension [8] or two dimensions, [9].In this paper, a modified model for a time-dependent heat diffusion model, [4] had been solved. The modification in the analytical model was on convection coefficient of gases at the inner surface of the barrel; it had been taken to exponentially decay at the discharge period instead of linear decreasing. On the other hand, the analytical model this time was validated using an experimental measurement not simulation model. The experimental work was carried on 7.62X51 mm General **P**urpose **M**achine **G**un's barrel using a highly sensitive thermocouple type k with only 0.5 mm thickness.

2. Modified Analytical Model

The heat diffusion equation in radial direction with no heat generated from the barrel elements and constant heat conductivity is as shown in equation (1), [10]:

IOP Conf. Series: Materials Science and Engineering **973** (2020) 012032 doi:10.1088/1757-899X/973/1/012032

$$\frac{\partial^2 T}{\partial r^2}+\frac{1}{r}\frac{\partial T}{\partial r}=\frac{1}{\alpha}\frac{\partial T}{\partial t} \tag{1}$$

Where; T is the barrel temperature in [k], r is the barrel radius in [m], α is the thermal diffusivity of the weapon barrel material in $[m^2/s]$ and t is time in [s].
The inner surface boundary condition, according to Newton's law of cooling is taken as follows, [11]

$$k\left|\frac{\partial T}{\partial r}\right|_{r=R_i}=h_f(T-T_f) \tag{2}$$

Where; k is the thermal conductivity of the weapon barrel material in[W/m k], $R_i$is the inner surface of the barrel, h_f is the flow- may be gases or air- convection coefficient in $[W/m^2\ k]$ and $T_f$is the flow temperature in [k].
Similarly, the outer surface boundary condition is as in equation (3), [11]

$$k\left|\frac{\partial T}{\partial r}\right|_{r=R_o}=h_\infty(T_\infty-T) \tag{3}$$

Where; $R_o$is the outer surface of the barrel, $h_\infty$is the ambient convection coefficient of air in $[W/m^2\ k]$ and $T_\infty$is the ambient temperature in [k].
The initial condition at time t =0 defines the temperature of the barrel before the firing:

$$T(r,t)_{t=0}=T_o \tag{4}$$

As the solution of equations (1-4) is too difficult, the solution is performed in two steps. The first step starts from the imitation of the cartridge till the end of gases discharge from the barrel and this the heating period. On the other hand, the second step ends before the initiation of the next cartridge and this period named as cooling period. Moreover, the coefficient of convection of gases and temperature of gases at the first stage usually is taken as an average value. The modification was on the convection coefficient of the heating period. The convection coefficient of gases$(h_{gas})$from the initiation of the cartridge till the point of the projectile leaves the muzzle is determined by, [12]:

$$h_{gas}=h_\infty+\lambda_n\bar{\rho}C_p\bar{v} \tag{5}$$

Where;$h_{gas}$is the convection coefficient of gases,$\lambda_n$is a dimension less constant which is approximated by$\lambda_n=[13.2+4\log(200*R_i)]^2$, $(\bar{\rho})$is the mean gases density and it simply calculated by dividing the propellant burnt mass by the inner volume of the barrel till the projectile base, $C_p$is the specific heat and it is considered to equals 1050 [J/kg k]and $(\bar{v})$ is the mean velocity of gases and it assumed to be linearly distributed inside the barrel. It equals the projectile velocity$(\bar{v}=v_p)$, from one end and equals the breech velocity$(\bar{v}=v_B=0)$ from the other end. $(\text{So, } \bar{v}=0.5*v_p)$.
The convection coefficient of gases in eq (5) is valid till the projectile leaves the barrel but, after the projectile leaves the barrel, the convection coefficient of gases$(h_{gas})$is not considered to linearly decay as [4] but it considered to exponentially decay according to Bravin's law, [6]as follows:

IOP Conf. Series: Materials Science and Engineering **973** (2020) 012032 doi:10.1088/1757-899X/973/1/012032

$$h_{gas} = h_{gas_M} e^{-At} \tag{6}$$

Where; h_{gas_M} is the convection coefficient of gases at muzzle section, A is the exponent of AAPG or the discharge period of powder gases from the barrel.

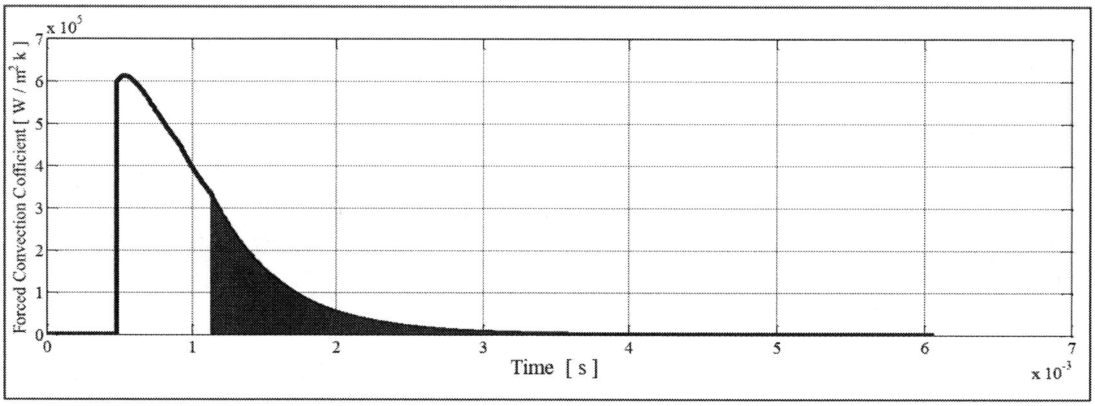

Figure 1. Gases convection coefficient using equation (6) for a single shot at the inner surface, 402 mm from the muzzle section

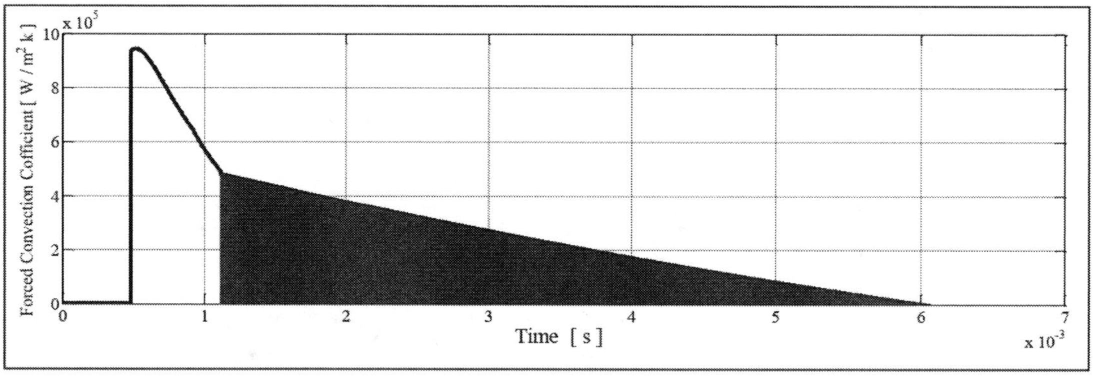

Figure 2. Gases convection coefficient using previous model, [4] for a single shot at the inner surface, 402 mm from the muzzle section

Figure 1 shows the convection coefficient of gases(h_{gas})using equation (6)during the firing time interval due to single shot at the inner surface, 402 mm away from the muzzle section and the exponentially decrease of it at the discharge period is clearly represented by the shaded area. It had a maximum value about $7x10^5$[W/m^2 k] and the average value about $4.5x10^4$[W/m^2 k]. On the other hand, figure 2 shows the gases convection coefficient using previous model, [4]for a single shot at the same section of figure 1, it had an average value about $2.6X10^5$[W/m^2 k].Using the algebraic solution of the system of equations (1-6) using the same technique of [4].

3. Experimental measurements

The experimental work was carried on General **P**urpose **M**achine **G**un's barrel. The GPMG was the best choice for measurement as it has a high rate of fire besides the outer surface of its barrel is almost totally uncovered with weapon casing and it is easy to measure at different stations. The measurement was carried using a highly sensitive thermocouple type k with only 0.5 mm thickness. The measurement had been performed at 2 stations at the outer surface. The two stations of measurement were selected at 140 (mm) and 402 (mm) from the muzzle. The selection of the two stations related to show the effect of increasing of the heating period and the effect of barrel thickness. As one station is near to the muzzle which had only6 mm thickness and the other station is near to the chamber which had a 15 mm thickness.

3.1. Description of experimental setup

The experimental setup as in figure 3 consists of:

- Test rig (1).
- GPMG 7.62x51 (mm) (2).
- Two thermocouples Type k made of Chrome-Aluminium with 0.5mm diameter (3).
- Data Acquisition system with a card with 64 ports and sampling rate 200 kHz maximum(4)
- PC-computer.

Figure 3.GPMG and its fixation.

3.2. Test procedure

The two thermocouples (2) are firmly fixed at two different stations at the outer surface of the GPMG's barrel which is elastically mounted on the test rig (1). Although the DAC was able to take a highly sampling rate, the phenomena of heating takes a very short times about 6 (ms) only. So, after each experiment a 60 (s) period of time is taken as a relaxation time. The first experiment was firing 3 successive shots in automatic trigger and 60 (s) as a relaxation time. For another continuous shot, experiment is repeated after the barrel was reached the room temperature. The second experiment was firing 10 successive shots and 60 (s) as a relaxation time. Before performing the third experiment the barrel is left till it reached the room temperature. The third experiment was firing 15 successive shots and 60 (s) as a relaxation time after the barrel was reached the room temperature.

4. Comparison of results

In order to compare the results of the analytical model with the experimental work, the analytical model was carried out at the same two sections of the experimental work 402 mm and 140 mm from the muzzle section. Also, the initial temperature of the barrel is considered as the same initial temperature of the experiment and changed from one experiment to another one.

In figure 4 and 5 show the outer surface temperature of the barrel for firing 3 successive shots and take 60 (s) as a relaxation time at 402 mm and 140 mm far from the muzzle section respectively. It is shown that the temperature in the two figures had the same trend as it increase till it reach about 301 [K] and 300 [K] respectively almost for the analytical model and the experimental measurement, then, it decay with time till the end of the relaxation period.

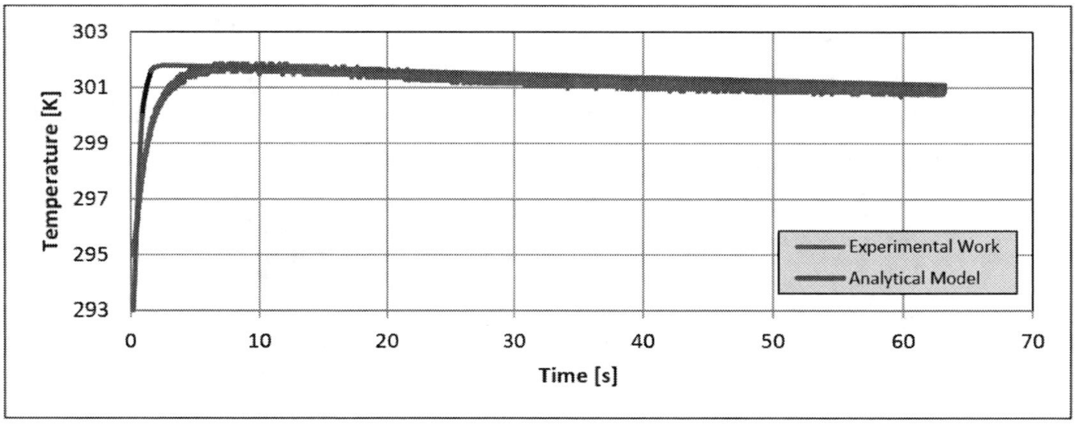

Figure 4. Transient temperature-time history of at the outer surface of the barrel for 3 shots, 402 mm from the muzzle section.

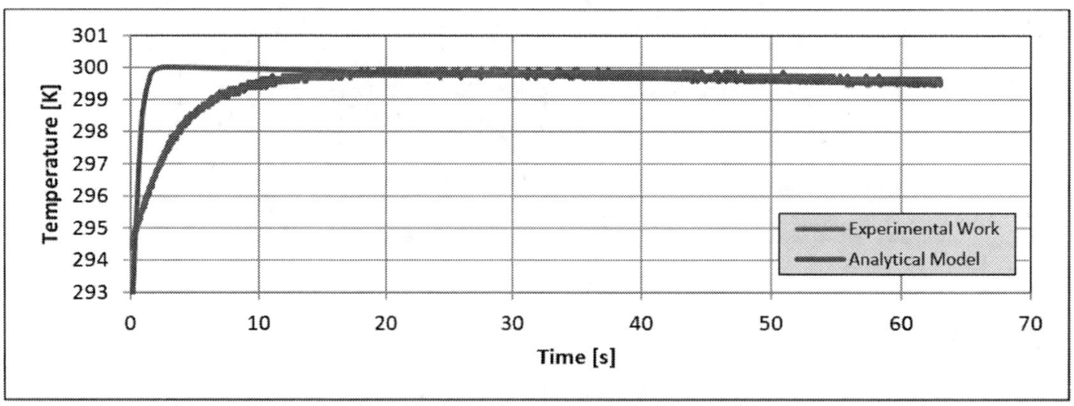

Figure 5. Transient temperature-time history of at the outer surface of the barrel for 3 shots, 140 mm from the muzzle section.

IOP Conf. Series: Materials Science and Engineering **973** (2020) 012032 doi:10.1088/1757-899X/973/1/012032

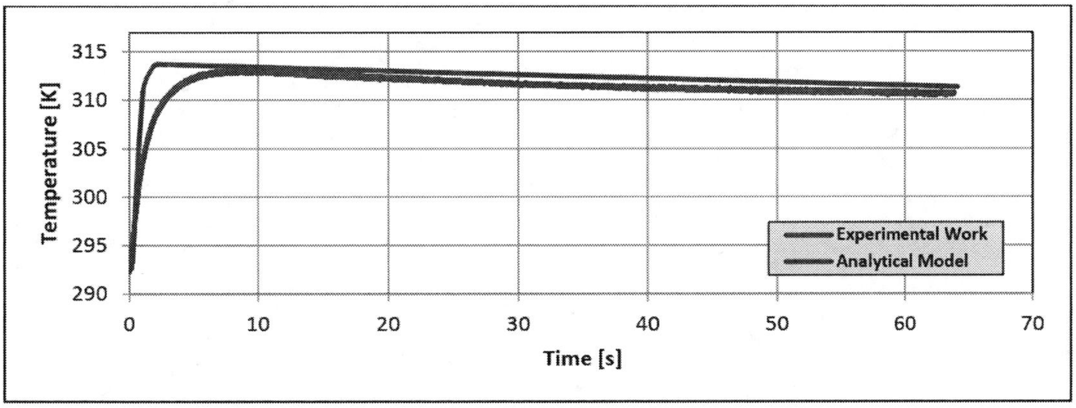

Figure 6. Transient temperature-time history of at the outer surface of the barrel for 10 shots, 402 mm from the muzzle section.

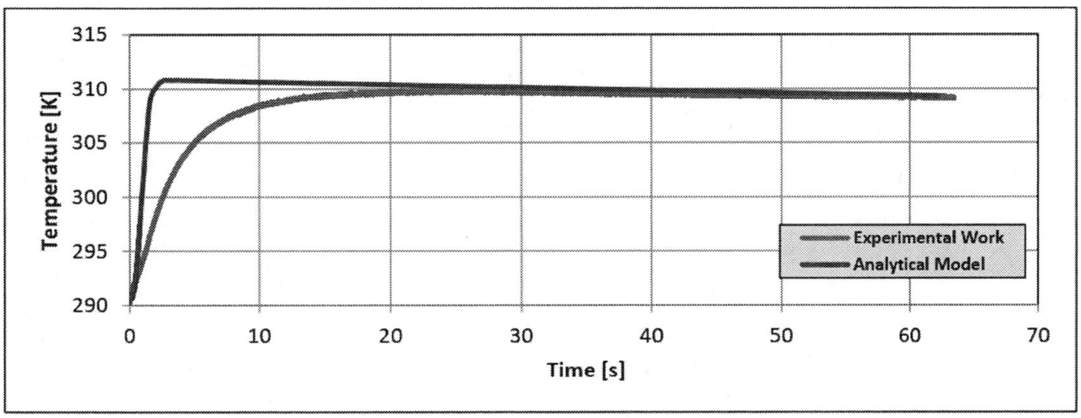

Figure 7.Transient temperature-time history of at the outer surface of the barrel for 10 shots, 140 mm from the muzzle section.

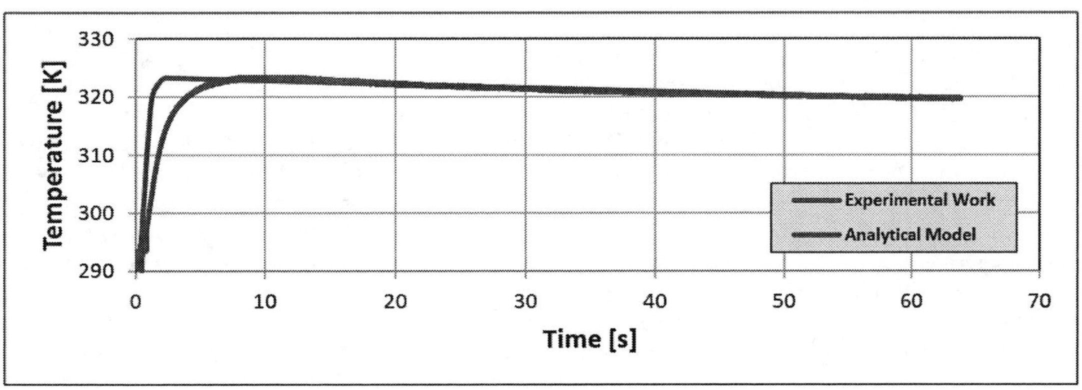

Figure 8. Transient temperature-time history of at the outer surface of the barrel for 15 shots, 402 mm from the muzzle section.

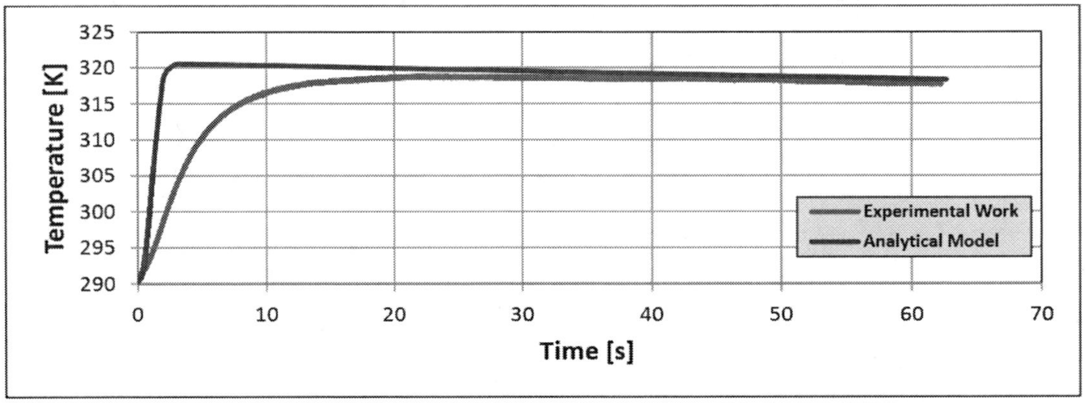

Figure 9. Transient temperature-time history of at the outer surface of the barrel for 15 shots, 140 mm from the muzzle section.

Figure 4 and 5 show the outer surface temperature of the barrel for firing 3 successive shots and take 60 (s) as a relaxation time at 402 mm and 140 mm from the muzzle section respectively. The temperature at the outer surface increases to 302 [K] and 300 [K], respectively. Figure 6 and 7 show the outer surface temperature of the barrel for firing 10 successive shots and take 60 (s) as a relaxation time at 402 mm and 140 mm from the muzzle section respectively. The temperature at the outer surface increases to 313 [K] and 311 [K], respectively. Figure 8 and 9 show the outer surface temperature of the barrel for firing 15 successive shots and take 60 (s) as a relaxation time at 402 mm and 140 mm from the muzzle section respectively. The temperature at the outer surface increases to 302 [K] and 300 [K], respectively. It is clear that the temperature in the two figures increase till it reach about 324 [K] and 321 [K] respectively. In each figure at the beginning of firing there is a difference between the analytical model and the experimental measurement, then, it show a good agreement between the two models at the relaxation period. The difference between the two models may be as a result of the very short time of the heating phenomena as it takes about 6 (ms) only as mentioned before. Moreover, although, the DAC was able to take a highly sampling rate, was not **a** real time results. But, when the DAC take some time with almost constant phenomena no heating impulse gives a good agreement with the analytical model.

5. Conclusion

In this paper, a modified analytical model had been used to investigate the transient temperature of the barrel thickness due to firing. The modification was on the convection coefficient of gases at the heating period especially after the projectile leaves the barrel till the end of discharge of powder gases. The convection coefficient of gases at this period is assumed to exponentially decay according to Bravin's law. On the other hand, the measurement had been performed at 2 stations at the outer surface of PMG's barrel. The two stations of measurement were selected at 140 (mm) and 402 (mm) from the muzzle. The selection of the two stations related to show the effect of increasing of the heating period and the effect of barrel thickness. As one station is near to the muzzle which had only 6 mm thickness and the other station is near to the chamber which had a 15 mm thickness. The analytical model show a good agreement with the experimental work, but there are a difference between them at the beginning. The difference mainly as a result of the very short time of the heating phenomena as it takes about 6 (ms) only, although the DAC was able to take a highly sampling rate and a highly sensitive thermocouple had been used. The temperature difference between the analytical solution and the experimental work is less than 3%. But, due to relaxation time the analytical model gives a good agreement with the experimental work and that declares the necessity of the relaxation period to enhance the measurement results due to the short period of the phenomena. Moreover, that declares the positive effect of the change of inner surface boundary conditions during heating period.

IOP Conf. Series: Materials Science and Engineering **973** (2020) 012032 doi:10.1088/1757-899X/973/1/012032

6. References

[1] Swardt R R and Andrews T D 2006 Stress Analysis of Autofrettaged Midwall Cooled Compound Gun Tubes *Journal of pressure vessel technology*. **128** (2) pp 201-207.

[2] Carslaw H and Jaeger J 1939 A problem in conduction of heat in Mathematical *Proceedings of the Cambridge Philosophical Society* pp 394-404.

[3] Carslaw H and Jaeger J 1959 *Heat in solids* **vol1**(Clarendon Press, Oxford)

[4] Mohamed G, Abdelsalam O, Guirgis S and Aboel Khair M 2017Solution of Heat Transfer Problem for Thick Walled Automatic Weapon Barrel Subjected to Continuous Firing *ASAT* 2017 **2**.

[5] Gu Y and Neal D L O1995 An analytical solution to transient heat conduction in a composite region with a cylindrical heat source *Journal of solar energy engineering* **117** pp 242-248.

[6] Mulholland G and Cobble M 1972 Diffusion through composite media *International Journal of Heat and Mass Transfer* **15** pp 147-160.

[7] Salt H 1983 Transient conduction in a two-dimensional composite slab—I. Theoretical development of temperature modes International *journal of heat and mass transfer* **26** pp. 1611-1616.

[8] De Monte F 2002 An analytic approach to the unsteady heat conduction processes in one-dimensional composite media International *Journal of Heat and Mass Transfer* 45 pp 1333-1343.

[9] Jain P K and Singh S 2009 Analytical solution to transient asymmetric heat conduction in a multilayer annulus *Journal of Heat Transfer* **131** p 011304.

[10] Daiajh H *Hand book of methematical formula and intgrals*.

[11] Bergman, TL, et al 2011 *Fundamentals of heat and mass transfer* (John Wiley & Sons).

[12] Hill R Dand ConnerJ M 2012 Transient heat transfer model of machine gun barrels *Materials and Manufacturing Processes* **27** pp 840-845.

IOP Conf. Series: Materials Science and Engineering **973** (2020) 012033 doi:10.1088/1757-899X/973/1/012033

Multi-fidelity drag prediction for base bleed projectile

M MAziz[1], AZ Ibrahim[1,2], M Y M Ahmed[1] and A M Riad[1]
[1] Mechanical Engineering Branch, Military Technical College, Kobri El-Kobba, Cairo, Egypt.
[2] Corresponding author, Email: **a.ibrahim@mtc.edu.eg.**

Abstract. A computational study is performed to predict the drag reduction for a base bleed projectile. 2-D and 3-D simulations are conducted to describe the flow past a projectile with base bleed. The calculated computational results are compared with measurements of total drag coefficients based on published results for 155mm K307 projectiles with live/dummy base bleed grain. In addition, the mean drag coefficient measurements based on live firing tests for a five K307 projectiles with live BB grain at different Mach numbers are used to validate results obtained from the computational results. Differences between the drag coefficient calculated from computational and published live firing measurements of the drag coefficient for K307 are less than 5.2% and 6.8% for projectile with dummy and live BB, respectively. On the other hand, the maximum deviation between computational results and real measurements from the live firing tests of live BB projectiles is 2.75%.

1. Introduction

Since the upper hand in battle for those with highest range of ammunition, most countries are rushing to find various ways to increase the range of their munitions. The range for artillery projectiles can be extended by many methods [1]; increasing muzzle velocity of the projectile, boosted projectile by using rocket assisted or by improved ballistics which includes drag reduction or using sub-calibre projectile. The drag reduction can be obtained by two different ways. First, pressure drag reduction through improving projectile nose shape or by using projectile with long ogive part. Second, base drag reduction by enhanced boattail shape or increasing the base pressure to overcome the formed wake behind the projectile base. The most common method of increasing base pressure is using a base bleed (BB) unit conjugated to the projectile with boat tail as depicted in figure 1.

Figure 1. A boattail projectile with base bleed (BB).

IOP Conf. Series: Materials Science and Engineering **973** (2020) 012033 doi:10.1088/1757-899X/973/1/012033

The base bleed (BB) unit is formed from a propellant grain which is considered as gas generator. This grain may have many shapes such as slots, tubular, star and wagon wheel. The slotted split tubular grain shape is the most famous configuration it is used in 155 mm projectiles such as M864 K310 and K307 projectiles. The burning of base bleed works to increase the base pressure in the wake (under pressure zone) behind the projectile. This wake is generated when the separation of the turbulent boundary layer from the body at the projectile base occurs. The downstream flow passes by a strong expansion at the base corner forming a free shear which separates the outer in viscid flow forming large Primary Recirculation Region (PRR) downstream of the projectile base where mass is drawn from the recirculation region. Therefore, the pressure on the blunt base is reduced. When the base bleed is activated, the position of primary recirculation region shifts downstream due to the injected gases from the orifice of base bleed while a part of the gases escapes from bleeding section creating the so called Second Recirculation Region (SRR). The base pressure behind the projectile increases due to the mass added from the gases injected from the base bleed. Figure 2 shows a schematic drawing of a flow past a projectile with/out base bleed.

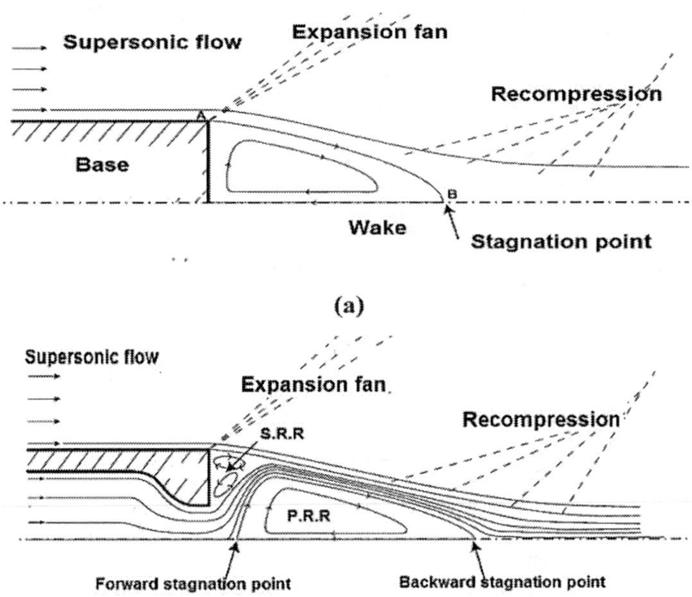

(b)

Figure 2. Scheme of a supersonic flow behind a blunt base (a) without BB (b) with BB [2].

Researches continue to this day on base drag reduction. Which can be obtained by two different ways. One of them is to increase the base pressure using active devices [3, 4] and /or passive devices [5-7] whereas, the other one is concerned with base drag reduction via boattailing [8]. There have been many trials to improve the ballistic performance of base bleed projectiles. The effect of the post-combustion of the base bleed grain on the base drag reduction was studied [9]. Also, it is confirmed that controlling the burning rate of the base bleed grain is an important factor to increase projectile range in the addition to the grain aging [10, 11]. Some studies examined different turbulent models to estimate drag coefficients of slender body with spin and base bleed [12]. Other studies proved that the nozzle pressure ratio of the base bleed unit has an influence on the drag reduction [13].

During developing projectiles with base bleed, it's to accurately estimate magnitude of drag reduction upon using base bleed. Drag on conventional projectiles can be estimated using engineering empirical methods computational techniques, wind tunnel tests, or live firings. However, for

IOP Conf. Series: Materials Science and Engineering **973** (2020) 012033 doi:10.1088/1757-899X/973/1/012033

projectiles with base bleed, no engineering empirical methods are available while wind tunnel test setup can be extremely complicated. Live firing is a reliable high-fidelity approach to estimate drag, however, it is expensive, time consuming, inflexible, and subject to design, production, and environmental uncertainties. Computational techniques based on CFD approach can be implemented to yield accurate prediction of drag on projectiles with base bleed. Being of a lower fidelity, computations need to be validated by comparing them with live firing counterparts.

A current research conducted by the authors is intended to examine novel designs of base bleed units based on computational simulations; validating the computational approach a necessary pre requisite.

The present study is carried out to a base bleed projectile model K307 to compare the calculations obtained from computational simulations with live firing measurements of drag coefficients at zero incidence. In addition, the computational approach is validated against available published firing results on the same projectile with and without base bleed unit. Based on results of this study, reliability of computational approach is assessed for further activities within the running research.

The remainder of the present paper is organized as follows. Details of the case study are explained next followed by study approach. Results are then presented and discussed and the paper finalized with the main conclusions.

2. Case Study

In this work, K307 projectile of 155 mm calibre with base bleed is studied and is manufactured especially for this purpose. Figure 3 shows the relative dimensions of the used projectile in terms of the calibre. A part from the 0.28D base bleed orifice, projectile dimensions are identical to original one. The mass of the projectile is about 46.5kg while the mass of the base bleed unit is 1.3kg.

This unit is consumed in 30seconds which implies that the flow rate of gases out of the base bleed unit is 0.045 kg/sec.

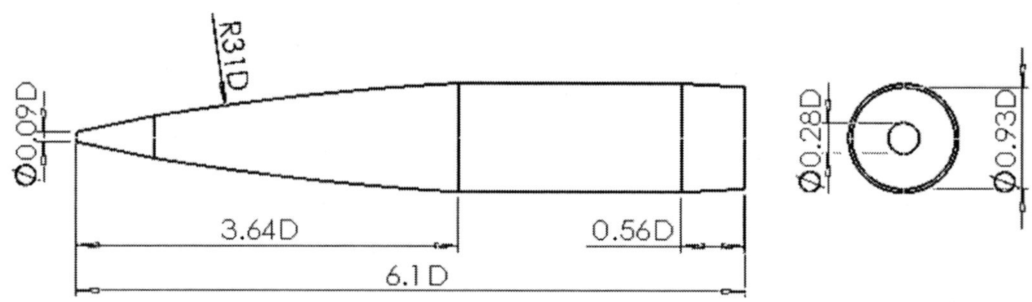

Figure 3. Relative dimensions of the projectile model in concern.

3. Methodology

3.1. Live firing experiments

FiveroundsK307 projectiles with live BB are prepared to check their performance along trajectory in a certified Shooting Range facility. A 155mm caliber gun is used in the firing tests while a system of tracking consisting of a Mono-Pulse Doppler Antenna is used to capture the projectile position during flight on its trajectory, relative to the gun.

The trajectory parameters are calculated using a software installed to a computer interfaced with the tracking radar system. Inputs to the software include projectile mass, initial temperature, meteorological data, and firing angles. For the five rounds of the present study, firing angle is fixed at

45 degrees. The outputs of this system calculated by data acquisition and processing units are trajectory parameters including the variation of range, altitude, velocity with time till impact. Therefore, the drag coefficient C_D can be estimated from the radar tracking data using many schemes as reported in Ref. [14].

3.2. Computational Fluid Dynamics CFD approach
Numerical simulations are conducted to estimate the drag coefficient on the projectile at zero incidence and different free stream Mach number values. Due to symmetry of both projectile and zero-incidence free stream, a one-eighth of the full 3D domain is constructed as shown in figure4. Boundary definitions are also shown.

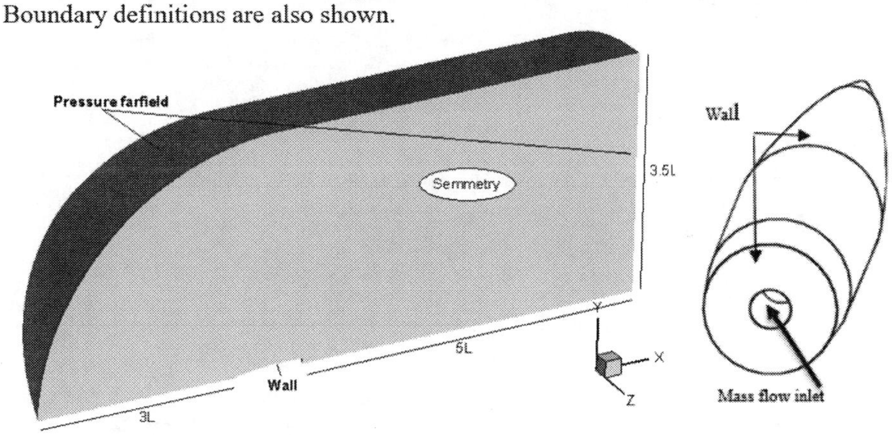

Figure 4. Computational domain and boundary condition.

The pressure far field boundary is set to the uniform upstream flow. The BB exit is defined as mass flow inlet with velocity satisfying the gases flow rate. The inlet flow conditions depend on the time and the position of the projectile along its trajectory [15]. The symmetric boundary is applied to the two side planes of the domain. The adiabatic no slip conditions are set to the solid projectile surfaces. Mass flow inlet boundary is defined at the base bleed orifice exit. The domain is extended in upstream up to 3L, 3.5L and 5L, respectively; L is the projectile length, lateral and downstream direction.

The domain is discretized into a structured grid with enhanced resolutions at the areas of interest, figure5.

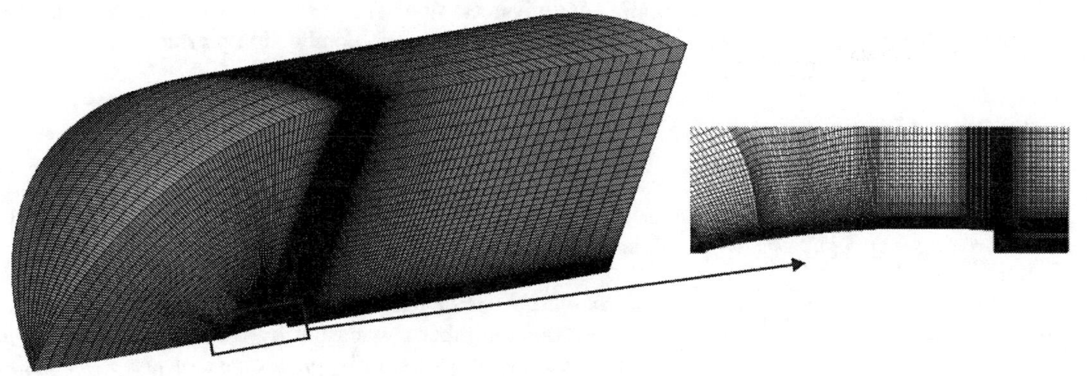

Figure 5. Structured 3-D grid (one-eighth domain).

Grid sensitivity study is done using 2-D version of the structured grids with different resolutions varying from 54300 cells to 85000 cells. Figure 6 shows the evolution of drag coefficient on the projectile unit grid size normalized by the size of the course one. The free stream for grid sensitivity check corresponds to Mach 1.5. A grid with 72300 cells is adopted since no significant improvement is attained with further increase in cell counts. This grid has 280 cells was distributed around the body (115 on ogive, 30 on cylindrical body, 50 on boattail and 85 on base), with about 140 cells upstream and above the model length and 200 cells downstream. Finally, the corresponding 3-D grid includes (714500 cells).

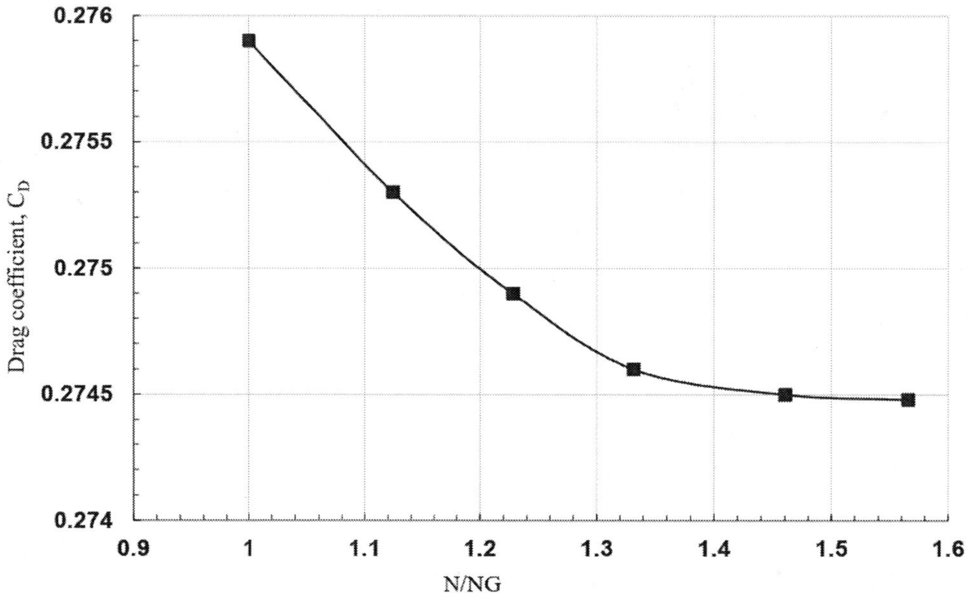

Figure 6. Drag coefficient versus normalized number of cells at M=1.5.

The implicit pressure base scheme is used to solve the system of differential equations. The second order upwind Green-Gauss node based scheme is also used in discretizing the spatial dependent variables in RANS equations. Coupled scheme is applied for pressure-velocity coupling. The turbulent model used in this work is K-epsilon Realizable Enhanced Wall Treatment. Air as an ideal gas is used as the working fluid for free stream flow. For base bleed gas flow, combustion products of base bleed change are defined with chemical reaction enabled [16].

4. Results and Discussion

4.1. Results of the live firing experiments

Figure7 aggregates all drag coefficients for all rounds over the range of flight Mach number. The mean variation is also plotted. All values are listed in table 1.

4.1.1. Comparison of the Computational simulations with published firing results. The live firing measurements of total drag coefficient against Mach number carried out by Hwang and Kim [17] is chosen to validate the corresponding results of computational simulations in case of projectile with a dummy and active BB unit.

Figure 8 compares the published experimental measurements, computational results of the total drag coefficient at different Mach numbers for projectile K307 with dummy and live base bleed. Good agreement is noticed between the CFD results of the drag coefficients with the published experimental measurements of the drag coefficient with difference less than 5.2% and 6.8% for dummy and live BB respectively. This is better shown in table 2.

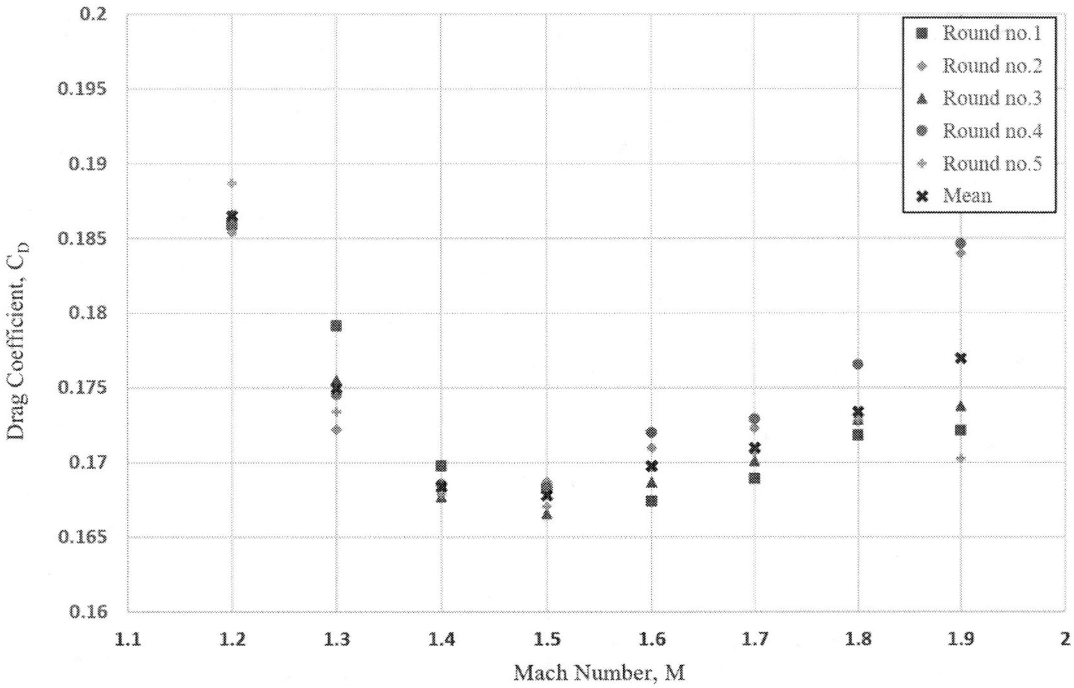

Figure 7. Drag coefficients versus Mach number for the live fired rounds.

Table 1. The measured drag coefficient for all rounds at all Mach number values.

	1.2	1.3	1.4	1.5	1.6	1.7	1.8	1.9
1	0.1859	0.1792	0.1698	0.1682	0.1674	0.1689	0.1718	0.1722
2	0.1854	0.1722	0.1681	0.1687	0.171	0.1723	0.1728	0.1839
3	0.1866	0.1755	0.1677	0.1665	0.1687	0.1701	0.1728	0.1737
4	0.1859	0.1745	0.1685	0.1683	0.1720	0.1729	0.1765	0.1846
5	0.1887	0.1734	0.1678	0.1670	0.1697	0.1706	0.1729	0.1702
Mean	0.1865	0.1750	0.1684	0.1678	0.1698	0.1710	0.1734	0.1770

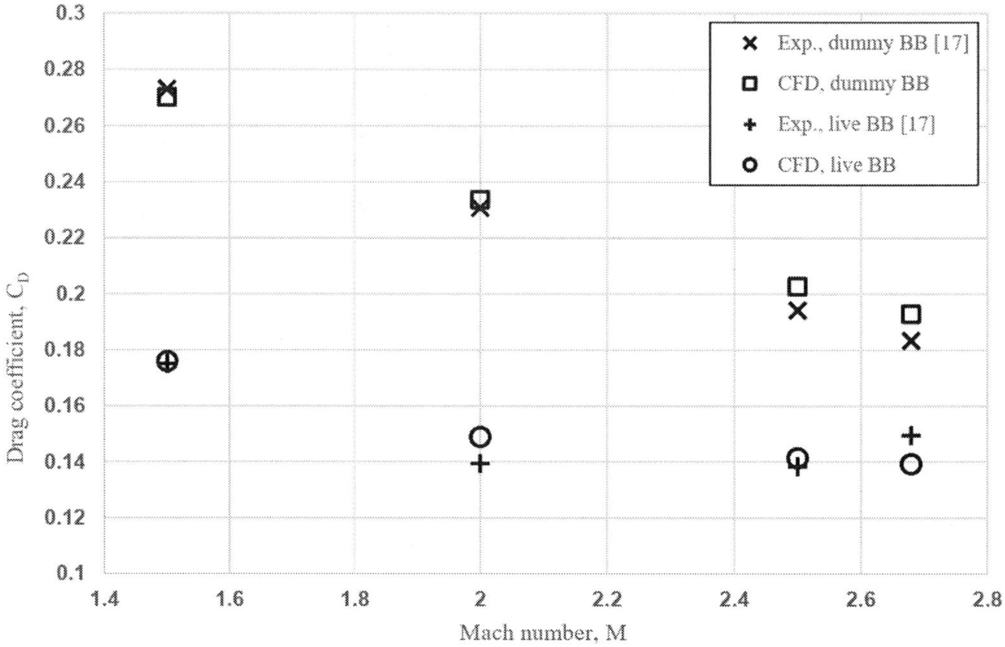

Figure 8. Experimental and computational drag coefficients versus Mach number for K307 with dummy and live BB.

Table 2. Experimental and computational drag coefficients result for K307 with dummy and live BB.

M	With dummy BB			With live BB		
	Exp. [17]	CFD	Difference (%)	Exp. [17]	CFD	Difference (%)
1.5	0.2731	0.2701	-1.1	0.175	0.1760	0.6
2	0.2307	0.2336	1.2	0.1394	0.1489	6.8
2.5	0.1941	0.2027	4.4	0.1384	0.1414	2.1
2.68	0.1833	0.1928	5.2	0.1495	0.1396	-6.8

The impact of live BB is cleared from the previous table; we can notice that the drag reduction is about 18% to 40% when live BB used. The role of base bleed can be better understood by exploring the flow field features of the base of the projectile. Based on computational reproduction of the work of Hwang and Kim [17].

4.1.2. Comparison of the Computational simulation with the live firing experiments. Finally, the total drag coefficient measurements obtained from live firing test for five projectiles K307 with live base bleed grain are used to the computational results. Figure 9shows the computational results and the mean values of the experimental measurements of the total drag coefficient versus Mach number. The deviation between the computational results and the experimental measurements does not exceed

IOP Conf. Series: Materials Science and Engineering **973** (2020) 012033 doi:10.1088/1757-899X/973/1/012033

2.75%. Table 3 lists drag coefficient of the computational and the mean drag coefficients for the experimental work.

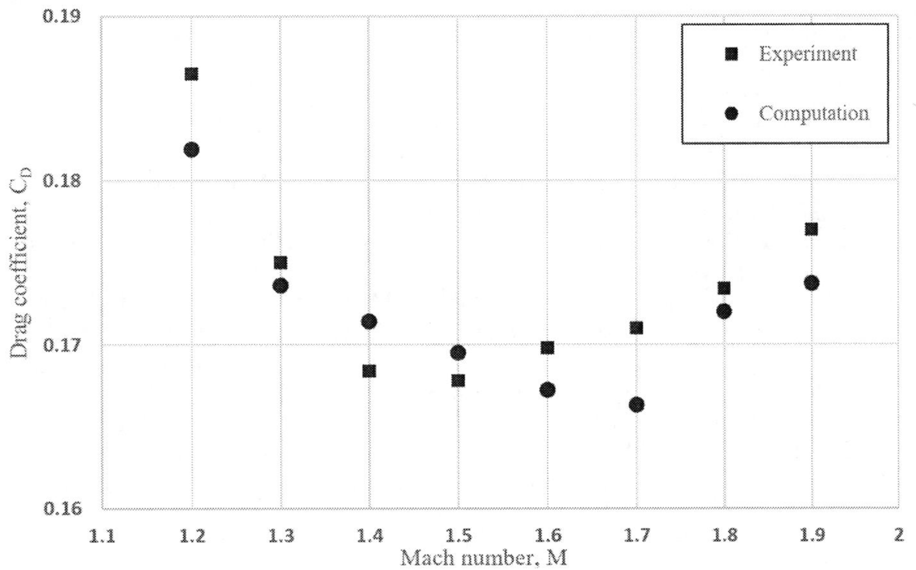

Figure 9. Experimental and computational drag coefficients versus Mach number (for projectile with live BB).

Table 3. Experimental and computational drag coefficients for a K307 with live BB.

M	Experimental (Mean value)	CFD	Difference (%)
1.2	0.1865	0.1819	-2.46
1.3	0.1750	0.1736	-0.8
1.4	0.1684	0.1714	1.78
1.5	0.1678	0.1695	1.01
1.6	0.1698	0.1672	-1.53
1.7	0.1710	0.1663	-2.75
1.8	0.1734	0.172	-0.81
1.9	0.1770	0.1737	-1.86

5. Conclusions

- The objective of the present paper was to check the validity of computational simulation in predicting the drag on projectiles with base bleed. Simulations on K307 projectile were compared with published and the live firing results.
- The main conclusion of this paper is the acceptable drag coefficient values calculated by CFD compared with real firing test measurements. The used computational method could be applied for predicting the drag coefficients of developed projectiles with and without base bleed.

IOP Conf. Series: Materials Science and Engineering **973** (2020) 012033 doi:10.1088/1757-899X/973/1/012033

References

[1] Gurnners N E, Andersson K andHellgren R 1987 *Aeronautics and Astronautics Progress*16(1) no 1 pp 537-61
[2] Kadic S S, Zecevic B, Terzic J and Catovic A 2012 *New Trends in Research of Energetic Materials* (Czech Republic: University of pardubice) part II pp 790-99
[3] Yu Wj, Yu Yg and Ni B 2014 *Defence Technology* **10**(3) pp 279-84
[4] Kubberud N and Ye I J 2011 *26th Int. Sympo. on Ballistics* (Miami: Florida)
[5] Ibrahim A Z and Filippone A2007 *Aircraft* **44**(6) pp 1865-76
[6] El-Awwad E, Ibrahim A Z,El-Sebaee A and Riad A2012 *15th Int. Conf. on Applied Mechanics and Mechanical Engineering*
[7] Suliman M A, Mahmoud O K, Al-Sanabawy M A and Abdel-Hamid O E 2009 *13th Int. Conf. on Aerospace Sciences & Aviation Technology* (Cairo Egypt: ASAT-13-FM-05)
[8] Cummings R M, Yang H T and Oh Y H 1995 *Computers & Fluids* **24**(4) pp 487-507
[9] Xiaochun X and Yonggang Y J 2016 *Applied Thermal Engineering* **109** pp 238-50
[10] Lingke Z Ruyuan Tand Zhuwei Z 2017 *Aerospace Science and Technology* **62** pp 31-35
[11] Zhang L K and Zheng X Y2018*Defence Technology* **14**(5) pp 422-25
[12] Péreza N, Velascob FJS, García JR, Otón RA, López A, Moratilla D, Rey F and Laso A2017 *Aerospace Science and Technology* **67** pp 126-40
[13] Longze M and Yonggang Y 2019 *Applied Thermal Engineering* **148** pp 502-15
[14] Dutta G, Singhal A, Kushari A and Ghosh A 2008 *Defence Science* **58**(3) pp 377-89
[15] Abou-Elela HA, Ibrahim AZ, Mahmoud OK and Abdel-Hamid OE 2013 *15th Int. Conf.on Aerospace Sciences & Aviation Technology* (Cairo Egypt: ASAT-15-207-AE)
[16] Marzouk O A and Huckaby E D 2010*Engineering applications of computational fluid mechanics* **4**(3) pp 331-56
[17] Hwang JS and Kim CK 1996*16th Int. Sympo. on Ballistics* (California; San Francisco) **2** pp 235-44

IOP Conf. Series: Materials Science and Engineering **973** (2020) 012034 doi:10.1088/1757-899X/973/1/012034

Experimental modal testing of a honeycomb sandwich plate

A Aborehab[1], M Kamel[2], A F Nemnem[3] and M Kassem[3]
[1]Space Technology Center, Cairo, Egypt
[2]Associate Professor, AircraftMechanicsDepartement, MilitaryTechnicalCollege, Cairo, Egypt
[3]Assistant Professor, AircraftMechanicsDepartement, MilitaryTechnicalCollege, Cairo, Egypt
ali.aborehab@gmail.com

Abstract. Honeycomb structures have been widely used in several applications and industries in the past decades. Such structures have a significant role in aerospace engineering especially in building up the satellite structures. Honeycomb structures provide key benefits represented in high specific strength, high specific stiffness, and superior dimensional stability. Detailed finite element modelling of such structures is a great challenge as it involves high computational expense. Thus, equivalent modelling is mandatory. The detailed and equivalent finite element models of a honeycomb plate are introduced intensively throughout analyzing its dynamic behaviour using the modal analysis module in ANSYS workbench software. The equivalent modelling is carried out via the three-layered sandwich theory and leads to the calculation of the first four natural frequencies and the related mode shapes. An experimental modal testing of the honeycomb sandwich plate is implemented for the sake of validating the computational results. A good agreement is obtained when comparing both experimental and computational results with mean error not exceeding 5%. Finally, a parametric study is performed to relate the modal analysis results with different boundary conditions cases. The results show that the model can be a reliable basis for satellite honeycomb sandwich structures design.

1. Introduction

In the past decades, honeycomb sandwich structures were used extensively in many applications, especially aerospace engineering [1]. Such structures are composed of two stiff facing sheets bonded to a more flexible honeycomb core. They comprise excellent mechanical properties represented in their high strength, stiffness and low mass. However, the detailed finite element modelling (FEM) of these structures is still a great challenge due to complexity and high computational time. Many researchers in the past years tried to use different equivalent modelling approaches to overcome the aforementioned difficulties.

In 2001, Xia et al. [2] introduced three modeling theories that can be used for equivalent modeling of sandwich structures. These theories are the sandwich theory, the equivalent plate theory, and the honeycomb plate theory. The authors validated the results of the modal analysis using the three theories by comparing them with analytical work.

In 2012, A. Boudjemai et al. [3] proposed the usage of the equivalent plate theory in the FEM of a satellite honeycomb sandwich structure. They carried out modal testing in order to validate the numerical results. They received a good agreement between experimental and numerical modal results.

In 2014, Jiang et al. [4] employed the sandwich theory during the computation of natural frequencies and their related mode shapes. They performed a dynamic testing to assess the use of sandwich theory in modeling sandwich structures with acceptable accuracy.

In 2017, Fu et al. [5] presented the dynamic response of a spacecraft solar blanket using the sandwich theory. They compared the results with analytical work to ensure the availability of using such theory in modeling sandwich structures.

Recently, sun et al. [6] implemented both modal testing and analysis of a honeycomb sandwich plate. Based upon results, they performed a FEM updating so as to efficiently estimate the accurate equivalent properties of the honeycomb core.

In this research, the modeling of honeycomb sandwich plate is introduced extensively via modal analysis module in ANSYS workbench using free-free boundary conditions. Detailed FEM is carried out; such model exhibits a high computational time. Consequently, an equivalent FEM is built, using the three-layered sandwich theory, to overcome such problem. Both detailed and equivalent results are validated using experimental modal testing results, where the first four natural frequencies are measured and compared to those computed numerically.

2. Detailed FEM of a honeycomb sandwich plate

Such type of modelling relies on representing all the geometric details of the utilized honeycomb core. These details includes the foil thickness (t), the cell vertical length (a), the cell inclined length (b), and the cell angle (θ) as shown in figure 1.

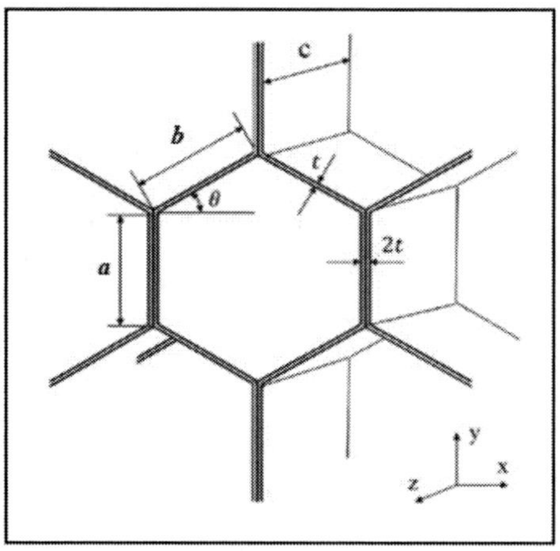

Figure 1. Honeycomb core geometry.

The modal analysis is implemented on a honeycomb sandwich plate utilizing free-free boundary conditions. The mentioned plate dimensions are 400mm x 400mm x 13 mm. the plate thickness includes 2 mm thickness for each facing sheet and 9 mm for the honeycomb core thickness. Both facing sheets and core are fabricated of Aluminum alloys. The geometric and material specifications of the selected core are shown intable 1 and table 2 respectively.

Table 1. Geometric properties of the core

Type	Material	Cell length (a=b)	Cell thickness (t)	Cell angle (θ)
Regular hexagon	AL-3003	4 mm	0.04 mm	30^0

IOP Conf. Series: Materials Science and Engineering **973** (2020) 012034 doi:10.1088/1757-899X/973/1/012034

Table 2. Core material specifications

Young's modulus (E_c)	Shear modulus (G_c)	Density (ρ)
70000 MPa	27000 MPa	2700 kg/m^3

The facing sheets are fabricated of AL-6061-T6, their material properties are displayed in table 3.

Table 3. Material properties of the facing sheets.

Young's modulus (E_c)	Shear modulus (G_c)	Density (ρ)
71000 MPa	27000 MPa	2770 kg/m^3

The detailed geometry of the core is performed via computer-aided design (CAD) software and then exported to ANSYS workbench "Design Modeler section" using "parasolid.x_t" format as shown in figure 2.

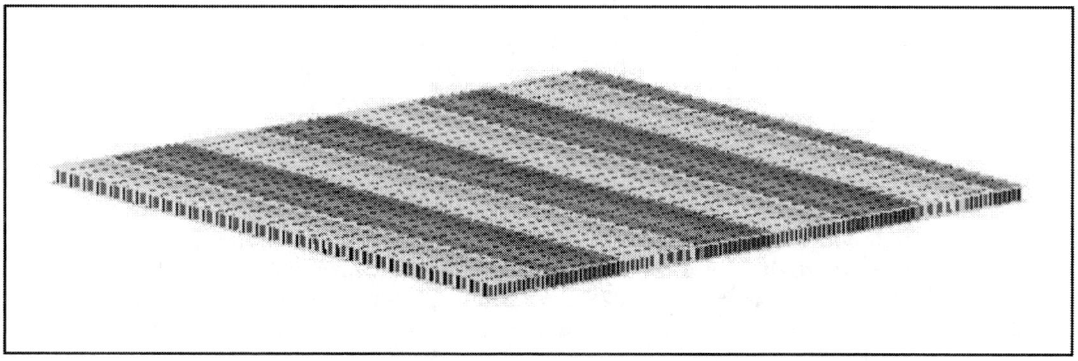

Figure 2. Imported detailed honeycomb sandwich plate.

Both facing sheets and core are meshed via shell finite elements "shell 181". Such element is a 4 node element with 6 degrees of freedom at each node. The Free-Free boundary conditions are employed to compute the four natural frequencies and the corresponding mode shapes.Meshing is performed using a fine mesh of element size 3 mm, where results converge and there is no more need to reduce the element size as shown in figure 3.Table 4 shows the total number of elements and nodes.

Table 4. Detailed model elements and nodes.

Element size, mm	Number of elements	Number of nodes
3	310643	249735

It can be noted that the CAD and FE detailed modeling is a time-consuming process and needs a computers of high computational resources.

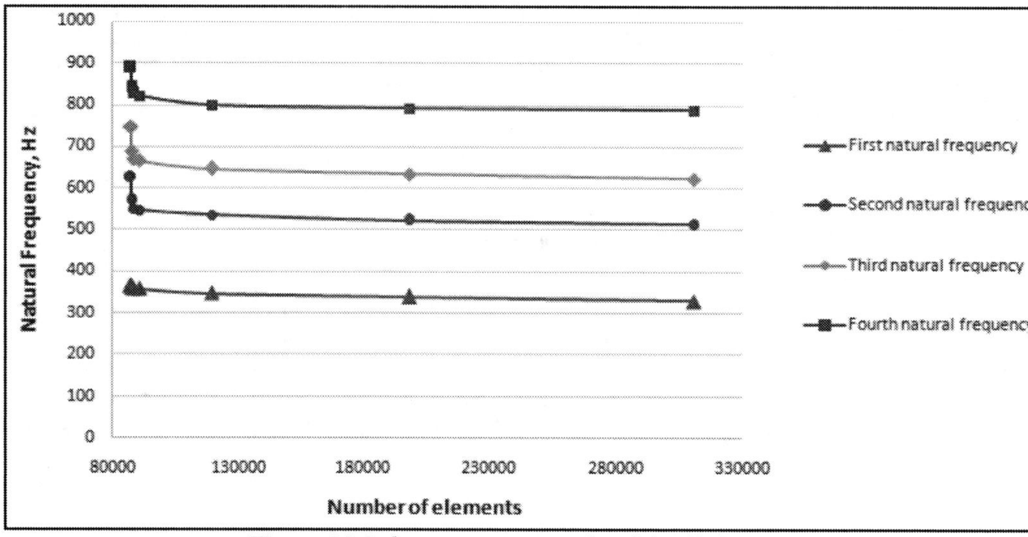

Figure 3. Mesh convergence study of detailed FEM.

3. Equivalent FEM of a honeycomb plate

The equivalent modelling helps in solving the problems encountered during the detailed modelling of honeycomb sandwich structures. In this research, the three-layered sandwich theory is used in the honeycomb sandwich plate modal analysis using free-free boundary conditions. The main difference between sandwich theory and detailed modelling is the way of representing the honeycomb core. In this theory, the core is dealt with as homogeneous continuum with orthotropic material properties. Thus, sandwich theory needs to calculate the equivalent properties of the core. The calculation of these properties depends on the geometric properties of the core in addition to the material properties of core material as shown in table 1 and table 2.

In this research, Gibson and Ashby formulas [7], with minor modification [8, 9] , are utilized to calculate the core equivalent properties as follows:

$$\rho_{eq} = \rho_c \frac{8}{3\sqrt{3}}\left(\frac{t}{\mathrm{a}}\right) \tag{1}$$

$$E_x = E_y = \frac{4}{\sqrt{3}}\left(\frac{t}{\mathrm{a}}\right)^3 E_c \tag{2}$$

$$G_{xy} = \frac{4}{5\sqrt{3}}\left(\frac{t}{\mathrm{a}}\right)^3 E_c \tag{3}$$

$$\upsilon_{xy} = \upsilon_{yx} = 1 \tag{4}$$

$$E_z = \frac{8}{3\sqrt{3}}\left(\frac{t}{\mathrm{a}}\right) E_c \tag{5}$$

$$\upsilon_{xz} = \upsilon_{yz} \approx 0 \tag{6}$$

$$G_{xz} = \frac{1}{\sqrt{3}}\left(\frac{t}{a}\right) G_c \tag{7}$$

$$\frac{\sqrt{3}}{2}\left(\frac{t}{a}\right) G_c \leq G_{yz} \leq \frac{5}{3\sqrt{3}}\left(\frac{t}{a}\right) G_c \tag{8}$$

$$G_{yz} = \frac{\sqrt{3}}{2}\left(\frac{t}{a}\right) G_c + \frac{0.787}{({}^{c}/_{a})}\left(\frac{5}{3\sqrt{3}}\left(\frac{t}{a}\right) G_c - \frac{\sqrt{3}}{2}\left(\frac{t}{a}\right) G_c\right) \tag{9}$$

In order to apply the sandwich theory in the finite element modeling using ANSYS software, shell-volume-shell (SVS) approach is employed[4, 6, 10], where the facing sheets are meshed using solid finite elements "solid 186", while the core is meshed using shell finite elements "shell 181".Meshing is performed using a fine mesh of element size 10 mm, where results converge and there is no more need to reduce the element size as shown in figure 4.Table 5shows both elements and nodes total number.

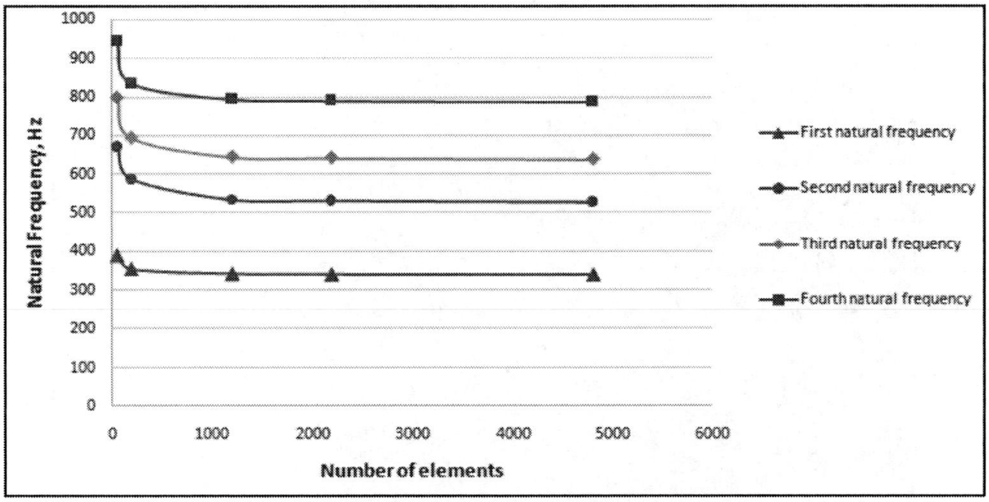

Figure 4. Mesh convergence study of Equivalent FEM.

Table 5. Equivalent model elements and nodes.

Element size, mm	Elements number	Nodes number
10	9604	32200

Based upon the aforementioned formulas, the equivalent properties of the utilized honeycomb core are estimated as shown intable 6.

Table 6. Equivalent properties of the core.

$\rho_{eq.}$	41.56 kg/m^3	E_z	1077.7 MPa

IOP Conf. Series: Materials Science and Engineering **973** (2020) 012034 doi:10.1088/1757-899X/973/1/012034

$E_x = E_y$	0.1616 MPa	G_{xz}	155.89 MPa
G_{xy}	0.097 MPa	G_{yz}	242.9 MPa
υ_{xy}	1	$\upsilon_{xz} = \upsilon_{yz}$	0

4. Experimental modal testing

Modal testing is carried out so as to validate the computational results of both detailed and equivalent modelling. The Free-Free boundary conditions are simulated by hanging the sandwich plate using stainless steel wires. The test configuration is shown in figure 5.

The test includes the following devices:

- Structure excitation using a hammer that comprises a force transducer.
- Response sensing elements using two accelerometers attached to the plate using special kind of wax. The location of such accelerometers is determined using numerical modal analysis results in the way that helps to extract the four mode shapes. This location is selected at the points of maximum deformations (corners and sides mid-distance).
- Analyzer for interpreting hammer force and accelerometers response.
- Laptop with Labview software for controlling the whole test process.

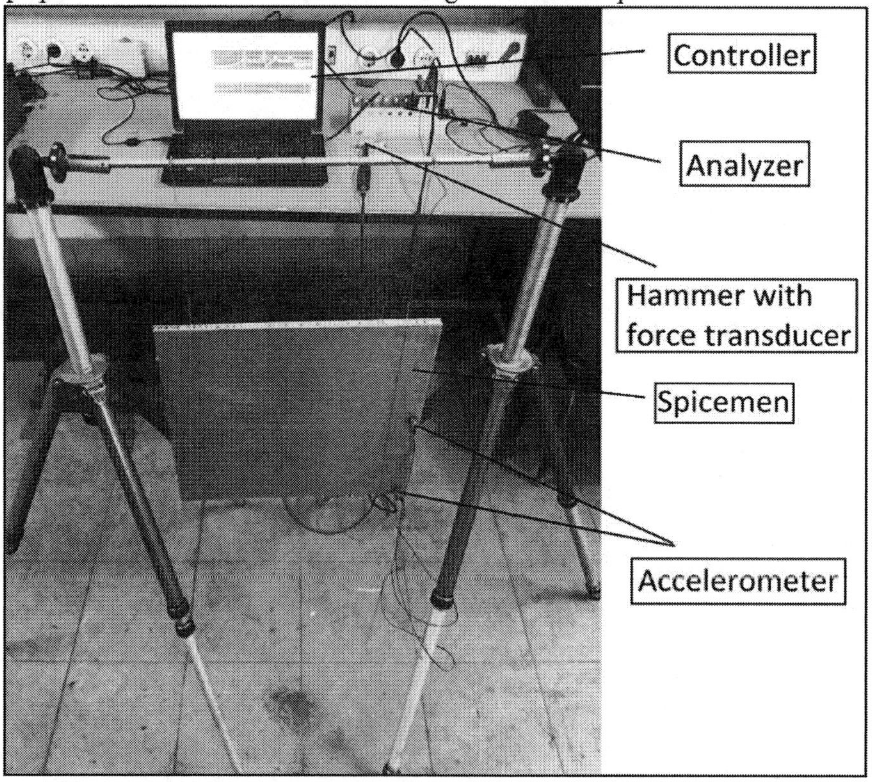

Figure 5. Modal testing configuration.

Eight excitation points are decided at the four corners, and the four mid points of each side. The test procedure implies performing five consecutive hammer impacts at each excitation point in order to

IOP Conf. Series: Materials Science and Engineering **973** (2020) 012034 doi:10.1088/1757-899X/973/1/012034

take the mean readings. The frequency response function (FRF) is recorded and the four natural frequencies are measured as shown in figure 6.

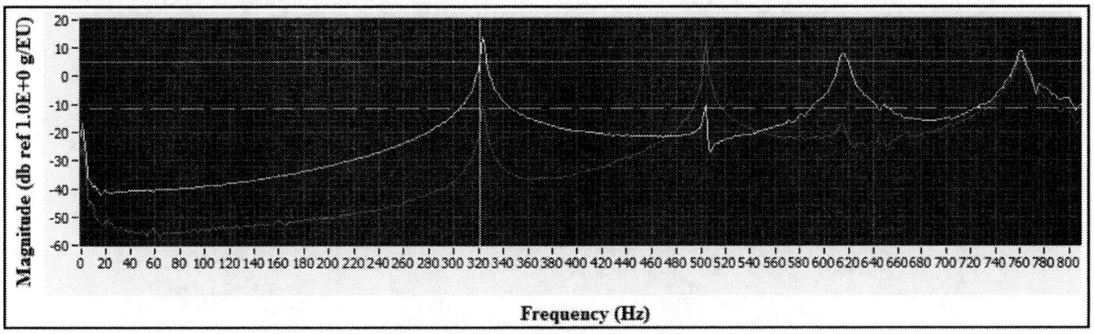

Figure 6. Modal testing FRF.

5. Results comparison and discussion

Figures 7 and 8 presents the numerically calculated natural frequencies and mode shapes using the detailed and equivalent modelling.

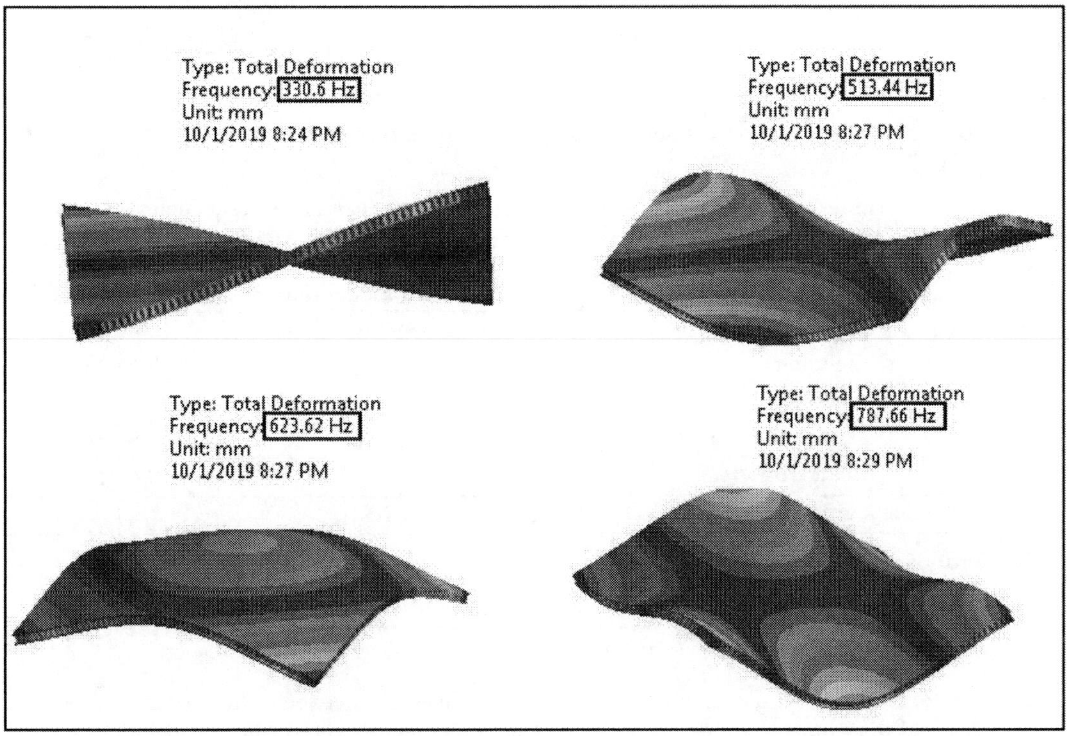

Figure 7. Detailed model results.

IOP Conf. Series: Materials Science and Engineering **973** (2020) 012034 doi:10.1088/1757-899X/973/1/012034

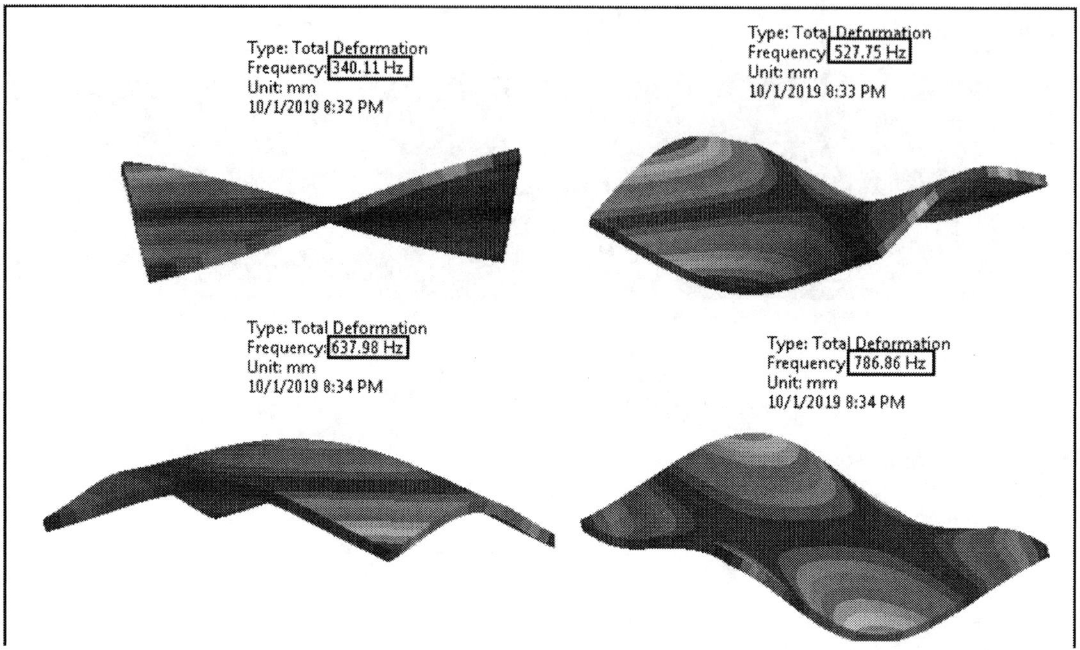

Figure 8. Equivalent model results.

The comparison of experimental and numerical results concerning the four natural frequencies is listed in table 7.

Table 7. Comparison of experimental and numerical modal frequencies.

Mode order	Mode Type	Experimental, Hz	Numerical, Hz	
			Detailed modelling	Equivalent modelling
First mode	Torsion	322	330.6	340.11
Second mode	Bending	504	513.44	527.75
Third mode	Bending	614	623.62	637.98
Fourth mode	Torsion	761	787.66	786.86
Mean error, %			2.403	4.41
Computational time, min.			210	1

The error of natural frequency value is calculated as follows:

$$error = \frac{|Numerical\ frequency - Experimental\ frequency|}{Experimental\ frequency} \times 100\% \qquad (10)$$

The previous results indicate clearly the good agreement between the experimental modal frequencies and the numerical ones using both types of modeling. Nevertheless, detailed modeling exhibit an extreme computational time and resources. Consequently, the use of equivalent modeling is mandatory for its acceptable accuracy and reasonable computational expense.

6. Boundary conditions parametric study

The main target of this study is to relate the natural frequenciesof the validated honeycomb sandwich plate with different boundary conditions cases including four sides clamped (CCCC), clamped-free-clamped-free (CFCF), and clamped-free-free-free (CFFF) cases. Equivalent model using SVS approach will be used during this study. Figure 9 shows the parametric study results.

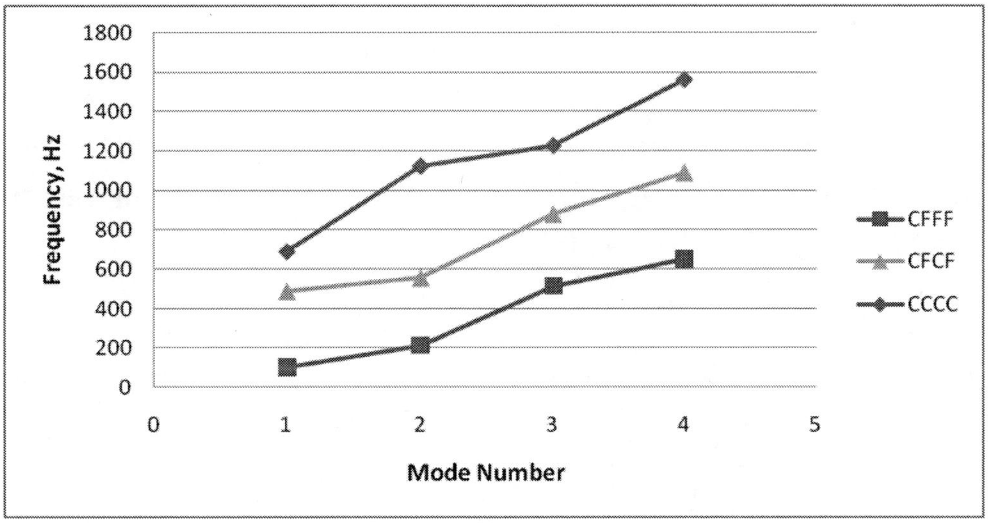

Figure 9. Parametric study results.

Conclusions

Detailed and equivalent FEM of honeycomb sandwich structures are thoroughly discussed in this article. Although detailed modelling is more precise, it undergoes a high computational time and computing resources. Hence, the utilizing of equivalent modelling in representing honeycomb sandwich structures is inescapable. The sandwich theory and its corresponding shell-volume-shell (SVS) approach achieve such modelling accurately with a mean error below 5% when compared with the experimental modal testing results. Moreover, a parametric study is carried out in order to relate the modal analysis results of the validated honeycomb plate with different boundary conditions. The results show that the proposed FEM can be a reliable basis for satellite honeycomb sandwich structures design.

References

[1] Aborehab A, Kassem M, Nemnem A F, Kamel M and Kamel H 2020 *Engineering Solid Mechanics*8 7-20
[2] Xia L, Jin X and Wang Y 2001*J Shanghai Jiao Tong Univ.* **37**pp999-1001
[3] BoudjemaiA,Amri R, Mankour A, Salem H, Bouanane M and Boutchicha D 2012 *Materials and Design***35**pp 266-275
[4] JiangD,Zhang D, Fei Q and Wu S 2014 *Finite Elements in Analysis and Design***90** 84-92
[5] Fu J,Luo H, Wang W and Wang Z 2017 *IEEE 3rd Information Technology and Mechatronics Engineering Conf.*
[6] Sun W and Cheng W 2017 *Structural and Multidisciplinary Optimization***55**pp121-139
[7] Gibson L and Ashby M 1997 *Cellular Solids: Structure and Properties*(Cambridge University Press)
[8] Grediac M 1993 *Int. J. of Solids and Structures***30** pp 1777-88
[9] Sorohan S, Sandu M, Sandu A and Constantinescu D 2015 *32nd DANUBIA ADRIA SYMP. on Advanced in Experimental Mechanics*
[10] Catapano A and Montemurro M 2014 *Joint Conf. on Mechanical, Design Engineering & Advanced manufacturing*

IOP Conf. Series: Materials Science and Engineering **973** (2020) 012035 doi:10.1088/1757-899X/973/1/012035

A new artificial neural network model integrated with a cat swarm optimization algorithm for predicting the emitted noise during axial piston pump operation

A H Elsheikh[1], MAbd Elaziz[2], HA Babikir[3], D Wu[3] andYLiu[3]
[1] Production Engineering and Mechanical Design Dept., Tanta University, Egypt.
[2]Department of Mathematics, Faculty of Science, Zagazig University, Zagazig, Egypt.
[3] School of Mechanical Science and Engineering, Huazhong University of Science and Technology, Wuhan, China.
ammar_elsheikh@f-eng.tanta.edu.eg

Abstract. This study presents a new artificial intelligence based methodology to predict the emitted noise of an Axial Piston Pump (APP). The suggested method depends on augmentation of conventional Artificial Neural Network (ANN) via integration with Cat Swarm Optimization (CSO). CSO is used to obtain the optimal structure of ANN. The training and testing of the approach were accomplished using experimental data sets considering six system operating pressures (0.5, 1.0, 1.5, 2.0, 2.5, and 3.0 MPa) and five speed levels (600, 900, 1200, 1500, and 1800 rpm). Two valve seat materials were investigated: polyetheretherket one (PEEK) and 316L stainless steel. A reasonable agreement was observed between the predicted results obtained by the developed method and the experimental data.

1. Introduction

APPs have been shown promising application in hydraulic systems that used to induce hydraulic power from mechanical power due to their advantages over other commonly used pumps such as their high operating pressure and speed.

However, APP has a main drawback as when it operates at higher pressure and displacements, high noise levels and severe vibration are generated[1]. APPs are classified into two main types based on the port type: plate or valve. The APP with port valve has a good abrasion resistance as it is not affected by water contamination which gives it an advantage to be operated under severe conditions in abrasive mediums such as silt-laden water. In a typical APP with port valve, two valves control the water flow: the inflow valve by which water enters the pump and the discharge valve by which water exits the pump. Therefore, APPs with port valve are the optimal choice for hydraulic systems that contain open hydraulic circle in which working fluid has excessive contamination. Moreover, this type of pump averts the use of extra filtration devices to clean the flowing water. In APP with port valve, inlet/outlet flow is controlled using port valves. These valves have a noteworthy effect on the life span, emitted noise, and volumetric efficiency of the pump. Xu et al. [2] proposed a methodology based on vibro-acoustic analyses to optimize the housing of an APP. A finite element model of the cover and housing has been developed to execute the modal analysis. Then, a topology optimization is carried out to obtain the optimal structure of the housing that reduces the frequency response function. The average level of sound pressure is decreased by about 2 dB(A) at operating pressure of 250 bar using the optimized housing. Pan et al. [3] investigated the transmission paths and radiated noise source of

IOP Conf. Series: Materials Science and Engineering **973** (2020) 012035 doi:10.1088/1757-899X/973/1/012035

an APP. A finite element model of the APP shell was developed to execute the modal analysis. The vibration performance as well as the excitation forces generated during operation and transmitted to the cover was analysed. The noise transmission path and the main noise source were determined. The sound pressure level was reduced by 1.26 dB (A) at 300 bar discharge pressure. An effort to reduce the sharpness of an APP with a reinforced shell with casting stiffeners has been conducted by Zhang et al. [4]. The sharpness of the APP was decreased by about 6.2% when reinforced shell is used compared with that of the reinforced shell-free APP at 21 MPa discharge pressure. A new design method of valve plate of APP to reduce the emitted noise resulted from the fluid flow and structure vibration has been proposed by Xu et al. [5]. The proposed design method has succeeded to reduce the flow ripple and eliminate the pressure undershoot and overshoot. Zhang et al. [6] proposed a new deposition technique to reduce the generated noise of an APP. This technique depends on depositing a 3 mm layer of a visco-elastic damping material on the external surface of the APP to increase the natural frequency of the APP by about 6%, which consequently decreases the modal displacement as well as the induced noise and vibration. The average level of sound pressure is reduced by about 6.3 dB(A) at 21 MPa discharge pressure. Yin et al.[7]proposed a reliability model to investigate the effects of valve-port on the emitted noise of an APP. The fluctuation in the contact pressure and rotational speed impairs the reliability of the pump. Therefore, it is recommended to preserve the speed of APPs at a constant value.

Recently, artificial intelligence based methods such as ANN and different metaheuristic optimization algorithms and their hybrid combinations have been used to optimize and predict the performance of different mechanical systems [8-15]. Ye et al. [16] used genetic algorithm, as a metaheuristic optimization tool, to minimize the noise of an APP which used as an objective function. A dynamic pump model was established considering the phenomena of cavitation and air-release. A comprehensive parametric investigation was carried out to analyse the effects of the valve plate design parameters on the induced noise. The average level of sound pressure is reduced by about 1.6 dB(A) at 15 MPa discharge pressure. Karkoub et al.[17]developed a feed-forward ANN model to predict the steady-state and dynamic behaviour of an APP. The pump speed was used as the input of the ANN, while the pressure was used as the output. Huayong et al. [18] used aback propagation ANN to obtain the emitted noise of APP for four materials of port plates. The ANN inputs were the operating pressure, flow rate, oil bulk modulus, and oil temperature, while the output was the noise level.

Based on the abovementioned survey, most of previous studies carried on APPs investigated dynamics of moving parts, fluid dynamics, application of visco-elastic damping material, valve plate optimization, and housing optimization. Few studies have been investigated the effects of port valve material on the pump lifespan, pump efficiency, and generated noise. There are two main materials used in port valve fabrication: polymeric and metallic. Polymeric port valves have superior sealing characteristics and high machinability compared to metallic ones. On the other hand, metallic port valves are recommended for severe operating conditions with high pressure due to their superior mechanical properties such as strength and rigidity. However, precision fabrication techniques is required to enhance the sealing capabilities of metallic port valves which results in increasing the cost of the pumps with metallic port valves.

2. Objective and approach

The main aim of this study is to develop an accurate model to predict the generated noise of an APP. Experiments were conducted on an APP experimental setup under different operating pressures and speeds. Two different valve seat materials were considered: PEEK and 316L. A modified version of ANN was proposed in which CSO was used to optimize the ANN structure. The performance of the proposed ANN-CSO model was compared with that of conventional ANN model using different statistical criteria.

3. Experimental setup

An APP with a port valve is used in this study. This pump is composed of two main components: driving components (case, shaft, swash plate, slipper, and bearing) and flowing components (discharge valve, inflow valve, piston, and cylinder). The valves used in this APP are disc-type poppet valves. This type of valve could be utilized to minimize pressure loss in comparison to ball and cone types. Due to the high salinity of seawater, different pump parts should be made of cost-effective materials with excellent anticorrosion characteristics, suchpolymers, alloy steel, and titanium alloys. PEEK is a well-known polymeric material which has good mechanical properties such as high modulus of elasticity, high tensile strength, creep resistance, and corrosion resistance. Titanium alloys have excellent corrosion resistance, low density, and outstanding strength. Stainless steels alloys have enhanced cavitation erosion resistance. In the current experimental setup,TC4 titanium alloy has been chosen as valve spool material for both of discharge and inflow valves. Meanwhile, PEEK and 316L stainless steel have been chosen as valve seat materials to figure out their role on the generated noise.
A schematic diagramand a photograph of the experimental setup are presented in figure 1. The inlet pressure was 0.01 MPa and adjusted via a water tank and a loading valve. The flow rate was controlled by adjusting the motor speed by an inverter. The pressure of the inlet line was measured by a pressure gauge with accuracy of ± 0.25% of full scale and a measuring range of 0–4 MPa. The flow rate was measured using a flow meter device with accuracy of ± 0.5% of full scale and measuring range of 0–50 L/min. The level of the generated noise was measured by a sound level meter with accuracy of ±0.5 dB which located at 1 m from the pump. The noise is measured according to A-weighted sound pressure level technique. In the current study, the effects of valve seat material (316L and PEEK), motor speed (600, 900, 1200, 1500, and 1800 rpm), and pressure (0.5, 1, 1.5, 2, 2.5 and 3 MPa) on the level of the emitted noise are investigated.

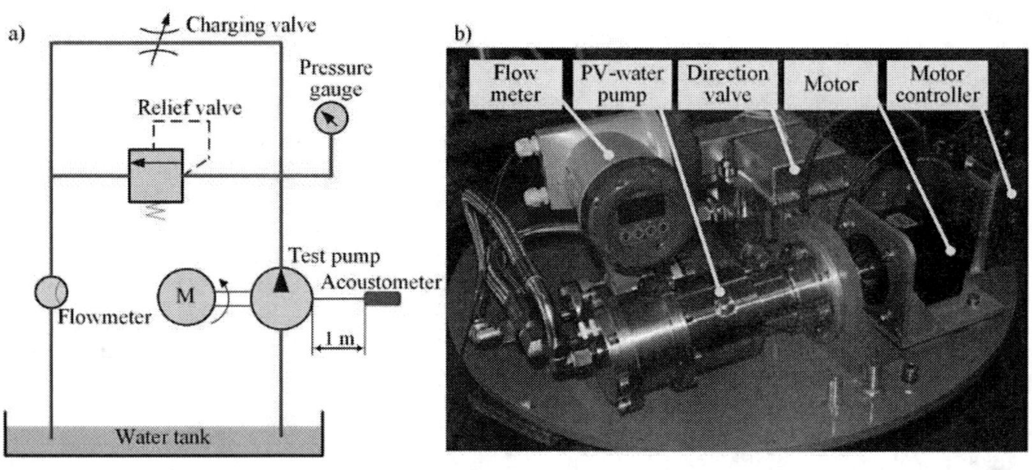

Figure 1. a) A schematic diagram; b) A photograph of the experimental arrangement[19].

4. The proposed approach

The proposed approach depends on improving the performance of the ANN through integration with CSO as a metaheuristic optimization method to find the optimal structure of ANN. ANN is intelligent computing tool which simulate the learning behaviour of the human brain. The performance of ANNs models is highly affected by the learning process. Multilayer perceptron (MLP) is a feed forward neural networks (FFNNs) which utilizes a supervised learning algorithm. MLP is composed of three main layers (figure 2 (a)): the input layer in which the input factors enter the network, the hidden layer in which data is processed, and the output layer. The neurons in different layers of MLP are organized in a one-directional manner. These interconnections are represented by weights defined by real numbers in the interval [-1.1]. The following equation can be used to describe each layer in an MLP:

$$o_i^{(\ell)} = \varphi(u_i^{(\ell)}) = \varphi\left(\sum_{j=1}^{n_\ell} o_j^{(\ell-1)} w_{j,i}^{(\ell)} + w_{0,i}^{(\ell)}\right), \quad 1 \le \ell \le L \qquad (1)$$

Wherein $\varphi(.)$ denotes the activation function used in the layer. The commonly used activation functions are classified as linear and non-linear (tangent hyperbolic). The linear activation functions are recommended for input and output layers, while the non-linear activation function is recommended for the hidden layers. The indexℓdenotes the layer in a network of L hidden and output layers, $n_\ell$is the number of neurons of layer ℓ, o_i^ℓ denotes the ineuron output in the layer ℓ, $w_{j,i}^{(\ell)}$, $1 \le \ell \le n_{\ell-1}$ are the weights of the interconnection of neuron i of layer ℓ with the neurons j of the layer $\ell - 1$, and $w_{0,i}^{(\ell)}$ is the neuron i bias of the layer. The output vectors of the layer ℓ =0 and layer $\ell = L$ are $O^{(0)} = x$and $O^{(L)} = y$, respectively. These vectors have lengths of n_0 and n_L and represent the input and the output vector of the network, respectively.

CSO algorithm is a swarm technique which simulates the behaviour of cats. Thisbehaviour can be categorized into two modes, the seeking and tracing. The CSO algorithm starts by defining its parameters then generating a set of N solutions with dimensionD.The next step is to compute the objective function for each cat and assign the mode for each cat (seeking, and tracing). The cats update their positions based on the mode of each cat. Then the best cat is determined based on the best fitness value. The previous steps are performed until the stop conditions met. The cats in the seeking mode, construct Seeking Memory Pool (SMP) copies (which called candidates) of the current cat. Then, the value of Self Position Consideration (SPC) parameter is used to determine either all the candidate solutions are updated randomly when SPC is false, or one of the candidate solutions doesn't updated and the rest are updated when the SPC is true. The next step in the seeking mode is to update the selected dimension based on the value of the Seeking Range of the selected Dimension (SRD). Therefore, the fitness function of each cat is computed then candidate is selected to exchange the original structure according to the probability Pro_i that defined as the following equation:

$$Pro_i = \frac{|F_i - F_{max}|}{F_{max} - F_{min}} \qquad (2)$$

Where F_{max}, F_{min} are the maximum and minimum value of the fitness for all candidates, respectively. The F_i is the fitness of i-th candidate. In the tracing mode, the CSO begins by computing the fitness function for each cat then determine the best solution. Then the velocity of each cat is updated using the following equation:

$$v_i = v_{i,d}^t + R \times C_1 \times (x_{b,d}^t - x_{i,d}^t), d = 1,2, \ldots, D \qquad (3)$$

Thereafter, the position of i-th cat is changed using the following equation

$$x_i = x_i + v_i \qquad (4)$$

where the $v_{i,d}^t$ represents the velocity of the i-th cat at the dimension j during the iteration t. $x_{b,d}^t$is the best position at the t-th iteration, while C and R are constant and random number respectively. The final steps of the CSO are given in Algorithm 1.

Algorithm 1: Cat Swarm optimization

1) Generate a set of N cats and determine the initialize position and velocity for each of them
2) Randomly select $N \times MR$ cats and put them into tracking mode, and the rest in seeking mode.
3) For each cat update its position based on its mode (seeking or tracking).

4) Assess the cats based on fitness value and determine the best cat which has the best value of fitness.
5) Check if the terminal conditions are met or not Return the best cat.

In general, the proposed hybrid ANN-CSO approach starts by creating the structure of the ANN and creating a random population which contains N solutions. Each generated solution defines the neurons weights of the ANN. To assess the accuracy of the obtained solutions, the RMSE is computed according to the following formula:

$$\mathrm{Fit}_i = \text{Root mean square error (RMSE)} = \sqrt{\frac{\sum_{j=1}^{N_s}(y_j - \hat{y}_j)}{N}} \quad (5)$$

Where $\hat{\mathbf{y}}_i$ and $\mathbf{y}_i$ are the predicted value and its corresponding target value, respectively. The following stage is to obtain the best solution with the minimum RMSE.

5. Results and discussions

Based on the conducted experiments, APP with valve seats made of PEEK has a lower noise level compared with that with 316L valve seats as shown in figure 2 (b). Moreover, the generated noise is increased by increasing the speed and/or the pressure. Experimental data is used to train the ANN models; in which 70% of the data are used to train the models and 30% are used to test the model validity. The results obtained by CSO-ANN showed a reasonable agreement with the experimental data compared with that obtained by ANN model. Different statistical criteria are used to assess the performance of the proposed models such as coefficient of determination R^2, mean absolute error (MAE), mean relative error (MRE), Root mean square error (RMSE), overall index (OI), and coefficient of residual mass (CRM). The lower values of RMSE, CRM, MRE, and MAE and approaching of R2, OI values to 1 indicate a better performance of the model. To assure the stability of the proposed approach, it was executed 20 times and the average value was used as the final output. The present results revealed the outperform of the ANN-CSO model over ANN model to predict the emitted noise of APP with different valve seat materials for both of training and test processes as tabulated in table 1. Figure 2 (c,d) shows the QQ-plot of the predicted and the target values of the emitted noise. A reasonable agreement between the predicted and experimental datasets could be realized for all the examined materials of the valve seat and for all investigated models. However, ANN-CSO has a better agreement compared with ANN, which indicates the ability of CSO to enhance the accuracy of ANN via determining the optimal ANN structure.

Table 1. Statistical coefficient for performance estimating ANN and ANN-CSO.

	PEEK				316L			
	Training		Test		Training		Test	
	ANN	ANN-CSO	ANN	ANN-CSO	ANN	ANN-CSO	ANN	ANN-CSO
R^2	0.999549	0.999978	0.998534	0.999955	0.999544	0.999993712	0.998745	0.999726
RMSE	1.48957	0.123499	2.827062	0.49768	1.558615	0.18296066	2.708818	1.266416
MRE	4.13E-05	2.89E-06	0.01156	-0.00024	0.000108	6.64012E-06	0.011173	0.006027
MAE	1.223408	0.050684	2.603294	0.374134	1.317489	0.101587302	2.262728	0.991515
OI	0.935835	0.996733	0.831525	0.98421	0.934328	0.995220697	0.848476	0.950006
CRM	8.94E-12	2.22E-14	0.011807	0.00043	1.09E-11	1.48997E-16	0.010489	0.004939

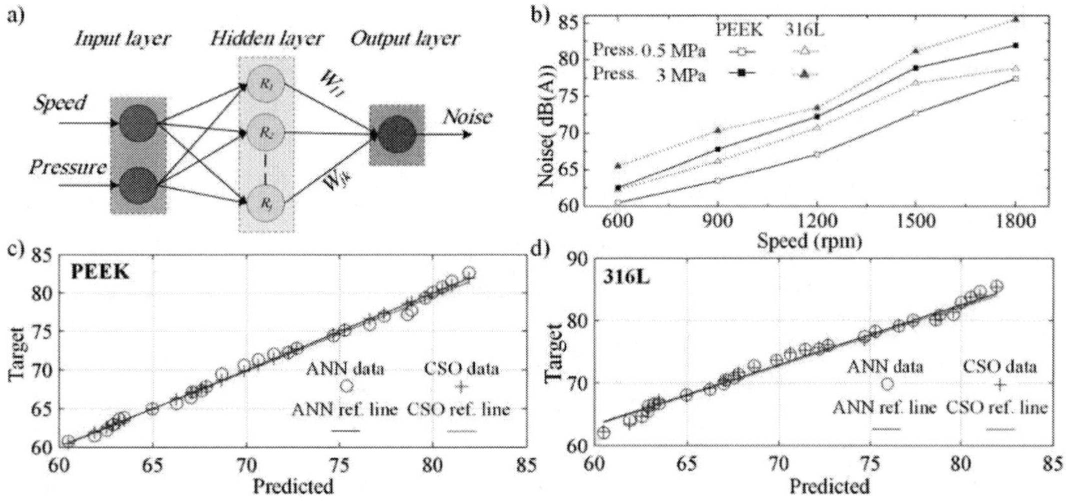

Figure 2. a) MLP ANN; b) Measured noise under the investigated operating conditions for PEEK and 316L as valve seat materials; c) QQ plot for PEEK; d) QQ plot for 316L.

6. Conclusion

A hybrid ANN-CSO model was proposed to predict the emitted noise of an APP for two different valve seat materials (PEEK and 316L) under different operating conditions. The CSO was utilized to determine the ideal ANN structure. ANN-CSO results were compared with conventional ANN model. Both models were trained using experimental data measured during experiments carried out on an APP experimental setup and then tested using another experimental data that have not been used in the training stage. The inputs of the models were speed (600-1800 rpm), and operating pressure (0.5- 3.0 MPa); while the output of the models was the emitted noise. ANN-CSO showed a better performance compared with conventional ANN. Moreover, to reduce the emitted noise of APP it is suggested to use PEEK as a valve seat material and decrease operating speed and pressure.

Reference

[1] Schuhler G, Jourani A, Bouvier S and Perrochat J M 2018 Efficacy of coatings and thermochemical treatments to improve wear resistance of axial piston pumps *Tribology International* **126** pp 376-85

[2] Xu B, Ye S and Zhang J 2016 Numerical and experimental studies on housing optimization for noise reduction of an axial piston pump *Applied Acoustics* **110** pp 43-52

[3] Pan Y, Li Y, Huang M, Liao Y and Liang D 2018 Noise source identification and transmission path optimisation for noise reduction of an axial piston pump *Applied Acoustics* **130** pp 283-92

[4] Zhang J, Xia S, Ye S, Xu B, Zhu S, Xiang J and Tang H 2018 Experimental investigation on the sharpness reduction of an axial piston pump with reinforced shell *Applied Acoustics* **142** pp 36-43

[5] Xu B, Sun Y-h, Zhang J-h, Sun T and Mao Z-b 2015 A new design method for the transition region of the valve plate for an axial piston pump *Journal of Zhejiang University-SCIENCE A* **16** pp 229-40

[6] Zhang J, Xia S, Ye S, Xu B, Song W, Zhu S, Tang H and Xiang J 2018 Experimental investigation on the noise reduction of an axial piston pump using free-layer damping material treatment *Applied Acoustics* **139** pp 1-7

[7] Yin F, Nie S, Ji H and Huang Y 2018 Non-probabilistic reliability analysis and design optimization for valve-port plate pair of seawater hydraulic pump for underwater apparatus *Ocean Engineering* **163** pp 337-47

[8] Oliva D, Elaziz M A, Elsheikh A H and Ewees A A 2019 A review on meta-heuristics methods for estimating parameters of solar cells *Journal of Power Sources* **435** 126683

[9] Elsheikh A H and Abd Elaziz M 2019 Review on applications of particle swarm optimization in solar energy systems *International Journal of Environmental Science and Technology* **16** pp 1159-70

[10] Babikir H A, Elaziz M A, Elsheikh A H, Showaib E A, Elhadary M, Wu D and Liu Y 2019 Noise prediction of axial piston pump based on different valve materials using a modified artificial neural network model *Alexandria Engineering Journal*

[11] Elaziz M A, Elsheikh A H and Sharshir S W 2019 Improved prediction of oscillatory heat transfer coefficient for a thermoacoustic heat exchanger using modified adaptive neuro-fuzzy inference system *International Journal of Refrigeration* **102** pp47-54

[12] Shehabeldeen T A, Elaziz M A, Elsheikh A H and Zhou J 2019 Modeling of friction stir welding process using adaptive neuro-fuzzy inference system integrated with harris hawks optimizer *Journal of Materials Research and Technology*

[13] Elsheikh A H, Deng W and Showaib E A 2019 Improving laser cutting quality of polymethylmethacrylate sheet: experimental investigation and optimization *Journal of Materials Research and Technology*

[14] Essa F A, Abd Elaziz M and Elsheikh A H 2020 An enhanced productivity prediction model of active solar still using artificial neural network and Harris Hawks optimizer *Applied Thermal Engineering* **170** 115020

[15] Elsheikh A H, Sharshir S W, Ismail A S, Sathyamurthy R, Abdelhamid T, Edreis E M A, Kabeel A E and Haiou Z 2020 An artificial neural network based approach for prediction the thermal conductivity of nanofluids *SN Applied Sciences* **2** 235

[16] Ye S-G, Zhang J-H and Xu B 2018 Noise Reduction of an Axial Piston Pump by Valve Plate Optimization *Chinese Journal of Mechanical Engineering* **31** 57

[17] Karkoub M A, Gad O E and Rabie M G 1999 Predicting axial piston pump performance using neural networks *Mechanism and Machine Theory* **34** pp1211-26

[18] Yang J, Xu B and Yang H 2006 Noise identification for hydraulic axial piston pumpbased on artificial neural networks *Chinese Journal of Mechanical Engineering(English Edition)* **19** pp 120-3

[19] Wu D, Liu Y, Li D, Zhao X and Li C 2017 Effect of materials on the noise of a water hydraulic pump used in submersible *Ocean Engineering* **131** pp 107-13

IOP Conf. Series: Materials Science and Engineering **973** (2020) 012036 doi:10.1088/1757-899X/973/1/012036

Passive vibration damping of acantilever plate

A Mohamed [1], A Hassan[1], A A Omer[1]
[1] Mechanical engineering department, MTC, Cairo, Egypt
anwer_mohamed@mtc.edu.eg

Abstract. Vibration control of mechanical systems divided into two types active and passive, this case study illustrates other approach of passive control of the cantilever plate by using double layers of the viscoelastic material (Dyad 606) one in the upper face of the Al substrate (5051) and the second one in the lower face. This layers are affected by axial uniform distributed force by applying tension load, the damping level depend on the value of this axial force where the vertical component of this force consistently the other way of the vibrating element that reduce vibration amplitude in this direction by about (95%) of the original amplitude without damping layers. Computational analysis using ANSYS program are used firstly to verify that the results from the software are compatible with that obtained from analytical and experimental evaluation by comparing the values of natural frequencies. Then using ANSYS model to calculate the reduction occurs in the lateral vibration with changing of the axial loads on the viscoelastic plates. The benefits of this model are enhancing the reliability and reducing the cost.

1. Introduction

Various techniques are used in damping process of the undesirable vibrations that may cause damaging of the structures and machines. Damping systems are divided into two main types active and passive [1], at the first, active vibration control uses many kinds of sensors and actuators to generate forces in the right direction and time to sustain the unwanted frequencies that cause vibration[2], this type of vibration control have higher cost and low reliability[3]. On the other hand, passive vibration control doesn't use any sensor or actuator that require external power units and depend only on the damping properties of the used material[4].
There are two main types of passive damping, the first one is unconstrained layer damping (UCLD)[5] by adding a completely bonded damping layer on the surface of the vibrating structure,as shown in figure 1 but this method cannot be used in many mechanical structures due to lower damping ratios[6].The second type In 1959, Kerwin, et al [7] first did a fundamental work in what is called passive constrained layer damping (PCLD), where the damping layer putted in between of the base material layer and constrained

layer that may be made of the same material of the base layer or different material as shown in figure2This

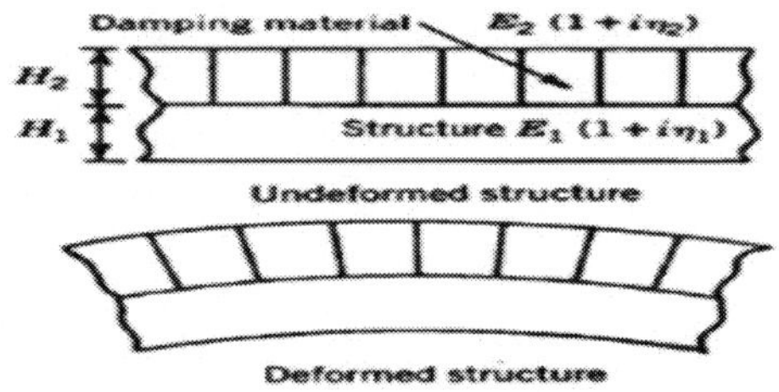

Figure .1 Un constrained layer damping.

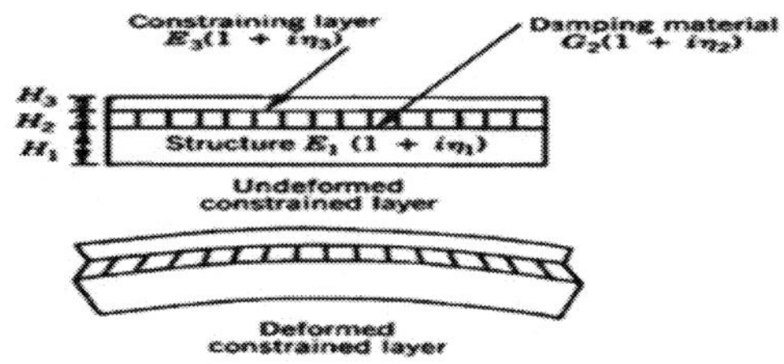

Figure .2 Constrained layer damping.

type of damping reduces the efficiency of the system and increasing the net cost[8].
In our work we study other approach of damping system that used two layers of damping material one of them on the upper surface of the base vibrating element and the second one on the lower side to achieve higher damping characteristic, low cost, simplicity and enhanced reliability[9]
Damping material having different types according to the level of vibration energy dissipation or loss factor (η) [10that classified into:

- Conventional materials (that having low loss factor).
- Composite materials, which exhibited higher viscoelastic properties than conventional materials and having superior specific stiffness and strength[11].
- Viscoelastic materials, for which the correlation between stress and strain depend on time so that the behavior of this type of material depend on the loading history and that have highest loss factor comparing with the composite materials[12].

In our damping system we use viscoelastic material type (Dayd 606) in both damping layers

2. Calculating of natural frequency

The natural frequency is the frequency at which a system tends to oscillate in the absence of any driving or damping force. when an object vibrates at frequency equivalents s to natural frequency, its vibration amplitude increases significantly which could lead to irreparable damage. Determination of natural frequency could be estimated with several methods.

2.1.Numerical invistagation

The process of developing the finite element model starts in the "Engineering Data section" in ANSYS workbench package version R15.0 [14], by assigning the mechanical properties of all materials used in building the FEM, The materials used are discussed in details in the next section.

2.1.1.Mechanical properties of FEM materials. An aluminum (AL 5051) rectangular plate and polymer (dyad 606)
Table 1 shows the mechanical and geometrical properties of the plain plate andthe damping material used in the (PTLD) treatment

Table 1. Mechanical and geometrical properties of the plain plate.

Layer	Thickness (mm)	Length (mm)	Width (mm)	Density (k_g/m^3)	Posion Ratio	Young's Modulus (Pa)
Aluminum alloy	1.2	300	50	2700	0.33	70×10^9
Damping material	0.5	300	60	1120	0.49	20×10^6

2.1.2. Meshing of the plate FEM .Meshing process is started by applying mesh tool to the developed 3-D model using fine mesh. The number of elements is decided after a convergence study made for the ANSYS model using mesh sensitivity analysis. The convergence isobtained at 5 mm element size.
The element type is "solid 186" – 20 node homogenous structural solid, figure 3 shows the solid 186 element configuration. It is a higher order 3-D 20-node solid element that exhibits quadratic displacement behavior. The element is defined by 20 nodes having three degrees of freedom per node: translations in the nodal x, y, and z directions. The element supports plasticity, hyper elasticity, creep, stress stiffening, large deflection, and large strain capabilities[15]. The geometry, node locations, and the element coordinate system for this element are shown in figure 4

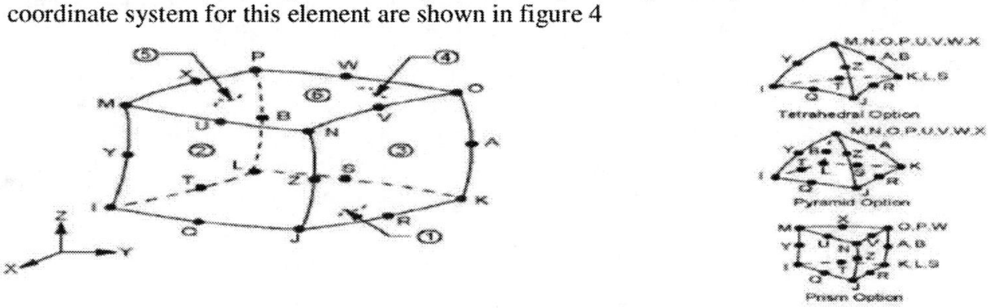

Figure. 3 Solid 186 element configurations.

Table 2. Number of elements and nodes used for the plate mesh.

MODEL TYPE	NUMBER OF ELEMENT	NUMBER OF NODES
The plate	720	5403
The plate with polymer	2160	16209

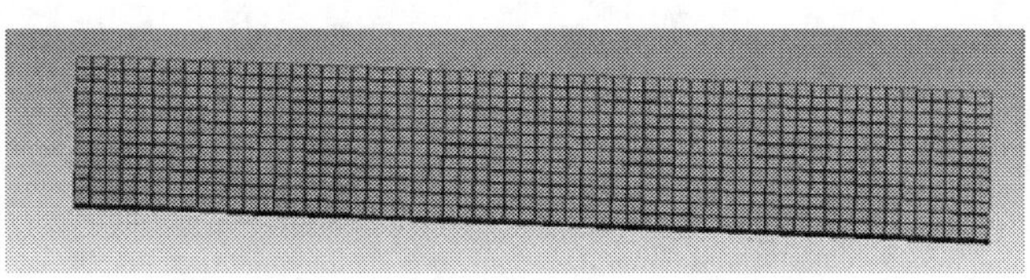

Figure. 4 The plate FEM with mesh.

2.1.3.Static analysis. Static structural analysis tool is performed to apply the steady inertia loads such as plate gravity, and used to determine the plate displacements, stresses, strains, etc… a cantilever boundary conditions are applied to the FE model. Figure5Shows the plate tip displacement due to gravitational load

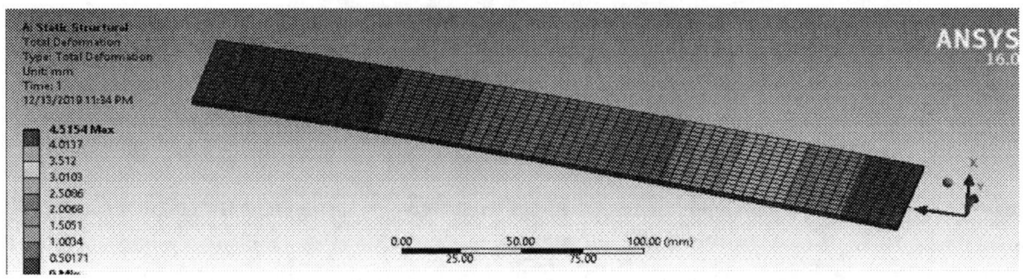

Figure. 5 The plate tip displacement due to gravitational load.

2.1.4.Modal analysis

Modal analysis is used to determine the vibration characteristics (natural frequencies and mode shapes) of the cantilever plate and the plate with polymer. It considered as a starting point for more detailed dynamic analysis, such as transient dynamic analysis, harmonic response analysis, or spectral analysis.

When frequencies of dynamic load match one of natural frequencies, then resonance takes place, which is considered from the main critical parameter of dynamic structure design. Mode shapes hereinafter called modes, and natural frequencies (NF) of the plate and the plate with polymer are obtained by modal analysis tool. The plate and the plate with polymer mode shape are shown in figures 6, 7, 8, 9. The values of natural frequencies of the plain plate and the plate with polymer layers are tabulated in Table.3

Table 3. The plate and the plate with polymer mode shape.

Mode Shape	Mode Shape Type	Plain Plate	Plate with Polymer
1	Bending	9.7975	8.7
2	Bending	70.267	60.3

IOP Conf. Series: Materials Science and Engineering **973** (2020) 012036 doi:10.1088/1757-899X/973/1/012036

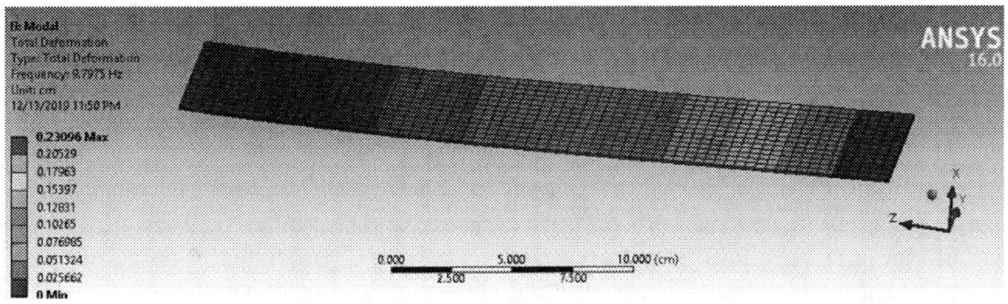

Figure. 6 First mode shape of plain plate.

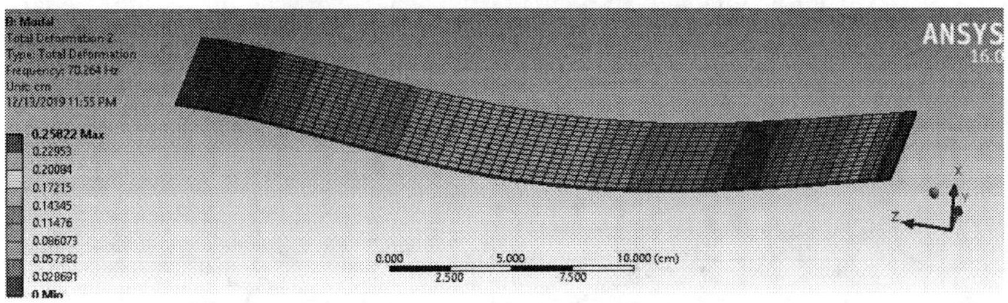

Figure. 7 Second mode shape of plain plate.

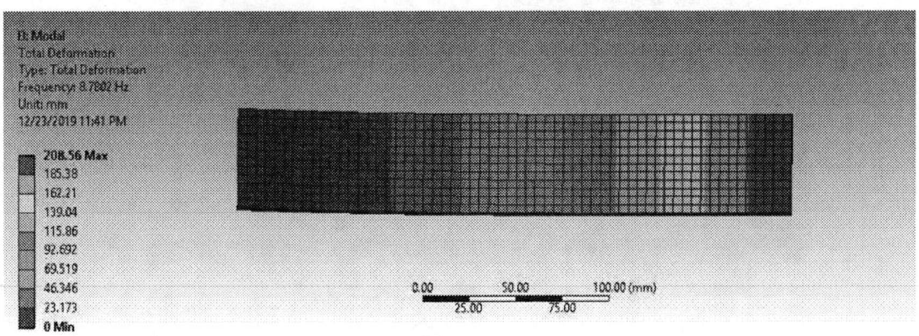

Figure. 8 First mode shape of plate with polymer.

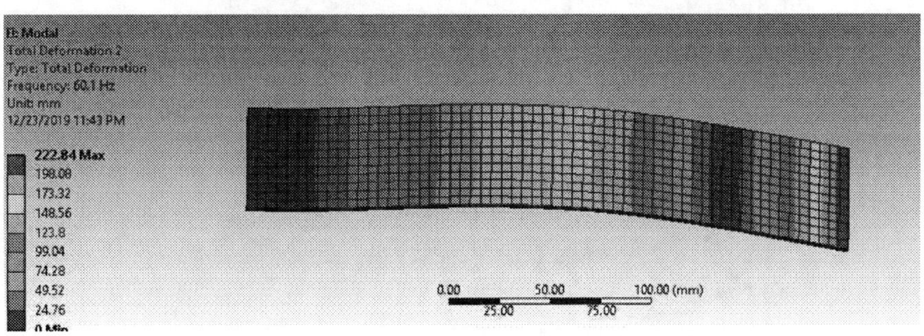

Figure.9 Second mode shape of plate with polymer.

2.2. Experimentalinvestigations

The experimental verification of finite element analysis is conducted. The experimental system layout is shown in figure 10. In this experiment, the plate is clamped from one end onto a shaker 4808 with fixation support, and the second end is free.

Two accelerometers type PCB PIEZOTRONICS are attached to the plate by a suitable adhesive, one is fixed on the plate free end to measure the output response, and the other is fixed on the shaker fixation support to measure the input excitation. The shaker B&k type 4808is driven by power amplifier B&k type 2712, which amplifies the sine sweep, signal (0-200) Hz coming from waveform generator. Chasses of NI compact DAQ-9188 is used for data acquisition system and PC with Lab Viewprogram is used to acquire and analyze the data. The most suitable technique used for characterizing the plate material properties in the frequency range (from 1 up to 1000 Hz) is the frequency response technique[13]. FRF results are processed by comparing the measured acceleration values taken by the accelerometer mounted on the plate free end (output) with the measured acceleration results taken from the accelerometer mounted on the shaker support (input). To obtain exact experimental results, ten repeated measurements are recorded for each experimental set up and the average values are considered as shown in figure 11.

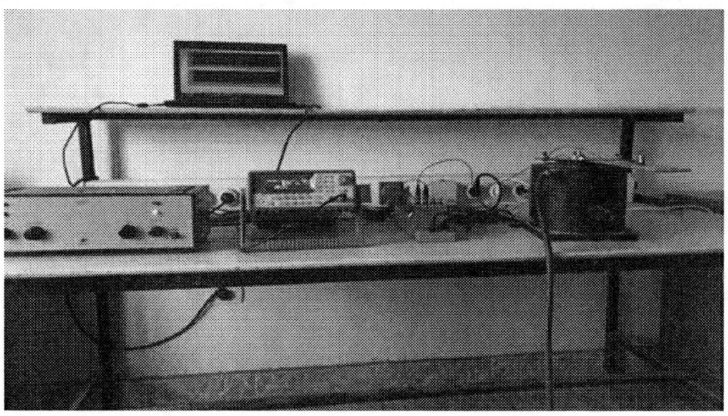

Figure.10 The experimental setup

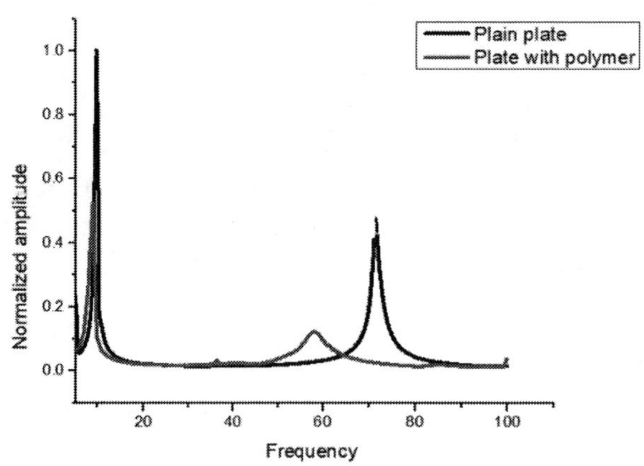

Figure.11 First and second natural frequency with and without polymer

IOP Conf. Series: Materials Science and Engineering **973** (2020) 012036 doi:10.1088/1757-899X/973/1/012036

3. Finite element modeling verification

The theoretical results are used to verify the FE model by conducting a comparison between the plainplate and the plate with polymer numericallyand experimentally. The comparison between the resonant frequencies of the numerical simulation and the experimental result are shown in table 4.

The results are in good agreement with an 4 % maximum error. The contrast among FE and exploratory outcomes can be credited to the accompanying:

- Distinction in material properties among FE and actual plate.
- Difference in boundary conditions; in FE model, the end is fixed, where in testing it is clamped and there may be still some relative freedom.
- -Difference in the portrayal of the accelerometer; in FE model, it is a concentrated mass, where in testing it is stuck to the plate.
- -Effect of environmental conditions during tests (temperature...etc.), which are not accounted for in simulation.
- In the FE model, just the shaker impact is signified (the environmental vibrations are not represented).

Table 4. The plate resonant frequencies(Hz)(numerically/experimentally)

MODE SHAPE	THE PLAIN PLATE		THE PLATE WITH POLYMER	
	Numerical	Experiment	Numerical	Experiment
1	9.7975	9.89	8.7	8.72
2	70.264	71.1	60.1	59.8

4.Casestudy

The main purpose of this technique is to illustrate numerically the fundamentals associated with damping the vibration of plate using a new method of passive control layer damping treatment. The developed new model is based on the idea of generating uniformly distributed forces on two layers of the damping materialexperiencing cyclic motionis shown in figure 12. The lateral components of the exerting forces always affecting in direction opposing the angle of rotation of plate. These lateral components have a complex form due to the inherent damping of the elastic layer bonded to the structure fibers. It will be demonstratednumerically that a more than 95% reduction in the peak of the fundamental bending mode of a cantilevered plate can be obtained by virtue of this new technique.

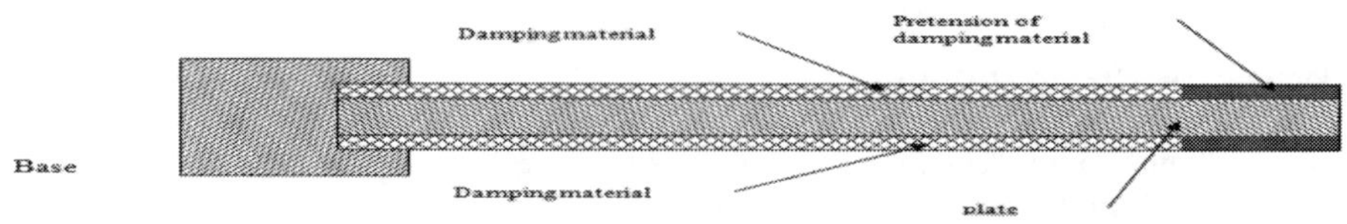

Figure. 12 Plain plate with Pretension Layer assembly.

4.1 Harmonic response analysis

A harmonic response tool is applied to study the plate FRF for the first two bending modes. It is carried out for the plate with /without the damping material. A transverse excitation force of 1 N with sine sweep signal varies from 0 to 200 Hz is applied at the plate structure. FRF of a structure otherwise called a transfer function is a ratio of the output signal to the input signal in frequency domain. Figure13 shows the amplitude of the plain plate and plate with polymer at different resonant frequencies or called amplitude FRF of the plate and the plate with polymer, hereinafter called the plate and the plate with polymer FRF.

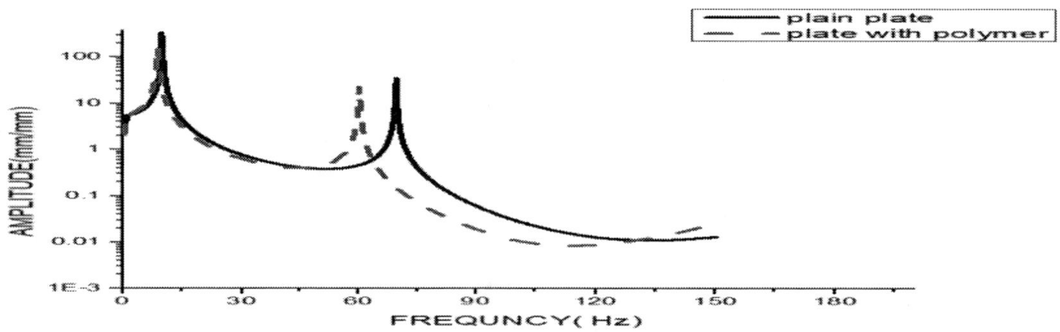

Figure. 13 The plain plate and the plate with polymer (FRF).

5. Results and discussion

The response of the generated exerting force on the reaction of the Plate/PTLD system is assessed. The attenuation percentage is calculated for different3Pretension3forces values, where, the comparison is made with respect to the response of the plain plate as a fixed reference.The attenuation percentageis determined from the accompanying equation.

$$\Delta_n^t(\omega_n, L) = \frac{|\alpha'(\omega_n, L)| - |\alpha(\omega_n, L)|}{|\alpha(\omega_n, L)|} 100\% \text{ [16]}$$

The performance of the system without the effect of the forces should be study, at this moment, the system response is that response of the Plate/UCLD system, and the damping produced is due to the direct shear strains in the damping layers, in addition to the damping of the base plate. Figure 13 gives the comparison between the Plate with polymer and the corresponding plain Plate.

Figure14 gives the comparison between the magnitudes of the frequency response functions of the system at different values of the Pretensionforce within the elastic range of the damping material. These figures indicate the high damping efficiency of the new approach. Table 5.gives the relation between the value of the Pretensionforce and the attenuation percentage of the new model at 1st resonance.

Table 5. The attenuation percentage at different Pretension force.

Pretension force [N]	Attenuation ratio[%]
0	52.3%
20	61.9%
30	69.3%
40	74.3%
50	86.8%
70	91.2%
100	95.2%

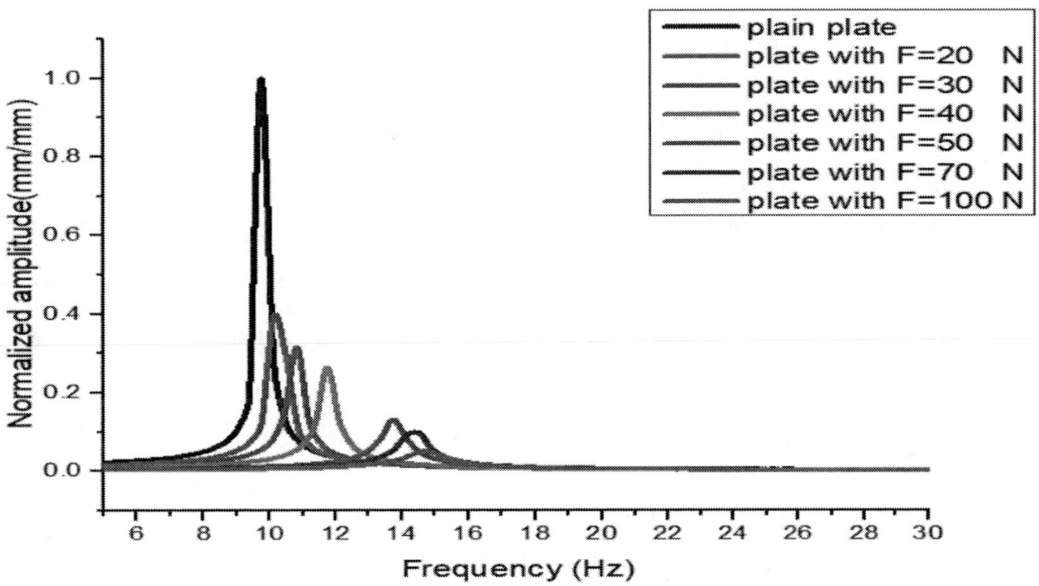

Figure. 14 Plain plate compare to plate with polymer and pretension polymer.

6. Conclusion

A new passive control model used for suppression of the lateral vibrations of a flexible base excited cantilevered Plate is illustrated in the present paper, This new approach, which is called pre-tensioned layer damping (PTLD), is a sort of artificial damping techniques, which is based on the theory of energy dissipation from vibrating systems. The devolved model has a simple, reliable and inexpensive solution

IOP Conf. Series: Materials Science and Engineering **973** (2020) 012036 doi:10.1088/1757-899X/973/1/012036

for the Plate /PTLD system. At which the major advance in the attenuation of the vibration amplitude has been discussed. the damping ratio is enhanced without the usage of complications control devices that used in case of Active Constrain Layer Damping (ACLD) It is noticed that the higher the elastic Pretensionforces in the pre-tensioned layer damping material that increase the attenuation percentage and the damping efficiency.

References

[1] Liang, C-C, et al.2001*The free vibration analysis of submerged cantilever plates.* Ocean Engineering,. **28**(9) pp 1225-1245.
[2] Baz A and J Ro 1993 *Finite element modeling and performance of active constrained layer damping.* in *Ninth VPI & SU Conference on Dynamics & Control of Large Structures.*
[3] Baz Aand J Ro1995 *Performance characteristics of active constrained layer damping.* Shock and Vibration, **2**(1) p. 33-42.
[4] Tong X and Zhao X 2018 *Passive vibration control of the SCOLE beam system.* Structural Control and Health Monitoring, **25**(8) pp e2204.
[5] Johnson CD1995 *Design of passive damping systems.*
[6] RoyPand Ganesan N1996*Dynamic studies on beams with unconstrained layer damping treatment.* Journal of Sound and Vibration, **195**(3) pp 417-427.
[7] Ross D, Kerwin E and Ungar E1960 *Damping of flexural vibrations by means of viscoelastic laminae, Structural Damping, Ch 3, J Ruzicka ed,* Pergamon Press, New York.
[8] ArisakaT, Saegusa T and KajiwaraI 2018 *Active vibration control device and design method therefor,* Google Patents.
[9] SherifH, Abd-Elwahab M and Omer A2003 *New Approach for Vibration Control of Cantilevered Structures.* in *International Conference on Aerospace Sciences and Aviation Technology.* The Military Technical College.
[10] Cai C et al2002 *Modeling of material damping properties in ANSYS.* in *CADFEM Users' Meeting & ANSYS Conference.*
[11] Treviso A, et al2015 *Damping in composite materials: Properties and models.* Composites Part B: Engineering, **78** pp 144-152.
[12] Nakra B1998 *Vibration control in machines and structures using viscoelastic damping.* Journal of Sound and Vibration, **211**(3) pp 449-466.
[13] HujarePP and Sahasrabudhe A D 2014*Experimental investigation of damping performance of viscoelastic material using constrained layer damping treatment.* Procedia Materials Science, **5**pp 726-733.
[14] Inc., A., *ANSYS Workbench User Guide*2012.
[15] Yadav D, Sharma A, and Shivhare V2015 *Free vibration analysis of isotropic plate with stiffeners using finite element method.* Engineering Solid Mechanics, **3**(3) pp 167-176.
[16] Abd-Elwahab M and Sherif H A2006 *Pre-tensioned layer damping as a new approach for vibration control of elastic beams.* Journal of vibration and acoustics. **128**(3) pp 338-346.

Modelling and experimental investigation of high – pressure common rail diesel injection system

W Niklawy[1], M Shahin[2], MI Amin[3] and AElmaihy[3]
[1] PhD candidate, Mechanical Power Engineering Department, Military Technical College, Cairo, Egypt.
[2] Mechanical Engineering Department, School of Engineering, Badr University, Badr, Cairo, Egypt.
[3] Mechanical Power Engineering Department, Military Technical College, Cairo, Egypt
niklawywaleed@gmail.com

Abstract. The high-pressure Common Rail (HPCR) injection system was originally introduced for diesel engines to both reduce pollutant emissions and enhancement of performance. HPCR separates fuel pressurization and injection processes from each other. The high injection pressure generated by the common rail system provides better atomisation and evaporation of fuel spray, resulting in improved air inlet and fuel jet mixing, which is advantageous for lowering soot emission. In this paper, a mathematical common rail injection system model has been presented. A Simulink/Matlab code was developed to execute this simulation. This work does not only seek to validate the presented numerical model but to have more insight for understanding the overall common rail injection system diesel engine performance under different operating conditions. Some simulation results are illustrated to highlight modelling capability.The engine used is an HCCI turbocharged diesel engine, 2776 cc, 4-stroke, and 4-cylinder, water-cooled with overhead valve mechanism. The common rail pressure, fuel consumption, start and duration of each injection through one engine cycle are measured at various engine speed and loads. The measured common rail fuel pressure and consumption are used to validate the simulation results. The findings of the simulation show good consistency with the experimental results. At last, some simulation results, which highlight the modelling capability, are illustrated at certain values of engine speed and load.

1. Introduction

Nowadays the HPCRS is the state-of-the-art technology used in most of the recently manufactured diesel engines. It came out as a way out for simultaneous improving diesel engine emissions with which the manufactures could reach the emission standards and allowing noise and performance to be improved. It provides more flexibility than any other used injection system [1-3]. In HPRCS, The pressure inside the common-rail is considered as the main fuel-injection measurement parameter. It also determines the injection -fuel pressure. The stabilization and excessive reaction of the common-rail pressure would affect all the engine regimes such as starting. Idling and acceleration[4]. Considering the preceding circumstances, the performance prediction of common rail equipment is of real significance in the design of the automotive engine to better understand the performance of diesel engine equipped with common rail, and the mechanisms that potentially can contribute to improved fuel consumption and reduced emissions through numerical simulations and experimental work. Numerical simulations of the fuel injection process become more and more essential as computational power especially with the enormous development of computation recourses contrary to prohibitively expensive and time-consuming experiments. The numerical simulation results after validation can be extended and used alone to predict some of the engine performance parameters.

Currently there are many computational and numerical methods for modeling and simulating the injectionsystem[5]. Different models can be developed by identification based on experimental data, whether linear or nonlinear. Automotive applications of such models do not consider modeling the injection system alone but mainly the whole engine modeling. Nicolao et al [6] designed a nonlinear idle speed regulator by using an integrated approach to identify a NARX (Nonlinear Auto Regressive Exogenous) model of an internal combustion engine. Maione et al [7], concluded that the nonlinear identification approach is a promising modeling technique for the injection system too.

Computational Fluid Dynamics (CFD) has evolved as a consistent and cost-effective tool that is now regularly utilized in the perfection of new or enhanced engine designs. While CFD methods are a practical way to explore the performance of an engine at a particular operating condition, they may need high computational recourses when used as an engine optimization tool, where numerous operating points may have to be evaluated. Therefore, engine optimization demands a special computationally efficient algorithm [8]. To investigate the effect of the injection system on the combustion process, it is frequent to combine the CFD analyses with the experimental study [9]. This approach can get validated results of different phenomena within the cylinder such as temperature distribution, fuel evaporation process, flame zones, and spray cloud profile and emission concentrations [10] .With such investigation of combustion and spray processes, it could be promising to achieve the desired emission formation reduction [11] and to have more insight on the effect different fuel injection strategies on the combustion process [12]

Chiavola and Giulianelli [13] , Mulemane et al [14] and Lino et al [15] used the Advanced Modeling Environment for Simulation environment (AMESIM) [16] to simulate the common rail injection system, which is a design package that permits the exploration of systems by using different models of distinct elements ready-made or self-made component libraries. Each injection system component has a mathematical model, including different effects of frictions and leakages. The aim of these models is to investigate system dynamics, to conduct parametric analysis and system components geometric optimization, to predict and confirm the impact of operational circumstances on the injection process. Even so, even with good evaluation capabilities, they could not be considered as proper approaches for control as they do not provide any mathematical representation of the process dynamics.

Phenomenological models, such as lumped parameters or one-dimensional model [4,17-19] are based on simple schemes. They take the full benefit of the experimental measurements that, usually, are represented by comprehensive quantities and are capable to catch the fundamental aspects of the phenomenology [9]. So, they may be considered as the optimum models which compromise between the validity of model results and the required computational recourses since.

Lino et al [20] proposed a model for diesel engine Common Rail injection system, appropriate for control purposes. It is based on a lumped parameter representation. It has the capability to be modified to other injection systems with the alike structural design, only by identification of satisfactory geometrical parameters. They concluded that their model accuracy is adequate to estimate the system performance to check the controller performances. The obtained results were validated with those obtained through the AMESIM.Due to the lack of experimental data, they considered that the AMESIM model presented in[21] is a reliable demonstration of the real injection system as it accurately predicts the system dynamics.

Wang et al [4] developed a mathematical modeling approach and common-rail regulation of common-rail HPCRIS. The proposed system that comprises three components of the Sub-system high pressure pump, common rail subsystem, and injector subsystem was done in Matlab/Simu- link. The model validation was conducted by comparing the model results with those obtained by

IOP Conf. Series: Materials Science and Engineering **973** (2020) 012037 doi:10.1088/1757-899X/973/1/012037

AMESim.As noted from literature most of the models presented were validated either with AMESim [4,13,15,20] or with injection rate test bench [14,18,19].

In this paper a mathematical lumped parameter model is built for a common rail injection system. The proposed HPCRS includes four sub-systems namely; high pressure pump, common rail, injector, and injections timing. To The model is implemented in Matlab/Simulink.

The experimental work does not only used for validating the presented numerical model but also to have more insight for understanding the overall engine performance using a common rail injection system under different loads and speeds. The experimental work was conducted on a complete rig for testing common rail turbocharged HCCI diesel engines. The test rig includes the engine and all the instrumentation necessary for measuring and recording a macro as well as micro-operating parameters.

The novelty of this work is the use of measured in-cylinder pressure obtained from the test rig instead of using constant values as in [4,20]. The multiple injection strategy are employed in this work .The injections start times and durations are expressed in the model as a functions of load and engine speed by data regression obtained from experimental measurements. Finally, the numerical results are validated with experimental results obtained from a complete diesel engine test rig instead of using AMESim injection rate test bench results.

The paper organization as follows. A brief description of the test rig experimental set up in the next section. a detailed description of the model is illustrated in section 3. Section 4 sets outthe validation of the model and to prove the performance of the model. The capability of the model is introduced in section 4. In section 5, the conclusion and future work is outlined

2. The Test Facility (Experimental Setup)

Experimental investigations were conducted on a transport diesel engine of type VM Motri with an open chamber. Four in-line cylinders water-cooled direct injection diesel engine with over-head valve gear and "common rail" fuel injection system. (Available in the laboratory of mechanical power and energy at the Military technical college) The main engine design parameters are listed in Table 1. The test rig includes the engine and all the instrumentation necessary for measuring and recording the operating parameters. An on-line data acquisition system is furnished to improve the speed and accuracy of data collection and recording. is used. Table 1displays engine specifications.

Table 1.Engine specifications.

Description	Specification
Engine	R2816K5A
Bore	94 mm
Stroke	100 mm
weight	270 kg
Compression Ratio	17.5
Number of Cylinders	4
Displacement	2776 cc
Max power	90 kW (120H.P.) @ 3800 RPM
Peak Torque	400 N·m @ 1800 RPM
Fuel primary pump	Electric pump
Fuel System	Direct Fuel Injection Common Rail System
High-Pressure Fuel Pump Injection	Radial Piston Pump (3 pistons) CP3 2nd Generation
Timing System	Valve Belt Driven DOHC Overhead Camshafts

2.1 Instrumentations

The parameters measured are classified into three main groups. a -External parameters: Engine speed, engine load, fuel consumption, airflow, crank angle, ambient conditions and real-time. B-Internal parameters: pressure inside the cylinder, inlet and exhaust manifolds and fuel line,andturbocharger speed. C-Safety parameters: Lubricating oil pressure and temperature, and cooling water, temperature. Figure 1 gives a general scheme showing numbered locations where important pickups and transducers are positioned. A list of these locations and the corresponding measured parameter used in this study at each is given below is given in table 2.

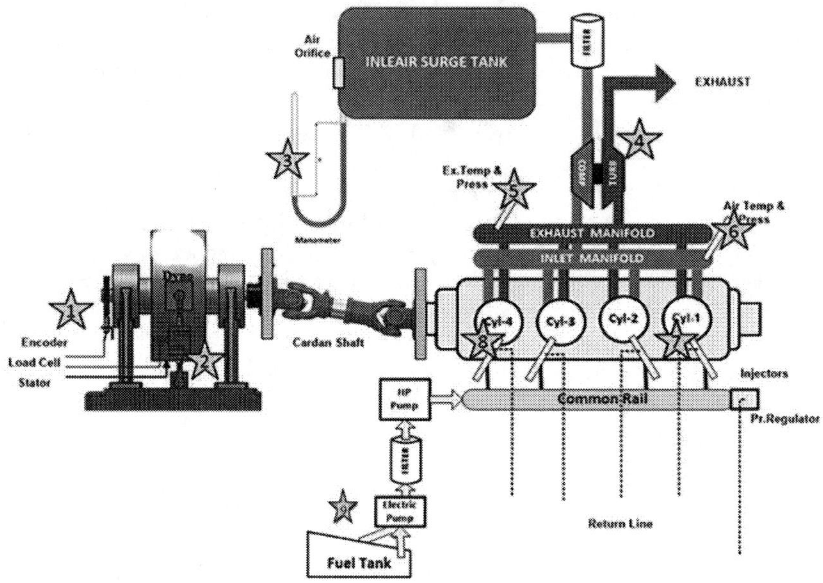

Figure 1. Test Facility Scheme

Table 2. A list of measured parameters and the corresponding locations relevant to figure 1

Location	Measured Parameters	Instrument	Accuracy
1	Engine Speed & Crank Angle	Incremental Encoder (WD GI 58B)	+ 7.5 % of the pulse width
2	External load (Torque)	S-type load cell	Combined error ≤+0.05%
7	Cylinder pressure	piezoelectric transducer type PCB (model no. 112B11)	<2.0 % FS
8	Fuel pressure sensor	piezoelectric transducer type (Kistler PN 6278)	<0.2 % FS
9	Fuel Consumption	Calibrated Bowel	NA

Engine external loading was carried out by an ELZE /Heenan hydraulic dynamometer. The fluid used was water with which the maximum braking power could reach 170 kW at 4000 rpm. The engine and dynamometer shafts were directly coupled through a cardan shaft. An S-type load cell (strain gauge type, max capacity 500kg) is inserted under the dynamometer torque reaction arm. The load cell is completed with a bridge amplifier that is excited (12VDC) from a regulated power supply. The whole setup is calibrated by applying known weights on the cell and observing the net voltage output.

A water-cooled piezoelectric transducer (type PCB model no. 112B11) is used for measuring cylinder pressure. The charge amplifier used is a PCB type, with the capability of statically holding the output charge for calibration processes.

IOP Conf. Series: Materials Science and Engineering **973** (2020) 012037 doi:10.1088/1757-899X/973/1/012037

A piezoelectric transducer type (Kistler PN 6278) and charge amplifier type (Kistler SN 284625) is used to measure fuel line pressure. This setup is capable of measuring pressure up to 3000 bars. The transducer is directly mounted on the high-pressure fuel line connecting the common rail to injector no.1. The transducer, amplifier, and cabling were calibrated together using a deadweight tester

Engine fuel consumption at steady-state operation is evaluated by recording the time of consumption of a certain volume of fuel. This old method remains particularly valuable today because of the high accuracy achieved. The measuring device consists of a glass flask of 250 cc volumes, an auxiliary fuel tank, a control cock, and a stopwatch.

An incremental digital quadrature encoder (type WDG 58B-360-ABN-G24-K3) is used for engine speed measurement. The encoder gives 2 trains of pulses (A and B), each has 360 pulses per encoder shaft revolution. The two trains are phased by ¼ pulses, and a third index train (N) with one pulse per each revolution is also produced. The Encoder is mounted on a special bracket fixed at the free end of the dynamometer shaft, figure 1.

2.2 Test procedure

Measured parameters included in this paper are engine speed, engine power (calculated from the dynamometer reading), fuel consumption, fuel line pressure, and cylinder pressure. The measurements were carried out at a fixed speed and engine load. The test procedure is required to fulfill the following main tasks in the prescribed order: first, pre-operational checks and preparations. Then, running the engine and sustaining the operating conditions as requested,(speed and load). Finally. Safety monitoring, e.g. overheating over-speed and overload.

3. Modeling

HPCR aims to deliver fuel with adequate high-pressure according to the engine operating condition. The system structure is indicated in figure 2. It consists of a high-pressure pump, a common rail, and injectors with their electronic control units. The fuel flow pressure is supplied from the tank to the pump with high pressure, where the fuel pressure goes up to the optimal high pressure. The high-pressure pump supply fuel with high-pressure to the rail, which distributes the fuel to each injector which in turn injects fuel with high-pressure to the cylinder. ECU determines the pressure of fuel based on the operational condition of the engine according to various engine sensor signals. HPCRS has a multifaceted injection process. The performance of the fuel injection system is affected by many factors, such as various components fuel leakage, fuel pipe elastic deformation due to high pressure, high temperature fuel compressibility and loss of high pressure and flow rate when the fuel flows through a variable cross-section etc.

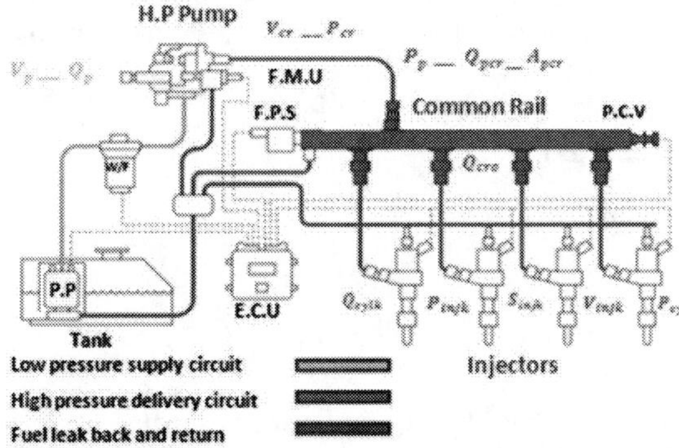

Figure 2.Common Rail Layout.

IOP Conf. Series: Materials Science and Engineering **973** (2020) 012037 doi:10.1088/1757-899X/973/1/012037

Some simplified assumptions are given base in system characteristics: the temperature variation of fuel is ignored throughout operations to ensure that system condition can be defined by its pressure; neglecting any pipe flow fluid dynamic phenomena, constant pressure of low-pressure pump is set [4, 20] but the in-cylinder pressure was taken from the experimental measurement in correspondence with engine load and speed. The HPCR model which contains four sub-models is constructed on the basis of the continuity equation, momentum equation and Newton's motion law. The compressibility of the fuel demonstrated by the elasticity bulk modulusE is

$$E = -\frac{dP}{dV/V} = \frac{dP}{d\rho/\rho} \quad (1)$$

E is calculated as s function of fuel pressure P [22,23], in this work E is taken as 32500 bars from From (1), the fuel pressure time derivative can be expressed as

$$\frac{dP}{dt} = -\frac{E}{V} \cdot \frac{dV}{dt} \quad (2)$$

Where:

VIs the compartment instantaneous volume $\frac{dV}{dt}$ express the time rate of volume variation caused by moving boundaries and the difference between inlet and the outlet flows. So equation (2) can be rewritten as

$$\frac{dP}{dt} = -\frac{E}{V}\left(\frac{dV_i}{dt} - Q_{inlet} + Q_{outlet}\right) \quad (3)$$

Q : The volumetric flow rate

$_{inlet\ and\ outlet}$: are for intake and outtake flows respectively

$\frac{dV_i}{dt}$: The time rate of instantaneous volume due to moving boundaries,

$\frac{dP}{dt}$: The time rate of high-pressure pump, common rail pipe and injector.

Equation (3) is the fundamental equation of pressure dynamics in every control volume.

Excluding the high-pressure pump, all boundaries in HPCRIS are fixed (i.e.$\frac{dV_i}{dt} = 0$).

Using the basic conservation lawQ_{inlet} and Q_{outlet} can be written as

$$Q = sign(\Delta P).c_d.A.\left(\frac{2|\Delta P|}{\rho}\right)^{1/2} \quad (4)$$

Where sign (ΔP) is a sign function which accounts for the direction of flow,

C_d Coefficient of discharge

A Orifice sectional area

ρ Fuel density

ΔP Fuel pressure differential over the orifice.

3.1 High-pressure pump (HPP)

It is composed of three plunger pump connected on one shaft with 120° phase angle. The pump speed depends on the engine speed as the pump is driven by the camshaft. It is linked to the low and high pressure circuits by a small orifice and a delivery check valve with a conical seat respectively. From equation (3) the fuel pressure time derivative is

$$\frac{dP_p}{dt} = -\frac{E}{V_{pi}}\left(\frac{dV_{pi}}{dt} - Q_p + Q_{pcr} + Q_{pl}\right) \quad (5)$$

Where, the subscripts denote the following

p for HPP

i for instantaneous

pcr for HPP outlet fuel flow to the common rail

pl for fuel leakage flow

The instantaneous volume of the HPP V_{pi} is expressed according to the camshaft motion by

$$\frac{dV_{pi}}{dt} = -A_p\omega\frac{dh_p}{d\theta} \quad (6)$$

A_p Sectional area of HPP plunger, $A_p = \frac{\pi}{4} d_p^2$
d_p Diameter of HPP plunger,
h_p HPP plunger displacement obtained from the kinematics of the plunger as a nonlinear function of the angular position and speed of the camshaft.
ω Angular camshaft speed,
θ Camshaft angular position

From Eq. (4) the HPP Fuel outlet flow to the Common Rail Q_{pcr} is given by

$$Q_{pcr} = sign(P_p - P_{cr}) c_{dcr} A_{pcr} \left(\frac{2}{\rho} |P_p - P_{cr}|\right)^{1/2} \quad (7)$$

Where, the subscripts CR denote common rail and p for pump

c_{dcr} Discharge coefficient which hanges with the pressure ratio in steady operational condition of the diesel engine. In this paper, it is set as a constant [4, 20].
A_{pcr} Area of the delivery valve connecting the pump and the high-pressure circuit,

$$\text{Sign}(P_p - P_{cr}) = \begin{Bmatrix} 0 & if P_p \le P_{cr} \\ 1 & if P_p > P_{cr} \end{Bmatrix} \quad (8)$$

The sign function determines the flow direction, when$P_p \le P_{cr}$, the delivery check valve (directional valve) between the HPP and CR closes to prevent the fuel flow from the rail to the HPP.
The HPP leakage flow Q_{pl} can be constant [4, 20][4,20].
So combining (5)–(8), Eq. (5) can be rephrased as

$$\frac{dP_p}{dt} = \frac{E}{V_{pi}}\left(A_{pcr}\omega \frac{dh_p}{d\theta} + Q_p - sign(P_p - P_{cr}) c_{dcr} A_{pcr} \left(\frac{2}{\rho}|P_p - P_{cr}|\right)^{1/2} - Q_{pl}\right) \quad (9)$$

3.2 Common rail subsystem

The rail pressure equilibrium is obtained by considering the relation between the inlet flow of the high-pressure pump and the outlet flows of the injector. Based on (3), the fuel pressure time derivative of common-rail can be modeled as

$$\frac{dP_{pcr}}{dt} = \frac{E}{V_{cr}}(Q_{pcr} - Q_{cro} - Q_{crb}) \quad (10)$$

Where, the subscripts denote the following
cr Common rail pipe
cro Common rail pipe outlet flow rate that is equal to the injectors he inlet fuel flow rate
crb Return flow, which is considered to be constant[4]

$$Q_{cro} = \sum_1^4 Q_{crok} \quad (11)$$

k injector number , k=1,2,3 and 4.
Q_{crok} Injection flow of each injector

$$Q_{crok} = sign(P_{pcr} - P_{injk}) c_{dinjk} A_{crik} \left(\frac{2}{\rho}|P_{cr} - P_{injk}|\right)^{1/2} \quad (12)$$

Where the subscript $injk$ is for the kth injector
c_{dinjk} Discharge coefficient of the orifice between common rail and kth injector,
A_{crik}Inlet port cross-sectional region in the kth injector.
Hence, Eq. (10) can be rewritable as

$$\frac{dP_{cr}}{dt} = \frac{E}{V_{cr}}\left(Q_{pcr} - \sum_1^4 sign(P_{cr} - P_{injk}) C_{dinjk} A_{crik} \left(\frac{2}{\rho}|P_{cr} - P_{injk}|\right)^{1/2}\right) \quad (13)$$

3.3 Injector subsystem

The common rail deliveries four injectors, one for each cylinder of the engine. Following from (3), the fuel pressure dynamics of the injector is shown as

$$\frac{dP_{injk}}{dt} = \frac{E}{V_{injk}}\left(Q_{crok} - Q_{cylk}\right) \quad (14)$$

Q_{cylk}fuel injection flow from the kth injector to a cylinder shown as

$$Q_{cylk} = sign(P_{injk} - P_{cylk})E_{tk}c_{dik}A_{injk}\left(\frac{2}{\rho}|P_{injk} - P_{cylk}|\right)^{1/2} \quad (15)$$

Where the subscript $_{cylk}$ is for the kth cylinder

c_{dik}Discharge coefficient of the injector nozzle

A_{injk} Orifice area of the nozzle of the injector.

E_{tk} Square signal that equals to1 during injections and zeroes otherwise. It represents the time interval at which the injection outlet orifice is open. The injection duration $\theta_{injection,\ i}$ or $dt_{injection,\ i}$in this work is related to the engine speed and load. Where $\theta_{injection,\ i}$ is the crank angle duration in degrees on which the injector orifice is open and $dt_{injection,\ i}$ is the time duration in [ms] on which the injector orifice is open.

Substitutes equations (12) and (15) in equation (14) ,results in

$$\frac{dP_{injk}}{dt} = \frac{E}{V_{injk}}\left(Q_{crok} - sign(P_{injk} - P_{cylk})E_{tk}\mu_{ik}S_{injk}\left(\frac{2}{\rho}|P_{injk} - P_{cylk}|\right)^{1/2}\right) \quad (16)$$

3.4 Injection timing subsystem

The Electronic Control Unit (ECU) activates each fuel injector 4 times per cycle.The fresh charge is internally prepared by injecting fuel twice, the first just before and the second just after the Top Dead Center at the beginning of the suction stroke. This early injection using the same high-pressure CR injection system allows for fuel heating and distribution and gives more time for fuel vaporization and mixing with air. The other two injections are the pilot and main injections that take place before and after the TDC at the start of expansion stroke The following subscripts are used to substitute the subscript i in $\theta_{injection,\ i}$ or $dt_{injection,\ i}$

p is for pilot injection
M is for main injection
1 pre is for first preparation injection
2 pre is for second preparation injection

There are two preparations injections ($\theta_{injection,\ 1\ pre}$ or$dt_{injection,\ 1\ pre}$)and($\theta_{injection,\ 2\ pre}$ or$dt_{injection,\ 2\ pre}$), pilotinjection ($\theta_{injection,\ p}$ or$dt_{injection,\ p}$) and main injection ($\theta_{injection,\ M}$ or$dt_{injection,\ M}$).

The start and the duration of each injection are obtained by data regression of the experimental data concerning the measurement of the triggering signals of the injector

The pilot injection is just before the Top Dead Centre (TDC) at the end of compression stroke. The small amount of fuel injected triggers the combustion of the fuel already present in the charge. Nearly 2 ms later, the main injection takes place.

The start of pilot injection TS $_p$ is calculated from the TDC at the end of compression stroke related to the engine speed and load by the equation

$$TS_p\ [ms] = 0.001017\times n + 0.00505\times load - 0.32 \quad (17)$$

The pilot injection duration $dt_{injection,\ p}$ is calculated as a function of engine speed and load from the following equation

$$dt_{injection,\ p}\ [ms] = (5.88\times10^{-5}\times n + 0.003222\times Load + 0.223048)\times (0.00015\times n + 0.002194\times Load + 0.75495) \quad (18)$$

so the end of pilot injection can be determined.

The start of main injection TS $_M$ is 2 ms later from the end of pilot injection

The main injection duration $dt_{injection,\ M}$ is calculated as a function of engine speed and load from the following equation

$$dt_{injection,\ M}\ [ms] = ((0.00004\times n + 0.002\times Load + 0.58))\times (0.00015\times n + 0.002194\times Load + 0.75495) \quad (19)$$

IOP Conf. Series: Materials Science and Engineering **973** (2020) 012037 doi:10.1088/1757-899X/973/1/012037

Two more short injections are timed just before the end of the exhaust stroke $\theta_{injection,\,1\,pre}$ and just after the start of the suction stroke $\theta_{injection,\,2\,pre}$. These to injection control the preparation of the fresh homogenous charge for the next cycle.

The Start of first preparation injection at the end of exhaust stroke of the previous engine cycle TS_{1pre} is adjusted 1.5 ms before the TDC.

$$TS_{1\,pre}\,[ms] = 1.5\ ms \qquad (20)$$

The first preparation injection duration $dt_{injection,\,1pre}$ equal 1 ms at all speeds and loads. So the end of first preparation injection can be determined as 0.5 before TDC.

The start of 2nd preparation injection $TS_{2\,pre}$ is 1 ms later from the end of first preparation injection i. e 0.5 after TDC at the start of suction stroke. The second preparation injection duration $dt_{injection,\,2pre\ equal}$ 1 ms at all speeds and loads.

The trigger signals before and after the TDC before the suction stroke, (1st and 2nd prepration injections) consist of multiples of very short successive triggers. To account for this disconiuity of the triggered signal a correction coefficient for the area of the injection nozzles during 1st and 2nd prepration injections is established as function of engine speed and load as

$$ACorr. = -0.0000495\times n + 0.001457\times Load + 0.19624 \qquad (21)$$

All time durations or starting times for injection can be converted to crank angle domain

$$\theta S_i\,[degree] = 6\times n\times T_{S\,i}\,[ms] \qquad (22)$$

$$\theta_{injection,\,i}\,[degree] = 6\times n\times dt_{injection,\,i}\,[ms] \qquad (23)$$

In the above equations engine speed is in rpm and engine load is in [N.m]

4. Results and Discussions

4.1 Validation for the simulation Results

The model has been conducted using the MATLAB/Simulink environment. The fourth-order Runge-Kutta integration method is adopted. The model parameters shown in table 3have been chosen according to the geometric specifications of the mechanical parts and experimental tests. The available experimental data, dimension of radial piston pump (bore-stroke), dimension of common rail, nozzle diameter of injector, number of the nozzle and speed of the engine, to evaluate the model validity, simulation outputs have been compared with experimental data obtained from a common rail injection diesel engine test rig described above. The available experimental data are the amount of fuel injected per cycle and the fuel line pressure at different engine loads.

Table 3. Model Input Parameters.

	Symbol	**Value**
Fuel density	ρ	870 kg/m^3
High pressure pump plunger diameter	d_p	0.65 cm
High pressure pump plunger stroke	Sp	0.7 cm
The geometric volume of High-Pressure Pump	V_g	0.696 cm^3
The volume of common rail	V_{cr}	20 cm^3
the volume of a single injector	V_{inik}	1.47 cm^3
Area of High-pressure Pump outlet port	Ap_{cr}	0.0707 cm^2
Area of the inlet port of a single injector.	A_{inik}	0.049 cm^2
Injector nozzle orifice diameter	d_n	0.015 cm
The total area of the injector nozzle orifices	A_{iniok}	0.000884 cm^2
The bulk modulus of fuel	E	3.25 Gpa
Discharge coefficients	C_{dcr}, C_{dinik} and C_{diniok}	0.64

Figure 3 shows a comparison between the measured and predicted common rail pressure at different speeds and loads. It is clear that the numerical simulation only predicted the average value of the fuel line pressure with relative error less than 8 % as shown in figure 4 which demonstrates the

IOP Conf. Series: Materials Science and Engineering **973** (2020) 012037 doi:10.1088/1757-899X/973/1/012037

relative error between the predicted, and the average value of the measured fuel line pressure and for different speeds and loads.

A comparison between the predicted and measured amounts of fuel injection during one cycle per one cylinder at different engine speeds and loads (30 points) is shown in figure 5. Figure6 shows the relative error of the predicted value compared to the measured value. It is shown that an acceptable agreement between the predicted and experimental results is obtained at all speeds and loads other than 0 and 50 N.m with relative error less than 9 %. The maximum relative error at load of 50 N.m is 13 % while a relative error of 21 % is obtained at 0 N.m load. The relatively large relative error at small loads (0 and 50 N.m) can be attributed to that the model does not simulate the common rail pressure wave fluctuations which affect the quantity of injected fuel. This deficiency combining with the small amount of fuel injected at small loads make the error is higher.

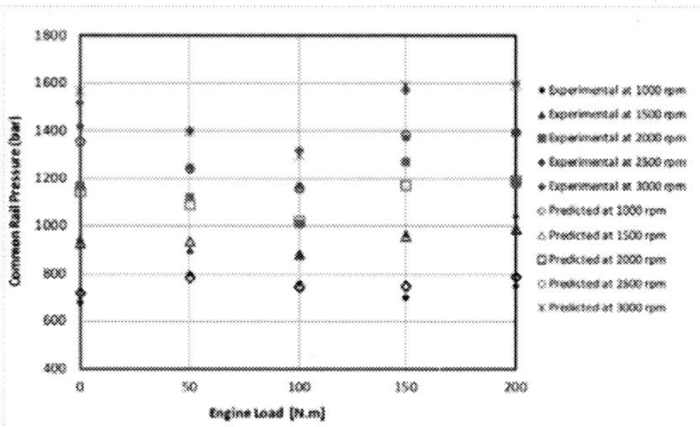

Figure 3. Experimental and predicted common rail fuel pressure at differentspeeds and loads.

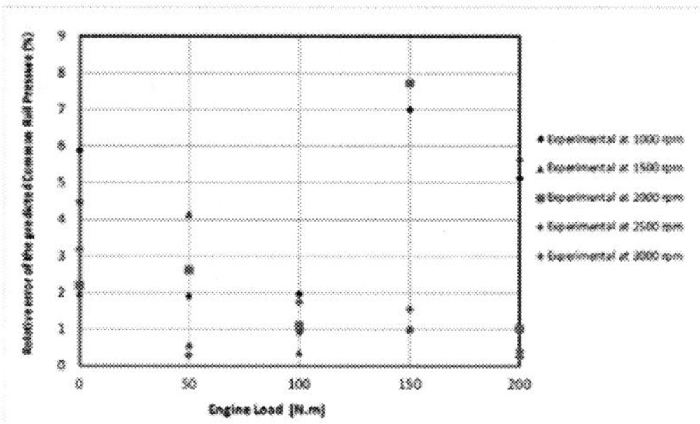

Figure 4. Relative error between the predicted and the average value of the measured fuel line pressure and for different speeds and load.

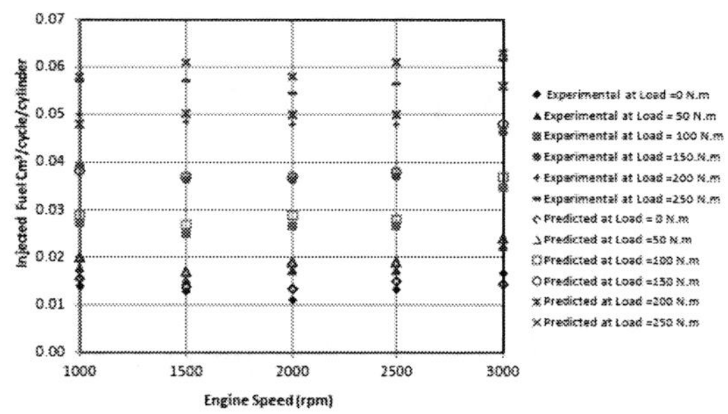

Figure 5. Experimental and predicted injected fuel at different engine speeds and different loads.

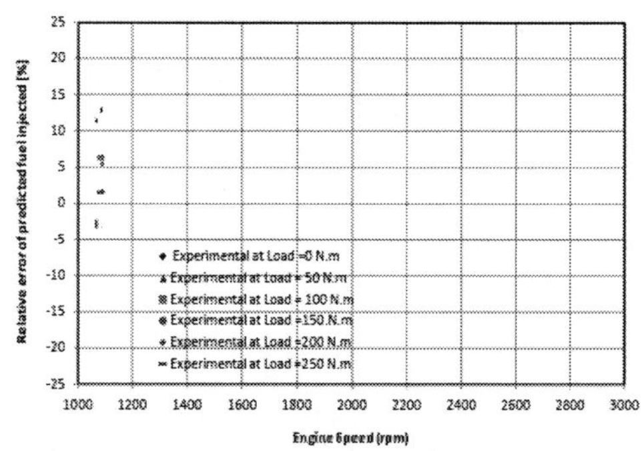

Figure 6. Relative error between the predicted and experimental injected fuel at different enginespeeds and loads.

4.2. Simulation Results

Some simulation results are demonstrated which highlight the modeling capability. Plunger displacement h_p is presented at 3000 rpm in figure 7. The used pump is basically a three-plunger pump derived by the engine crankshaft. The flow rate of the pump exit is regulated by the delivery valve between the high-pressure pump and common rail. As seen in the figure the alternative motion of each piston with the phase shift of 120° between each of them is presented and with the displacement of each piston, the flow rate varies, so from figure 8, the High Pressure Pump outlet flow rate is demonstrated

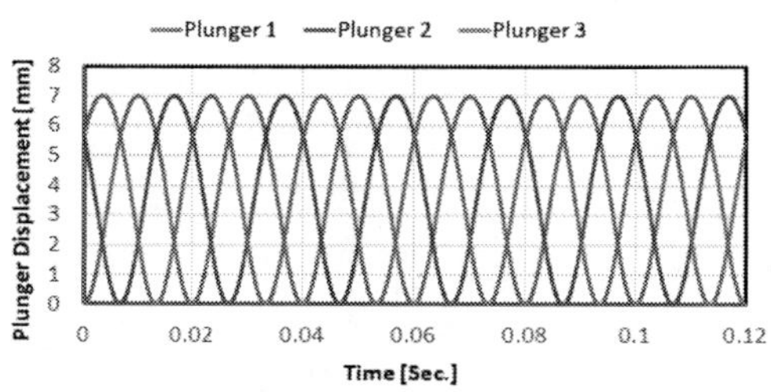

Figure 7. The displacementof the piston plunger.

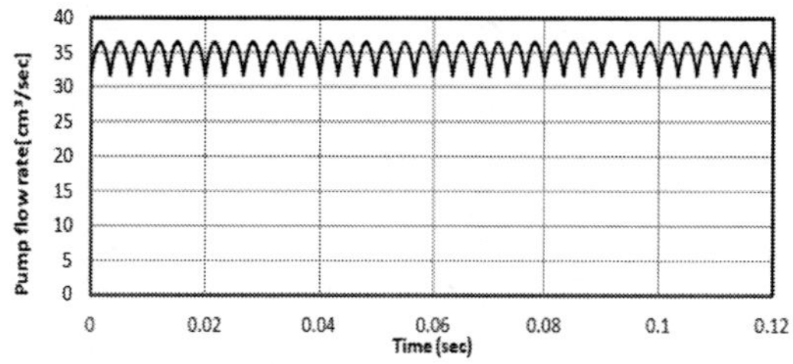

Figure 8. The volumetric high pressure pump flow rate.

Another simulation output is the common rail pressure at different engine speeds and loads. Figure. 9 illustrates the pressure of common rail predicted by Matlab/Simulink at an engine speed of 1500 rpm and load of 250 N.m for three successive cycles as an example. Common Rail pressure is almost constant at about an average value of 1022 bar with a standard deviation of 2.1 bars through the three cycle. Rail pressure fluctuations result from the variations of the fuel supplied from a high-pressure pump, and the quantity of the fuel output to the injectors. These small oscillations shown in figure 10 Occurs at time of injections for the four cylinders. The small oscillations may be attributed to the large Rail volume of 20 cm^3 which is sufficient to provide the desired injection quantity with small pressure variations. Also, the study of the dynamics of common rail pressure wave is not included in this simulation. Payri et al [24], concluded that A lumped model cannot model the rail real dynamics because of the inhomogeneous pressure alternations across the rail length due to the instabilities caused by the incoming flow from the high-pressure pump to rail and the out coming flow to the injectors. This event is mainly intermittent concerning the engine crank angle-scale due to the engine reciprocating motion [25-26]. On the other hand, the pulsation period will vary in the time-scale, if the engine speed varies with time and hence appropriate control of common rail pressure is complicated to realize [27-28].

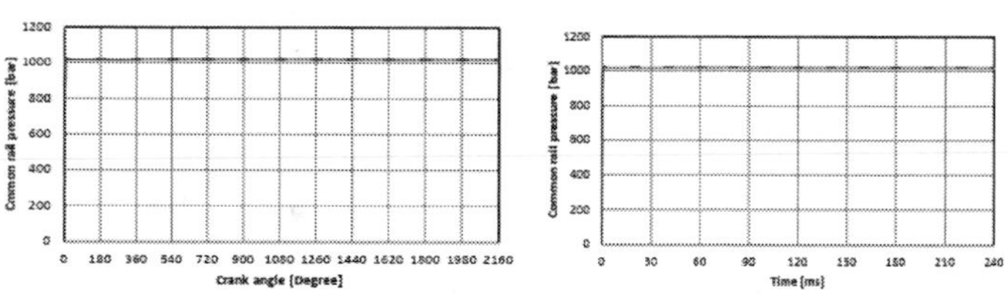

A B

Figure 9. The predicted common rail pressure for three successive cycles (A) in crank angle scale and (B) time scale

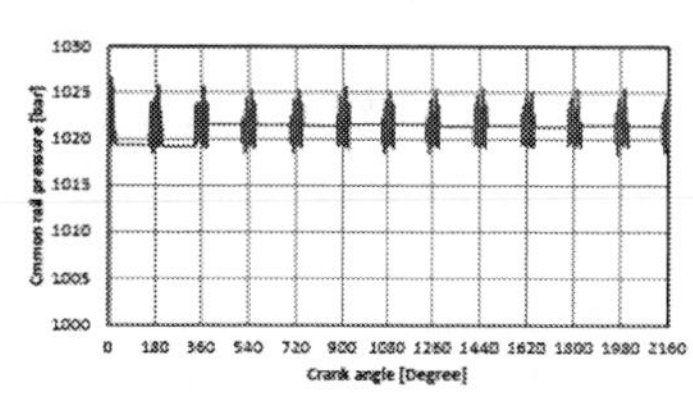

Figure 10. The predicted common rail pressure oscillation in crank angle scale

Figure 11 shows the injection pressure of cylinder no. 2 at engine speed of 1500 rpm and 250 N.m predicted by Matlab/Simulink. The high pressure from the rail-to-injectors pipes right up to the injector entry keeps equal or below the common rail pressure with the existing system layout (both pipes lengths and diameters) so the pressure remains below the durability border of the components. It is noticed that the basic pressure pulsation period (pressure falls due to the injection) is constant on the time-scale due to the constant speed of 1500 rpm on which the engine run so it will be invariant also concerning the engine rotational angle. It is expected to change if the engine speed is changed and this issue will be discussed later.

IOP Conf. Series: Materials Science and Engineering **973** (2020) 012037 doi:10.1088/1757-899X/973/1/012037

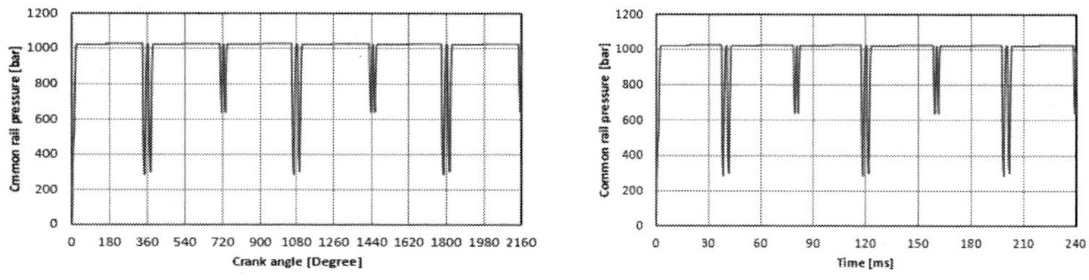

A B

Figure 11. The predicted injection pressure for three successive cycles (A) in crank angle scale and (B) time scale

Figure 12 demonstrates the injected mass flow rate in both the time-scale and crank angle-scale for cylinder no. 2 at engine speed of 1500 rpm and 250 N.m predicted by Matlab/Simulink. The fuel injector solenoid receives a control signal from the Electronic Control Units which stimulates the injector to introduce the fuel into the engine combustion chamber, results in the pressure in the injector to drop. The injection event is principally intermittent on the time-scale and the rotational angle-scale at constant engine speed irrespective of its value. The variation in the duration of the injection in the time scale due to the change in engine speed and load affects the magnitude of the decrease in the common rail and injection pressure [29-30]. This issue will be discussed later. In the rest of this chapter the results will be discussed over one engine cycle other than the first cycle to avoid the effect of transient pressure rise from zero time due to the set initial conditions. This issue will be studied in future work. The predicted injected flow rate profile for one engine cycle is shown in figure 13.

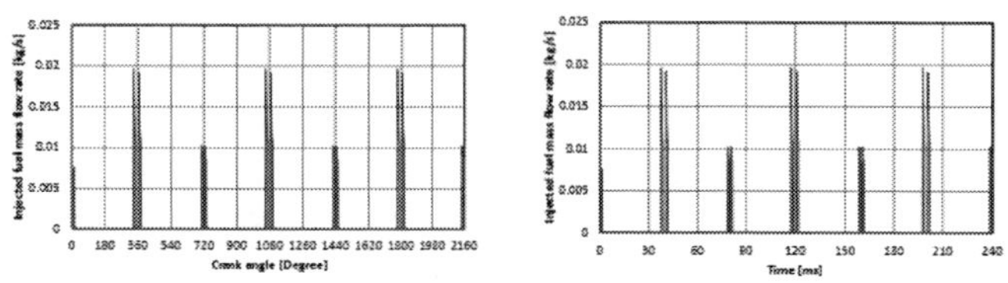

A B

Figure 12. The predicted injected fuel flow rate for three successive cycles (A) in crank angle scale and (B) time scale

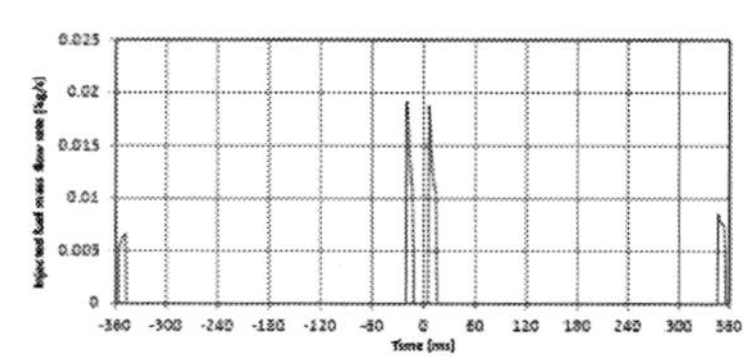

Figure 13. The predicted injected fuel flow rate profile for single cycle cycles in crank angle scale

5. Conclusions

This paper sets out a HPCRS mathematical model for diesel engines. Model equations are generated by using the physical laws governing the fluid dynamic and mechanical phenomena. The model parameters have been chosen based on the mechanical parts geometric specifications and experimental tests at different operating conditions. The measured in-cylinder pressure obtained from the test rig is used inside the model instead of using a constant value in-cylinder pressure as found in the literature. The injection start time and injection duration are expressed as a function of engine load and speed such that it simulates the change in injection start and duration with engine load and speed.

The proposed model is validated by comparing simulation results with experimental results carried out on diesel engine test rig. The comparison demonstrates the capability of the model to accurately predict the injection system behavior in different operating conditions, in terms of injected flows, and injection pressure.

IOP Conf. Series: Materials Science and Engineering **973** (2020) 012037 doi:10.1088/1757-899X/973/1/012037

6. References

[1] Kazutoshi M 1997 *SAE* **970753** 19-29.
[2] Pischinger F 1998 *J. Eng. Gas Turb. Power ASME* **120** pp 641-647.
[3] TennisonP J and Reitz R 2001 *J. Eng. Gas Turb. Power ASME* **123** pp 167-174.
[4] Wang H, Zheng Dand Tian Y 2016 *ISA Transactions* **63** pp 265-273.
[5] Allocca L, Belardini P, BertoliC and Corcione F 1992 *SAE* **920576**.
[6] De Nicolao G, Rossi C, Scattolini R and Suffritti M 1999 *Control Engineering Practice* **7** pp 1061-1069
[7] Maione B, Lino P and Rizzo A 2004 *WSEAS Transactions on Systems* Issue 5, **3** pp. 2164-2169.
[8] Tanner F and Srinivasan S 2009 *Applied Mathematical Modeling* **33** pp 1366–1385
[9] GuoC et al 2018 *Energy Conversion and Management* **160** pp 302–12.
[10] Petranović Z, Sjerić M, Taritaš I, Vujanović M and Kozarac D 2018 *Energy Conversion and Management* **171** pp 1–11.
[11] Zhao F, Yang W, Zhou D, Yu W, Li J and Tay K L 2017 *Fuel* **188** pp 382–9.
[12] Jurića F, Petranovićb Z, Vujanovića M, Katrašnikc T, Viharc R, Wangd X and Neven Duić N 2019 *Energy Conversion and Management* **185** pp 730–739
[13] Chiavola O and Giulianelli P 2001 *SAE Technical Paper* **2001-01-3183**.
[14] Mulemane A, Han J, Lu P,Yoon S and Lai M 2004 *SAE Technical Paper* **2004-01-0536**.
[15] Lino P, Maione B and Rizzo A 2007 *Applied Mathematical Modeling* **31** pp 1770–1784.
[16] IMAGINE SA 2000 *IMAGINE Technical Bulletin* 110, Roanne, France.
[17] Catania A, Dongiovanni C and Mittica A 1992 *J. Eng. Gas Turb. Power ASME* **114** pp 534-543.
[18] Catania A, Dongiovanni C and Mittica A, Negri C and Spessa E 1996, *SAE* **962020**.
[19] Catania A, Dongiovanni C and Mittica A, Negri C and Spessa E 1999, *J. Eng. Gas Turb. Power ASME* **121** pp 186-196.
[20] Lino P, Maione B, and Rizzo A 2005 *IEEE Conference on Emerging Technologies and Factory Automation.*
[21] Dinoi A 2002 Control Problems in Common Rail Diesel Injection Systems (in Italian), Master Theses, Bari.
[22] Kouremenos A, Rakopoulos C, Hountalas D and Kotsiopoulos P 1991 *Proc. ASME-WA Meeting, Atlanta, Georgia* AES pp. 91-102.
[23] Nikolic B, Kegl B, Markovic S and Melanija S 2012 *Thermal Science* **16** Suppl. 2, pp. S569-S579
[24] Payri F, Lujan J M, Guardiola G and Rizzoni G 2006 *Proc. IMechE-Part D: J Automob. Eng.* **220** pp 347–357.
[25] Chin Y and Coats F1986 *SAE* **860412**.
[26] Yurkovich S and Simpson M 1997 *Proc. Am. Control Conf* pp 278–283.
[27] Sun Z , Zhang Z and Tsao T 2009 *Control Lett.* **58 (6)** pp 452–460.
[28] Zhang Z and Sun Z 2010 *ASME J. Dynam. Syst. Measure. Control* **132 (011004)** pp 1–10.
[29] Coppo M Dongiovanni C and Negri C 2004 *J. Eng. Gas Turb. Power ASME* **126** pp 874–885.
[30] Ubertini S 2006 *J. Eng. Gas Turb. Power ASME* **128** pp 694–701.

IOP Conf. Series: Materials Science and Engineering **973** (2020) 012038 doi:10.1088/1757-899X/973/1/012038

Performance of homogeneous charge compression ignition (HCCI) engine with common rail fuel injection

W Niklawy[1], M Shahin[2], M I Amin[3] and A Elmaihy[3]
[1] PhD candidate, Mechanical Power Engineering Department, Military Technical College, Cairo, Egypt.
[2] Mechanical Engineering Department, School of Engineering, Badr University, Badar, Cairo, Egypt.
[3] Mechanical Power Engineering Department, Military Technical College, Cairo, Egypt
niklawywaleed@gmail.com

Abstract. Homogeneous Charge Compression Ignition (HCCI) engine is a newly introduced technology. However, controlling the HCCI timing of ignition and theheat release rate, at all operating regimes, present the biggest challenge due to the complicated and highly coupled combustion problems. The successful implementation of the common rail injection system in diesel engines has made this possible. The main key for this success is the capability to optimize the most important fuel injection parameters, namely timing, rate, and duration.This work experimentally investigates the performance of an HCCI Diesel engine with a Common Rail Direct Injection (DI) fuel system. The engine used is a turbocharged 2776 cc, 4-stroke, 4-cylinder, water-cooled with overhead valve mechanism. The engine performance is evaluated and presented at different loads and speeds. The Apparent Heat Release Rate during combustion is evaluated from the measured cylinder pressure-crank angle data. The pressure in the fuel line is measured at the entry to different fuel injectors to investigate the effect of pressure wave propagation in the fuel line on the fuel injection system.

1. Introduction

Although diesel engines have high thermal efficiencies than that for petrol engines, they release higher amounts of NOx, smoke and particulate matter (PM). To defeat these drawbacks, a relatively new combustion concept known as HCCI (Homogeneous charge compression ignition) is being explored all overt he world. HCCI engines can operate with reduced NOx, smoke and PM emissions simultaneously [1-3].HCCI was developed as a consequence of current progress in electronic engine control, variable valve timing and lift systems, so the implementation of the HCCI combustion concept in the manufacture of engines is now considered to be practical [4,5].HCC uses the advantages of spark ignition and compression ignition engines. They have better thermal efficiency than that for Spark Ignition engines and come up with the efficiency of a Compression Ignition engine [6-7].

Homogeneous Charge Compression Ignition (HCCI) uses a lean homogeneous air-fuel mixture as a means to decrease the combustion temperature. This mixture can be carried out by prolonging the ignition delay period with extremely advanced injection timing. Most of the fuel is injected, evaporated and mixed in the premixed mode of combustion, such that the mixture at the moment of ignition is near homogeneous. Therefore, the premixed mode of combustion is the predominant phase in HCCI [8]

In recent times, the multiple injection approach was useful in commercial common rail Direct Injection HCCI diesel engines to accomplish an organized air-fuel mixture and evade wall wetting. The common rail system provides higher injection pressure for better fuel spray atomization and evaporation, improved air entrainment and mixing in the fuel jet, which is advantageous for lowering soot emissions [9,10]. Several Direct Injection HCCI concepts have used advanced injection circumstances to sufficiently prolong the ignition delay period so that full fuel vaporization event preceding to combustion. The electronically controlled common rail injection permits diesel engines to vary the injection timing broadly and also gives the ability to achieve multi- injection strategies which result in improved charge mixing and homogeneity [11].

During HCCI engine running, a homogeneous fuel and air mixture is set throughout the compression process and the mixture auto ignition takes place by compression only at different in-cylinder positions when reaching the chemical activation energy. The combustion process is controlled entirely by chemical kinetics like Compression ignition engines [12] instead of ignition by a spark followed by a flame front propagation such as Spark Ignition engine [13]. The two-step heat release processes with low-temperature kinetic (cool) reactions occurring firstly followed by high-temperature kinetic reactions are noticed in HCCI engine combustion.

Many researchers have studied the effect of engine speed on the combustion behavior and performance characteristics of a HCCI engine. Kong et al. [14] investigated the consequence variations of ignition startmoment on HCCI engine combustion at different speeds. They concluded that the ignition start would be an efficient and practical required parameter to manage timing of ignition for the dynamic engine regimes. Iida et al. [15] concluded that the start of heat release is less sensitive to engine speed variations than that of compression ratio, suction air, and coolant water temperatures. Aroonsr isopon et al. [16] experimentally examined different reference fuels with an octane number of 70, which reveal significant heat release at low-temperature, at different engine speeds from 600 to 2000 rpm with constant suction air temperature. It was found that the low-temperature heat release period decreases as engine speed increases. Furthermore, decreasing low-temperature heat release duration causes shifting the high-temperature heat release period to later timings, consequently, a long low-temperature (cool) heat release period resulted in maintain high-temperature heat release. Iida et al. [17] used n-butane as a fuel to investigate the effect of inlet air temperature, compression ratio and engine speed on HCCI operating ranges in a research engine. they found that operation of HCCI at a relatively high engine speed was achieved of 2000 rpm, with a high compression ratio and a high temperature inlet air of 16.55 and 400 K, respectively. Szybist and Bunting [18] concluded that the amount of energy released during the low-temperature heat release interval increased suddenly as engine speed reduced. Besides, a nearly constant temperature rise which occurred during the low-temperature heat release period was noticed for all engine operating conditions. Garcia et al. [19] and Genzale et al. [20] reported that the angle of the start of combustion increases as the engine speed increases. Agarwal [21] concluded that the maximum pressure rise rate and coefficient of indicated mean effective pressure[IMEP] variation are the critical factors, that can be used to characterize HCCI operating range. Ebrahimi and Desmet [22] experimentally studied the effect of engine speed and cyclic variations in an HCCI engine.

Khandal et al [5] presented an exhaustive review on the performance characteristics, exhaust emission and combustion behavior of Common Rail Direct Injection and HCCI engines operated using with different fuels. Hadia et al [23] numerically investigated the combined effects on efficiency of the compression ratios and steam injection, characteristics of combustion and emissions of an HCCI engine. The contaminant emission is restricted by the dilution of the reactant by using steam injection. Combustion is performed by using engine modeling.

Calam et al [24] parametrically investigated the effects of compression ratio on HCCI combustion, performance, and emissions. The engine operational BSFC maps were the results of running the

IOP Conf. Series: Materials Science and Engineering **973** (2020) 012038 doi:10.1088/1757-899X/973/1/012038

engine at 169 experimental data points. The experiments were carried out at different equivalence ratios with predetermined intake air temperature (353 K) and by using RON20 and RON40 fuels.
El Shenawy et al [25] reported that one of the essential aspects to reduce engine emissions is how to prepare a homogenous charge for HCCI engines. One strategy is the external mixture formation, or port fuel injection (PFI). PFI used only for high volatile fuel such as alcohol [26-28].

The second is the in-cylinder mixture preparation, which can be classified according to injection timing early injection in which fuel is injected in the compression stroke and late injection. in which fuel injection is delayed behind the top dead center [29].

One of the main challenges that the HCCI combustion technique has to face is the occurrence of knocking at high loads due to high combustion pressure which causes the engine to knock. This may limit this technique commercially practical [30]. However, there are very few studies investigating the effects of the external load on HCCI engines parametrically at different engine speeds. The objective of this work experimentally investigates the performance of an HCCI Diesel engine with a Common Rail Direct Injection (DI) fuel system. The engine performance is evaluated and presented at different loads and speeds. The Apparent Heat Release Rate during combustion is evaluated from the measured cylinder pressure-crank angle data. The pressure in the fuel line is measured at the entry to different fuel injectors to investigate the effect of pressure wave propagation on the fuel line pressure at different operating conditions.

2.The Test Facility (Experimental Setup)

Experimental investigations were implemented on a transport diesel engine of type Mercedes-Benz with an open chamber. A four in-line cylinders, water-cooled direct injection diesel engine with over-head valve gear and “common rail” fuel injection system. (available in the laboratory of mechanical power and energy at the Military technical college) The main engine design parameters are listed in table 1. The test rig includes the engine and all the instrumentation necessary for measuring and recording the operating parameters. An on-line data acquisition system is furnished to improve the speed and accuracy of data collection and recording. is used. Detailed engine data are given in table 1.

Table 1. Engine data.

Description	Specification
Engine	R2816K5A
Bore	94 mm
Stroke	100 mm
weight	270 kg
Compression Ratio	17.5
Number of Cylinders	4
Displacement	2776 cc
Max power	90 kW (120H.P.) @ 3800 RPM
Peak Torque	400 N·m @ 1800 RPM
Fuel primary pump	Electric pump
Fuel System	Direct Fuel Injection Common Rail
High-Pressure Fuel	Radial Piston Pump (3 pistons)
Injection pump	CP3 2nd Generation Common Rail
Timing System	Valve Belt Driven DOHC Overhead

2.1 Instrumentations

The parameters measured are classified into three main groups. a -External parameters: Engine speed, engine load, fuel consumption, airflow, crank angle, ambient conditions and real-time. B-Internal parameters: pressure inside the cylinder, inlet and exhaust manifolds and fuel line,and turbocharger speed. C-Safety parameters: Lubricating oil pressure and temperature, and cooling water, temperature. Figure 1 gives a general scheme of installed test rig showing numbered locations where important pickups and transducers are positioned. A list of these locations and the corresponding measured parameter used in this study at each is given below is given in table 2.

Engine external loading was carried out by an ELZE /Heenan hydraulic dynamometer. The fluid used was water with which the maximum braking power could reach 170 kW at 4000 rpm. The engine and dynamometer shafts were directly coupled through a cardan shaft. An S-type load cell (strain gauge S type, max capacity 500kg) is inserted under the dynamometer torque reaction arm. The load cell is completed with a bridge amplifier that is excited (12VDC) from a regulated power supply. The whole setup is calibrated by applying known weights on the cell and observing the net voltage output.

A water-cooled piezoelectric transducer (type PCB model no. 112B11) is used for measuring cylinder pressure. The charge amplifier used is a PCB type, with the capability of statically holding the output charge for calibration processes. The transducer is mounted on the cylinder head above cylinder no.1 where suitable and enough space could be provided. The glow plug of this cylinder was machined from the inside to form a cavity. The sensor is mounted at the outer side of the cavity that is opened to the combustion chamber on the inner side

A piezoelectric transducer type (Kistler PN 6278) and charge amplifier type (Kistler SN 284625) is used to measure fuel line pressure. This setup is capable of measuring pressure up to 3000 bars. The transducer is directly mounted on the fuel line high pressure connecting the rail to an injector no.1. The transducer, amplifier, and cabling were calibrated together using the deadweight tester
Engine fuel consumption at steady-state operation is evaluated by recording the time of consumption of a certain volume of fuel. This old method remains particularly valuable today because of the high accuracy achieved. The measuring device consists of a glass flask of 250 cc volumes, an auxiliary fuel tank, a control cock, and a stopwatch.

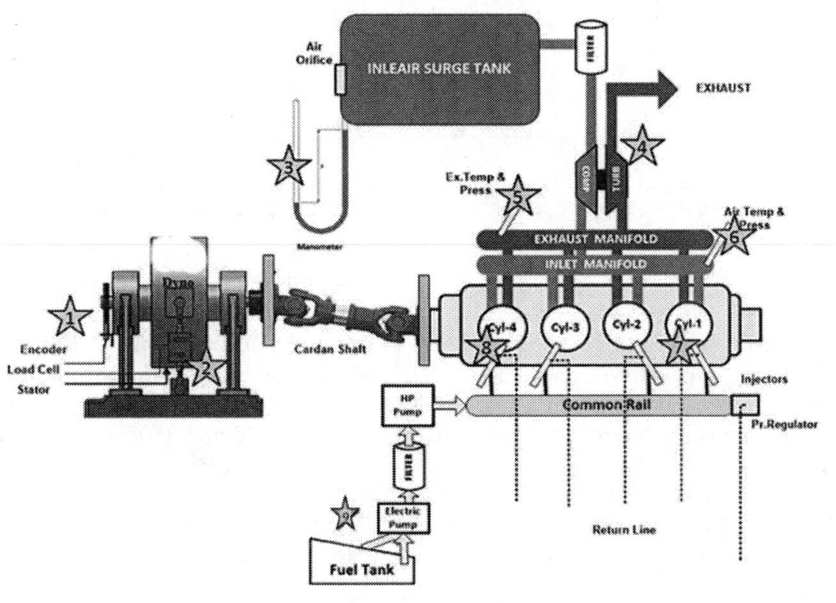

Figure 1. Scheme of the engine test facility.

An incremental digital quadrature encoder (type WDG 58B-360-ABN-G24-K3) is used for engine speed measurement. The encoder gives 2 trains of pulses (A and B), each has 360 pulses per encoder shaft revolution. The two trains are phased by ¼ pulse, and a third index train (N) with one pulse per each revolution is also produced. The Encoder is mounted on a special bracket fixed at the free end of the dynamometer shaft, figure 1.

IOP Conf. Series: Materials Science and Engineering **973** (2020) 012038 doi:10.1088/1757-899X/973/1/012038

A standard orifice (40 mm) diameter is mounted at the air surge tank entrance with a U-tube water manometer for measuring the air pressure drop across the orifice. The airflow rate is calculated from the orifice area and the manometer reading. This technique is considered as one of the most accurate methods for measuring the fluid flow rate.
The unsteady pressure in the exhaust manifold is measured using a strain gauge transducer (model PT124BG1/4) which comes complete with its amplifier). The transducer is positioned such as direct exposure to the exhaust blows from cylinder outlet ports are avoided.

Table 2.A list of measured parameters and the corresponding locations relevant to figure 1.

Location	Measured Parameters	Instrument	Accuracy
1	Engine Speed & Crank Angle	Incremental Encoder (WD GI 58B)	+ 7.5 % of the pulse width
2	External load (Torque)	S-type load cell	Combined error ≤+0.05%
3	Inlet air pressure drop across the orifice pressure	U tube manometer	NA
5	Exhaust Temperature	Thermocouple type K (Nicr-Ni)	+/- 2.2C or +/- .75%
	Exhaust Pressure	Strain gauge transducer model PT124BG1/4	0.5%FS, 1.0%FS
6	Inlet Air Temp	type K (Nicr-Ni)	+/- 2.2C or +/- .75%
	Inlet Air Pressure	strain gauge transducer model AST4000AV0015P6C0000	<± 0.5% BFSL
7	Cylinder pressure	piezoelectric transducer type PCB (model no. 112B11)	<2.0 % FS
8	Fuel pressure sensor	piezoelectric transducer type (Kistler PN 6278)	<0.2 % FS
9	Fuel Consumption	Calibrated Bowel	NA

A strain gauge pressure transducer (model AST4000AV0015P6C0000, complete with an amplifier, is used for measuring the pressure inside the intake manifold. Thermocouples type K (Nicr-Ni), are used to measure the exhaust gasses and inlet air temperature

2.2 Test procedure

Measured parameters included engine speed, engine power (calculated from the dynamometer reading), fuel consumption, inducted air mass flow rate, fuel line pressure, and cylinder pressure. The measurements were carried in the range from 1000 to 3000 rpm with 250 rpm increment. At each speed, the engine external load was adjusted to 0, 50, 100, 150, 200 and 250 N.m which corresponds to 0, 2.25, and 4.5, 6.75 and 9 bar brake mean effective pressure (BMEP).

The test procedure is required to fulfill the following main tasks in the prescribed order: first, pre-operational checks and preparations. Then, running the engine and sustaining the operating conditions as requested,(speed and load).Finally. Safety monitoring, e.g. overheating over-speed and overload.

3. Results and Discussions

3.1 In-cylinder and Fuel Line Pressures

In-cylinder pressure records investigation is the mainly efficient method to examine the engine combustion performance due to the direct effects of in-cylinder pressure history on the engine output power and combustion behavior characteristics. In this paper, in-cylinder pressure is recorded versus crank angle. Cycle to cycle variation was tested with the number of cycles. It was notice that the cycle to cycle variation is constant for the number of cycles more than 30 cycles. Average pressure values for 42 successive cycles is calculated to evade the cycle to cycle variations and to report for measurement uncertainty.

In HCCI engines The combustion is divided into two subsequent processes namely; low-temperature process (cold flame) and high-temperature process with relatively small ignition delay period which strongly depends on the energy released during the low-temperature process [31].Ebrahimi and Desmet [22] introduced some fundamental combustion parameters of HCCI engine as shown in figure2.$HRRH_{max}$ and $HRRL_{max}$ are the maximum rates of heat release of high and low-temperature processes (reaction). In the first stage, the following parameters are defined. LTRA and LTRE are the low-temperature reaction appearance and end which are the points of intersection of the horizontal axis and the tangent line of the point where the gradient is maximum on the rising side and falling sides of the curve respectively. LTRD is the low-temperature reaction duration which is the duration between LTRA and LTRE. HTRA and HTRE are the high-temperature reaction appearance and end which are the points of intersection of the horizontal axis and the tangent line of the point where the gradient is maximum on the rising side and falling sides of the curve respectively. HTRD is the high-temperature reaction duration which is the duration between LTRA and LTRE. but for the second-stage combustion. The zone of negative temperature coefficient (ZNCT) is the period between FRBT and ARHT is named

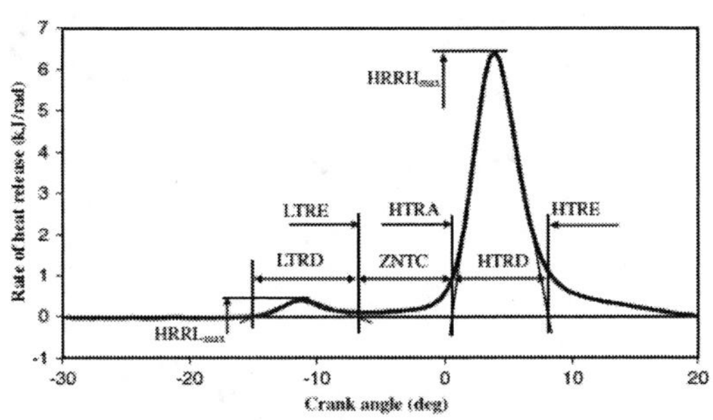

Figure 2. Definition of symbols and combustion parameters [22].

The apparent (net) heat release rate AHRR generally used in the literature is calculated by [32,33]

$$AHRR = \frac{dQ_{app}}{dt} = \frac{\gamma}{\gamma-1}pdv + \frac{1}{\gamma-1}vdp$$

Pressure-crank angle (P-φ) diagram provides information about combustion start, rate of pressure rise (pressure roughness), and maximum cylinder pressure.

The variation in in-cylinder pressure and apparent Heat Release rate are represented versus engine speed in figures 3 and 4.The speed changes from 1000 to 3000 rpm at a constant load of 200 N.m. It is noticed that there is more than one peak in P-φ history. This is a characteristic of multiple injection strategies of HCCI engines. The figures shows that the auto-ignition (during HCCI combustion) takes place in all cases. It is seen that heat is released in waves. These waves follow the injection pattern (pilot and main injections). The number of burning waves, however, is not the same in all cases, but changes with engine speed. It is directly obvious that as the engine speed increased the phasing of the main energy release shifts to later in the engine cycle. Also as the engine speed increases the magnitude of the low-temperature flame energy release becomes greater with fewer burning waves. The crank angle delay between the low temperature and the main flame energy release becomes longer for higher speeds. It is hypothesized that the HCCI auto ignition process is largely kinetically controlled, which is to say it is time-based [16]. The crank angle is directly proportional to the engine speed.Figure4shows that LTRD, ZNTC, and HTRD based on a crank angle are observed to increase with increasing the engine speed. As a result, the angle-scale interval of chemical reactions increases. The engine speed has a remarkable influence on LTRD in spite of the other parameters.

Figures 5 and 6 dedicate the in-cylinder pressure variation recorded versus the crank angle for an engine speed of 3000 rpm and different loads of 50, 100, 150 and 200 N.m. The high-pressure rise line is shifted lower crank angle as the load increases (amount of injected fuel increases). As the mixture becomes richer (higher engine load), the start of combustion shifts towards earlier times (crank angle) due to the presence of a satisfactory amount of fuel at higher temperatures and pressure which leads to an earlier auto-ignition of charge. The maximum pressure is shown to increase as engine load increases.

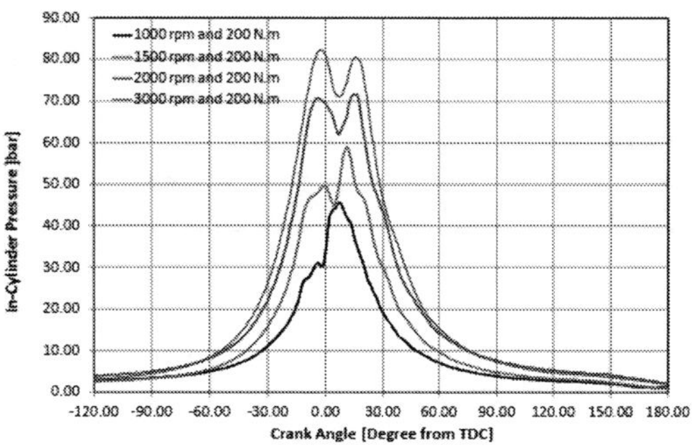

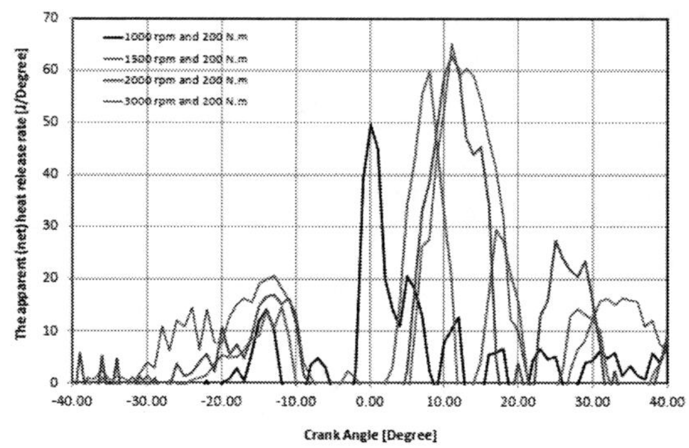

Figure 3. Cylinder pressure variation versus to crank angle for a constant load of 200 N.m and different engine speed.

Figure4. Apparent Heat Release rate versus to crank angle for a constant load of 200 N.m and different engine speed.

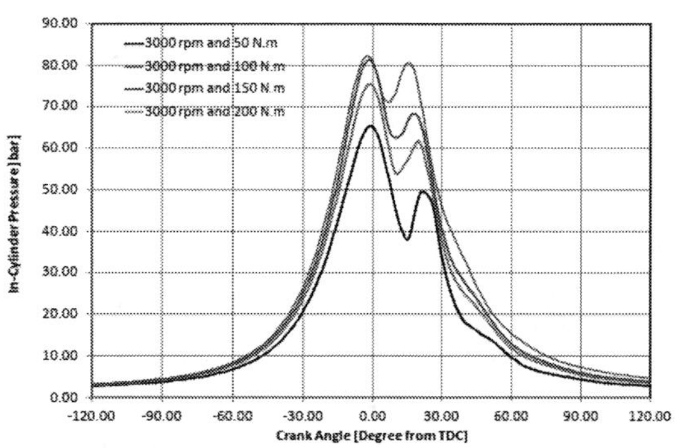

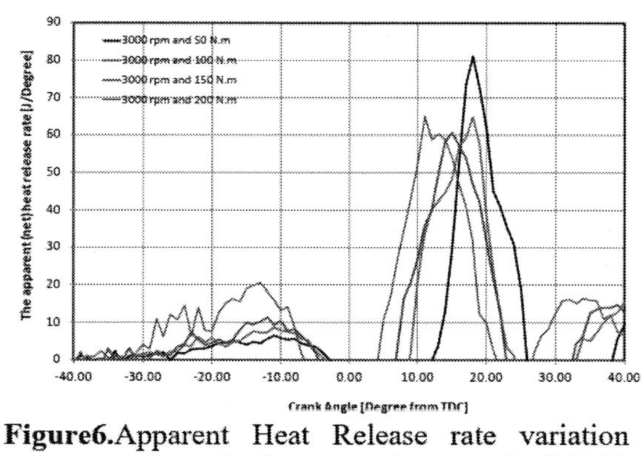

Figure 5.in-cylinder pressure variation versus crank angle for an engine speed of 3000 rpm and different loads.

Figure6. Apparent Heat Release rate variation versus crank angle for an engine speed of 3000 rpm and different loads.

Figure 3 and figure 5 indicate that the crank angle interval at any in-cylinder pressure becomes larger as engine speed and load increases so for the same injected fuel the maximum pressure and so the maximum temperature may be decreased at high loads due to the distribution of high-pressure values on larger crank angle duration. This effect may be attributed to multiple injection strategies used in HCCI engines.

The measured fuel line pressure showed a repetitive pattern which is related to the pressure wave initiated at the beginning of each injection, it depends on the lengths of fuel pipes and a common rail in addition to the fuel bulks' modulus at each condition. The disturbances in the pressure distribution, due to injector and pump flow, propagate back and forth inside of the rail, reflecting off of the ends.

Initial pressure distribution in the rail superimposes onto the pressure distribution that is generated by the next injection [34].Figure 7shows the fuel line pressure variation versus crank angle for an engine speed of 3000 rpm and different loads of 50, 150 and 200 N.m. The average value of fuel line pressure increased from 1408 bar at 50 N.m brake load to 1584 bar at 150 N.m then increased to 1610 bar at 200 N.m which may be attributed to higher wave amplitude encountered due to the increase of the injector opening duration which promote the building up of pressure wave and retard the reverse of wave travel direction which helps in damping of the wave generated at the moment of injector opening.

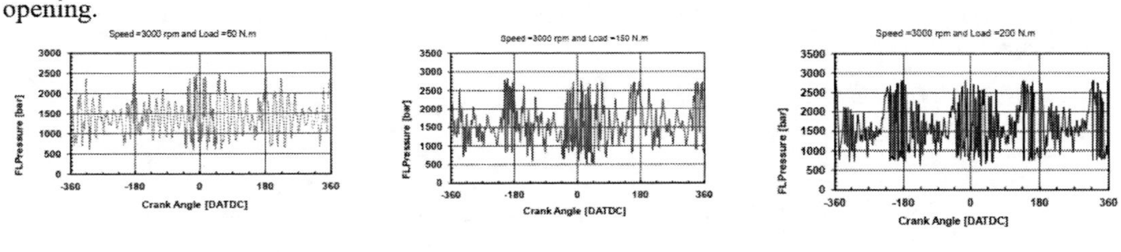

(a) Engine load of 50N.m (b)Engine load of 150N.m (c)Engine load of 200N.m

Figure 7. Fuel Line pressure variation versus crank angle for an engine speed of 2000 rpm and different loads.

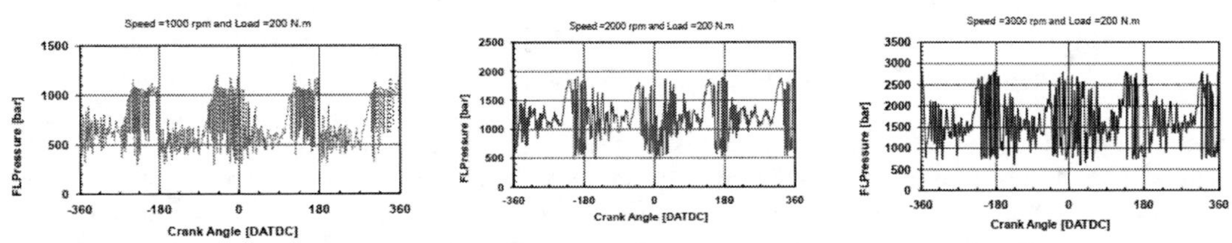

(a) Engine load of 50N.m (b)Engine load of 150N.m (c)Engine load of 200N.m

Figure 8. Fuel line pressure variation versus crank angle for a constant load of 200 N.m and different engine speed.

Figure 8 shows the fuel line pressure variation versus crank angle for a constant load of 200 N.m and different engine speeds of 1000, 2000 and 3000 rpm. The average value is found to increase from 750bar at 1000 rpm to 1150 bar at 2000 rpm to 1610 bar at 3000 rpm. This is may be attributed to the higher flow rate of the high-pressure pump at higher speed and the higher wave amplitude encountered due to the increased number of injectors opening at a higher speed. However, the fuel line pressure needs more analysis in the crank angle domain and time domain which is the matter of future work. The effect of engine speed on pressure fluctuation is more dominant than that the effect of engine load. Alzahabi[34] concluded that fluctuation in the fuel line pressure promotes the fluctuation in fuel delivery. The pressure fluctuation is not enough to cause emissions problems at engine part loads but is enough to cause problems at full load.

3.2 Engine Performance

Figures (8(a) to 13a) show the speed characteristics of the turbocharged common rail diesel engine at different speeds from 1000 to 3000 rpm. Performance curves are presented at 6 fixed external loads, namely 0.0, 50, 100, 150, 200 and 250 N.m corresponding to 0, 2.25,4.5, 6.75, 9 and 11.25 bar BMEP in preference of brake torque. The results at each load are plotted against engine speed.

For a sake of preference, the speed characteristics of a turbocharged diesel engine with a conventional direct injection system are presented in figures (8b-13b). The technical specifications of the conventional engine are attached in appendix A. Performance curves are presented at 5 fixed external loads, namely 0.0, 70, 140, 210, 200 and 280 N.m corresponding to 0, 1.56, 3.11, 4.67 and 6.22 bar BMEP in preference of brake torque [35].

The engine power shown in figure 8(a) is a direct function of engine speed and brake mean effective pressure (BMEP). This is clear from the linear dependence on speed drawn at fixed BMEP. Comparing figure 8(a) and figure 8(b) on can find that the maximum presented power is 79.57 kW at 3000 rpm (Specific power of 28.8 kW/liter swept volume) for Common rail injection engine and 82.38 kW at 2800 rpm (Specific power of 14.5 kW/liter swept volume) for the conventional injection diesel engine which means that the two engines produce almost the same power at the same speed but at completely different BMEP which is 11.25 bar for common rail and 6.22 bar for the conventional engines so it is clear that the specific power of common rail engines is much greater than that for conventional injection engines.

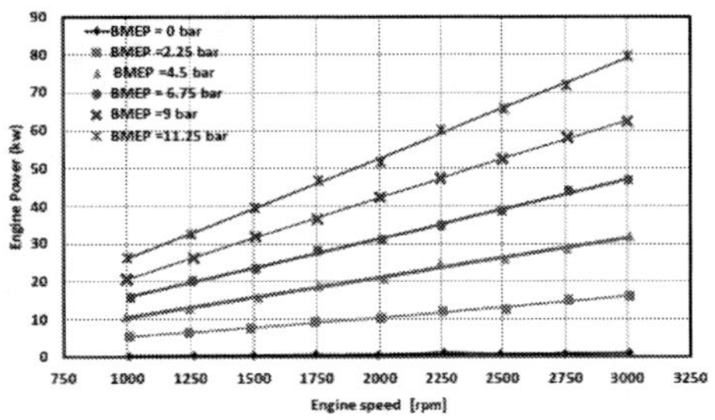

Figure 8(a). Engine power for diesel engine equipped with a common rail.

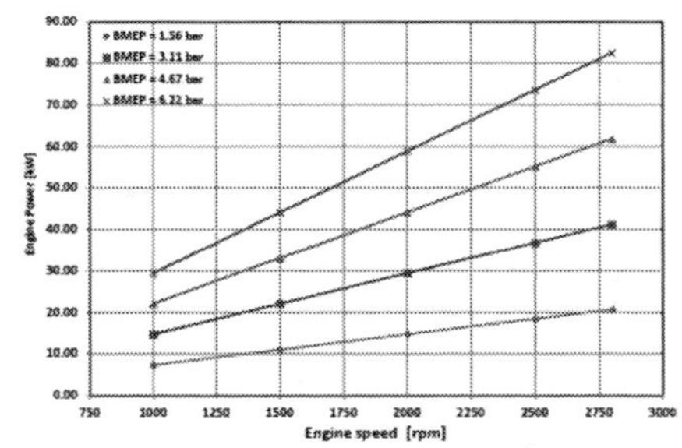

Figure 8b. Engine power for diesel engine equipped with a conventional injection system.

The equivalence ratio for the two engines is presented in figures 9(a) and 9(b). The mixture equivalence ratio increases with the brake mean effective pressure. This is due to the greater amount of fuel injected when higher loads are required. The effect of higher speed is attributed to better pump efficiency and fuel line expansion behavior. The figures reveal comparable results at the maximum load (BMEP). At lower external loads comparable results could be obtained at the same rations of $BMEP/BMEP_{max}$.

The brake specific fuel consumption (BSFC) presented in figure 10is shown to have an increasing trend with the decrease in engine speed due to the colder engine and bad charge mixing at low speeds. Then it is shown to increase again with the speed increase. This may be attributed to the higher friction losses at higher engine speed especially at part loads. The specific fuel consumption of the common rail injection engine (minimum value 210 gm/kW. hr at 2000 rpm and 11.25 bar BMEP) is much lower than that for the conventional engines (minimum value 292 gm/kW. hr at 2000 rpm and 6.22 bar BMEP). This because the displacement volume of the conventional engine is higher than that of the common rail injection engine and both working with the same fuel and equivalence ratios and the better utilization of fuel attained by high-pressure injection. The thermal efficiency is inversely proportional to the brake specific fuel consumption which is clear by comparisons of figures. 10 and 11.

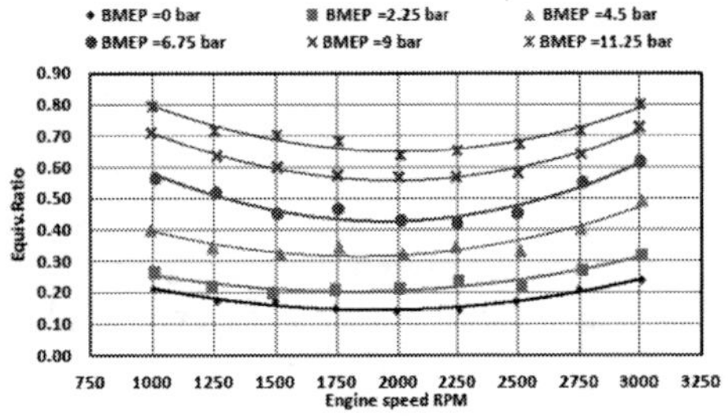

Figure 9(a). Equivalence ratio for diesel engine equipped with a common rail.

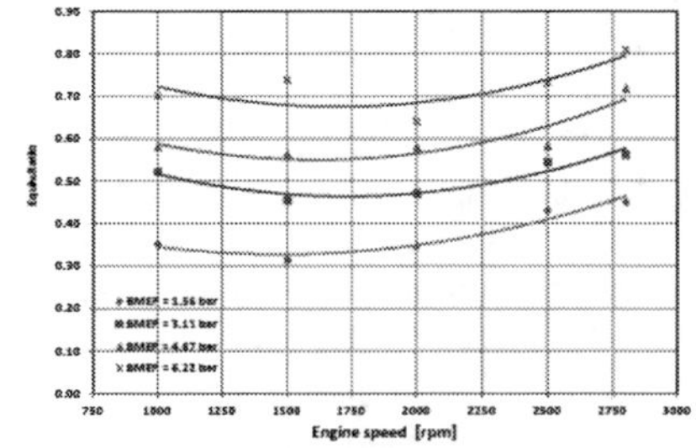

Figure 9(b). Equivalence ratio for diesel engine equipped with a conventional injection

IOP Conf. Series: Materials Science and Engineering **973** (2020) 012038 doi:10.1088/1757-899X/973/1/012038

system.

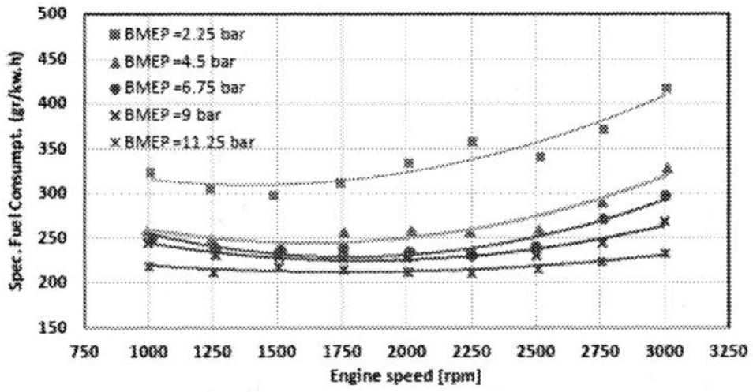

Figure10(a). BSFC for diesel engine equipped with a common rail injection system.

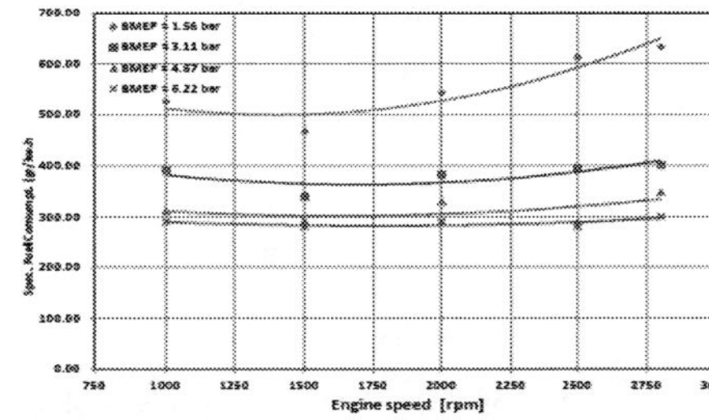

Figure 10(b). BSFC for diesel engine equipped with a common conventional injection system

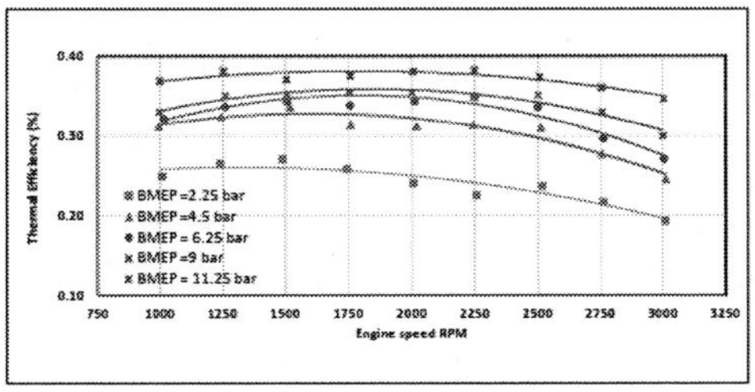

Figure 11(a). Thermal efficiency for diesel engine equipped with a common rail.

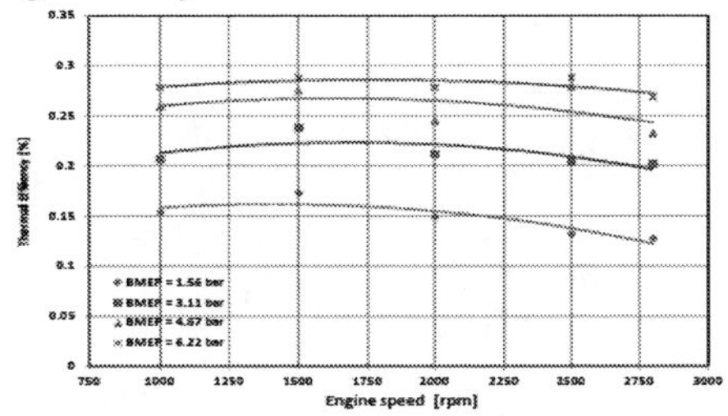

Figure 11(b). Thermal efficiency for diesel engine equipped with a conventional injection system.

Exhaust gas temperature shown in figure 12 is shown to increase with engine speed and load. The effect is explained by the less amount of heat transferred to cylinder walls as engine speed increases, and hence more heat is carried away with the exhaust. The effect of the increased load is a direct result of the more fuel injected and consequently more heat evolved during the combustion processes. The exhaust temperature of the common rail injection engine is much lower than that for the conventional engines. This may be attributed to the lean mixture that HCCI uses to lower the maximum combustion temperature resulting in reducing the engine exhaust temperature.

The volumetric efficiency calculated based on the air density achieved with turbocharging is shown in Figure 13. The volumetric efficiency of the common rail injection engine and conventional injection engine are almost similar except the engine speed at maximum volumetric efficiency which corresponds to the speed at which maximum torque occurs for both engines.

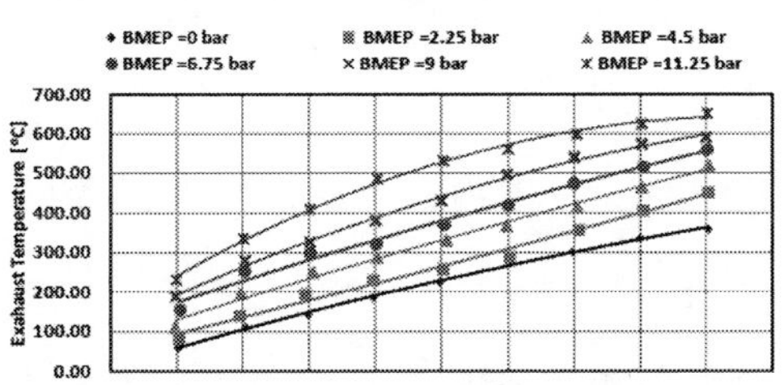

Figure 12(a). Exhaust temperature for diesel engine equipped with a common rail.

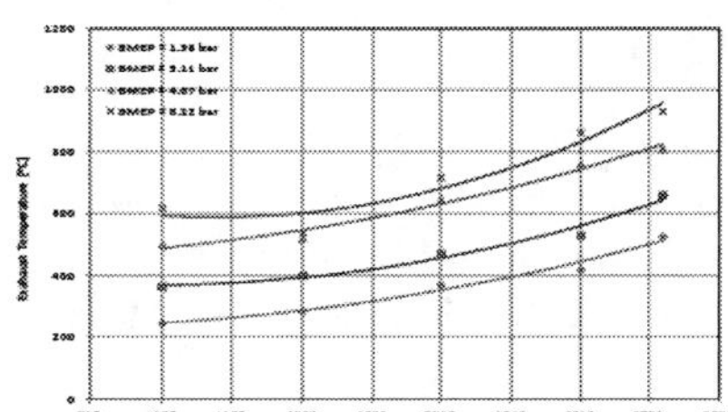

Figure 12(b). Exhaust temperature for diesel engine equipped with a conventional injection system.

IOP Conf. Series: Materials Science and Engineering **973** (2020) 012038 doi:10.1088/1757-899X/973/1/012038

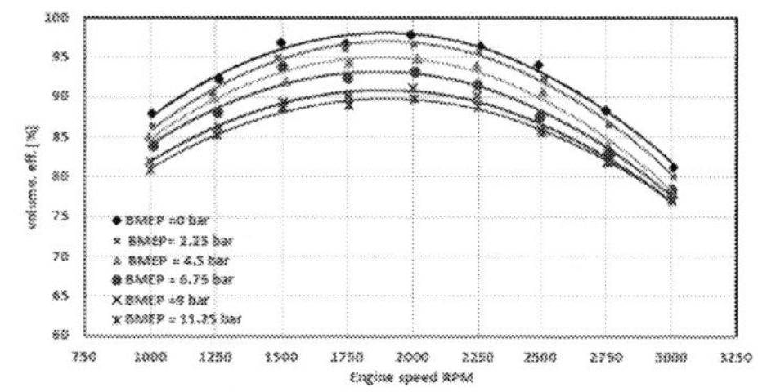

Figure 13(a). Volumetric efficiency for diesel engine equipped with a common rail injection system.

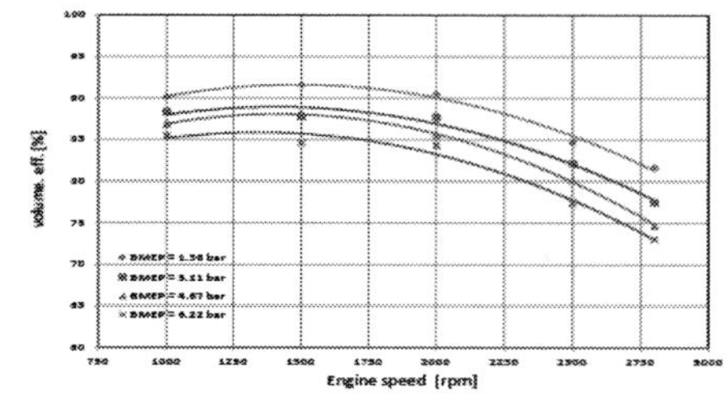

Figure 13(b). Volumetric efficiency for diesel engine equipped with a conventional injection system.

4. Conclusions

In the absence of studying the performance of HCCI engines at different loads, the objective of this work is to experimentally investigates the performance of an HCCI Diesel engine Common Rail Direct Injection (DI) fuel system. The engine performance is evaluated and presented at different loads and speeds.The analysis of in-cylinder pressure, heat release rate and fuel line pressure were performed to investigate the combustion behavior characteristics. The major conclusions from these investigations are listed below:

- The power density of the HCCI engine is much higher than of the conventional engine.
- The HCCI engine has higher thermal efficiency and lowers exhaust temperature than that for the conventional engine at all speed and loads
- The maximum in-cylinder pressure and temperature may be decreased at high loads due to the distribution of high-pressure values on larger crank angle duration. This effect may be attributed to multiple injection strategies used in HCCI engines. This may be advantageous in decreasing Nox or may give the chance to increase the engine compression ratio.
- LTRD, ZNTC, and HTRD based on the crank angle-scale increase with the increase of engine speed. So the chemical reaction intervals increase.
- As engine load increases (the mixture becomes richer), the start of combustion shifts towards BTDC side. Also as the engine speed increases the magnitude of the low-temperature flame energy release becomes greater with fewer burning waves. The crank angle delay between the low temperature and the main flame energy release becomes longer for higher speeds.
- The pressure wave amplitude generated in the high-pressure line due to the opening of the injector is shown to increase as engine speed and load increase. The effect of engine speed on pressure fluctuation is more dominant than that the effect of engine load. The fuel line pressure needs more analysis in crank angle domain and time domain which is the matter of future work

5. References

[1] Zhao H 2007 *HCCI and CAI engines for the automotive* industry (Cambridge England: Wood head Publishing Limited).
[2] Mathivanan K, Mallikarjuna J M and Ramesh A 2016 *Experimental Thermal and Fluid Science* 77 pp 337-346
[3] Yoon, S H, Kim H J and Park S 2018 *Applied Thermal Engineering* **144** pp 1081-1090.
[4] Han Z, Uludogan A, Hampson G and Reitz R 1996 *SAE* **960633.**
[5] Khandal S, Banapurmath N,Gaitonde V and Hiremath S 2017 *Renewable and Sustainable Energy Reviews* **70** pp 369–384
[6] Epping K, Aceves S, Bechtold R and Dec J 2002 *SAE Technical Paper* **2002-01-1923.**
[7] Agarwal K, SinghA P and Maurya R K 2017 *Progress in Energy and Combustion Science,* **61** pp 1-56.

[8] Han S, JaeheunKim J, and Bae C 2014 *Applied Energy* **119** pp 454–466
[9] Wang X, Huang Z, Zhang W, Kuti O and Nishida K 2011 *Applied Energy* **88** pp1620–1628
[10] Nishida K, Zhu J, Leng X and He Z 2017 *International J of Engine Research* **18**(1-2) pp 51–65.
[11] Nathan S, Mallikarjuna S and Ramesh A 2009 *SAE* **2009-26-028** .
[12] Najt P and Foster D 1983 SAE **830264**.
[13] Kong S and Reitz R 2002 *J. Eng. Gas Turb. Power ASME* **124** pp 702–7.
[14] Kong S, Marriott C, Reitz R and Christensen M 2001 *SAE* **2001-01-1026**.
[15] Iida M, Aroonsrisopon T, Hayashi M, Foster D and Martin J 2001 *SAE* **2001-01-1880/4278**
[16] Aroonsrisopon T, Foster DE, Morikawa K and Lida M 2002 *SAE* **2002-01-1924; 2002**.
[17] Iida M, Hayashi M, Foster D and Martin J 2003 *J. Eng. Gas Turb. Power ASME* **125** pp427–78.
[18] Szybist J and Bunting B 2005 *SAE* **2005-01-3723**.
[19] Garcia M, Aguilar F and Lencero T 2009 *Energy* **34** pp 159–71.
[20] Genzale C, Kong S and Reitz R 2008 *J. Eng. Gas Turb. Power ASME* **130** p 8.
[21] Maurya R and Agarwal A 2009 *SAE* **2009-01-1345**.
[22] Ebrahimi R and Bernard Desmet B 2010 *Fuel* **89** pp 2149–2156
[23] Hadia F, Wadhah S, Ammar H and Ahmed O 2017 *Case studies in thermal engineering* **10** pp 262-271.
[24] Alper Calam A, Solmaz H, Icingur Y and Yilmaz E 2018 *Energy* **168** pp 1208-1216
[25] Shenawy E, Elkelawy M, Bastawissi H A, Panchal H and Shams M 2019 *Fuel* **249** pp 277–285
[26] Ganesh D and Nagarajan G 2010 *Energy* **35(1)** pp148–57.
[27] Yu S, Li L and Zheng M 2019 *Fuel* **242** pp 243–54.
[28] Liu H, et al 2018 *Fuel* 231 pp 26–36.
[29] Lu X, Han D and Huang Z 2011 *Prog. Energy Combust Sci.* **37(6)** pp741–83.
[30] An Y et al 2018 *Energy* **158** pp181–91.
[31] Tanaka S, Ayala F, Keck J and Heywood J 2003 *Combustion and Flame* **132** pp 219–239
[32] Asad U, Zheng M 2008 *International Journal of Thermal Sciences* **47** pp 1688–1700
[33] Fathi M, Saray R K, Checkel M D 2010 *Fuel* **89** pp 2323–2330
[34] Alzahabi B and Schulz K 2008 *Int. Jnl. of Multiphysics* **2** Number 3
[35] Elmaihy A 1997 MSc. thesis, Military Technical Collage.

Appendix A: Engine Technical Data

Make and Model	Mercedes 352 seriesNatural aspiratedDiesel engine
Compression ratio	17: 1
No. of Strokes	4
No. of cylinder	6
Arranging	In-line
Cooling	Water
Bore	97 mm
Stroke	128 mm
Combustion chamber	Open type, Direct Injection
Camshaft	Sided
Speed range	800-2800 rpm
Maximum power	120 HP at 2800 rpm
Maximum torque	28 kp.m at 1600 rpm
Static injection	23 CA BTDC
Firing order	1 5 3 6 2 4 1
Min.compression pressure	20 bar at 150-200 rpm
Injector opening pressure	200 bar

IOP Conf. Series: Materials Science and Engineering **973** (2020) 012039 doi:10.1088/1757-899X/973/1/012039

Experimental investigation and numerical simulation of the thermosyphon heat pipe charged with R134a

S Y Zakaria[1], A G Ibrahim[2], A M Rashad[3]and S E El-shamarka[4]
[1] MSc Researcher, Egyptian Armed Forces, Czakareya@yahoo.com.
[2]Assoc. Prof., MTC, Cairo,agmibrahim@hotmail.com.
[3]Assoc. Prof., MTC, Cairo, rashad@mtc.edu.eg.
[4]Prof.,Former Commandant of MTC, Cairo,selshamarka@mtc.edu.eg.
czakareya@yahoo.com

Abstract: This work presents an experimental and numerical study of Two-Phase Closed Thermosyphon (TPCT) heat pipe filledwith R134a as a base fluid. For this purpose, a test rig was designed and developed toperform a series of tests on the TPCT heat pipe by applying different heat inputs at the evaporator section. The surface temperature along the TCPT heat pipe and temperature changes of the cooling water across the condenser section were measured. The influence of the TCPT heat pipe operating conditions on its thermal performance was reviewed. The considered parameters were; heat pipe transport capacity, conductance, both heat transfer coefficients of the evaporator and condenser sections and the thermal efficiency. Then, a CFD model was developed to investigate the two-phase flow and heat transfer mechanism during the transient and steady-state operation of the TPCT heat pipe. The results of the experimental work were used to validate the CFD model and acceptable agreements were noticed. The validated CFD model is utilized to predict features of the mass and heat transfer processes, as well as the nucleate pool boiling and the liquid film condensation phenomena during TPCT heat pipe operation.

1. Introduction

Over the past decades, the applications of the thermosyphon heat pipe into thermal engineering systems are known, primarily in air conditioning systems, waste heat recovery, electronics cooling, chemical engineering, heat exchangers, water heater, solar collectors and power generation. The Two-Phase Closed Thermosyphon (TPCT) heat pipes have high effective coefficient thermal conductivity; Which can be significance orders greater than those of highly conductive solid materials, such as copper[1]. It consists of a tube or enclosed pipe consisting of a material, which is suitable with the working fluid, without wick structure lined on the inside pipe wall as shown in figure 1. It is partlycharged with a working fluid according to a specific filling ratio and then bolted; the mass of working fluid is selected, so that the heat pipe comprises both vapor and liquid through the operating temperature range. The degradation and failures of TPCT heat pipe thermal performance can occur in the sealed container wall, if any of the parts are not compatible (including the working fluid). The TPCT heat pipes are considered as passive heat transfer devices and constructed of three main sections:

- The evaporator section where the heat load is applied.
- The adiabatic section which is in the middle of the TPCT heat pipe.
- The condenser section where the heat load is rejected.

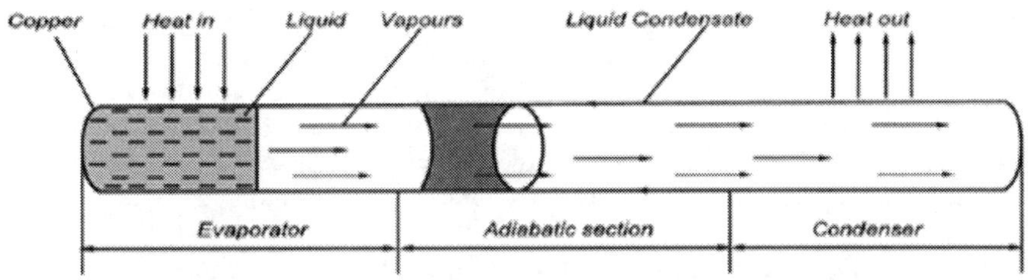

Figure 1. The construction for a TPCT heat pipe, [2].

The main advantages of TPCT heat pipe are low manufacturing cost, high heat fluxes transport and simple construction rather than the other heat pipes. For low temperature applications, different refrigerants like R404, R134a, R22, R410a and ammonia which are selected as working fluids with compatible metals of sealed shell as steel, aluminum and copper are used. For medium temperature applications, water is considered a compatible working fluid for temperatures from30 °Cto300 °C, with good compatibility with different metals counting stainless steel and copper. For high temperature applications, different organic fluids and liquid metals are selected as working fluids for operating temperature above 300 °C. As heat pipe principle of work depends on the working fluid phase change to transfer heat from heat source to heat sink or heat distribution, working fluid charged in heat pipe which has an important role to consolidate the heat pipe thermal performance[3]&[4]. Concerning the importance of TPCT heat pipe applications, several researches have been carried out to investigateits thermal performance with severaldesign parameters, base fluids, materials, geometries, filling ratios and different operating conditions, heat load, inclination, etc. .

Naphon et al.[5]presented an experimental study by using a fabricated heat pipe with 600 mm length and 15 mm diameter from straight copper tube. The heat pipe is tested with alcohol, de-ionic water and alcohol-titanium nanofluids. The reported work indicated the influence of heat pipe tilt angle, filling ratio of working fluid and concentrations of nanoparticles volume on the performance of heat pipe.Results showed that an enhancement in thethermal efficiency of heat pipe with utilizing nanofluids as working fluids compared to use of based fluids.

Tsai et al. [6]compared the common steady-states test with a novel dynamic test method to investigate; shapes of heat pipes, filling ratios and bending angle showing their effect on performance. Experimental results indicated that the heat pipe performance would deteriorate considerably due to deformation of heat pipes. The operation limitations were increased due to larger filling ratios.

Attia and El-Assal[7]presented experimental study on heat pipe filled with methyl alcohol and water to examine heat pipe thermal performance at different filling ratios, and solution of propylene glycol and water were tested to investigate the influence of using surfactant at two concentrations.The results showed that using water as working fluid indicates much better thermal performance than using methyl alcohol. The filling ratio 30% was optimum for most tested working fluids. Using surfactant with water was much better than using pure water.

Wang et al. [8]presented an experimental work to examine an inclined miniature mesh heat pipe charged with water-CuOnanofluid as working fluid. The aim of this workwas to show the effects of operating temperature and the inclination angle on heat pipe performance. Results showed a considerableinfluence on heat pipe thermal performance charged with water-CuOnanofluiddue to inclination angle.The best results were obtained at 45° inclination angle of heat pipe.

Peyghambarzadeh et al [9]reported an experimental work to study a multi-evaporator heat pipe charged with three working fluids (methanol, water and ethanol). Low heat flux inputs are applied at evaporator section (up to 2500 W/m^2), a water coolant is circulating at constant temperature across condenser section with three levels 15, 25 and 35 °C. Experimental results showedthat evaporator heat transfer coefficient was increased with increasing heat flux input. It was noticed for methanol at high heat flux input, the heat transfer coefficient was degraded due to dryout effect at evaporator section.

An improvement in heat transfer coefficients were acquired for ethanol and water compared to methanol anddegradationin the thermal resistance of heat pipe with increasing the inclination angle.

Aniket and Ravindra[10]focused on operating parameters effect on TPCT heat pipe performance parameters such as heat load, coolant mass flow rate, aspect ratio, inclination angle and filling ratio. The novel heat transfer techniques were illustrated such as CFD modeling and analysis, resurfacing and use of ultrasonic wave. This study indicated the need of mathematical modeling of thermosyphon thermal operation with new efficient and minimum global worming potential refrigerants.

Fadhl et al. [11]presented a numerical simulation model of a thermosyphon heat pipe charged with water as base fluid using the volume of fluid (VOF) method in FLUENT ANSYS. Thismodel simulated the two-phaseflow, mass and heat transfer processes over heat pipe operation. The numerical model wasverified using experimental data at the same condition, and a good agreement was observed.

Abou-Ziyan et al [12]Noticed during their studies that the adiabatic mean temperature is very similar to the saturation temperature ofworking fluid. EL-Genk and Huang [13]reportedthe saturation temperature at the end of adiabatic section. Those results were utilized for computation the evaporator and condenser heattransfercoefficients.

Noie[14]investigated experimentally a vertical copperTPCT heat pipe on steady-state condition of total length 980 mm, outside diameter 32 mm and thickness 3.5 mm charged with distilled water. According to input heat rates from 100 to 900 W, filling ratio from 0.3 to 0.9, and the evaporator lengths for aspect ratios7.45, 9.8, and 11.8. The wall surface temperature of the TPCT heat pipe, the applied heat to evaporatorsection and rejected heat from condenser section were defined, also. A comparison of experimental evaporation heat transfer coefficient with current correlations was presented. The results showed the optimum filling ratio for a specific aspect ratio, that the TPCT heat pipe is operating at its best.

Wang et al.[15]performed a CFD model to stimulate the geyser boiling conception in a TPCT heat pipe. The commonly used Lee modelis enhanced by presenting the regard of superheat to raise the predicted performance. CFD simulation of heat transfer process was achieved using both original and enhanced Lee model using VOF technique.These results were evaluated and compared to those from experiments. From experimental results, the cycle time of 80-100s is occurred at 400 mL/min at (95°C) flow rate of heating water and 300 mL/min at (30°C) of cooling water. The conformable modeling resulted in agreement with the experimental data at same conditions proposed that the enhanced Lee model has a better expectation performance.

Alizadehdakhelet al.[16]carried out a numerical model using VOF method to investigate the vapor/liquid two phases interaction. Experiments were performed at severalworking conditions. The predicted wall temperatures gradient from the numerical model showed a good agreement with those from experimental tests results. This concluded that the usage of CFD as a proposed technique to simulate and clarify the two-phase flow and heat transfer process in heat pipes.

Asmaie et al.[17]performed CFD model of TPCT heat pipe filled with two base fluids deionized water and CuO/Water nanofluid. The results showed that, the ability of thermosyphon to transport heat load is increased about 46 % higher with usage of nanofluid as operating fluid than that of water, Moreover the volume concentration of 1 wt. % was found as the best concentration, and a reduction in thermosyphon heat pipe wall temperature was observed with increasing the nanofluid concentration in base fluid.

As yet, there isneeding of new studies and researches to explain the heat transfer features of TPCTheat pipe to improve its thermal performance in different thermal engineering applications. This study investigates, experimentally and numerically, the thermal performance of a TPCT heat pipe. Different performance parameters, which have a substantial effect on application of TPCT heat pipe in engineering systems, namely; heat transport capacity, heattransfer coefficients, thermal conductance and overall thermal efficiency are reported under different heat loads.

IOP Conf. Series: Materials Science and Engineering **973** (2020) 012039 doi:10.1088/1757-899X/973/1/012039

This article is organized as follows. The of heat pipe performance indices are summarized in the next section, the general design of the experimental test stand and experimental results are described. Then, the development of numerical model and validation of simulation results are presented. Finally, the main findings of present work conclude the paper.

2. Performance parameters

The considered parameters utilized to estimate the thermal performance of the TPCT heat pipe are summarized as follows:

2.1. Heat transport capacity

The heat transport capacitywas specified by determining the mass flow rate, temperature changes of coolant water through the condenser section.

$$\dot{Q}_{out} = \dot{m}\, C\, (T_{out} - T_{in}) \qquad (1)$$

2.2. Heat transfer coefficients

At the evaporatorsection, from experimental measurements theevaporatorheattransfer coefficient is definedby the equation:

$$h_{ev} = \frac{\dot{Q}_{in}}{A_{ev}\ (T_{ev} - \tilde{T}_{ad})} \qquad (2)$$

At the condenser section, from experimental measurements the condenser heat transfer coefficient is defined by the equation:

$$h_{con} = \frac{\dot{Q}_{in}}{A_{con}\ (\tilde{T}_{ad} - T_{con})} \qquad (3)$$

2.3. Thermal conductance

The thermal conductance represents the ability of thermosyphon to transport input heat. It is computed using the equation:

$$c = \frac{\dot{Q}_{in}}{(T_{ev} - T_{con})} \qquad (4)$$

2.4. The overall thermal efficiency

The overall thermosyphon thermal efficiency is computed using the equation:

$$\eta_{th} = \frac{\dot{Q}_{out}}{\dot{Q}_{in}} \qquad (5)$$

3. Experimental work

3.1. Experimental set-up

The experimental system was built to measure the surface temperature profiles of the TPCT heat pipeat variable operating conditions. Figure 2presentsthe scheme of the experimental arrangement forthetested TPCT heat pipe in vertical position. The tested TPCT heat pipe is made from standard red copper tube. The tube has 1400 mm total length, 20 mminner diameter and 1 mm wall thickness. The investigated heat pipe is filled with R134a. Closure of the heat pipe is made by soldering brass cap at both ends. The TPCT heat pipe is consisted of three main sections; evaporator, adiabatic and condenser sections. The specifications of the tested thermosyphon are listed in table 1.

The evaporator section is located in lower end of the pipe; the thermal power is supplied by mean of a wire heater of Nickel – Chrome. The electrical resistance (3.334 ohm/m) is uniformly coiled the evaporator section surface such that to guarantee it is not close to thermocouples. The input electric power to electric tape is measured in each experiment and adjusted via a Variac resistance. The thermosyphon evaporator section is insulated by a high-density wool glass isolator to minimize heat losses during experiments. The evaporator section cap end is insulated. The temperature at the surface of wool glass isolator is measured to determine the heat losses by natural convection at the evaporator section. The evaporator wall surface temperature is measured by means of three calibrated (NiCr - Ni) K – type thermocouples. The thermocouples are embedded on the evaporator outside wall surface and spaced on specific locations on the evaporator circumference along its length.

Similarly, the adiabatic section in-between evaporator and condenser sections is insulated by high density wool glass insulator and its wall temperature surface is measured by two spaced calibrated (NiCr - Ni) K - type thermocouples.

In the top end of the pipe the condenser section is located. A glass cooling water jacket of length 350 mm, 40 mm inside diameter and 2 mm thickness insulated by high density wool glass is mounted around the condenser section overall length as the cooling of condenser is accomplished by circulating water through the glass jacket. The temperatures of water inlet and outlet are measured by two calibrated (NiCr - Ni) K - type thermocouples. The water mass flow is adjusted and measured at each experiment. To ensure steady flow rate of water to the condenser cooling water jacket during experiments, an overhead tank is used. Also, the condenser wall surface temperature is measured by three spaced calibrated (NiCr - Ni) K - type thermocouples.

All the reading measurements of the eight thermocouples that are embedded on the outside wall surface of three sections of the thermosyphon are logged by lab view program through two analog data acquisition cards for thermocouples. Figure 3shows the main components of the tested thermosyphon, while figure 4shows the locations of thermocouples along the thermosyphon.

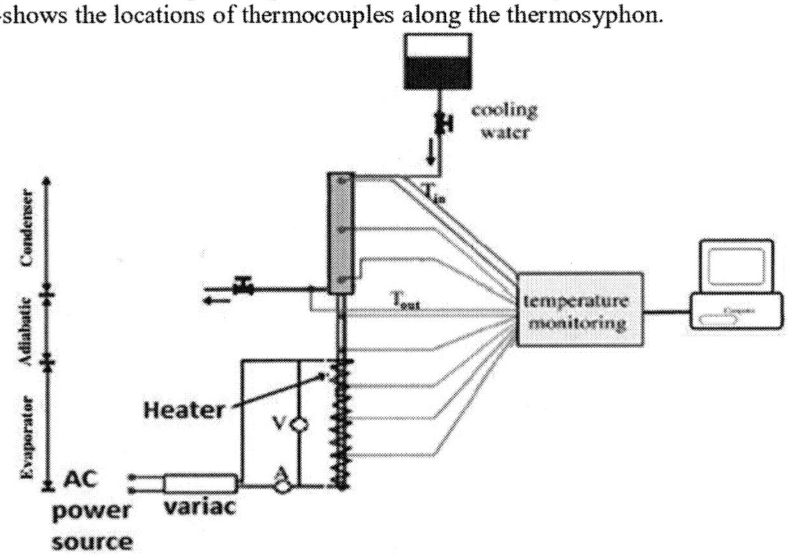

Figure 2. A scheme of the experimental arrangement.

Table 1. The specifications of the tested thermosyphon.

Material	Red copper
Total length	1400 mm
Evaporator length	550 mm
Adiabatic length	500 mm
Condenser length	350 mm
Outside diameter	22 mm
Wall thickness	1 mm
Filling ratio	100%
Working fluid	R134a
Working fluid temperature range	- 40:100 °C

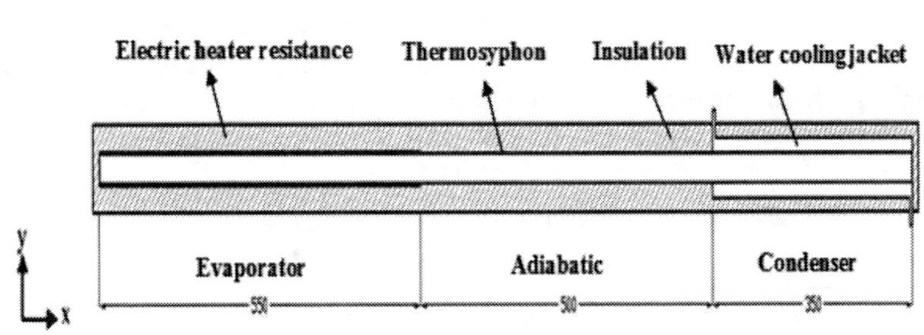

Figure 3. A longitudinal cross section of thermosyphon heat pip.

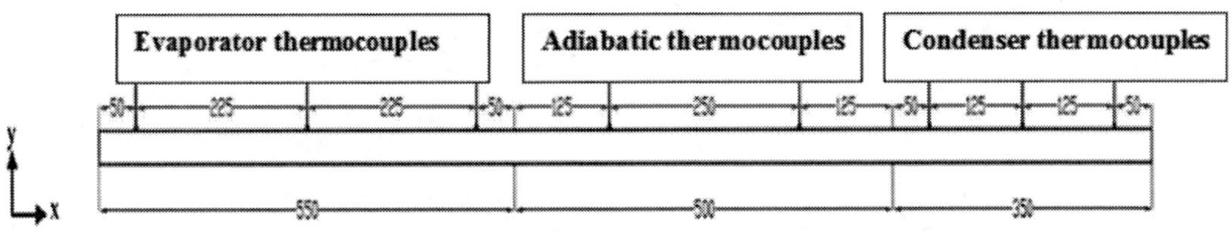

Figure 4. Locations of thermocouples along the thermosyphon heat pipe.

3.2. Test procedure

The water flow rate was adjusted to the required value at the beginning of each steady-state test, and then the electrical heat load was set to the evaporatorsection. A constant working pressure was achieved by balancing the rate of heat load added to the evaporator section to heat rejected from the condenser section. After that, the equipment waspermitted to stabilize about 20 min before recording any readings. By using the two analog data acquisition cards,all temperatures from the thermocouples were registered. As soon as, a steady state was established within (15 - 20) min, thermocouples temperature and power readings were recorded. For different power inputs from 30.4 to125.4 W, the process was repeated. This range was investigated in order to ensure reliableperformance of the TPCT heat pipe within a prolonged workingperiod. To ensure that there was no degradation of the TPCT heat pipeperformance was occurredalong operation, the 2-h observing period was considered.

3.3. Experimental results

The experimental study of the performance characteristics for a TPCT heat pipe, according to different heat loads to evaporator section, includes; thermosyphon transient response, output heat at condenser section, surface temperature profilethroughout the thermosyphon, heat transfer coefficient at both evaporator and condenser sections,conductance of thermosyphon and overall thermal efficiency. The experimental results are presented in next sections.

3.3.1 Heat transport capacity. The experimental technique is imitated by adjustment of the cooling water flow rate across the condenser section, to accomplish a reasonable temperature variation across it during whole test period using overhead tank, then applying the desired input power to the warred electric heater along evaporator surface. The applied heat load to evaporator section is increasedsteadily, to begin a new test. All test parameters are evaluated and registered once the steady-state is established. Figure 5shows output heat from condenser section as a function of applied input heat to evaporator section.The output heat rises with increasing input heat.

3.3.2 Surface temperature. The surface temperatures at the different sections of TPCT heat pipe are shown in figure 6 for different heat loads from 30.4 to 125.4 W. The temperature over the outer surface of the TPCT heat pipe was recorded at eight different locations.The surface temperature

profile according to each constant input heat load is obtained at the steady state condition and it was recorded for 2 h to grantee the stabilization thermal operating cycle of the heat pipe.
As depicted from figure 6the surface temperature in the thermosyphon sections is almost uniform; the difference in temperature is within (3-10) °C.

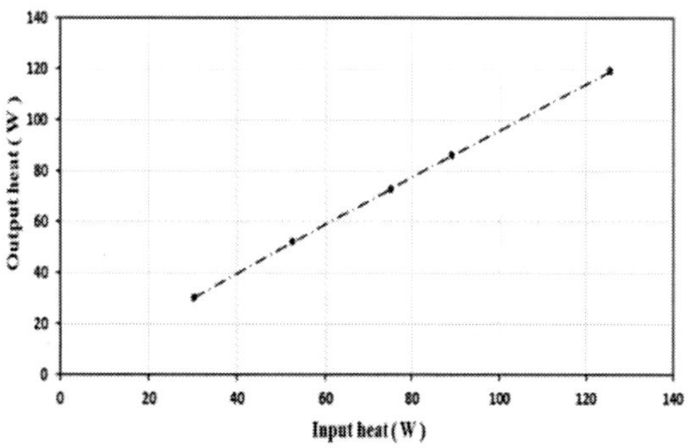

Figure 5. Heat transport capacity according different heat inputs.

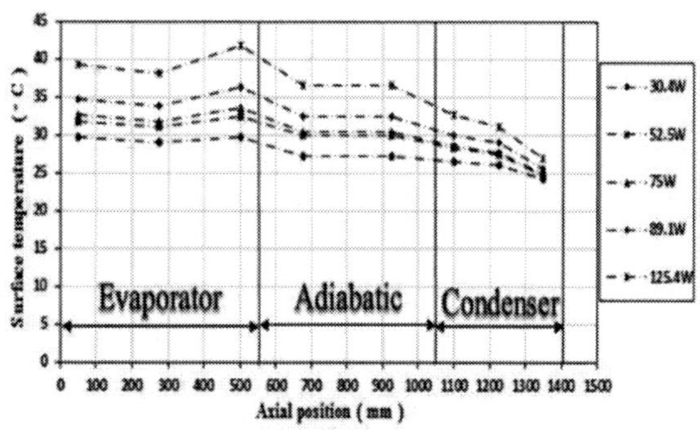

Figure 6. Surface temperature profiles along the thermosyphon.

3.3.3 Heat transfer coefficients. Figure7 and figure 8show the evaporator and condenser sections heat transfer coefficients respectively against the input-heat at evaporator section. As shown in these figures, at constant working pressure, the heat transfer coefficients of the evaporator and condenser increase by increasing the input heat into the evaporator section.

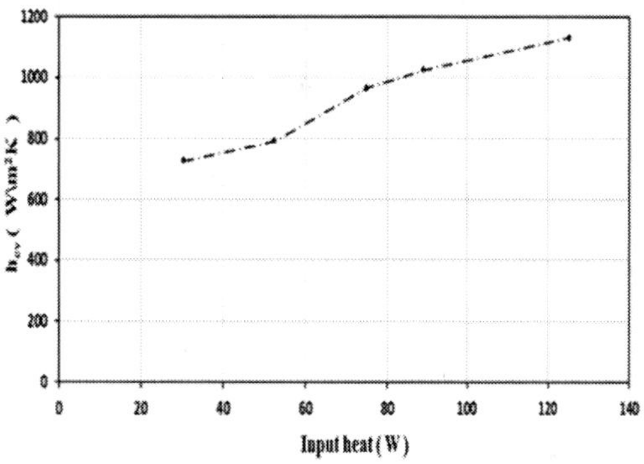

Figure 7. Evaporator heat transfer coefficient according to several heats input.

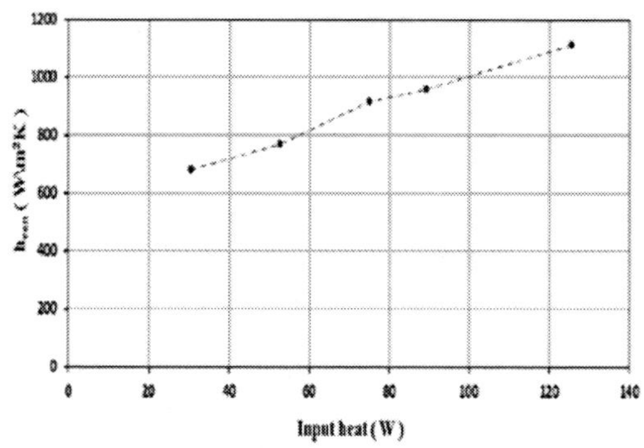

Figure 8. Condenser heat transfer coefficient according to several heats input.

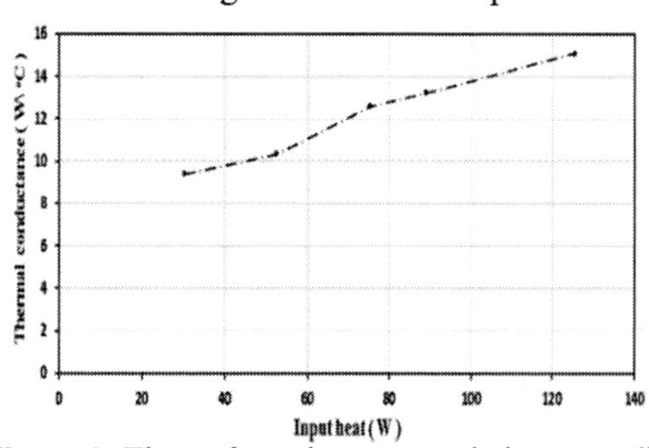

Figure 9. Thermal conductance variation according to different heat inputs.

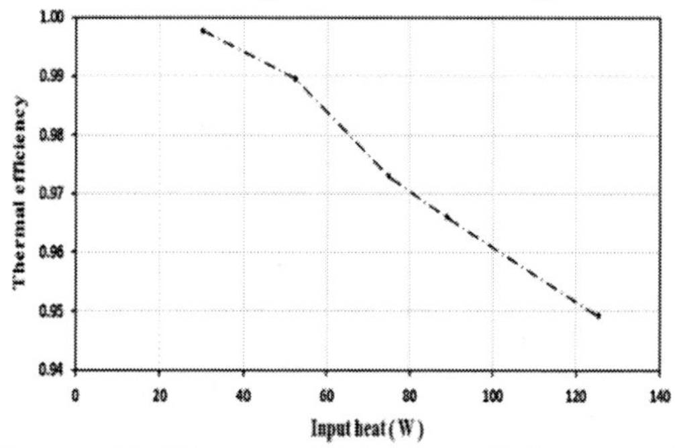

Figure 10. Thermal efficiency variation according to different heat inputs.

3.3.4 Thermal conductance. Figure 9 shows the thermosyphon thermal conductance variation with different heat inputs at evaporator section. It increases with increasing input heat at evaporator section, showing the ability of thermosyphon to transport input heat.

3.3.5 Thermal efficiency. Figure 10 shows the overall thermosyphon thermal efficiency according to differentheatinputs at evaporator section. It was shown that thermosyphon thermal efficiency decreases with increase input heat at evaporatorsection.

4. Numerical simulation

In the current work, the numerical simulation of TPCT heat pipe is performed by CFD approach. The commercial package ANSYS FLUENT 18.1 [18] and the VOF method were applied for the modeling of TPCT heat pipe. Throughout the functioning of TPCT heat pipe,the phase changes continuously from liquid to vapor phase and vice versa due to pool boiling inside the evaporator section and film condensation inside the condenser section, respectively. The VOF method defines the volume fraction in the cells of the simulated domain for each phase (liquid / vapor). To simulate phase change processes, customized mass and energy source terms to the offeredNavier-Stokes equations in the ANSYS FLUENT 18.1 kit were employed.

For the evaporation process, to compute mass quantity released from the liquid phase and the mass quantity additional to the vapor phase, two mass source terms were used. Similarly for the condensation process, two mass sources specifying the vapor mass transferquantity to the liquid phase were applied.

For heat flow, only one source term for two phases was employed in the evaporation or condensation processes. The heat transfer was defined by multiplying one mass source term by the latent heat of evaporation. Additionally, instead of liquid or vapor temperature a mixture temperature was presented, because of the VOF method combines parameters such as temperature and velocity with the mixture phase, not with a specified one.

4.1. Computational domain

In order to simulate two phase heat transfer process in tested TPCT heat pipe, a two-dimensional model was produced. The total length 1400 mm, 22 mm andwall thickness 1 mm of closed copper tube were applied to the model geometry. The lengths of the evaporator, adiabatic and condenser sections were represented as 550,500 and 350 mm, respectively.

4.2. Mesh generation

The two-dimensional geometry is meshed using GAMBIT's grid generation software. The evaporator section contains 26,400 cells, the adiabatic section contains 24,000 cells and the condenser section contains 16,800 cells. As a result, 67,200 Quad cells are generated for liquid regions. Nearby to the TPCT heat pipe inner walls, fifteen layers of cells wereutilized to take the thin liquid film that obtains near the walls. The size of first grid was 0.01 mm with growth ratio 1.2, as shown in figure 11.

Various mesh sizes were applied to test grid independence as shown in figure 12. The mean surface temperatures of the evaporator ($\widetilde{T_{ev}}$), adiabatic ($\widetilde{T_{ad}}$) and condenser ($\widetilde{T_{con}}$) sections for various sizes of mesh were compared and shown in 'Table 2' for input heat load 30.4 W. It was observed that nearly the similarmean surface temperature differences at different sections were acquired for various mesh sizes. Hence, the mesh size of 67,200 Quad, Map cells is chosen for the modeling study.

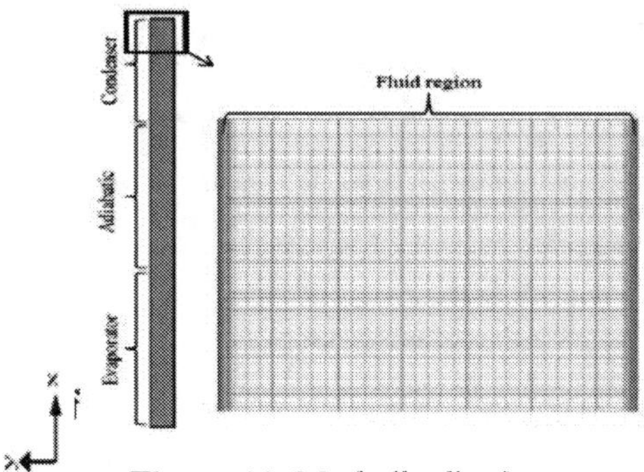

Figure 11. Mesh distribution.

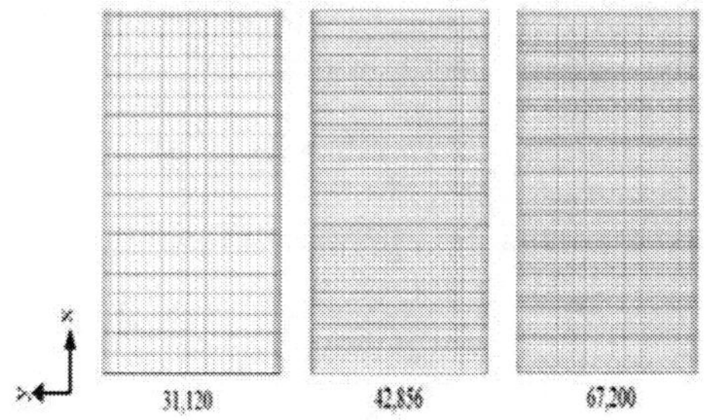

Figure 12. A section of the computational meshes.

Table 2. Grid independence results for the TPCT heat pipe charged with R134a for heat input of 30.4 W.

Mesh size	31,120	42,856	67,200
$\widetilde{T_{ev}}$ °C	29.85	29.75	29.77
$\widetilde{T_{ad}}$ °C	25.22	25.22	25.22
$\widetilde{T_{con}}$ °C	24.02	24.02	24.02

4.3.Boundary conditions

A constant heat flux at the evaporator wall, a zero-heat flux at adiabatic wall, the upper and lower caps of the TPCT heat pipe assuming these regions are insulated and a convection heat transfer coefficient at condenser wall, were defined as boundary conditions.Ano-slip boundary condition is defined at the inner walls of the TPCT heat pipe.It is consideredthat the filling ratio 100%, i.e.FR = 1.0, for R134a charged the TPCT heat pipe. By using the mesh adaption ability in ANSYS FLUENT 18.1wasutilizedto describethe filling ratio at the evaporator section.

Referring to experimental results, the temperature variation along evaporator and condenser sections was restricted to 10 deg. Hence, the thermo-physical property of R134a is supposed to be temperature-independent to decrease thecomputationaltime.These properties are shown in 'Table 3'. The density of the liquid phase (ρ_l) is defined as:

$$\rho_l(\mathrm{T}) = \sum_{\mathrm{i}=0}^{\mathrm{n}=4} \mathrm{C_i} \cdot \mathrm{T^i} \quad (6)$$

The influence of surface tension (σ_{lv}) over the interface between the liquid-vapor phases is comprised in the model by applying the following correlation:

$$\sigma_{lv}(T) = \sum_{\mathrm{i}=0}^{\mathrm{n}=3} \mathrm{B_i} \cdot \mathrm{T^i} \quad (7)$$

Where $\mathrm{C_i}$ and $\mathrm{B_i}$ are shown in table 4.

Table 3. Thermo-physical properties of R134a, [19].

Physical property	Value	Unit
Latent heat of evaporation	177.79	kJ/kg
Density of vapor phase	32.35	kg/m^3
Specific heat of liquid phase	1.4246	kJ/kg.K
Specific heat of vapor phase	1.0316	kJ/kg.K
Thermal conductivity of liquid phase	0.081134	W/m.K
Thermal conductivity of vapor phase	0.013825	W/m.K
Viscosity of liquid phase	1.9489×10^{-04}	kg/m.s
Viscosity of vapor phase	1.1693×10^{-05}	kg/m.s
Molecular weight	102.03	kg/kmol
Critical temperature	374.21	K
Critical pressure	40593	kPa

Table 4. Density and surface tension correlations for R134a, [19].

		Density (ρ_l)		
C_0	C_1	C_2	C_3	C_4
3952.801	- 25.9914	0.09482	- 1.290x10^{-04}	0
		Surface tension (σ_{lv})		
B_0	B_1	B_2	B_3	-
0.04929	- 8.34x10^{-05}	- 3.95x10^{-07}	7.071x10^{-10}	-

4.4. Solution procedure and convergence criterion

With a time step 0.001 s,a transient simulation is performed to model the dynamic behavior of the liquid-vapor flow and the evaporation and the condensation processes. This time step was designated according to the courant number, representing the ratio of the time step to the fluid time fluid needs to move through a cell. In this CFD model, the interaction between two phases was detected by solving the continuity equation to determine each phase volume fraction, and the energy equation was solved to observe the temperature through the computational domain. The flow is assumed to be laminar. FLUENT offers several algorithms for pressure-velocity coupling. The SIMPLE algorithm for pressure-velocity coupling and a first-order upwind scheme for the definition of momentum and energy were chosen in the model.

When the scaling residuals are $10^{-5}$for volume fraction and velocity components and $10^{-6}$for temperature variable, the numerical calculation is converged. Through the CFD computational domain the governing equations are solved such that variables of the governing equations are defined by the entity of the particular phase in each cell in the computational domain. The primary and secondary phases are described as vapor and liquid phases, respectively. In the material properties format the saturation temperature and the latent heat of evaporation are defined.

4.5. CFD simulation results

4.5.1 Heat transfer process. The operating conditions were set to saturation values (25 °C)at the start of heating process. At the evaporator section a constant heat flux is defined, evaporating heat transfer keeps on the wall of the evaporator section. The generated vapor moves across the adiabatic section to reach the condenser section due to that a high temperature region shows in the condenser section. The region near the condenser interior wall has a lower temperature than the central region that causes vapor condensing over the interior wall of the condenser section. After 287 s the steady state is achieved as the temperature appearedintothe TPCT heat pipeseems uniform, figure 13shows temperature distribution contour inside TPCT heat pipe.

4.5.2 Evaporation process. During the evaporation the pool boiling process occurs within the evaporator section was visualized. Figure 14illustrates the volume fraction contour distribution within the evaporator section,the blue color represents liquid phase with value 0 of the vapor volume fraction and the red color represents the vapor phase with value 1of the vapor volume fraction.

A localized natural convection flows can be seen at the lower half of the evaporator regionbecause of a minor rise of the working fluid saturation pressure / temperature. The adjacent liquid film to the wall starts to boil, the neighboring liquid film to the wall begins to boil therefore essential radii are surpassed at the local nucleation places so that persistent nucleation takes place. Vapor bubbles begin to form at these places where the liquid boils. Then, the isolated vapor bubbles develop and float to the top of the liquid pool before decomposition and release of their vapor quantity. Through this cycle, the volume fraction of the liquid phase reduces while the volume fraction of the vapor phase increases in the region of the evaporator.

4.5.3 Condensation process. Following to the liquid pool boiling process, the saturated vapor movesacross the adiabatic regionreaches the condenser region. As the saturated vapor contacts the

IOP Conf. Series: Materials Science and Engineering **973** (2020) 012039 doi:10.1088/1757-899X/973/1/012039

condenser's surface, the vapor condenses over the inner cold wall of the condenser region developing a liquid film. Figure 14 illustrates the condensate film configuration.

Under gravity effect this liquid film will drop down across the adiabatic region, the liquid pool in the evaporator region will be filled by a continued thin liquid film. During the above cycle, the latent heat of evaporation that was stored in the evaporator region is extracted by the vapor and released due to phase change process occurred at the inner wall of the condenser region.

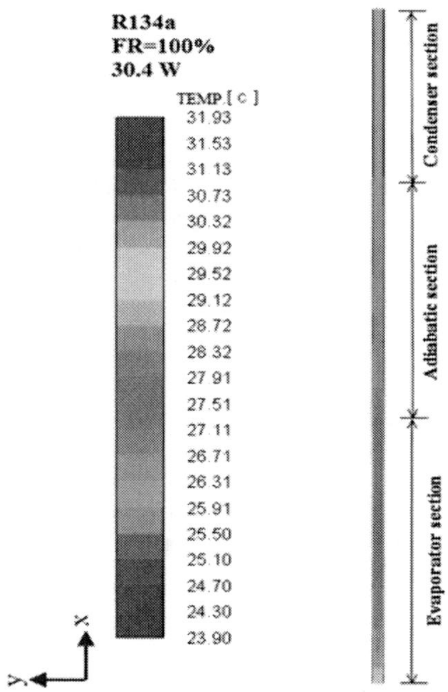

Figure 13. Temperature distribution contour for R134a filled thermosyphon at steady state.

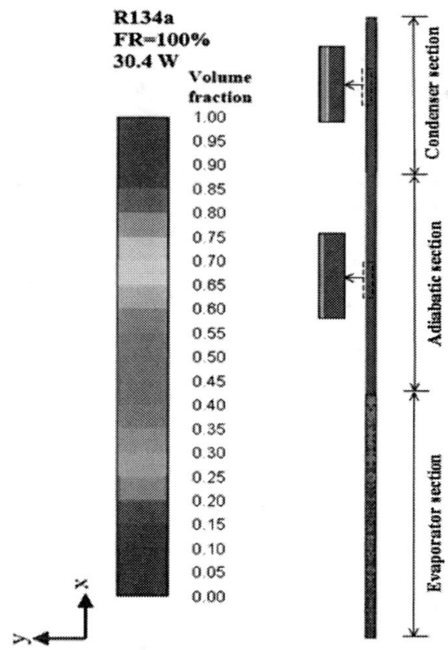

Figure 14. Volume fraction contour for R134a filled thermosyphon at steady state.

5. Validation of CFD simulation results

Figure 15shows the experimental and CFD simulation temperature distribution profiles over the R134a charged the TPCT heat pipe, for applied heat load 30.4 W. As depicted in figure 15 the simulation results presentedan acceptable agreement with the experimental data. The simulation results of temperature profile werereviewed with the experimental data by defining the average relative error (*ARE*). As a result of this,*ARE*of the evaporator, adiabatic and condenser average wall temperatures are 1.01 %, 7.51 % and 6.08 %, separately, table 5.

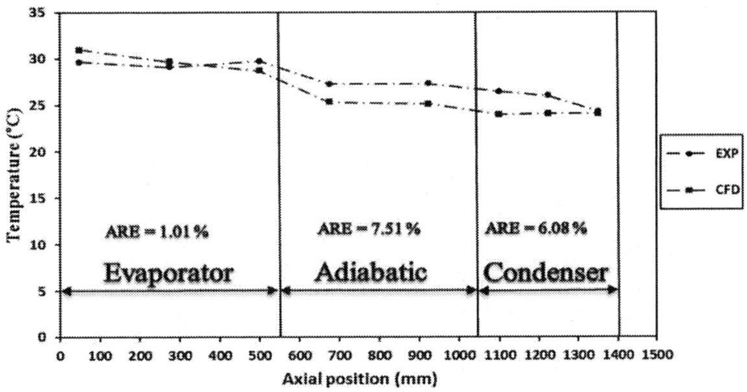

Figure 15. Temperature distribution profiles for experiment and CFD simulation.

Table 5. Comparison between experimental data and CFD simulation for heat input 30.4 W.

Section	Position	T_{EXP}	T_{CFD}	$\Delta T_{EXP-CFD}$	RE	$\widetilde{T_{EXP}}$	$\widetilde{T_{CFD}}$	$\Delta\widetilde{T_{EXP-CFD}}$	ARE
		°C	°C	°C	%	°C	°C	°C	%
Evaporator	T_{e1}	29.62	30.93	1.31	4.43	29.48	29.77	0.3	1.01
	T_{e2}	29.11	29.69	0.58	2.00				
	T_{e3}	29.70	28.70	1.00	3.36				
Adiabatic	T_{a1}	27.25	25.33	1.92	7.05	27.27	25.22	2.05	7.51
	T_{a2}	27.29	25.11	2.18	7.97				
Condenser	T_{c1}	26.43	23.97	2.46	9.32	25.57	24.02	1.55	6.08
	T_{c2}	25.99	24.04	1.95	7.50				
	T_{c3}	24.30	24.05	0.25	1.02				

6. Conclusions

The present study reports a numerical simulation of TPCT heat pipe filled with R134a. The advanced CFD model considered the nucleate pool boiling and the liquid film condensationprocesses occurred during operation of the TPCT heat pipe. The CFD model results are validated with the experimental data at the same working conditions. The comparison between experimental and numerical results indicated the good ability of the CFD model. The capabilities of validated CFD model will be used to derive the limitations that can restrict the transportation of heat pipe fluid in satellite application.

References

[1] Chang YW, Cheng CH, Wang JC and Chen SL 2008 *Energy Conv. and Mgmt.***49**3398-404
[2] Huminic G, Huminic A, Morjan I and Dumitrache F 2011, *Int. J. of Heat and Mass Transfer***54**656-61
[3] Liu ZH and Li YY. 2012*Int. J. of Heat and Mass Transfer***55**6786-97
[4] Liu ZH, Li YY and Bao R 2011*Int. J. of Thermal Sciences***50**558-68
[5] NaphonP, AssadamongkolP and BorirakT 2008*Int. Comm. in Heat and Mass Transfer***35**1316-19
[6] Tsai TE, Wu GW, Chang CC, Shih WP and Chen SL 2010 *Int. J. of Heat and Mass Transfer***53**4567-78
[7] Attia A A and El-Assal B T 2012*Heat Pipe Science and Technology, An Int. J.*335-51
[8] Wang PY, Chen XJ, Liu ZH and Liu YP 2012 *Thermochimica Acta***539**100-08
[9] PeyghambarzadehS, Shahpouri S, Aslanzadeh N and Rahimnejad M 2013 *Ain Shams Engineering J.*4855-61
[10] AniketD P and Ravindra B Y 2012 *Int. J. of Emerging Technology and Advanced Engineering*2202-06

[11] Fadhl B, Wrobel L C and Jouhara H 2013 *App. Therm. Eng.* **60** 122-31
[12] Abou-ZiyanH, Helali A, Fatouh M and El-Nasr M A 2001 *App.Therm.Eng.* **21** 813-30
[13] El-GenkM Sand Lianmin H 1993 *Int. J. of Heat and Mass Transfer* **36** 3823-30
[14] Noie S 2005 *App. Therm. Eng.* **25** 495-506
[15] Wang X, Wang Y, Chen H, and Zhu Y 2018 *Int. J. of Heat and Mass Transfer* **121** 703-14.
[16] Alizadehdakhel A, Rahimi M and Alsairafi A A 2010 *Int. Comm. in Heat and Mass Transfer* **37** 312-18
[17] Asmaie L, Haghshenasfard M, Mehrabani-Zeinabad A and Esfahany M N 2013 *Heat and Mass Transfer* **49** 667-78
[18] Ansys FLUENT Theory Guide *2017* (Release 18.0). *Multiphase Flows. ANSYS, Inc.* 715-46
[19] Lemmon E, Huber M and McLinden M 2010 *NIST Standard Reference Database 23, Reference Fluid Thermodynamic and Transport Properties (REFPROP)* ver 9.0 National Institute of Standards and Technology.

Nomenclature

A	Area (m^2)
C	Water specific heat (J/kg.°C)
c	Conductance (W/ °C)
h	Heat transfer coefficient (W/m^2.°C)
$\dot{Q}$	Heat rate (W)
T	Temperature (°C)
$\dot{m}$	Mass flow rate (kg/sec)

Greek symbols

η_{th}	Thermal efficiency
α	Volume fraction
ρ	Density (kg/m^3)
σ	Surface tension coefficient (N/m)

Superscripts

$\sim$	Average

Subscripts

ev	Evaporator
ad	Adiabatic
con	Condenser
in	Inlet
out	Outlet
l	Liquid phase
v	Vapor phase

IOP Conf. Series: Materials Science and Engineering **973** (2020) 012040 doi:10.1088/1757-899X/973/1/012040

Aerodynamic performance improvement using a micro-cylinder as a passive flow control around the S809 airfoil

W Mostafa[1,*], A Abdelsamie[2,3], MMohamed[2,4], DThévenin[3] andM. Sedrak[2].
[1]Egyptian Russian University
[2]Lab. of Fluid Mechanics, Helwan University, Cairo, Egypt
[3]Lab. of Fluid Dynamics & Technical Flows, University of Magdeburg "Otto von Guericke", Germany
[4]Mechanical Engineering Dept, College of Engineering and Islamic Architecture, Umm Al-Qura University, Makkah, Saudi Arabia
[*]Corresponding author:eng.wafaamostafa@gmail.com

Abstract. Aerodynamic performance improvement of airfoilsis the first step towards enhancement ofthe wind turbine performance in electricity generation and energy conversion inrenewable energy applications. The flow behavioraround wind turbine blades profilecan be improved by introducing active and/or passive flow controls. This work numerically describes the impact of adding micro-cylinder, as a passive flow control around S809 airfoil, on aerodynamic performance under various operating conditions. A suitable combination of flow analysis and optimization technique has been used in the current work.The numerical simulation has been performed using ANSYSFluent 18.2 software. The airfoil was numerically analysed in flow atReynolds number of 10^6; aerodynamic coefficients (lift and drag coefficients) at different angle of attacks were validated with the experimental data reported by Somers in NREL. The Response Surface Method (RSM) is applied to obtain the optimum position of micro-cylinder to achieve maximum lift to drag ratio. It has been found that the total aerodynamic forces are sensitive to the location of the micro-cylinder. A significant enhancement of lift to drag ratio can be achieved by adding micro-cylinder in front of S809 airfoil especially at high Reynolds number.

1. Introduction

Active and passive flow controls are the two different techniques that are used to control the flow state. The active flow control is relatively restricted, as it needs an additional controller, actuators and power supply to control theflow. On the other side, passive flow control is relatively simple and cheap.Therefore, the passive flow control is more applicable in practical applications but on a narrow range of operating conditions [1,2]. Several publications investigated the impact of different passive technique on the aerodynamic performance of wind turbine blades, such as adding micro-cylinder [1,2], micro-tab [3], vertex generators, gurney flaps[4,5, 6], tubercle leading edge[7, 8, 9], split blade [10], slot [11, 12], and rod vortex generators [13]. Wang et al. 2018 discussed in detail the effect of adding micro-cylinder with different diameters and positions in front of the NREL Phase VI wind turbine blade. It was found that (1) it possibleto suppress the flow separation on wind turbine blades, (2) the blade torque was increased by 27.3%when the micro-cylinder is close to the blade leading edge point, with smaller vertical distance = $2.3\times10^{-2}$and diameter = 6.78×10^{-3} (dimensions in this paper normalized by chord). In the work of Luo et al. 2018, a micro-cylinder was added near to the leading

IOP Conf. Series: Materials Science and Engineering **973** (2020) 012040 doi:10.1088/1757-899X/973/1/012040

edge of a stalled airfoil (NACA 0012) inflow at Re= 6 x10^6;The Reynolds-averaged Navier–Stokes equations(RANS)simulations were performed to determine the optimal control parameter. Luo et al. 2018 were able to capture the transient vortical structures in the massive separation region by relying on Delayed Detached Eddy Simulation (DDES) for the controlled and uncontrolled cases. Overall, adding micro-cylinder improved lift coefficient and lift-to-drag ratio that can be achieved after stall. RANS and DDES results indicated that the heavy stall can be effectively delayed; the extent of large separation region, on the airfoil suction surface, can also be diminished[2].
In the currentwork, a micro-cylinder is selected as the control element and proposed to be installed around the airfoil leading edge.The S809 airfoil is chosen for simulation. This thick airfoil (21% thickness) has been designed for a horizontal axis wind turbine and tested experimentally[14]. The main objectives, that have been achieved for this airfoil comparing to the other availableairfoils, are maximizingthe lift, minimizing the profile drag, and reducing the sensitivity to roughness. Practically, the airfoil-S809 is using now for some horizontal axis wind turbine rotor;this was tested experimentally in NREL phase II (3-bladed HAWT)[15] and NREL phase VI[16] (2-bladed HAWT).This airfoil is popular for validation of numerical solutions[1, 6, 12, 17, 18].Here, the simulation will be carried out using ANSYSFluent 18.2 to investigate the impact of the location and diameter of the micro-cylinder. Response Surface Method (RSM) is employed in this simulation as the optimization tool.

This paper is organized as follows: in Section 2 the numerical approaches and governing equations are reviewed. The results are discussed in Section 3. Summary and conclusion come at the end.

2. Numerical approaches and governing equations

In this study, the simulations were carried out using ANSYSFluent 18.2 for incompressible flow. In this package, the Navier-Stokes equation is solved to resolve the fluid flow. In order to consider the turbulence effect, Reynolds-Averaged Navier-Stokes equations (RANS) andshear-stress transport (SST) k-ω turbulence model are used. These equations can be reviewed as follows:
(1) Continuity equation

$$\frac{\partial(\overline{\bar{u}_i + u_i'})}{\partial x_i} = \frac{\partial(\bar{\bar{u}}_i + \overline{u_i'})}{\partial x_i} = 0, \qquad \frac{\partial(\bar{u}_i)}{\partial x_i} = 0, \qquad \frac{\partial(\overline{u_i'})}{\partial x_i} = 0 \tag{1}$$

(2) Momentum equation

$$\frac{\partial}{\partial t}\rho(\bar{u}_i) + \frac{\partial}{\partial x_j}\rho(\bar{u}_i\bar{u}_j) = -\frac{\partial(\bar{p})}{\partial x_i} + \mu\frac{\partial}{\partial x_j}\left(\frac{\partial \bar{u}_i}{\partial x_j} - \rho(\overline{u_i'u_j'})\right) + \rho(\bar{F}_i) \tag{2}$$

In equation (1) and (2), u_j' is the turbulence fluctuation in the velocity component, $-\rho(\overline{u_i'u_j'})$term represents the product of turbulence fluctuations in the velocity components; it is known as the turbulent shear or Reynolds stresses ($\tau_{ij}' = -\rho\overline{u_i'u_j'}$) which are unknown. The Reynolds stresses cannot be calculated directly but have to be modeled. This is commonly referred to as a closure problem (See Ferziger and Peric[19]). Therefore, in 1877 the Boussines approximation was introduced, resulting in the eddy viscosity or eddy diffusion model.

$$\overline{\tau_{ij}} = \mu_t\left(\frac{\partial \bar{u}_i}{\partial x_j} + \frac{\partial \bar{u}_j}{\partial x_j}\right) - \frac{2}{3}\rho k\delta_{ij} \tag{3}$$

where μ_t is the eddy (turbulent) viscosity which has the same dimension as the molecular viscosity(μ), it isn't a fluid property but rather a turbulence property and k is the turbulent kinetic energy $k = \frac{1}{2}\left[\overline{u'}^2 + \overline{v'}^2 + \overline{w'}^2\right]$ .The problem has been transformed from one of finding the individual Reynolds stresses to one of determining the eddy (turbulent) viscosity μ_t.Shear-stress transport (SST) k-ω turbulence model is a type of hybrid model, combining two models in order to calculate flow in the near-wall region. It uses a standard k-ε model to obtain flow properties in the free-stream (turbulent) flow region far from the wall, while using a modified k-ε model near the wall using the turbulence frequency ω as a second variable instead of turbulent kinetic energy dissipation term ε, which is expressed as $\omega = \varepsilon/k$ [s^{-1}]. The transport equation for turbulent kinetic energy k for the k-ω model can be reviewed as follows[19]:

$$\frac{\partial(\rho k)}{\partial t} + \frac{\partial(\rho k \bar{u}_j)}{\partial x_j} = \frac{\partial}{\partial x_j}\left[\left(\mu + \frac{\mu_t}{\sigma_k}\right)\frac{\partial(k)}{\partial x_j}\right] + P_k - \beta^* \rho k \omega, \quad (4)$$

$$\frac{\partial(\rho \omega)}{\partial t} + \frac{\partial(\rho \omega \bar{u}_j)}{\partial x_j} = \frac{\partial}{\partial x_j}\left[\left(\mu + \frac{\mu_t}{\sigma_\omega^*}\right)\frac{\partial(\omega)}{\partial x_j}\right] + \alpha \frac{\omega}{k} P_k - \rho \beta \omega^2. \quad (5)$$

Equations [4] and [5] can be summarized as follows,

$$\begin{pmatrix}\text{Rateofchange}\\ \text{of}\\ k/\omega\end{pmatrix} + \begin{pmatrix}\text{Transportof}\\ k/\omega\\ \text{byconvection}\end{pmatrix} = \begin{pmatrix}\text{Transportof}\\ k/\omega\\ \text{bydiffusion}\end{pmatrix} + \begin{pmatrix}\text{Rateof}\\ \text{production}\\ \text{of}\\ k/\omega\end{pmatrix} - \begin{pmatrix}\text{Rateofdestruction}\\ \text{of}\\ k/\omega\end{pmatrix},$$

In this model, the eddy viscosity is expressed as $\mu_t = \rho\frac{k}{\omega}$, $\alpha = \frac{5}{9}$, $\beta = 0.075$, $\beta^* = 0.09$, $\sigma_k^* = 2$, andσ_ω^* = 2 [14].

3. Results and discussions

In this section, the impact of adding micro-cylinder around the S809 airfoil and optimizationfor its location and diameter will be discussed.

3.1. Aerodynamic analysis around S809 using CFD

As it usual in standard computational fluid dynamic (CFD) works, the first step it to perform the mesh independent study. For that, three different meshes were examined to ensure that the simulation results are independent of grid size. Table 1 lists three different mesh sizes with the number of elements 53650, 60000, and 75254. Figure 1 shows the mesh topology and its mesh refinement around S809 airfoil. To achieve good precision the thickness of the first layer around airfoil should be $y^+<5$ and it will be more accurate if y^+approaches 1 [20].The goal is achieved by setting the first layer thickness = 2×10^{-5}m.The mesh sensitivity analysis is presented infigures.2 and 3. Where figure 2 shows the behavior of lift coefficient versus the simulation iterationat different mesh size; and figure 3 illustrate the trend of lift coefficient versus the angle of attack (AOA) at different mesh size. As it can be observed from these figures, the results become independent on the mesh size≥ 60000,therefore, so this mesh size (60000)is selectedas a default case for the current simulations. The domain dimensions at the upstream and downstream are 20 and 40 times chord length as shown in figure 1 [21].In these simulations, the boundary conditions areas follows: (1) left boundary is velocity inlet, (2) upper and lower boundaryis symmetry, (3) the right boundary is pressure outlet.In this work, the numerical approaches that have been used in the Ansys Fluent 18.2 can be summarized as follows: 1) the pressure-velocity coupling scheme (coupled), and (2) (2) Least squares (cell-based) is used for spatial discretization, (3) second-order scheme for pressure equation, (4) second-order upwind scheme for the momentum equation, and (5) second-order upwind scheme for turbulent kinetic energy and specific dissipation rate equations.

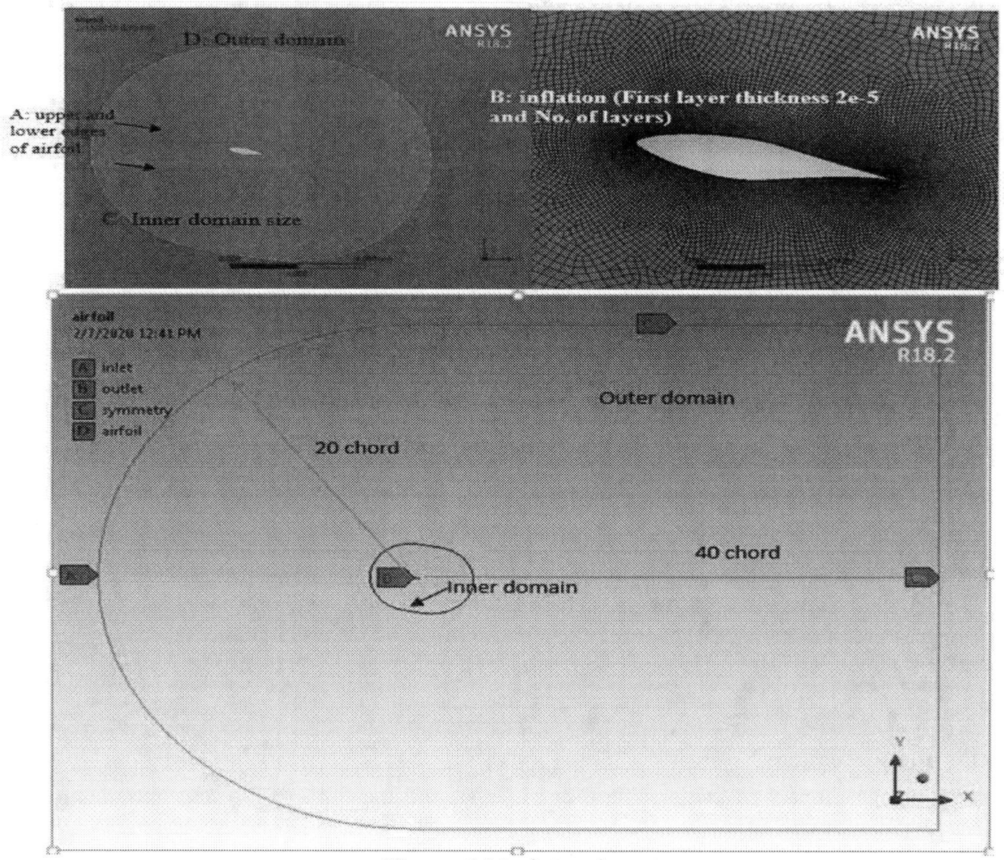

Figure 1. Mesh topology.

Table 1. Mesh Refinement.

Mesh size	A	B	C	D
53650	180	60	0.3	0.6
60000	200	70	0.3	0.6
75254	200	70	0.2	0.4

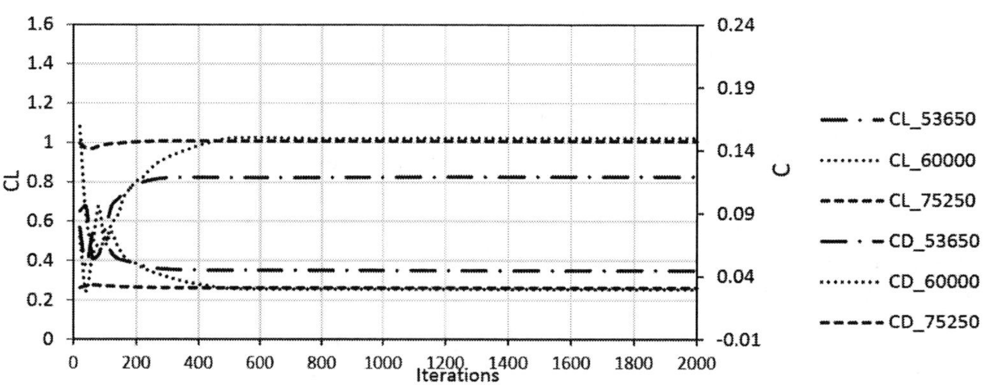

Figure 2. Lift and drag coefficient convergence history at AOA = 11.2°.

Validation with experimental results [21]is performed at Reynolds number of 10^6 and Mach number of 0.044. Different angles of attacks are selected for validation, which are AOA= 0.0°, 2.1°, 4.1°, 6.1°, 8.2°, 10.1°, and 11.2°. Figure 3 shows that our numerical results for lift and drag coefficient agree well with the experimental values.

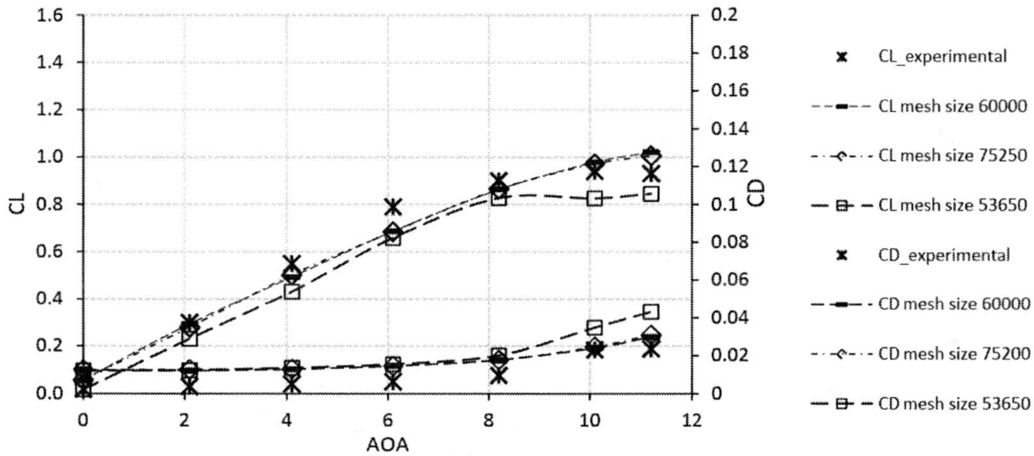

Figure 3. Lift and drag coefficient validation with experimental at Re=10^6.

In order to achieve our design point (The optimum location of the micro-cylinder and the optimum diameter),first the optimum angle of attack of the airfoil S809 that satisfied the target of maximum lift to drag ratio at certain design point is obtained. Thenthe optimum location and dimension of micro-cylinder at the selected design point is examined.

Response surface optimization feature impeded inANSYSpackage will be considered asthe optimization tool used in this study. Sparse grid initialization is selected to proceed the step of the Design of Experiments (DOE) which is responsible of collecting a representative set of data relating to a process, to calculate a response surface and then executing the optimizationusing Multi-Objective Genetic Algorithm (MOGA) which supports all types of input parameters.

*3.1.1.Design point analysis around S809 airfoil (Re=7.2×10^5).*In this section our goal is to obtain the optimum angle of attack (AOA) at Re=7.2×10^5 (corresponding velocity V=10.5 m/s). The input parameter for the optimization algorism is the angle of attack while the output parameter is the lift to drag ratio. The optimization criteria is the AOA at which maximum lift to drag ratio achieved. The maximum and minimum values of angle of attack for step of design of experiment is 0° and 14° respectively.
Design of experiment samples are illustrated in table 2, and the refinement points created by response surface are tabulated in table 3. MOGA optimization technique then applied. It has been observed that the resultant value of the optimum angle of attack that satisfied maximum lift to drag ratio is 6.65° as shown in table 4.

Table 2. Design of experiment (sparse grid initialization).

Name	P9 - AOA (degree)	P10 - cd-op	P11 - cl-op	P12 - lift-to-drag-ratio-op
1	7	0.017	0.760	45.121
2	0	0.013	0.044	3.420
3	14	0.058	0.977	16.720

Table 3. Refinement points created by response surface (sparse grid).

Name	P9 - AOA (degree)	P10 - cd-op	P11 - cl-op	P12 - lift-to-drag-ratio-op
1	3.500	0.015	0.420	28.579
2	10.500	0.029	0.961	33.487
3	8.750	0.017	0.953	55.022
4	12.250	0.041	0.990	23.914
5	7.875	0.018	0.832	46.214
6	9.625	0.023	0.937	40.429
7	1.750	0.013	0.236	17.808
8	5.250	0.015	0.595	39.355

Table 4. Optimum Lift to drag ratio after verification (MOGA).

P9 - AOA	P10 - cd-op	P11 - cl-op	P12 - lift-to-drag-ratio-op
6.65	0.016	0.722	46.412
7.77	0.018	0.823	45.849
8.85	0.020	0.899	44.321

3.2. Optimum location and diameter of micro-cylinder around S809 airfoil (Re=720000)

From the previous section, the optimum angle of attack that satisfied maximum lift to drag ratio is calculated. Then the next step is to optimize the micro-cylinder location and diameter as it will be discussed in the next section. The following study focus mainly on two positions of the micro cylinder: (1) micro-cylinder is just above the upper surface of the airfoil, and (2) micro-cylinder is front of the S809 airfoil leading edge. In the following section the general setting of micro-cylinder will be illustrated.

3.2.1. General setting for different optimization methods. As the first test, the micro-cylinder is added and the number of cells = 66824 cell, AOA = 6.65°(optimum one)corresponding to wind speed = 10.5 m/s as shown in table 4 and figure 4. The input parameters are (1) dx which is the horizontal distance between micro-cylinder and blade leading edge, (2) dy which isthe vertical distance between micro-cylinder and blade leading edge, and (3) D which is the diameter of cylinder all dimensions are normalized by chord length in this paper chord length = 1m . The output parameters are (1) lift and drag coefficients, and (2) lift to drag ratio. The optimization technique is response surface optimization method as stated before. In this optimization, Latin hypercube sampling design is chosen for Design of Experiments (DOE); Response Surface method is Kriging; and optimization technique is MOGA.

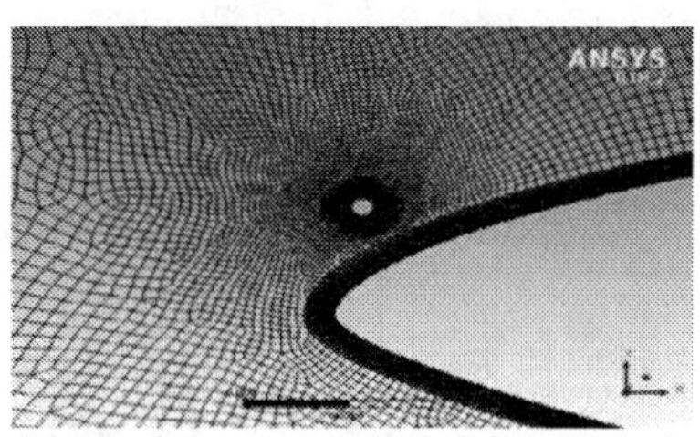

Figure 4. Grid topology around airfoil with micro-cylinder.

3.2.2.Optimization of micro cylinder location above the upper surface of S809 airfoil. In this section, micro-cylinder is located above the upper surface of the S809 airfoil to illustrate the effect of adding it on the values of lift to drag ratio. The range of horizontal distance from leading edge is dx=0 - 0.05, vertical distances from leading edge of the airfoil is dy = 0.054 - 0.08 and diameter of micro-cylinder is D = 0.008 - 0.02 as illustrate in figure 5.

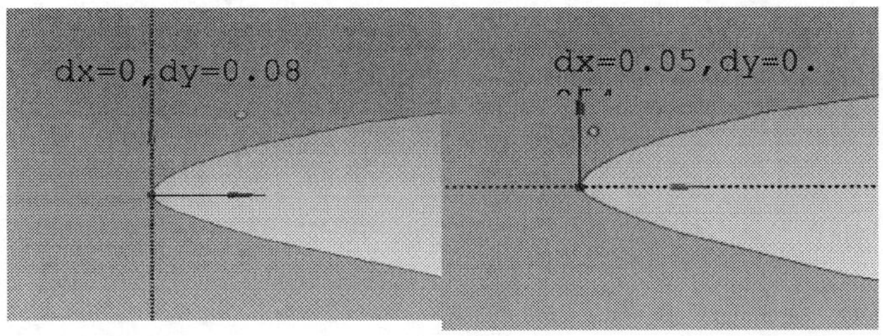

Figure 5. Samples of addingthe micro-cylinder above upper surface of airfoil.

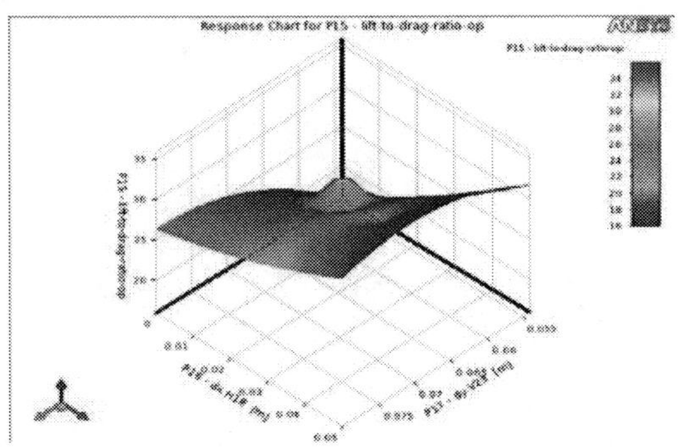

Figure 6..Response chart of lift to drag ratio versus dx and dy.

Kriging method is selected to create the response surface as shown in figure 6. This method is non-parametric, meaning that the prediction will depend on all existing data points. Thus the method could be slow in case of usinglarge amount of data. It is also slow in fitting the data due to the calculation of pair-wise distances among data points. However, this method automatically fits through all data points.
By using MOGA optimizationmethod, the location of micro-cylinder with AOA =6.65 found to be at dx = 0.05, dy = 0.056, and D = 0.008.

*3.2.3 Off-design charts in case of existence of micro-cylinder above upper surface.*At optimized location of micro-cylinder off design charts was created to predict the flow behavior. The variation of the lift coefficient, drag coefficient, and the lift to drag ratio versus angle of attack with and without the micro-cylinder are shown in figures 7, 8, and 9 respectively.

IOP Conf. Series: Materials Science and Engineering **973** (2020) 012040 doi:10.1088/1757-899X/973/1/012040

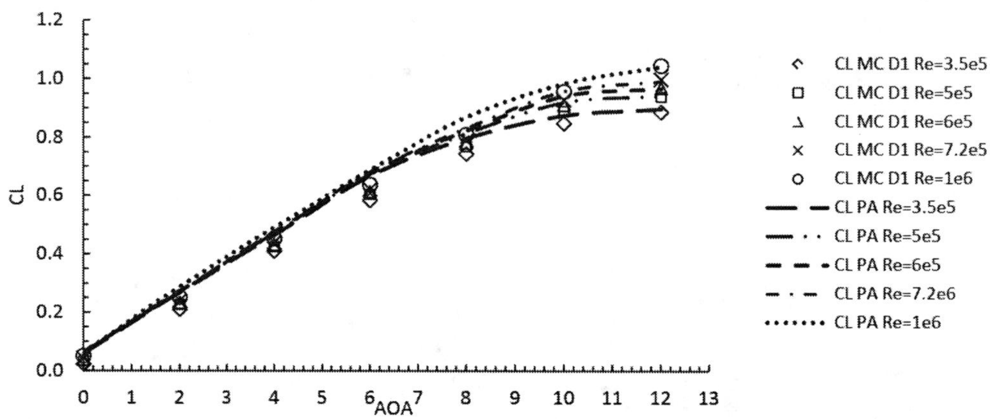

Figure 7. Lift coefficient versus AOA. Lines representairfoilwithout the micro-cylinder. Symbols are used to represent the cases with adding the micro-cylinder.

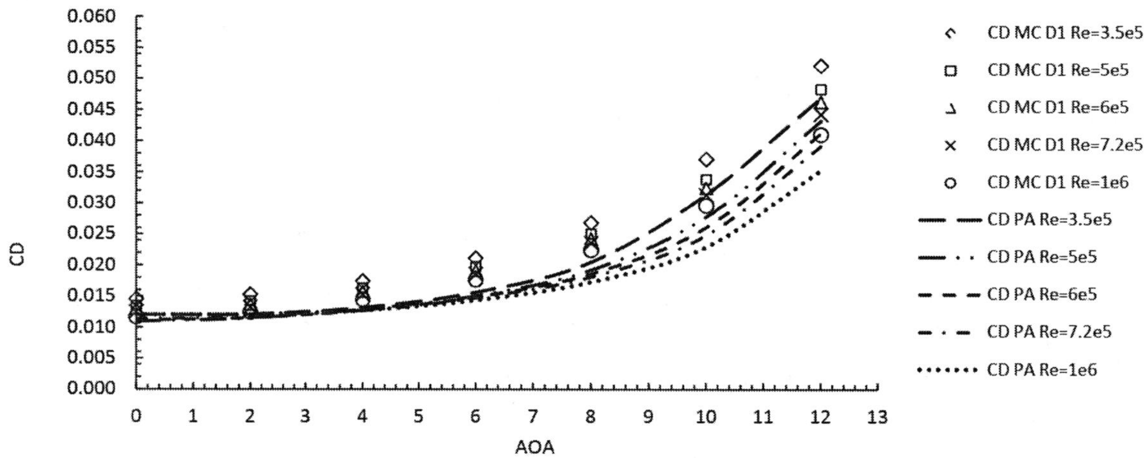

Figure 8. Drag coefficient versus AOA.Lines representairfoilwithout the micro-cylinder. Symbols are used to represent the cases with adding the micro-cylinder.

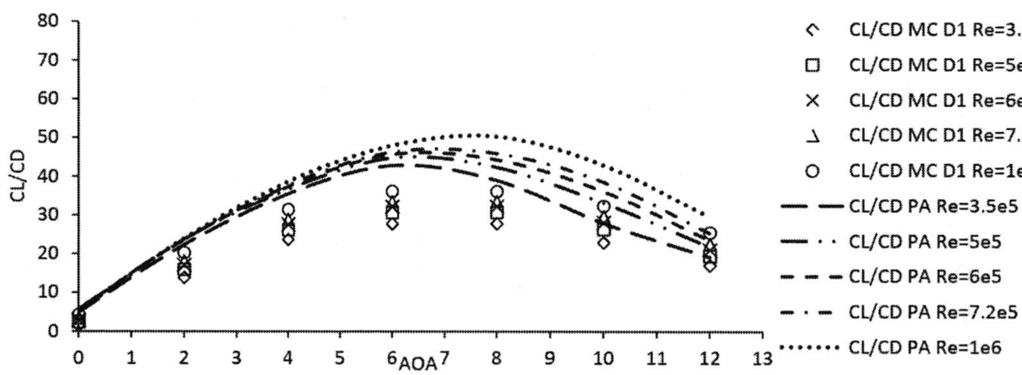

Figure 9. Lift to drag ratio versus AOA. Lines representairfoilwithout the micro-cylinder. Symbols are used to represent the cases with adding the micro-cylinder.

It is clear from these figures (7, 8, and 9) thatadding the micro-cylinder above the upper surface of S809 airfoil, slightly decreases the lift coefficient as shown in figure 7 while the drag coefficient is increased as shown in figure 8.Consequently, the lift to drag ratio is decreased as shown in figure 9. So it is concluded that the adding the micro-cylinder above upper surface of the airfoil will affect negatively on the aerodynamic performance.

3.2.4 Optimization of micro cylinder location in front of the S809 airfoil leading edge. The micro-cylinder now is located in front of the S809 airfoil leading edge to examine the impact on the values of lift to drag ratio. The range of horizontal dx = 0.01-0.06, and vertical distances dy = 0 - 0.06 , from leading edge of the airfoil and diameter of micro-cylinder D = 0.002 – 0.02 , examples of distances are illustrate in figure10.

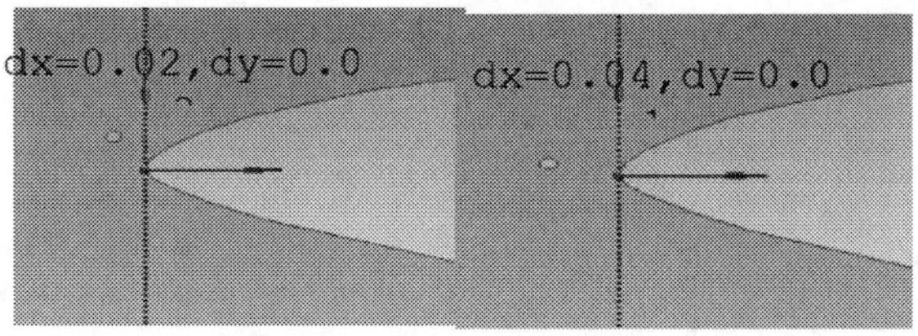

Figure 10. Samples of addingthe micro-cylinder in front of the S809 airfoil leading edge.

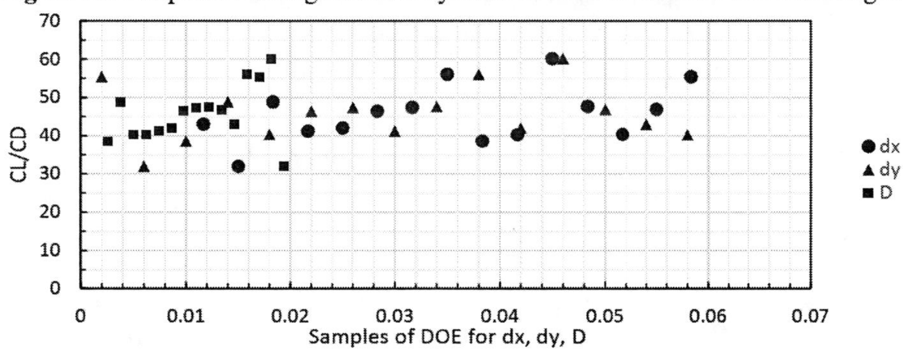

Figure 11. Lift to drag ratio versus input parameters (Design of experiments LHS).

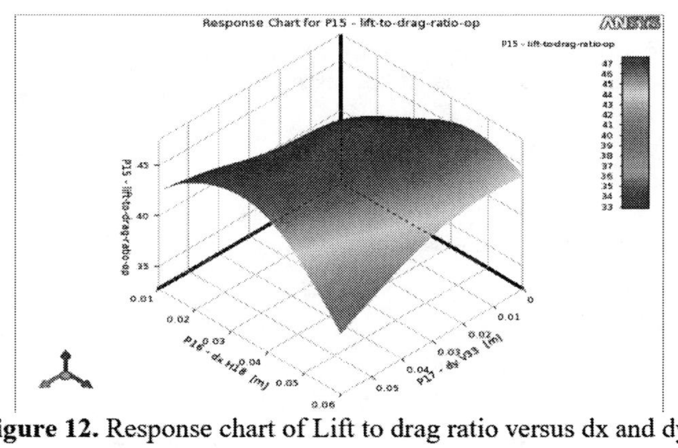

Figure 12. Response chart of Lift to drag ratio versus dx and dy.

Latin hypercube sampling design are shown in figure 11 and the created response Surface is shown in figure 12.It has been observed that the values of lift to drag ratio in figure 12 are greater than that

IOP Conf. Series: Materials Science and Engineering **973** (2020) 012040 doi:10.1088/1757-899X/973/1/012040

infigure 6 (case with adding the micro-cylinder above the airfoil). From MOGA optimization tool it is found that the optimum location is dx = 0.05, dy = 0.05, D = 0.02.

*3.2.5.Off-design Charts in case of existence of micro-cylinder in front of the S809 airfoil leading edge.*It is clear from the above figures (11- 14) thatwhen adding micro-cylinder in front of the S809 airfoil leading edge, the lift coefficient is slightly decreased as shown in figure 13 while the drag coefficient is highly decreased as shown in figure 14. As a consequence, the lift to drag ratio is increased as shown in figure 15. Therefore, it can be concluded that the adding of micro-cylinder in front of the S809 airfoil leading edge will affect positively on the values of lift to drag ratio and consequently the aerodynamic performance.

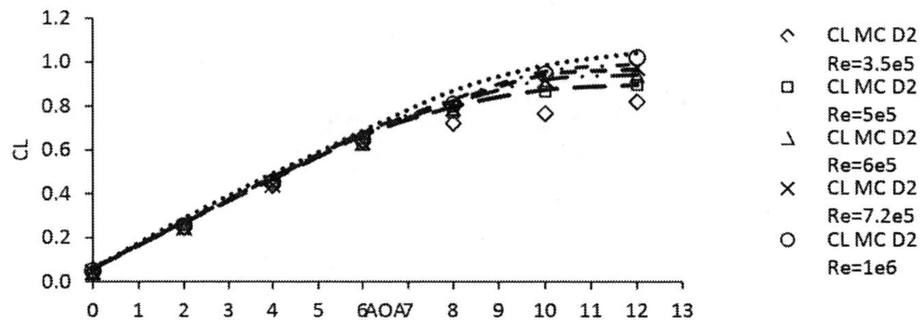

Figure 13. Lift coefficient versus AOA .Lines representairfoilwithout the micro-cylinder. Symbols are used to represent the cases with adding the micro-cylinder.

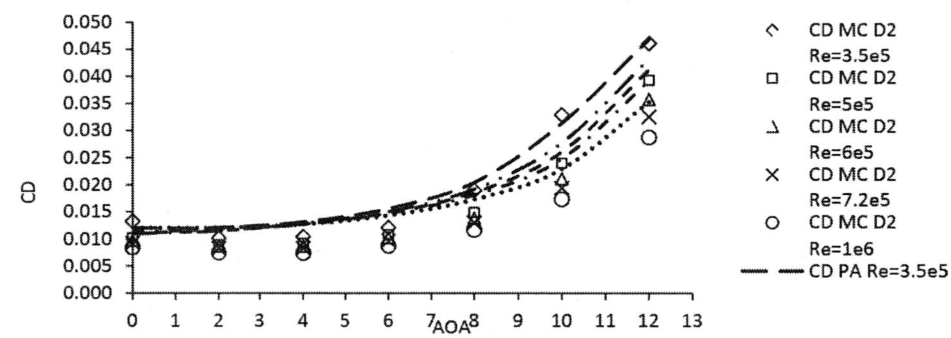

Figure 14. Drag coefficient versus AOA. Lines representairfoilwithout the micro-cylinder. Symbolsare used to represent the cases with adding the micro-cylinder.

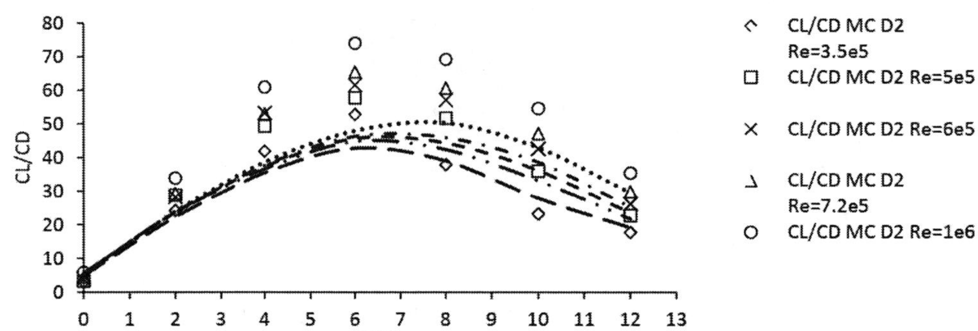

Figure 15. Lift to drag ratio versus AOA. Lines representairfoilwithout the micro-cylinder. Symbols are used to represent the cases with adding the micro-cylinder.

IOP Conf. Series: Materials Science and Engineering **973** (2020) 012040 doi:10.1088/1757-899X/973/1/012040

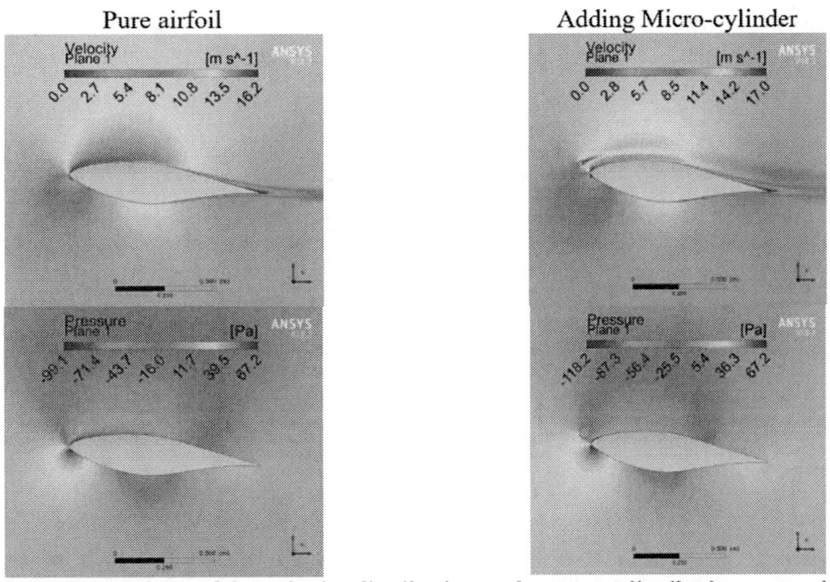

Figure 16. Comparison of the velocity distribution and pressure distribution respectively around the controlled and uncontrolled S809 airfoils at angle of attack 6.1°

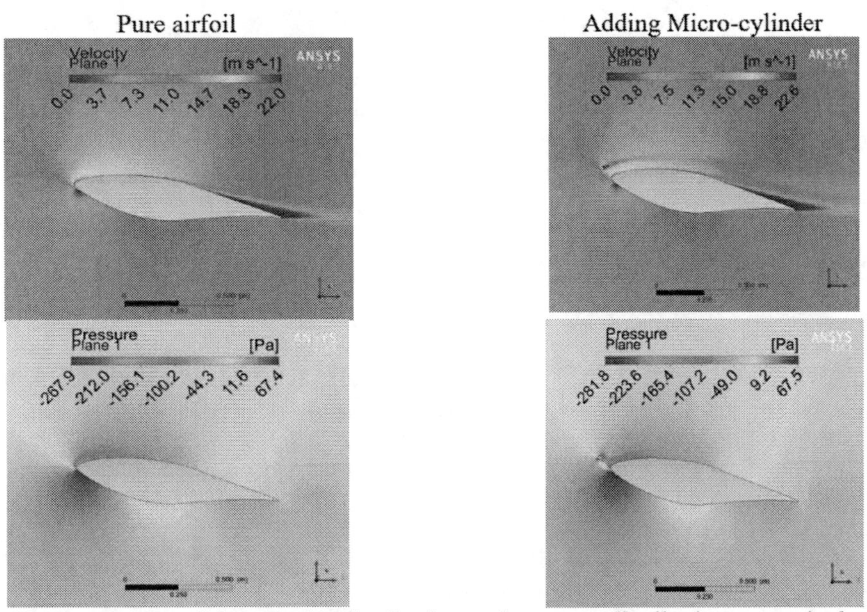

Figure 17. Comparison of the velocity distribution and pressure distribution respectively around the controlled and uncontrolled S809 airfoils at angle of attack 10.1°.

The presence of the micro-cylinder affects the velocity and pressure distribution around the S809 airfoil,the change in velocity and pressure distribution at the angle of attack 6.1° is greater than that at the angle of attack 10.1° as shown in figures 16 and 17. These agree with the results are shown in figure 15, where the values of lift to drag ratio has a peak value at angles of attack around

6.65°.Therefore, the micro-cylinder has a direct impact on the total aerodynamic forces and leads to a decrease in the drag coefficient and highly increase in the lift to drag ratio as shown in figures 14, 15.

4. Conclusion

A passive flow control strategy for the S809airfoil was investigated inthis study. The effect of installing a micro-cylinder near the S809 airfoil was numerically investigated, with a focus on thetotal aerodynamic forces and flow fields at angle of attacks ranging from 0 to 12 .Thediameter and the location of the micro-cylinder were theinput parametersto be considered in this investigation and lift to drag ratio was the output parameter. Response surface optimization tool was the method used to calculate the input parameter and achieving the target of maximum lift to drag ratio. The main results from this study is that adding micro-cylinder in front of the leading edge increases the lift to drag ratio especially at Reynolds numbers above $5\times10^5$and angles of attack from 4^0 to 10^0, while the effect is slightly less at low velocities, Re = 3.5×10^5 .

Reference

[1] Wang Y, Li G, Shen S, Huang D and Zheng Z 2018 Investigation on aerodynamic performance of horizontal axis wind turbine by setting micro-cylinder in front of the blade leading edge. *Energy,* **143**, pp.1107-1124

[2] Luo D, Huang D and Sun X 2017 Passive flow control of a stalled airfoil using a microcylinder. *Journal of Wind Engineering and Industrial Aerodynamics*, **170**, pp.256-273.

[3] Ebrahimi A and Movahhedi M 2018. Wind turbine power improvement utilizing passive flow control with microtab. *Energy*, **150**, pp.575-582

[4] Fernandez-Gamiz U, Zulueta E, Boyano A, Ansoategui I and Uriarte I 2017Five megawatt wind turbine power output improvements by passive flow control devices. *Energies*, **10(6)**, p.742

[5] Holst D, Pechlivanoglou G, Wegner F, Nayeri C N and Paschereit C O 2017Potential of Retrofit Passive Flow Control for Small Horizontal Axis Wind Turbines. *Journal of Engineering for Gas Turbines and Power*, **139(3)**, pp.1-8

[6] Wang H, Zhang B, Qiu Q and Xu X 2017 Flow control on the NREL S809 wind turbine airfoil using vortex generators. *Energy*, **118**, pp.1210-1221

[7] Aftab S M A and Ahmad K A 2017 CFD study on NACA 4415 airfoil implementing spherical and sinusoidal Tubercle Leading Edge. *PloS one*, **12(11)**,pp.1-27

[8] Asli M, Mashhadi Gholamali B and Mesgarpour Tousi A 2015 Numerical analysis of wind turbine airfoil aerodynamic performance with leading edge bump. *Mathematical Problems in Engineering*, **2015**

[9] Huang G Y, Shiah Y C, Bai C J and Chong W T 2015 Experimental study of the protuberance effect on the blade performance of a small horizontal axis wind turbine. *Journal of Wind Engineering and Industrial Aerodynamics*, **147**, pp.202-211

[10] Moshfeghi M, Shams S and Hur N 2017Aerodynamic performance enhancement analysis of horizontal axis wind turbines using a passive flow control method via split blade. *Journal of Wind Engineering and Industrial Aerodynamics*, **167**, pp.148-159

[11] Semwal M, Tripathi R, Sati N, Rana V and Rawat K S 2016 Drag reduction of wind turbine blade to enhance aerodynamic performance: A numerical study. *International Journal of Scientific & Engineering Research*,**7(12)**, pp. 199–203

[12] Xie Y, Chen J, Qu H, Xie G, Zhang D and Moshfeghi M 2013 Numerical and experimental investigation on the flow separation control of S809 airfoil with slot. *Mathematical Problems in Engineering*, **2013**

[13] Ostachowicz W, McGugan M, Schröder-Hinrichs J U and Luczak M eds2016 *MARE-WINT: new materials and reliability in offshore wind turbine technology*. Springer.

[14] Somers D M 1997 *Design and experimental results for the S809 airfoil* (No. NREL/SR-440-6918). National Renewable Energy Lab., Golden, CO (United States).

[15] Butterfield C P, Musial W P, Scott G N and Simms D A 1992 *NREL Combined Experimental Final Report--Phase II* (No. NREL/TP-442-4807). National Renewable Energy Lab.(NREL), Golden, CO (United States).

[16] Lindenburg C 2003*Investigation into Rotor Blade Aerodynamics Analysis of the stationary measurements on the UAE phase-VI rotor in the NASA-Ames wind tunnel.* Energy research Centre of the Netherlands (ECN) Wind Energy publication, ECN-C--03-025.

[17] Moshfeghi M, Song Y J and Xie Y H 2012Effects of near-wall grid spacing on SST-K-ω model using NREL Phase VI horizontal axis wind turbine. *Journal of Wind Engineering and Industrial Aerodynamics*, **107**, pp.94-105

[18] Wang Y, Zheng X, Hu R and Wang P 2016 Effects of Leading Edge Defect on the Aerodynamic and Flow Characteristics of an S809 Airfoil. *PloS one*, **11(9)**, pp. 1-17

[19] Ferziger J H and Peric M 2002*Computational Methods for Fluid Dynamics*Third, Rev Edition, Springer, 2002.

[20] El Khchine Y and Sriti M 2017 Boundary Layer and Amplified Grid Effects on Aerodynamic Performances of S809 Airfoil for Horizontal Axis Wind Turbine (HAWT). *Journal of Engineering Science & Technology*, **12(11)**, pp.3011-3022

[21] He Y and Agarwal R K 2014 Shape optimization of NREL S809 airfoil for wind turbine blades using a multiobjective genetic algorithm. *International Journal of Aerospace Engineering*, **2014**

Appendix :TableA. Coordinates and characteristics of NREL airfoil

X/C	Y/C	X/C	Y/C	X/C	Y/C
1	0	0.171409	0.078687	0.193738	-0.08357
0.996203	0.000487	0.136174	0.07058	0.231914	-0.09244
0.98519	0.002373	0.104263	0.061697	0.271438	-0.09991
0.967844	0.00596	0.076035	0.052224	0.311968	-0.10528
0.945073	0.011024	0.051823	0.042352	0.35337	-0.10818
0.917488	0.017033	0.03191	0.032299	0.395329	-0.10801
0.885293	0.023458	0.01659	0.02229	0.438273	-0.10455
0.848455	0.03028	0.006026	0.012615	0.48192	-0.09735
0.80747	0.037766	0.000658	0.003723	0.527928	-0.08657
0.763042	0.045974	0.000204	0.001942	0.576211	-0.07398
0.715952	0.054872	0	-0.00002	0.626092	-0.06064
0.667064	0.064353	0.000213	-0.00179	0.676744	-0.04744
0.617331	0.074214	0.001045	-0.00348	0.727211	-0.0351
0.56783	0.084095	0.001208	-0.00372	0.776432	-0.0242
0.519832	0.093268	0.002398	-0.00527	0.823285	-0.01516
0.474243	0.099392	0.009313	-0.0115	0.86663	-0.0082
0.428461	0.10176	0.02323	-0.0204	0.905365	-0.00336
0.382612	0.10184	0.04232	-0.03027	0.938474	-0.00049
0.33726	0.10007	0.065877	-0.04082	0.965086	0.000743
0.29297	0.096703	0.093426	-0.05192	0.984478	0.000775
0.250247	0.091908	0.124111	-0.06308	0.996141	0.00029
0.209576	0.085851	0.157653	-0.07373	1	0

IOP Conf. Series: Materials Science and Engineering **973** (2020) 012041 doi:10.1088/1757-899X/973/1/012041

Assistive exoskeleton hand glove using soft pneumatic actuators: design optimization

M N El-Agroudy[1], M I Awad[1], A Rehan[1] and S A Maged[1]
[1] Mechatronics Department, Faculty of Engineering, Ain Shams University
shady.maged@eng.asu.edu.eg

Abstract. An assistive exoskeleton hand glove can be the solution to people who suffer from impairment in hand functionality. Soft pneumatic actuators (SPA) are used to actuate such glove. This paper focus on optimizing the design of the soft actuator. Finite element (FE) model is linked with the design optimization tool to find the optimal dimensions and materials of the SPA that give maximum displacement of the actuator with acceptable stresses on the actuator's body. Performance analysis shows the relation between different parameters and how each affect the performance of the actuator. Results show the dimensions and materials needed to have maximum actuator displacement without being subjected to high stresses.

1. Introduction

Our hands are of great importance to our daily activities, whether you're a 10-year-old or even 60 years old. Its functionality is irreplaceable. But this functionality can be impaired due to strokes or accidents. Thus, rehabilitation is now needed to restore the lost functionality of the hand. Rehabilitation and physical therapy require the patient to go to a clinic or hospital for up to 6 months to perform specific tasks with certain difficulty that is directly proportional to the improvement in the patient's performance. Such process can be expensive on the long run and time consuming. Also, the need of an assistive device still exists to assist the patient in performing the everyday tasks. Thus, an exoskeleton hand glove is needed.

The glove is designed so that it's actuated with soft pneumatic actuators (SPA). The actuators will tend to move in any direction when inflated. So, the motion must be restricted to give the desired motion. For example, if the actuator is to bend, we need to restrict the other degrees of freedom. Thus, an inextensible layer is added so that the actuator will not be able to elongate. Also, Fibres in both clockwise and anti-clockwise directions are added to restrict the twisting motion. Therefore, the actuator can only bend.

In last decade, many researchers start to model and optimize the parameters of the SPA that is used in the exoskeleton hand glove, like in [1]-[4]. Zhongkui Wang and Shinichi Hirai modelled, simulated and optimized the performance of a soft pneumatic gripper for handling deformable objects [1]. In [2], Gundula Runge, Jan Peters and Annika Raatz modelled and optimized the design of SPAs using genetic algorithms. Also, Panagiotis Polygerinos et al. modelled, fabricated and controlled SPAs that are made of elastomeric and inextensible materials [3]. Gabriel Dämmer et al. presented, in [4], the design and shape optimization of polyjet bellows actuators.

In this manuscript, the SPA is modelled and simulated. The model is linked with a design optimization tool to find the optimal design parameters that give maximum displacement of the actuator without being subjected to high stresses.

This paper is organized as follows: Section 2 presents the finite element model of the soft pneumatic actuator. Section 3 demonstrates the optimization of the design. The optimization results and discussion will be presented in section 4. Finally, concluding remarks will be drawn in Section 5.

2. FE modelling of the SPA

The actuator is designed as a tube with a semi-circular cross-section. The actuator is modelled via a FE modelling and simulation tool called Abaqus [5]. First, the tube is modelled as 3d-deformable part with inner radius of 6.35 *mm* and outer radius of 7.35 *mm* [6]. The length of the tube is 95 *mm* but the cavity inside it is only 85 *mm* in length (The tube is solid from each side by a length of 5 *mm* and the rest is cavity), as shown in figure 1. Then, the cables are created by creating another part and running our python script which will draw our cables. Two scripts exist for clockwise and anticlockwise cables. A set of the cables and an instance are added to the assembly. The cables prevent the SPA from twisting. Figures 2 and 3 show both clockwise and anticlockwise cables, respectively. An outer layer of silicon is added around the cables with thickness of 1 *mm*, as shown in figure 4. Datum planes are created to partition the tube so that the lower 0.5 mm of the tube's lower face can act as a strain-limiting material. The strain-limiting layer prevents the SPA from elongation. Since the SPA now can neither twist nor elongate, the SPA will bend when pressure is applied.

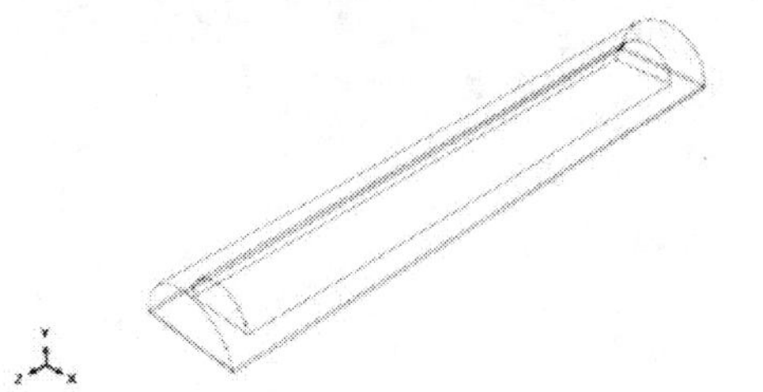

Figure 1. Tube model isometric view.

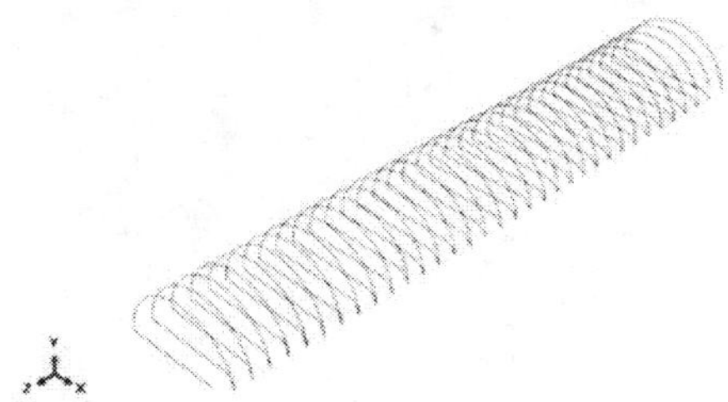

Figure 2. Clockwise cable isometric view.

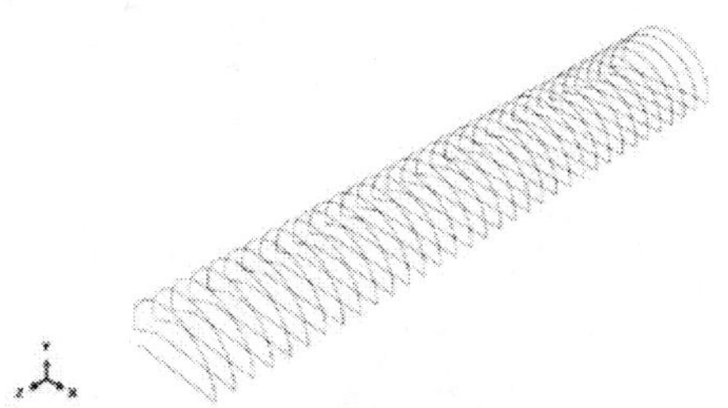

Figure 3. Anti-clockwise cable isometric view.

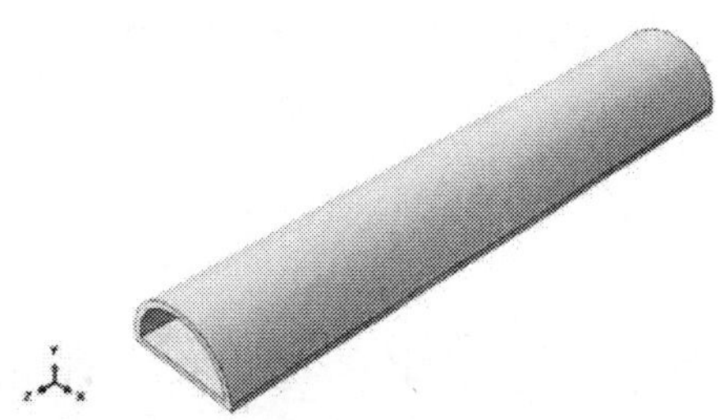

Figure 4. Outer layer isometric view.

The next step is to create the different materials used in the actuator and assign each material to each of the parts. The materials are dragon-skin 20 [7] for the actuator's upper part, dragon-skin 10 [8] for the outer layer. Fiberglass for the strain-limiting material, Kevlar for the cables. Dragon-skin 20 and Dragon-skin 10 are both modelled as hyperelastic neo-hooke materials. The material is defined with two parameters; C10 and D1, which are temperature dependant parameters, related to shear modulus (mu) and bulk modulus (kappa), where C10 = mu/2 and D1 = 2/kappa [10]. Dragon-skin 20 and Dragon-skin 10 are modelled with C10 coefficients of 0.0316485 and 0.0425 respectively, and D1 coefficients of 0 for both. Fiberglass and Kevlar are modelled as elastic materials with Young's modulus of 72 GPa and 31067 MPa and Poisson's ratio of 0.297 and 0.36, respectively.

Outer and inner surface of the actuator are then defined. Then, a step is created in which we will apply the pressure load. A static general step is selected with minimum increment size of 0.01 and maximum number of increments of 1000. The boundary conditions are then defined by imposing zero displacement on one face of the actuator. Also, an amplitude is created to control the rate at which the pressure load is applied. The amplitude is selected to be equally spaced. Next, load is created of type

pressure and applied to the inner surface that we already defined. The magnitude of pressure is of value 0.03 MPa. This value can be changed anytime to view how it affects the actuator force.
The next step is meshing. After meshing, the cables and the elastomeric tube need to be connected to each other in some way, and this connection is modelled as a 'tie constraint', which ties the cables to the surface of the actuator. Another tie constraint is created to connect the outer layer with the tube.
Finally, a job is created and submitted for analysis. Results are viewed and the actuator is functioning as shown in figure 5. The actuator bends and the displacement of any node can be viewed as xy-graph, as shown in figure 6. A node is selected at the actuator's face and the following graph shows its displacement.

Figure 5. Simulation results and bending of the actuator.

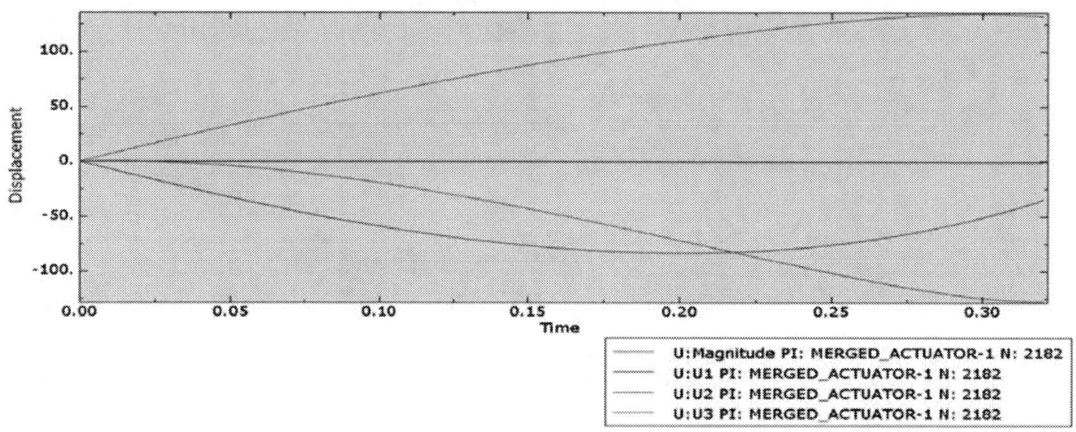

Figure 6. Displacement graph of the actuator.

3. Design optimization of the SPA

After simulation, the maximum stress and the maximum displacement of the SPA are now available. Optimization of the design of the SPA is investigated to reach a higher displacement with acceptable maximum stress. Isight (Dassault System, MA) software tool is used for the optimization problem. Abaqus is added as component. The Abaqus FE job is loaded to select input and output parameters. Neighbourhood cultivation genetic algorithm (NCGA) is selected as the optimization technique [9]. This is a multi-objective exploratory technique. Input parameters (i.e. dimensions, materials, etc.) are results from the optimization and inputs to the Abaqus job. Output parameters (i.e. maximum displacement, maximum stress, etc.) are outputs from the Abaqus job and inputs to the optimization task. The optimization process loop is shown in figure 7.

IOP Conf. Series: Materials Science and Engineering **973** (2020) 012041 doi:10.1088/1757-899X/973/1/012041

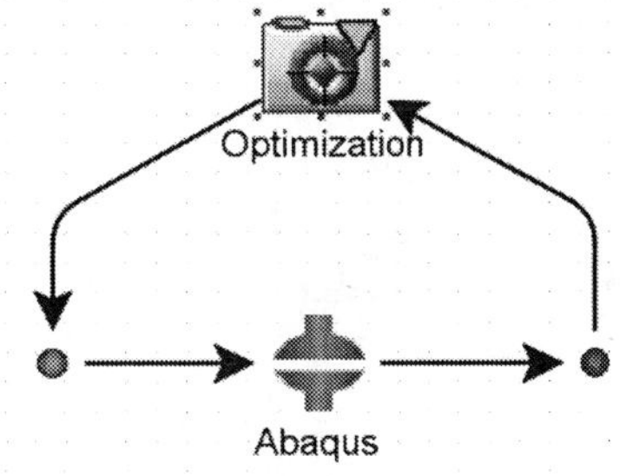

Figure 7. Optimization process loop.

3.1. Dimensional Optimization

The dimensions of the actuator are selected as input parameters. Inner radius and the solid extruded part at the tube's tip are selected and given their allowed values. By changing the solid extruded part's length, the cavity length will change. Inner radius allowed values belong to a discrete set from 4.05 *mm* to 6.35 *mm* with 0.1 *mm* increment, as shown in figure 8. The solid extruded part length allowed values are defined the same way but from 3 *mm* to 5 *mm* with 0.1 *mm* increment. Thus, the cavity length will change from 85 mm to 87 mm with 0.1 mm increment, as shown in figure 9.

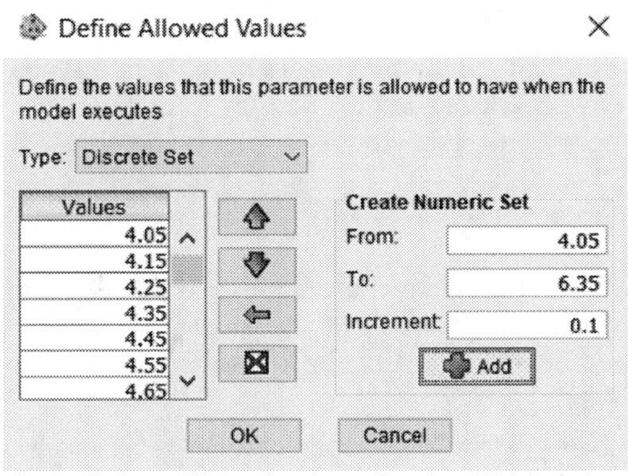

Figure 8. Defining allowed values for the SPA's inner radius.

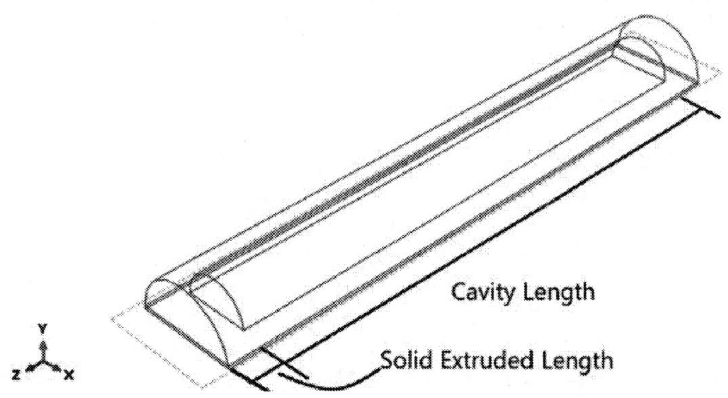

Figure 9. The relation between solid extruded length and cavity length.

IOP Conf. Series: Materials Science and Engineering **973** (2020) 012041 doi:10.1088/1757-899X/973/1/012041

The objectives are then specified. The first objective is for the displacement of the most displaced point of the SPA to be maximum. The second objective is for the maximum stress at any point on the actuator to target a value of 450 MPa, as shown in figure 10.

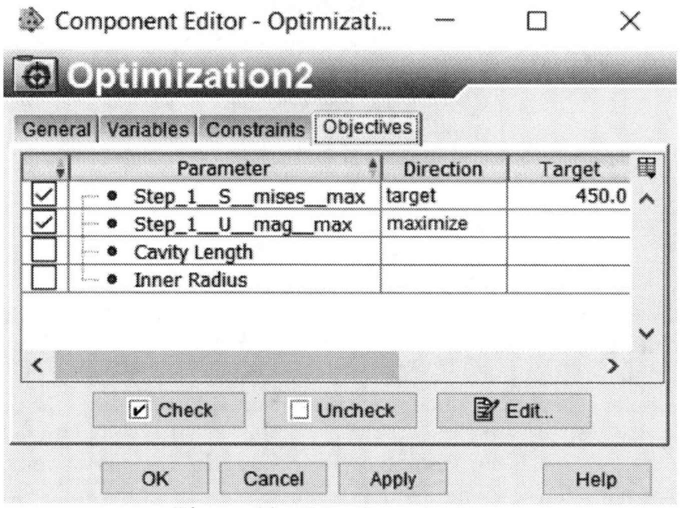

Figure 10. Objective selection.

Another model is also run but with input parameters as inner radius and outer layer thickness. The inner radius has the same allowed values and the outer layer thickness has allowed values from 0.5 mm to 5 mm with 0.5 mm increment. The output parameters and objective functions are the same as the previous model.

3.2. Material optimization

The material of the outer layer and inner layer are optimized by selecting the C10 coefficients of both materials as input parameters. The allowed values are 0.0425 for Dragon-Skin 10, 0.031648 for Dragon-Skin 20, 0.03 for EcoFlex-50 and 0.12 for Elastosil. The objective function is the same as in dimensional optimization model.

4. Optimization results

4.1. Dimensional optimization results

For the first model with inner radius and cavity length as input parameters, the number of iterations and computation time are shown in Table 1.

Table 1. Number of iterations and computation time for first model.

Number of iterations	Computation time
201	33 hours

Isight iterated the parameter for 201 times. The best cases results are in light blue and the optimum result is in green. The optimum solution is an inner radius of 6.25 *mm* and solid excluded length of 4 *mm* which corresponds to a cavity length of 86 *mm*. Figure 11 shows a part of the results' Table. It's noticed that the optimum solution corresponds to a maximum displacement of 100.58 and maximum stress of 470.33 Mpa. The best cases are the blue points and the optimum case is the green point. The solid green line in the maximum stress graph represents the target stress value of 450 MPa. Figure 12 shows the optimization results for each parameter.

IOP Conf. Series: Materials Science and Engineering **973** (2020) 012041 doi:10.1088/1757-899X/973/1/012041

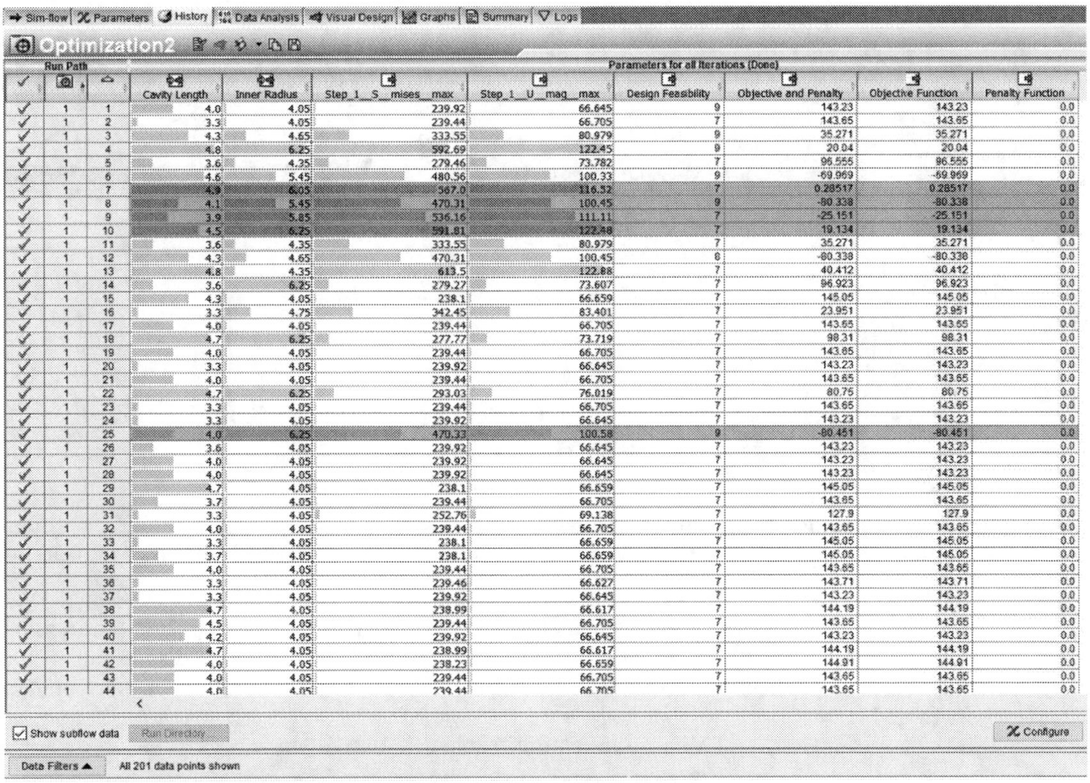

			Cavity Length	Inner Radius	Step_1__S__mises__max	Step_1__U__mag__max	Design Feasibility	Objective and Penalty	Objective Function	Penalty Function
✓	1	1	4.0	4.05	239.92	66.645	9	143.23	143.23	0.0
✓	1	2	3.3	4.05	239.44	66.705	7	143.65	143.65	0.0
✓	1	3	4.3	4.65	333.55	80.979	9	35.271	35.271	0.0
✓	1	4	4.8	6.25	592.69	122.45	9	20.04	20.04	0.0
✓	1	5	3.6	4.35	279.46	73.782	7	96.555	96.555	0.0
✓	1	6	4.6	5.45	480.56	100.33	9	-69.969	-69.969	0.0
✓	1	7	4.9	6.05	567.0	116.52	7	0.28517	0.28517	0.0
✓	1	8	4.1	5.45	470.31	100.45	9	-80.338	-80.338	0.0
✓	1	9	3.9	5.85	536.16	111.11	7	-25.151	-25.151	0.0
✓	1	10	4.5	6.25	591.81	122.48	7	19.134	19.134	0.0
✓	1	11	3.6	4.35	333.55	80.979	7	35.271	35.271	0.0
✓	1	12	4.3	4.65	470.31	100.45	8	-80.338	-80.338	0.0
✓	1	13	4.8	4.35	613.5	122.88	7	40.412	40.412	0.0
✓	1	14	3.6	6.25	279.27	73.607	7	96.923	96.923	0.0
✓	1	15	4.3	4.05	238.1	66.659	7	145.05	145.05	0.0
✓	1	16	3.3	4.75	342.45	83.401	7	23.951	23.951	0.0
✓	1	17	4.0	4.05	239.44	66.705	7	143.65	143.65	0.0
✓	1	18	4.7	6.25	277.77	73.719	7	98.31	98.31	0.0
✓	1	19	4.0	4.05	239.44	66.705	7	143.65	143.65	0.0
✓	1	20	3.3	4.05	239.92	66.645	7	143.23	143.23	0.0
✓	1	21	4.0	4.05	239.44	66.705	7	143.65	143.65	0.0
✓	1	22	4.7	6.25	293.03	76.019	7	80.75	80.75	0.0
✓	1	23	3.3	4.05	239.44	66.705	7	143.65	143.65	0.0
✓	1	24	3.3	4.05	239.92	66.645	7	143.23	143.23	0.0
✓	1	25	4.0	6.25	470.33	100.58	9	-80.451	-80.451	0.0
✓	1	26	3.6	4.05	239.92	66.645	7	143.23	143.23	0.0
✓	1	27	4.0	4.05	239.92	66.645	7	143.23	143.23	0.0
✓	1	28	4.0	4.05	239.92	66.645	7	143.23	143.23	0.0
✓	1	29	4.7	4.05	238.1	66.659	7	145.05	145.05	0.0
✓	1	30	3.7	4.05	239.44	66.705	7	143.65	143.65	0.0
✓	1	31	3.3	4.05	252.76	69.138	7	127.9	127.9	0.0
✓	1	32	4.0	4.05	239.44	66.705	7	143.65	143.65	0.0
✓	1	33	3.3	4.05	238.1	66.659	7	145.05	145.05	0.0
✓	1	34	3.7	4.05	238.1	66.659	7	145.05	145.05	0.0
✓	1	35	4.0	4.05	239.44	66.705	7	143.65	143.65	0.0
✓	1	36	3.3	4.05	239.46	66.627	7	143.71	143.71	0.0
✓	1	37	3.3	4.05	239.92	66.645	7	143.23	143.23	0.0
✓	1	38	4.7	4.05	238.99	66.617	7	144.19	144.19	0.0
✓	1	39	4.5	4.05	239.44	66.705	7	143.65	143.65	0.0
✓	1	40	4.2	4.05	239.92	66.645	7	143.23	143.23	0.0
✓	1	41	4.7	4.05	238.99	66.617	7	144.19	144.19	0.0
✓	1	42	4.0	4.05	238.23	66.659	7	144.91	144.91	0.0
✓	1	43	4.0	4.05	239.44	66.705	7	143.65	143.65	0.0
✓	1	44	4.0	4.05	239.44	66.705	7	143.65	143.65	0.0

Figure 11. Optimization results for the first model.

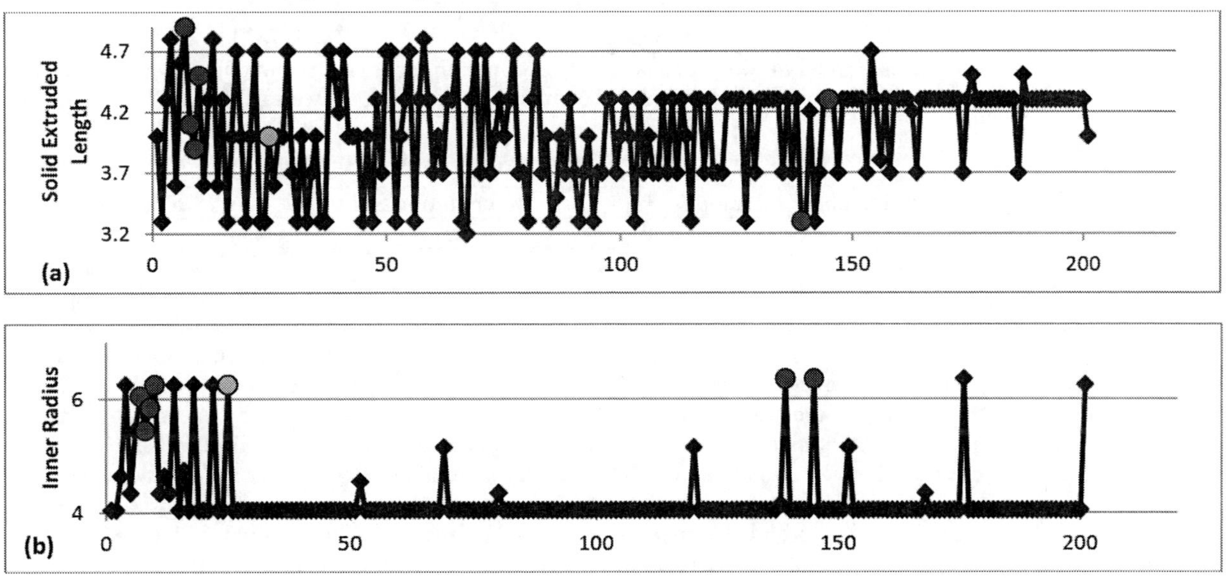

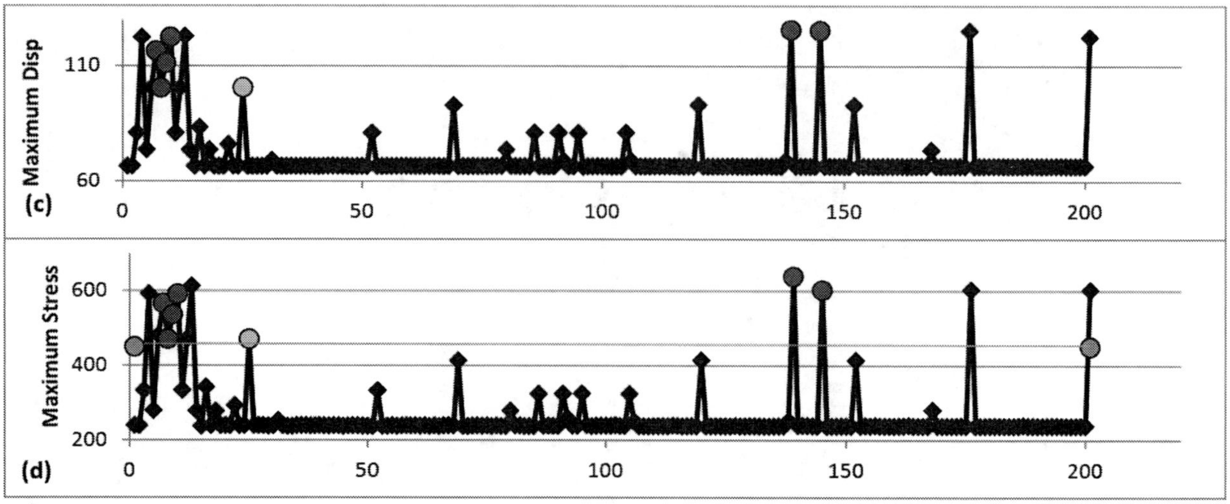

Figure 12. Optimization results of the first model. (a) shows the extruded length values for all points, (b) shows the inner radius values, (c) shows the maximum displacement and (d) shows the maximum stress values.

The linear correlation values between input and output parameters are also extracted. The solid extruded length has low positive correlation with maximum stress and maximum displacement. The inner radius has high positive correlation with both maximum stress and maximum displacement, as shown in Table 2.

Table 2. Correlation Table of first model.

	Maximum Stress	Maximum Displacement
Solid extruded Length	0.114523212	0.115576479
Inner Radius	0.843821803	0.847783115

As for the second model for dimensional optimization with inner radius and outer layer thickness as input parameters, the number of iterations and computation time are shown in Table 3.

Table 3. Number of iterations and computation time for second model.

Number of iterations	Computation time
63	47 hours

The increased computation time is due to the increase in the complexity of the model when the outer layer thickness is changed. The optimum design case is inner radius 5.15 *mm* and outer layer thickness of 0.5 *mm*. The corresponding maximum displacement and maximum stress are 66.679 and 433.53 MPa, respectively. Figure 13 shows the optimization process results for each parameter.

It's noted that one of the best cases is point 45 which the same outer layer thickness of 0.5 *mm* and an inner radius of 6.25 *mm* which has the same radius obtained from the first model. The linear correlation values between input and output parameters are also extracted. The inner radius has high positive correlation with maximum stress and low positive correlation with maximum displacement. The outer layer thickness and the maximum stress has nearly no correlation and the outer layer thickness and the maximum displacement have high positive correlation, as shown in Table 4.

4.2. Material optimization results

The material optimization process with the inner layer and outer layer materials as input parameters is iterated. Number of iterations and computation time are shown in Table 5.

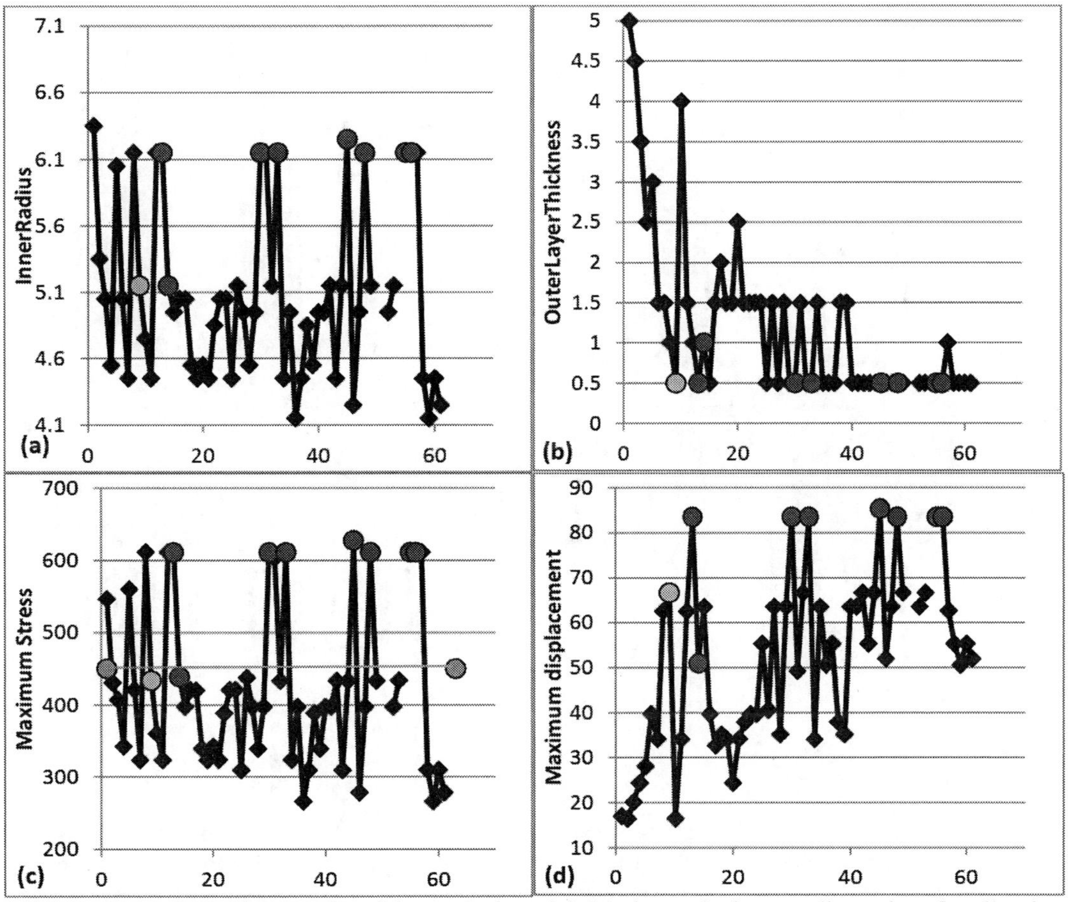

Figure 13. Optimization results of the second model. (a) shows the inner radius values for all points, (b) shows the outer layer thickness values, (c) shows the maximum stress and (d) shows the maximum displacement values.

Table 4. Correlation Table of second model.

	Maximum Stress	Maximum Displacement
Inner Radius	0.991336239	0.443522174
Outer Layer Thickness	-5.47E-04	-0.813641802

Table 5. Number of iterations and computation time for third model.

Number of iterations	Computation time
48	29 hours

The model is to select between the C10 coefficient of 4 materials: Dragon-skin 20, dragon-skin 10, EcoFlex 50 and Elastosil. The optimum design case is dragon-skin 20 for the inner layer material and EcoFlex 50 for the outer layer material. This case corresponds to a maximum displacement of 61.332 and maximum stress of 473.02 MPa. Figure 14 shows the optimization process results.

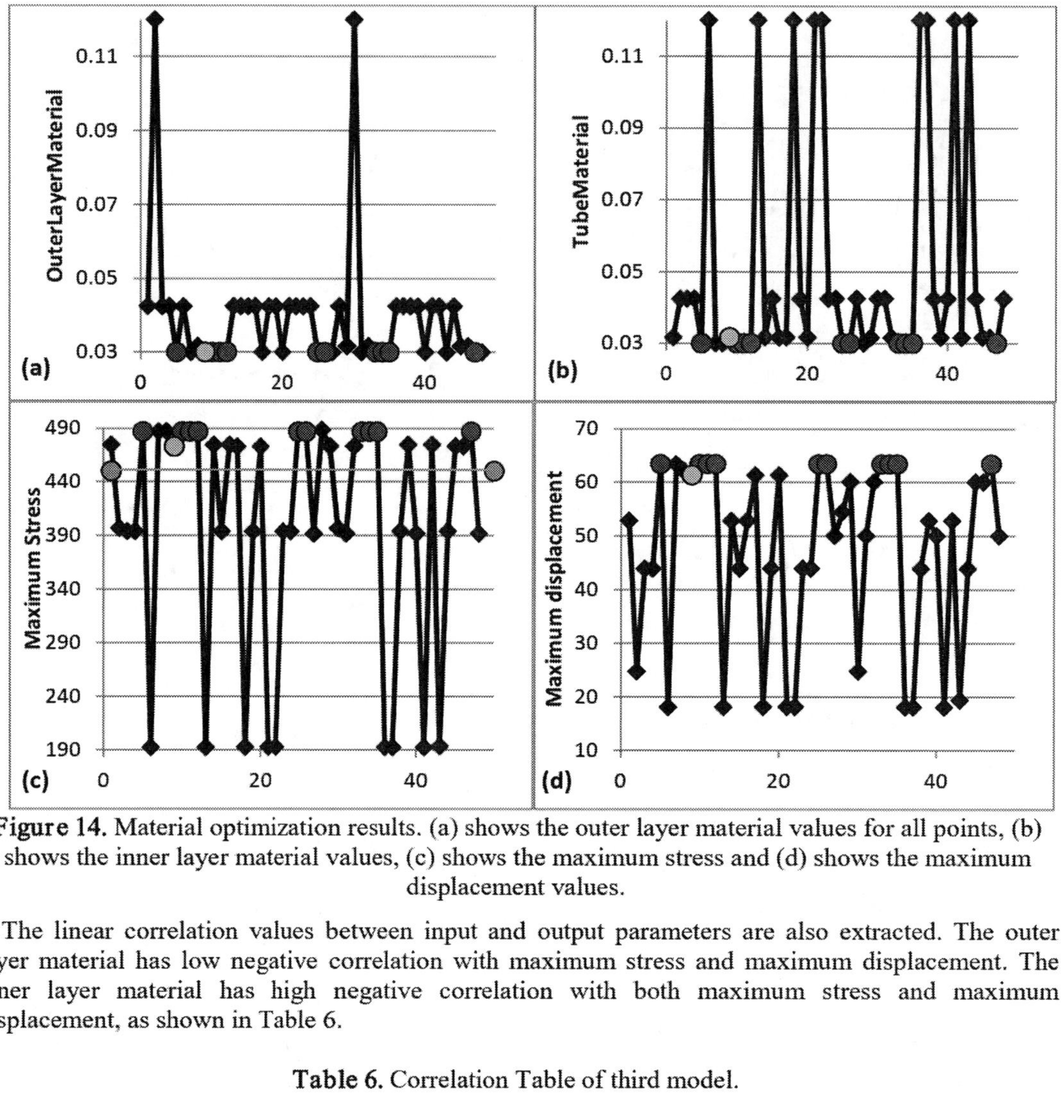

Figure 14. Material optimization results. (a) shows the outer layer material values for all points, (b) shows the inner layer material values, (c) shows the maximum stress and (d) shows the maximum displacement values.

The linear correlation values between input and output parameters are also extracted. The outer layer material has low negative correlation with maximum stress and maximum displacement. The inner layer material has high negative correlation with both maximum stress and maximum displacement, as shown in Table 6.

Table 6. Correlation Table of third model.

	Maximum Stress	Maximum Displacement
Outer Layer Material	-0.176361891	-0.475591571
Inner Layer Material	-0.978232335	-0.897922165

5. Conclusion

The design of the soft actuator can be optimized using an optimization tool that iterates the FE model several times until the optimum solution is reached. Isight optimization tool is used and three models runs are performed. Dimensional optimization results in an optimum inner radius of 6.25 mm, cavity length of 84 mm and outer layer thickness of 0.5 mm. Material optimization shows that the optimum inner layer material is dragon-skin 20 and that of outer layer is EcoFlex-50.

IOP Conf. Series: Materials Science and Engineering **973** (2020) 012041 doi:10.1088/1757-899X/973/1/012041

References

[1] Wang Z and Hirai S 2018 Geometry and Material Optimization of a Soft Pneumatic Gripper for Handling Deformable Object *IEEE International Conference on Robotics and Biomimetics (ROBIO)*, pp. 612-617 (Kuala Lumpur Malaysia)

[2] Runge G, Peters J and Raatz A 2017 *Design optimization of soft pneumatic actuators using genetic algorithms IEEE International Conference on Robotics and Biomimetics (ROBIO)* pp. 393-400 (Macau).

[3] Polygerinos P et al. June 2015 Modeling of Soft Fiber-Reinforced Bending Actuators *IEEE Transactions on Robotics*, vol. 31, no. 3, pp. 778-789

[4] Gabriel Dämmer, Sven Gablenz, Alexander Hildebrandt, and Zoltan Major 2018 *Design and shape optimization of PolyJet bellows actuators* IEEE International Conference on Soft Robotics (RoboSoft)Livorno pp. 282-287

[5] Dassault Systèmes Simulia Corp 2019 https://www.intrinsys.com/software/simulia/abaqus/why

[6] Holland D. P., Berndt S, Herman M, Walsh C 2018 Soft Roboticsvol. 5 no. 2 pp. 119-121 *Growing the Soft Robotics Community Through Knowledge-Sharing Initiatives*

[7] Smooth-on, Inc. 2019 *https://www.smooth-on.com/products/dragon-skin-20/*

[8] Smooth-on, Inc. 2019 *https://www.smooth-on.com/products/dragon-skin-10-medium/*

[9] Hiroyasu, Tomoyuki & Miki Mitsunoric 2003 *NCGA: Neighborhood Cultivation Genetic Algorithm for Multi-Objective Optimization Problems*

[10] Dassault Systèmes Simulia Corp. 2017 *https://abaqus docs.mit.edu/2017/English/SIMACAEMATRefMap /simamat-c-hyperelastic.htm*

IOP Conf. Series: Materials Science and Engineering **973** (2020) 012042 doi:10.1088/1757-899X/973/1/012042

Design of gain scheduled fuzzy PID controller for multi-link robot manipulators

Maya Emarah
Assistant Professor, The High Institute of Engineering, City of Science and Culture, 6 of October City, Giza, Egypt.
Maya.emarah2007@Gmail.com, Maya_m_e@Yahoo.com.

Abstract. Gain Scheduling (GS) controller is one of the most popular approaches to nonlinear control design. Commonly, Gain Scheduling (GS) controllers have better performance than robust ones. In this paper, the structure of a proposed gain scheduled PID fuzzy controller (GS-PIDF) designed and analyzed. The Fuzzy PID controller gains weighted through gain scheduling designed system to grantee both the stability and the best performance of a multi links robot manipulator system. The overall designed controller system is applied to two links robot manipulator system, which is subject to modelling using Lagrangian dynamics method and simulation vie MatLab and Simulink package to minimize the vibration and acquire best performance.

1. Introduction

Gain scheduling is widely technique applied for nonlinear control strategy. It has been used in many applications. Historically, gain-scheduling method applied with flight control, gas turbine control, robotics and under actuated mechanical systems [1, 2, 4 and 8].

Most techniques of gain scheduling control design are based on finding approximation linearized system at a number of selected operating points. The control parameters were then calculated based on different criteria at each operating points [1, 2, 4, 6 and 9].

The control parameters were applied to the nonlinear system under conditions of slow performance, near to the operating point and with or without stability grantee. Most old studies put several strategies to overcome the disadvantages of the gain scheduling method; e.g. working close to the operating point, slow change between operating points, freeze the system between the operating points and even global shut down the system if show unacceptable performance [1, 2, 4, 6 and 8].

In this paper, a suggested strategy of gain scheduling combined with PID fuzzy controller applied to nonlinear two links robot arm is designed. The whole operating range of the robot arm is divided into several local operating points. Linearized dynamic model of the robot arm is calculated at each operating point. The control parameters range at each operating point is calculated using Routh stability criteria. The control gains for all operating points are tabulated and interpolated in between the operating point to cover the whole operating range. The gain-scheduled system is designed to calculate and feed the control gains to the PID fuzzy control at any operating point to force the robot arm to follow the input trajectory to the system.

The paper is organized as follows. In section two, modelling of mechanical systems with equation of motion and linearized model are represented. Section three provides the design control strategy and

IOP Conf. Series: Materials Science and Engineering **973** (2020) 012042 doi:10.1088/1757-899X/973/1/012042

steps. In section four, simulated results of the gain scheduling PID fuzzy controller with two links robot arm are represented. In section five, the conclusions are summarized.

2. Modelling of Mechanical Systems

Generally, the mathematical models for multi links robot manipulators are highly nonlinear; the mathematical models of such systems usually consist of a set of linear or nonlinear differential equations derived by using some form of estimation and simulation. The dynamic models of multi links robot manipulators may depend not only on their structure but also on their internal characteristics [7 and 9].

2.1. Equation of Motion

The two link robot manipulator is a nonlinear time invariant system. The general form of the Lagrangian dynamics equation of the system is given by:

$$M(x)\ddot{x} + c(x,\dot{x})\dot{x} + G(x) = Lu(t) \quad (1)$$

$$M(x)\ddot{x} + N(x) = Lu(t) \quad (2)$$

Where, $\boldsymbol{x} = [\boldsymbol{\theta_1}, \boldsymbol{\theta_2}, \dot{\boldsymbol{\theta}}_1, \dot{\boldsymbol{\theta}}_2]^T$ the state variables of the system which taken to be the angle and angular velocity of each link, u(t) is the applied torque to the links, M(x) include the inertia matrix and N(x) is a vector including Cariolis and Centrifugal$\boldsymbol{c(x,\dot{x})}$, gravity forces $\boldsymbol{G(x)}$and damping effect.

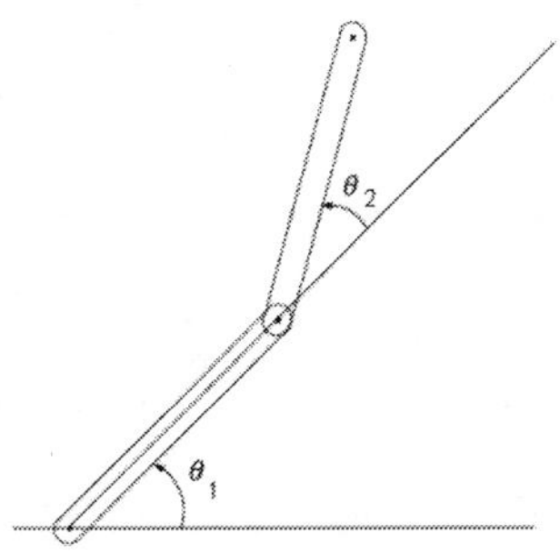

Figure 1. Two link robot arm.

The system is designed according to the following assumptions; the actuator inertia of each link is neglected and the mass of each link is concentrated as lamped mass at the end of the link.
Equation (1) is written in the classical *mass-spring-damper* form, as follows;

$$\begin{bmatrix} a_1 & a_2c_{12} \\ a_2c_{12} & a_3 \end{bmatrix}\begin{bmatrix} \ddot{\theta}_1 \\ \ddot{\theta}_2 \end{bmatrix} + \begin{bmatrix} B_1 & a_2s_{12}\dot{\theta}_2 \\ -a_2s_{12}\dot{\theta}_1 & B_2 \end{bmatrix}\begin{bmatrix} \dot{\theta}_1 \\ \dot{\theta}_2 \end{bmatrix} + \begin{bmatrix} -a_4s_1 \\ -a_5s_2 \end{bmatrix} = \begin{bmatrix} T_1 \\ T_2 \end{bmatrix} \quad (3)$$

Where,

$$\begin{aligned} s_1 &= \sin(\theta_1) \\ s_{12} &= \sin(\theta_1 - \theta_2) \\ c_{12} &= \cos(\theta_1 - \theta_2) \\ a_1 &= (m_1 + m_2)l_1^2 \\ a_2 &= m_2l_1l_2 \\ a_3 &= m_2l_1^2 \\ a_4 &= (m_1 + m_2)gl_1 \\ a_5 &= m_2gl_2 \end{aligned} \quad (4)$$

2.2. Linearized Model of Two Links Robot Arm

In order to drive the linearized model of the system the mathematical equations of the system (3-4) approximated near to each operating points representing the whole range of required trajectory the robot will take.

Equation (5) represents the linear model of two-link robot arm:

$$\begin{bmatrix} a_1 & 0 \\ 0 & a_3 \end{bmatrix}\begin{bmatrix} \ddot{\theta}_1 \\ \ddot{\theta}_2 \end{bmatrix} + \begin{bmatrix} B_1 & 0 \\ 0 & B_2 \end{bmatrix}\begin{bmatrix} \dot{\theta}_1 \\ \dot{\theta}_2 \end{bmatrix} + \begin{bmatrix} -a_4 c_1 \\ -a_5 c_2 \end{bmatrix} = \begin{bmatrix} T_1 \\ T_2 \end{bmatrix} \quad (5)$$

Where;
$c_1 = \cos(\theta_1)$
$c_2 = \cos(\theta_2)$
B_i is the damping coefficient at each hinge.
T_i is the torque at each hinge.

2.3. Case Study

The physical parameters of the two links robot arm designed as shown in Table 1:

Table 1. Specification of the two links robot arm.

Link	**1st Link**		**2nd Link**	
The link length; l_i	l_1=0.5	[m]	l_2=0.5	[m]
Link mass; m_i	m_1= 5	[kg]	m_2= 4.5	[kg]
Damping coefficient at each hinge; B_i	□□= 20	[kgm^2/s]	□□= 20	[kgm^2/s]

3. Controller Design

The gain scheduled PID fuzzy (GS-PIDF) controller system is designed according the suggested steps:

- The whole operating range of the robot arm is divided into small interval operating points, $\theta_i = 0^o \to 90^o, \Delta\theta_i = 10^o$
- The nonlinear time invariant mathematical model of the two links robot arm islinearized at the selected operating points.
- Each linearized model is applied to classical PID close loop control system.
- Using Routh stability Criteria for each operating point to calculate the range of each gain.

Where, each link has:

$$\begin{aligned} k_{Pi\,min} &< k_{Pi} < k_{Pi\,max} \\ k_{Ii\,min} &< k_{Ii} < k_{Ii\,max} \\ k_{di\,min} &< k_{di} < k_{di\,max} \end{aligned} \quad (6)$$

Where;
i = number of the links.
k_{pi} = Proportional gain,
k_{Ii} = Integral gain,
k_{di} = Derivative gain.

- The selected gain of each operating point is calculated as the norm of the maximum and minimum values.
- The selected gain is applied to the PID fuzzy controller system at each operating point to perform the whole GS-PIDF controller system [3 and 8].

The designed control system shown in the following block diagram, figure 2

IOP Conf. Series: Materials Science and Engineering **973** (2020) 012042 doi:10.1088/1757-899X/973/1/012042

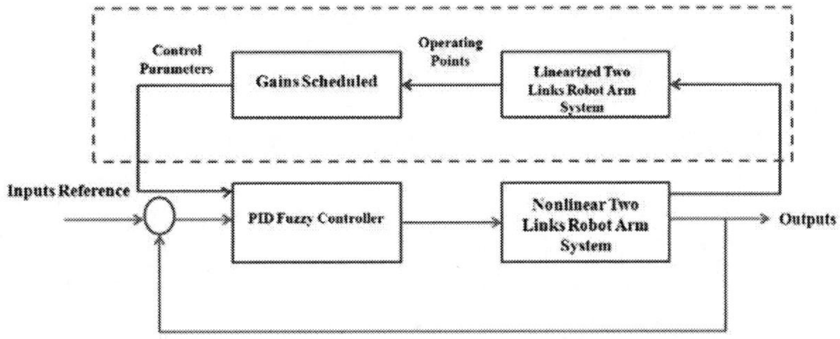

Figure 2. The GS-PID Fuzzy Controller System of Two Links Robot Arm.

4. Simulation Results

The dynamic model of two links robot arm in Equation (3) is simulated using MatLab Simulink Package using the physical parameters given in table 1. The gains variation for each link at the selected operating points is shown in figure 3. The variation of P and I gains k_{p1}, k_{p2}, k_{i1}, k_{i2} for both links are the same for each operating point in figure 3[a-b]. While, the D gains for each link k_{d1}, k_{d2} show large deviation at each operating point in figure 3[c].

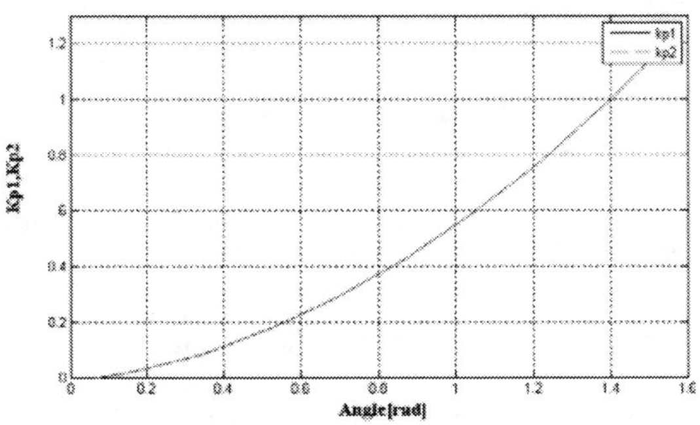

[a] The Change of P Gains with Operating Points

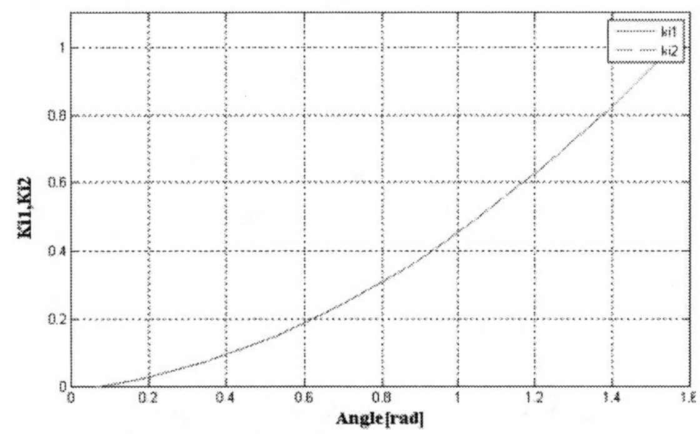

[b] The Change of I Gains with Operating Points

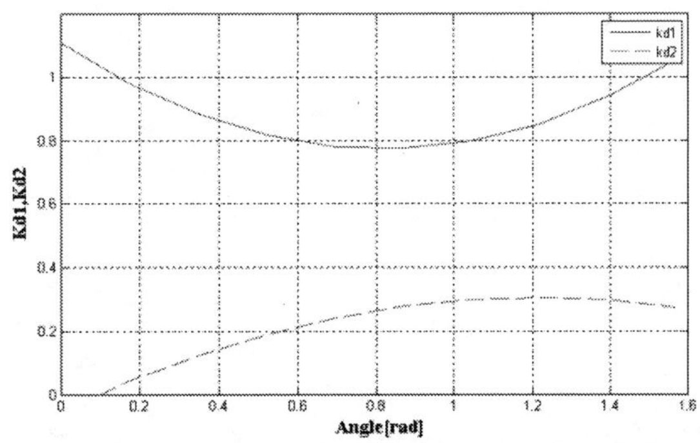

[c] The Change of D Gains with Operating Points

Figure 3. The gains variation for each link at the selected operating points.

In this study, for each link two fuzzy logic controller systems were designed; PI fuzzy logic controller has inputs: the position error (e), the integration of the error (Ie) and PD fuzzy controller has inputs: the position error (e), change in the error (de) and one output is the change of torque (du). Input and

IOP Conf. Series: Materials Science and Engineering **973** (2020) 012042 doi:10.1088/1757-899X/973/1/012042

output fuzzy members functions are symmetric and triangle member functions were used in membership functions [3, 5, and 6].

The performance of the GS-PIDF controller for unit step input is shown in figure 4. The maximum overshoot of [21% - 27%] at [0.5 – 1 sec] and zero steady state error.

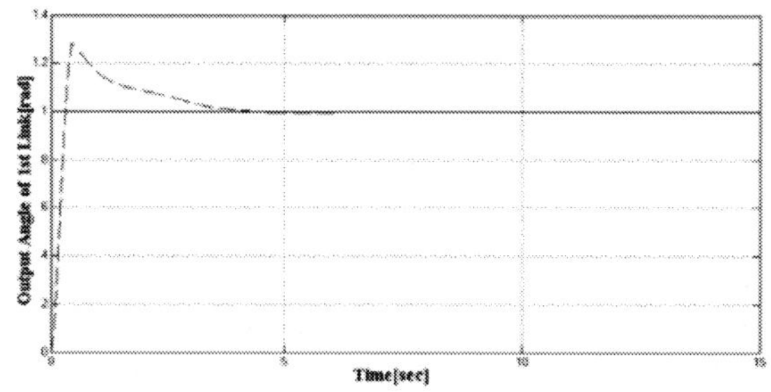

[a] 1st Link

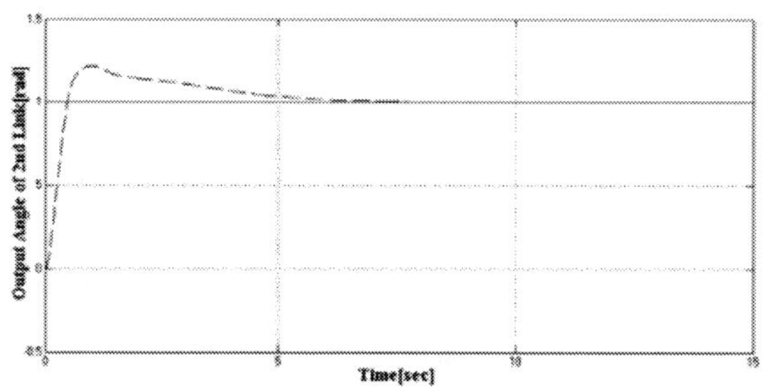

[b] 2nd Link

Figure 4. The Performance of the GS-PIDF Controller for Unit Step Input.

The GS-PIDF controller is applied to the two links robot arm with different trajectories. The controller system successes to force the robot arm to flow the given trajectory with minimum error as shown in figure 5.

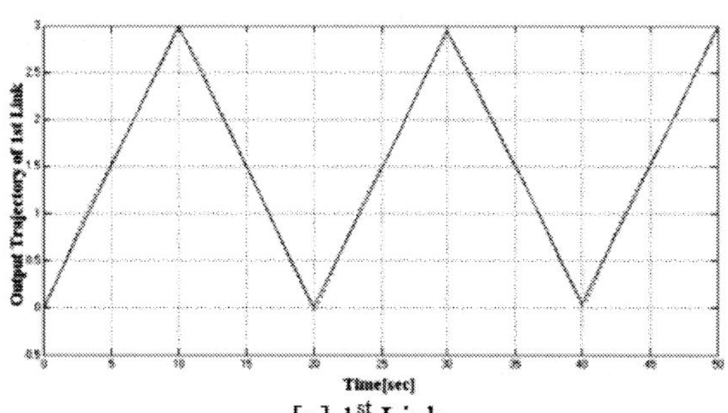

[a] 1st Link

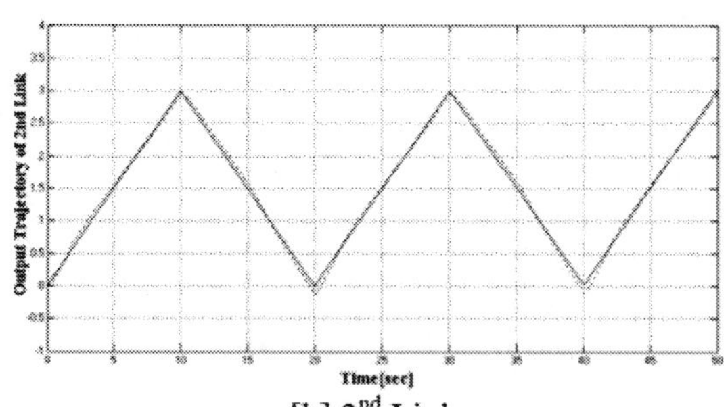

[b] 2nd Link

IOP Conf. Series: Materials Science and Engineering **973** (2020) 012042 doi:10.1088/1757-899X/973/1/012042

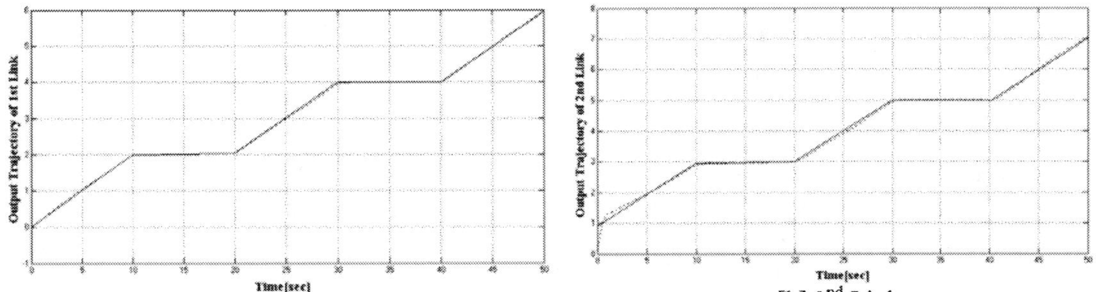

[a] 1st Link [b] 2nd Link

Figure 5. The Performance of the GS-PIDF Controller for Different Input Trajectories.

5. Conclusions

In this paper, brief acknowledge of the gain scheduled control systems was introduced. Then a suggested control strategy was designed based on linearized the nonlinear dynamic model of two links robot arm at selected operating points. The strategy is based on grantee the stability of the system at each operating point by calculating the PID gains range at each operating point using Routh stability criteria. Also, these grantees the stability performance of the system in between any two operating points. The norm PID gains at each operating point were the applied to a PID fuzzy controller to force the robot arm to follow the selected trajectory. The simulation designed for the system used MatLab and Simulink package. The control strategy shows acceptable performance of the robot arm with different given trajectories.

References

[1] Leith D J and Leithead W E 2010 Survey of gain-scheduling analysis and design *International Journal of Control* **73(11)** pp 1001-1025

[2] Ilka A 2015 *Gain-Scheduled Controller Design* (Bratislava : Slovak University of Technology Faculty of Electrical Engineering and Information Technology, Institute of Robotics and Cybernetics Ph. D. Dissertation).

[3] Jantzen 2007 *Foundations of Fuzzy* Control (John Wiley & Sons, Ltd.)

[4] Bıdıklı B 2018 *A Self-Tuning PID Control Method for Multi-Input-Multi- Output Nonlinear Systems* (İzmir Turkey : Izmir Katip Çelebi University School of Engineering and Architecture Department of Mechatronics Engineering Ph. D. Dissertation)

[5] Bedoud K, Rachedi M, Bahi T and Lakel R 2015 *Adaptive Fuzzy Gain Scheduling of PI Controller for control of the Wind Energy Conversion Systems* (Euro-Mediterranean Institute for Sustainable Development, Elsevier Ltd).

[6] Ahmed H and Rajoriya A 2016 Hybrid of Sliding Mode Control and Fuzzy Gain Scheduling PID Control using Fuzzy Supervisory Switched System for DC Motor Speed Control System *International Journal of Grid and Distributed Computing* **9** No. 5.

[7] Tarmizi M R 2014 *Design and modeling of 6 DOF revolute robot using fuzzy PID controller* (Malaysia : University Of Hussein Faculty of Electrical & Electronic Engineering Msc. Dissertation).

[8] Jadhav V K and Gawale N 2018 Gain Scheduling of PID Controller for Level Control System & Comparative Analysis of PID Controllers International *Journal of Advanced Research in Electrical, Electronics and Instrumentation Engineering.*

[9] Goos J 2016 *Modelling and identification of Linear Parameter-Varying systems*(Brussels Belgium: Vrije Universiteit Brussel Faculty of Engineering Department of Fundamental Electricity and Instrumentation Ph.D. Dissertation) .

IOP Conf. Series: Materials Science and Engineering **973** (2020) 012043 doi:10.1088/1757-899X/973/1/012043

Design of bio-inspired muscle sarcomere structure using a hybrid hydrogel-IPMC actuator

H M Hamdi[1], SAAbdelwahhab[2], MIAwad[3]and FATolbah[3]
[1] Design and Production Engineering Department, Faculty of Engineering, Ain Shams University, Cairo, Egypt.
[2] Production Technology Department, Faculty of Industrial Education, Helwan University, Cairo, Egypt.
[3] Mechatronics Engineering Department, Faculty of Engineering, Ain Shams University, Cairo, Egypt.
hend.hamdi@eng.asu.edu.eg

Abstract. Biological muscle is considered a powerful actuator due to its flexibility, lightweight, and efficiency. The building unit of a muscle, sarcomere, and the study of its energy balance cycle is considered in researches due to its importance to mimic the micro-level muscle structure to improve the artificial muscle performance. In this work, a new design of a linear actuator based on the sarcomere behavior is developed. The design is inspired by studying the four steps adenosine triphosphate (ATP) hydrolysis cycle, which is the main source of the required energy for sarcomere contraction. A new developed hybrid hydrogel-polymeric material actuator is designed in this paper using a combination between Ionic Polymeric Metallic Composites (IPMC) and hydrogel to behave like the sarcomere. This new actuator proposes an autonomous cycle using the effect oscillatory Belousov–Zhabotinsky (BZ) reaction. The physical model is proposed, and the mathematical model of the actuator is derived and formulated and identified using MATLAB/ Simulink.

1. Introduction

Bioinspired actuators have been recently intensified to cover many fields, as the continuous improvement of automated systems requires efficient energy utilization [1], where the conventional actuators have weakness in force to weight ratio [2], [3]. The natural muscle is considered an important source of inspiration for its multi-tasking motion, flexible design, and efficient energy utilization. [2].Natural muscle has many models such as Hill model, in which the muscle is studied as three element mechanical system, and Huxley model, in which the muscle is studied on the micro level and the main building unit- sarcomere- behavior is considered. The previously mentioned models are widely used, and it was found that studying the muscle behaviour on the micro-level, and simulating the sarcomere behavior gives better and smoother results, so this study focuses on Huxley's model. Then it is required to understand the four-states Adenosine Tri Phosphate (ATP) chemical cycle, which is the actuating sarcomere contraction cycle.

The evolution of Electro Active Polymers (EAP) helps to fulfil the bioinspired actuators designs. They give a large mechanical response as stretching, bending, buckling, and volumetric change under the effect of many types of physical stimulators [4]. EAP is divided into two main types, electronic and ionic. Ionic Polymeric Metal Composites (IPMC) is a class of ionic polymers [5], which exerts a

IOP Conf. Series: Materials Science and Engineering **973** (2020) 012043 doi:10.1088/1757-899X/973/1/012043

large bending value when excited by a low electric voltage. This privilege was utilized in [6] to simulate the muscle protein motion in the sarcomere. This design is extended in this work adding a chemically stimulated hydrogel terminal that can shrink, expand, and move to different binding sites. The proposed hybrid material actuator simulatesthe four-state ATP actuation cycle of the sarcomere.

2. Artificial Muscles and Their Materials

2.1 Natural Muscle Contraction Cycle

Natural skeletal muscle is responsible for motion control. It is responsible for maintaining bones and joints stability and prevents excessive movements which may cause damage or deformation[7].The sarcomere is the main actuating unit of striated muscle. It is important to study its construction and its contraction mechanism to understand the muscle behaviour.

The striated muscle consists of myofibrils shown in figure 1 (a). It consists of periodically repeated structures called sarcomeres. Each sarcomere consists of coiled protein filaments, the thin chain called actin and the thick one called myosin. Which are the main motive elements of the muscle contraction [8].

The study of the energy balance of muscle contraction cycle needs to go through the micro level of muscle and brief description of actin-myosin cycle of the sarcomere. It was proven that the trajectories of force and muscle length along time were smoother and more relevant. Studies shows that the microscopic level of modeling is more appropriate[9].

The muscle contraction cycle is directly related to the ATP hydrolysis cycle which is the main source of the required energy. In the absence of calcium, there is no activities done by any of both sarcomere motor proteins, actin and myosin. As Ca++ binds the actin sites, the cycle starts. The ATP cycle was first illustrated by Lymn and Taylor as follows:

$$ATP \rightleftharpoons ADP + Pi + energy \tag{1}$$

where ADP is the responsible for myosin head attachment to the actin and cross bridge formation. The resulting energy causes the pull the actin to move forward[10].Then, the ATP hydrolysis enables the myosin head to start the power-stroke, causes the slipping between the actin filaments and myosin head.[11].

The one- direction myosin head movement from a binding site along the actin to another is divided into 4 main steps:Disengagement between actin and myosin and a new ATP molecule arrives, preparing for the new step forward which called recovery stroke, Engagement between actin an myosin at the actin binding sites, The myosin head moves forward pulling the actin filament using the released energy coming from ATP hydrolysis[12]. The coupling between the mechanical and chemical cycles of the motor proteins is illustrated in figure 2, showing the power and return stroke synchronized with the ATP hydrolysis[13].

(a) The myofibril configuration and the action zones during (b) A single sarcomere unit.

Figure 1. Natural Sarcomere Details.

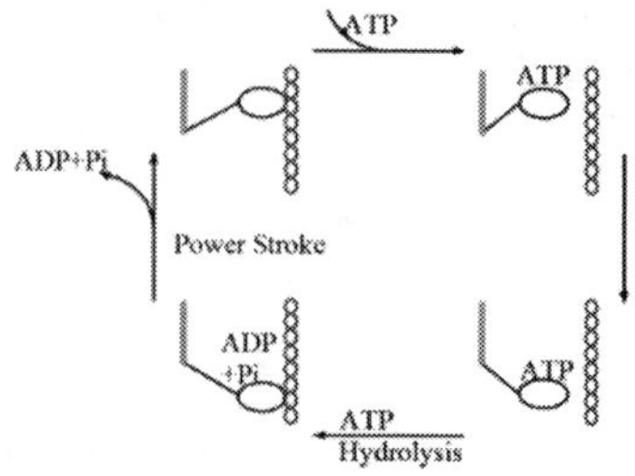

Figure 2. The coupling between the mechanical and chemical states during the ATP hydrolysis 4 steps cycle.

All the previous work agreed with the presence of four mechanical states resulting from the ATP hydrolysis reaction. Here we will focus on the actuation cycle of the sarcomere. The objective of this work is to propose a new sarcomere-like actuator simulating the previously mentioned four sates cycle of the muscle contraction in order to be simulated, aiming to optimize the energy required to perform the cycle.

2.2 *Artificial Muscles Designs and Materials*
Artificial muscles are widely used in the bio-robots field. Many designs were developed to resemble thenatural muscle behavior efficiently. The mechanical motors and pneumatic and hydraulic systems early designs of those biomimetic actuators but the large size and the need of hydraulic pumps caused low system efficiency regarding to size [15] [16].

Although there are many types of biomimetic actuators were invented to act as the natural muscle, electroactive polymers (EAP) were found to be the closest type to the actual performance, since they have close range of actuation force, same nature in deflection, and suitable mechanical work cycle needed to give the required chemical cycle steps [2]. They are superior to SMA because they have lower density and higher response and resilience. The two basic classes are electronic EAP and ionic EAP, the second is preferred duo to lower required activation voltage and higher resultant displacement. Hybrid material actuator is proposed in [17] to behave like myosin unit during power stroke and return, since the behaviour of each material results in different deflection under activation.
In current research, the designs in [6] and [17] are incorporated to propose a hybrid bio-inspired four states cycle actuator that can attach a sequence of binding sites. The IPMC is electrically actuated and acts as myosin lever arm, and attached with a chemically actuated hydrogel terminal which acts as myosin head. The self-oscillating chemical reaction enables the actuation cycle to be repeated to perform muscle like contraction.

2.3 *Hydrogels and Swelling Reaction*
Hydrogels are elastic materials that have increasing volume (or swelling) if it is actuated by stimulus, which can be chemical, electrical, magnetic, or thermal, which makes them suitable to be used in biomimetic actuators design. Chemically stimulated gels- even by oscillating reaction or varying pH- give more powerful results rather than using electrical stimulation [14].It is performed by immersing gel into water or solvent or changing the chemical environment, which results in cation and anion exchange [15]. The oscillatory Belousov–Zhabotinsky (BZ) reaction was utilized by Yoshida in [16] is one of the chemical stimulation methods for the hydrogels, generating an autonomous cycling of oxidation and reduction needed for hydrogel swelling. It possesses better performance than the pH stimulation since it doesn't need any external control for the surrounding reactors needed to change the solution status[17]. This internal stimulation cyclic was applied to propose a chemo-mechanical actuator using the effect of generated chemical wave inside the gel to provide motion. Further, the generated osmosis pressure of this propagated wave was used to actuate a chemical robot that can provide motion to an object [18].

IOP Conf. Series: Materials Science and Engineering **973** (2020) 012043 doi:10.1088/1757-899X/973/1/012043

The beating behavior of the gel is due to the existing metal ions- ruthenium Ru- in the polymer network working as a catalyst. The gel swells while Ru oxidation (Rubpy3) and de-swells while the reduction (Rubpy2). This repeating cycle which is driven by the BZ reaction provides autonomous controllable motion as a kind of chemical energy transformation to mechanical energy used in biorobots design [19].

The abovementioned behavior is utilized in the developed actuator as a beating head generates an autonomous voltage signal. This voltage signal results in cyclic IPMC strip bending. The beating gel head turn on the operating switch while swelling and the switch is turned off while deswelling.

2.4 Ionic Polymeric Metallic Composites (IPMC's)

IPMC's are commonly used in the field of soft actuator because of their ability to transform electrical energy into mechanical energy in a form of strip bending when the electric filed is applied. Further, it shows a relatively large bending displacement under the effect of low voltage[20]. It consists of Nafion membrane, plated with two electrodes on each side [21].

The shape of IPMC is a polymeric cantilever membrane contains fixed anions on the polymer chain, free cations, and water molecules. When a voltage is applied as shown in figure 4, the hydrated cations are attached to the negative electrode side, causes osmosis pressure results in overall membrane deformation, mainly bending displacement [22], [23].

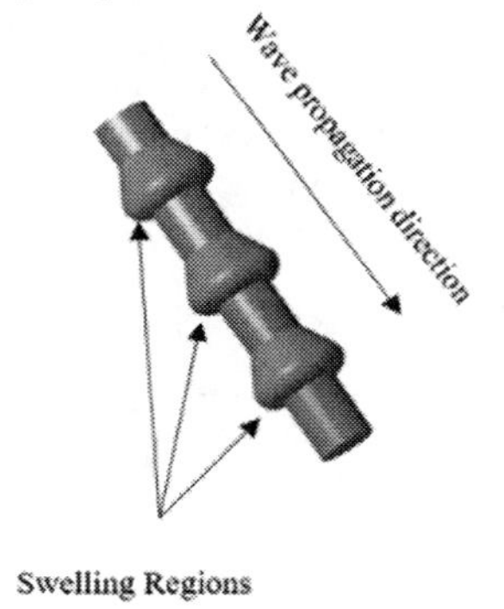

Figure 3. The chemical wave generated by the BZ reaction and the resulting swelling shape in the hydrogel.

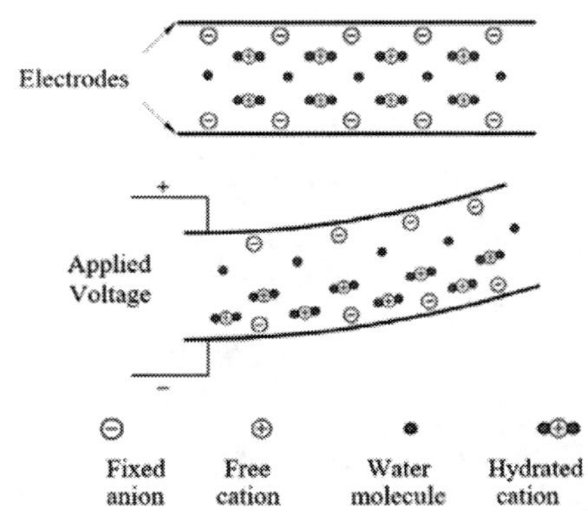

Figure 4. The effect of electrical activation of the IPMC membrane by applying voltage on the surface electrode and the resulting bending.

This previously described electromechanical behavior of the IPMC material was analyzed to determine the cantilever tip displacement and resultant force needed for actuation. The applied voltage on the IPMC strip is due to the hydrogel head expansion which results in sequence of pulses with variable duration controlled by changing the initial concentration of the acids as escribed in the following section.

3. Proposed Sarcomere-Like Actuator Mechanical Model

The presented actuator is a new sarcomere-like actuator simulating the previously mentioned four sates cycle of the muscle contraction. The proposed actuator is adopted from [6]. The design is extended to fulfil the proposed autonomous cyclic motion requirements by adding a hydrogel terminal as an actuator head. The gel terminal is affected by the cyclic oxidation-reduction resulting from the Belousov–Zhabotinsky (BZ) reaction, and the wave propagates along its edge. The operating sequence of the actuator is presented in figure 5.

The main actuating element of the proposed actuator is an IPMC strip ended with a cyclic swelling hydrogel head, which controls an on-off switch during swelling-deswelling cycle. The switching point is located inside the slider so that it is in contact with the gel wave at the middle of the cycle as illustrated in figure 6 enabling it to go forward and return by strip bending as a result of electrical activation of the bottom electrodes. The gel heads are loosely positioned into a slider slots, supplied with the acidic solution.

At the period of contact between the gel wave region and the slider connection point, an electric voltage signal is given to the electrode attached to the IPMC strip terminal, which starts bending in turn. When the wave travels away from the switching point, the voltage signal is removed letting the IPMC flapper strip to return to the idle position. Since the gel head is attached to the slider, it performs forward stroke during bending and then it stays at the new position. During the return stroke the gel head is loose and the IPMC strip returns to initial position. This cycle can be repeated enabling the slider to move steps forward instead of on-off movement and fluctuating between two points.

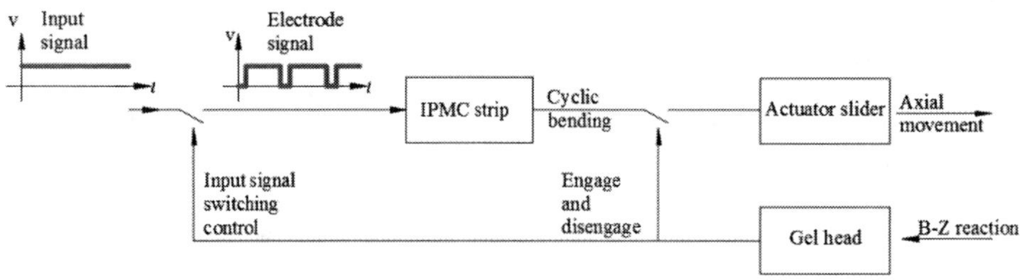

Figure 5. The operating sequence of the proposed hybrid material actuator.

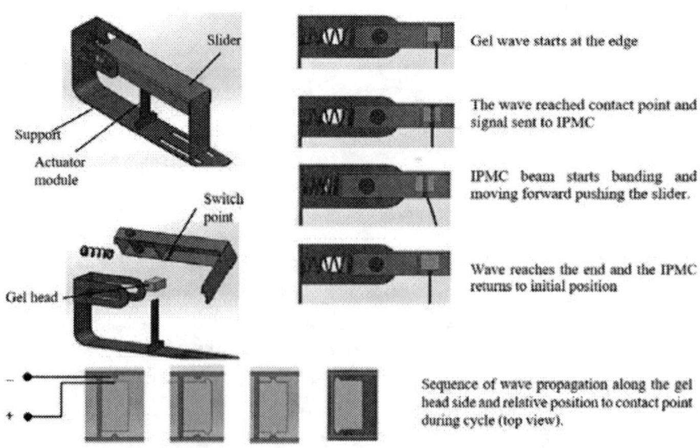

Figure 6. The proposed actuator mechanical model.

IOP Conf. Series: Materials Science and Engineering **973** (2020) 012043 doi:10.1088/1757-899X/973/1/012043

4. Mathematical Modeling

4.1 Modeling of Hydrogel Head Swelling Wave

The mathematical model of the BZ chemical reaction is presented in the following equations (2) to (4).

$$\frac{du}{dt} = F(u, v, \phi) \tag{2}$$

$$\frac{dv}{dt} = G(u, v, \phi) \tag{3}$$

$$F(u, v, \phi) = (1-\phi)^2 u - u^2 - (1-\phi) f v [u - q(1-\phi)^2/u + q(1-\phi)^2] \tag{4}$$

$$G(u, v, \phi) = \varepsilon[(1-\phi)^2 u - (1-\phi) v] \tag{5}$$

where u is the stimulus component ($HBrO_3$), v is the oxidized ion of the metal in polymer chain (Ru^{3+}), , F, G are the functions of reactants rate, f,q and $\varepsilon$are dimensionless reaction parameters, and $\phi$is the resultant volumetric fraction which is responsible of swelling and deswelling [24][19].The resulting increase in gel volumetric fraction causes the head width to expand and then increase in length. This gel swelling travelling wave can be modelled as a sine wave mentioned in equation (8) to describe the change in length with time and position [16].

$$Z_{ox}(x, t) = cos^2 \pi (x - t) \tag{6}$$

This function is dependent on wave periodic time and wavelength, it can be described by the following dimensionless mathematical presentation:

$$1 = 2\int_0^1 Z_{ox}(x, t)dx = 2\int_0^1 Z_{ox}(x, t)dt \tag{7}$$

Assuming there is no residual gel length at the edge, the resulting gel mechanical wave can be modelled as follows;

$$L(t) = 2\int_0^{l_o} Z_{ox}(x, t)dx \tag{8}$$

where l_o is the average value of gel length can be calculated as follows;

$$l_o = \int_0^1 L(t)dt \tag{9}$$

The resultant wave propagation surface along the gel head edge and the resultant start and end times for the input voltage signal will be discussed in section 5.

4.2 Electromechanical Model of IPMC:

The mathematical model of the IPMC is derived from its electromechanical nature. The electrical part of the model considers the IPMC as series of*R-C* elements, as it can store discharge[25].When the electrical activation is started by the input voltage, a discharge rate is created depends on the electrical impedance of the connected layers. Figure 7 shows the electrical IPMC model, where R represents the actuator membrane DC resistance, R_i and C_i represent the ith resistance and capacitance of the electrode element i respectively [26]. These circuit elements values estimation is evaluated according to physical considerations based on IPMC geometry as follows;

$$R_i = 2\rho_i b/Lw, \ C_i = \varepsilon_i Lw/2b \tag{10}$$

where ρ_i is electric the resistivity of segment *I* material, $\varepsilon_i$is electric permittivity of element *i*,*L*,*w* and *b* are the segment length, width and thickness.

To determine the discharge *Q(t)*, produced by the input voltage *v(t)*, the generalized equation of the electrical model is formed according to Kirchhoff's current law at segment i loop;

IOP Conf. Series: Materials Science and Engineering **973** (2020) 012043 doi:10.1088/1757-899X/973/1/012043

$$C_i v_i = \frac{v_i - v_{i+1}}{R_{i+1}} - \frac{v_{i+1} - v_{i+2}}{R_{i+2}} \tag{11}$$

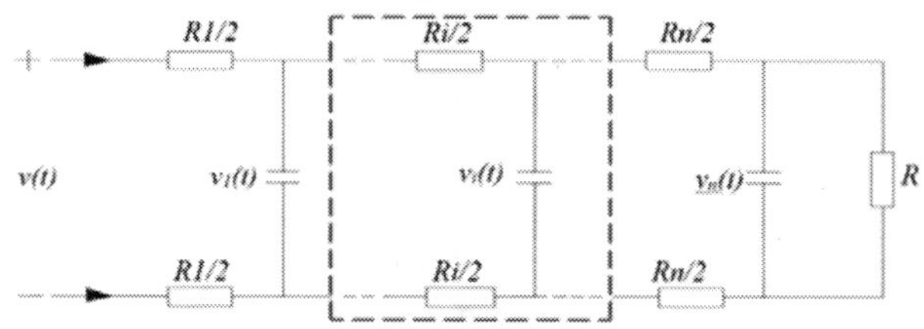

Figure 7. The electrical R-C circuit model of the IPMC.

When this model has low frequency working conditions, it is convenient to reduce the model complexity and consider the number of elements n=2 [27]. Then, relation between the input voltage *v(t)* and the absorbed current *i(t)* can be modelled in the Laplace domain as follows;

$$\frac{I(s)}{V(s)} = \frac{as^2+bs+1}{cs^2+ds+e} \tag{12}$$

Then, the discharge can be estimated from the transfer function;

$$\frac{Q(s)}{V(s)} = \frac{as+b}{cs^2+ds+e} \tag{13}$$

where,

$$\begin{aligned} a &= RR_2C_1C_2 \\ b &= RC_1 + R_2C_1 + RC_2 \\ c &= RR_1R_2C_1C_2 \\ d &= R_1R_2C_1 + RR_1C_1 + RR_1C_2 + RR_2C_2 \\ e &= R + R_1 + R_2 \end{aligned} \tag{14}$$

This discharge causes a stress distribution across the cross section of the IPMC beam, which in turns crates a banding moment. The IPMC beam deflects, figure 8 due to bending moment applied and its moves forward with a displacement δ[28].

The beam bending displacement δ with no load results from the rearrangement of water molecule inside the beam- is determined by equation (15),

$$\delta(t) = k_v W(t) = 4k_v Q_i(t) \tag{15}$$

where k_v is deformation coefficient, $W(t)$ is water molecules concentrations[29].
Then we take the Laplace transformation for equation (15) and get,

$$\delta(s) = k_v W(s) = 4k_v Q_i(s) \tag{16}$$

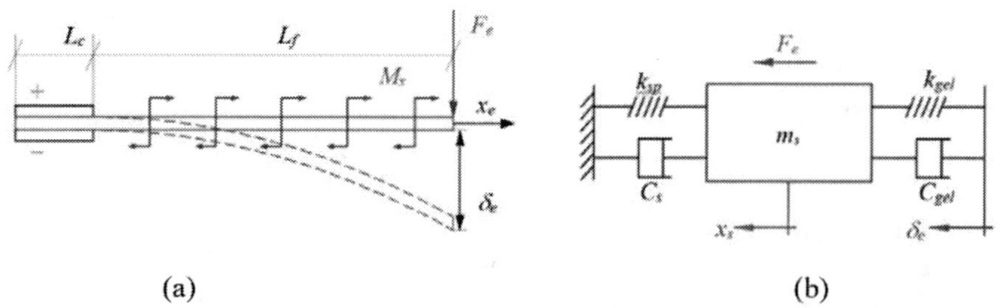

(a) (b)

Figure 8. Motion analysis of the mechanical elements (a) IPMC strip (b) Slider.

IOP Conf. Series: Materials Science and Engineering **973** (2020) 012043 doi:10.1088/1757-899X/973/1/012043

Substitute from (13) into (16) we get,

$$\delta(s) = 4k_v \frac{as+b}{cs^2+ds+e} V_i(s) \tag{17}$$

To estimate the IPMC tip force, the relationship between the tip deflection δ_x and distributed moment generated along the stripM_x is given by;

$$\delta_x(t) = M_x(t).x(l_f - \tfrac{x}{2})/EI \tag{18}$$

So, the total induced bending moment$M_l(t)$ is given by;

$$M_l(t) = \int M_x(t)dx = F_e(t).L_f \tag{19}$$

Then the tip force $F_e(t)$can be estimated as follows;

$$\delta(t) = \int \frac{M_l(t).l_f^2}{2EI} dt = \frac{F_e(t).l_f^3}{3EI} \tag{20}$$

$$\delta(s) = \frac{M(s).l_f^2}{2EIs} = \frac{F_e(s).L_f^3}{3EI} \tag{21}$$

Then, F_e canbe calculated as follows;

$$F_e(s) = \frac{3EI}{L_f^3} \delta(s) \tag{22}$$

As the slider connected to the IPMC beam end, considering the tip force as an input to the slider, then the slider equation of motion can be formulated as follows;

$$\dot{x} = \boldsymbol{A}x + \boldsymbol{B}u,\ y = \boldsymbol{C}x + \boldsymbol{D}u \tag{23}$$

where $x$is the state vector, y is the output slider motion x_s,$\boldsymbol{A} = \begin{bmatrix} 0 & 1 \\ -\frac{k_{eq}}{m_s} & -\frac{D_{eq}}{m_s} \end{bmatrix}$, $\boldsymbol{B} = \begin{bmatrix} 0 \\ 1 \end{bmatrix}$,$\boldsymbol{C}$=[1 0], $\boldsymbol{D}$=1,$u = F_e$, $m_s$is the slider mass,$k_{eq}$is the spring stiffness,D_{eq} is the damping coefficient due to sliding along the support guideways.

5. Simulation and Results

In order to analyse the proposed actuator performance, the wave propagation behavior is analysed to calculate the voltage activation and deactivation time. Times of start and end of voltage signal determines the duration of IPMC input signal.

To determine the wave propagation profile with time and edge length, equation (6) is solved and results in the dimensionless presentation shown in figure 9. The gel head edge at which the wave is created has a width of 5 mm. The slider switch point position is adjusted at the middle of gel head and its width is set so that the duration time is 0.25% of each cycle.

Figure 10 shows the chemical wave profile as it reaches and leaves the contact point, which is. The cycle time of the wave depends on the malonic acid concentration in the reaction beginning, the cycle time in this work is set to 60 seconds. It results in t1=15 sec. and t2= 30 sec. the resultant voltage input pulses for one operating cycle is illustrated and then it is simulated with signal builder block inside the Simulink model.

IOP Conf. Series: Materials Science and Engineering **973** (2020) 012043 doi:10.1088/1757-899X/973/1/012043

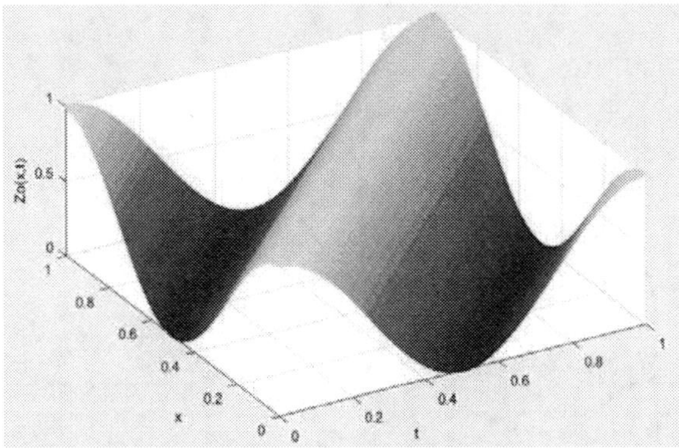

Figure 9. Chemical wave propagation along the gel head edge.

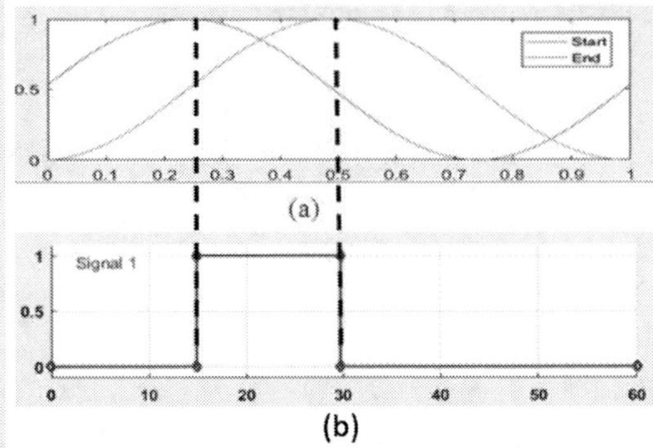

Figure 10. (a) Wave profiles at signal start and end and (b) The resultant voltage signal for one cycle.

The previously estimated IPMC mathematical model is built on MATLAB Simulink as shown in figure 11 with the input voltage signal results from the previous stage. A pulse of voltage of 1 vols is given to the system simulation as an input signal at the previously determine times. The model is estimated for an IPMC strip deflection is adopted from [30], which addresses the IPMC deflection due to input voltage at no load at the tip. The transfer function parameters (a, b, c, d, e) are estimated using system identification toolbox with one zero and two poles to match the estimated model configuration. To estimate our actuator performance, the mathematical model is extended to find the resultant IPMC tip force and slider displacement relative to the input voltage signal. The model parameters are estimated based on the mechanical and electrical properties of the moving parts given in Table 1.

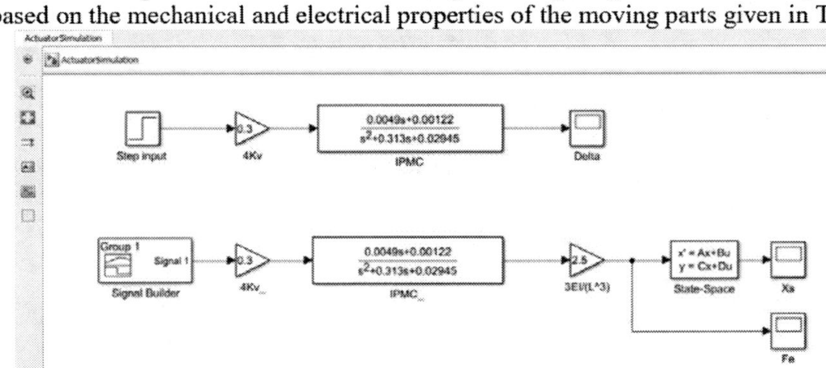

Figure 11. MATLAB Simulink model of the proposed actuator.

Table 1. Model mechanical parameters

Parameter	Value
Nafion Density *(kg/m³)*	2100
Electrode Density *(kg/m³)*	19300
E (Pa)	$8*10^7$
B (m)	0.000032
*L*w (m2)*	0.03*0.005
m_s *(kg)*	0.002
k_v	0.075
k_{eq}*(N/m)*	0.01
D_{eq}*(N.s/m)*	0.01

IOP Conf. Series: Materials Science and Engineering **973** (2020) 012043 doi:10.1088/1757-899X/973/1/012043

The model simulation was run on two steps, the first step was done to validate the estimated model on the adopted data using a step input signal of 3 volts and the step response is analyzed. Afterwards, the estimated model of the mechanical and electrical model parts is built combined with signal builder to simulate the resultant voltage signal to study the actuator performance criteria for one cycle, which are the slider displacement and actuator force.

The resultant step response of the identified model is shown in figure 12. It was analysed and found to have a good match with the original model step response. Then it was convenient to use the resultant voltage deflection transfer function given by equation (24);

$$\frac{\delta(s)}{V(s)} = \frac{0.3(0.0049s+0.00122)}{s^2+0.313s+0.02945} \quad (24)$$

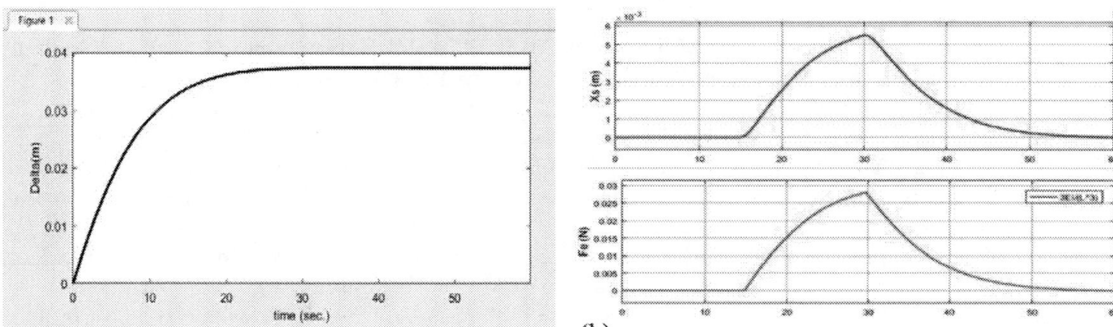

(a) (b)

Figure 12. (a) Unit step response of the IPMC tip deflection for step input 1 volt based on identified data. (b) Slider Displacement and and Actuator Force.

Based on the abovementioned results, it can be concluded that, for input signal of 1 volt for duration time of 15 seconds, the resultant actuator slider axial movement is 5 mm, and the maximum tip force is 0.028 N. These performance measurements result from one actuating module (one IPMC strip connected with gel head) illustrated in the design model.

6. Conclusion

Inspired by the four-step cycle of the natural sarcomere, this work presents an autonomous, hybrid material, single unitartificial sarcomere. The beating hydrogel head is powered by the BZ chemical reaction, the chemical wave propagation which is responsible for voltage signal creation is estimated based on the nature of the wave inside the gel. The IPMC strip is analysed by an electromechanical model to determine the proposed actuator characteristics. This work focused on the study of one operating cycle of one actuating module. The proposed design has the ability to repeat the cycle without any external signal because of cyclic nature of BZ reaction, which is considered a privilege enables the artificial actuators to act in natural behavior to improve their performance. The resultant actuation force is relatively small, so the mechanical design is proposed so that it has extendable configuration to add more modules in order to enhance the resultant response, actuation force and displacement.

References

[1] Nomura M, Volkenburgh E and Mizunami M 2016 Bioinspired Actuators and Sensors *Cambridge University Press.*

[2] Abdelwahab S 2015 Design and Development of a Motion Control System Based on Human Like Actuator and IP *Ain Shams University.*

[3] Peng C, Yin Y, Hong H, Zhang J, and Chen X 2017 Bio-inspired design methodology of sensor-actuator-structure integrated system for artificial muscle using SMA *Procedia CIRP* **65** p 299.

[4] Hilber W 2016 Stimulus-active polymer actuators for next-generation microfluidic devices *Appl. Phys. A Mater. Sci. Process* **122**(8) p 1-39.

[5] Bar-cohen Y 2002 Electroactive Polymers as Artificial Muscles: A Review Introduction **39**(6).

[6] Tolbah F, Abdelhameed M, Awad M and Abdelwahab S 2014 Modelling and simulation of a new bioinspired muscle actuator *5th Int. Work. Res. Educ. Mechatronics.*

[7] Seeley C, Rod, Russo, Andy, Regan, Jennifer and VanPutte 2014 Anatomy & Physiology *McGraw-Hill.*

[8] Williams W 2011 Huxley's model of muscle contraction with compliance *Journal of Elasticity* **105**(1–2) p 365–380.

[9] Jovanovic K, Vranic J and Miljkovic N 2015 Hill's and Huxley's muscle models - tools for simulations in biomechanics *Serbian J. Electr. Eng* **12**(1) p 53–67.

[10] Lymn R and Taylor E 1971 Mechanism of Adenosine Triphosphate Hydrolysis by Actomyosin *Biochemistry* **10**(25) p 4617–24.

[11] Chen J, Zhang X, Lin S, Wang H and Gu L 2015 Multiscale modeling of skeletal muscle active contraction in relation to mechanochemical coupling of molecular motors *Micromachines* **6**(7) p 902–14.

[12] Chen J, Zhang X and Wang H 2015 Research on Biomechanical Model of Muscle Fiber Based on Four-State Operating Mechanism of Molecular Motors *IEEE –NEMS* p 529–32.

[13] Kull F and Endow S 2013 Force generation by kinesin and myosin cytoskeletal motor proteins *Journal of Cell Science* **126** (1) p 9–19.

[14] Calvert P 2008 Gel Sensors and Actuators *MRS Bull* **33**(3) p 207–12.

[15] Ballhause D and Wallmersperger T 2008 Coupled chemo-electro-mechanical finite element simulation of hydrogels : I Chemical stimulation (*IOP Science Smart Mater. Struct* p 1-10.

[16] Yoshida R, Kokufuta E and Yamaguchi T 1999 Beating polymer gels coupled with a nonlinear chemical reaction Beating polymer gels coupled with a nonlinear chemical reaction *Choas* **9**(2) p 260-66.

[17] Yoshida R and Ueki T 2014 Evolution of self-oscillating polymer gels as autonomous polymer systems (*Nature Publishing Group* **6** p 1–14.

[18] Maeda S, Hashimoto S and Yoshida R 2004 Design of chemo-mechanical actuator using self-oscillating gel *IEEE International Conference on Robotics and Biomimetics* p 474–79.

[19] Yashin V, Levitan S and Balazs A 2015 Modeling the entrainment of self-oscillating gels to periodic mechanical deformation *Chaos: An Interdisciplinary Journal of Nonlinear Science* **25**(6) p 1-9.

[20] Vokoun D, He Q, Heller L, Yu M and Dai Z 2015 Modeling of IPMC Cantilever' Displacements and Blocking Forces *J. Bionic Eng* **12**(1) p 142–51.

[21] Kazem B and Khawwaf J 2016 Estimation Bending Deflection in an Ionic Polymer Metal Composite (IPMC) Using Artificial Neural Network Model (*Jordan Journal of Mechanical and Industrial Engineering* **10**(2) p 123–31.

[22] Kim K *et al.* 2016 Electromechanically Active Polymers *London UK- Springer.*

[23] Ruiz S 2013 3D Modeling and Design Optimization of Rod Shaped Ionic Polymer Metal Composite Actuator *University of Nevada, Las Vegas.*

[24] Yashin V and Balazs A 2007 Theoretical and computational modeling of self-oscillating polymer gels Theoretical and computational modeling of self-oscillating polymer gels *Journal of Chemical Physics* **126** p 1-17.

[25] Moeinkhah H, Rezaeepazhand J and Akbarzadeh A 2013 Analytical dynamic modeling of a cantilever IPMC actuator based on a distributed electrical circuit *Smart Mater. Struct.* **22**(5) p 1-10.

[26] Wang Z, He B, Liu X, and Wang Q 2017 Development and modeling of a new ionogel based actuator (*J. Intell. Mater. Syst. Struct.* **28**(15) p 2036–50.

[27] Dougal R, Gao L and Liu S 2004 Ultracapacitor model with automatic order selection and capacity scaling for dynamic system simulation (*J. Power Sources* **126**(1–2) p 250–57.

[28] Shi L, Guo S and Asaka K 2012 Modeling and experiments of IPMC actuators for the position precision of underwater legged microrobots *IEEE Int. Conf. Autom. Logist. ICAL* p. 415–20.

[29] Gong Y, Fan J, Tang C and Tsui C 2011 Numerical simulation of dynamic electro-mechanical response of ionic polymer-metal composites *J. Bionic Eng* **8**(3) p 263–72.

[30] Gutta S, Trabia M and Yim W 2007 Modeling of ionic polymer metal composite actuator *ASME International Mechanical Engineering Congress and Exposition.*

Performance investigation and control parameters choice for sliding mode control of coupled tanks system

M A Louis[1], M R Roman[2], O E Mahmoud[2] and M F Sedrak[2]
1 Faculty of engineering, Egyptian Russian University, Cairo, Egypt
2 Faculty of engineering at mataria, Helwan University, Cairo, Egypt
Corresponding author's e-mail address:mina-Louis@hotmail.com

Abstract. In several food processing and chemical industries, liquid is pumped and kept in interrelating coupled tanks. However, automatic regulation of the liquid level and flow control between these tanks is a challenging problem because of the complexity and high non linearity of such system. This paper deals with the liquid level control of two horizontal coupled tanks system. A comprehensive comparative study is made for most popular sliding mode control (SMC) algorithms found in literature, namely Proportional-Derivative Sliding Mode Control (PD-SMC), Proportional-Integral-Derivative SMC (PID-SMC), Fractional Order SMC and finally dynamic SMC. Special emphasis is put on the effect of the sensor noise on the controller performance. Simulated experiments including robustness to variation in plant parameters and step input disturbances are made. Control algorithms parameters are selected to optimize designed performance indices by using MATLAB optimization toolbox. Simulation results reveal that dynamic SMC is superior to other control algorithms in the presence of sensor noise and has a significant reduction in the actuator chattering phenomenon.

1. Introduction

The liquid level and flowing control between two or more tanks are essential applications in several industrial areas such as food processing, water distribution and paper industry. However, a coupled tank system is an interesting control problem due to complexity and high non linearity.An exact model besides a proper control strategy are very important so as to maintain the required tank level under uncertainties and disturbances. Numerous control tuning approaches of PID controller were used because of its structure simplicity and parameters adjustment easiness. [1], [2]. Yet, whenever requiring a good tracking with high precision of liquid level, traditional tuning methods of PID controllers failed providing an appropriate performance to level control. Moreover, most of the PID controllers were designed using lower order linearized process models [3], [4], [5] and [6]. Because of the model order reduction, this leads to additional parametric uncertainty.Nevertheless, not many PID controllers are designed on the basis of higher order model[7], [8]. However, the robustness problem still unsolved, if not considering the robustness measure in designing process.

Conventional PID was compared with fuzzy control[9]. For the nonlinear quadruple tank system, nonlinear Artificial Neural Network (ANN) control was designed [10]. Also, for coupled tank system, backstepping controllers and Adaptive controllers are implemented [11], [12] and [13]. Observer-based back stepping controller waspresented[14]. For regulating the coupled tank systemlevel, a robust decentralized PI controller was introduced[15]. A fractional order PI (FOPI) controller for coupled tank system was introduced[16] and [17]. A comparison between cascaded FOPD, FOPI and integer order PD (IOPD), integer order PI (IOPI)controllers for coupled tank system was presented [18].For

IOP Conf. Series: Materials Science and Engineering **973** (2020) 012044 doi:10.1088/1757-899X/973/1/012044

nonlinear processes control, an evaluation for multi-model FO controller is made, and was compared with multi-model IO controller [19]. Fuzzy FOPID controller for controlling thelevel of liquid of spherical coupled tanks system was introduced[20].FOID controller for theproblem of level control in two input two output liquid level system was proposed [21].

SMC has several attractive features as better disturbance rejection capabilities, good transient performance, and faster response. Mainly, in face of the uncertainties, SMC laws are inherently more robust[22], [23]. Variable structure systems (VSS) with sliding modes design and systems analysiswerestudied[24], [25]. A fuzzy SMCwith nonlinear sliding surface waspresentedfor coupled tanks system [26], such thata fuzzy logic controller was usedfor improving the chattering. An adaptive fuzzy SMCwas proposed for coupled tanks system [27]. Aneuro-fuzzy-SMCusing a nonlinear sliding surface was developed for coupled tanks system in order to smooth the switching signal[28]. A PI-SMC and a backstepping PI-SMC for a quadruple tank system are presented [29].

A static SMC design for a coupled tank system was introduced[30]. To reduce the chattering problem,two different dynamic SMC algorithms werealso proposed [30]. In a quadruple tank system,feedback linearization combinedwith SMC algorithmwas applied [31]. A second order SMC algorithm was presented[32], [33].For four coupled tank system using higher order SMC, a robust observer based controller was proposed [34].A Feed-forward adaptive second order SMC is presented to reduce the disturbance effect on the coupled tank system [35]. A chattering free SMCwas proposed to realize level position control of a coupled tank system[36]. A nonlinear SMC with varying boundary layer was presented to improve the tracking performance of a coupled tank system against various uncertainties [37]. Fuzzy FOSMC is proposed to find a chattering-free robust method for a coupled tank system [38]. SMC for quadrupletank withmulti input multi output process with time delay compensation was introduced [39].

In many industrial circumstances, sensor signal is interfered with noise. This may be caused by many reasons such as, long wires connections, close proximity to other electrical equipment, etc... In such cases the ability of the closed-loop controller to manage these noises is crucial. In many researches that are found in literature, SMC is reported as a robust faster response control that can deal with nonlinear uncertain systems. However, sensor noise effect on the SMC chattering phenomenon has not been investigated.

In this paper, a comprehensive comparative study for four control algorithms; namely PD-SMC, PID-SMC, fractional-SMC and dynamic-SMC are conducted. The comparison in the performances of the four algorithms was done by using Simulink-MATLAB. The novelty in this paper is the investigation done about the effect of the sensor noise on the performance of these famous SMC algorithms, Specially on the chattering phenomenon. The paper is organized as follows. The coupled tank system model presented in Section 2. The Controller design is proposed in Section 3. The simulation results are discussed and shown in Section 4. Finally, the conclusion in section 5.

2. The coupled tank system model

Considering the horizontal coupled tanks shown in figure 1, the system governing equations can be expressed as,

$$C\frac{dh_1}{dt} = q - q_1,\ C\frac{dh_2}{dt} = q_1 - q_2 \tag{1}$$

where, $q(t)$ is the inlet flow rate in (m^3/sec), $q_1(t)$ is the flow rate from tank 1 to tank 2 in (m^3/sec), $h_1(t)$ is the level in the first tank in (m), $h_2(t)$ is the level in the second tank in (m), $q_2(t)$ is the flow rate out of tank 2 in (m^3/sec) and C is the cross-section area of tank 1 and tank 2 in (m^2).

The intermediate and outlet flow rates q_1 and q_2 can be expressed as,

$$q_1 = c_{12}\sqrt{2g(h_1 - h_2)} \quad \text{for } h_1 > h_2,\ q_2 = c_2\sqrt{2gh_2} \quad \text{for } h_2 > 0 \tag{2}$$

where c_{12} is the coupling orifice area in (m^2), c_2 is the outlet orifice area in (m^2), and g is the gravitational constant in (m^2/sec). In the coupled tanks system, the fluid flow rate (q) into tank 1 will be positive because the water pumped only into the tank by the pump. Therefore, the inflow rate will be

$$q \geq 0 \tag{3}$$

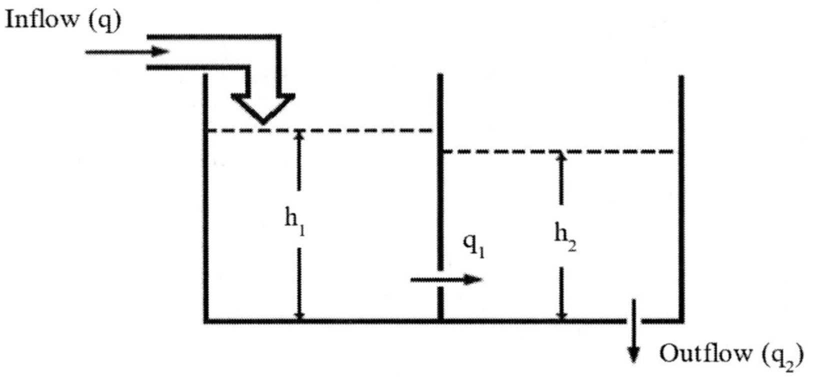

Figure 1. Horizontal coupled tanks problem.

$$\dot{h}_1 = -\frac{c_{12}}{A}\sqrt{2g|h_1-h_2|}sgn(h_1-h_2)+\frac{q}{A}$$
$$\dot{h}_2 = \frac{c_{12}}{A}\sqrt{2g|h_1-h_2|}sgn(h_1-h_2)-\frac{c_2}{A}\sqrt{2gh_2} \quad (4)$$

At equilibrium, for constant liquid level desired point, the derivatives must be zero, i.e.,
$$\dot{h}_1 = \dot{h}_2 = 0, \quad (5)$$
therefore,
$$-\frac{c_{12}}{C}\sqrt{2g\,|h_1-h_2|}\,sgn(h_1-h_2)+\frac{Q}{C}=0, \frac{c_{12}}{C}\sqrt{2g\,|h_1-h_2|}\,sgn(h_1-h_2)-\frac{c_2}{C}\sqrt{2gh_2}=0 \quad (6)$$
where Q is the equilibrium inlet flow rate. From equation (6) and to justify the restriction in equation (3) on the inlet flow rate,$sgn(h_1-h_2)$ should be positive.

Considering $z_1 = h_2 > 0,\ z_2 = h_1 - h_2 > 0,\ \mathbf{Z} = \begin{bmatrix} z_1 \\ z_2 \end{bmatrix}$, u = q(t)

Also let $a_1 = \frac{c_2\sqrt{2g}}{C}$ & $a_2 = \frac{c_{12}\sqrt{2g}}{C}$

thus, the dynamic model can be written as,
$$\dot{z}_1 = -a_1\sqrt{z_1} + a_2\sqrt{z_2}\ ,\dot{z}_2 = a_1\sqrt{z_1} - 2a_2\sqrt{z_2} + \frac{u}{C}, y = z_1 \quad (7)$$
where$z_1 = h_2$ is considered the system output.

The control algorithm objective is regulating the output $y(t) = z_1(t) = h_2(t)$ to a required value h_{2d}. As can be seen, the coupled tanks system dynamic model is complex and highly nonlinear. Therefore, a transformation will be defined so that the dynamic model in equation (7) can be converted to a form that will facilitates the control design.

Considering the state vector $\boldsymbol{x} = \begin{bmatrix} x_1 \\ x_2 \end{bmatrix}$ and defining the transformation $\boldsymbol{x} = T(\boldsymbol{z})$ such that,
$$x_1 = z_1, x_2 = -a_1\sqrt{z_1} + a_2\sqrt{z_2} \quad (8)$$
The inverse transformation $\boldsymbol{z} = T^{-1}(\boldsymbol{x})$ is such
$$z_1 = x_1, z_2 = \left(\frac{a_1\sqrt{x_1}+x_2}{a_2}\right)^2 \quad (9)$$
The dynamic model in equation (7) then can be written as,
$$\dot{x}_1 = x_2, \dot{x}_2 = \frac{a_1a_2}{2}\left(\frac{\sqrt{z_1}}{\sqrt{z_2}} - \frac{\sqrt{z_2}}{\sqrt{z_1}}\right) + \frac{a_1^2}{2} - a_2^2 + \frac{a_2}{2c}\frac{1}{\sqrt{z_2}}u \quad (10)$$
where the values of $z_1$and$z_2$are function of x_1 and x_2 as given by equation (9).

Thus, the system dynamic model can be written as,

$$\dot{x}_1 = x_2,\ \dot{x}_2 = f + \phi u, y = x_1 \tag{11}$$

where,

$$f = \frac{a_1 a_2}{2}\left(\frac{\sqrt{z_1}}{\sqrt{z_2}} - \frac{\sqrt{z_2}}{\sqrt{z_1}}\right) + \frac{a_1^2}{2} - a_2^2,\ \phi = \frac{a_2}{2c}\frac{1}{\sqrt{z_2}} \tag{12}$$

3. Controller design

3.1. Classical sliding mode controller (SMC)

SMC is a robust nonlinear control algorithm which is basing on Lyapunov method, where an nth order uncertain and nonlinear system is converted to a 1st order system. SMC has many features such as: the relative simplicity in the design, robustness to process dynamic characteristics and external disturbances. It is known that robust solution is provided by SMC to the control problem; hence it allows coping with the changes in plant without noteworthy deterioration of the performance. Even that SMC is fundamentally a discontinuous control result, however, it is clear that the control to be designed must drive the trajectory to the switching surface and once it has been reached is preserved on this surface. The problem of the control is making the system respond to track a desired and specified trajectory using a SMC. The procedure will be introduced by means of the next Single input single output system [22]:

$$x^{(n)} = f(x) + \phi(x)u, \tag{13}$$

such that x is a state vector, u is a scalar input, n is the order of the system and f and ϕ are nonlinear functions of the states. The objective of the control is that the state must follow a required vector state trajectory $x_d(t)$. Considering the surface S(t) in the state-variable form:

$$S(x,t) = \left(\frac{d}{dt} + \lambda\right)^{n-1} \tilde{x} \tag{14}$$

such that $\tilde{x} = x - x_d$ represents the error, while $\lambda$is a positive constant. Considering Lyapunov function as $V = \frac{1}{2}S^2$,selection of the control law should be in done in a way that reduces the distance to this surfacein (14) along with all state trajectories of the system (sliding condition). In other words,

$$\frac{1}{2}\frac{d}{dt}S^2 \le -\eta|S|, \tag{15}$$

such that η is a positive constant. The whole state trajectories were obviously enhanced and are getting nearer to the sliding surface in finite time, and for all future times, staying on the surface. As soon as settling the system behavior on the surface, it's said that the sliding mode ($\dot{S} = 0$) took place. When the surface is touched by the initial state for the first time, that time will be as follows

$$t_{reach} \le \frac{S(t=0)}{\eta} \tag{16}$$

Therefore, under SMC, an ideal motion consists of two phases; reaching and sliding, such that the motion is limited to the sliding surface during these phases [22]. The designing method of the sliding mode consists of two steps. The first step involving the design of a switching function $S = 0$, such that the design specifications have been satisfied by the sliding motion. The second step was the concerning of the description of the sliding mode by the control law which will be selected, so that its existence and reaching conditions are satisfied[40]. The next subsections explain the designing of switching surfaces for a coupled tank using four alternative methods. Firstly, presenting sliding surfaces based on linear compensation networks PD and PID. Then, the fractional form of one of these networks $PI^{\lambda}D^{\mu}$ is used so as to obtain the sliding surfaces. Finally, a dynamic sliding mode controller is presented.

3.1.1. Sliding surface through a PD controller. Assume H is the required system output constant value; $h_2(t)$, then defining the error

$$e = (h_2 - H) = (z_1 - H) \tag{17}$$

Therefore, obtaining PD-SMC as

$S = K_d\dot{e} + K_p e = K_d\dot{z}_1 + K_p(z_1 - H)$(18)

Differentiate both sides of equation (18) with respect to time, yields

$\dot{S} = K_d\ddot{z_1} + K_p\dot{z_1}$(19)

$\dot{S} = K_d\left((-a_1\dot{z_1}/2\sqrt{z_1}) + (a_2\dot{z_2}/2\sqrt{z_2})\right) + K_p\dot{z}_1$(20)

Substituting from (7) into (20) leads to

$\dot{S} = K_d\left[\frac{a_1^2-2a_2^2}{2} + \left(\frac{a_1a_2}{2}\right)\left(\frac{\sqrt{z_1}}{\sqrt{z_2}} - \frac{\sqrt{z_2}}{\sqrt{z_1}}\right) + \left(\frac{a_2}{2C\sqrt{z_2}}\right)u\right] + K_p(-a_1\sqrt{z_1} + a_2\sqrt{z_2})$(21)

To satisfy Lyapunov stability criterion, we introduce

$\dot{S} = -K\,\mathrm{sgn}(S)$(22)

$$\text{where} \,\mathrm{sgn}(S) = \begin{cases} +1, & \text{if } S > 0, \\ 0, & \text{if } S = 0, \\ -1, & \text{if } S < 0, \end{cases}$$

substituting in (21)we get

$-K\,\mathrm{sgn}(S) = K_d\left[\frac{a_1^2-2a_2^2}{2} + \left(\frac{a_1a_2}{2}\right)\left(\frac{\sqrt{z_1}}{\sqrt{z_2}} - \frac{\sqrt{z_2}}{\sqrt{z_1}}\right) + \left(\frac{a_2}{2C\sqrt{z_2}}\right)u\right] + K_p(-a_1\sqrt{z_1} + a_2\sqrt{z_2})$(23)

solving for u we get

$u = \left(\frac{2C\sqrt{z_2}}{a_2}\right)\left[-\frac{a_1^2}{2} + a_2^2 - \left(\frac{a_1a_2}{2}\right)\left(\frac{\sqrt{z_1}}{\sqrt{z_2}} - \frac{\sqrt{z_2}}{\sqrt{z_1}}\right)\right] - (K_pK_d)(-a_1\sqrt{z_1} + a_2\sqrt{z_2}) - (K/K_d)\,\mathrm{sgn}(S)$(24)

Using the control law in the above equation, the system states are now reaching to the hyperplane. The error vectors are forced to tend to zero asymptotically. At the same time, the required value H will be asymptotically converged by the height $h_2(t)$. Therefore, the SMC assures the output asymptotic convergence to the desired value.

The switching function control scheme suffers from the chattering problem. In the control signal, the sign function is included causing the chattering. This means that the control could switch its value at any instant and with nearly zero-time delay. The chattering could be reduced with the usage of a saturation function [22], equation (25).

$$K\,\mathrm{sat}(S/\Delta) = \begin{cases} +1, & \text{if } (S/\Delta) \geq 1, \\ S/\Delta, & \text{if} -1 < S < 1, \\ -1, & \text{if } (S/\Delta) \leq 1, \end{cases} \tag{25}$$

such that K is a switch gain with positive value, Δ is the boundary layer width and rewriting equation (24)using the saturation function we get

$u = \left(\frac{2C\sqrt{z_2}}{a_2}\right)\left[-\frac{a_1^2}{2} + a_2^2 - \left(\frac{a_1a_2}{2}\right)\left(\frac{\sqrt{z_1}}{\sqrt{z_2}} - \frac{\sqrt{z_2}}{\sqrt{z_1}}\right) - \frac{(K_p}{K_d}\right)(-a_1\sqrt{z_1} + a_2\sqrt{z_2}) - \left(\frac{K}{K_d}\right)\mathrm{sat}\left(\frac{S}{\Delta}\right)$(26)

3.1.2 Sliding surface through a PID controller. The PID controller helps getting system output to the desired value, in a short time, with minimal overshoot, and with little error [41]. It is also the most adopted controller in the industry due to the its low cost and simplicity.

A PID-SMC is designed as follows:

$S = K_p e + K_I\int e\,dt + K_d\dot{e} = K_p(z_1 - H) + K_I\int(z_1 - H)dt + K_d\dot{z}_1$(27)

Differentiate equation (27) with respect to time, results in

$\dot{S} = K_p\dot{z_1} + K_I(z_1 - H) + K_d\ddot{z_1}$(28)

Substituting (7)in (28),yields

$\dot{S} = K_p(-a_1\sqrt{z_1} + a_2\sqrt{z_2}) + K_I(z_1 - H) + K_d[\frac{a_1^2-2a_2^2}{2} + \left(\frac{a_1a_2}{2}\right)\left(\frac{\sqrt{z_1}}{\sqrt{z_2}} - \frac{\sqrt{z_2}}{\sqrt{z_1}}\right) + \left(\frac{a_2}{2C\sqrt{z_2}}\right)u]$ (29)

Same as the procedure in (24), when $\dot{S}$is forced to zero, the control signal yields

$$u = \left(\frac{2C\sqrt{z_2}}{a_2}\right)\left[-\left(\frac{K_P}{K_d}\right)(-a_1\sqrt{z_1} + a_2\sqrt{z_2}) - \left(\frac{K_I}{K_d}\right)(z_1 - H) - \frac{a_1^2}{2} + a_2^2 - \left(\frac{a_1a_2}{2}\right)\left(\frac{\sqrt{z_1}}{\sqrt{z_2}} - \frac{\sqrt{z_2}}{\sqrt{z_1}}\right)\right] - \left(\frac{K}{K_d}\right)\mathrm{sat}\left(\frac{S}{\Delta}\right)(30)$$

3.1.3 Sliding surfaces through fractional $PI^{\lambda}D^{\mu}$ controller. As many systems in real life are described using fractional order differential equations, the notion of fractional order controllers or processes were the subject of considerable researches [42], [43], [44], [45] and [46]. A popularization of the PID

controller is presented, namely; the $PI^{\lambda}D^{\mu}$ controller, such that λ and μ are the indices [46]. Obviously, the PID controller is a special case of FOPID. The expansion of fractional order of derivative and integral terms could provide much more flexibility in PID controller design.
A $PI^{\lambda}D^{\mu}$-SMC is as follow:

$$\begin{aligned} S &= K_p e + K_I D^{-\lambda} e + K_d D^{\mu} e \\ &= K_p(z_1 - H) + K_I D^{-\lambda}(z_1 - H) + K_d D^{\mu}(z_1 - H) \end{aligned} \quad (31)$$

such that D is a general fundamental operator which denotes the fractional order differentiator and K_p, K_I, K_d, λ and μ are the design parameters which have to be determined. Taking the time derivative for both side of equation (31), results in

$$\begin{aligned} \dot{S} &= K_p \dot{z}_1 + K_I D^{1-\lambda}(z_1 - H) + K_d D^{1+\mu}(z_1 - H) \\ &= K_p \dot{z}_1 + K_I D^{1-\lambda}(z_1 - H) + K_d D^{\mu-1}\ddot{z}_1 \end{aligned} \quad (32)$$

Using the defined variable in (7) resulting in

$$\dot{S} = K_p \dot{z}_1 + K_I D^{1-\lambda}(z_1 - H) + K_d D^{\mu-1}\left(\left(-\frac{c_1 \dot{z}_1}{2\sqrt{z_1}}\right) + \left(\frac{c_2 \dot{z}_2}{2\sqrt{z_2}}\right)\right) (33)$$

$$\dot{S} = K_p(-a_1\sqrt{z_1} + a_2\sqrt{z_2}) + K_I D^{1-\lambda}(z_1 - H) + K_d D^{\mu-1}\left[\frac{a_1^2 - 2a_2^2}{2} + \left(\frac{a_1 a_2}{2}\right)\left(\frac{\sqrt{z_1}}{\sqrt{z_2}} - \frac{\sqrt{z_2}}{\sqrt{z_1}}\right) + \left(\frac{a_2}{2C\sqrt{z_2}}\right)u\right] (34)$$

Forcing $\dot{S} = 0$, provides:

$$K_d D^{\mu-1}\left(\left(\frac{a_2}{2C\sqrt{z_2}}\right)u\right) = -K_p(-a_1\sqrt{z_1} + a_2\sqrt{z_2}) - K_I D^{1-\lambda}(z_1 - H) - K_d D^{\mu-1}\left[\frac{a_1^2 - 2a_2^2}{2} + \left(\frac{a_1 a_2}{2}\right)\left(\frac{\sqrt{z_1}}{\sqrt{z_2}} - \frac{\sqrt{z_2}}{\sqrt{z_1}}\right)\right] \quad (35)$$

Then obtaining an equivalent control signal as

$$u = \left(\frac{2C\sqrt{z_2}}{a_2}\right)\left[\left(\frac{-K_p}{K_d}\right)D^{1-\mu}(-a_1\sqrt{z_1} + a_2\sqrt{z_2}) + \left(-\frac{K_I}{K_d}\right)D^{2-\lambda-\mu}(z_1 - H) - \frac{a_1^2 - 2a_2^2}{2} - \left(\frac{a_1 a_2}{2}\right)\left(\frac{\sqrt{z_1}}{\sqrt{z_2}} - \frac{\sqrt{z_2}}{\sqrt{z_1}}\right)\right] - \left(\frac{D^{1-\mu} K sat\left(\frac{S}{\Delta}\right)}{K_d}\right) \quad (36)$$

3.1.4 Dynamic sliding mode controller. To decrease the chattering because of the static-SMC (PD-SMC), a dynamic SMC is introduced [30]. Let α_1, α_2 be positive scalers, an input-dependent sliding surface S is defined as,

$$S = \ddot{x}_1 + \alpha_1 \dot{x}_1 + \alpha_2(z_1 - H) (37)$$

where x_1 is defined by the alternate system model defined by equations (11) and (12). Substituting (11) and (12) into (37) we get

$$S = \left(\frac{a_1 a_2}{2}\right)\left(\frac{\sqrt{z_1}}{\sqrt{z_2}} - \frac{\sqrt{z_2}}{\sqrt{z_1}}\right) + \frac{a_1^2}{2} - a_2^2 + \frac{a_2}{2C}\frac{1}{\sqrt{z_2}}u + \alpha_1(-a_1\sqrt{z_1} + a_2\sqrt{z_2}) + \alpha_2(z_1 - H) (38)$$

Taking the derivative of equation (37) with respect to time using equations (8), (10) we obtain

$$\dot{S} = \dddot{x}_1 + \alpha_1 \ddot{x}_1 + \alpha_2 \dot{x}_1$$

$$= f_1 + \alpha_1\left(\frac{a_1 a_2}{2}\left(\frac{\sqrt{z_1}}{\sqrt{z_2}} - \frac{\sqrt{z_2}}{\sqrt{z_1}}\right) + \frac{a_1^2}{2} - a_2^2 + \frac{a_2}{2C}\frac{1}{\sqrt{z_2}}u\right) + \alpha_2(-a_1\sqrt{z_1} + a_2\sqrt{z_2}) + \frac{a_2}{2C}\frac{1}{\sqrt{z_2}}\dot{u} - \frac{a_2}{4C}\frac{1}{\sqrt{z_2^3}}\left(a_1\sqrt{z_1} - 2a_2\sqrt{z_2} + \frac{1}{C}u\right)u \ (39)$$

Satisfying Lyapunov stability criterion by equating (39) to, $\dot{S} = -Ksat(S/\Delta)$, from which we get,

IOP Conf. Series: Materials Science and Engineering **973** (2020) 012044 doi:10.1088/1757-899X/973/1/012044

$$\dot{u} = \frac{-2C\sqrt{z_2}}{a_2}\left[f_1 + \alpha_1\left(\frac{a_1 a_2}{2}\left(\frac{\sqrt{z_1}}{\sqrt{z_2}} - \frac{\sqrt{z_2}}{\sqrt{z_1}}\right) + \frac{a_1^2}{2} - a_2^2 + \frac{a_2}{2C}\frac{1}{\sqrt{z_2}}u\right) + \alpha_2(-a_1\sqrt{z_1} + a_2\sqrt{z_2}) + K\mathrm{sat}\left(\frac{S}{\Delta}\right)\right] + \frac{1}{2z_2}\left(a_1\sqrt{z_1} - 2a_2\sqrt{z_2} + \frac{1}{C}u\right)u \quad (40)$$

where

$$f_1 = \frac{-a_1 a_2(z_1+z_2)}{4\sqrt{(z_1 z_2)^3}} * \left(a_1 z_1^{\frac{3}{2}} - 2a_2 z_1\sqrt{z_2} + \frac{1}{C}z_1 u + a_1 z_2\sqrt{z_1} - a_2 z_2^{\frac{3}{2}}\right) \quad (41)$$

The state trajectories accompanying the unforced discontinuous dynamics equation (22) show a finite time reachability to zero from any given initial condition providing that the constant K is positive. Because S is driven to zero in finite time, the output $y = z_1 = h_2$ is governed after such a finite time, by the second-order dynamics $\ddot{y} + \alpha_1\dot{y} + \alpha_2(y - H) = 0$. Hence, the required value H will be asymptotically converged by the output y(t), because α_1 and α_2 are positive scalars. Then, the dynamic SMC given by equations (40) and (41) assures the asymptotic convergence of the output y(t) $= z_1(t) = h_2(t)$ to its required value H.

4. Simulation Results and Discussion

Each control algorithm has between four to seven tuning parameters. Selecting controller parameters by using the trial and error method that was used in many past researches to achieve certain response requirements is not practical. In this research, MATLAB optimization toolbox is used in order to find the optimal values of the controller parameters to achieve the specified response. The optimization toolbox provides functions for finding parameters that minimize or maximize objectives while satisfying certain constraints. The toolbox can perform design optimization tasks, including parameter estimation and parameter tuning. The optimization method used in order to find the optimal parameters values is the gradient descent method. Figure 2 describes the constraints used for the optimization in all proposed algorithms. The rise time (90%) is 35 seconds, the settling time (5%) is 100 seconds, overshoot is 8% and undershoot is 5%.

To ensure finding the best controller parameters, two objective functions are defined. The first is the integral of square of the error between the liquid level, h_2,and the desired height, h_{2d}. This function will be called error index.

$$Q_1 = \int (h_2 - h_{2d})^2 \,.\, dt \quad (42)$$

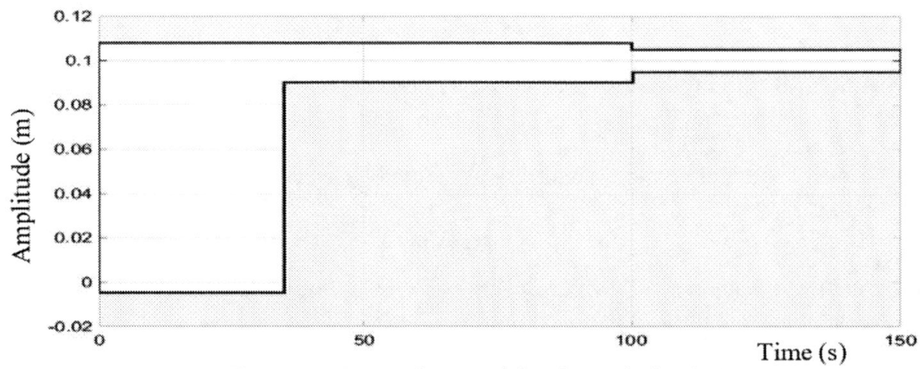

Figure 2. Constraints used for the optimization.

The second objective function is the integral of the square of time derivative of the control signal, u. This function will be called chattering index. It is described as

$$Q_2 = \int \left(\frac{du}{dt}\right)^2 .\, dt \quad (43)$$

This function decreases as the chattering rate of the control signal decreases. Minimizing both functions while keeping the signal inside the constraints described above ensure the best performance of the controller. The optimization algorithm searches for the best combinations of the controller parameters to satisfy this target. The range for searching for each controller parameter is described in table 1:

Table 1. Searching range for each controller parameter.

Coefficient	Range	Coefficient	Range
K	0→∞	μ	0→1
Δ	0→∞	Λ	0→1
K_p	0→∞	α_1	0→∞
K_I	0→∞	α_2	0→∞
K_d	0→∞		

4.1 Step response

The coupled tank system is subjected to a step input of 0.1 meters. Table 2 shows the tuning parameters found by the optimization algorithm for each controller to fit the response into the constraints shown in figure 4 and to satisfy both objective functions given by (42) and (43).Figures 3 to 6 show the response of the four control algorithms to this disturbance. As can be seen from the figures, PD-SMC, PID-SMC and Dynamic-SMC reached the required level with zero steady state error, while the fractional PID-SMC has a steady state error of 0.0002 m. PD-SMC has the minimum overshoot of 1.6% while the PID-SMC has the maximum overshoot of 4.9%. The four algorithms almost do not have any chattering as can be seen from figure 3-b to figure 6-b. A comparison between the rise time, percentage overshoot, steady state error, error index, and the settling time (5% criterion) for the four controllers is shown in table 3. As can be seen from the table that PD-SMC has the minimum error index.

Table 2. Tuning parameters for each controller.

Parameter	PD-SMC	PID-SMC	Fractional-PID	Dynamic-SMC
K	1.935	10.17	2.59	1.036
Δ	0.832	47.213	0.509	0.406
K_p	8.55	1.828	14.765	-
K_I	-	$1.0*10^{-5}$	$1.416*10^{-6}$	-
K_d	50.865	1.099	78.653	-
λ	-	-	0.999	-
μ	-	-	0.965	-
α_1	-	-	-	0.233
α_2	-	-	-	0.012

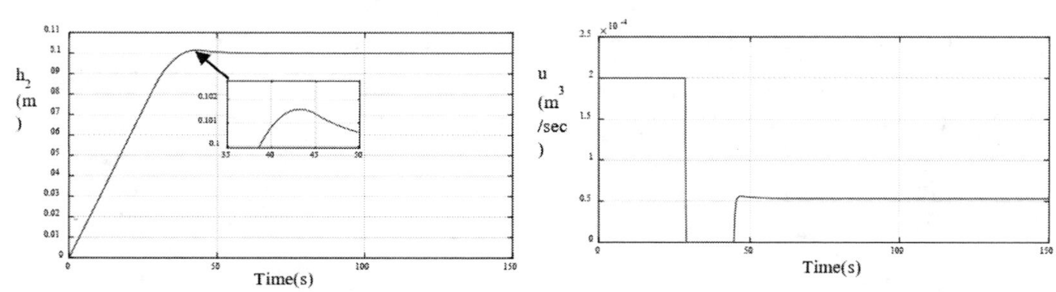

(a) (b)

Figure 3. PD-SMC (a) Step response (b) Control signal.

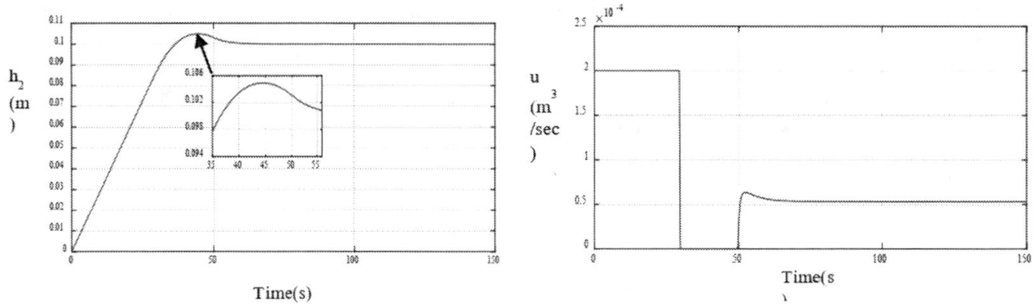

(a) (b)

Figure 4. PID-SMC (a) Step response (b) Control signal.

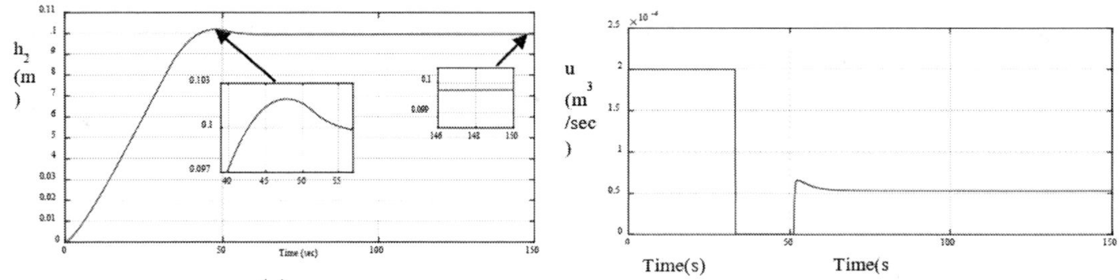

(a) (b)

Figure 5. FractionalPID-SMC (a) Step response (b) Control signal.

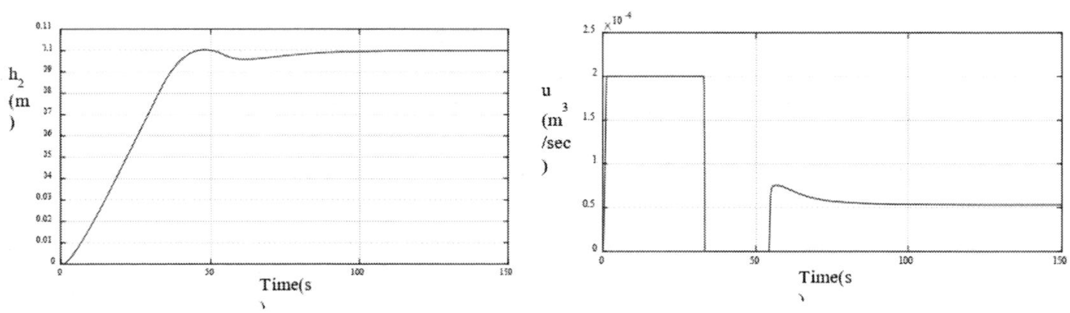

(a) (b)

Figure 6. Dynamic-SMC (a) Step response (b) Control signal.

Table 3. Comparison between the step response of the four controllers.

Controller	Rise time (s)	Overshoot (%)	Steady state error (m)	Error index (m^2.s)	Settling time (s)
PD-SMC	28	1.6	Zero	0.1159	34.2
PID-SMC	27.8	4.9	Zero	0.1162	33.5
Fractional-PID	29.8	1.9	0.0002	0.1488	38.5
Dynamic-SMC	30	-	Zero	0.1573	39.8

4.2 Input disturbance

An input disturbance step signal of -0.0001 m^3/s (50% of the full scale of the pump) is added to the control signal (u) when the running time in the simulation is 100 seconds. Figures 7 to 10 describe the response of the four controllers to that disturbance. As can be seen from the figures, the four controllers almost reach the desired value but all have steady state error. The dynamic-SMC has the lowest steady state error of 0.0001 m while the PD-SMC and PID-SMC have the largest steady state error of about 0.0007 m.Also, the four algorithms almost do not cause any chattering in the control signal,figures 7-b to 10-b. A comparison in the values of the steady state error, the error index and the settling time (5% criterion) is shown in table 4.

Table 4. Comparison between the four controllers after a disturbance signal is applied.

Controller	Steady state error (m)	Error index (m^2.s)	Settling time (s)
PD-SMC	0.0007	0.11596	34.2
PID-SMC	0.0007	0.11626	33.5
Fractional-PID	0.0005	0.1488	38.5
Dynamic-SMC	0.0001	0.1573	39.8

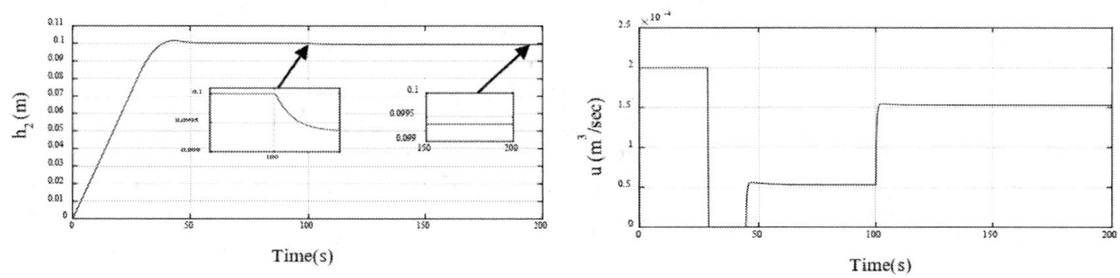

(a) (b)

Figure 7. PD-SMC with a step disturbance in the control signal (a) Response (b) Control signal.

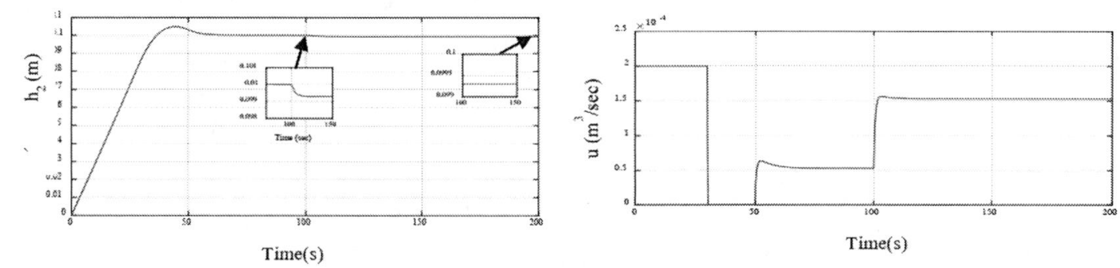

(a) (b)

Figure 8. PID-SMC with a step disturbance in the control signal (a) Response (b) Control signal.

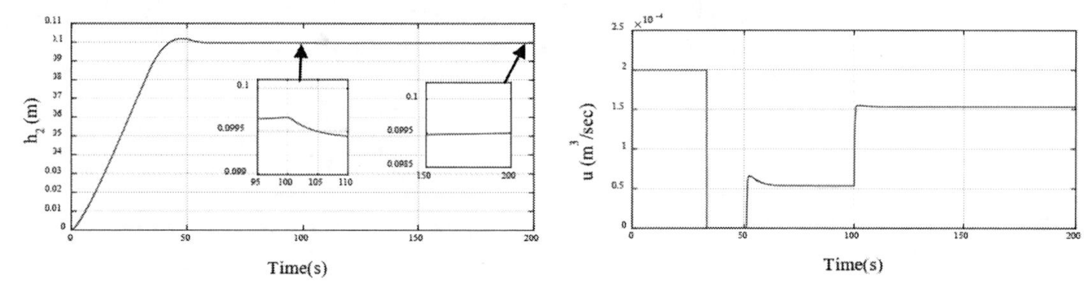

(a) (b)

Figure 9. FractionalPID-SMC with a step disturbance in the control signal (a) Response (b) Control signal.

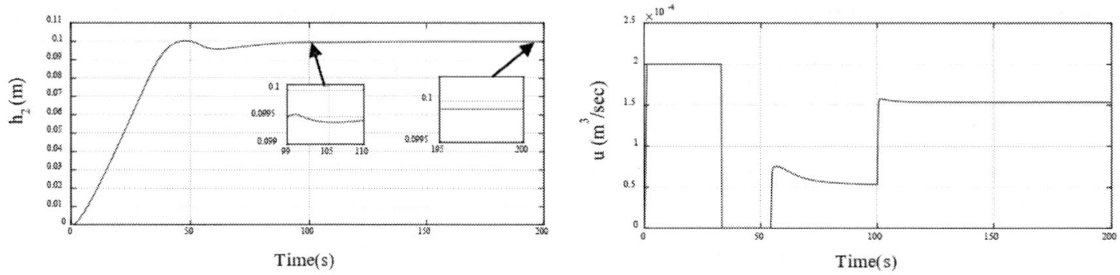

(a) (b)

Figure 10. Dynamic-SMC with a step disturbance in the control signal (a) Response (b) Control signal.

4.3 Robustness test: 25% variation in plant parameters

To investigate the robustness of the proposed control algorithms, a 25% variation in the plant parameters, a_1, a_2 and c is applied. As can be seen from figures 11 to 14, all of the four algorithms reached the required value and still the fractional order PID has the highest steady state error.Also, the four algorithms almost do not have any chattering,figures 11-b to 14-b. However, as can be noticed, the dynamic-SMC response has an overshoot of 24.2%. A comparison between the rise time, percentage overshoot, steady state error, the error index, and the settling time (5% criterion) for the four controllers is given in table 5. As can be seen from the table, PD-SMC has the minimum error index of 0.1342 m^2.s and the minimum rise time of 34.7 s while both PD-SMC and PID-SMC have the minimum steady state error of 0.0002 m.

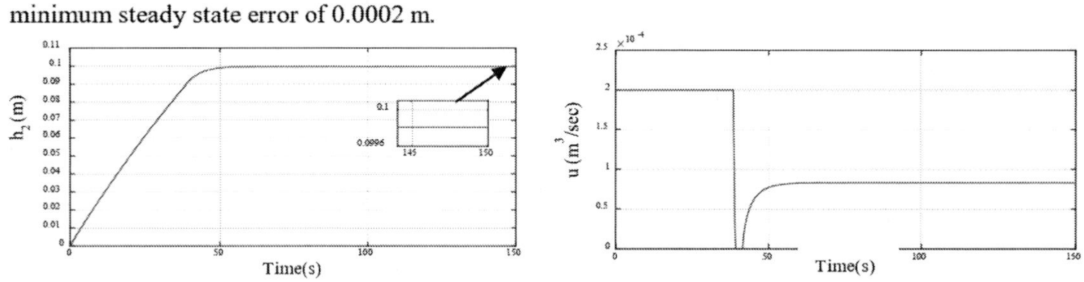

(a) (b)

Figure 11.PD-SMC with 25% variation in the plant parameters (a) Response (b) Control Signal.

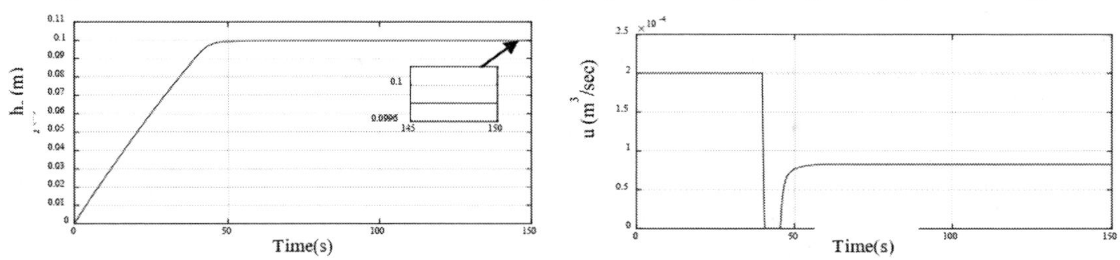

(a) (b)

Figure 12.PID-SMC with 25% variation in the plant parameters (a) Response (b) Control Signal.

IOP Conf. Series: Materials Science and Engineering **973** (2020) 012044 doi:10.1088/1757-899X/973/1/012044

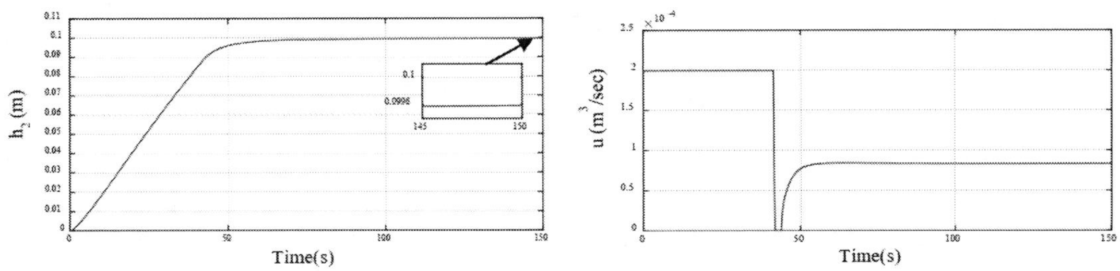

(a) (b)

Figure 13. Fractional PID-SMC with 25% variation in the plant parameters (a) Response (b) Control Signal.

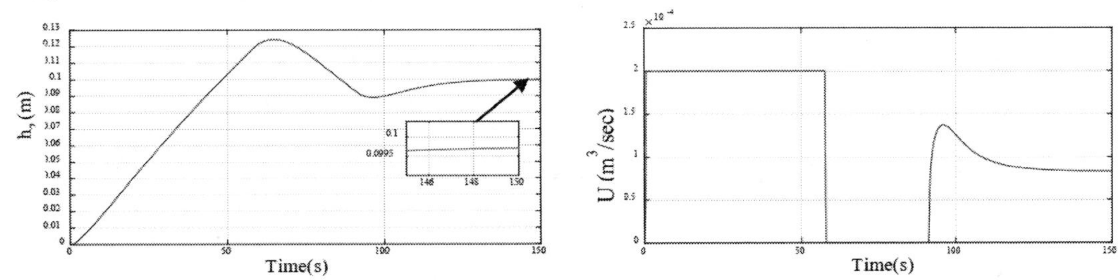

(a) (b)

Figure 14. Dynamic-SMC with 25% variation in the plant parameters (a) Response (b) Control Signal.

Table 5. Comparison between the four controllers in the presence of 25% variation in the plant parameters.

Controller	Rise time (s)	Overshoot (%)	Steady state error (m)	Error index (m^2.s)	Settling time (s)
PD-SMC	34.7	-	0.0002	0.1342	42.1
PID-SMC	35.5	-	0.0002	0.1374	42
Fractional-PID	36.9	-	0.0004	0.166	58
Dynamic-SMC	36.7	24.2%	0.0003	0.1829	122.3

4.4 Sensor noise

In order to test the controllers' sensitivity to sensor noise, a gaussian white noise with signal to noise ratio (SNR) ranges from 19 dB to 23 dB is added to the level sensor, figure 15. The time response to this test showed a high chattering level for all control algorithms. It worth nothing that, the controller parameters in all algorithms have been optimized to reduce the error signal and chattering in the absence of sensor noise. Therefore, a new run of the optimization algorithm is done in order to get the optimized controller parameters in the presence of sensor noise, see table 6.

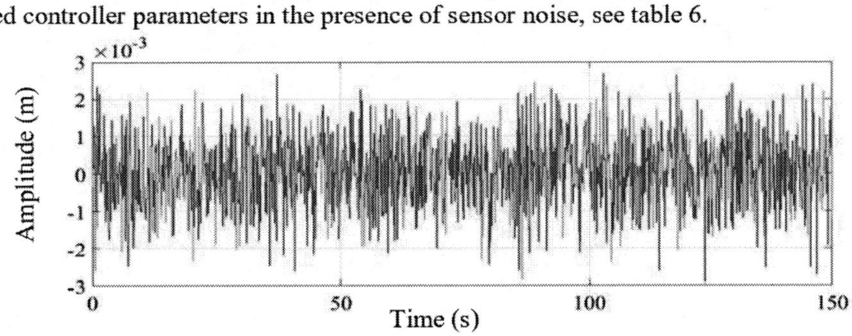

Figure 15. Gaussian white noise added to the level sensor 2.

IOP Conf. Series: Materials Science and Engineering **973** (2020) 012044 doi:10.1088/1757-899X/973/1/012044

Table 6 New tuning parameters in the presence of sensor noise.

Parameter	PD-SMC	PID-SMC	Fractional-PID	Dynamic-SMC
K	$7.281*10^6$	$2.148*10^{12}$	$7.901*10^3$	0.146
Δ	$4.366*10^7$	$9.600*10^{14}$	12.0	2.179
K_p	$3.259*10^6$	$6.494*10^7$	$1.194*10^{10}$	-
K_I	-	$6.182*10^6$	$5.786*10^9$	-
K_d	$2.042*10^7$	$3.890*10^8$	$4.215*10^{10}$	-
λ	-	-	0.9999	-
μ	-	-	0.9179	-
α_1	-	-	-	0.3398
α_2	-	-	-	0.0493

Using the new tuning parameters, the chattering in the control signal is decreased significantly in comparison with that using the old ones. However, the fractional-SMC still suffers from heavy chattering in its control signal, while the dynamic-SMC has the least chattering, figures 16-b to 19-b. Also, as can be seen from figures 16 to 19, the largest steady state error was found in PID-SMC, while the fractional-PID has the maximum overshoot percentage

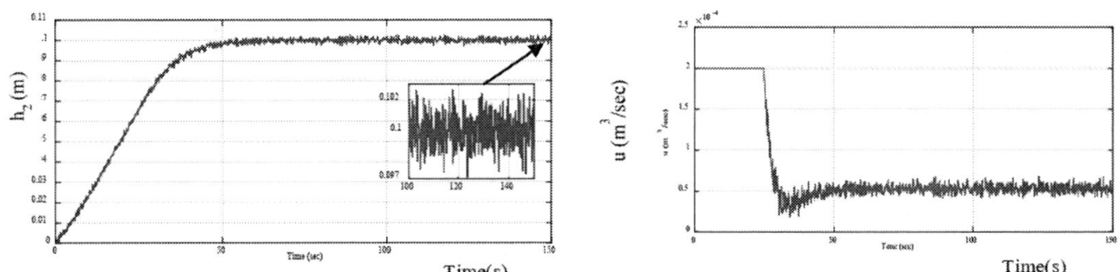

(a) (b)

Figure 16.PD-SMC with sensor noise (parameters retuned) (a) Response (b) Control Signal.

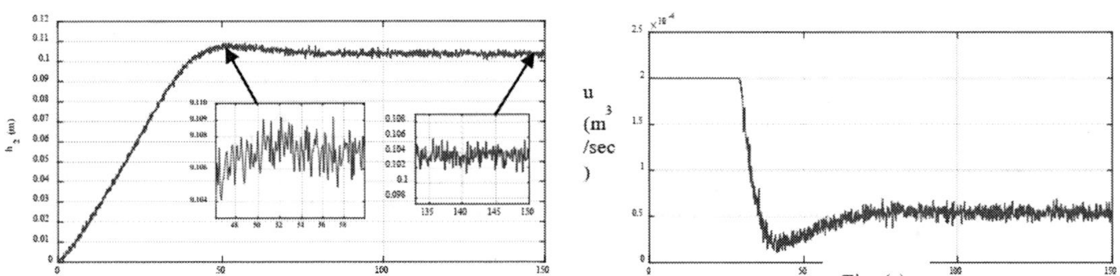

(a) (b)

Figure 17.PID-SMC with sensor noise (parameters retuned) (a) Response (b) Control Signal.

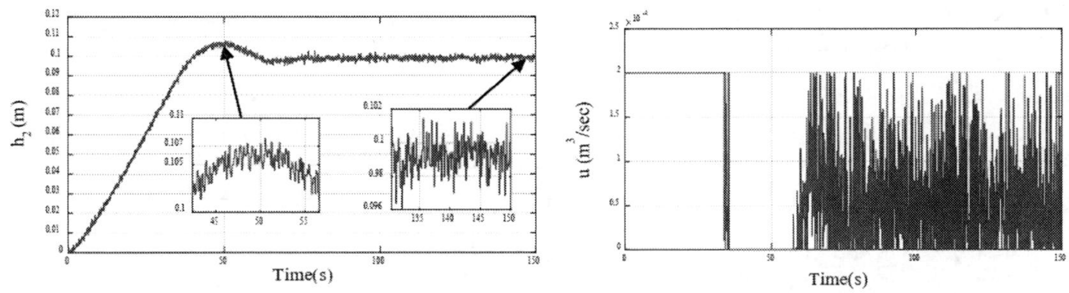

(a) (b)

Figure 18. FractionalPID-SMC with sensor noise (parameters retuned) (a) Response (b) Control Signal.

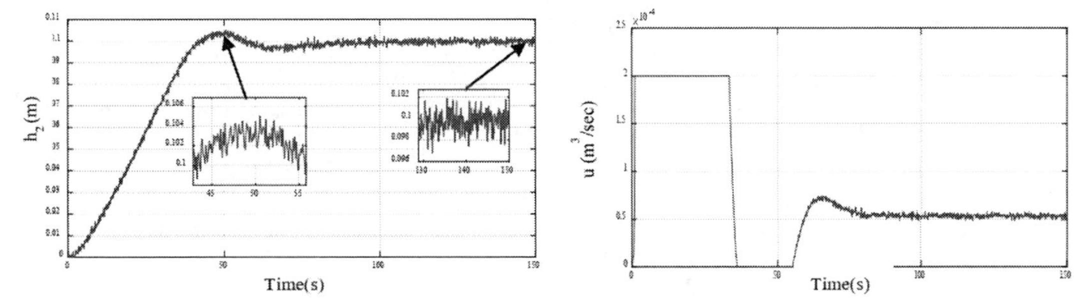

(a) (b)

Figure 19. Dynamic-SMC with sensor noise (parameters retuned) (a) Response (b) Control Signal.

A comparison between the rise time, percentage overshoot, steady state error, the chattering index, and the settling time (5% criterion) for the four controllers is shown in table 7. As can be seen from the table, PD-SMC has in general a better performance in comparison to the other algorithms irrespective of the chattering level. However, in accordance to the application, if the chattering level of the actuator is considered, dynamic-SMC is superior to the other controllers.

Table 7. Comparison between the four controllers in the presence of sensor noise.

Controller	Rise time (s)	Overshoot (%)	Steady state error (m)	Chattering Index	Settling time (s)
PD-SMC	31.4	-	0.0015	$9.999*10^{-7}$	45
PID-SMC	29.8	9.2	0.0035	$8.2296*10^{-7}$	>150
Fractional-PID	29.8	7.8	0.002	0.0038	54.8
Dynamic-SMC	30.7	5.0	0.0016	$1.999*10^{-7}$	75.8

5. Conclusion

In this paper, the performance of four SMC algorithms are investigated for the control of the liquid level in horizontal coupled tank system. The performances of the four controllers are studied when they are exposed to step input, a step input disturbance, a 25% variation in the plant parameters and finally in the presence of sensor noise. The simulation results indicate that the behaviors of the four proposed control algorithms have no significant differences when the step input and input disturbance are applied. In case of plant parameters variation (25%), the PD-SMC, PID-SMC and fractional-SMC give better response than dynamic-SMC. However, the dynamic-SMC has a significant reduction in the chattering level compared to the other three controllers when sensor noise is added. Based on the results presented in this study, the dynamic-SMC is highly recommended in case of sensor used in the feedback control system has a significant noise level. This can significantly protect the actuator from damage and increase its life span.

6. References

[1] Ko C C, Chen B M, Chen J, Zhuang Y and Tan K C 2001 Development of a web-based laboratory for control experiments on a coupled tank apparatus *IEEE Trans. Educ.* **44** 76–86
[2] Ang K H, Chong G and Li Y 2005 PID control system analysis, design, and technology *IEEE Trans. Control Syst. Technol.* **13** 559–76
[3] Astrom K J and Hagglund T 1995 *PID Controllers:Theory,Design,and Tuning-2nd Edition* (Instr. Society of America)
[4] Wang Q G, Lee T H, Fung H W, Bi Q and Zhang Y 1999 PID tuning for improved performance *IEEE Trans. Control Syst. Technol.* **7** 457–65
[5] Tavakoli S, Griffin I and Fleming P J 2006 Tuning of decentralised PI (PID) controllers for TITO processes *Control Eng. Pract.* **14** 1069–80
[6] Maghade D K and Patre B M 2012 Decentralized PI/PID controllers based on gain and phase margin specifications for TITO processes *ISA Trans.* **51** 550–8
[7] Wang Q G, Zhang Z, Astrom K J and Chek L S 2009 Guaranteed dominant pole placement with PID controllers *J. Process Control* **19** 349–52
[8] Malwatkar G M, Sonawane S H and Waghmare L M 2009 Tuning PID controllers for higher-order oscillatory systems with improved performance *ISA Trans.* **48** 347–53
[9] Kaur A 2012 Comparison between Conventional PID and Fuzzy Logic Controller for Liquid Flow Control□: Performance Evaluation of Fuzzy Logic and PID Controller by Using MATLAB / Simulink *Ijitee* **1** 84–8
[10] Khalid U, Shah Y A, Qamar S, Gohar W, Riaz R and Shah W A 2014 Flow and level control of coupled four tanks system using artificial neural network *Am. J. Comput. Commun. Control.* **1** 30–5
[11] Cartes D and Wu L 2005 Experimental evaluation of adaptive three-tank level control *ISA Trans.* **44** 283–93
[12] Pan H, Wong H, Kapila V and de Queiroz M S 2005 Experimental validation of a nonlinear backstepping liquid level controller for a state coupled two tank system *Control Eng. Pract.* **13** 27–40
[13] Kangwanrat S, Tipsuwannaporn V and Numsomran A 2010 Design of PI controller using MRAC techniques for coupled-tanks process *ICCAS 2010 - Int. Conf. Control. Autom. Syst.* **1** 485–90
[14] Gouta H, Said S H, Barhoumi N and M'Sahli F 2015 Observer-Based Backstepping controller for a state-coupled two-tank system *IETE J. Res.* **61** 259–68
[15] Mahapatro S R, Subudhi B and Ghosh S 2019 Design and experimental realization of a robust decentralized PI controller for a coupled tank system *ISA Trans.* **89** 158–68
[16] Roy P and Roy B K 2016 Fractional order PI control applied to level control in coupled two tank MIMO system with experimental validation *Control Eng. Pract.* **48** 119–35
[17] Roy P, Kar B and Roy B K 2017 Fractional order PI-PD control of liquid level in coupled two tank system and its experimental validation *Asian J. Control*
[18] Kar B and Roy P 2018 A Comparative study between cascaded FOPI–FOPD and IOPI–IOPD controllers applied to a level control problem in a coupled tank system *J. Control. Autom. Electr. Syst.* **29** 340–9
[19] Prasad G M, Adithya A and Rao A S 2019 Design of multi model fractional controllers for nonlinear systems: an experimental investigation *Computer Aided Chemical Engineering* **46** (Elsevier) pp 1423–8
[20] Jegatheesh A and Kumar C A 2020 Novel fuzzy fractional order PID controller for non linear interacting coupled spherical tank system for level process *Microprocess. Microsyst.* **72** 102948
[21] Gurumurthy G and Das D K 2020 An FO − I^{λ} $D^{1-\lambda}$ controller design and realization for inverted decoupled Two Input Two Output-Liquid Level System *Int. J. Dyn. Control* 1-14
[22] Slotine J J and Li W 1991 *Applied NonLinear Control_part1* (Englewood Cliffs, New Jersy:

Prentice hall)
[23] Żak S H 2003 *Systems and Control* (New York: Oxford University Press)
[24] Utkin V 1977 Variable structure systems with sliding modes *IEEE Trans. Automat. Contr.* **22** 212–22
[25] Hung J Y, Gao W and Hung J C 1993 Variable structure control: a survey *IEEE Trans. Ind. Electron.* **40** 2–22
[26] Boubakir A, Boudjema F, Boubakir C andLabiod S 2008 A fuzzy sliding mode controller using nonlinear sliding surface applied to the coupled tanks system *Int. J. Fuzzy Syst.* **10** 112–8
[27] Mahapatro S R, Subudhi B and Ghosh S 2019 Design and real time implementation of an adaptive fuzzy sliding mode controller for a coupled tank system *Int. J. Numer. Model. Electron. Networks, Devices Fields* **32** e2485
[28] Boubakir A, Boudjema F and Labiod S 2009 A neuro-fuzzy-sliding mode controller using nonlinear sliding surface applied to the coupled tanks system *Int. J. Autom. Comput.* **6** 72–80
[29] Aksu I O and Coban R 2019 Sliding mode PI control with backstepping approach for MIMO nonlinear cross-coupled tank systems *Int. J. Robust Nonlinear Control* **29** 1854–71
[30] Almutairi N B and Zribi M 2006 Sliding mode control of coupled tanks *Mechatronics* **16** 427–41
[31] Biswas P P, Srivastava R, Ray S and Samanta A N 2009 Sliding mode control of quadruple tank process *Mechatronics* **19** 548–61
[32] Khan M K and Spurgeon S K 2006 Robust MIMO water level control in interconnected twin-tanks using second order sliding mode control *Control Eng. Pract.* **14** 375–86
[33] Ding S, Park J H and Chen C-C 2020 Second-order sliding mode controller design with output constraint *Automatica* **112** 108704
[34] Kashyap R, Jaggi N and Pratap B 2019 Higher-order Sliding-Mode Observer based robust controller design for four-tank coupled system *2019 3rd International conference on Electronics, Communication and Aerospace Technology (ICECA)* (IEEE) pp 832–7
[35] Dumlu A and Ayten K K 2019 A combined control approach for industrial process systems using feed-forward and adaptive action based on second order sliding mode controller design *Trans. Inst. Meas. Control* **41** 1160–71
[36] Derdiyok A and Başçi A 2013 The application of chattering-free sliding mode controller in coupled tank liquid-level control system *Korean J. Chem. Eng.* **30** 540–5
[37] Benayache R, Chrifi-Alaoui L, Bussy P and Castelain J M 2009 Design and implementation of Sliding Mode Controller with varying boundary layer for a coupled tanks system *17th Mediterranean Conference on Control and Automation* (IEEE) pp 1215–20
[38] Delavari H, Ghaderi R, Ranjbar A and Momani S 2010 Fuzzy fractional order sliding mode controller for nonlinear systems *Commun. Nonlinear Sci. Numer. Simul.* **15** 963–78
[39] Shah D H and Patel D M 2019 Design of sliding mode control for quadruple-tank MIMO process with time delay compensation *J. Process Control* **76** 46–61
[40] Calderón A J, Vinagre B M and Feliu V 2006 Fractional order control strategies for power electronic buck converters *Signal Processing* **86** 2803–19
[41] Piazzi A and Visioli A 2006 A noncausal approach for PID control *J. Process Control* **16** 831–43
[42] Birbil Ş I and Fang S C 2003 An electromagnetism-like mechanism for global optimization *J. Glob. Optim.* **25** 263–82
[43] Birbil Ş I, Fang S C and Sheu R L 2004 On the convergence of a population-based global optimization algorithm *J. Glob. Optim.* **30** 301–18
[44] Cao J and Cao B 2006 Design of fractional order controllers based on particle swarm optimization *1ST IEEE Conference on Industrial Electronics and Applications* (IEEE) pp 1–6
[45] Chang P C, Chen S H and Fan C Y 2009 A hybrid electromagnetism-like algorithm for single machine scheduling problem *Expert Syst. Appl.* **36** 1259–67
[46] Wu P, Yang W H and Wei N C 2004 An electromagnetism algorithm of neural network analysis—an application to textile retail operation *J. Chinese Inst. Ind. Eng.* **21** 59–67

IOP Conf. Series: Materials Science and Engineering **973** (2020) 012045 doi:10.1088/1757-899X/973/1/012045

Effect of applying lean maintenance in oil and gas fields

M Nasr Hassan [1], A F Barakat[2] and A S Sobh[3]

[1]Section Head Mechanical Engineer, Gas Affairs Department, Ministry of Petroleum and Mining Resources, Nasr City, Cairo, Egypt.

[2] Professor, Doctor, Mechanical Engineering Department, Faculty of Engineering, Helwan University, Helwan City, Cairo, Egyptand Professor, Doctor, Faculty of Engineering, Ahram Canadian University (ACU), 6th of October City, Giza, Egypt.

[3] Doctor, Mechanical Engineering Department, Faculty of Engineering, Helwan University, Helwan City, Cairo, Egypt.

Email:eng.mohamed.nasr.h@gmail.com

Abstract. Oil and Gas Sector are facing a huge increase and delay in Operations & Maintenance tasks for all facilities beside continuous demand from management to improve the productivity of wells and the efficiency of facilities. The top management decides to use Lean Maintenance as a new tool, technique and methodology to improve and modify the current operations, production and maintenance systems as done before by many researchers in the oil and gas sector. This improvement will be done through three real case studies by integrating Building Information Modelling with lean concepts, monitoring Key Performance Indicators within a modern Computerized Maintenance Management Software and finally using some LM tools such as Total Productive Maintenance, Value Stream Mapping and 5S Methodology to increase productivity, efficiency, and quality of the output production and services. The results of implementation since July 2017 till June 2018 show average increase in crude oil production by 6.72% per day in addition to overcome natural decline of oil wells, reduced Technician's total Preventive Maintenance tasks by 52.63% and transfer the less difficult inspections to Operator to be responsible for making checks to all parameters of all equipment and to enable Technicians to perform other maintenance tasks, Saving about $724,770 in Water Injection Pumps Operation's Expense. Last but not least, accelerate the workflow of production and maintenance processes by 28% compared with the previous financial year.

1. Introduction

Currently, companies carry out planned routine maintenance, inspections, replacement, cleaning, painting, defect correction, etc. These set of actions are what we in general called maintenance activities which are considered as one of the basics of financial management [1]. The Word Lean meaning perfect use of manufacturing resources by cutting the Non- Value Adding (NVA) activities or wasting in product or service that doesn't add value to the customer's need [2].Integrating lean thinking in maintenance is known as Lean Maintenance (LM) [3]. Lean Maintenance is necessary for the success of a Lean manufacturer, and it provides a complete approach to the function of maintenance [4, 5].

The paper Objective is to create the methodology of implementing lean Maintenance in Oil & Gas Fields and re-design the Maintaining System based on the concepts of Lean Maintenance. This paper is applied at Qarun Petroleum Company (QPC), which is one of the crude oil production companies and related to the Egyptian General Petroleum Corporation (EGPC). Nowadays, the company is facing a huge increase and delay of operations & maintenance for all facilities at QPC fields. So, the upper management decides to use new LM techniques and tools to improve the current operations & production system in order to meet EGPC demands. By creating two case studies similar to the published papers in Oil &Gas Sector in addition to new innovation.

Integrating Building Information Modelling (BIM) with lean concepts especially for the upper management in head office, in addition to using the CMMS software to monitor KPIs to get the required data for equipment per each field. The data include (Location, Tag, Status if it is running or in Stand-by, Availability, Reliability, Utilization, Online Hours, Up-set Hours, PM Hours, Shut-in Hours, comment if any) for implementing Lean Maintenance Techniques in the Case Study. Although using the most usable LM tools such as Total Productive Maintenance (TPM), Value Stream Mapping (VSM) and 5 S Methodology in implementation.

The primary objective of any maintenance function is to see that machines and equipment are maintained in a way that enables a plant to operate with the lowest unit cost consistent with the safety and well-being of the workers. Maintenance function becomes a significant contributor towards to achieve strategic objectives of an organization in today's competitive markets [6]. The maintenance process is to serve the production facilities of high productivity, it comprises planned and unplanned actions carried out to retain a physical asset to the acceptable operating conditions [7]. Maintenance aims at increasing the value of the reliability, safety, availability, and quality of an asset (e.g. production plant, equipment, or building) with acceptable economical costs [8].

The cost of maintenance activities ranges from 15 to 70% of the total production costs [6]. This is the second largest part, after energy costs, of the operational budget [5]. In the United States, the estimated cost of maintenance increased from $200 billion in 1979 to $600 billion in 1989 [9]. Maintenance activities account for an average of 28 percent of the total cost of finished goods. One of the reasons for such a significant portion of maintenance of the total operating cost is that the machinery has become highly automated and technologically very complex. For example, modern operating systems depend on sensor-driven management systems that provide alerts, alarms, and indicators. Consequently, maintenance costs are expected to be even higher in the future [10].

Generally, the maintenance costs are proportional to the downtime (DT). The DT is the time interval when the equipment/system is down for maintenance until it is back to the normal working conditions. The increased DT is caused by the non-value added (NVA) activities or wastes within the entire maintenance process. One of waste elimination strategies is the application of lean thinking in all activities between suppliers and customers. Integrating lean thinking in maintenance is known as lean maintenance (LM) [3]. Some researchers emphasized that LM is prerequisite for the success of a lean manufacturer as it provides a holistic approach to the function of maintenance [5].

The lean integration in any process is carried through adopting lean principles which begins with specifying the customer value [4]. In the maintenance environment, any maintenance service could be considered as a final intangible product. The service is provided to a customer, which, in this case, could be assumed as an asset (e.g. production line). Therefore, it is essential to identify the value from the asset perspective, which can improve its availability and reliability through efficient maintenance. Then, mapping the maintenance value stream, which fundamentally consists of all the collective activities to deliver the maintenance service. Later, improving the maintenance value stream by abolishing the waste, which assists in minimizing the lead-time [5].

The Results show that the implementation of lean maintenance in O&G fields increases productivity, Profit, improves quality, decreases wastes, cycles times, Inventory, raw materials, costs and number of Operators. This paper will be useful in developing a more generic approach to design a lean environment. Last but not least Applying KPIs, CMMS software, BIM and Lean Maintenance Tools, are recommended to be implemented in all Oil & Gas Companies, to get the same improvement.

IOP Conf. Series: Materials Science and Engineering **973** (2020) 012045 doi:10.1088/1757-899X/973/1/012045

2. Research methodology

This study employs a complete literature Survey in order to explore the effect of applying lean Maintenance in Oil & Gas Fields. Then, extracting and synthesizing the selected resources to group and analysis the suitable Lean tools and concepts. Finally, two real-word cases studies similar to the published papers in Oil and Gas Sector in addition to a new innovation case study is chosen to verify the integrated approach. The intensive case studies are conducted and focused on productivity improvement activities at the Oil & Gas Company. This research focuses scenarios simulation and role-playing which guided by seasoned managers and brainstorms were used to sort out the Oil & Gas industry maintenance process and collect the primary problems that relate to the process. The case studies were focused on the process of corrective maintenance. In addition, the discussion was conducted not only reside on the past implementation, but also focus on the future plans and developments via regular seminars.

The research criteria depend on reviewing the title and then the abstract which saves the researcher time and efforts. The following criteria has been considered to include/ exclude the articles:

- Articles published between 2000 and 2020, as the origin of LM, raise in 2001.
- Search for articles published in scientific journals or conference proceedings.
- Ensure that selected articles contain at least one of the below search keywords in the title or abstract; (Maintenance, Lean Maintenance, Lean TPM, Lean Principles, Lean Maintenance Tools, Oil & Gas).
- Search well in online databases like Taylor & Francis, EBSCO, Emerald, IEEE, Inderscience Publishers, ProQuest, Sage, Science Direct, and Springer Link.

3. Lean, Lean Tools and Lean Maintenance

3.1. Lean

The word lean means the efficient use of the available resources by cutting the NVA activities or wastes [2]. Lean represents a collection of tools that work together synergistically to create a streamlined, high-quality system that produces finished products at the same place as the customer demand [11]. Waste in lean is defined as any activities that add cost to a product/service without adding values from a customer's perspective. In recent decades, the lean system is gaining momentum across different industrial sectors. It has been originally started as Toyota production system (TPS) which describes the manufacturing philosophy of Toyota Motor Corporation [12, 26]. This lean system has been successfully extended to service industries, such as maintenance service, airlines, restaurants, public sector, education, food, and hospitals.Researchers stating that lean principles can be transferred to any organization [5, 13].

3.2. Lean tools

The lean tools are representing the lean principles for the implementation process. Reducing the NVA within maintenance activities can be accomplished through implementing Lean tools. To achieve LM improvement effectively, key lean tools such as VSM, 5S, and visual management need to be employed [5]. A comprehensive lean tool developed for maintenance activities within an organization include 5S, TPM, OEE, Kaizen, Poka-Yoke, process activity mapping, Kanban, computer managed maintenance system (CMMS), enterprise asset management (EAM) system, Value Stream Mapping (VSM), mapping, inventory management, visual management, seven wastes, single minute exchange of die (SMED) and Takt time [15-18].

3.3. Lean Maintenance

LM term was coined in the last decade of the twentieth century. Some researchers define LM as a proactive maintenance operation employing planned and scheduled maintenance activities through TPM practices using maintenance strategies developed through application of Reliability-Centered maintenance (RCM) decision logic and practiced by empowered (self-directed) action teams [14]. LM

IOP Conf. Series: Materials Science and Engineering **973** (2020) 012045 doi:10.1088/1757-899X/973/1/012045

generates a desirable outcome by minimizing consumption of inputs. LM represents adopting lean principles into the maintenance, repair, and overhaul (MRO) operations. It could reduce unscheduled DT through optimizing maintenance support activities and maintenance overhead. [15, 19].

4. Case Study (I): Re-organize Operations Dept. Using Building Information Model (BIM)

The highly critical nature of Oil and Gas (O&G) activities leads to the high demand for smooth, uninterrupted and safe running of O&G facilities. Operations and Maintenance (O&M) departments in O&G Sector are facing many problems during repair or during any activities, as Maintenance Staff usually used original blueprints & as-built drawing and most of those data need an upgrade by time. In addition to the difficulty of transmission of data between departments which cause low productivity, labor wastage, and high cost [20].

Building Information Model (BIM) is a 3D information model for Operations & Maintenance Process coordination. Recent years, many types of research paid close attention to cooperation between BIM and lean concepts but little has published on lean concepts and BIM in O&M management. In this case study, we will combine BIM with Lean concepts to improve the reliability of flows and waste estimation. The case study extracted from an actual case study in O&G Sector with little modifications due to the change in organization structure from one company to another but both of case studies is facing the same problems [20-22].

The case study focused on the process of corrective maintenance of Qarun Petroleum Company (QPC) and the discussion was conducted not only reside on the past implementation, but also focus on future plans and developments via regular seminars.

4.1. Old Organization Process & Phases:

Based on the QPC old organization, there were two departments carry out the corrective maintenance work. Operations Dept. is responsible for the facilities operation management, quality assurance (QA) and quality control (QC) under the supervision of two head General Managers (Engineering & Fields). Projects Dept. is responsible for implementing the constructions and repairs required by the Operations Department. The workers divided into three levels: Managers, Engineers, and Operators.

Oil & Gas industry mainly composed of a large number of pipelines & Equipment and the production is highly relying on them. The uninterrupted production process is the key factor for the performance and success of Oil & Gas plants. There are a series of problems can be pointed out in Oil & Gas plants, such as (Time consuming or insufficient facilities procurement - a Large incidence of no value-adding activities in the maintenance process, resulting in long duration and not enough time for generating maintenance solutions - Maintenance requirements between different departments are not all identified).

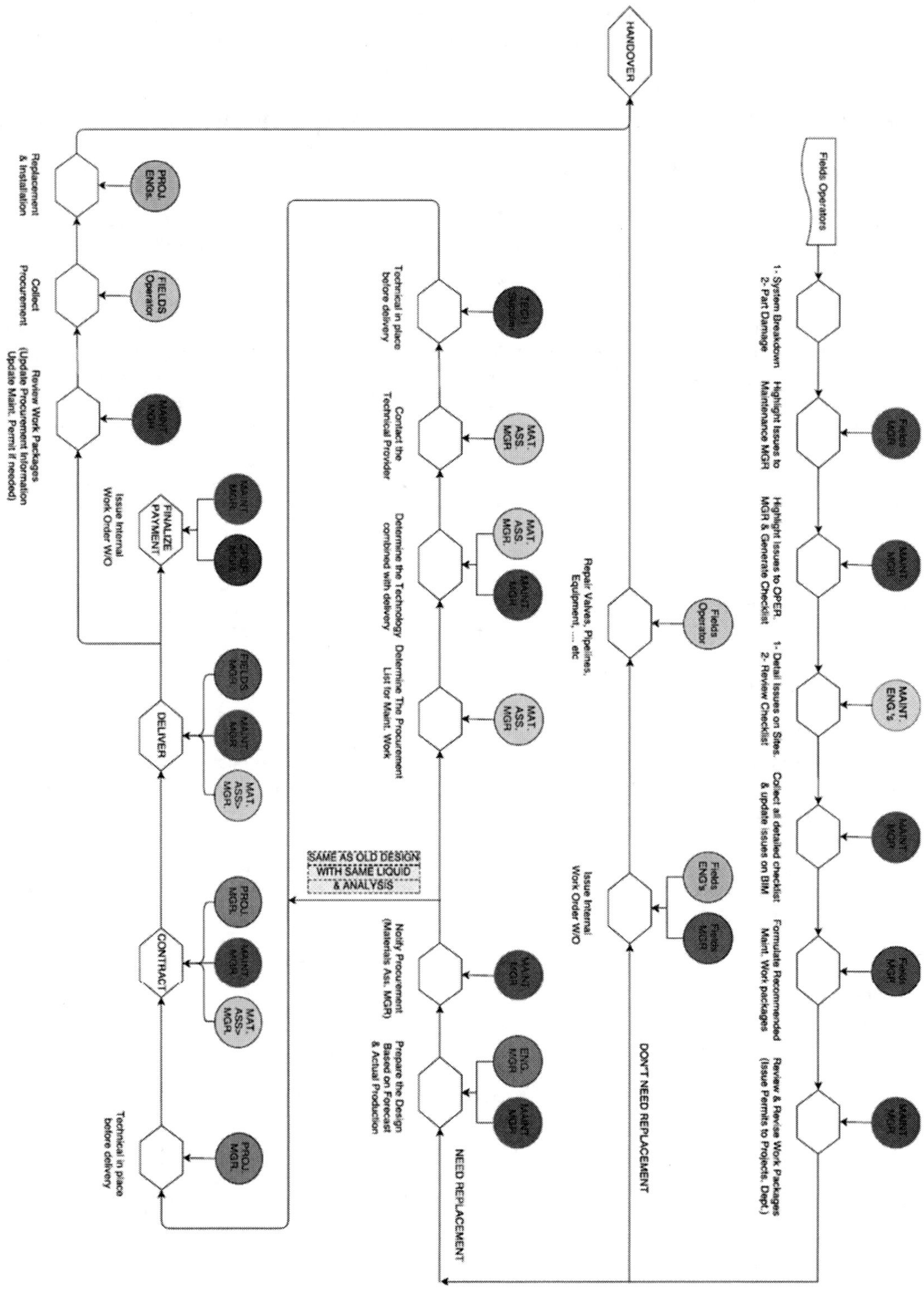

Figure 1. Optimized maintenance road within integration of BIM and Lean Concept.

IOP Conf. Series: Materials Science and Engineering **973** (2020) 012045 doi:10.1088/1757-899X/973/1/012045

4.2. New Organization After Using Building Information Modelling (BIM):

The main target of applying BIM in Maintenance process is to create a complete database for facilities (e.g. location, maintenance status, product information, and process), also to guarantee the complete control of information management (information integration and sharing in different departments and workers), in addition to allow actions of automatically collect and update the maintenance information [20, 25].

A new BIM was built and re-organized the Operations Dept. by issue two sub-department under the Operations General Manager to handle the Maintenance & Materials issues besides the Engineering & Projects Departments all over QPC fields. The new BIM enabled the workflow to start from system breakdown warning till handover as shown above in figure 1 and as explained in the following steps:

- The system breakdown warning and part damage warning problems sent by Field's Operators to Fields General Managers to highlight the urgent issues to Maintenance General Managers.
- The Maintenance General Manager create and generate the checklist to Operations General Managers based on priority from high to low urgency.
- Maintenance Engineers start to collect the details of issues in sites and review again the checklist with Maintenance General Manager to update in the new BIM model created based on the revised checklist.
- Maintenance General Managers Formulate Recommended Maintenance Work Packages after review & revise with Fields General Managers.
- In case of do not need replacement, Fields issue internal work orders W/O to concerned dept. to repair valves, pipelines, equipment through Fields Operators and then handover it for process operation.
- In case of need replacement, another maintenance roadmap taken between different department to facilitate and minimize the cycle of study of each issue and eliminate the wasted time taken per each department.
- Maintenance & Engineering Managers prepare the design based on Forecast and actual production to review if the replacement will need an upgrade or not.
- In case of an upgrade is required during the replacement. Maintenance & Engineering Dept. prepare the procurement list for maintenance work and export it to Materials Ass. General Manager to check the latest technology in the market with all Technical suppliers and vendors, in addition, to review the delivery of list before shipment to match with projects time.
- Maintenance Managers in both cases of upgrade or not will issue permits to the Projects Department to coordinate with the contractors who will proceed with the installation and replacement under the supervision of Projects, Maintenance and Materials. To finalize the project in time and handover it to Operations for process and pay all bending invoices on this project.
- In this maintenance workflow, including the following: procurement, delivery, permission, installation, and handover, all the participants shared work-related information such as how is the work going on and gave feedback for updating the BIM model, which allowed the other workers to real-time tracking and adjust the work plan in time [20].

4.3. Results of Implementation

- The results of applying this method increased workflow of maintenance and production processes by 28% since applying it in July 2017 till June 2018 compared to the previous financial year as shown in figure 2.

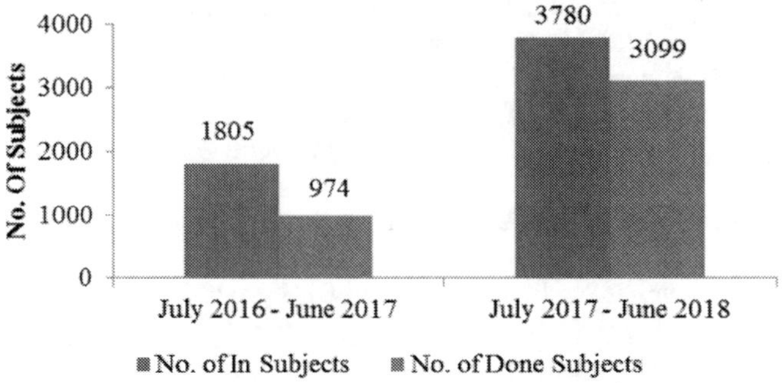

Figure 2. No. of Subjects (Inand Done).

- This case study demonstrated that the integration of BIM and lean concepts had strong application prospects in Oil & Gas maintenance, and the maintenance process was primarily satisfied with efficiency improvement.
- BIM mainly used for the inspection, assessment, repair and 3D visualization.
- Lean concepts focus on reducing the variability, reduce non-value adding work and increase the process transparency.
- The implementation of lean thinking and BIM in the process of maintenance enabled complete control of information management for entire works by focusing on complete process and by transferring data information flow from 2D-based information to 3D virtual models.
- Increased process transparency by providing an accurate model for all facilities, all the information introduced in the model was real-time tracking and in time adjusting which allow workers from different positions to communicate and coordinate in the whole process.

The completion of further studies for the development of a more comprehensive BIM lean platform will enable the maintenance process in the Oil & Gas industry to adopt this emerging technologies and management thinking for actual field practice [20].

5. Case Study (II): Improve Maintenance Service Using TPM & 5S Methodology

Qarun Petroleum Company "QPC" is one of the biggest Oil Production Company in Egypt, QPC has mainly four large fields (Qarun, Karama, EBAH & Beni-Suef) with total production 40,000 Barrel Per Day and each one of those fields has its own organization & structure and also has its shipping facilities, Oil production wells, workshops, and warehouse. Each facility has storage tanks, shipping pumps, motors and control room. The maintenance of those facilities was carried by Maintenance Dept. Engineers and technicians based on preventive maintenance plan while the supervision of facilities is done by production Dept. Engineers and operators.

QPC Always looking for improve the maintenance service that is currently provided, this study aims to investigate and intervene in the points where it is possible to improve the current system of maintenance management. It is also intended to start the implementation of new techniques for industrial maintenance, namely the Total Productive Maintenance and the 5S Methodology. The case study extracted from an actual case study in O&G Sector with little modifications due to the change of equipment's type & structure from one company to another but both of case studies are facing the same problems and both of them are focusing to use the TPM & 5S methodology in implementation [1, 23].

This case study will focus on the implementation of Autonomous Maintenance which is one of the most important pillars of TPM. The application of autonomous maintenance aims to reduce the major stops of the machines/equipment for repairs, thus increasing their availability. This method allows achieving more appropriate and efficient management of human resources. It is through self-

IOP Conf. Series: Materials Science and Engineering **973** (2020) 012045 doi:10.1088/1757-899X/973/1/012045

maintaining activities that begins the process of training the operators to make them responsible for their workplace and equipment.

5.1. Applying lean concepts using TPM & 5S methodology:

A first approach to the autonomous maintenance was to implement it to the equipment associated with the shipping facility which requires more maintenance and stronger control by the operator and equipment whose proper functioning is critical to the continuity of the process. The maintenance is guided through documents that serve as a guide to the operators when performing maintenance [11, 17].

Before the implementation, we should know that the operators had not at his charge any maintenance activities, except for some visual inspection tasks. Equally, the shipping facility operators had no maintenance responsibility, while there was a person exclusively to maintain this division. With the introduction of TPM techniques, it was possible to migrate the most basic maintenance tasks to the operators and free the technician for other maintenance tasks. It is also necessary to assess the tasks to be delegated to the operator, as well as evaluating the equipment manufacturer's recommendation.

Implementation of the 5S methodology in maintenance activities is done by eliminating any kind of materials that no longer used or obstruct access to materials and equipment [27]. Pack the materials and equipment of the maintenance services by creating labels for all component and by improving the identification of the workshop components. Clear all the areas of the maintenance service by removing floor garbage. Standardize all equipment service by creating quality standards, cleanliness standards, safety standards & standards of working procedures. Finally, respect the change made in the maintenance service.

First, QPC has to select and sort all types of equipment used inside the shipping facility which has a critical effect on production. Some of the equipment which needs criticality evaluation is Oil Storage Tank, Valve, Strainers, Booster Pump, HPS pump, Motors, and controllers. There are several methods that let us analyze this need, one of those is IPINZA Method.

After estimating the scores for all equipment by summing grades as shown in table 1, we can select the type of maintenance to be applied to the equipment according to the criticality of each one (Optional: Score (0 to 6) – Convenient: Score (6 to 13) – Important: Score (13 to 19) – Critical: Score (19-22)). In order to guide the technician in their inspections, we were created maintenance procedures for each machine. These documents contain a list of activities that should be undertaken by the operator or maintenance technician, aimed to ensure the proper functioning and cleaning of equipment.

After selecting Horizontal Pumping System (HPS) which has the largest score results in IPINZA method, the list of preventive routine maintenance was sorted and reclassified between the Technician and Operator to migrate the most basic maintenance tasks to the operators and free the technician for other maintenance tasks (Monthly, 3 Months and 6 Months) such as inspection, oil change, clean, grease, visual check of mechanical parts. In addition to the standards and internal procedures created called "TPM TAG". These documents are no more than labels that inform you whether the equipment has been inspected and what tasks were undertaken during the inspection.

5.2. Results of implementation

- The implementation of this method reduced total PM tasks done by Technicians for HPS Pump by 52.63% and transfer the less difficult inspections to Operator to be responsible for making checks to all parameters of all equipment and to enable Technicians to perform other maintenance tasks. As shown in figure 3.

IOP Conf. Series: Materials Science and Engineering **973** (2020) 012045 doi:10.1088/1757-899X/973/1/012045

Table 1. Scoreboard IPINZA Method and Assessment.

	Description	Answer	Score	Oil Storage Tank (Eq-01)	Valve (Eq-02)	Strainers (Eq-3)	Booster Pump (Eq-4)	HPS Pump (Eq-5)	Motor (Eq-6)	Controller (Eq-7)
1	Effect in production	Stop	4	4	4			4		
		Slow Down	2			2	2		2	2
		No effect	0							
2	Economical Value of the Equipment	High	4	4				4	4	
		Medium	2				2			2
		Low	1		1	1				
3	Consequences of Failure Machine	Yes	2	2	2		2	2	2	2
		No	0			0				
4	Consequences of Failure of Process	Yes	3	3	3			3	3	3
		No	0			0	0			
5	Consequences of Failure to Staff	Dangerous	1						1	1
		Not Dangerous	0	0	0	0	0	0		
6	Dependence on Logistics	Foreign	2				2	2		1
		Local	0	0	0	0			0	
7	Manpower	Subcontract	2					2		
		Own	0	0	0	0	0		0	0
8	Reliability	High	1	1	1	1	1	1	1	1
		Low	0							
9	Repair Difficulty	High	1	1				1	1	1
		Low	0		0	0	0			
10	Flexibility and Redundancy	Simple	2							
		By-Pass	1	1			1	1		1
		Double	0		0	0			0	
	Total Score			16	11	4	10	20	14	14
				Important	Convenient	Optional	Convenient	Critical	Important	Important

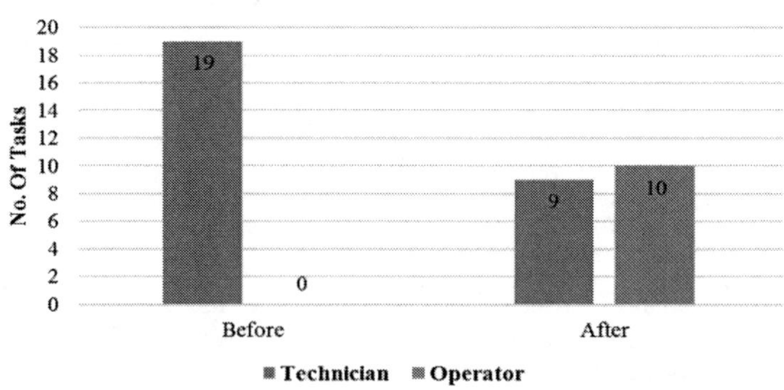

Figure 3. Classification of Preventive Maintenance Tasks for HPS Pumps Before & After Implementation LM Tool.

- With the application of this methodology, companies can improve not only the costs and efficiency but also the quality, flexibility, and innovation.
- The new guidelines employee motivation will increase and it will contribute to the increase in discipline and safety levels and
- The standardization of procedures and tasks is now company policy.
- This autonomous maintenance policy generates a reduction in waste such as the motion of technical maintenance, as well as the allocation of a technician to a specific activity in 100%.

6. Case Study (III): Check the availability, Reliability & Utilization of each equipment by Integrating Key Performance Indicators with Computerized Maintenance Management System

Qarun Petroleum Company "QPC" have a Computerized Maintenance Management System (CMMS), this system consists of two software (GMS) Global Maintenance System and (MMS) Materials Management System. The GMS is used by fields Engineers & Technicians and responsible for issuing the work orders for each department based on the PM Plan of each equipment according to the maintenance instruction book. While MMS is the main general software used by Operations & Office engineers to review the Materials Requisition (MR) & purchase orders (PO) from ordering till receiving in Fields warehouse. CMMS can reinforce reporting and analysis capabilities[24].

Accessibility and accuracy of information can provide more reliable decisions in CMMS because of closer working relationships between maintenance and production. So, the Planning Department is responsible to review and supervise the update of that information with each department to ensure that each item is up to date and each preventive maintenance "PM" is done in time [25].

Generally, the CMMS have all required features like Work Order Management, Planning Function, Scheduling Function, Budget/Cost Function, Spares Management and Key Performance Indicators.CMMS software prices vary depending on method of deployment, number of users, and level of functionality and customizations. For a web-hosted CMMS, small businesses with only one or two users might pay $50 to $100 per month, while mid-sized companies will probably pay several hundred dollars. Companies can also purchase an off-the-shelf CMMS without customizations for $750 to $1000.For customized on-premise solutions, vendors typically charge a base fee for the software and a separate fee for each user. Small companies might pay between $1,500 and $10,000, while larger businesses with more comprehensive needs can spend $10,000 to $40,000.These prices don't include support and software upgrades, which can run about 20% of the total purchase price each year.

This case study is a new innovation method done based on the integration of Key Performance Indicators within the CMMS system to check the availability, Reliability & Utilization of each equipment from 2012 to 2017. By getting the required data for each equipment which exist in Qarun & Karama Fields. The case study focused on the Weatherford Pumps & National Oil Well Pumps “NOW” (Triplex Reciprocating Pumps), those pumps are in operations since 2008 at Qarun & Karama Fields. Weatherford & NOW pumps have a lot of downtime due to repeated problems Compared to current new Water Flood “WF” Horizontal Pumping System "HPS" pumps (REDA & WOODGROUP). The Selection of those pumps was done based on the results of Key Performance Indicators: “Mean Time Between Failure” (MTBF), “Mean Time to Repair” (MTTR) and “Mean Time Between Repair” (MTBR) during financial year (July 2017 till June 2018). After reviewing the total Existing Water Flood Pumps (Weatherford & NOW) Skids at QPC we found that 50 % of those skids are down.

Also, we found that the Operational Costs of one Weatherford pump per year exceed the price of purchasing a new HPS pump by $144,954, it means that the payback for HPS Pump will be less than one year and the replacement of five skids per one financial year will save about $724,770 in addition to the availability, Reliability & Utilization of HPS Pumps will be better than Weatherford Pumps. In addition to the Operations Cost of HPS pump per year is very low comparing to the Operations costs of Weatherford pumps.Those Problems cause QPC a lot of costs especially in spare parts replacement during the repair & Overhauling. In addition to the operational cost of Diesel consumption, Field

Service Representative "FSR" Site visit and delay of supply, while in other side comparing to HPS Pumps the availability & Reliability, Replacement, and Repair of those are perfect as shown in figure 4.

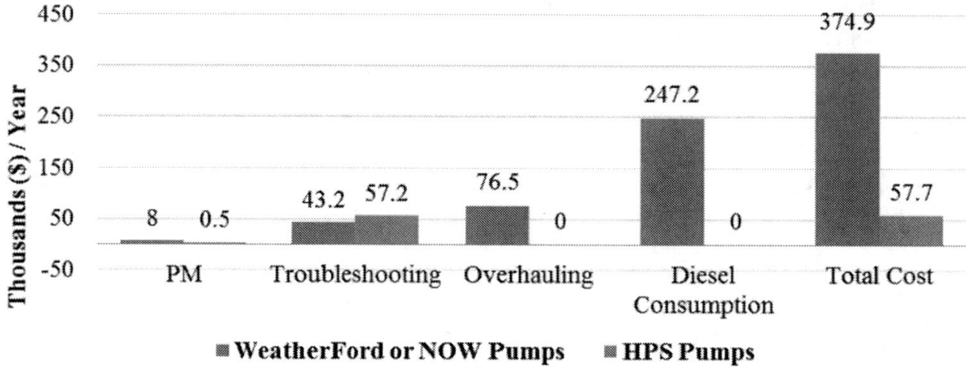

Figure 4. Cost Comparison between Triplex Pumps and HPS Pumps/ One year/ One Skid.

6.1. Results of Implementation

- The Results of the new innovation Case study demonstrated that the integration of Key Performance Indicators within the CMMS system had strong application prospects in Oil & Gas maintenance.
- The Operational Costs of one Weatherford pump per year exceed the price of purchasing new HPS pump by $144,954 it means that the payback for HPS Pump will be less than one year and the replacement of five skids per one financial year will save about $724,770 in addition to the availability, Reliability & Utilization of HPS Pumps will be better than Weatherford Pumps.
- Weatherford & National pumps must be replaced by another HPS pumps due to the low cost of operation of HPS pumps compared to Weatherford & National pumps.

7. Conclusions

The implementation of Lean Concepts for maintenance in Oil and Gas Sectors will bring great economic benefits and powerful techniques support for the equipment maintenance reform. Applying Lean Maintenance strategy also can ensure a high degree of utilization, reliability, and availability of facilities especially in a continuous production process. In addition to increasing productivity, Profit, improving quality, decreasing wastes, cycling times, Inventory, costs and number of Operators.

Propose three case studies using lean maintenance tools, two case studies similar to the published papers in Oil &Gas Sector: Integrating Building Information Modelling (BIM) with lean concepts and Improve Maintenance Service Using (TPM) & (5S) Methodology. In addition to a new innovation case study by monitoring Key Performance Indicators (KPIs) within a modern Computerized Maintenance Management Software (CMMS). The results of case studies from July 2017 till June 2018 is as follows:

- An average increase in crude oil production by 6.72% per day besides to overcome the natural decline of oil wells.
- Reduced Technician's total Preventive Maintenance (PM) tasks by 52.63% and transfer the less difficult inspections to Operator to enable Technicians to perform other maintenance tasks.
- Saving about $724,770 in Water Injection Pumps Operation's Expense and accelerate the workflow of maintenance and production processes by 28% comparing to the previous year.

IOP Conf. Series: Materials Science and Engineering **973** (2020) 012045 doi:10.1088/1757-899X/973/1/012045

Last but not least Applying KPIs, CMMS software, BIM and Lean Maintenance Tools, are recommended to be implemented in all Oil & Gas Companies, to get the same improvement.

References

[1] Tiago Palmeira, Helena Navas, Teresa Morgado 2015 Lean Maintenance Management Activities in An Oil Terminal: Case Study *6th International Conference on Mechanics and Materials in Design* Paper Ref.: 5579.

[2] Carrasqueira M, Machado V C 2008 Strategic logistics: Re-designing companies in accordance with Lean Principles *International Journal of Management Science and Engineering Management* **3** pp 294–302.

[3] Baluch N, Abdullah C S Mohtar S 2012 TPM and lean maintenance – A critical review *Interdisciplinary Journal of Contemporary Research In Business* **4** pp 850–857.

[4] Bhasin S 2015 *Lean management beyond manufacturing*(New York: Springer International Publishing).

[5] Sherif Mostafa, Sang-Heon Lee, Jantanee Dumrak, Nicholas Chileshe, Hassan Soltan 2015 Lean thinking for a maintenance process *Production & Manufacturing Research: An Open Access Journal* **3:1** pp 236-272.

[6] Fraser K 2014 Facilities management: The strategic selection of a maintenance system *Journal of Facilities Management* **12** pp 18–37.

[7] Faccio M, Persona A, Sgarbossa F, Zanin G 2014 Industrial maintenance policy development: A quantitative framework *International Journal of Production Economics* **147** pp 85–93.

[8] Márquez C A 2007 *The maintenance management framework: Models and methods for complex systems maintenance*(London: Springer-Verlag).

[9] Bevilacqua M, Braglia M 2000 The analytic hierarchy process applied to maintenance strategy selection *Reliability Engineering & System Safety* **70** pp 71–83.

[10] Tinga T 2013 *Principles of loads and failure mechanisms*(London: Springer-Verlag) pp 161–186.

[11] Shah R, Ward P T 2007 Defining and developing measures of lean production *Journal of Operations Management* **25** pp 785–805.

[12] Melton T 2005 The Benefits of Lean Manufacturing What Lean Thinking has to Offer to Process Industries *Chemical Engineering Research and Design* **83(A6)** pp 662–673.

[13] Demeter K, Matyusz Z 2011 The impact of lean practices on inventory turnover *International Journal of Production Economics* **133** pp 154–163.

[14] Smith, Ricky e Hawkins, Bruce 2004*Lean maintenance: reduce costs, improve quality, and increase market share* (s.l. : Elsevier Butterworth–Heinemann).

[15] Danilo Ribamar Sá Ribeiro, Fernando Antônio Forcellini, Milton Pereira2020 Maintenance's Value Stream Mapping As A Useful Tool For The Implementation Of Lean Maintenance: Case Study In An Injection's Sector Of A Thermoplastics Company *Journal Of Lean Systems* **5** 1 pp 47-66.

[16] Soltan H, Mostafa S 2014 Leanness and agility within maintenance process *International Journal of Engineering Research and Technology* **3** pp 553–555.

[17] Okhovat M A, Ariffin M K A M, Nehzati T, Hosseini S A 2012 Development of world class manufacturing framework by using six-sigma, total productive maintenance and lean *Scientific Research and Essays* 7 pp 4230–4241.

[18] Patrício Cordeiro, J C Sá, A Pata, M Gonçalves, G Santos and F J G Silva 2020 The Impact of Lean Tools on Safety—Case Study *Occupational and Environmental Safety and Health II Studies in Systems, Decision and Control* **277**.

[19] Tendayi T G, Fourie C J 2013 The combined AHP-QFD approach and its use in lean maintenance*25th Annual Southern African Institute of Industrial Engineering Conference* 554.

IOP Conf. Series: Materials Science and Engineering **973** (2020) 012045 doi:10.1088/1757-899X/973/1/012045

[20] Wenchi Shou, Xiangyu Wang, Jun Wang, Lei Hou and Martijn Truijens 2014 Integration of BIM and Lean Concepts to Improve Maintenance Efficiency: A Case Study *Computing In Civil And Building Engineering* ASCE 2014.

[21] Salla Marttonen-Arola, David Baglee 2019 Assessing the information waste in maintenance management processes *Journal of Quality in Maintenance Engineering* ISSN: 1355-2511.

[22] Horváthová J, Mokrišováa M, Suhányiová A and Suhányi L 2015 Selection of key performance indicators of chosen industry and their application in formation of Creditworthy model *Procedia Economics and Finance* **34** pp 360-367.

[23] Tomé Pombal, Luís Pinto Ferreira, J C Sá, Maria Teresa Pereira, F J G Silva 2019 Implementation of Lean Methodologies in the Management of Consumable Materials in the Maintenance Workshops of an Industrial Company *29th International Conference on Flexible Automation and Intelligent ManufacturingProcedia Manufacturing* **38** 975–982.

[24] Ahmed En-Nhaili, Anwar Meddaoui and Driss Bouami 2016 Effectiveness improvement approach basing on OEE and lean maintenance tools *Int. J. Process Management and Benchmarking* **62**.

[25] Becerik-Gerber, B., Jazizadeh, F., Li, N. and Calis, G. (2012). "Application Areas and Data Requirements for BIM-Enabled Facilities Management *Journal of Construction Engineering and Management* **138(3)** pp431-442.

[26] G F L Pinto, F J G Silva, R D S . Campilho, R B Casais, A J Fernandes, A Baptista 2019 Continuous improvement in maintenance: a case study in the automotive industry involving Lean tools *Procedia Manufacturing* **38** pp1582–1591.

[27] Börger A, Alfaro J, León P 2019 Use of the Lean Methodology to Reduce Truck Repair Time: A Case Study *New Knowledge in Information Systems and Technologies* pp 655–665.

A new integrated product and process development model

AAyman[1], A BAlhamaki[2] and M H Abdellatif[3]
[1]MSc student, Mechanical Department, Faculty of Engineering, Ain shams university
[2] Assoc. Prof. Mechanical Department, Faculty of Engineering, Ain shams university
[3] Prof. Mechanical Department, Faculty of Engineering, Ain shams university
eng.alshimaa.asu@gmail.com

Abstract. Currently, the demand for more robust, economic, and quality products has increased the complexity in the integrated process development and product design model. In the knowledge economy century, all nations are competing to innovate new technologies to develop a new system that produces goods in less time and cost with compatible quality. Hence, a new IPPD model is needed to offer support and information/ technology sharing during the entire product life cycle. The new IPPD model proposed that integrate all domains shall affect the product development. The model has a friendly computerized interface supported with internet enablers that offer support to the user to collect and share data easily. This model contains 6 domains. The Business Environment is the first domain, which analyzing the pre-feasibility of the product creation. The second domain is the market analysis and segmentation taking into consideration customer needs and competition. Managerial readiness of the product developing company is the third domain, including all managerial functions and the necessity of implementing the organizational change techniques if necessary. The fourth domain is Technological creation and deals with product concepts and robust quality design. Production/Launch is the fifth domain which consists of production processes and quality techniques to assure producing with the required quality and price levels. Finally, the sixth domain deals with the vital information of product distribution, maintenance and after-market service. The model is easily adapted to the industry need

1. Introduction

The product realization process and information integration are a major concern and prerequisite for competition in today's global market. Integrated Product and Process Development (IPPD) is associated with information integration to innovate and manage a new product/platform or improving an old one, through establishing a systematic and standardized system that applies innovation and improvements within particular steps. Information integration and cooperation are the basic aspects of a successful integrated model. Researches in business success among different organization and products have shown that number of factors are important to create a competitive product.

To establish a successful, and effective product and process development model not only avails to control the production process effectively but also to integrate all available information and inputs that will trigger product improvement.

Our proposed model encompasses these mutual relationships between the external forces drive the business, the market forecasting, and the organizational aspects with the production process technicality. Prior IPPD models were linear, such that the activities done in series with no integration, and not considering the international changes in developing products, neglecting its complexity and

IOP Conf. Series: Materials Science and Engineering **973** (2020) 012046 doi:10.1088/1757-899X/973/1/012046

related needs for innovative ideas, concepts and new technologies in order to face competition and realizing the speed to market[1], the integration also allowed us to consider aspects have been neglected such as business environment, market, company strategy management. The global market nowadaysdepends on information integration, and to cope with it the product development system used must be integrated as wellSystematic, Coherence, flexible, and innovative. The concept of Integrated Product Development Process (IPPD) establishes an overview on each theme, as well as their relation to each other.

2.Methodology

Research articles and papers from scholarly journals, conferences, and reviewing references and books. The author studied the IPPD existing models, knowing each model's advantages and disadvantages and limitations then build a new integrated model that has a set of new domainsand tools.

2.1. Integrated Product and Process Development

I.P.P.D provides a systematic approach to product development that achieves timely collaboration of relevant stakeholders through the product life cycle to better satisfy customer needs.

2.2 PESTEL Tool

The PESTEL tool is the most common approach for considering the external business environment [2]. PESTEL refers to P forpolitical, E for economical, S for social, T for technological, L for legal, and E is for ecological.In the new model,it shall assist the user understood and configure the external impact on the business, and this tool is especially useful when starting a new business or making a new platform or introducing a new market.

2.3 SWOT Analysis

It is a tool used for strategic planning, it can be used effectively to build competitive strategies[3]. SWOT stands for strengths, weaknesses, opportunities, and threats.

3. Theory: The New Model

In Our Proposed Model the domains emanate from each other and the activities are not done in series, they are integrated together and allowing the necessary feedback. This model has 6 domains Each domain contains several activities, and the final model will help the user to understand the surroundings and their needs. The domains are:

- The Business environment
- The Market environment
- Managerial readiness
- Technological creation
- Production/launch
- Post Launch.

This paper focuses on the first three domains, as shown in figure 1.

3.1 Business Environment Domain

It is the external factors that affect the business. There is an unbroken relationship between the organization and the business environment, thus every business organization needs to react with its environment.

The organization's success or failure principally depends on the effectiveness of its interaction with the surrounding environment [4] explained that various restrictions are decreed in all organizations by the environment.

The readiness of the business environment will facilitate the decision of investing and the success of the project as shown in figure 2, and it contains 3 subdomains, PESTEL tool (external environment) as it is the most common approach for consideringthe external business environment. Resources evaluating and internal environment.

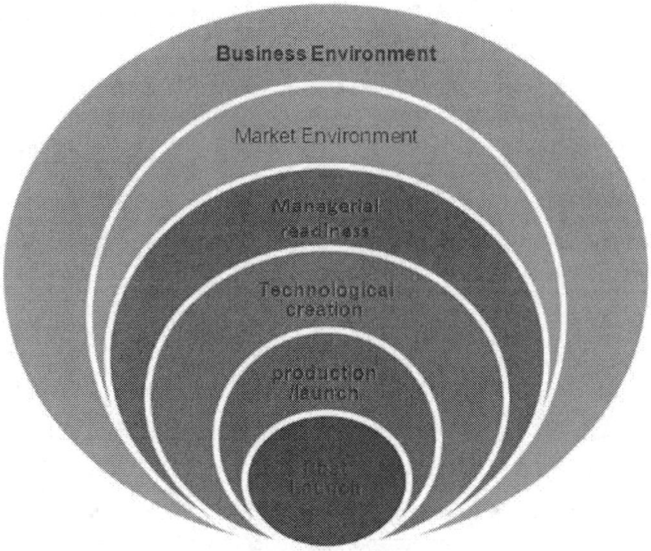

Figure1. New Proposed model six domainsoutline.

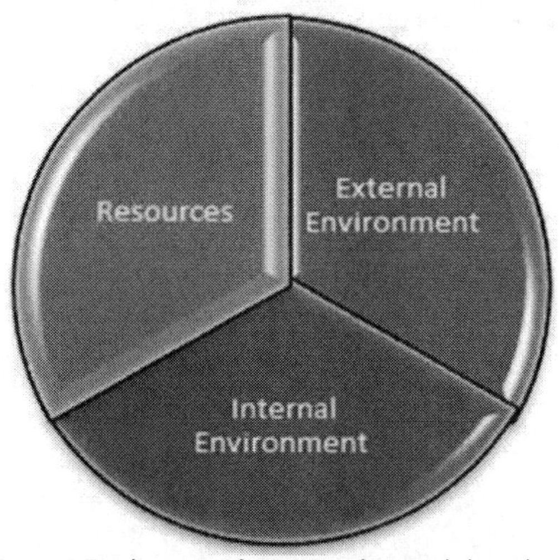

Figure2. Business environment three subdomains.

3.2 Market Environment Domain

Marketing term refers to the group of consumers or organizations that is interested in the product and has the resources to purchase it. The Market is permitted by regulations to acquire the product. In this domain, market research takes place to analyze the market need, know the demand, potential investors,Supplier integration[5],study the competitors and their strength and weakness points, also make the initial marketing plan with the help of the gathered information, finally collect the potential customers' needs and main concerns.

Market domain contains 7 subdomains so that market environment shall be covered from all aspects.

3.3Managerial readiness

The third domain in the model helps in managerial readiness of the product developing company including all managerial functions. This domain is also necessary for implementing the new vision and organizational techniques if necessary.It contains 3 subdomains planning and company strategy[6]. Planning contains 3 steps as shown in figure 3.

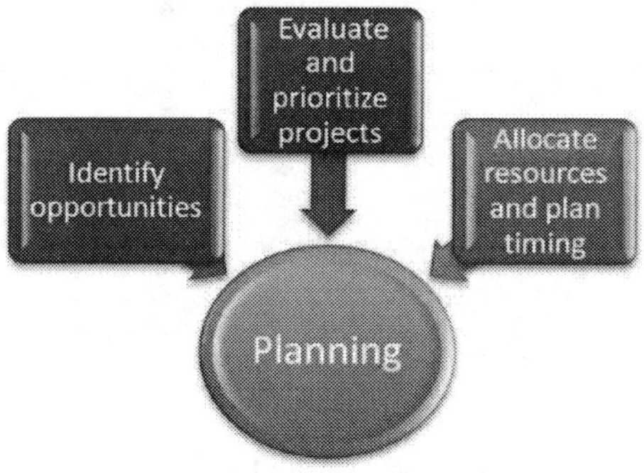

Figure 3. Planningsteps (help users to set up project goals).

4. Application

Our proposed model is consistent withthe requirements of the economic century so we designed a user-friendly interface that interact and support the users need. The interface is supported by internet enablers and data entry pagesto help the users in their research and the final decision[7]. The new model interface shown in figure 4 supported with internet enablers and programmed sheets which will help the users to enter the data into the program as well as lead them to follow the right procedure.
As shown in figure 4 the model interface including the domains and subdomains icons, save and save as. By pressing any icon, another sheet or chart or internet linkwill be opened in certain procedures.

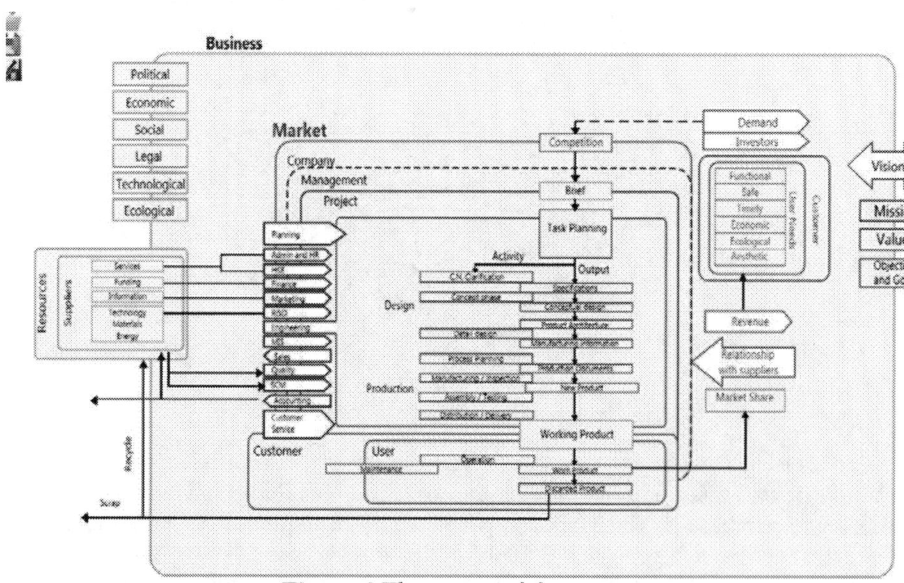

Figure 4. The new model program.

IOP Conf. Series: Materials Science and Engineering **973** (2020) 012046 doi:10.1088/1757-899X/973/1/012046

At first, the business domain, which include the PESTEL tool that contains of 6 external factors each factor listed inside the parameters effect on it and some linked to internet enablers,as shown in figure 5.Example,If the user press on Tariffs, the user will be transferred to the Egyptian ministry of finance / Tariffs section. So that the user can know all what is need and be aware of any changes happened in the tariffs.

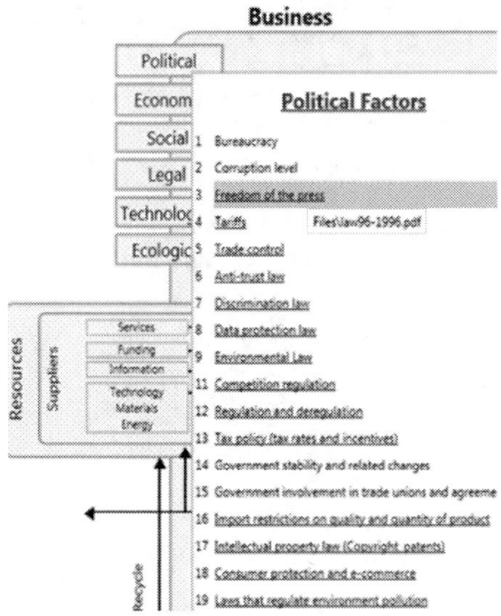

Figure 5. PESTEL tool drop-down menu.

After studding those parameters, a questioner for each factor must be done as shown in figure 6.

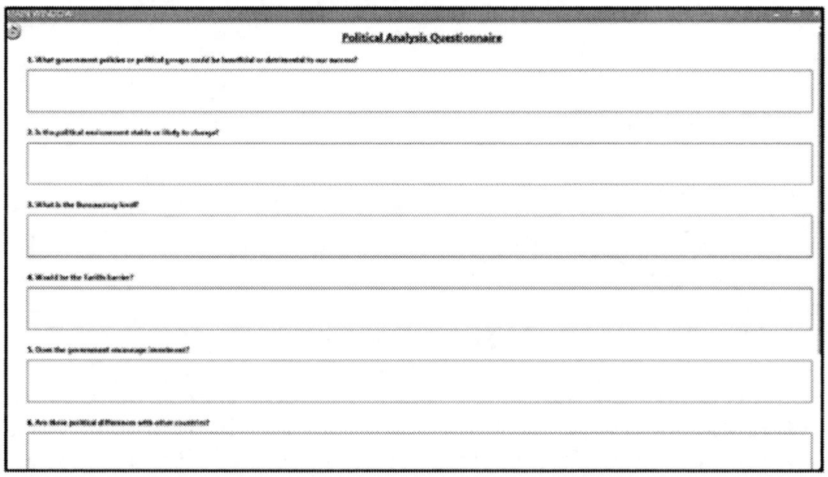

Figure 6.Political analysis questioner.

After studying the PESTEL factors and answering the questioner, the user press Business icon then a new sheet appears in it (SWOT/PESTEL analysis) table and the user has to rate each slot from 0 to 10 and the programe will have coloured those slots, green for go and red for risk. This step should help the user in his decision, figure 7.

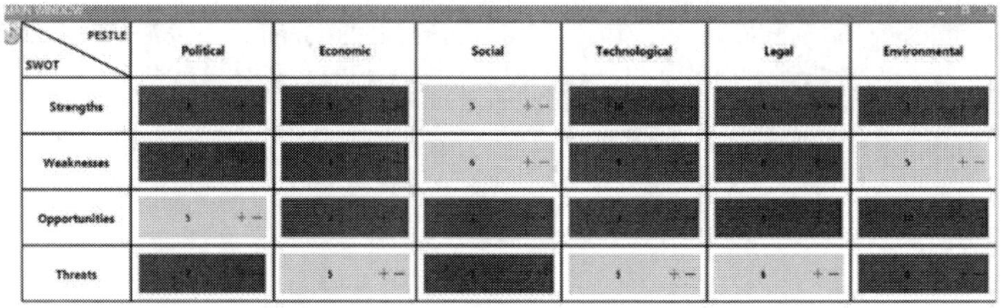

Figure 7.SWOT/PESTEL analysis.

Now, the internal environment section, which contains of 4 parts that the enterprise should create, they are vision, mission, values, objectives and clear goals and millstone[8].As in figure 8.

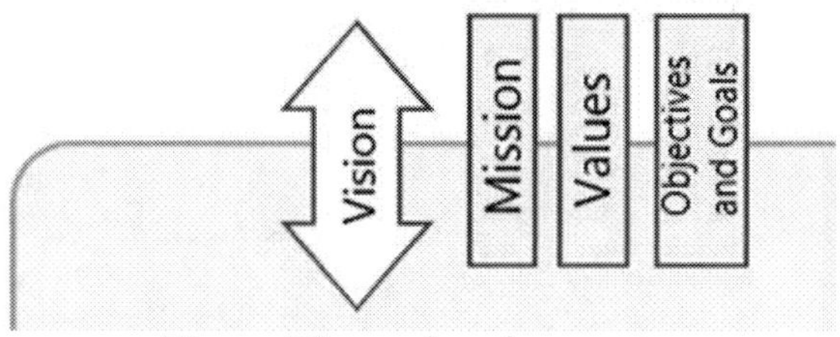

Figure 8.Internal environment.

Each slot linked to another page that allowed the user to write down the data.For example, mission is shown in figure 9.

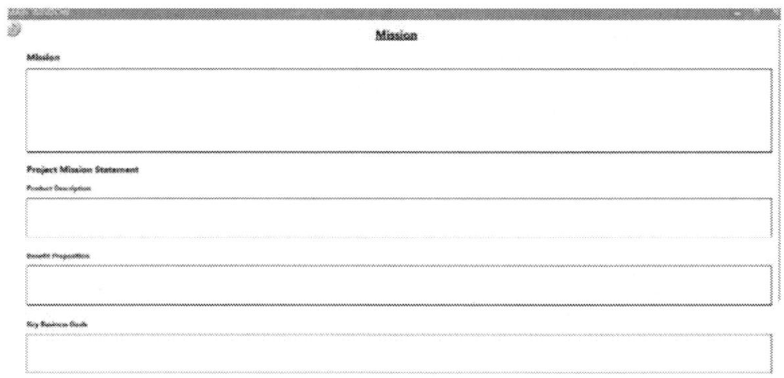

Figure 9. Mission statement window.

The 3rd subdomain is resources availability, and it 15 questions about the raw material, energy resources, and the services availability. Theuser shall study the availability with the help of internet enablers then rank them from 0 to 5. A pie chart will appear showing the availability percent, as shown in figure 10.

IOP Conf. Series: Materials Science and Engineering **973** (2020) 012046 doi:10.1088/1757-899X/973/1/012046

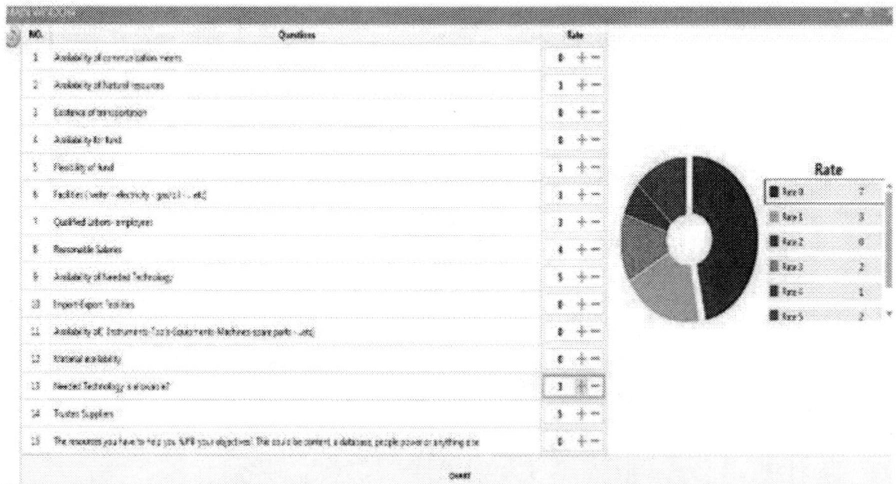

Figure10. Resources and services availability.

The 2nd domain "market domain" contains of 7 subdomains as shown in figure 11. Each subdomain contains activities. If user press any icon another window will open that contains helping internet enablers and links, also contains a questioner that user can answer it using the links or via other methods.

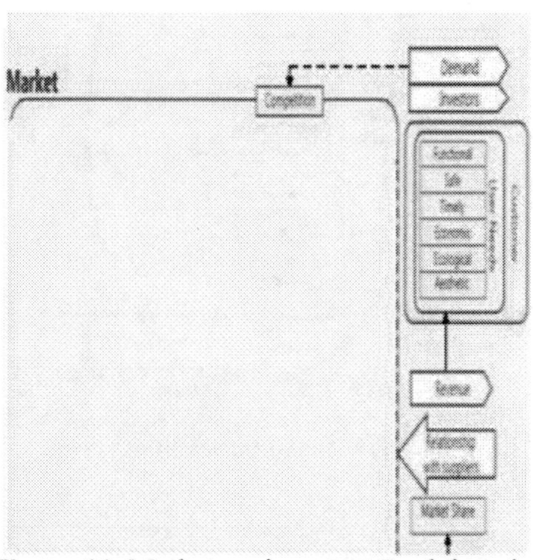

Figure 11. Market environment 7 subdomains

For example, potential investors subdomain that shown in figure 12 and figure 13.The 3Rd domain is Managerial readiness, it contains 2 subdomains,Planning and Organizational Structure.

IOP Conf. Series: Materials Science and Engineering **973** (2020) 012046 doi:10.1088/1757-899X/973/1/012046

Figure 12.potential investors window that contains a questioner and helping internet enablers

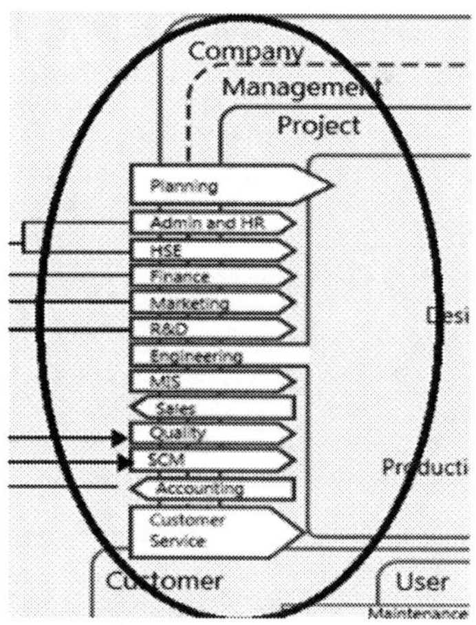

Figure 13. Managerial readiness domain

For example, planning subdomain that contains 3 steps as in figure 14

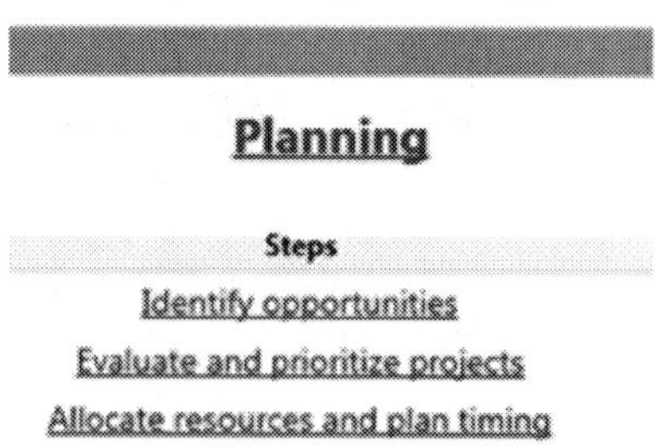

Figure14.planning subdomains window

IOP Conf. Series: Materials Science and Engineering **973** (2020) 012046 doi:10.1088/1757-899X/973/1/012046

Allocate resources for example the user allocates the facilities and resources for each quad as in figure 15.

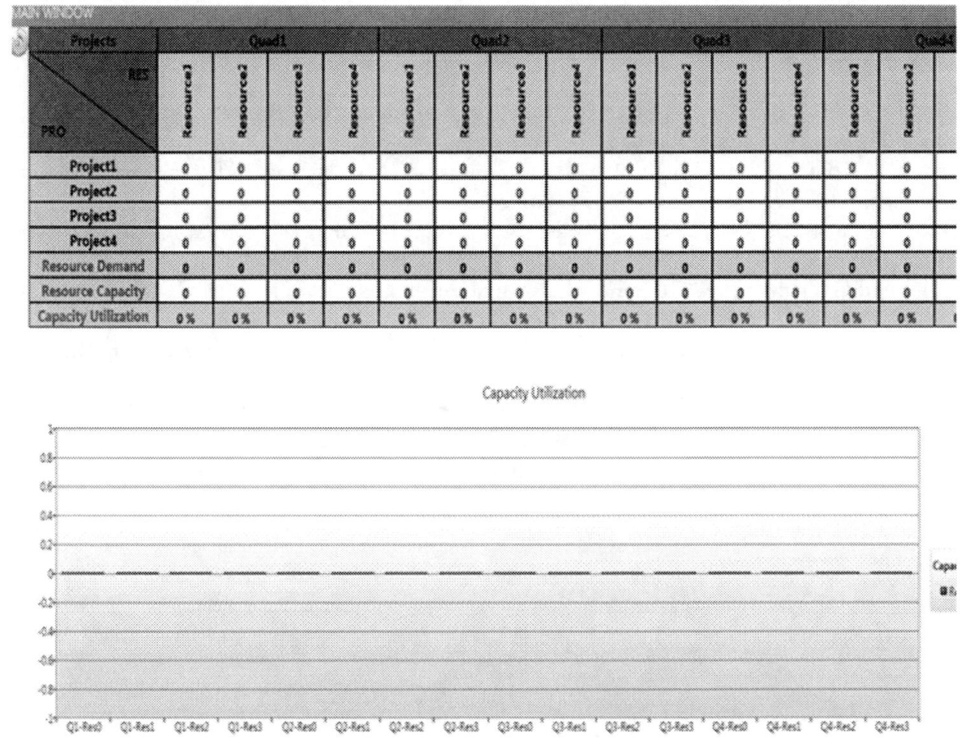

Figure 15. Allocate resources table

After finishing the first three domains a decision must be taken to continue in this project or not.If the project will be continued, then proceed in the rest of the model domains.The New IPPD Model has a friendly interface that the users can deal with easily and smoothly, the user can save as many as projects he wants.This model will help in managing and development of the entire process.

5. Conclusion

The findings of this paper are:

1) A common challenge in the product, process modelling, and tried to simulate the information integration era.
2) Our findings have important implications for entrepreneurs and managers who want to introduce a new product, improve old one, or even enter a new market.
3) Our proposed model encompasses all the domains that will affect the project success, taking into consideration the variation between industries segments and their needs, the model is Taylor made such that it is flexible and can be adapted.
4) By using this model Entrepreneurs will be able to decide when and where they will establish their projects, as they will scan the business environment and forecast the market need, turning those needs to a reliable design with competitive characteristics. Such information will have impact on the entrepreneur's decision.
5) Finally, our study also highlights two important aspects:

IOP Conf. Series: Materials Science and Engineering **973** (2020) 012046 doi:10.1088/1757-899X/973/1/012046

- The importance of simulate the reality and let the model be integrated with activates emanate from each other not done in series
- The importance of take into consideration, the external forces that drive the business and emphasize the criticality of the information integration within the organization and how to manage them.

References

[1] ROBERT G.COOPER 1983 A Process Model for Industrial New Product Development *IEEE Transactions on Eng. Manag***30**(no. 1)
[2] Abhishek G 2013 Environment & PEST Analysis: An Approach to External Business Environment ,*Int. J. Modern Soc. Sci.,* **2**(1) pp 34-43.
[3] EmetG ,Merba T 2017 SWOT Analysis: A Theoretical Review , *The Journal of Int. Soc. Research* **10**Issue 5.
[4] P. Kotler and G. Armstrong2010 Principles of marketing **13**.
[5] Kenneth J. Petersena, Robert B. Handfieldb and Gary L. Ragatzc 2005 Supplier integration into new product development: coordinating product, process and supply chain design,*Journal of Operations Management.***23**pp371-388.
[6] Karl T. Ulrich, Steven D. Eppinger 2012 *Product Design and Development***5**pp53-70.
[7] YingkuiGu, Juanjuan Liu and Weidong Wu 2006 Integrated Product and Process Development Mode Based On Models Coupling the 6th*World Congress on Intelligent Control and Automation.*
[8] Abbie Griffin and Stephen Somermeyer 2007 *The PDMA Tool Book 3 for New Product Development.* **3** pp 367-369.

IOP Conf. Series: Materials Science and Engineering **973** (2020) 012047 doi:10.1088/1757-899X/973/1/012047

Ultrasonic-assisted drilling of nickel-based super alloy in conel 601: An experimental study

A M Abdelaziz[1,3], H Youssef [1], M Al-Makky[1] and H El-Hofy[1,2]
[1] Production Engineering Department, Faculty of Engineering, Alexandria University, Alexandria, Egypt
[2] Industrial and Manufacturing Engineering Department (IME), School of Innovative Design Engineering Egypt-Japan University for Science and Technology, Alexandria, Egypt
[3] Corresponding author,E-mail: ahmedmustafa@alexu.edu.eg

Abstract. In this study, a comparison between ultrasonic-assisted drilling (UAD) and conventional drilling (CD) is presented under different feed rates using thrust force, torque, and hole geometrical errors as output responses. The experiments were done on plates of Inconel 601 (nickel-based superalloy), which is classified as a difficult-to-cut alloy due to its high Nickel content (60%), high hardness (43 HRC), and low thermal conductivity. The experiments were performed using DMG Mori Ultrasonic 20 linear, which is equipped with ultrasonic tool holderoscillating at 20 kHz with 7 μm amplitude. A coated carbide single margin twist drill had been used in the experiments. Full factorial design of experiments approach was employed, and the results had been statistically analyzed to find the most significant factor affecting the process responses. The results showed that the ultrasonic assistance had reduced the thrust force, and torque compared to conventional drilling (CD). Also, a reduction in holecylindricity error was detectedduring UAD, whichimprovesthe hole quality. In case of UAD, twist drills did not suffer from a physical wear, however notch wear was observed in CD drills. Chip morphology was also studied. Short segmented chips were obtained when using UAD which improved chip evacuation and reduced the chance of chip jamming in the drill flutes.

1. Introduction

Since the invention of the aircraft gas turbine, a steady increasingdemand for materials showing higher mechanical properties at elevated temperatures appeared. This requirement drove the motivation to develop high temperature alloys which has resulted in many useful alloys to meet a variety of applications. One family of these alloys is Ni-based superalloys. They exhibit high mechanical strength, ductility, creep strength at high working temperatures, high fatigue strength, and also an excellent resistance to oxidation and corrosion even at elevated operating temperatures[1].

On contrast, from the machining point of view, these superior properties made the cutting of such alloys more complicated and,consequently,they were classified as difficult-to-cut alloys. During machining, a major portion of superalloys strength is preserved due to their high temperature properties. Even though superalloys are not very hard at room temperature (the hardness range being about 25-40 HRC)[2], they create difficulties during machining due to their tendency for work hardening. The work hardening quickly occursand results in high cutting forces, burr formation and notch wear at the cutting edges during machining. Also, the poor thermal conductivity of superalloys generates high concentrated temperature at the tool tip, which decreases the tool life[1, 3].

As a result of the low machinability of superalloys, it was crucial to find new techniques to machine these difficult-to-cut alloys instead of conventional ones. Ultrasonic-assisted drilling (UAD); using high frequency and low amplitude oscillations, is a promising technology in drilling superalloys[4]. In this study, one of these Ni-based superalloys (Inconel 601) was investigated to evaluate the effect of UAD technique on the process performance compared with conventional drilling (CD).

UAD process is similar in concept to the CD in many aspects; it uses conventional twist drill as a cutting tool, which rotates and feeds axially in z-direction, however in UAD ultrasonic oscillations are superimposed in the direction of feed as shown in Figure1. The main challenge in this process is how to build a system to generate the ultrasonic oscillations, work in the resonance mode, and transmit the oscillations to the cutting tool without any harmful effect on machine tool spindle. Three proposed designs were introduced. First one, through the design of a non-rotary ultrasonic tool holder and give the rotational speed to the workpiece[5, 6]. Secondly, design of a rotary tool holder and transmit theelectrical signal from power generator to the piezoelectric ceramics via two different techniques; using slip rings[7-12], or inductive transmission system[13-15]. Last technique is by transmitting oscillations to the workpiece not cutting tool[16-19].

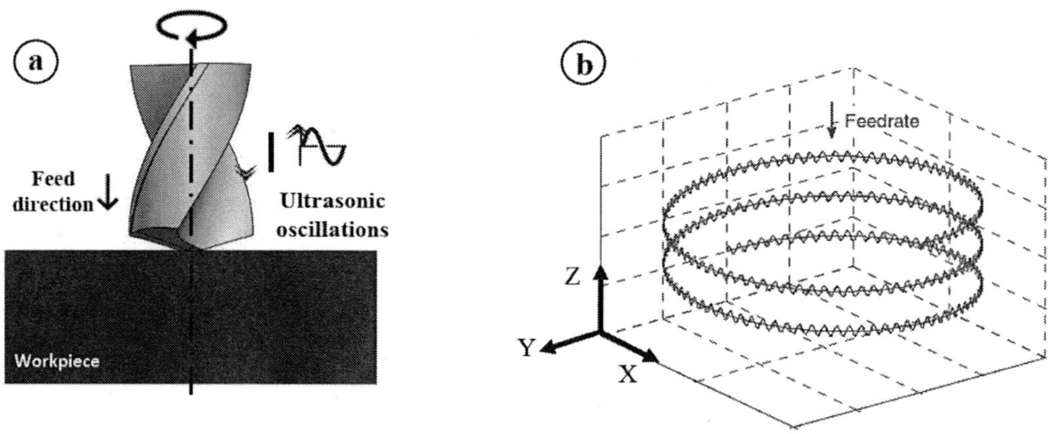

Figure 1. (a) Schematic drawing of ultrasonic-assisted drilling operation (b) the trajectory of any point on the cutting edges of twist drill during UAD.

Ultrasonic assistance (UA) improves hole quality alongall drilling stages from the beginning to the end of hole. In the first stage, chisel edge begins to contact with workpiece, and ultrasonic oscillations help in drill quick penetration; plays as a centering operation, so they reduce tool skidding (displacement) and more location accuracy could be obtained[20]. After that the main cutting edges start to penetrate into the workpiece. Due to UA, chips become more segmented, which facilitate chip evacuation from drill flutes[5, 21]. Chip segmentation helped in cutting forces reduction. Azghandi, et al. [21] concluded that UAD showed a significant decrease in thrust force of around 60% compared to CD when cutting high alloyed steel (X20Cr13). Furthermore, with increasing the oscillation amplitude, the drilling force was slightly decreased. Also, tool wear was significantly reduced when UAD was applied, which increased tool life[6, 21, 22]. UAD improved hole geometrical quality, as it was found that UAD provided better hole circularity, cylindricity and reduced hole surface roughness drastically [23, 24]. At hole exit, UAD playeda crucial role in burrs elimination, because the last layer in workpiece is cut not punched as in case of CD. So, a decrease in burrs height and width was observed[6, 17, 25].

2. Experimental Procedure

In this study, the Inconel 601 (Ni-based superalloy) workpiece material was supplied as plates of 25 mm width, 200 mm long, and 6.45 mm thick. The chemical composition of Inconel 601 as analyzed by WDXRF Spectrometer is provided in Table 1. The hardness measurement value was 43 HRC.

Table 1. Chemical composition of Inconel 601.

C	Si	S	Cr	Ni	Al	Ti	Co	Cu	Mo	Nb
0.06	0.20	0.003	**22.80**	**59.20**	1.23	0.38	0.08	0.11	0.39	0.14

All experiments were performed on DMG Mori Ultrasonic 20 linear machining center which is specially designed for ultrasonic-assisted machining operations and equipped with distinctive HSK ultrasonic toolholder. The toolholder contains the piezoelectric crystals as the oscillators and the electric signal is transmitted to the oscillators via wear-free inductive system, which enables the full compatibility to the automatic tool changers as shown in figure 2. The ultrasonic system has 300-Watt power generator with frequency range from 17.5 to 48 kHz. In order to avoid any oscillations from being transferred back to the spindle, the connection point between the transducer and the tool holder is in the nodal plane, where the oscillation is zero. The cutting tool used in this study was a single-margin AlTiN-coated carbide twist drill with a 6-mm diameter, supplied from Kyocera-SGS tools. The drills had 140° point angle and 30° helix angle with a full length of 63 mm. It was recommended to minimize the tool overhang, in order to minimize the radial vibrations which might affect the drilled hole quality. All experiments were performed using water-based emulsion cutting fluid with a concentration of 10%. figure 3 shows the twist drill fixed to the ultrasonic tool holder and the external cutting fluid tubes are targeted to drill tip.

Figure 2. ultrasonic actuator system consisting of (1) HSK interface, (2) inductive signal transmission system, (3) piezoelectric transducer, (4) tool interface, (5) vibration-free mounting flange, (6) ER-11 collet, and (7) twist drill.

Figure 3. Drill fixation to the US-toolholder.

The experiments were designed to study the effect of changing feed rates on the responses with the application of ultrasonic oscillations (UAD) at 20 kHz and amplitude of 7 µm and without (CD).The frequency and amplitude of oscillation are fixed in all experiments. The tool manufacturer recommended to use 10 m/min surface cutting speed (rotational speed: 530 rpm) todrill Inconel 601, so it was fixed in all experiments. Feed rate values used in the present investigation were 0.020, 0.035, 0.050 mm/rev. Number of test runs was 6with 2 replications as shown in table 2.

Table 2. Test runs of the design matrix.

Drilling Technique	Feed rates		
	0.020 mm/rev.	0.035 mm/rev.	0.050 mm/rev.
CD	1	2	3
UAD	4	5	6

Three different responses were recorded to assess the performance of ultrasonic and conventional drilling while changing feed rates. The responses measured were; thrust force, torque, and cylindricity error.For the measurements of thrust force and torque, a Kistler dynamometer type 9272 was used with charge amplifier type 5070A connected to the computer with USB cable as shown in figure4. A special fixture was designed to clamp the workpieces to the dynamometer based on dynamometer manufacturer recommendations to ensure measurements accuracy and precision (figure5). Hole geometrical errors were measured using CNC Coordinate Measuring Machine (Model: Mitutoyo CRYSTA-Apex S 544) with a resolution of 1 μm.

Figure 4. Kistler hardware system used in torque and force measurements

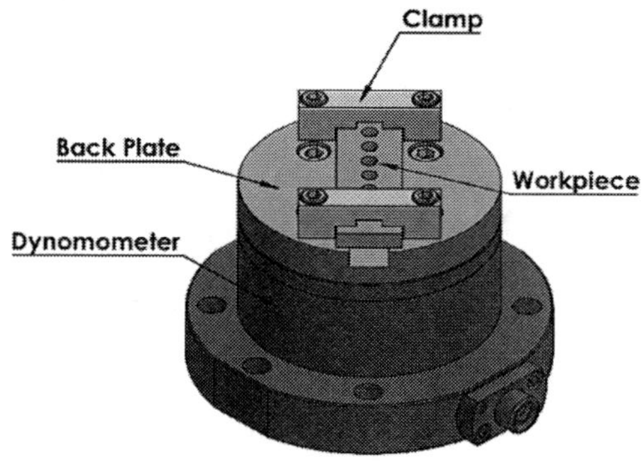

Figure 5. Fixture designed for workpiece clamping and fixed to the dynamometer with 4 bolts.

3. Results and discussion

3.1. Thrust force and torque

Drilling of Inconel 601 was a challenging process, due to its high hardness, and high mechanical properties, which may appear in thrust force and torque plots in figure 6.UAD shows a clear reduction in the values of thrust force and torque for the same cutting conditions. As a result of UAD, an average reduction of 18.5% and 20% in thrust force and torque values were observed respectively.

Figure 6 shows the plots of thrust force and torque at the three selected feed rates. Each plot represents a comparison between CD and UAD values against drilling time. The dotted lines represent the position of twist drill through the drilling operation. The mean values of thrust force and torque had been obtained from the steady cutting zones in these plots, which are located between the two dotted lines (from drilling depth of 0.8 mm to 6.4 mm).All torque plots demonstrated that the torque values of CD are saddling the torque values of UAD, whereas the plots of the thrust force in UAD are located below those of CD. As a result of the superimposed ultrasonic oscillations on the drilling operation in UAD, the thrust force on the first 0.8 mm depth in workpiece is less than in CD through all feed rates. The ultrasonic oscillations work as a hammering process, so they guide the tool during drilling, which reduce the tool skidding (displacement) and, therefore, reduce the thrust force. After the depth of 0.8 mm, the main cutting edges of the drill is totally engaged in the workpiece. Consequently, the effect of ultrasonic impact action begins which leads to more segmented chips than that in CD. Segmented chips are easier to evacuate from the drill flutes which reduce the possibility of chip jamming, and tool breakage. Also, increasing in tool life and improving the hole quality can be realized.

Regarding the torque plots, it is noticed that the amplitude (distance between peak and valley) of UAD plot is larger than that of CD due to the tool separation process as the drill leaves and engages again into the workpiece during the ultrasonic oscillations. At the depth of 6.45 mm, where the chisel edge reaches the bottom surface of workpiece, while the main cutting edges are still in engagement, thrust force and torque begin to decrease to their minimum values.

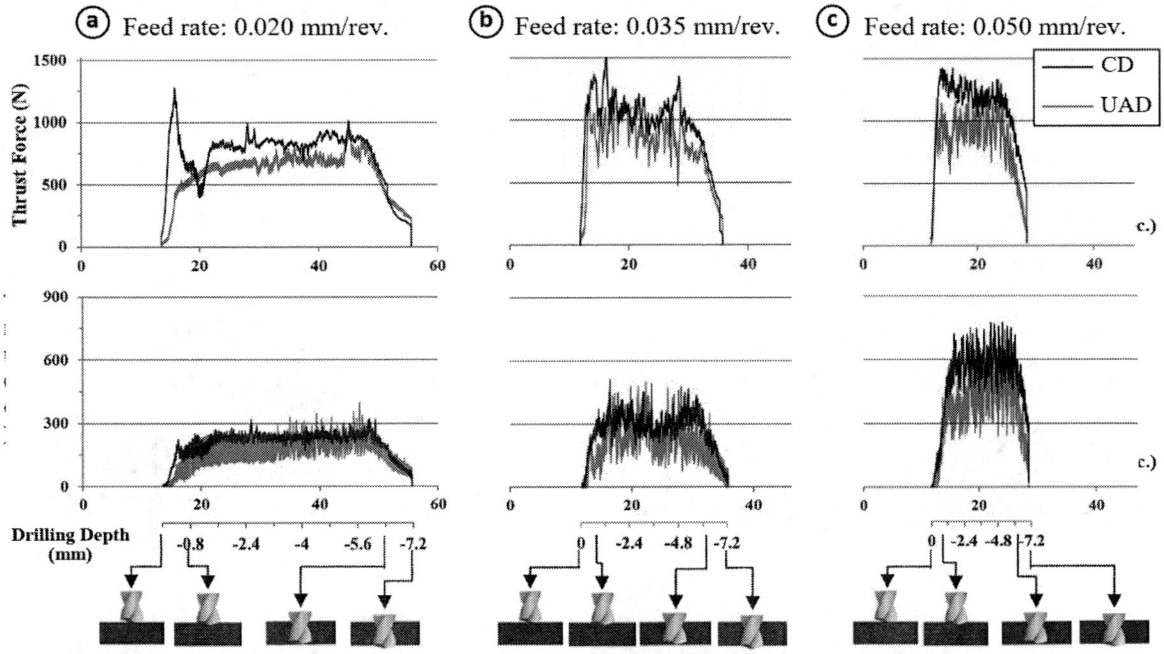

Figure 6. Thrust force and torque plots vs time and drilling depth for different feed rates (a) 0.02 mm/rev. (b) 0.035 mm/rev. (c) 0.05 mm/rev. NB: the dotted lines illustrate the position of twist drill during cutting.

Figure 7 shows that the mean values of thrust force and torque are increased with increasing feed rates for the two techniques CD and UAD. However, the rate of increase of thrust force and torque with increasing the feed rate in UAD is not equivalent to that in CD. Figure 7(b). shows that the mean torque value for feed rate of 0.05 mm/rev. is significantly higher than that at low and medium feed rates.

In most metals and alloys, the last layers of workpiece at hole exit are usually transformed to burrs. However, in case of Inconel 601, the last layers did not form burrs, but were pierced to form a cup as shown in figure 8. This phenomenon occurred during both CD and UAD processes, which can be attributed to the high shear strength of Inconel 601, so the cup was pierced as a result of the thrust force at the end of cut. The indentation lines formed due to ultrasonic oscillations were clearly noticeable on the cup at UAD as shown in figure 8, while the cup from CD had only the usual circular machining marks (lays).

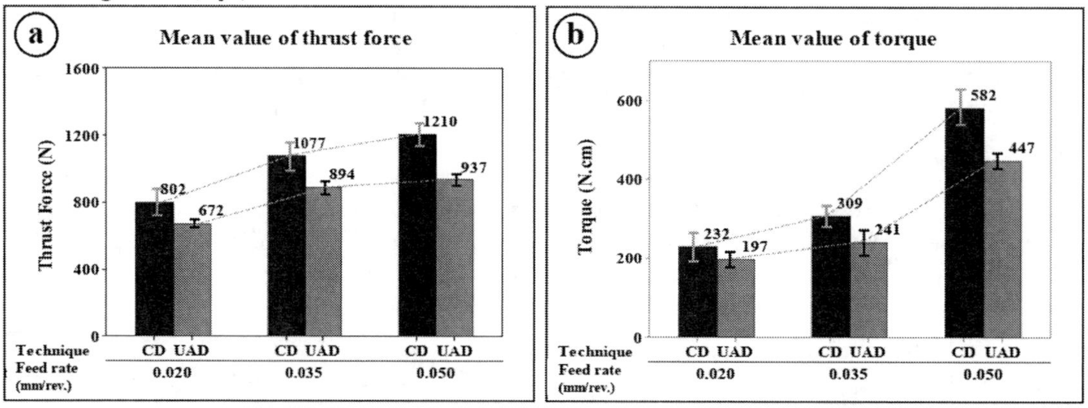

Figure 7. Bar chart of the mean values of (a) thrust force and (b) torque during the drilling of Inconel 601 using conventional drilling (CD) and ultrasonic-assisted drilling (UAD) with different feed rates (N=530 rpm).

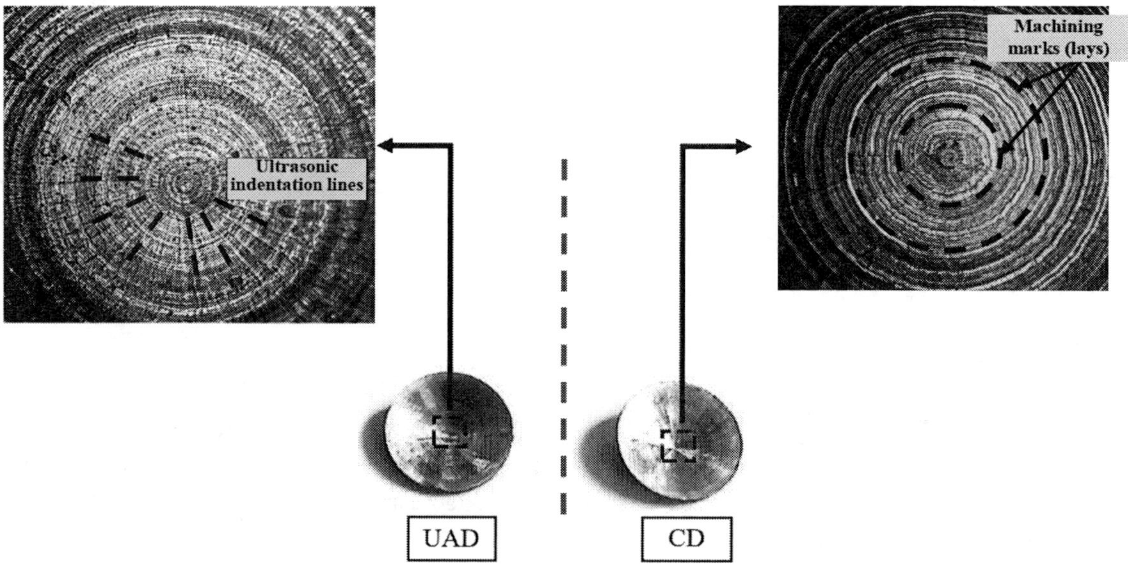

Figure 8. The cup formed due to piercing at the end of cut in UAD and CD.

In order to define the most significant factor affecting the responses, ANOVA approach was used for both thrust force and torque. The analysis of thrust force demonstrated that the two factors; drilling technique (A) and feed rate (B) are significant, but the feed rate change is slightly more effective than drilling technique as shown in figure 9. On the other hand, the effect of feed rate factor is more significant in case of torque response as demonstrated in figure 10(a). Also, in figure 10 (b,c) the main effects plot and the interaction plot are shown respectively. The effect of interaction factor (AB) is minimum. So, in order to reduce thrust force and torque during the drilling of Inconel 601, lower feed rates should be selected.

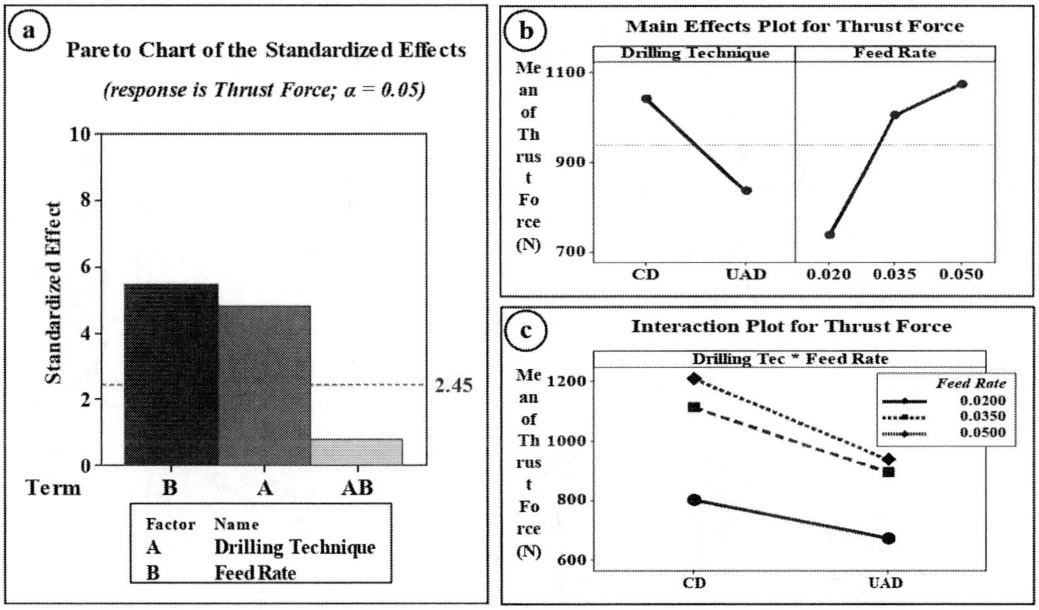

Figure 9. ANOVA results for thrust force (a) pareto chart for the significant factors, (b) main effects plot, (c) Interaction plot.

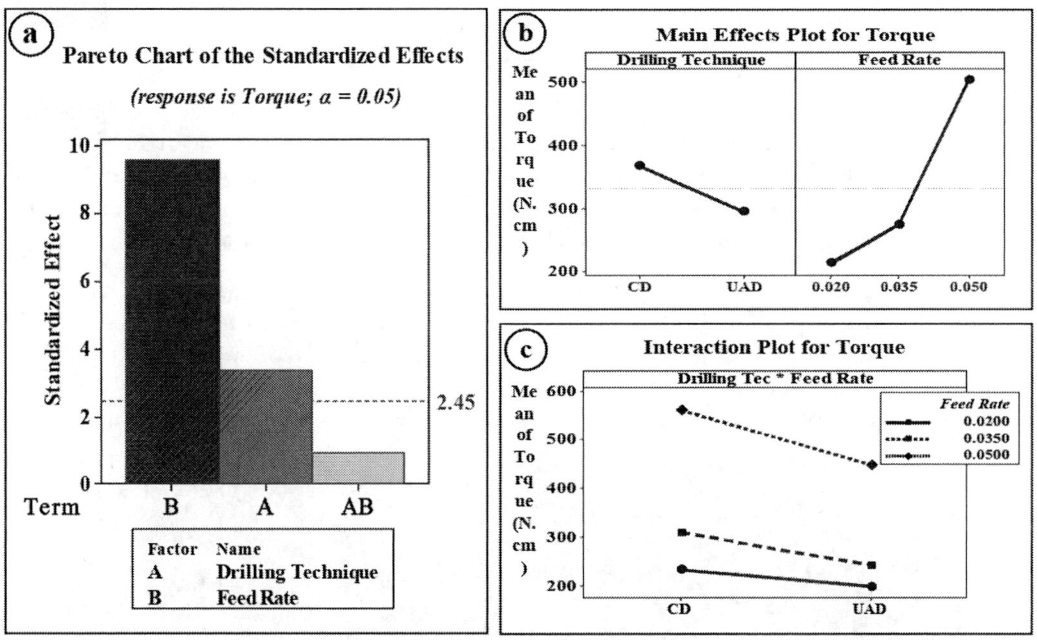

Figure 10. ANOVA results for torque (a) pareto chart for the significant factors, (b) main effects plot, (c) Interaction plot.

3.2. Hole quality

In order to evaluate the hole quality, the cylindricity error of the produced holes was measured using a CNC CMM by probing 40 points randomly through the hole surface. Then, the least square cylinder method had been employed to calculate the cylindricity error. Figure 11 shows the bar chart of mean values of cylindricity error for the test matrix. Using UAD, cylindricity error was reduced significantly than that in CD by 33%. Due to force and torque reduction in UAD process, the drill cutting edges are less affected by the radial displacement (tool skidding). Additionally, the axial ultrasonic oscillation helps in tool alignment through the entire depth of hole, which reduces the cylindricity error and, therefore, improves the hole quality.

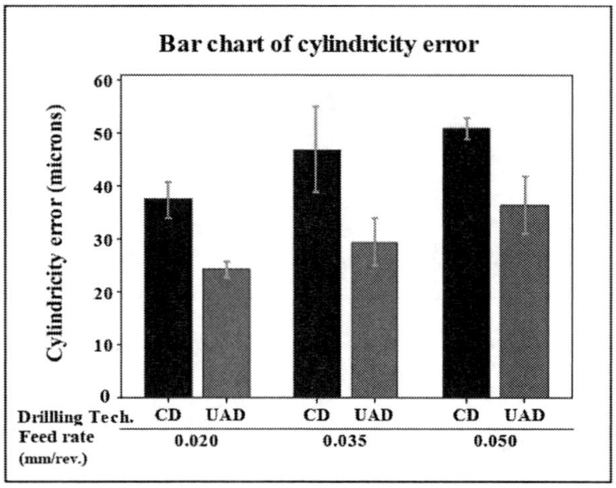

Figure 11. Bar chart of drilled hole cylindricity error at different feed rates using CD and UAD using constant surface cutting speed 10 m/min.

3.3. Tool wear

In order to investigate the tool wear, A MEJI microscope equipped with a digital camera was employed. The experiments were designed to use a new drill every 6 CD experiments and a new one for every 6 UAD experiments. After analysis, the drill used for UAD experiments did not suffer from wear, only scratches in the coating layer (figure 12-a),while no physical wear occurs to the tool.

On the other hand, CD tool suffered from a notch wear as shown in figure 12(b), which lead to BUE adhesion in this location. Tool-workpiece separation process due to the ultrasonic oscillation helped in the dissipation of heat from the cutting zone and the reduction of the pressure (stress) on the cutting edges, which lead to minimize the chance of tool chipping as compared toCD. Furthermore, it helped in the renewal of the cutting fluid with consequent reduction of tool temperature and tool wear.

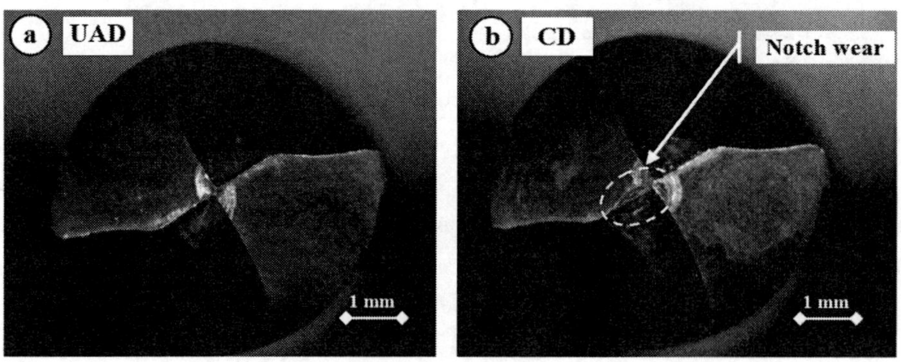

Figure 12. Tools wear after drilling using (a) UAD, (b) CD.

IOP Conf. Series: Materials Science and Engineering **973** (2020) 012047 doi:10.1088/1757-899X/973/1/012047

3.4. Chip morphology

As mentioned before, one of the reasons that makes UAD more effective is the chip segmentation. As shown in figure 13(a), segmented chips were obtained while cutting with UAD, however short continuous chips were also noticed in UAD in the beginning of cut due to the material ductility. On contrary, chips obtained in CD were long continuous chips as shown in figure 13(b). Long continuous chips might cause severe chip jamming inside the narrow helical flutes and edge chipping, leading to early tool wear and breakage, poor hole quality and decreased machining productivity.

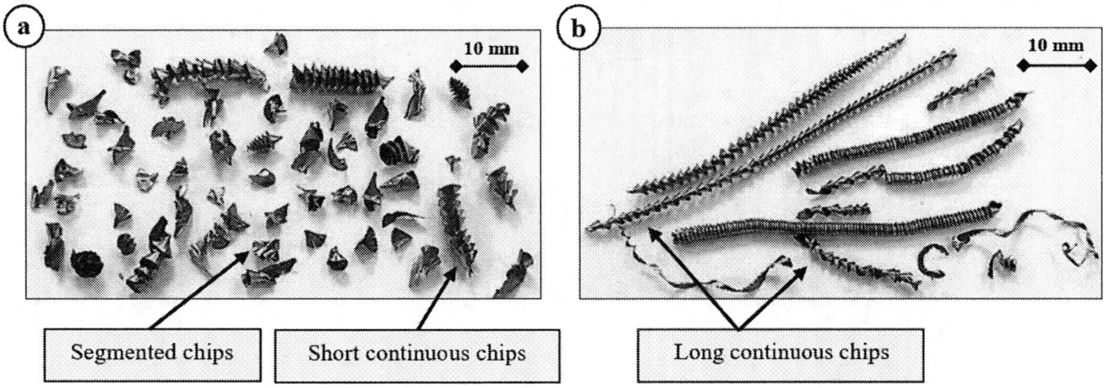

Figure 13. Chips morpholgy during (a) UAD process, (b) CD process.

4. Conclusions

In this study, experimental investigations on drilling of Ni-based superalloy (Inconel 601) were performed using conventional drilling (CD) and ultrasonic-assisted drilling (UAD). A set of experiments was designed using AlTiN-coated carbide drills to explore their effects on thrust force, torque, hole quality, and tool wear. It can be concluded that;

- When UAD was applied, thrust force decreased by 18.5% in average, compared with that in CD.
- At higher feed rates, thrust force increased in UAD and CD, however the rate of increase in thrust force was not steep as in case of CD.
- Torque is also reduced during UAD by 20% in average,compared with that in CD.
- In case of thrust force, ANOVA demonstrated that the two factors; feed rate and drilling techniques were significant and almost have the same effect.
- However, in case of torque, the feed rate factor was more significant than the drilling technique.
- UAD improves drilled hole quality as hole cylindricity error was reduced by 33% than that in CD.
- Notch wear and BUE adhesion was observed in CD, while in UAD the drill wear was minimum.
- Segmented chips were obtained while cutting with UAD, but in CD long continuous chips were observed.
- The indentation lines formed due to ultrasonic oscillations were clearly noticeable on the pierced cup at UAD.

IOP Conf. Series: Materials Science and Engineering **973** (2020) 012047 doi:10.1088/1757-899X/973/1/012047

References

[1] Ezugwu E O, Wang Z M and Machado A R 1999 The machinability of nickel-based alloys: a review*Journal of Materials Processing Technology* **86** pp 1-16.
[2] Polkowska A *et al.* 2019 Microstructure and Hardness Evolution in Haynes 282 Nickel-Based Superalloy During Multi-variant Aging Heat Treatment *Journal of Materials Engineering and Performance* **28** pp 3844-51.
[3] Jindal P C, Santhanam A T, Schleinkofer U and Shuster A F 1999 Performance of PVD TiN, TiCN, and TiAlN coated cemented carbide tools in turning *International Journal of Refractory Metals and Hard Materials* **17** pp 163-70.
[4] Feucht F, Ketelaer J, Wolff A, Mori M and Fujishima M 2014 Latest machining technologies of hard-to-cut materials by ultrasonic machine tool *Procedia CIRP* **14** pp 148-52.
[5] Chen S, Zou P, Tian Y, Duan J and Wang W 2019 Study on modal analysis and chip breaking mechanism of Inconel 718 by ultrasonic vibration-assisted drilling *The International Journal of Advanced Manufacturing Technology.*
[6] Azarhoushang B and Akbari J 2007 Ultrasonic-assisted drilling of Inconel 738-LC*International Journal of Machine Tools and Manufacture* **47** pp 1027-33.
[7] Lotfi M and Amini S 2019 Effect of longitudinally intermittent movement of cutting tool in drilling of AISI 1045 steel: A three-dimensional numerical simulation," *Proceedings of the Institution of Mechanical Engineers, Part C: Journal of Mechanical Engineering Science* **233** pp . 4081-90.
[8] Babitsky V, Astashev V and Meadows A 2007 Vibration excitation and energy transfer during ultrasonically assisted drilling *Journal of sound and vibration* **308** pp 805-14.
[9] Chu N-H, Nguyen V-D and Ngo Q-H 2019 Machinability enhancements of ultrasonic-assisted deep drilling of aluminum alloys*Machining Science and Technology* pp 1-24.
[10] Li X F, Dong Z G, Kang R K, Wang Y D, Liu J T and Zhang Y 2016 Comparison of Thrust Force in Ultrasonic Assisted Drilling and Conventional Drilling of Aluminum Alloy in *Materials Science Forum*, pp 38-43.
[11] Sanda A, Arriola I, Navas V G, Bengoetxea I and Gonzalo O 2016 Ultrasonically assisted drilling of carbon fibre reinforced plastics and Ti6Al4V *Journal of Manufacturing Processes* **22m**pp 169-76.
[12] Baghlani V, Mehbudi P, Akbari J, Nezhad E Z, Sarhan D and Hamouda A 2016 An optimization technique on ultrasonic and cutting parameters for drilling and deep drilling of nickel-based high-strength Inconel 738LC superalloy with deeper and higher hole quality*The International Journal of Advanced Manufacturing Technology* **82** pp 877-88.
[13] Shao Z, Jiang X, Li Z, Geng D, Li S and Zhang D 2019 Feasibility study on ultrasonic-assisted drilling of CFRP/Ti stacks by single-shot under dry condition *The International Journal of Advanced Manufacturing Technology.*
[14] Dong S, Liao W,Zheng K, and Ma W 2019 Investigation on thrust force in rotary ultrasonic drilling of CFRP/aluminum stacks*Proceedings of the Institution of Mechanical Engineers, Part C: Journal of Mechanical Engineering Science.*
[15] Liang W, Xu J, Ren W, Liu Q, Wang X and Yu H 2019 Study on the influence of tool point angle on ultrasonic vibration–assisted drilling of titanium alloy *The International Journal of Advanced Manufacturing Technology* pp 1-14.
[16] Chang S and Bone G 2009 Thrust force model for vibration-assisted drilling of aluminum 6061-T6 *International Journal of Machine Tools and Manufacture* **49** pp 1070-76.
[17] Chang S and Bone G 2005 Burr size reduction in drilling by ultrasonic assistance *Robotics and computer-integrated manufacturing* **21** pp 442-450.
[18] Pujana J, Rivero A, Celaya A and de Lacalle L 2009 Analysis of ultrasonic-assisted drilling of Ti6Al4V *International Journal of Machine Tools and Manufacture* **49**, p 500-08.
[19] Azghandi B and Razfar M 2011 An experimental study on the effects of ultrasonic assisted drilling on chip characteristics and tool life in *Advanced Materials Research* pp 351-56.

IOP Conf. Series: Materials Science and Engineering **973** (2020) 012047 doi:10.1088/1757-899X/973/1/012047

[20] Baghlani V, Mehbudi P, Akbari J and Sohrabi M 2013 Ultrasonic Assisted Deep Drilling of Inconel 738LC Superalloy *Procedia CIRP* **6** pp 571-76.
[21] Azghandi B, Kadivar M and Razfar M 2016 An Experimental Study on Cutting Forces in Ultrasonic Assisted Drilling *Procedia CIRP* **46** pp 563-66.
[22] Dahnel A, Ascroft Hand Barnes S 2016 The effect of varying cutting speeds on tool wear during conventional and ultrasonic assisted drilling (UAD) of carbon fibre composite (CFC) and titanium alloy stacks*Procedia CIRP* **46** pp 420-23.
[23] Mandegari Mand Behbahani S 2013 Experimental Analysis of a Novel Rotary Ultrasonic Assisted Drilling (RUAD) Machine *Materials and Manufacturing Processes* **28** pp 481-487.
[24] Moghaddas M, Short M, Wiley N, Yi A and Graff K 2018 Performance of an ultrasonic-assisted drilling module *The International Journal of Advanced Manufacturing Technology* **94** pp 3019-28.
[25] Li Z, Zhang D, Jiang X, Qin W and Geng D 2017 Study on rotary ultrasonic-assisted drilling of titanium alloys (Ti6Al4V) using 8-facet drill under no cooling condition *The International Journal of Advanced Manufacturing Technology* **90** pp 3249-64.

Acknowledgment
The authors sincerely acknowledge Prof. HelmiAttia, Adjunct Professor of Mechanical Engineering, McGill University, for his prompt response and full dedication in the supply of cutting tools related to this research.

IOP Conf. Series: Materials Science and Engineering **973** (2020) 012048 doi:10.1088/1757-899X/973/1/012048

Computational fluid dynamic model for machining using minimum quantity coolant

W Abdelfattah[1,2], H Hegab[2], A Mohany[1] and H A Kishawy[1]
[1]Faculty of Engineering and Applied Science, University of Ontario Institute of Technology, Oshawa L1H 7K4, Canada.
[2] Faculty of Engineering, Cairo University, Giza 12613, Egypt.
Corresponding author. E-mail address: waleed.ahmed3@ontariotechu.net

Abstract. The cooling applications during machining has significant effects on the production costs, surface quality, and the mechanical properties of the final product. In conventional flood cooling, a large amount of continuous cooling fluid is usually used, and that increases the cost of the product as well as the harmful effects on the environment and the machining operator. This study focuses on simulating alternative cooling system, called minimum quantity coolant (MQC), which used an optimal flow rate compared to classical flood cooling. The cooling fluid is directly provided to the cutting edge through the insert holder. In the current work, a computational fluid dynamic (CFD) model has been developed to study the effects of the cooling fluid velocity on the accessibility of coolant to the chip-tool interface area under using various types of cooling fluids. Three types of coolant are used (i.e. water, mineral oil, and nano-fluid). The results of the proposed CFD model have been classified into two phases. The first phase obtains the coolant accessibility percentage into the chip-tool interface (MA%) with different coolant velocities (i.e., 0.5, 1, 1.5, and 2 m/s) for the three studied coolants. The second phase discusses the heat transfer effectiveness for the employed coolants with different inlet velocities since it is an important aspect, especially when machining hard-to-cut materials. It was found that increasing the coolant velocity would increase the coolant accessibility percentage into the chip-tool interface area. However, no significant effect has been found after 1.5 m/s for all employed coolants.

1.Introduction

Difficult-to-cut materials arelargelyused in several engineering applications, including and not limited to army, automotive, and aerospace industries. That is because of their superior properties,such ashigh corrosion resistance and high strength-to-weight ratio [1]. The main problem associated with machining hard to cut materials is the change in the tool geometry (i.e., wear) within a relatively short time because of high temperature levels. Thus, the tool performance indicator is directly depending on the ability of the cutting insert to maintain a good tool life (i.e., nearly the same geometry) at this severe temperature. The excessive temperature occurs due to the fact that most of the hard to cut materials have a low thermal conductivity which leads to concentrate the generated heat in the cutting zone. As a result, the tool wear rate is highly increased and also the mechanical characteristics of workpiece material are influenced [2].The main reasons behind using the cutting coolant during the machining process are to reduce the temperature in the cutting zone and reduce friction between the tool and the chip (see figure 1) by providing a lubricant effect. Flood cooling is the commonly used technique in order to improve the machinability of difficult-to-cut materials; however, it has severe environmental and health impacts. Various studies showed that minimum quantity coolant (MQC) is a potential alternative to flood cooling that minimizes the amount of cutting fluid supplied to the cutting

IOP Conf. Series: Materials Science and Engineering **973** (2020) 012048 doi:10.1088/1757-899X/973/1/012048

zone [3, 4, 5]. MQC is an approach in which an optimal amount of cutting fluid is forced to penetrate the cutting zone by means of compressed coolant. It provides optimum amount of lubricant through a system consists of a compressor, base fluid container, tunings, flow control system and spray nozzles.

The direct measurements of the coolant accessibility to the chip-tool interface area or tool cutting edge is still a challenge due to the dynamic nature of cutting processes as well as the small contact area during chip formation. In such cases, finite element analysis and simulation could provide an acceptable prediction of different process characteristics such as coolant accessibility to the tool-chip interface, cutting temperature, and residual stresses under different machining conditions in a virtual environment.

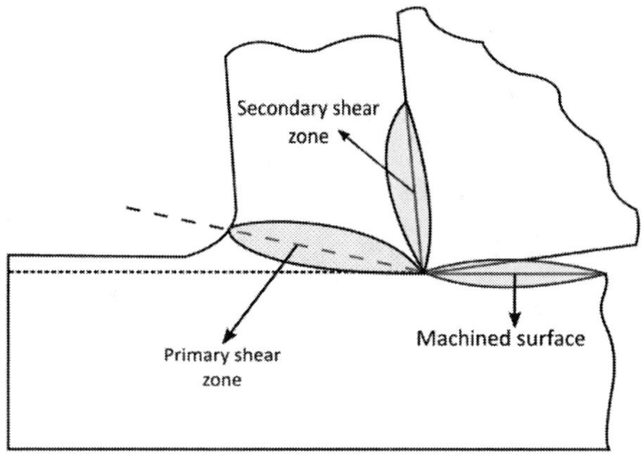

Figure 1. Heat generation zones in machining processes.

Due to the complexity of the studied domain, developing an analytical model expressing the two-phase flow interaction (coolant with surrounded air), and coolant flow in all possible directions is still a challenge. Thus, using the numerical approach can be valid for studying and analyzing such cases. In the current work, a computational fluid dynamic model (two-phase flow) has been employed using the ANSYS-FLUENT module in order to simulate and analyze the coolant accessibility and behavior into the tool-chip interface area when using MQC system. Three base fluids have been used under MQC systems: water, mineral oil, and proposed nano-fluid oil. The coolant speed effect on the accessibility into the tool-chip interface area and heat transfer performance have been presented and discussed for all employed fluids. Also, another objective is studying the coolant type effect on the induced heat transfer coefficient.

2.Proposed CFD Model

A computational Fluid Dynamics (CFD) modelis developed to simulate the MQC volume,which is bounded by the tool flank face and the workpiece newly generated surface,as shown in figure 2.

2.1. Domain Geometry, Materials, and Meshing

For the simulation purposes, the current model is followed the same geometry and operation conditions of Hegab et al. [6] work as follows:

- Three types of coolant are used (i.e. water, mineral oil, and nano-fluid). The thermophysical properties of the employed coolant are listed in table 1. The used nano-fluid is based on vegetable oil with dispersed multi-walled-carbon-nano-tubes (4 wt.%).
- The used workpiece is Ti-6Al-4V alloy (bar), which has a diameter of 50 mm and a length of 200 mm. The mechanical, thermal, and physical properties of the used alloy are provided in table 2 [7].

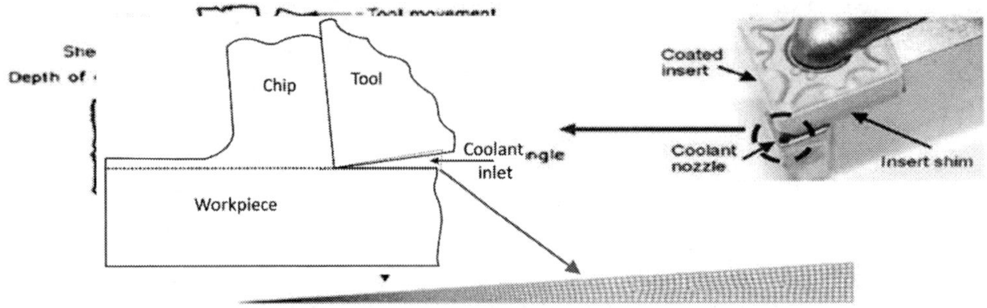

Figure 2. The developed domain used in the CFD simulation (coolant effect in tool-chip interface).

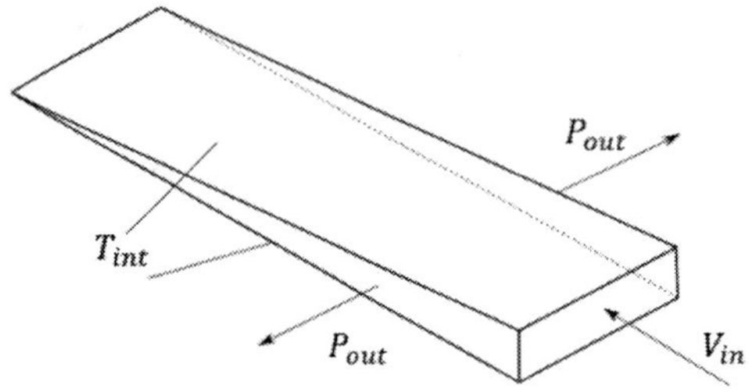

Figure 3. Model boundary conditions.

- The cutting conditions are; cutting velocity of 0.5 m/s, feed rate of 0.15 mm/rev, and depth of cut of 0.1 mm
- The temperature of the chip-tool interface is assumed to be about 600°C
- Throughout the proposed model, in order to accurately define the outlet boundary conditions of the chip-tool interface sides, a three-dimensional effect has been considered.

Figure 3 shows the model boundary conditions. Mist inlet velocity is definedtothe right side, whilea zero gauge pressure outlet is assigned to the model sides. Constant interface temperature is consideredat the top and bottom (i.e. the tool flank face and the newly generated workpiece surface). To consider the workpiece motion, the bottom of the modelis defined as a moving wall.

Generally, quadrilateral elements with an average size of $1\mu m$ have been used in the meshing construction and laminar steady-state pressure-based solver was used. Then, mesh dependent study was performed in order to confirm that the results are not depending on the mesh size. This is done by comparing the current results with more fine mesh size results.

Table 1. Thermo-physical properties of the employed coolants.

	Coolant Types		
	Oil	Water	Nano-fluid
Density (Kg/m^3)	820	1000	930
Specific-heat (J/Kg. K)	1670	4182	1963.304
Thermal-conductivity (w/m. K)	0.162	0.6	1.191
Viscosity (kg/m. s)	0.015	0.001	0.07154

IOP Conf. Series: Materials Science and Engineering **973** (2020) 012048 doi:10.1088/1757-899X/973/1/012048

Table 2. Properties of Ti-6Al-4V alloy (work-piece material).

Chemical composition (wt.%)	Ti:Bal, Al:5.5-6.75%, V:3.5-4.5%, C:0.1%,Fe:0.3%, O:0.2%, N:0.05%, H:0.0125%
Density (Kg/m^3)	4470
Young's modulus (GPa)	114
Poisson's ratio	0.3
Tensile strength (MPa)	895
Yield strength @ 0.2% offset (MPa)	828
Elongation (%)	10
Hardness (Rc)	30-34
Specific Heat (J/kg °C)	560
Thermal Conductivity (W/m °C)	7.2
Melting Point (°C)	1649

2.2. Governing Equations

Navier-Stokes equations are considered in the proposed model to simulate the flow of the coolant-air interaction in the tool-chip interface area using MQC system attached in the tool holder. Equations for the conservation of mass, the linear momentum, the energy, and the volume fraction are described, respectively, as shown in equation (1-6) [8].These equations must be solved simultaneously in order to obtain velocity components, temperature, pressure, and volume fraction of the coolant.

$$\nabla \cdot \left(\rho \vec{\upsilon} \right) = 0 \tag{1}$$

$$\rho \left(U \frac{\partial U}{\partial x} \nabla \cdot \left(\rho \vec{\upsilon} \right) = 0 + V \frac{\partial U}{\partial y} \right) = -\frac{\partial p}{\partial x} + \rho g_x + \mu \left(\frac{\partial^2 U}{\partial x^2} + \frac{\partial^2 U}{\partial y^2} \right) \tag{2}$$

$$\rho \left(U \frac{\partial V}{\partial x} + V \frac{\partial V}{\partial y} \right) = -\frac{\partial p}{\partial y} + \rho g_y + \mu \left(\frac{\partial^2 V}{\partial x^2} + \frac{\partial^2 V}{\partial y^2} \right) \tag{3}$$

$$\nabla.(\vec{\upsilon}(\rho E + p)) = \nabla.\left(k \nabla T + (\tau.\vec{\upsilon}) \right) + S_h \tag{4}$$

$$\rho = \alpha \rho_l + (1 - \alpha) \rho_g \tag{5}$$

$$\nabla \cdot \left(\alpha \vec{\upsilon} \right) + \nabla \cdot \left(\alpha (1 - \alpha) \vec{\upsilon} \right) = 0 \tag{6}$$

3.Results and Discussions

In order to confirm the solution convergence, the solution is continued until the difference between two successive iterations in each flow conservation equation (i.e., residuals) becomes less than 10^{-3} for the continuity equation and 10^{-4} for all other equations [9].

The results of the proposed CFD model have been classified into two phases. The first phase obtains the coolant accessibility percentage into the chip-tool interface (MA%) with different coolant velocities (i.e. 0.5, 1, 1.5, and 2 m/s) for the three studied coolants. The second phase discusses the heat transfer effectiveness for the employed coolants with different inlet velocities since it is an important aspect, especially when machining hard-to-cut materials. The results of the coolant accessibility percentage (phase I) are provided as shown in figures (4-5). As can be seen in these figures, the water-based coolant offers better results than oil and nano-fluid-based coolants. It is mainly due to its viscosity (see Table 1). Also, the lowest accessibility percentage has been obtained when using nano-fluid-based coolant, and it is also attributed to its viscosity because of the nano-tubes presence in the base oil. It can be also stated that increasing the nano-fluid dispersion stability could effectively decrease the resultant viscosity as has been mentioned in previous research works. Also, it

IOP Conf. Series: Materials Science and Engineering **973** (2020) 012048 doi:10.1088/1757-899X/973/1/012048

was clearly found in figure 4 that increasing the coolant velocity would increase the coolant accessibility percentage into the chip-tool interface area. However, no significant effect has been found after 1.5 m/s for all employed coolants.

Regarding the second phase analysis, the average heat transfer coefficient has been used to express the used coolant heat transfer effectiveness. The results (see figure 6) showed that the water-based coolant has offered the best performance. Also, the nano-fluid based coolant has shown better results than mineral oil, and that is mainly attributed to the superior thermal properties (higher thermal conductivity and heat convection coefficients) of the added nano-additives. Also, increasing the coolant velocity showed a noticeable increase in the average heat transfer coefficient. Although water-based coolant offers better results for both studied characteristics (i.e., accessibility percentage & average heat transfer coefficient), water-based coolant is an undesirable approach for cutting hard-to-cut materials due to its poor tribological and chemical characteristics (e.g., flash point, tool wear resistance, and corrosion resistance).

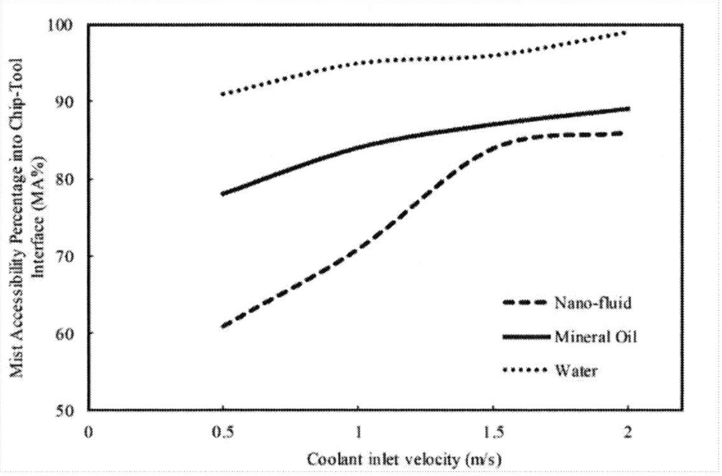

Figure 4. Results of coolant accessibility percentage with different coolant velocities.

Conclusion

Flood coolant is widely used to dissipate the generated heat, which occurs during the machining processes; however, it is not a sustainable technique because it has high harmful effects on the environmental and operators' health as well. Besides, using a large amount of coolant increases the machining cost by approximate 10-15 %. In this paper, the utilization of different alternative techniquesis studied, where the optimal coolant can be directly fed to the cutting zone through a nozzle in the tool holder located below the insert flank face. CFD model for this technique is developed in order to study the accessibility distance and heat transfer coefficient of three kinds of coolant: water, mineral oil, and nano-fluid. The results show the water-based coolant has the best accessibility distance due to its low viscosity, which leads to an increase in the heat transfer coefficient. However, water-based coolant cannot be used as it has poor tribology aspects. The mineral oil coolant shows a higher accessibility distance than the nano-fluid. However, the nano-fluid provides a better heat transfer coefficient even at its short accessibility distance. This can be explained by the superior thermal properties of the added nano-additives. Future works need to be done in order to study the effect of higher inlet pressure on the accessibility distance of the nano-fluid, as the heat transfer coefficient is found to be mainly affected by the accessibility distance.

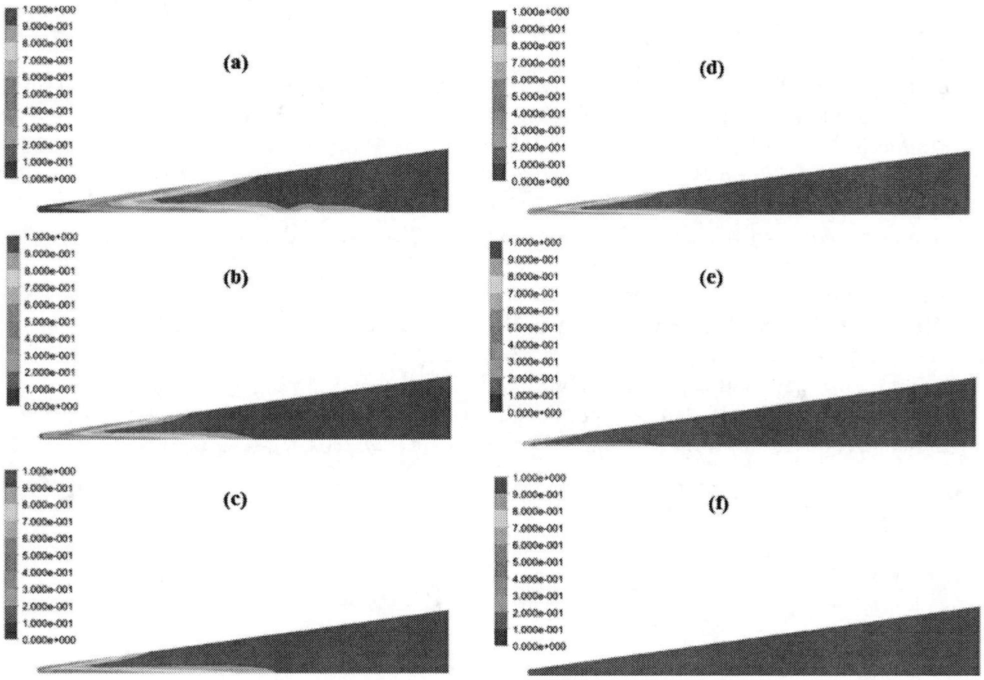

Figure 5. Contours of coolant volume fraction for some studied cases; (a) nano-fluid at 1 m/s, (b) nano-fluid at 2 m/s, (c) mineral oil at 1 m/s, (d) mineral oil at 2 m/s, and (e) water at 1 m/s, (f) water at 2 m/s.

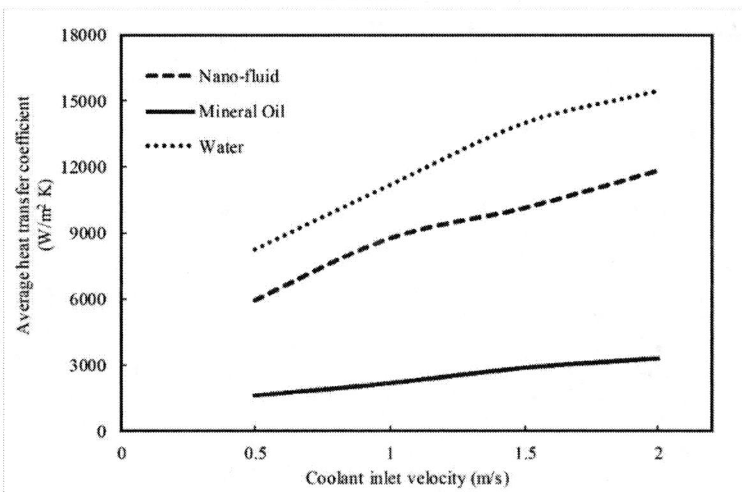

Figure 6. Results of average heat transfer coefficient with different coolant velocities.

References

[1] Vereschaka A, Vereshchaka A, Sharipov O, Bublikov J and Sitnikov N 2016 Edge Cutting Tools for Machining of Hard-to-Cut Materials *Materials Science Forum* **857** pp 213-20.

[2] Ezugwu E 2005 Key improvements in the machining of difficult-to-cut aerospace superalloys*International Journal of Machine Tools and Manufacture* **45**(12-13) pp 1353-67.

[3] Dhar N ,Kamruzzaman M and Ahmed M 2006 Effect of minimum quantity lubrication

(MQL) on tool wear and surface roughness in turning AISI-4340 steel*Journal of materials processing technology* **172**(2)p 299-04.

[4] Liao Y , Liao C and Lin H 2017 Study of oil-water ratio and flow rate of MQL fluid in high speed milling of Inconel 718*International Journal of Precision Engineering and Manufacturing* **18**(2) pp 257-62.

[5] Hegab H, Abdelfattah W, Rahnamayan S, Mohany A and Kishawy H 2018 Multi-objective optimization during machining Ti-6Al-4V using nano-fluids.

[6] Hegab H, Kishawy H, Umer U and Mohany A 2019 A model for machining with nano-additives based minimum quantity lubrication*The International Journal of Advanced Manufacturing Technology*p pp 1-16.

[7] Welsch G, Boyer R and Collings, E W 1993 Materials properties handbook: titanium alloys*ASM international.*

[8] Kundu P, Cohen I and Dowling D 2008. Fluid Mechanics*ELSEVIER*, Oxford.

[9] El-Rahman A, Abdelfattah W and Fouad M 2017 A 3D investigation of thermoacoustic fields in a square stack. *International Journal of Heat and Mass Transfer* **108** pp 292-300.

IOP Conf. Series: Materials Science and Engineering **973** (2020) 012049 doi:10.1088/1757-899X/973/1/012049

Optimizing ladle-refining performance during treating special steel melts for aviation technology

S A Elghazaly[1], K KGyula[2] and W S Elghazaly[3]
[1] Steel Department,Central Metallurgical R&D Institute, P.O.Box 87Cairo, Egypt
[2] Institute of Metallurgy, Miskolc University, H35015, Hungary.
[3] Steel Department, Central Metallurgical R&D Institute, P.O.Box 87 Cairo, Egypt
saiedelghazaly@yahoo.com

Abstract. The ever increasing demands of consumers towards superior quality steel products urged steelmakers to consider the implementation of various relatively new metallurgical technologies into the classical operations. The necessity to produce high quality steels led to the use of two stage steel production processes, one is the primary steelmaking and the other stage is the outside furnace treatment through which many metallurgical functions can be achieved like degassing, stirring and inclusion removal, Inclusion modification, desulphurization, deoxidation, decarburization, heating and alloying. In this research a trial was done to optimize the performance and usage of 30tons-ladle refining system during production of X65-pipeline steel as final product-Aluminium killed steel melts through controlling the mass flow contour using optimized modelling, optimum usage of Al2O3 and Sulphide modifiers and enhancing removal of non-metallic inclusions by altering their morphologies and hence floatation speed. Assessment of fine clusters of inclusions in the final steel product has been industrially correlated to the cleanness of melt before refining, to the slag composition and to the parameters of materials flow rates as well as their effects on the mechanical properties of final X65-steel product. Scanning-EM+DX analyzing unit and metallurgical microscopes were used to emphasis qualitatively and quantitatively the characters of non-metallic inclusions.

1. Introduction

The cleanness of different steels is governed by numerous standards; however, it differs according to the applications of the final product. Products having minor cross-sectional dimension, such as thin strip steel, where the presence of critically sized inclusions has increasing effects on product performance. Inclusion particle size distribution is frequently used to provide characteristic description of steel cleanliness. A comprehensive cleanliness evaluation also requires complementary information on non-metallic inclusions found such as quantity, spatial distribution, type, and morphology. With the growing demand for high quality materials, the cleanliness of different steel types has become very important [1-3]. High strength low alloy (HSLA) steels offer improved weldability, and superior mechanical and corrosion-resistant properties compared to mild or low carbon steel at a minor price premium Load-bearing HSLA steel found in structural applications is susceptible to fatigue failure initiated by non-metallic inclusions. [4-6]

The great challenge faced by steelmakers, at the present, is the lack of reliable cleanliness assessment methods to quickly evaluate the quality of steel products at low cost. The cleanliness assessments performed traditionally using metallurgical microscopy, chart comparison, and visual inspection of semi-finished products no longer gives adequate feedback in regard to micro-cleanliness. Image analysis is a powerful tool in assessing steel cleanliness with improved efficiency and accuracy. However, it has been reported[7-8] that

with automated image analysis using optical microscopy there exists a common problem of potential erroneous detection of defects arising from inadequate sample preparation.
The use of inert gas stirring is considered to be a reliable and widespread ladle metallurgy operation in steel production. Argon stirring has an important role in the control of impurities in steels that have been heavily deoxidized and micro- alloyed. In Al-deoxidized HSLA steels, the presence of the classic Al_2O_3 type inclusions is unfortunately unavoidable. [9-10] These impurities cause difficulties either with castability because of occasionally occurring nozzle clogging, or with the ability to meet the ever-stricter requirements for cleanliness. [11-12] There are some ways to avoid these ladle metallurgy problems. However; due to the need for very time-consuming material tests and analysis, these solutions require extremely precise production control (which is challenging even in the case of a well-equipped steel mill), and the technological conditions involved makes quality control difficult to handle.

2. Aim of Research

In this investigation a series of heats were evaluated to optimize the performance and usage of ladle refining system during production of HSLA-X65 steel as final product- Aluminium killed steel melts through controlling the mass flow contour using optimized modelling, optimum usage of Al2O3 and Sulphide modifiers and enhancing removal of non-metallic inclusions by altering their morphologies and hence floatation speed. Assessment of fine clusters of inclusions in the final steel product has been industrially emphasized to the cleanness of melt before refining.

3. Melting and Refining

Several heat Logs of HSLA steel grade X-65 has been taken into consideration for further analysis. All heats were melted in 100-130 tons basic-Oxygen steel furnace, deoxidized before tapping using 1000kg burnt lime-200kg Flourspar-900kg FeMn-80kg Al blocks. To ensure deepdeoxidation some deoxidants were also added during tapping to accomplish lowest oxygen activity of about 75ppm,then outside furnace degassed, micro-alloyed and CaSi-Al wire treated in ladle furnace (LF) technology to minimize and optimize count and morphology of non-metallic inclusions respectively. The overall technology flow diagram for steel melting, desulphurization and refining in (LF) is shown as in Figure (1). The average steel composition in final coils was as in Table1.

Table 1.Average values of chemical analysis for X-65 HSLA steel Coils.

Heats	C	Mn	Si	V	Nb	Al	S	P
x-65	0.2-0.25	1.5-1.6	0.5-0.6	0.08	0.09	0.04-0.06	0.007	0.015

3.1 Steel Deoxidation

After oxidation end enhancing melting using gaseous Oxygen, the amount of [o] in the steel melt increased to maximum equilibrium values. It is well known that the maximum solubility of oxygen in liquid iron at the eutectic of 1527°C is about 0.16%[3], while the oxygen solubility in solid iron, at temperature slightly below its melting point, approaches zero.Accordingly, and upon solidification, majority of dissolved oxygen will precipitate as FeO, SiO2 and MnO inclusions. However the presence of alloying elements such as carbon can influence the dissolved oxygen content as clear from: [wt%C] • [wt%O] = ~0.0023 [2] at 0.6%C maximum. This means that in order to prevent blowhole formation, porous cast product, or precipitation of FeO inclusions in sizeable quantities, liquid steel must be deoxidized prior to casting.

3.2Deoxidation with Aluminium

Aluminium is one of the most effective deoxidizers used for steel deoxidation. In aluminium deoxidized steel, there are generally two species of deoxidation products: solid (FeO-Al2O3) spinel, and solid (Al2O3) corundum. Among the two deoxidation products, corundum is the dominant species found in steel. Corundum phase is characterized by having unique faceted shapes and relative smaller diameter as single particle and during deoxidation, follows dendritic growth. For steels deoxidized solely with aluminium, α-Al2O3 products are formed; clusters of these particles tend to remain as inclusions in steel. Corundum inclusions, usually having the particle size of 1 to 5 μm, have a tendency to agglomerate upon colliding with one another in order to lower the overall contact area with molten steel and therefore effectively stabilize the entire unit by minimizing the surface energy.

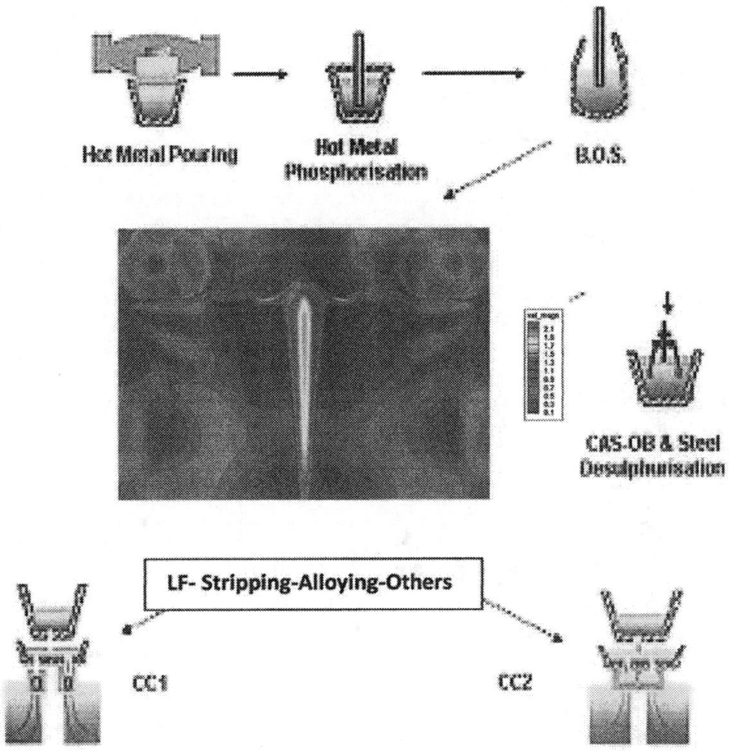

Figure 1. Melting and Refining technologies used in HSLA steel production.

4. Morphology of Nonmetallic Inclusions

After complete deoxidation of the molten steel (in furnace or in the tapping ladle) tremendous amounts of nonmetallic were detected, one of them the harmful Al2O3 clusters that deteriorate the continues casting operation through nozzle clogging and the mechanical and rolling conditions of the steel. Calcium has a strong affinity to oxygen and could potentially be utilized as steel deoxidizer and modifier for Al2O3. The challenge, however, lies in the following properties of calcium: low boiling point (1439°C), limited solubility in steel (0.032% Ca at 1600°C), and high vapor pressure at 1600°C (1.81atm).[10] Due to these reasons, it is rather difficult to introduce calcium to molten steel in its metallic form, and it is usually added as various iron-containing Ca-Si alloys. The primary deoxidation products are therefore calcium silicates, which may also contain other oxides. When combinations of Ca and Al or Mn/Si deoxidation are carried out, the primary deoxidation products can be modified to oxides with lower activity and hence improve the removal of dissolved oxygen. By converting the solid alumina inclusions to liquid calcium aluminates, the extent of deoxidation can be improved from 8-10ppm O to 1ppm O in Al-killed steel (0.05% Al).[9] With a CaO:Al2O3 ratio of 12:7, calcium

IOP Conf. Series: Materials Science and Engineering **973** (2020) 012049 doi:10.1088/1757-899X/973/1/012049

treated Al2O3 can reach a melting point of 1360°C at the CaO-Al2O3 eutectic (Figure 2-7) and therefore exists in the liquid state at steelmaking temperatures. Moreover, there exist five modifications of calcium aluminates as indicated in Figure 2-7; 12CaO•7Al2O3, 3CaO•Al2O3 and CaO•Al2O3 are liquid, while CaO•2Al2O3 and CaO•6Al2O3 are solid at steelmaking temperatures. Instead of agglomerating, in alumina inclusions, liquid calcium aluminates will coalesce upon contact due to better wetting with liquid steel and will not easily attach onto refractory surfaces. Hence, solid deoxidation products can also be calcium treated so that the steel casting process is clogging-free.

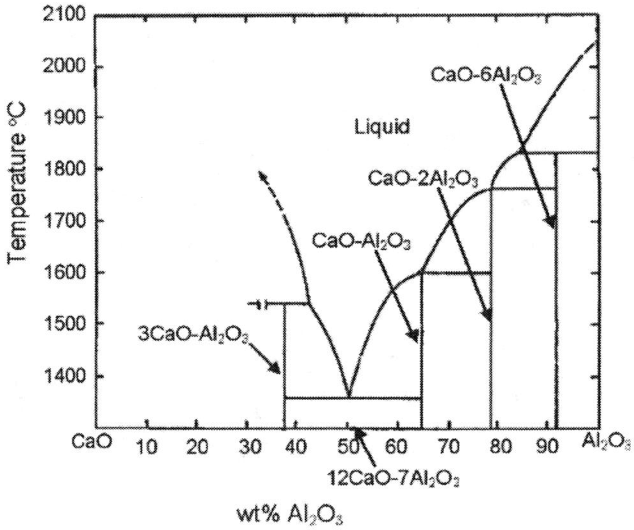

Figure 2. CaO-Al2O3 equilibrium phase diagram [10].

The modification of inclusions and removing them to the slag layer was accomplished in Ladle-Furnace (LF), in which cored calcium-aluminum wire feeding and melt stirring with Argon gas was used. The maximum gas flow rate was modeled to obtain the optimum turbulence between molten metal and slag layer as shown in Figure 3. It is evident that if an abrupt increase in gas flow rate was exercised, the molten metal can be spilled over the slag layer and then atmospheric oxidation is happened unfortunately again.

To achieve optimized inclusion count and morphology after Ca-cored wire feeding as shown in Figure 4, moderate bath stirring and the technological parameters projected in Table 2 must be used during melting and refining steel melts. The overall outcome revealed that optimal and satisfactory Al2O3 inclusion removal from the molten steel could be accomplished in ladle-Furnace at [O] and [Al] levels reached 2-3ppm and 0.05-0.06% respectively. This was only possible at the technological values given in table 2.

However the overall modifications occurred during Ca-treatment of liquid steel in LF can be summarized as in Figure 5, where Calcium Aluminates centered the inclusions and CaS precipitated as a rim around the inclusions. Meanwhile, MnS precipitated as scattered globules and cuboids but reversion of [Mn] and [Si] can also happened. The floatation of even Ca-Al2O3 minute inclusions (1-5 microns) to the slag layer was possible leaving behind a clean steel melts. The assurance of intimate final refining slag analyses can give answer for what is the meaning of Optimized Ladle-Furnace treatment. Table 3 shows the overall analyses of the slag after LF processing.

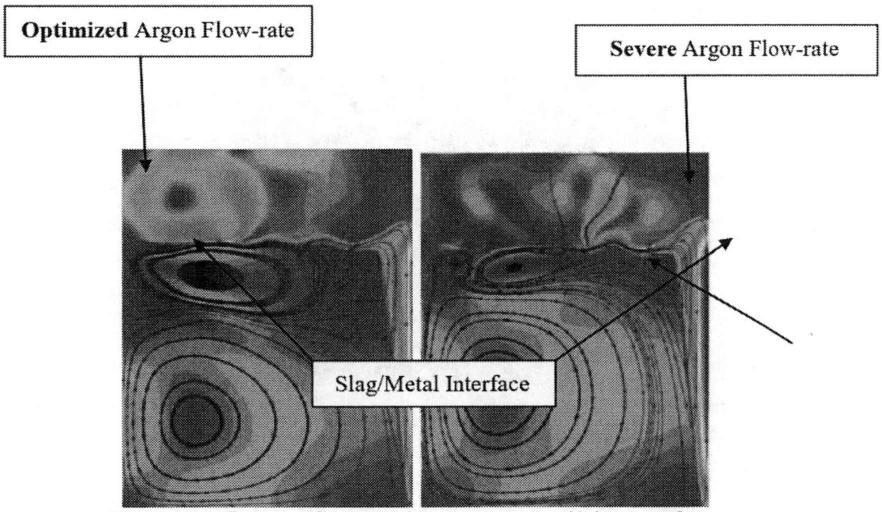

Figure 3. Influence of Argon Flow rate on Metal/Slag Interface.

Table 2. Technological parameters of experimental heats produced at mild stirring.

Total Al kg	CaSi-Al wire kg	Argon used m3x100	Blowing time min	[O],.ppm	[Al], %
210	55	31	15	2.4	0.055
245	53	81	25	4.7	0.063
230	57	72	27	4.0	0.057
225	55	67	15	3.6	0.062
210	21	54	19	3.2	0.060
280	34	58	35	3.5	0.061
230	12	53	21	3.0	0.057
210	16	45	21	2.6	0.052
265	52	50	25	3.1	0.053

Argon flow rate Nm^3/min	1000	2000	3000	4000
Depth of molten steel [m]	2.85	2.85	2.85	3.12
Bulge height [m]	0.21	0.24	0.3	0.39
velocity [m/s]	2.0	2.1	2.2	2.4

5. Optimized Argon Stirring

The removal of minute inclusions depends to a great extent not only on the intensity of Argon stirring but also on the time of blowing, meanwhile care must be taken on the molten steel temperature, not to cool down after long time gas stirring. The dependency between floatation of nonmetallic inclusions (1-50 microns) and the parameters of argon stirring in LF can be controlled according to the turbidity occurred during gas purging and the time required for such operation. However, and according to our gained experience to reduce melt temperature about 4000 Nm^3/min of Argon must be blown, to homogenize steel bath about 3000 Nm^3/min, to enhance desulfurization process 2000 Nm^3/min, and about 1500-2000 Nm^3/min can be used for removing impurities.

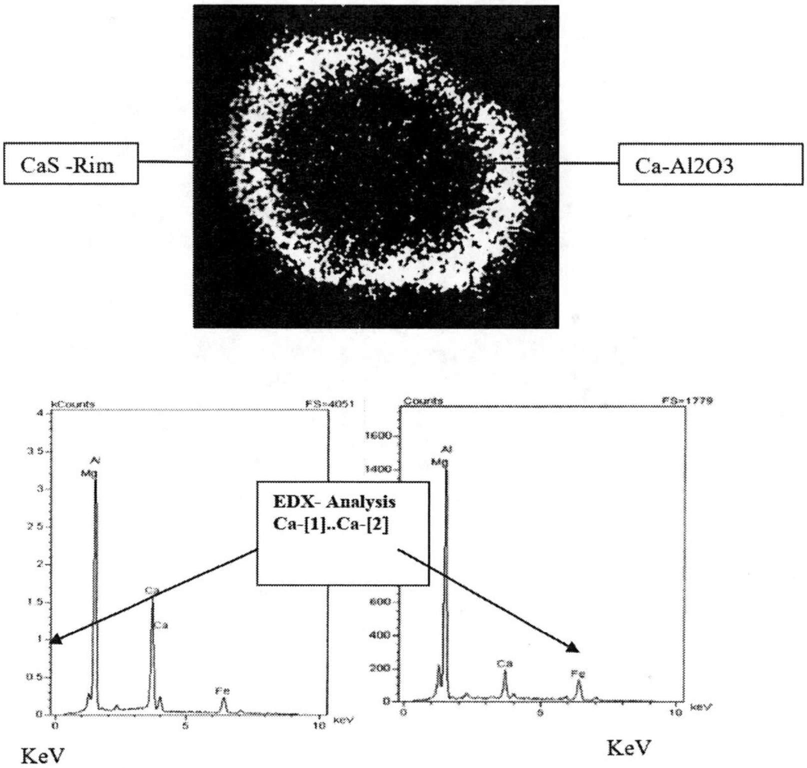

Figure 4. Morphology of modified Al2O3 inclusion and its EDX-composition ratios.

The effect of low to high gas flow rate at a given time of purging (15minutes) was investigated through gathering the changes occurred in the final liquid slag and metal analyses such as for (Al_2O_3), (FeO), (SiO_2), [O] and [Al] after treatment in LF. The results can be easily followed as in Figure 6. It is evident that a moderate decrease in (FeO)% along moderate argon blowing (500-1500 Nm3/min), reaching minimal content (4%) at 2000-2500 Nm3/min. However, (FeO) content is drastically increased at higher stirring rates. This is due to the increased bulge height causing direct contact between molten steel and atmospheric air. It is also evident that at increased stirring intensity the activity of oxygen[O] in molten steel and content of Alumina (Al_2O_3) are increased to reach (6-10ppm) and (40-60%) respectively. The optimum LF working zone at satisfactory concentrations of (FeO), (Al_2O_3) and [O] can be obtained at stirring intensity between 2000-2500Nm3/min.

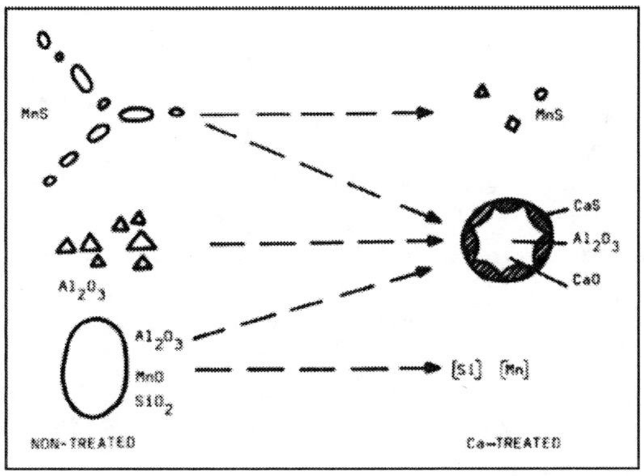

Figure 5. Modification of composition and morphology of inclusions.

Table 3. Overall analyses of slags after LF process Al-Wire Feed50-60kg + Argon Stirring.

(CaO) %	(FeO) %	(Al_2O_3) %	B	(CaO) %	(FeO) %	(Al_2O_3) %	B
57.78	2.4	25.81	7.23	55.9	2.8	24.8	7.2
58.22	2.24	28.22	13.04	60.8	2.3	25	7.3
56.30	3.45	25.04	6.59	61.3	4.5	24	11.4
62.30	3.78	24.10	12.35	61.4	4.1	25	13
61.17	4.46	23.19	11.53	63.5	3.2	24.3	12.7

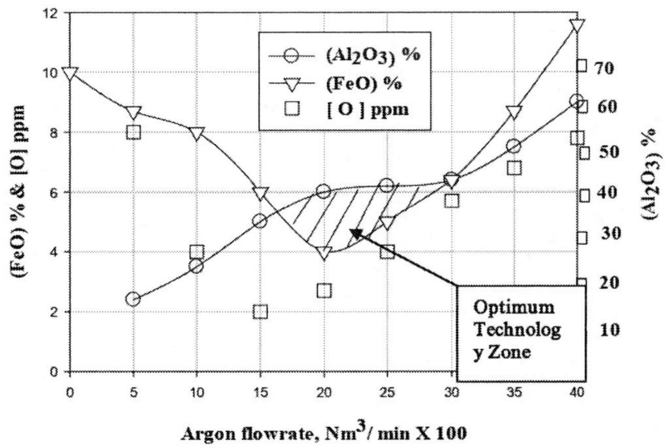

Figure 6. Stirring intensity versus (FeO), (Al2O3) slag contents and [O] activity in steel.

IOP Conf. Series: Materials Science and Engineering **973** (2020) 012049 doi:10.1088/1757-899X/973/1/012049

6. Conclusion

1) Optimized Ladle-Furnace treatment was accomplished not only through adequate bottom stirring modeling but also through industrial implementation and data base establishment.
2) The maximum gas flow rate was modeled to obtain the optimum turbulence between molten metal and slag. At abrupt increase in gas flow rate, the molten metal can be spilled over the slag layer and then atmospheric oxidation is happened unfortunately again.
3) Optimal and satisfactory Al2O3 inclusion removal from the molten steel could be accomplished in ladle-Furnace at [O] and [Al] levels reached 2-3ppm and 0.05-0.06% respectively.
4) floatation of nonmetallic inclusions (10-50 microns) depends on the projected parameters of argon stirring in LF.
5) To achieve optimized inclusion count and morphology after Ca-cored wire feed, moderate bath stirring and the optimum technological parameters zone has to be followed.

References

[1] Millman S and Lee KY 2004 Belgium, *IISI Study on Clean Steel* pp 409-442
[2] Jacobi H. and Wunnenberg K 2004 Belgium, IISI Study on Clean Steel, International Iron and Steel Institute pp 307-309
[3] Zhang L , Thomas BG and Cai K 2002 Warrendale, PAEvaluation and Control of Steel Cleanliness – Review 85th Steelmaking Conference Proceedings, ISS-AIME pp 431-452
[4] HSLA Steels 2015conference proceedings, the Chinese Society for Metals ISBN: 978-1-119-22330-6
[5] Krishnadev MR et al 2009Strong Tough HSLA Steels Via Processing and Heat Treating of Cu-Ni-Nb and Cu-Ti-B Compositions Journal of Engineering, Mat. Technology **3 vol.103**
[6] Park J.H and Todoroki H 2010, ISIJ Int. **13 50** pp 33–46.
[7] Schamber F 2009 Introduction to Automated Particle Analysis by Focused Electron Beam, Fischer Corporation
[8] Van Ende M.A, Guo M and Jung IH 2013 ISIJ Int. 53 pp 1974–82.
[9] Bilenicki Mand Cwudzinski A 2015 ISIJ Int. 60 Issue 1 pp 257-262.
[10] Zhang L and Thomas B G2005 Application of computational Fluid Dynamics to Steel Refining and Casting in Fourth International Conference on CFD in the Oil and Gas Process Industries pp 1-9
[11] Cong Li, Jian-guo Li and Jia-cheng Ji 2017 Mechanism to remove oxide inclusions from molten aluminium by solid fluxes refining methodFoundry, China pp 233-243 DOI:10.1007/S41230-017
[12] Jowsa J and Cwudziński A 2015 Numerical modelling of metal/flux interface in a Continuous casting mould Archives of Metallurgy and Materials, 4 **60** pp 2905-2912

Acknowledgments

The Authors would like to acknowledge assistance and encouragement from many colleagues both working with central metallurgical research and development institute-Egypt and with the institute of metallurgy at Miskolc university-Hungary.

AUTHOR INDEX

Abdalla, H M 1, 50
AbdAllah, A. 194
Abdel-Kader, M A 348
Abdelaziz, A M. 522
Abdeldayem, M A 146
Abdelfattah, W. 533
Abdelghany, A A 276
Abdelhamed, M 237
Abdellatif, M H. 226, 512
Abdelnaby, M A 257
Abdelsalam, O 361
Abdelsamie, A 442
Abdelwahhab, SA 471
Abo-Elnor, M E 146
Aborehab, A. 378
Abu-El-Yazied, T G. 298
Abu-Okail, M. 206
Adel, WM 338
Ahmed, H M 288
Ahmed, H 172, 184
Ahmed, M Y M 369
Ahmed, M Y 250
Ahmed, M. 116
Al-Makky, M 522
Al-Sanabawy, M A 19, 36
Alhamaki, A B 512
Ali, S. 257
Aly, M H. 100
Amin, M I 417
Amin, MI 404
Ata, W G. 237
Attai, YA 338
Attia, T A. 159
Awad, M I. 455
Awad, MI. 471
Ayman, A 512
Azab, N A 226
Aziz, M M. 369
Azzam, B 194
Babikir, HA 387
Badawy, A 276
Barakat, A F. 499
Bassioni, G. 226
Belal, H. 1, 19, 36, 50
Bondok, M M 137
Darwesh, O 172
Edriys, I I 326
El Maksoud, R M Abd 76
El-Agroudy, M N 455
El-Desouky, M A 267
El-Gindy, M 116
El-Hadek, M A. 137
El-Hofy, H. 522
El-Naggar, M 1
El-Sallamy, W 194
El-Shabasy, A B. 159, 305
El-Shamarka, S E 429
El-Sheikh, M A 137
El-Wazery, M S. 267
Elaziz, M Abd 387
Elenien, K F Abo 226
Elghazaly, S A. 540
Elghazaly, W S. 540
Elkshen, M 19
Elmaihy, A 404, 417
Elsheikh, A H. 387
Elsheikh, AH. 317
Elshimy, M. 91
Emarah, Maya 465
Farag, SS 250
Fattouh, M. 326
Gadallah, N 206
Ghanem, M 361
Ghoneim, A M 100
Gyula, K K. 540
Habaka, M A. 250
Hamdi, H M 471
Hannora, A E. 288
Hasabelnabi, S. 184
Hassan, A 394
Hassan, H A 298, 305
Hassan, HA 159
Hassan, M Nasr 499
Hegab, H 533
Hegazy, M M 276
Hegazy, M. 36
Ibrahim, A G 429
Ibrahim, A Z. 348
Ibrahim, AZ. 369
Ibrahim, S A. 288
Ibrahim, S. 91
Ismail, M A 338
Kamal, A 317
Kamel, Hisham. 129
Kamel, M 378
Kassem, M 378

Khafagy, A Z 237
Khalil, A 184
Khalil, M........ 250
Kishawy, H A 533
Kotb, A T........ 76
Lang, H 116
Liu, Y........ 387
Lotfy, H M........ 159
Louis, M A........ 483
Mabrouk, M H 146
Mabrouk, OM 267
Magdy, A M 305
Maged, S A........ 455
Mahmoud, O E 100, 483
Mahmoud, T S 217
Makady, A M........ 298
Makled, A 36
Malalasekera, W 91
Masoud, R........ 326
Mohamed, A 394
Mohamed, M 442
Mohammed, R K 217
Mohammed, S S........ 217
Mohany, A 533
Mostafa, W 442
Nawar, M A 76
Nemnem, A F 378
Niklawy, W........ 404, 417
NourEldin, A F 338
Omer, A A 394
Osman, T 194
Rashad, A M........ 429
Rehan, A 455
Riad, A M 348, 369
Roman, M R 483
Sabry, I 206
Said, A E........ 298
Salah, A 184
Salama, A 50
Salem, A M........ 237
Sedrak, M F 483
Sedrak, M........ 442
Seleman, M. M 288
Shahin, M 404, 417
Shawky, Omar 63
Shiba, M S 257
Showaib, E........ 317
Sobh, A S........ 499
Thévenin, D 442
Tolbah, FA........ 471
Wu, D 387
Youssef, H 522
Zakaria, S Y 429

Institute of Physics
Dirac House, Temple Back
Bristol BS1 6BE UK

ISSN: 1757-8981
ISBN 978-1-7138-4792-2